HANDBOOK
OF CURRENT
SCIENCE &
TECHNOLOGY

HANDBOOK
OF CURRENT
SCIENCE &
TECHNOLOGY

BRYAN BUNCH

GALE

DETROIT • NEW YORK • TORONTO • LONDON

Written by BRYAN BUNCH

Gale Research Staff

James A. Edwards, *Project Coordinator*

Nicole Beatty, Erika Berry, Christine Jeryan, Jacqueline Longe, Zoran Minderović, Donna Olendorf, and Pamela Proffitt, *Assisting Editors*

Mary Beth Trimper, *Production Director*
Shanna Heilveil, *Production Assistant*

Cynthia Baldwin, *Product Design Manager*
Michelle DiMercurio, *Art Director*

ISBN 0-8103-9552-5

Printed in the United States of America.

10 9 8 7 6 5 4 3 2 1

Contents

ASTRONOMY AND SPACE

BIOLOGY

CHEMISTRY

EARTH SCIENCE

PHYSICS

TECHNOLOGY

APPENDICES

INDEX

HOW TO USE THIS BOOK

The Handbook of Current Science and Technology is designed to provide easy access to major developments in science and technology that occurred during the mid-1990s. It also presents background for interpreting these events and serves as an up-to-date handbook of basic scientific information. Data included have been chosen to cover facts that are frequently needed while reading or writing. The same data also form an essential aid to understanding new science devices, experiments, and theories wherever encountered.

As the following plan for the book indicates, *The Handbook of Current Science and Technology* (or *Current Science*, for short) is designed so that it can be read from start to finish—perhaps skimming or skipping some of the tabular material—to obtain an overview of science and technology in the mid-1990s. Many will prefer to treat *Current Science* like a magazine, however, browsing among the articles that are of particular interest. In either case, the structure and content of the book make it a reference work to which one can return again and again.

Plan of the Book

The book begins with an introduction describing the current state of scientific enterprise. Following that, the main branches of science and technology are discussed in alphabetical order, although some groupings are used as well. There is no strict separation of technology from science. Some technological issues, such as telescopes for astronomy or applications of chemistry, are treated in conjunction with the related science.

- Anthropology and archaeology are grouped—anthropology here is largely physical anthropology of human ancestors.
- Archaeology is split into Old World, New World, and some general issues that apply across the board.
- Astronomy is grouped with space science and technology.
- Biology focuses mostly on whole organisms with the exception of a general discussion of genetics and development.
- Chemistry is primarily inorganic, although carbon compounds are a central element of the chapter.

- The several earth sciences are grouped together, including paleontology, oceanography, and geology.
- Physics combines both theory and applications.
- Technology includes computer advances and other so-called "high-tech" issues.

Within each subject area, the same basic plan is followed:

State of the Subject.   Each subject-area section begins with an article on the state of the subject in the mid-1990s.

Timetable of the Subject.   This is followed by a historical timetable through 1992, consisting of short entries describing major developments in the subject area in chronological order. The timetables are designed both for easy reference and so that the reader can develop a feeling for how a subject has evolved.

Sections.   The bulk of the book consists of articles on developments in science and technology that took place in the period from 1992 through 1995. The individual articles are grouped into several sections on a given theme. For example, the **Physics** chapter contains three main sections after the timetable: "Particle and Nuclear Physics" updates particle physics experiments and theory and describes progress in nuclear fusion; "Basic Theories" discusses experimental confirmation of some theories, developments in quantum theory, and the most advanced ideas of theorists; and "New Devices in Physics" is concerned with progress in particle accelerators and in superconductivity.

Articles.   Within each section, individual topics are covered in one or more articles. For example, "Transportation and Communications Technology" contains separate articles on modern developments in transportation, focusing on trains and tunnels; new developments in communicating by telephone, radio, and television; and finally a review of communication through time instead of space—which is called "data storage and retrieval." This last is one of the frequent "UPDATE" articles, which are intended to summarize many new developments in a rapidly changing field.

References.   Articles conclude with a listing of periodical references and suggestions for additional reading. These listings are limited to magazines that are commonly available in general libraries in the United States. Foreign periodicals are excluded.

Set-Off Background Material.   An important feature within many articles is background material that is set off from the body of the article. The main focus of this book is on current science and technology, but some readers may not have the prerequisite knowledge to understand new technological developments. Sometimes a simple glossary is all that is needed to clarify an article or a table; other times it may be a paragraph or two that provides a brief history of a key development.

Science periodicals or yearbooks, which constitute the most accessible source of current scientific and technological information outside of this book, either omit such background material or incorporate it into the body of articles. *The Handbook of Current Science and Technology* is unique in allowing the reader a choice.

Tables, Charts, and Lists of Useful Data.   Most tables that have to do with a particular subject matter are grouped at the end of the subject sections. Sometimes, when it is particularly useful with regard to a specific article, a table is included in or immediately follows the article. Every effort has been made to include in considerable detail the data that might be most useful.

Who Needs This Book

Everyone who regularly uses current scientific data that span several fields or that cover a particular subject needs *The Handbook of Current Science and Technology*. Among those people are science writers and reporters, science teachers, and investors in new technology. This group also includes any others who have to know details of recent developments in science.

Science as an International Enterprise

Every effort has been made to include not only the discoveries and developments of science around the world, but also the viewpoints of scientists in various nations. A work published in English in the United States tends to have a North American point of view. This is compounded in science and technology because, despite many problems with science education in the United States, a great many of the advances in current science take place in the United States. Scientists and institutions are identified by nationality throughout the book to help the reader from outside the United States keep track of the progress of international science.

The State of Science

THE STATE OF SCIENCE TODAY

Scientific enterprise in the mid-1990s continues to increase its power to provide a clearer understanding of the universe, and to affect the lives of people around the world. Only those living in rain forests or other fringe environments remain outside the scope of its influence, and that number is receding rapidly.

Growth of Knowledge

Scientific enterprise as a method for trying to understand the universe is traditionally considered to have started with the Ionian philosophers of 600 B.C. Since that inception, growth in science has proceeded rapidly, along a steeper-than-exponential curve. What do I mean, "steeper-than-exponential"? A pure exponential curve has a constant doubling time; the quantity being measured doubles regularly every n years. For scientific knowledge, growth has been so fast that the doubling time itself may have a halving time. That is, it seems to take half the time for knowledge to double as the previous amount of time required. If it first doubled in n years, it will double again in $n/2$ years or so.

Since late in the nineteenth century, scientists themselves have suggested that this kind of expansion cannot continue. In many instances of various types of growth, scientists observe what biologists call the reproduction curve for a growing population. A reproduction curve begins as an exponential curve and then flattens out as the resources on which the population depends are depleted. The reproduction curve eventually becomes an elongated S. Many scientists expect that the growth of knowledge will follow such a reproduction curve in the long run. The unanswerable question is when the flattening of the curve will commence.

Despite nineteenth-century pessimism, and more recent caution over the possibility of further scientific progress, there seems to be no serious depletion of the underlying impetus for the growth of science: ignorance. If we have reached the flattening out of gains in knowledge, it is not yet apparent.

In May of 1994, the Santa Fe Institute in New Mexico hosted a workshop on this topic called "Limits to Scientific Knowledge." The theme was whether or not there were any limits to what science can determine. Mathematicians have already demonstrated that there are many important problems in mathematics that in effect cannot ever be solved. Is this true of physics or biology as well?

There has been some progress toward answering this question. For example, Gregory Chaitin, an IBM mathematician, has proved that it is not possible to know for all theories whether or not they are the "best" in the sense of being the simplest description of reality. The same proof establishes that it is impossible to show in general that apparently random data have a pattern that we cannot find, even if the pattern is there.

As mathematicians have learned, however, the key words are "for all" and "in general." As long ago as the beginning of the nineteenth century, mathematicians proved that there was no set of rules that could solve every equation involving x to a power higher than four, yet individual equations of this type are routinely solved by high-school students. This is because they are *specific* equations, not *all* equations of this type or equations in general. Furthermore, if one is willing to accept approximations that are as good as one cares to make them (that is, to as many decimal places as you want to measure), all equations in general can be solved. Similarly, Chaitin suggests that abandoning infinitely precise numbers would get around the limitations he has discovered.

However, other limitations also exist. Problems that can be solved for simple systems become intractable for systems only slightly more intricate. In this case, chaos theory has had some success at finding patterns. But it may be that real problems of complex molecules, interacting galaxies, and living environments are too complicated even for chaos theory or for approximate solutions. Perhaps the day will come when the object of scientific enterprise will be only that of adding an additional decimal place to what we already know. However, in the 1990s, that day does not seem to have arrived.

Science Education

In the United States, the world's leader in science for the past 50 years, there continues to be considerable concern over whether young people are attracted to science in sufficient numbers, as well as whether they are obtaining the necessary foundation in science education. This is especially true with regard to women and minorities. One study, released in 1993, showed that three-quarters of all scholarships from the National Academy for Science, Space, and Technology (awarded on the basis of scores on a standardized test) are given to boys. Such a statistic is suggestive of the disparity between males and females in science preparation.

At the end of 1992, the U.S. National Research Council issued standards for fourth, ninth, and twelfth grades that they hoped would lead to improved science education throughout elementary and secondary education. In October of 1993, the American Association for the Advancement of Science tried its hand at the idea with *Benchmarks for Science Literacy*. After several delays, even more definitive standards were published in a preliminary form on December 1, 1994, by the U.S. National Academy of Sciences (USNAS). This publication covers content from kindergarten to twelfth grade, divided into three levels (kindergarten through fourth grade, fifth through eighth grades, and ninth through twelfth grades). Emphasis is on the teaching of concepts and not vocabulary, with advice on testing as well. The standards were then sent to some 200 focus groups for further criticism and refinement. Although scheduled for completion by the end of 1995, it is taking longer to reach consensus.

Had the USNAS standards for science education been put into practice earlier, the results of a 1994 survey exploring the science knowledge of 1,225 adults might have been less distressing. Only 21% of the adults got scores of 60 or better on a test of relatively well-known science facts, such as the number of planets in the solar system (65% did not know the number to be nine). The scores might have been skewed downward, however, by the religious convictions of the test takers. Many of the test items were based on scientific theories which are considered heretical by certain fundamentalist Christian groups, such as the evolution of living things, and changes in Earth's geology.

Despite all the concern, or perhaps because of it, United States school-age children tested better in science and mathematics in 1992, the last year for which data from the National Assessment of Educational Progress (NAEP) have been assembled, than they have since the early 1970s when the NAEP was started. Testing in science and mathematics is repeated every two years on samples of children ages nine, thirteen, and seventeen. Results of the 1994 round of testing should be available in 1996.

School is not the only place one can learn about science. At one time, many a scientist was self-taught from newspapers, magazines, and especially books. Books continue to be essential for the basics, but the main sources for information on current science are periodical publications. Yet, throughout the late 1980s and the early 1990s, there has been a steady decline in the number of periodicals written about science for the nonscientist. In the United States, between 1988 (the peak year for science sections in newspapers) and 1992, 53% of the 95 newspaper science sections were dropped, and another 22% reduced the amount of space devoted to science. Popular science magazines have had a tough time surviving into the 1990s as well, frequently changing owners because of poor profits, and often resorting to more sensational news stories to try to sell more copies.

Where public institutions stumble, private ones—in the ideal American system, at least—are expected to come to the rescue. For example, the Howard Hughes Medical Institute, a private foundation founded in 1953 to promote science education, has given $290 million between 1988 and 1994 to a total of 213 institutions for such purposes as improving teaching and laboratory equipment, as well as for outreach programs to secondary schools, women, and minorities.

International Science

Despite legitimate worries, the United States has continued to dominate the international scene in both science and, to a somewhat lesser extent, technology. The ongoing rise of European unity and of the nations of the Pacific rim may result, however, in a change in the location of the active center of international science away from the United States. A 1992 U.S. National Science Board report, "Science and Engineering Indicators," sounded the warning when it reported that funding for science in the United States began falling in 1989 and has continued to fall, while such funding for science has increased in other nations. In 1992, Japan, Germany, France, Italy, Sweden, and the United Kingdom together funded a third more basic research than the total funding in the United States for similar science projects. Currently, Japan leads the world in percentage of gross national product spent on research. Japan's research, however, tends to be product-oriented, because the majority of its funding comes from industry.

Meanwhile, the former Soviet Union, which has many brilliant scientists, continues somewhat out of the mainstream, but for a different reason than before. When communism was dominant, the totalitarian government tried to control science and the movements of scientists. Now that democracy of a sort is in place, the controls on science come largely from disorganization and poverty. Figures quoted for salaries of scientists in Russia vary, but are typically less than $50 a month on prevailing currency exchange rates. Russian scientists have had to supplement such low wages one way or another to stay alive.

Many of the best Russian scientists have simply left Russia for the West, in particular the United States. According to one survey of Russian scientists, four out of five would like to leave the country and work in the West. Another survey showed that of the best mathematicians in Russia—mathematics has long been a Russian specialty—four out of five have already left. Physicists and biologists are also in demand in the West. Many top Russian scientists are Jews, and many of these Russian Jewish scientists have emigrated to Israel. After emigration to Israel from the East became relatively easy, within three years the number of scientists living in Israel doubled.

Soon after the collapse of communism, the budget of the Russian Academy of Sciences was cut by 60% in terms of actual buying power. Some Western corporations or governments took advantage of the new circumstances by hiring whole Russian laboratories. For example, Sun Microsystems hired a computer-design laboratory of about 50 Russian scientists and technicians for salaries only slightly higher than their Russian incomes. The U.S. Department of Energy obtained the services of a group of more than 100 Russian fusion scientists at an average salary of about $900 a year. Similarly, AT&T's Bell Labs and Corning Inc. have each hired Russian groups of more than 100 scientists at similarly low salaries. The Russians hired by Bell Labs and Corning specialized in fiber optics and other glasses. Bell Labs liked their first deal so well that they soon hired another group of 27 Russian semiconductor researchers.

Another source of outside help to Russia has been direct Western aid, primarily fellowships from the Soros Foundation. Still, because many Russian scientists have not been able to emigrate or to obtain salaries or supplements from the West, many have turned to other professions such as banking or even manual labor to supplement their incomes.

Changes in Funding

Government support for science in the United States, especially for big science, has eroded greatly in the early 1990s, as it has to a somewhat lesser extent in other Western nations. The United States national laboratories, no longer fueled by cold-war concerns, have reduced their staffing levels. There are fewer jobs for scientists in colleges and universities than there are graduating scientists—sometimes hundreds of applicants apply for a single opening in a desirable institution. In 1992, there were physics job openings for fewer than 60% of the United States graduates in physics.

Industry has greatly reduced the number of jobs in pure science, and to some degree in applied science as well—the National Science Foundation reported that funding by industry for basic research peaked in 1989 at about $79 billion, and since then has dropped. Among the corporations discontinuing or downsizing research efforts have been such giants as AT&T, IBM, and Westinghouse. The decline would be even steeper in the United States, if it were not for an increase in research funding by foreign-owned industries. The pharmaceutical industry (not exactly in financial trouble—pharmaceutical profits are higher than profits in most other industries) laid off 27,000 people in 1993, part of a trend toward downsizing that affected all industries in the United States in the early 1990s. The College Placement Council, which includes 402 colleges as members, found that the number of job offers to recent graduates in aerospace, petroleum geology, and pharmaceuticals had decreased significantly between 1988 and 1993.

United States government spending on science measured in constant dollars fell 7% between 1988 and 1994. Furthermore, the aim of funding has changed from largely a defense-oriented set of goals to goals for improving the economy or personal health. In the universities, the percentage of funds spent on research that came directly from the federal government dropped consistently, falling from 72% in 1968, to 61% in 1988, to 56% in 1993. About 20% of the total is spent on basic research.

Despite the declines in funding and the shifts in emphasis, much of the funding for basic science in the United States still comes from government, chiefly through the National Science Foundation and the National Institutes of Health (NIH). In 1993, Harold Varmus (a prominent and distinguished scientist) became the NIH's new director. Varmus quickly moved to review the way that the NIH reviews grants. Among the changes instituted are quicker elimination of grant applications that are not likely to be approved, a change in the amount of financial documentation needed early in the review process, and better use of electronic networks to handle grant submissions. Such changes, and possibly others, are intended to make more money available for science by cutting the amount of money spent on writing grant proposals.

Just as in science education, research funding by wealthy individuals and foundations financed by such individuals has tended to fill in for the lack of funding by the government and industry. In the early 1990s, the largest contributions to environmental research in the United States were from an individual, Edward P. Bass, best known for his creation of and support for the Biosphere 2 experiment. Other notable contributors to science research include the late David Packard, the founder of Hewlett Packard, who funded ocean research through his Monterey Bay Aquarium Research Institute ($13 million through 1993) and the W. M. Keck Foundation of Los Angeles ($144 million for the Keck telescopes on Hawaii).

In 1993, the amount of money granted to basic scientific research in the United States, according to the National Science Foundation, could be broken down as follows:

	Federal Government	*Industry*	*Universities*	*Nonprofits*
Money granted	62.6%	17.8%	13.6%	6%
Money spent	12.6%	20.0%	57.9%	9.5%

Legal Position of Scientific Testimony

The early 1990s were marked by a U.S. Supreme Court decision that is expected to affect all scientific expert testimony for years to come. From 1975 to 1993, the U.S. Court System operated on the basis of two quite different rules for acceptability of expert testimony by scientists. The older standard (known as the Frye rule from a 1923 case concerning an early form of lie detection) required that a theory be generally accepted by the scientific community prior to allowing such a theory as the basis of court testimony. In practice, this meant that many courts deemed as admissible only that scientific information which had already been published in a refereed publication. Since 1975, there has been a competing standard, enacted by the U.S. Congress, which is known as the Federal Rules of Evidence. These rules have been interpreted to mean that testimony can be accepted from anyone identified as a scientist, such as a person holding a master's degree or higher academic degree in a scientific discipline.

Trial lawyers hoping to win settlements for medical incompetence or other forms of liability prefer the looser Federal Rules of Evidence. Industries and medical doctors that are often on the receiving end of lawsuits prefer the Frye rule. State courts have felt free to follow whichever rule suited them. Federal courts have been required by law to follow the Federal Rules of Evidence, but this requirement did not prevent a federal judge in California from rejecting expert testimony in a product liability case because the statistical method used for interpreting the data was not the standard one used by most scientists. In effect, this California court ignored the Federal Rules of Evidence.

In this case, the plaintiff was able to win a rehearing by the U.S. Supreme Court, whose decision in mid-1993 struck a middle ground. The new rule, announced for the case of Daubert v. Merrell Dow Pharmaceuticals, requires the trial judge to ascertain if the scientific theory proposed as evidence has been developed using true scientific method, meaning that questions such as the following must be answered affirmatively: Can the results be reproduced? Can the hypothesis be shown to be false by a suitable experiment? Has such an experiment been carried out? Is the amount of data sufficient to justify the conclusion? Is the error rate low enough to exclude different conclusions? These and other questions regarding the method used to obtain scientific testimony in a case must be answered by the judge before admitting the testimony. While this puts a great burden on the judge, most people who have studied the issues feel that the new criterion will be workable and an improvement over the confusion of the past.

Miscellaneous Trends

About 8% of doctoral degrees in science in the United States in the early 1990s were earned by women; about 1% were earned by African Americans. In physics, nearly half of the degrees granted by United States institutions were awarded to citizens of other nations.

In 1994, at least one scientist was fined for deliberately sabotaging another's research, although the motive in that case was not clear.

Concerns of Specific Disciplines

Many major trends in science tend to fall more heavily on one discipline than on science as a whole. Special concerns that affect particular branches of science will be discussed later in the book, in articles on "the state of" each discipline.

(Periodical References and Additional Reading: *New York Times* 1-14-92, p C1; *New York Times* 2-4-92, p C1; *New York Times* 2-21-92, p A1; *New York Times* 3-3-92, p A1; *New York Times* 3-15-92, p IV-3; *New York Times* 6-14-92, p D1; *New York Times* 6-30-92, p C1; *New York Times* 7-18-92, p IV-3; *New York Times* 8-13-92, p D1; *New York Times* 10-3-92, p IV-4; *New York Times* 10-20-92, p D4; *New York Times* 11-11-92, p A21; *New York Times* 12-7-92, p B12; *New York Times* 1-2-93, p A1; *New York Times* 6-30-93, p A12; *New York Times* 7-15-93, p A25; *New York Times* 11-171-93, p B7; *New York Times* 11-21-93, p IV-18; *New York Times* 2-20-94, p A1; *New York Times* 4-21-94, p D23; *New York Times* 7-10-94, p IV-1; *Science* 7-22-94, p 467; *Science* 8-26-94, p 1167; *New York Times* 8-31-94, p A16; *Scientific American* 9-94, p 72; *Science* 9-16-94, p 1648; *New York Times* 10-4-94, p A17; *Science* 12-9-94, p 1637)

FRAUD REMAINS AN ISSUE

Scientific misconduct continued to be an issue throughout the early 1990s. As competition has increased in science, not only between nations but also between individual scientists, the attraction of fraud has increased. In an earlier era, when the most that could be gained from successful scientific fraud was the knowledge of having put something over on fellow scientists, fakery was probably rare. With the strengthened link between career advancement and the production of new science came an increased possibility that the unscrupulous might use less-than-honest means to improve their fortunes. At least, that's how a lot of people perceived things at the start of the 1990s, resulting in investigations and the positioning of new watchdogs for the scientific community.

Various allegations of fraud in the 1980s continued to make news throughout the early 1990s. In 1991, there were about six seminars presented on the subject of scientific fraud, including three run by the NIH; this is just one measure of scientists' concern about the issue. Constant changes and recommendations for change in the mechanisms used to investigate and punish misconduct—as scientists prefer to call fraud—kept the issue alive into the mid-1990s. The *American Scientist* published a survey in 1993 reporting that 43% of students and 50% of faculty members knew of misconduct by colleagues, including 6-9% who knew of scientific misconduct by faculty members. The study, which obtained replies from 2,600 of 4,000 scientists and graduate students, was conducted by Judith P. Swazey of the Acadia Institute in Maine. Originally, the institute submitted their results to *Science*, but the paper was not accepted on the grounds that it was too long, dealt with perceptions of misconduct rather than evidence of misconduct, and was poorly written. *Science* has argued on its editorial pages that scientific misconduct is rare.

The Imanishi-Kari Affair

A notorious fraud allegation from 1986 continued to make headlines in the early 1990s, and was never completely resolved. This case ended with the alleged perpetrator still claiming innocence even after her principal sponsor had retracted the original paper and various groups had declared her guilty. Throughout the affair, controversy surrounded the U.S. National Institutes of Health's Office of Scientific Integrity and its successor, the Public Health Service Office of Research Integrity, which had become the main arbitrators of the problem. Along the way, a Nobel Laureate was sufficiently discredited by the affair that he had to resign a major post. To appreciate fully the extent and twists and turns of this tale, it is presented as a timetable.

TIMETABLE OF A FRAUD INVESTIGATION

1970	David Baltimore is one of two scientists who independently discover reverse transcriptase, the enzyme that is the key to genetic engineering.
1975	Baltimore receives a share of the Nobel Prize in physiology or medicine for his work on reverse transcriptase.
1983	About this time Baltimore, as director of the Whitehead Institute for Biomedical Research at the Massachusetts Institute of Technology (MIT), begins collaborating with various other biologists on a study of transgenic mice.
1985	
June 20-22	According to biologist Thereza Imanishi-Kari, she performs critical subcloning of genes from transgenic mice at this time while at the MIT Center for Cancer Research, leading to results reported in 1986. She files for an NIH grant to proceed with the work and the grant is awarded.
	Postdoctoral student Margot O'Toole begins working with, and disagreeing with, Imanishi-Kari at MIT.
1986	
April 25	A paper by David Weaver, Imanishi-Kari, Moema Reis, Baltimore, and two others ("Altered repertoire of endogenous immunoglobulin gene expression in transgenic mice containing a rearranged Mu heavy chain gene") is published in *Cell*. It claims foreign genes transplanted into a line of mice can indirectly change the antibodies produced by the mice's own genes.
May	O'Toole discovers that 17 pages of supporting data from Imanishi-Kari's work contradict claims in the April 25 paper, and reports this to Henry Wortis, her senior advisor at Tufts University, but Wortis finds no evidence of misconduct after a brief investigation with two other researchers.
June 6	O'Toole prepares a formal list of her objections to the *Cell* paper for an investigation into her charges by Herman Eisen of MIT. This investigation is inconclusive, finding minor errors, but no misconduct.
June 16	According to O'Toole, she and Imanishi-Kari tell Baltimore, Weaver, and

Eisen that some of the data in the paper are false. Baltimore says that does not call for retraction of the paper, or even an erratum.

O'Toole abandons her career in science, resigning from MIT.

1988

January	The NIH appoints a panel of three immunologists to review the dispute between O'Toole and Imanishi-Kari. Later, two of the panelists are accused of having ties to Baltimore.
April 11	Congressman Ted Weiss (D-NY) of the House Government Operations Subcommittee holds the first hearing on "scientific fraud and misconduct" to mention the Imanishi-Kari affair.
April 12	Congressman John Dingell (D-MI) of the Subcommittee on Oversight and Investigations of the Energy and Commerce Committee conducts a first hearing on the Imanishi-Kari matter.
June	The NIH appoints an expert panel to review Imanishi-Kari's research and O'Toole's charges.
August	Dingell asks the U.S. Secret Service to examine Imanishi-Kari's scientific notebooks for evidence of fraud.

1989

January 31	The NIH panel investigating the O'Toole charges finds "no evidence of fraud," but concludes that the research contains serious errors.
March	The NIH establishes an Office of Scientific Integrity (OSI) as a branch of the Public Health Service (PHS).
May 4	At a hearing of the Dingell subcommittee, Baltimore attacks the subcommittee for the way it is conducting the investigation, and defends Imanishi-Kari.
May	OSI begins operation. Suzanne Hadley is deputy director and will supervise the Imanishi-Kari matter.

1990

April 30	Baltimore is interviewed by OSI teams and says that incorrect data in Imanishi-Kari's notebooks are not "fraud" because the data was not published.
About this time	After four years of being unable to get a job in science, O'Toole is hired by the Genetics Institute.
May 14	U.S. Secret Service agents testify at the third Dingell hearing that notes purporting to be Imanishi-Kari's work of June 20-22, 1985, are falsified.
June 29	News that the NIH failed to renew funding for Imanishi-Kari as early as April 1990 is published in *Science*.
July 1	Baltimore accepts the office of president at Rockefeller University.
December	Federal judge Barbara Crabbe rules that OSI guidelines for misconduct were not drawn up in accordance with United States law, since they were not open to public comment.

1991

January	U.S. Health and Human Services (HHS) Secretary Louis Sullivan appoints an NIH committee to look into the operations of OSI.

March 20	A report from OSI concludes that Imanishi-Kari fabricated data, and calls O'Toole a hero. Baltimore and coauthors—except Imanishi-Kari—ask *Cell* to withdraw 1986 paper, which it agrees to do on March 22.
	A National Academy of Sciences (NAS) 25-member panel, chaired by Edward E. David, issues a draft report calling for an independent body to prepare standards for and to monitor scientific misconduct.
April	Bernadine Healy becomes director of NIH.
May 3	Baltimore's 14-page apology to O'Toole is revealed to the press.
May 15	Imanishi-Kari calls the whole affair a witch-hunt that arose from O'Toole's misunderstanding of her research notes.
June 13	The NIH Office of Scientific Integrity Review publishes the rules under which OSI operates for public comment.
June	A group of 143 scientists, many of them eminent immunologists, complain that OSI has denied basic rights to Imanishi-Kari.
July	Hadley, who has supervised the Imanishi-Kari investigation, is removed from the case by NIH director Healy.
August	Anthony Cerami, a professor of biochemistry who opposed Baltimore's appointment at Rockefeller University, leaves the University, taking his 30-person team to the Picowar Institute for Medical Research in Manhasset, New York.
September	PHS receives thousands of letters, including a joint letter from various university associations, attacking OSI rule-making procedures.
October	Nobel Prize winner Gerald M. Edelman, who led those forces at Rockefeller University opposed to hiring Baltimore, announces that he and his lab are moving to the Scripps Research Institute.
	David Rockefeller gives $20 million to Rockefeller University to indicate his satisfaction with Baltimore as its president.
November	The NIH advisory committee appointed by Secretary Sullivan in January urges OSI to tighten its definition of misconduct and to allow scientists under investigation to defend themselves.
December 2	Baltimore announces his November resignation as president of Rockefeller University, because of the "climate of unhappiness" created by the Imanishi-Kari investigation. He remains on the faculty as head of his own laboratory.
1992	
March	The U.S. Federal Bureau of Investigation (FBI) is brought in to look for leaks in the OSI and interviews Suzanne Hadley on March 11.
April 22	Another panel of the NAS chaired by Edward E. David Jr. states that while government has the right to investigate plagiarism, data fabrication, and falsification, it prefers creation of a new Scientific Integrity Advisory Board responsible to the scientific community and research institutions. The panel also comments on official PHS and National Science Foundation (NSF) definitions and calls for dropping language which labels as misconduct "other serious deviations from accepted research practices." The panel reports that only about 30 of

some 200 accusations of scientific misconduct reported to the government in 1989-91 resulted in actual confirmation of misconduct.

May 19 — MIT announces that Baltimore will return to work there in the spring of 1994, this time as a professor, rather than getting his old job back as director of the Whitehead Institute for Biomedical Research.

June — The NIH Office of Scientific Integrity is replaced with a new Office of Research Integrity (ORI) within HHS. The office becomes a part of the PHS.

July — United States Attorney Richard Bennet announces that the Department of Justice will not prosecute Imanishi-Kari as a criminal. Baltimore immediately announces his confidence in the results of the original paper, in effect reversing his retraction of the spring of 1991.

August 15 — Jules Hallum resigns as director of the new ORI, saying it is too big and too dominated by lawyers to be effective.

October 23 — *Science* reports that the PHS plans to set new rules, including a new definition for "scientific misconduct," for the ORI, and to establish specific criteria under which a scientist may be accused.

November — The ORI agrees to permit scientists to appeal adverse decisions to a board of five lawyers known as the Research Integrity Adjudications Panel. This panel was originally established by HHS in 1973 to resolve disputes over Medicare and Medicaid contracts and had reviewed scientific disbarment cases since 1987.

1993

January 4 — Most of the glass plates used by the U.S. Secret Service as evidence of fraud in the Imanishi-Kari case are found smashed as a result of various postal accidents while they were being shipped back and forth between prosecution and defense attorneys and expert witnesses.

January 29 — Donald E. Buzzelli, a senior scientist for the Office of Inspector General at the NSF, writes in *Science* to defend a definition of misconduct that includes the disputed phrase "other serious deviations from accepted research practices."

May 4 — The NIH announces that on May 10 it will reassign Walter Steward and Ned Feder, two of the principal attackers on Baltimore and Imanishi-Kari, to duties unconnected with fraud investigation, saying that the two investigators had begun to exceed their mandate.

April 15 — Two papers in *Cell* coauthored by Imanishi-Kari claim to support the original research for which she was accused of fraud.

June — The first attempt at an appeal to the Research Integrity Adjudications Panel quickly shows that the rules of law are different from the rules of the ORI.

November — In the year that it has been the appeals board for the ORI, the Adjudications Panel has sided with the scientists and rejected the ORI position in every case taken to it (six in all). Faced with this record, the ORI drops its case against Robert Gallo (accused originally of stealing the HIV virus from French researchers, although by 1993 the charges had been lowered to relatively minor misdeeds connected with the virus). The ORI notes that 13 of its 22 misconduct judgments were not appealed by the scientists involved. In three other cases the scientists first appealed, and then withdrew their appeal before the board acted.

1994

February

Early in the month, the NAS makes a joint statement with the U.S. National Academy of Engineering and the Institute of Medicine, recommending that all United States agencies adopt a common standard of scientific misconduct, but that research institutions adopt broader standards that go beyond any legal requirements. The organizations issue a written statement complaining that their proposed guidelines for misconduct have not been adopted. Officers of the NAS say that their organization should take over misconduct investigations.

May

Alan Price of the ORI announces that the office is prepared to drop the phrase "practices that seriously deviate from those that are commonly accepted within the scientific community" from its definition of scientific fraud. The new definition will emphasize deliberate misrepresentation of data or analysis.

June 20

A 12-member commission chaired by Kenneth Ryan of Harvard Medical School begins meeting at the start of a two-year program intended to make recommendations to Health and Human Services and to the U.S. Congress on how to enforce research integrity. Panel members indicate that they will go far beyond simply trying to "fix" the ORI, and will make recommendations for fundamental changes in the way scientific misconduct should be handled in the future.

November 25

The ORI issues a final report alleging 19 counts of misconduct by Imanishi-Kari, accompanied by a letter to Imanishi-Kari summarizing its conclusions: "Acts of deliberate misrepresentation, falsification and fabrication described, constitute a pattern of conduct extending over a four-year period that establishes a lack of integrity and honesty on your part." The report also concludes that she falsified data in her original grant application in 1985. HHS says it plans to bar Imanishi-Kari from U.S. funding for a period of 10 years. Imanishi-Kari announces that she will appeal. The report also includes new allegations that Moema Reis, a co-author of the original *Cell* paper, also fabricated data. (Reis took a sick leave from MIT shortly after the *Cell* paper was published, and has never returned.)

December

Tufts University reports that Imanishi-Kari is "conclud[ing] her service as a Tufts faculty member," although she will remain a research associate at Tufts until July 1995. Should she win her appeal, she can reapply for tenure.

When called to the attention of investigative bodies, most accusations of misconduct are rejected after study. Of the first 110 investigations completed by OSI, only 16 cases, or one-seventh of the total, were ruled to be fraud. Furthermore, not all OSI investigations concern fraud at all. About 40% of the investigations so far have focused on other forms of misconduct, such as corruption of the peer review system, or theft of intellectual property.

(Periodical References and Additional Reading: *New York Times* 3-14-92, p 11; *New York Times* 4-23-92, p A20; *New York Times* 5-19-92, p C5; *New York Times* 5-5-93, p A21; *New York Times* 11-12-93, p A22; *Science* 6-18-93, pp. 1714 & 1715; *Science* 1-7-94, p 20; *New York Times* 2-6-94, p I-23; *Science* 5-27-94, p 1243; *Science* 6-24-94, p 1841; *New York Times* 11-27-94, p I-37; *Science* 12-2-94, p 1468; *Science* 12-16-94, p 1811)

THE END OF BIG SCIENCE IN THE UNITED STATES?

As the bulk of scientific funding comes increasingly from governments rather than other institutions, the critical issue becomes what kinds of scientific endeavors should be funded. The issue of "big science" or "small science," problematic since the success of the Manhattan Project during World War II, has become increasingly important to scientists in search of funds. The 1990s saw a continuation of the debate regarding the relative importance of big and little science to scientific enterprise, and to the world in general.

"Big science" is experimental work that requires enormously expensive high-technology equipment, tended by scores or perhaps hundreds of scientists. The classic model for big science is the large particle accelerator, such as the once-planned $8.4 billion Superconducting Supercollider (SSC). A $40 billion space station is certainly "big," although some would say that the planned U.S. space station is not science at all, just a display of technology. Large astronomical observatories on the ground and in space are probably "big" since they use expensive telescopes, even though most of the actual experiments utilizing such equipment are carried out by individuals or by small groups. The Human Genome Project (*see* "Genome Project UPDATE," p 366) is considered big science for the opposite reason—the required equipment is not especially big or expensive, but the number of experimenters collaborating is enormous and therefore the cost of the project is great. There are also borderline cases. Large-scale programs of ocean drilling are almost "big," while superdeep boreholes are probably "big."

Such categorizations have become important because the 1980s and the early 1990s have been rife with complaints by proponents of small science that big science is taking most of the money without producing much in the way of return. At the same time, proponents of individual projects are claiming that vast sums must be spent on their particular project because it is the next important step or the next logical step in science.

Small science advocates complain that the evaluation procedures for large, expensive projects are flawed and that once the projects are initially funded, it is almost impossible to get rid of them. Evaluation of small projects is traditionally conducted by peers—people whose own projects in the same field are fighting for the same funds in about the same amounts. The peers are thought to be much more stringent when evaluating a project that is directly competing with their own projects for scarce funds. Therefore, it is expected that peer-approved projects are particularly good projects. Furthermore, the peers are scientists who know the field in detail, making it difficult to submit half-baked or poorly designed concepts. On the other hand, very expensive projects are evaluated by politicians, people with no vested interest in or depth of knowledge about the proposal seeking funding.

Once a large project is approved, it has its own momentum. No one wants to waste the first few billion dollars allocated because the idea is not working out. Small projects run through their smaller investments completely in a few years. If the concept for a small project was no good, it no longer matters financially because all the money appropriated has already been spent and no one is going to spend any more money on what proved to be a dumb idea.

Scientists, of course, are concerned about the economic benefits big-science projects have for their own workforce. Loss of the SSC, for example, resulted in more than a thousand scientists losing their jobs.

Let's look at the record. Here is a summary of what seem at this time to be the greatest accomplishments in science during the 1980s and early 1990s categorized as "big" and "small."

"BIG"	*"SMALL"*
1980	
Voyager reaches Saturn	Quantum Hall effect discovered
	K/T iridium layer found
	Inflationary model of universe developed
	Scanning tunneling microscope invented
	Genetic engineering made possible
1981	
	Catalytic action of RNA discovered
	Quadratic-sieve method of factoring developed
1982	
Soviets soft-land space probes on Venus	Bell's theorem confirmed by experiment
Artificial heart employed	
1983	
IRAS finds evidence of planets forming around stars	Dinosaur fossils near found near Arctic Circle
CERN (*Centre Européen de Recherche Nucléaire*) finds W and Z particles	Development of each cell of *Caenorhabditis elegans* mapped
	Polymerase chain reaction invented
	Gene markers for diseases found
1984	
	Apes and chimps found to have nearly identical genes
	Soft-bodied fossil deposits found at Chenjiang, China
1985	
	Buckminsterfullerene discovered
1986	
Comet Halley investigated	Gene for muscular dystrophy discovered
	High-temperature superconductors discovered
1987	
Supernova 1987A detected by optical and neutrino telescopes	Mitochondrial genes used to find African "Eve"
	Gravitational lenses observed

"BIG"	*"SMALL"*
1988	
Voyager reaches Neptune	Whale with legs found
1989	
1990	
COBE shows background radiation is smooth	Gene for neurofibromatosis discovered
	Gene for maleness located
Hubble Space Telescope is launched, but found to be flawed	Impact crater at Chicxulub, Mexico, that was probable cause of K/T mass extinction, found
1991	
Magellan orbits Venus	Gene therapy begins
Radar studies made of Mercury	
1992	
	First fossil of *Ardipithecus ramidus* found
	Possible planet found around neutron star
1993	
Hubble Space Telescope successfully repaired	Andrew Wiles proves Fermat's last theorem
1994	
Elements 110 and 111 created in laboratory	Chauvet Cave found in France

Recent Developments in the United States

Even before the economy-minded Republican majority took control of the U.S. Congress at the start of 1995, congressional approval of funds for big-science projects was dwindling rapidly. Between the late 1960s—the glory days of the race to put humans on the moon—and the early 1990s, congressional appropriations for science research fell from 5.2% to 1.7% of the budget. Congress, for example, voted to kill funds for the SSC in October 1993. Congress had also reduced the scope of the *Freedom* space station that June, and came within one vote of halting that project entirely. The head of the House appropriations subcommittee that controls spending on space research called for taking grants away from "curiosity-driven research" and spending the money on "strategic research" that would have clear-cut economic benefits.

This retreat does not mean the end of big science in the United States, however. New projects continue to be planned, and sometimes executed.

A new proposal from the National Science Foundation (NSF) would replace relatively small science projects at the South Pole with a new $250-million polar station that would ensure that all subsequent research at the pole would become "big." The station would be built over an eight-year period. Although most polar research up to now has consisted of various small projects, the total NSF budget for the South Pole has been running at nearly $200 million a year, and the cost of the new construction would be added to that—although one NSF official said that the need was so great that he would prefer to see regular funding cut, if necessary, rather than cut funding for the new station.

One of the primary objectives of the new polar station is to develop a site for new telescopes that would take advantage of both the extremely low humidity at the South

Pole, as well as a location that would give an unparalleled view of the sky above the Southern Hemisphere. For different reasons, the South Pole is already the site of a $1 million neutrino telescope, and a big-science $50 million neutrino telescope has been proposed for the polar station as well, although that project is far from being even a detailed plan (*see* "Neutrino Astronomy Comes of Age," p 136).

If everything is approved, the new South Pole facility would become the fourth project segregated financially by the NSF to protect big science from criticism by small science claims that big science is taking all the money. The other three segregated projects are: a planned $150 million millimeter-range radio telescope; the twin eight-meter optical telescopes called GEMINI (half the cost of which—$88 million—is to be put in by NSF with the rest to come from outside the United States); and the laser interferometry gravity wave observatory (LIGO). LIGO was originally expected to require $250 million in order to build twin facilities in Livingstone, Louisiana, and Hanford, Washington. Revised estimates at the end of 1994, however, reached $297 million to build it and $68 million to operate it for four years.

The NSF is not the only U.S. government agency struggling with big-science issues. The National Oceanic and Atmospheric Administration (NOAA) wants to spend $1.9 billion on new research vessels; in turn, some members of Congress want to eliminate NOAA altogether, folding its more popular programs into other agencies and dropping the rest. NOAA ships now survey fishery resources, police marine mammal protection, research climate trends, and map the ocean bottom. All these functions are imperiled by the deteriorating ships that are currently used, which in 1995 had an average age exceeding 30 years. Five-sixths of the fleet was built in 1970 or earlier. One-fourth of the fleet of 24 was not operational in 1994.

At the National Aeronautics and Space Agency (NASA) the ax has fallen on its big-science projects, such as a mission to Pluto and a planned $250 million orbiting telescope called the *Far Ultraviolet Spectroscopic Explorer* (*FUSE*). FUSE had been approved in 1989 for launch in 2000, but was scrapped in 1994. Instead, NASA has promised scientists a host of smaller missions that will accomplish many of the same goals.

The U.S. Department of Energy (DOE) is also having trouble obtaining funding for big-science projects. At the end of 1994, it was not clear whether the $2.9 billion Advanced Neutron Source, a powerful nuclear reactor, would be in the next administration budget, much less approved by a Republican-dominated Congress bent on saving money. Many DOE big-science projects, including the Intense Pulsed Neutron Source, the Advanced Light Source, as well as the Continuous Electron Beam Accelerator Facility and the Advanced Photon Source, both scheduled for completion in 1995, do not have enough financial support for full-time operation. A last-minute compromise kept the Tokamak Physics Experiment (*see* "Fusion Progress At Last," p 639) and the International Thermonuclear Experimental Reactor in the budget at the end of 1994, although the scope of both was reduced and the Tokamak eventually shut down.

Meanwhile in Europe

The European answer to the SSC, the Large Hadron Collider (LHC), was approved in a squeaker by CERN on December 16, 1994. The anticipated cost is $2 billion, making the particle accelerator definitely big science. Plans call for construction in an existing 17 mi (27 km) tunnel in France and Switzerland, the site of the present CERN large accelerators. If all goes well, the project should be fully operational in 2008.

(Periodical References and Additional Reading: *New York Times* 6-28-93, p A13; *New York Times* 2-20-94, p A1; *New York Times* 6-29-94, p A17; *Science* 6-24-94, p 1836; *Science* 7-8-94, p 175-176; *Science* 9-25-94, p 1801; *Science* 10-21-94, p 356; *Science* 11-11-94, p 963; *Science* 11-18-94, p 1147; *Science* 11-25-94, p 1314; *Science* 6-6-95, p 26)

FERMAT'S "LAST THEOREM" PROVED

Puzzle buffs and mathematics amateurs know a few unproved conjectures from pure mathematics, but only one has become so much a part of the general knowledge of educated citizens that it is regularly referred to in graffiti, cartoons, and other popular media. The reference is generally to a remark made along with its original statement: "I have discovered a truly remarkable proof, which this margin is too small to contain." It is generally known as Fermat's last theorem. In 1994, after a near success in 1993 that made headlines all over the world, Princeton mathematician Andrew J. Wiles discovered a truly marvelous proof of Fermat's theorem. The margin would have been far too small to contain Wiles' proof, since the proof runs more than 150 pages.

The theorem itself, unlike many important results of more recent mathematics, is easily stated: There are no counting numbers (positive integers) a, b, c, and n for which $c^n = a^n + b^n$ is true if n is greater than 2.

Pierre Fermat, who first stated the theorem sometime in the seventeenth century, probably did not possess the marvelous proof he claimed, although that conjecture is likely to be even harder to prove than his "last theorem." Wiles' proof is based upon important results in modern mathematics that were not known in Fermat's time. The methods available to Fermat had been tried by great mathematicians and unschooled amateurs alike for about 300 years with only small progress toward a proof. Mathematicians who have studied Fermat's other work believe that he probably thought briefly that he had a proof, but soon realized that he was in error, and therefore did not communicate his result or his theorem to any of his many correspondents. Yet there remains the small possibility that some relatively simple proof exists. For one thing, Wiles' proof shows that the conjecture is correct.

Wiles first announced his results on June 23, 1993, and at that time it appeared that he probably had achieved the goal of showing that $c^n = a^n + b^n$ is impossible for every whole number n greater than 2 if a, b, and c are also positive integers. But when Wiles began to put the lengthy proof into shape for publication, he found a gap in the chain of reasoning, which he announced in December 1993. The gap came in the use of a mathematical bookkeeping system device recently invented by Russian mathematician Viktor A. Kolyvagin, which requires the construction of an entity called an Euler system. The Euler system had not been correctly constructed.

Although the gap, near the end of the proof, did not seem to be necessarily fatal, many mathematicians told reporters that they thought it might take years either to resolve the existing proof or to find a way around the problem by using a different approach. Wiles got some help, however, from a former student, Richard L. Taylor, who suggested a bridge that was similar to an approach Wiles had tried much earlier and discarded. Wiles used Taylor's idea and solved the problem. The final proof actually consists of two papers, one by Wiles alone called "Modular Elliptic Curves and Fermat's Last Theorem," and a second paper by Wiles and Taylor covering a new subtheorem needed to fill the gap. The revised theorem is shorter and somewhat more direct than the previous one with the flawed Euler system.

The course of a mathematical proof is seldom absolutely direct, however, and Wiles' proof is typical in this regard. When an infinity of possibilities is involved, as in the Fermat theorem, mathematicians from Euclid's day to the present have taken refuge in the "law of the excluded middle." This axiom states that if there are only two possibilities and one of them leads to a contradiction, then the other must be true. Wiles knew that earlier mathematicians had established that assuming Fermat's theorem to be false implies the existence of a particular curve. A famous conjecture, known as the Taniyama-Shimura conjecture, implies among other things that such a curve cannot exist. Thus, a proof that the Taniyama-Shimura conjecture is true—or even just the part that says the curve is impossible—implies that assuming Fermat's theorem to be false leads to an unacceptable contradiction. Therefore, the Fermat theorem is true if the Taniyama-Shimura conjecture is true. As Sherlock Holmes told Watson: When you have eliminated the impossible, what remains must be the answer. Wiles eliminated the impossible curve, so Fermat must have been right.

For the future course of mathematics, the proof that Fermat's last theorem is true is not very important. The truth of the Taniyama-Shimura conjecture, however, and the methods used in Wiles' proof of part of it, are very much at the heart of modern mathematics.

Background

Pierre Fermat [French: 1601-1665], a lawyer and civil servant who lived in Toulouse in southern France, was among the greatest of mathematicians, although technically an amateur. His originality went far beyond the "last theorem." Fermat independently invented analytic geometry and the basic elements of differential and integral calculus, although these ideas were not published in his lifetime. Fermat's correspondence with Blaise Pascal is considered the beginning of organized probability theory. In the theory of whole numbers, sometimes called higher arithmetic, Fermat made many discoveries that remain at the heart of number theory today.

The "last theorem" is a conjecture from number theory. Its roots are in one of the earliest known results of pure mathematics: the existence of whole-number triples that can be the lengths of the sides of a right triangle. These triples are closely related to another very early result of mathematics—the Pythagorean theorem.

In the earliest known civilizations—Egypt, Mesopotamia, and China—mathematicians used one form or another of what we call today the Pythagorean theorem: In a right triangle, the square of the side opposite the right angle (the longest of the three sides) is the same number as the sum of the squares of the other two sides. Traditionally, the three sides are labeled a, b, and c, with c always the longest, so the theorem can be stated as the equation $c^2 = a^2 + b^2$. Ancient

Egyptians used the whole-number solution to this equation $a = 3$, $b = 4$, and $c = 5$ to form right triangles of rope that they used in surveying and probably in building the pyramids.

The Egyptians probably knew that $5^2 = 3^2 + 4^2$ ($25 = 9 + 16$) and perhaps other small-number solutions, such as $a = 5$, $b = 12$, and $c = 13$. Scholars in Mesopotamia about 1500 B.C. listed on a clay tablet 38 whole-number solutions to $c^2 = a^2 + b^2$, including $a = 119$, $b = 120$, and $c = 169$ and even larger examples. All such solutions to $c^2 = a^2 + b^2$ are known today as Pythagorean triples.

Given the antiquity of Pythagorean triples, it may seem surprising that it was not until the seventeenth century that a mathematician began to contemplate what today seems like the obvious generalization, $c^n = a^n + b^n$, where n is some specified whole number greater than 2. On the other hand, the modern notation for variables and exponents was not commonly in use until the seventeenth century, so before that time it was less obvious how to generalize Pythagorean triples. Fermat got the idea for the "last theorem" from a somewhat different source. He had been reading the first modern edition of the *Arithmetica* of Diophantus, published in Greek with a Latin translation in 1621. One problem given and solved by Diophantus was to find two small squares with sides of rational length whose area is the same as that of a given square with integral sides. Next to that problem in his copy of the book Fermat added in the margin (in Latin), "In contrast, it is impossible to divide a cube into two cubes, or a fourth power into two fourth powers, or in general any power beyond the square into two powers of the same degree. I have discovered a truly remarkable proof, which this margin is too small to contain." Fermat's copy of *Arithmetica* was filled with similar generalizations, often including a sketch of the proof.

Fermat's son published the annotated edition in 1670, making the notes generally available some five years after his father's death. Within a few years one of Fermat's correspondents who knew Fermat's methods was able to use them to show that the conjecture was correct for $n = 4$; that is, it is impossible to divide a fourth power into two fourth powers, or, equivalently, there are no integral solutions for $c^4 = a^4 + b^4$. A hundred years after the first publication of Fermat's note, Leonhard Euler offered new proofs for $n = 4$ and proved the case $n = 3$ for the first time, using methods that also were similar to those of Fermat. More than 50 years later the case $n = 5$ was added. About that time Sophie Germain [French: 1776-1831] give a general theorem that enabled other nineteenth-century mathematicians to prove many special cases. But no one before Wiles could prove that the Fermat theorem is true for the infinity of whole-number exponents greater than 2.

(Periodical References and Additional Reading: *Discover* 1-94, p 61; *Science* 11-4-95, p 725; *Science News* 11-5-94, p 295; *Scientific American* 2-95, p 16)

Anthropology and Archaeology

THE STATE OF PALEOANTHROPOLOGY

If, as Alexander Pope has stated, "the proper study of mankind is man," then anthropology is the most proper of sciences. Yet in some ways, it seems that anthropology is the least proper science: it is not clear exactly what it is, and anthropologists are famous for feuds that exceed those of any other branch of science.

Because what we call anthropology is really a collection of barely related disciplines, it is difficult to find a completely accurate name for the topic of the following section of this book—that is, the anthropology part of the chapter "Anthropology and Archaeology." Strictly speaking, "physical anthropology" is any study of the human body in general, including studies of the body dimensions or other characteristics of people living in particular cultural groups today. On the other hand, there is another discipline, which might be considered a *social* science, called "cultural anthropology" which is primarily concerned with how different groups of people live, and includes study of customs and language. The section that follows is not really about either of these topics, although it is closer to physical anthropology. More frequently these days, people studying the topics which will be included here call themselves paleoanthropologists. The prefix *paleo-* is used to mean "ancient," as in *paleolithic*, a formal name for the Old (Ancient) Stone Age. A strict definition of paleoanthropology, however, is "the study of ancestors of *Homo sapiens*." Nevertheless, many "paleoanthropologists" today find themselves dealing with early representatives of our own species.

The section that follows is almost entirely about the physical remains of the ancestors of humans, which is truly physical paleoanthropology; or about very early humans and how they lived, which might be called cultural paleoanthropology. In some cases, the archaeology or paleoarcheology concerned is also discussed. A more specific description would be that "anthropology" in this book is largely evolutionary physical anthropology, as well as the archaeology of the period before the Agricultural Revolution. In this latter context, we discuss the remarkable new cave art found in the early 1990s (*see* "Two New Caves," p 52).

Background

Understanding of the various relatives of modern human beings and how they are interconnected is continually changing. Currently, there is a developing consensus among paleoanthropologists regarding the details of who is related to whom among the earliest ancestors of humans. Surprisingly, the most controversy surrounds later periods (*see* "The Species Problem," p 35).

Taxonomists often use the article *the* to indicate that there is a single species extant, while referring in the plural to members of a genus that contains more than one species. Among rodents, for example, they speak of the capybara (one species) and mice (very many species of the mouse tribe). Following this convention for primates means that one refers to the orangutan (one species) and to chimpanzees and gibbons (more than one species). Since there is only a single species of human today, it is proper to speak of "the human," rather than "humans," although the phrase falls oddly on the ear. In the following discussion, however, use of the taxonomists' convention helps keep the record straight, and also helps in viewing the human as a species.

The human is one species in an order of mammals called **primates** (*see* "Very Early Primates," p 29). Primates more closely related to the human are called **hominoids**, and include gibbons; the orangutan; the gorilla (or gorillas—evidence from mitochondrial DNA suggests that the West African lowland gorilla is a separate species from the eastern population); chimpanzees (perhaps two full-size species on the basis of DNA, as well as the clearly differentiated pygmy chimpanzee); and the human, and their closest ancestors. Until recently, our relatives that are not gibbons or great apes (the orangutan, the gorilla, and chimpanzees) were classified along with us as **hominids**, while the great apes were classified as **pongids**. Evidence from DNA studies and other recent work suggests that the gorilla and chimpanzees are more closely related to the human than either gibbons or the orangutan. Currently, many anthropologists identify three families: for living species, these are Hylobatidae for gibbons, Pongidae for the orangutan, and Hominidae for the human, the gorilla, and chimpanzees. Thus, it is necessary to specify "the hominid great apes" to mean just the gorilla and chimpanzees.

Although there is only a single species of human alive today, in the past there are thought to have been at least two others. In the old classification, the Hominidae, or hominids, had two main branches over the course of time: australopithecines and humans. In the new system there would either be three branches, to account for the gorilla and chimpanzees, or four branches—humans, australopithecines (including *Ardipithecus*), hominid great apes, and a genus called *Paranthropus* that is lumped with the australopithecines in the three-branch scheme.

The East Side Story

The origin of the hominid line, according to no less of an authority than Charles Darwin, was in Africa (although *see* "Out of Africa, but When?" p 47). Darwin's basis for this hypothesis was the close resemblance between human anatomy and that of chimpanzees and the gorilla. Today, DNA studies have shown that the relationship is even closer than Darwin knew. From the genetic point of view, the pygmy chimpanzee and the human are essentially the same animal. The slight differences in the genes suggest that humans and chimps diverged only a short time ago as evolutionary periods are counted—variously estimated between 15 and 3 million years ago, with the most popular estimate being a little less than 8 million years ago.

In 1982, Yves Coppens of the College of France in Paris related this apparent time of divergence to big geological events in Africa. According to Coppens, the Rift Valley of Africa formed about 8 million years ago. At the same time, the assemblage of plants and animals that constitute the Ethiopian savanna also formed to the east of the Rift, mainly as a result of dry conditions caused by the mountains along the western rim of the Rift Valley. Among that assemblage were the australopithecines, which eventually evolved into humans. In Coppens' theory, the common ancestor of both the australopithecines and the chimpanzees inhabited all of Africa. When the Rift Valley formed, it geographically separated that common ancestor into two populations. The population on the drier eastern side of the Valley evolved into australopithecines, while the population in the wetter forests of the west side became chimpanzees. Coppens has somewhat whimsically named his theory "the east side story."

According to standard geological theory, the rifting in Africa started much earlier than 8 million years ago, near the beginning of the Miocene epoch, which itself started about 23.5 million years ago. But the rifting process, caused as part of the African tectonic plate tried to separate from the rest of the African plate, continued throughout the Miocene epoch, which did not end until 5.2 million years ago. By then the Rift was completely formed.

Geologists also agree that a major effect of the rifting has been the drying out of the part of Africa to the east of the Rift. Thus, although Coppens' east side theory may overemphasize the suddenness of the Rifting event, the broad outlines of the idea are a plausible explanation of the origin of the hominids.

THE HUMAN AND ANCESTORS

The first column below gives scientific and common names (with *man* used only because of traditional terminology; modern paleoanthropologists shun the use of *man* to mean "human"), along with the time and place in which the animal is known to have flourished. The abbreviation *mya* means "million years ago."

Purgatorius c. 60 mya Africa	The earliest known primate, a toothy creature somewhat like a lemur of today, but with a longer snout (but *see* "Very Early Primates," p 29).
Catopithecus browni c. 40 mya Egypt	The earliest known representative of the higher primates, which include the Old World monkeys, apes, and humans.
Aegyptopithecus c. 30 mya Egypt	This monkey-like creature, about the size of a large domesticated cat, may be the earliest known ancestor of the hominoids. First fossils found in 1965.
Proconsul: Three known species c. 20 mya East Africa	Oddly named after Consul, a popular chimpanzee in the London Zoo (*Proconsul* means "before Consul"), these apes, the first major discovery of Louis and Mary Leakey (in 1948), are generally recognized as the ancestors of all hominoids. Lacking a tail, like modern apes, *Proconsul* had a larger brain than monkeys of comparable size.
Afropithecus c. 18 mya Kenya	In some respects, *Afropithecus*, discovered in the mid 1980s, seems like a better candidate for an ancestor of hominoids than does *Proconsul*.

Kenyapithecus c.14 to 11 mya Kenya	A large ape that was somewhat more "advanced" than either *Afropithecus* or *Proconsul*. The first hominoid to have teeth covered with thick enamel.
Otavipithecus namibiensis c. 13 mya Namibia	The first hominoid known from southwestern Africa.
Dryopithecus c. 12 to 10 mya France, Spain, and Hungary	A small ape not much different from *Proconsul*, but identified in 1992 as the common ancestor of all hominids, including the hominid great apes, the australopithecines, and *Homo*.
Ouranopithecus (also known as **Graecopithecus**) c. 12 to 8 mya Greece	Thought by some to be the most promising candidate among the Miocene apes to be a direct ancestor of the hominids.
Sivapithecus c. 10 mya India, Pakistan, and Turkey	Once believed to be a direct ancestor of modern humans, *Sivapithecus* ("Shiva's ape," from its discovery in Hindu India) is now considered to be a close relative of the orangutan.
Gorilla gorilla, gorilla c. 9 mya to present Equatorial Africa	Three subspecies exist today, although research by Maryellen Ruvolo of Harvard University, reported in the September 13, 1994, *Proceedings of the National Academy of Sciences*, suggests that the West African group is actually a separate species. The gorilla is thought to have split from the line that leads to modern humans after the orangutan and before the chimpanzee.
Pan—two, possibly three species: **P. troglodytes**, common chimpanzee **P. paniscus**, pygmy chimpanzee, or bonobo c. 7 mya to present Sub-Saharan Africa	Several studies of proteins and DNA suggest that the chimpanzee, especially *P. paniscus*, is the closest living relative of humans. Phillip Morin and coworkers from the University of California, Davis, utilized chimpanzee hair to perform DNA studies which imply that *P. troglodytes* is actually two different species, found in different parts of Africa. Traditionally, the West African branch has been classed as the subspecies *P. troglodytes verus*.
Ardipithecus ramidus c. 4.4 mya Ethiopia	Discovered in 1994 at Aramis, Ethiopia, this woodland-dweller is one candidate for the direct ancestor of *A. afarensis*, and is very close to the common ancestor of the hominid great apes and the human line. (*See also* "New Root for Human Family Tree," p 32.)
Australopithecus anamensis c. 4.2 to 3.9 mya Northern Kenya (Lake Turkana)	This group is represented by 21 fossils found by Maeve Leakey and Alan Walker in 1994 and announced in 1995. *A. anamensis* seems to have walked on two legs like later hominids.
Australopithecus afarensis "Lucy" and "The First Family" c. 4 mya to 3 mya Ethiopia and East Africa	*A. afarensis* is thought to be on the direct line to humans. In addition to the famous skeleton of "Lucy" and the collection known as the First Family—all from Ethiopia and dating from about 3.3 mya—a trail of footprints, left in volcanic ash at Laetoli, Tanzania, and discovered in 1978, demonstrated that *A. afarensis* walked upright, unlike any of the great apes. In the January 1994 *Geology* geochronologist Robert C. Walter announced that his argon-argon redating of some of the famous australopithecine fossils shows that Lucy lived about 3.18 million years ago, and the First Family was slightly older, having lived about 3.2 million years ago.

Australopithecus
africanus
"The Taung Child"
c. 3 mya to 2.5 mya
South Africa

A. africanus was the first of the genus to be discovered, in 1924, although very few anthropologists accepted the species until after World War II. Like all australopithecines, except possibly *Ardipithecus ramidus*, *A. africanus* walked upright and, while large brained even for a primate, had a much smaller brain than that of members of *Homo* of the same general size. *A. africanus* is also thought to be on the direct line to humans.

Paranthropus or the robust line of **Australopithecus**
c. 2.5 mya to 1 mya
Eastern and southern Africa

Recently, there has been a tendency to reclassify several early species as *Paranthropus*, because they appear to be a separate branch of the human family that has become extinct without issue. Previously, the most common classification was *Australopithecus*, in recognition of their apparent descent (along with humans) from *A. afarensis* and probably *A. africanus*. Among the several species that have been recognized by one or more authorities are the South African *P. robustus* and possibly a *P. crassidens*; the East African *P. bosei* (sometimes known as *Zinjanthropus* or as "Nutcracker Man"); fragmentary remains from Ethiopia classed as *P. aethopicus*; and the earliest known, from Kenya, *P. walkeri* or perhaps also *P. aethopicus*.

Homo habilis
"Handy Man"
c. 2.4 mya
East Africa

Earliest known member of our own genus, *H. habilis* gets his nickname from tools found at sites that are both physically and chronologically close to sites where fossil *H. habilis* has been found. Recently, anthropologists have argued that *H. habilis* was a scavenger instead of a hunter, as had previously been believed.

Homo spp.
c. 2.4 to 1.6 mya
East Africa

Larger fossils found at sites with *H. habilis* are sometimes assumed to be a different species, which has been christened *H. rudolfensis*, although most anthropologists do not recognize this and believe instead that male *H. habilis* individuals were much larger than females (*see* "New Root for the Human Family Tree," p 32). A different *Homo* species is more widely accepted, based on a skull and a separate nearly complete skeleton from different sides of Lake Turkana in Kenya. The Lake Turkana specimens have been classified as *H. ergaster*, although a few anthropologists think that they are early representatives of the African branch of *H. erectus*. Previously, the fossils were classified as *H. habilis*.

Homo erectus
"Java Ape Man"
"Heidelberg Man"
"Peking Man"
c. 1.5 mya to .2 mya
Africa, Asia, Europe

Conservative anthropologists continue to lump all human fossils after *H. habilis* and before *H. sapiens*—if the Neandertals are counted as a subspecies of *H. sapiens*—into *H. erectus*. A new trend is to put fossils from Asia into one species, with different subspecies from Java and from China. African fossils from about 1.2 mya are considered a separate species, which may be the same as the species found in Europe. Since the first of the African-European species was "Heidelberg Man," found in 1908, the other species is sometimes labeled *H. heidelbergensis*. More commonly, however, the African fossils are given priority, and their name, *H. ergaster*, is used for the non-Asian group. (*See also* "The Species Problem," p 35.)

Homo sapiens neanderthalensis or *Homo neanderthalensis* "Neandertal Man" (previously "Neanderthal Man") c. 200,000 years to 35,000 years ago Europe, the Near East, Africa

The species problem is particularly acute for the Neandertals, and some experts, such as Erik Trinkaus, have almost abandoned the argument, saying that it is simply a technical problem. Everyone agrees that the Neandertals were anatomically and culturally distinct hominids who in some places lived side by side with archaic *H. sapiens*, but who disappeared abruptly about the same time as modern *H. sapiens* arrive in Europe.

Archaic *Homo sapiens* c. 360,000 years to 40,000 years ago Africa and the Near East at first; Asia and Australia somewhat later

Fragmentary fossils from the southern tip of Africa appear to be from *H. sapiens*, and are dated at 120,000 years in the past. Some think archaic *H. sapiens* appeared as early as 350,000 years ago in China (*see* "Out of Africa, But When?" p 47). The main reason to call these humans "archaic" is that they precede the well-known Cro Magnon fossils from Europe, and have some primitive features.

Homo sapiens "Cro Magnon Man" c. 36,000 years ago to the present Worldwide

Early Europeans who resemble modern people were first found at a site known as Cro Magnon. These humans completely displaced the Neandertals in Europe. Other parts of the Old World had already been settled by *H. sapiens*. Some anthropologists believe that *H. sapiens* came to the New World about the same time as the Cro Magnons came to Europe, but the more traditional view is that *H. sapiens* first came to the New World about 11,000 years ago.

(Periodical References and Additional Reading: *Science* 9-25-92, pp 1864 & 1929; *Science* 5-14-93, p 892; *Science News* 1-15-94, p 46; *Scientific American* 5-94, p 88; *New York Times* 5-17-94, p C1; *Science* 8-26-94, p 1173; *Science* 9-16-94, p 1661; *Nature* 1-5-95, p 15)

TIMETABLE OF PHYSICAL ANTHROPOLOGY TO 1992

1776   *De genera human variegate* ("On the natural varieties of mankind") by anthropologist Johann Friederich Blumenbach [German: 1752–1840] separates humans into five different races: Caucasian, Mongolian, American Indian, Malayan, and Ethiopian.

1790   Friederich Blumenbach's *Collectionis suae craniorum diversarum* gives descriptions of 60 human skulls.

1805   Friederich Blumenbach founds the science of anthropology with his *Handbuch der vergleichen Anatomie* ("Handbook of comparative anatomy").

1856   A skeleton found in a cave in the Neander valley near Düsseldorf (Germany) is the first known remains of what anthropologists now call a Neandertal; the Neandertals are either a vanished subspecies of *Homo sapiens* or a separate species, *H. neanderthalensis*.

1860   Edouard-Armand-Isidore Hippolyte Lartet [French: 1801–1871] about this time discovers a mammoth tooth engraved with a sketch of a mammoth, evidence that the engraver lived in a time when mammoths were extant.

1868   Workmen building a road in France discover five skeletons in a cave named Cro-Magnon. The skeletons appear to be about 35,000 years old, and the location gives rise to the common name Cro Magnon for *Homo sapiens* of the Old Stone Age.

1886   Additional Neandertal skeletons are discovered.

1894   Marie-Eugène François Thomas Dubois [Dutch: 1858–1940] publishes his discovery of fossil remains in Java of a precursor of *Homo sapiens*. Called *Pithecanthropus erectus* by Dubois, the species is now generally known as *Homo erectus*.

1895   Eugène Dubois brings to Europe the hominid fossils he has found in Java, where they are met with considerable skepticism by scientists and laypersons.

1908   On August 5, the Abbé Henri Breuil [French: 1877–1961] and Jean Bouyssonie and an assistant, Josef Bonneval, discover a nearly complete Neandertal skeleton at La Chapelle-aux-Saints, France. Remarkably, the skeleton seems to have been deliberately buried, and comes to be nicknamed *Le Viellard* ("Old Man").

In England, fossils are located in a gravel pit near Barkham Manor on Piltdown in Sussex. These later become the first part of the fraudulent "Piltdown Man."

1911   Pierre-Marcellin Boule [French: 1861–1942] publishes the first part of his reconstruction of the appearance of a Neandertal man, taken mostly from the skeleton of the *Viellard* of La Chapelle-aux-Saints. Completed in 1913, his inaccurate version becomes for many years the accepted way a "cave man" is pictured—stooped and brutish, with an oddly thrust-forward head that highlights a protruding face and heavy ridges above the eyes.

1912   Charles Dawson [English: 1864–1916] announces that he has found further fossils of Piltdown Man, which he identifies as the earliest Englishman. The unusual skull combines a modern braincase with the jaw of an ape. By 1953, however, the fossil is unmasked as a hoax.

1924   In South Africa, Raymond Arthur Dart [Australian: 1893–1988] is given the skull of the Taung child, the first recognized fossil of any australopithecine. Dart correctly observes that the Taung child was a hominid but not a human. He names the species *Australopithecus africanus*, and announces the discovery in 1925.

1927   Davidson Black [Canadian: 1884–1934] discovers a single hominid tooth that he names *Sinanthropus pekinensis*. This "Peking Man" is now most often viewed as a race of *H. erectus*.

1929   Pei Wenzhong and coworkers discover the partial skull of a specimen of what Davidson Black had christened *Sinanthropus* in a cave known as Choukoutien (Zhoukoudian), near Peking (Beijing).

1932   In India, G. Edward Lewis finds part of the jaw of a primitive primate that he names *Ramapithecus*. For the next 50 years, most anthropologists believe that *Ramapithecus* is the earliest known ancestor of *H. sapiens*. Instead, evidence later shows that *Ramapithecus* is the ancestor of the orangutan.

1935   German paleontologist G. H. R. "Ralph" von Koenigswald finds the first evidence—fossil teeth in a Hong Kong apothecary shop—of *Gigantopithecus*, the largest primate known.

1937   Franz Weidenreich [German: 1873–1948] discovers (between 1934 and the end of his career) 14 partial skulls, 11 jawbones, and multiple teeth and limb bones of Peking Man.

1952   In November, Kenneth Page Oakley [English: 1911–1981] announces that tests on the jawbone of the Piltdown Man reveal it to be an ape mandible doctored with a chemical to make the bone appear ancient.

1956   William Clouser Boyd [American: 1903–1983], assisted by his wife Lyle, develops a list of 13 races of *H. sapiens* based on blood groups. Unexpectedly, the Basques (a people inhabiting the western Pyrenees at the border of France and Spain) appear to be the remnant of an earlier European race.

1959   English anthropologist Louis Seymour Bazett Leakey [English: 1903–1972] uncovers fossil remains of an early hominid from about 1.75 million years ago, which he names *Zinjanthropus*. Later anthropologists usually classify the specimens as either *Paranthropus* or *Australopithecus bosei*, or as *P.* or *A. robustus*.

1960   Louis Leakey and Mary Leakey [English: 1913] find the first fossil remains of *Homo habilis*, or "Handy Man," in the Olduvai Gorge in Tanganyika (Tanzania). The first actual fossils are discovered by the Leakey's 19-year-old son Jonathan in November.

1963   Ice-age skeletons found in the Tomito cave near Cosenza, Italy, include that of a dwarf. Later analysis suggests that the 17-year-old dwarf was accepted by the hunter-gatherers despite the fact that he could not have contributed much to the community.

1967   Elwyn L. Simons announces his discovery of the 30-million-year-old skull of an ape, which he names *Aegyptopithecus*.

1974   A team led by Donald Johanson and Maurice Taieb discover "Lucy," 40% of the skeleton of an early hominid more than 3 million years old. Lucy is a representative of a species first identified by Johanson and Taieb and named *Australopithecus afarensis*.

1981   J. Desmond Clark and Timothy D. White discover more fossils of *A. afarensis*.

David Pilbeam determines that *Ramapithecus* is probably an ancestor of the orangutan, and not the first and earliest hominid.

1984   Charles G. Gibley and Jon E. Ahlquist use studies of DNA to conclude that humans are more closely related to chimpanzees than either chimpanzees or humans are related to other hominoids.

Timothy D. White and Donald Johanson locate 302 pieces of a female *Homo habilis*, later known as OH62, that is 1.8 million years old. Included are the first known limb bones from this species.

1988   French and Israeli scientists announce that fossils found in Israel, dated at 92,000 years old, are of *Homo sapiens*, which more than doubles the length of time this species was previously thought to have existed.

1991   Martin Pickford, working in Namibia with a French-American team, discovers part of a jaw of an early ape, later named *Otavipithecus namibiensis*, that may have lived 12 to 15 million years ago. It is the first known hominoid fossil from southwestern Africa.

NEWS ABOUT EARLY HUMANS AND OUR OTHER ANCESTORS

VERY EARLY PRIMATES

The study of the roots of the human branch of the class of mammals underwent considerable progress in the early 1990s, with further exploitation of the main site for early primate research in the Fayum Depression of Egypt and the discovery of additional rich sites in North America, in Algeria, and in southeastern China. One result is that the general picture of early primate evolution accepted just a decade ago has now been thrown into total confusion. Today, not only is the base of the primate family tree obscure, but it is also unclear as to whether the tree is planted in Africa or in Asia. Furthermore, the tree germinated much earlier than previously believed. Until recently, it was thought that primates originated about the time of the demise of the dinosaurs, 65 million years ago, but new evidence has pushed the date back to about 80 million years ago. This increases the time spent on primate evolution by 23%, and also increases the amount of unknown primate history by an even greater percentage. Taxonomists, looking at the newly emerging primate family tree, continue to find paradoxes that no rearrangement of branches can resolve, due to incompatible characteristics within each of the main branches.

Accepted primate history has been considerably revised in other ways as well. The 1990s began with the recognition that the presumed ancestor of the primates was not a primate at all, but instead was the ancestor of the colugo. Nicknamed the "flying lemur," the colugo is not a lemur and glides rather than flies; it is a primate cousin perhaps, but not a primate.

Another recognition of the early 1990s was that the anthropoid line, previously believed to have separated from the prosimian primates about 35 million years ago, is actually far older. Not only were older and more diverse fossils found in the Fayum Depression, but the dates for previously known fossils were shoved back by new geological studies of the region. As a result, the accepted date for the anthropoids was pushed back 20 million years. Yet many primate paleontologists felt that even an origin of 55 million years ago was likely to be too late, and that forthcoming evidence would push the date back even further. This expectation is based largely on the still-existing paucity of early primate fossils. Also, an earlier date would make it easier for primates to have spread—as they certainly did—from Africa to South America, since the two continents were joined together some 130 million years ago, separating slowly over a period of tens of millions of years after that, and still moving slowly apart today.

Home of the Primates

Near the village of Shanghuang in Jiangsu Province in China, a limestone quarry has uncovered four deposits of Eocene epoch fossils—that is, fossils from the period of 40 to 60 million years ago. Each deposit was, in Eocene times, a large crack or fissure caused by the same forces of water and erosion that produce caves and sinkholes in limestone-based landscapes (a topography known to geologists as karst, after a limestone district in Europe). The cracks were about 6 ft (2 m) wide before they filled with dirt, rocks, and bones. While such cracks fill, small animals make their homes in parts of them, or fall into the cracks, or their remains wash into the cracks. Once a dead animal is in the fissure, the hard parts are preserved away from forces that would scatter or destroy small animal bones or teeth left on the surface. Because early primates were all small, and also were apparently abundant in that part of Eocene China, the fissure deposits have become an important new source of early primate fossils. Since January, 1992, a team of K. Christopher Beard and Mary R. Dawson of the Carnegie Museum of Natural History in Pittsburgh, Pennsylvania, and Tao Qi, Banyue Wang, and Chuankuei Li of the Academia Sinica in Beijing, China, have been digging through the fissures and studying the fossils. They have concluded that most of the fossils in the four fissures formed about 45 million years ago. The team described their conclusions in the April 14, 1994 *Nature*.

While the fossils represent various small mammals, the main interest is in the primate remains. These remains are more diverse than those known from the rest of China for Eocene times. Two of the new species are adapids, a group of primates that either became extinct 5 million years ago without issue, or are related to modern lorises. One species is from another long-extinct group of tarsier-like primates, the omomyids, while another is the oldest tarsier ever found (by about 20 million years). The last new species is identified by the team as a basal anthropoid, although at least one prominent primate researcher thinks that it is really a sort of hedgehog, and not a primate at all. The finders have named their basal primate *Eosimia*, or "dawn simian," a name that is likely to stick whether or not the weight of later evidence is in favor of anthropoidality or hedgehogness. Because identification of all the fissure fossils is based on partial jawbones (or, in the case of the tarsier, on loose teeth) there is considerable opportunity for opinions to differ.

Meanwhile, another contender for the earliest anthropoid appeared. Isolated teeth found in Algeria were placed at the bottom rung of the anthropoid ladder by Marc Godinot of Université de Montpellier in France and Mohamed Mahboubi of Université d'Oran in Algeria. Because of the small size of the teeth and the location of their discovery, the species was given the name *Algeripithecus minutus*. The formation in which the teeth were found was dated at either the Middle Eocene (40 to 45 million years ago) or Early Eocene (46 to 50 million years ago). News of the fossil was published in the May 28, 1992 *Nature*.

In 1994, the noted primate researcher Elwyn L. Simons of Duke University, working with D. Tab Rasmussen of Washington University, also described his contribution to the new set of early primate remains. Simons found the complete skull of the first recognized member of what Simons and Rasmussen call, in the *Proceedings of the National Academy of Sciences*, the first known member of a whole new family of primates. Indeed, they put the creature, which they named *Plesiopithecus teras*, into a new superfamily. While *Plesiopithecus* was known only from a single lower-jaw fragment, the species was thought possibly to be an anthropoid—*Plesiopithecus* means "near to the ape." With the whole skull now known, it is clear that the large and startling teeth of *Plesiopithecus*—a giant lower front tooth combined with a dagger-like upper canine—make it a primate like no other. The other two known Eocene superfamilies of primates are the adapids and the omomyids. While there should be prosimians and primates from this time, none have been definitely identified, except possibly for *Eosimias* from China.

Although *Plesiopithecus* is younger than *Eosimias* by about 10 million years, most paleontologists still believe that early primate evolution began in Africa (home of *Plesiopithecus*) and spread to Asia (home of *Eosimias*). A key factor in the argument for either side is the number of different families or different genera found at any given site. One quarry in the Fayum Depression shows a great diversity, including four different families. At the genus level, Fayum quarry L-41 has four genera of anthropoids, and two of prosimians. The earlier deposits in China at Shanghuang are evidence in favor of an Asian origin for the primates, since they also claim four families: adapids, omomyids, a tarsier, and a "basal" anthropoid. The diversity argument appears to result in a tie between the two continents. Thus far, no one has discovered the type of fossil that would end all arguments: a skull, and possibly some

---

PRIMATE LANGUAGE

The primates are an ancient and diverse order of mammals with two principal suborders, known technically as Prosimii (**prosimians**) and Anthropoidea (**anthropoids** or **simians**). In this case, the technical language has a different meaning from common English usage. In laypersons' usage, anthropoid is an adjective that refers to similarity to or membership in the "great apes"—the gibbons, orangutan, chimpanzees, and gorillas. American primatologists, however, use *anthropoid* as a noun meaning any New World monkey, Old World monkey, ape, or hominid, including of course the human. English anthropologists are more likely to use *simian* (which in common usage is an adjective meaning resembling a monkey or an ape) to mean the same as *anthropoid*.

The situation is further confused by the name of the other order, which in nontechnical English is prosimian, meaning "before the simian." Thus, the name strongly suggests that the prosimian suborder is older than the anthropoid suborder, although the main reason for thinking this is that we humans are in the anthropoid suborder. Our usual human narcissism causes us to assume that the anthropoid suborder must surely be a higher, and therefore later, branch on the tree of evolution. The prosimians consist of five families: tree shrews, lemurs, the aye-aye, lorises, and tarsiers. The anthropoids also are grouped into five families: New World monkeys, marmosets and tamarins, Old World monkeys, gibbons, and apes and humans. But there is good reason to think that the tarsiers and the anthropoids share a common ancestor, meaning that the tarsier family of prosimians could not have come before the anthropoid suborder. The best evidence today, which everyone agrees is singularly incomplete, suggests that the anthropoid-tarsier line separated from the non-tarsier prosimian line sometime in the late Cretaceous period, perhaps about 75 million years ago, right after the emergence of the presumed common ancestor of all the primates. If this is indeed the case, then *prosimian* is a misnomer.

limb bones or a pelvis, from the "missing link" between the prosimians and the anthropoids. The continent on which such a fossil is found will likely be declared the original home of the primates.

(Periodical References and Additional Reading: *Nature* 5-28-92, p 324; *Science* 6-12-92, p 1516; *Nature* 5-20-93, p 223; *Nature* 4-14-94, p 587; *New York Times* 4-19-94, p C12; *Science* 10-28-94, p 541)

NEW ROOT FOR HUMAN FAMILY TREE

Since 1856, progress in understanding human evolution has depended largely, although not completely, on discovery and interpretation of extremely rare fossils. Early humans and their ancestors often failed to die in places where bones turned easily into stone. In addition, human hard parts are not so large or hard as those of some other creatures. Furthermore, the early population of our ancestors was not very large. Despite these impediments, the early and middle 1990s brought forth reports of several new fossils that may lead to important new understandings.

Out of the Woods

The most important new find came from Ethiopia, as have many others of the past two decades. Tim D. White of the University of Californian, Berkeley, Gen Suwa of the University of Tokyo, and Berhane Asfaw of Ethiopia discovered the earliest known species of australopithecine near the Aramis River in the Afar region of Ethiopia, an upstream tributary of the Awash River. The Awash basin was also the discovery site of the australopithecines previously believed to be the earliest. Because White, Suwa, and Asfaw believe that the new hominid is close in time to the split between chimpanzees and the human line, they originally named the species *Australopithecus ramidus*, using *ramid*, the Afar word for "root," as the base of the specific name. The generic name, *Australopithecus*, designates a group of three to six related species of extinct hominds that walked upright and had brains considerably smaller than those of the genus *Homo*. Some anthropologists suggested immediately, however, that further discoveries of *ramidus* fossils may suggest the need to name a new genus to contain them. This would be the case if the new fossils differ from *Australopithecus* species in significant ways. Indeed, later study resulted in placing *ramidus* in a new genus, *Ardipithecus*.

The *ramidus* fossils were found during the 1992-93 and 1993-94 seasons and reported in the September 22, 1994 *Nature*. Gen Suwa has the honor of having collected the first fossil of the new species, for on December 17, 1992, he recognized a single molar tooth lying among the pebbles on barren, desert shingle. Teeth of hominids are characteristic for each species, and Suwa immediately knew that he had made a major find, especially since it was found in a fossil-bearing region with many non-hominid species already known to be more than 4 million years old. Even so, he thought the tooth belonged to *Australopithecus afarensis* until a few days later when another researcher found a whole set of hominid lower teeth that clearly were more primitive than those of *A. afarensis*. Eventually, ten different searchers on the expedition in 1992 and 1993 found more teeth, jawbones, the base of a skull and other skull fragments, and a couple of arm bones from the newly found species. The fossils have been identified as the remains of 17 different individuals and dated at about 4.4 million years ago.

Each part of *Ardipithecus ramidus* has a story to tell the scientists. The teeth are like those of chimpanzees, but smaller in proportion to the jaw. In shape as well as size, they are between those of chimpanzees and those of *A. afarensis*, a species that would appear in the same region about 800,000 years after *A. ramidus*. The ramidus teeth are also in between those of chimpanzees and *afarensis* in enamel thickness, an important characteristic that differentiates hominid great apes from other hominids. The cranial base of the *A. ramidus* skull is shorter than that of chimpanzees, more like that of a human. The elbow is also shaped differently from those of chimpanzees, suggesting that the arm is more like that of a human than that of a knuckle-walking tree-dweller. Tim White said of the elbow, "This is not the arm of a knuckle-walker." However, since no leg, pelvic, or foot bones were found, it was not clear at first whether *A. ramidus* walked upright.

In November of 1994, White returned to the site. An Ethiopian member of his team, Yohannes Haile Selassie, found 90 fragments that together make up about 45% of the skeleton of an individual *ramidus*. These fragments do include leg, foot, and pelvic bones, so more answers will soon be forthcoming. In another development at about the same time, rumors circulated through the anthropological community that Maeve Leakey, who has previously made several significant discoveries, had also located the leg bone of a different human ancestor from about the same time as *A. ramidus* at a site near Lake Turkana in Kenya. This was confirmed in a report in the August 17, 1995 *Nature*. Leakey and coworkers named their new species *Australopithecus anamensis*.

The evidence found so far, however, suggests that *A. ramidus* is almost as close to the chimpanzee line as to the australopithecines. This confirms DNA and protein analysis that indicates that chimpanzees are the closest living relatives of the human. Indeed, a few researchers have suggested that *ramidus* is a chimpanzee ancestor instead of a human ancestor, primarily on the basis that its tooth enamel is more like that of chimpanzees than that of other australopithecines or humans.

If the split between the ape line and the human line is a million years later than most evidence suggests, and if *A. ramidus* developed a half million years earlier than the individuals who left the Aramis fossils, then *A. ramidus* could be the famous "missing link." Otherwise, there is still a million-and-a-half-year gap in which the "missing link" could have split off from the apes.

For most scientists, the biggest surprise in the Aramis finds was the environment in which both *A. ramidus* and *A. anamensis* lived. Giday Woldegabriel and Grant Heiken, both of Los Alamos National Laboratory in New Mexico, and coworkers dated the Aramis fossils based on layers of volcanic ash and glass that overlay and underlie the deposits. These researchers also observed that numerous monkey fossils, forest-dwelling antelope, petrified wood, and tree seeds demonstrate the woodland nature of the Aramis environment of 4.4 million years ago. The Lake Turkana home of *A. anamensis* was also wooded around 4 mya. Previously, most anthropologists had expected human ancestors to have evolved in a savanna environment, which is mainly grassland with few trees. A few anthropologists, however, such as Andrew Hill of Yale University in New Haven, Connecticut, feel that their theoretical expectations have been fulfilled by finding early ancestors in a woodland, rather than a savanna, setting.

Other Australopithecine Finds

After more than a decade of turmoil, Ethiopia is once again politically ready for fossil hunters. Anthropologists have returned, and have extended our knowledge in more ways than just the discovery of *Ardipithecus ramidus*. In 1990, Tim White located the jaw of a large male *Australopithecus afarensis* at Maka in the Middle Awash region, publishing an analysis of it and other Maka fossils in the November 18, 1993 *Nature*; and in 1991, Yoel Rak of Tel Aviv University in Israel found the first skull of *A. afarensis* at Hadar, in Ethiopia, describing it in the March 31, 1994 *Nature*. William H. Kimbel and Donald C. Johanson, on the same expedition with Rak, also found various limbs and jawbones of the species. The discovery of the skull (which had to be assembled from more than 200 pieces) and jawbones has helped support a theory previously espoused by White and by Donald C. Johanson, who stated that the male *A. afarensis* was much larger than the female. This theory is their attempt to explain the two sets of different-sized fossils from this period in East Africa. Mary D. Leakey has been the most prominent supporter of an alternative theory—that the larger fossils represented a different species from *A. afarensis*, most likely a species of *Homo*.

The question is whether *Australopithecus afarensis* is one sexually dimorphic species with large males and small females or should be reclassified as two different species, one large and one small. Even modern humans are statistically sexually dimorphic. Although many female *Homo sapiens* are larger than many males of the same species, the average male is 5-12% taller and 15-20% heavier than the average female in all modern human populations. If *A. afarensis* is a single species, however, the males were nearly twice as heavy as the females, which is a dimorphism more like that of *Gorilla gorilla* than that of *H. sapiens*. Specifically, Kimbel estimated that female *afarensis* individuals were about 3.5 to 4 ft (1.1 to 1.2 m) tall and weighed about 75 lb (35 kg), while males were at least 5 ft (1.5 m) tall and weighed more than 100 lb (50 kg).

Kimbel and White claim that the new fossils confirm the single-species theory, on the grounds that the fossils demonstrate the bones and teeth of the various *afarensis* specimens are too much alike, whether large or small, to be from different species. Thus, this particular species controversy is considered settled by most anthropologists, although not everyone is completely convinced.

The newly discovered jaw is right in the middle of the range of dates now known for *A. afarensis* fossils. The new skull, assembled from about 200 bone fragments, has the most recent *afarensis* date—3 million years old. Some very old *afarensis* remains were also found in the Middle Awash near Maka. These extend the *afarensis* line back to 3.9 million years ago.

And the Oldest Human

Although *Ardipithecus ramidus* is the oldest hominid known to be on the human family tree, it is not a human. It is not until 2 million years after *ramidus* that the first known members of *Homo* appear in the fossil record. In 1992, using a special variety of radioactive dating, a particular fossil was dated as the oldest known human. The 3 in (8 cm) long piece of a skull had actually been found in 1965 in the Lake Baringo Basin of Kenya by Australian geologist John Martyn. In the late 1980s, Andrew Hill of Yale University saw the cranium fragment at the National Museums of Kenya and immediately recognized it as belonging (most likely) to *Homo* and not to *Australopithecus*. At Hill's behest, Steven Ward of Northeastern Ohio University's College of Medicine obtained the date of 2.4 million years ago for the fossil bed in which the Baringo bone fragment was found, making it the earliest *Homo* fossil by about half a million years.

Although it is easy to assume that the Baringo fragment belonged to *H. habilis*, there is not enough of the fragment remaining to be certain. Indeed, the only reason that Hill can argue that the bit belongs to *Homo* instead of *Australopithecus* is that the size of the brain appears to be too large for an australopithecine of that period. On the other hand, many anthropologists believe that *Homo* dates from about 2.9 million years ago, allowing for the possibility of a species that comes before *H. habilis*. Thus, the skull fragment could even be the first-known fossil from a previously unknown species.

(Periodical References and Additional Reading: *New York Times* 2-20-92, p A14; *Science* 10-1-93, p 27; *Nature* 11-18-93, pp 207 & 261; *Science News* 11-20-93, p 324; *New York Times* 3-31-94, p A1; *Science* 4-1-94, p 34; *Science News* 4-2-94, p 212; *New York Times* 4-3-94, p IV-2; *Archaeology* 5/6-94, p 14; *New York Times* 8-26-94, p A20; *Science* 9-16-94, p 1661; *Nature* 9-22-94, pp 280, 306, & 330; *Science* 9-30-94, p 2011; *New York Times* 9-22-94, p A1; *Discover* 1-95, pp 37 & 38; *New York Times* 2-21-95, p C1)

THE SPECIES PROBLEM

The classical definition of *species* is a group of living organisms that breed with each other under natural conditions. While this simple concept is not truly that simple — for example, it fails to take into account organisms that reproduce without sexual breeding, such as bacteria, protists, many plants, and some vertebrates — it is a helpful way to describe what scientists mean by *species* for most modern animals and plants. The concept is more difficult to apply to extinct organisms, knowledge of whose breeding habits is based largely on analogy and speculation.

Anthropologists continually rearrange the species, and even the genus, of early hominids. So-called "lumpers" assign fossils to as few species as possible, while "splitters" tend to give each fossil a new specific or generic name, especially for those fossils discovered by the splitter. While it would be no surprise if most of the reclassification had been of the oldest hominids, some of the most difficult cases to settle have been the most recent—Neandertals and *Homo erectus*.

Passing the Subway Test, and More

The first truly early hominids to be recognized were the Neandertals, with skeletons first found in 1856 (three years before publication of Darwin's *Origin of Species*,

Background

Since 1735, scientific classification of plants and animals has been based upon the binomial, or "two-name," system of Linnaeus. Each single type of organism has a specific (adj.) name that tells its species (n.), while closely related species are assigned a generic (adj.) name, creating a somewhat artificial grouping called a genus (n.). Thus, the domestic dog, with all its variety, is considered a single species in the "dog genus," different from other doglike creatures, such as the wolves, coyote, and jackals. The genus of all these species is given the Latin name for dog, which is *Canis*—always in italics with an initial capital letter. Each then has its own specific name, which is *familiaris* for the domestic dog and *latrans* for the coyote. After a genus name is used in discussion, scientists abbreviate further mentions of the name with the initial capital letter. Thus, the first mention of the scientific name for the domestic dog is given as *Canis familiaris*, and later mentions become *C. familiaris*. Furthermore, the abbreviated form is used for other members of the genus, such as *C. lupus* for the gray wolf.

In theory, species are evolutionary units that persist as separate groups because the members of one species do not interbreed with other members of their genus. In practice, interbreeding within a genus often occurs. Throughout the United States, for example, coyotes and domestic dogs interbreed, while wolves, domestic dogs, and coyotes also interbreed in complicated patterns when there is opportunity to do so.

Biologists who classify animals use every possible means at their disposal to identify a species. These means range from color and size to behavioral traits to genetic analysis. As a result, the field of taxonomy (classification) is filled with differing, strongly held opinions. Some taxonomists have lumped the fox into the genus *Canis*, while others have split the jackals into two different genera (plural of genus). Some have even split the domestic dogs into two species—"man's best friend," *C. familiaris*, and the Australian wild version of a domestic dog, *C. dingo*.

although Darwin apparently did not know of the Neandertal finds at that time). By the mid-1990s, considerable attention had been focused on trying to determine Neandertal abilities, lifestyle, and relationship to people living today. Neandertals were physically well-adapted to living in ice-age conditions—stocky, with a respiratory tract designed for living in a cold, dry climate. Neandertals used similar tools to those of anatomically modern humans, and had similar customs, such as burial of the dead. Some anthropologists claim that, despite their stocky physique and heavy brow ridges, Neandertals pass "the subway test." That is, if a properly dressed Neandertal were riding on a New York subway, the other passengers would notice nothing unusual.

Despite the subway test, anthropologists still hotly debate whether Neandertals are our literal ancestors, or a branch of the hominid bush that was pruned after the most recent glacial advance. Champions of the ancestor concept include Milford Wolpoff of the University of Michigan, while the driven-to-extinction school is led by Christopher B. Stringer of the Natural History Museum in London.

People who believe interbreeding took place between Neandertal and anatomically modern humans find evidence in present-day bones of Europeans, evidence dismissed by the other school. David Frayer of the University of Kansas has suggested that comparisons of Neandertal bones with bones of Europeans from more than 300 years ago show more evidence of interbreeding than do comparisons between Neandertal bones and those of contemporary Europeans. People who think Neandertals did not or could not interbreed with our species get some support from mitochondrial DNA studies. Such studies point to an African origin for modern humans, and suggest no contribution from early European mitochondria. Neandertal-appearing skulls and bones from Sierra de Atapuerca in Spain (*see also* "Out of Africa, but When? " p 47) suggest that the Neandertals were becoming a separate entity as early as 300,000 years ago.

The Atapuerca skulls also provide another window on the species problem. Of the three skulls, one adult skull has the largest cranial capacity of any found from that time period, while the other adult skull has the smallest. Similar variability is demonstrated by all discernable traits from the Atapuerca fossils. This suggests that all the fossils from this time period, and probably from the ensuing Neandertal period, are within the range that should be called Neandertal.

One of the main problems with explaining the Neandertals is the apparent proximity in both time and space of Neandertals and anatomically modern humans in the Mount Carmel region of what is now northern Israel. Wolpoff believes that the two were living in contact with each other, just as different races do in the same region today, and that the period of togetherness lasted 50,000 years, plenty of time for interbreeding. As part of the evidence for this, Wolpoff points out that the cultures of the two groups, as established by archaeology, were essentially identical.

Against this idea of two "races" in the Mount Carmel region, Yoel Rak of Tel Aviv University points out that the fossilized skeleton of a baby found in the region already shows distinctive Neandertal traits, suggesting to Rak that the Neandertals of that time (50,000–60,000 years ago) were a separate species. Wolpoff thinks that the baby was too young to have definite traits of either Neandertals or modern humans.

Mysterious *Homo erectus*

If the Neandertals are members of our own species, as many anthropologists believe, the first fossilized remains of a human ancestor known to science are those of *Homo erectus*, first described in 1894. This species, or something very much like it, appears to have dominated the Old World for a million years, leaving behind not only large numbers of characteristic stone tools, but also hearths, remains of hunting expeditions, and even signs of constructed dwellings. Consequently, it has been easy to assume that *H. erectus* is well understood, and that the mysteries of human origins lie earlier. But it has become clear in the 1990s that *H. erectus*—or perhaps various other species confused with *erectus*—is among the least well understood and well documented of our ancestors.

For one thing, despite the apparent million-year reign over Africa, Asia, and Europe, there are comparatively few fossils attributed to erectus. Far more remains have been found of the earlier *H. habilis,* and there are many better preserved and more complete fossils of even the earliest known australopithecines.

Although the conventional view has been that *erectus* spent a million years in Africa before breaking forth into the remainder of the Old World, more *erectus* fossils have been excavated from Asia and Europe than from Africa. The best location for *erectus* fossils continues to be where they were first found, along the Solo River in Java. Nine

---

NAMING NAMES

Late in the nineteenth century, many people who had come to believe in evolution concluded that the precursors of humans had most likely lived in Asia. Eugène Dubois, who favored an Asian origin, managed to get to Indonesia in his quest for Asian "ape-men." Amazingly—since subsequent discoveries in Asia have been uncommon—he found fossils, which he named *Pithecanthropus erectus* ("erect ape-human"). Later anthropologists concluded that the fossils Dubois found should be classed in the same genus as modern humans, so the scientific name was changed to *Homo erectus*.

According to the rules of taxonomic naming, the specific name given by the original discoverer is eternal, whatever its flaws may be. But the generic name for an established genus may take over the originator's name if subsequent study warrants it. Dubois' "erect" was kept, but his "ape-human" lost out to plain "human." Thus, the species that many believe to be the most recent ancestor of modern humans has a name that literally means "erect human," suggesting incorrectly either that this was the first bipedal ancestor, or that the modern human might be less than erect.

Similar problems arise throughout the hominid line. Most commonly, hominids are grouped into anywhere from two to six genera—the maximum list currently includes *Homo, Australopithecus, Ardipithecus, Paranthropus, Pan,* and *Gorilla*. Fossil genera are most often classed as *Homo* or as *Australopithecus,* which means "southern ape-human." The name *Australopithecus* was coined by Raymond Dart because the first specimen was found in South Africa.

The genus name *Paranthropus* originated with Robert Broom in 1938, and means "next to man." The majority of later anthropologists moved *Paranthropus* into the *Australopithecus* genus, however, until a recent shift back to *Paranthropus*. Oddly, while two or more *Australopithecus* species are often considered to be on the evolutionary line to humans, the species Broom named *Paranthropus robustus*—usually called *Australopithecus robustus* until recently—is not generally thought to be close to humans ("next to man") at all. It is because *robustus* is not on the human line that many anthropologists now argue for restoring the *Paranthropus* name.

Meanwhile, a predecessor to Dart's "southern ape" was found in Ethiopia, which is not southern at all in reference to Africa. But, by the rules of taxonomic nomenclature, a new member of the same genus must abide by the name originally assigned to the genus. Even if *Australopithecus afarensis* had been discovered at the north pole, instead of the Afar valley of Ethiopia, it would still be called a "southern ape."

of the Solo fossils are skulls. Two *erectus* skulls, one of them the most complete ever found, were discovered near the Solo in 1993.

The complete *erectus* skull, found by local farmers who told University of Idaho paleoanthropologist Donald Tyler of its location, is also the oldest skull ever found. Tyler thinks the skull is between 1.1 and 1.4 million years old. It has a cranial capacity estimated at 856 cubic centimeters, near the low end of the range for *H. erectus*.

Two fossil craniums discovered in the Bodo sands of the Middle Awash Valley of Ethiopia and a fossil arm bone found at the same location in 1990 were dated in 1994 by J. Desmond Clark of the University of California, Berkeley. Clark estimated these remains to be from 600,000 years before present. The craniums, thought to be archaic *Homo sapiens*, had been dated previously at about 350,000 years before present. The new dates, combined with the discovery of typical hand axes, suggest that the fossils are actually of *H. erectus*.

The largest single group of *erectus* fossils found at one site consists of bones from 40 different individuals which were found in Zhoukoudian cave near Beijing, China. These fossils, collectively known as Peking Man, were lost during World War II. Another group of fossils and tools, sometimes assigned to *erectus*, has been found in geographically dispersed areas of Germany, France, Spain, and perhaps between the Caspian and Black seas in Georgia. But there are only two *erectus* sites in Africa—Lake Turkana in Kenya and Olduvai Gorge in Tanzania—and some think that neither is the same species as the fossils found in Asia. Increasingly, the fossils from Lake Turkana are classed as *Homo ergaster* ("working man"). In this view, *H. ergaster* traveled to Asia, where evolution into *H. erectus* took place.

Some lumpers may classify the European "*erectus*" as *H. ergaster*, while splitters place them in the old taxonomic class of *H. heidelbergensis*, named after a fossil found in a gravel pit near Heidelberg, Germany, which is often called the Mauer Mandible. The true lumpers, however, have maintained that the Mauer Mandible, like the African fossils, are all part of *H. erectus*, which would make erectus the only human on Earth for about a million and a half years or even longer.

(Periodical References and Additional Reading: *New York Times* 2-4-92, p C1; *Nature* 4-8-93, pp 501 & 534; *New York Times* 4-20-93, p C8; *Archaeology* 9/10-93, p 18; *New York Times* 11-18-93, p A9; *Science* 6-4-94, p 1907; *Archaeology* 7/8-94, p 14; *Discover* 9-94, p 80; *Discover* 1-94, p 39)

HANDS AND FEET

The human can be characterized among living animals by unique hands and feet, as well as by the well-known large brain. Anthropologists have sometimes hypothesized that the hands, so useful for handling tools, led to the growth of the brain. Another theory is that walking upright preceded development of the hand, because the hand needed to be free of locomotive duties before it could develop into its present form. In the past quarter of a century, fossils and other evidence have for the first time enabled scientists to establish objective evidence for or against such theories.

These Legs Are Made for Walking

Evidence that human ancestors (at least since *Australopithecus anamensis*) stood and walked upright is supported not just by study of the anatomy of ancient legs and feet, but also from study of fossil footprints. In 1976, famous fossil footprints in volcanic rock were found when paleontologist Andrew Hill, now of Yale University, was dodging a piece of playfully tossed elephant dung in Laetoli, Tanzania, and noticed animal and bird tracks preserved in rock. In 1977, subsequent investigations by Mary Leakey established the prints of australopithecines, identified by the date that they were left in wet volcanic ash, as being the prints of *A. afarensis*. The trail of prints by two or three australopithecines convincingly established that members of the species were bipedal at least part of the time. In 1993, Andrew Hill observed that the site was eroding and endangered by plants growing on the famous prints. The Getty Conservation Institute announced on December 15, 1994, that they were beginning a three-year project to preserve the prints for future study.

There has been speculation that bipedal locomotion occurred when the climate dried, causing one or more of the following: (1) apes preadapted for an upright posture by living mostly in trees were forced into life on the ground; (2) knuckle walkers (like the great apes) or quadrupeds (like baboons) had to stand up to see over the tall grass while hunting; (3) our ancestors who had become preadapted to bipedalism by life mostly in shallow water had to come out on the land when the waters dried up; (4) something else happened to cause our ancestors to stand up.

Some of the bones of *A. afarensis* reveal information about walking and lifestyle, but competing anthropologists read the fossil record slightly differently. Leslie C. Aiello of the University College in London thinks that the *afarensis* arm bones are "ideally suited to a creature which climbed in the trees but also walked on two legs when on the ground." Other researchers note that curved toe bones in *afarensis* could indicate tree climbing. Tim D. White and his coworkers from the Rift Valley Research Mission in Ethiopia write that a femur they found at Maka, as well as the pelvis, knee, and foot found earlier at Hadar, all show a leg "specialized for terrestrial bipedality," while lacking "the key arboreal adaptation" of very long arms. No one doubts that *afarensis* could climb trees when it wanted to—even modern humans can do that. The point of contention is how much time was spent in the trees, and how much on the ground.

Although some details of the walking style of early australopithecines may differ from current human walking style, it is important to note that behavior is rarely found in the fossil record. Consequently, one of the best ways available to gain insight into what might have happened with our early ancestors is to follow the observable actions of our close living relatives, the chimpanzees, which is what K. D. Hunt did in the late 1980s and early 1990s. Hunt carefully measured the amount of time that knuckle-walking chimpanzees stand or walk bipedally. The standing time and the walking time are markedly different. Chimps regularly stand up on two legs to feed, and they occasionally stand on two legs for other reasons, yet very rarely do they actually walk bipedally. They spend 20 times as much of their bipedal moments in feeding behavior as they do in walking behavior. They do not walk across open grasslands on two legs. Instead, they sometimes walk bipedally from one place under a tree where they are feeding standing up to another nearby place where they can also eat vertically.

Hunt used what he learned in studying chimpanzee postural behavior to analyze the probable occasions of bipedalism in early hominids. He believes that australopithecines were more like chimpanzees than like current day humans in this regard. Australopithecines probably walked only when they had to, in Hunt's view, but they probably fed upright more frequently than do present-day chimpanzees. Later hominids, such as *Homo erectus* or *Homo ergaster*, appear to have taken to regular walking.

A number of recent studies by P.E. Wheeler, including a 1993 paper published in the *Journal of Human Evolution*, suggest that the need to keep cool may help to explain why an australopithecine preadapted for walking would make bipedalism a regular part of its life after the climate became hotter and drier. A biped exposes less of its body to the heat of the sun (especially if it has hair on top of its head), and the nearly hairless skin elsewhere on the body releases heat easily. Wheeler studied heat absorption by bipedal and quadrupedal models, finding that the quadrupeds need two-thirds more water than the bipeds do. Wheeler also concludes that *H. erectus* was larger and taller than its ancestors because size and length are good ways to control heat.

Once the preadapted australopithecine or its ancestor begins to walk, the advantages to the practice, in Wheeler's theory, become obvious. Other savanna creatures—such as baboons, gazelles, and lions—would also benefit, it would seem, from assuming an upright posture, but they are not preadapted by earlier evolutionary changes.

Randall L. Susman of the State University of New York at Stony Brook demonstrated at the April 1994 meeting of the American Association of Physical Anthropologists that early australopithecines may have walked differently from modern humans— they "walked funny," in Susman's phrase. He reasoned that the 30% longer foot of *A. afarensis* would have resulted in a different stride. By adding "clown shoes" to humans, he was able to show that longer feet in modern humans necessitates a high-stepping stride that is more like trotting than walking. Donald C. Johanson of the Institute of Human Origins in Berkeley, California, who first identified *A. afarensis*, criticized the demonstration, pointing out that the early australopithecine leg was also different from the modern human leg in bone-length ratios. Thus, in Johanson's view, the longer foot of the australopithecine would have been used for normal walking.

Another study of early hominid walking, also reported at the April 1994 meeting, focused on the bone structure that houses the inner ear, which contains the organs of balance. C. Fred Spoor of the University of Liverpool in England hypothesized that balancing would be different in creatures that walk upright compared to those that do not, and Frans W. Zonneveld of the University of Utrecht in the Netherlands was enlisted to make x-ray studies of the skulls of humans, chimpanzees, gorillas, and orangutans to help evaluate this idea. The x-rays showed that human skulls contain larger semicircular canals (the balance structure itself) than those found in ape skulls. The same technique was then used to analyze various early hominids. The research was conducted too early to use a skull from *A. afarensis* (*see* "New Root for Human Family Tree," p 32), but 35 early specimens of *Homo* and *Australopithecus* were tested. Only the specimens identified as *H. erectus* had semicircular canals similar to those in *H. sapiens*. Thus the "erect" in the specific name may be justified after all. Presumably, the other species studied spent more time living in trees, or traveling by some means other than upright walking. However, at least one anatomist, Owen Lovejoy of Kent State University in Ohio, claims that the changes in semicircular canals are caused by expansion in brain size, and have nothing to do with walking.

Hands That Are Not All Thumbs

Randall L. Susman, of the clown-feet experiment, also analyzed how early hominds used their hands. He proposed in the September 9, 1994 issue of *Science* that a single bone in the hand could be used to identify primates whose hands were capable of making tools. That bone, the first metacarpal, is the one to which the thumb is attached. The ends of the other metacarpals are the knuckles, and similarly the prominent bump that corresponds to a knuckle at the base of the thumb is the end of the first metacarpal.

What Susman observed is that there are three hand muscles in humans, each attached to that first metacarpal, all of which are lacking in great apes. The attachments for these muscles, and the stresses they engender, produce a thicker and differently shaped bone from the corresponding metacarpal in apes. The human thumb is broader at the head, in relation to its length, than is the chimpanzee thumb. These additional muscles allow the human thumb greater strength than that of the apes, allowing humans to use a combination of the thumb and one or more fingers to grasp and hold small objects tightly. The fingers generate the majority of the apes' hand strength, making their hands most suitable for grasping larger objects, such as branches. In both apes and the human, the thumb is fully opposable—capable of touching each of the other fingertips—and therefore useful in manipulating small objects. Only the human, however, has additional muscle power strengthening the thumb.

Because the first metacarpal bone is easily preserved in the fossil record, Susman reasoned that it could be used to identify which hominids were capable of making tools. He was able to compare modern human and chimpanzee metacarpals with those from four earlier hominids. Corresponding with known manufacture of stone tools, *Homo erectus* from Swartkrans, South Africa, and Neandertals from Shanidar Cave in Iraq both left behind first metacarpal bones of the modern-human type, while the hand bones of *Australopithecus afarensis* from Ethiopia matched those of the chimpanzee specimens, which also included bonobos. There has never been any association between *A. afarensis* and stone tools.

The most problematic result of Susman's study has to do with the remaining specimen of an earlier hominid. The first metacarpal of a hand identified as belonging to *Paranthropus robustus* from Swartkrans is of the human type. Identification of this bone's species of origin does not appear to be as clear-cut to other anthropologists, however. Susman claims that the fossil is definitely *P. robustus*, based upon an unpublished analysis, while some other anthropologists think that the thumb bone could be from *H. erectus*, whose fossils are also found at Swartkrans.

If Susman is correct in ascribing a toolmaking hand to *Paranthropus*, the conclu-

Background

In 1962, J. R. Napier argued persuasively that human hands were anatomically different from those of the great apes, in that human hands can produce a "precision grasp" needed for toolmaking, while ape hands are better suited for hanging onto branches, using what Napier called a "power grasp." Because of its name, the precision grasp is usually thought of in terms of control, but for manufacture of tools, especially stone tools, the power of the thumb is more important than its precision.

When Napier was writing, most people believed that only humans or their immediate ancestors made tools. A couple of years before Napier's work, however, scientists had observed chimpanzees making tools. Subsequently, chimpanzees and the orangutan have often been found to make simple tools from branches or leaves. It is important to note, however, that of all living species, only humans chip stone to make tools.

Because chipped stone survives easily in the archaeological record, anthropologists have long known that early ancestors (thought to be a different species from modern humans) also made stone tools. There is no controversy about stone tools less than a million years old. Aside from the great apes, the only hominids on Earth since that time have been various species of *Homo*, and it is quite certain that individuals in the species *Homo* have made (and in some cultures continue to make) stone tools. Before a million years ago, the situation is less clear, especially with regard to simple pebble tools of the style called Oldowan. Indeed, when Louis and Mary Leakey first identified the Oldowan tools, they believed that the maker was a non-*Homo* species now most commonly called *Paranthropus bosei*, but known to the Leakeys as Zinj (for *Zinjanthropus*, an earlier genus designation). By coincidence, the discovery of the Oldowan tools occurred about the same time as Napier's hypothesis and the first scientific observation of chimpanzee toolmaking. Shortly after the Leakey's identification of Oldowan tools, the same team discovered fossils they identified as *Homo* in the same strata as the tools and as Zinj. The Leakeys even found hand bones that appeared capable of toolmaking, so they called their discovery *Homo habilis*, or "Handy Man." They then attributed toolmaking to Handy Man, not to Zinj.

sion is surprising, but not totally unexpected. Careful anthropologists have noted that the earliest stone tool industry, named the Oldowan, is associated with sites where both early humans and *Paranthropus* species left fossils. Although it has seemed likely that the comparatively large-brained *Homo habilis* made the tools found at these sites, there has never been any definitive proof that *Paranthropus* did not make the tools.

Susman's analysis of the significance of the thumb bone is widely recognized as a breakthrough. In the future, analyses of first metacarpals from *H. habilis*, *P. bosei*, and more definitively identified specimens of *P. robustus* should go a long way toward clearing up the role of the hand in toolmaking and in the development of a larger brain.

(Periodical References and Additional Reading: *Nature* 6-17-93, p 587; *Science News* 4-2-94, p 212; *Science News* 4-9-94, p 231; *Science* 4-15-94, p 350; *Science* 9-9-94, pp 540 & 1570;

New York Times 9-13-94, p C1; *Science News* 9-17-94, p 183; *New York Times* 12-16-94, p A30; *Discover* 1-95, p 42)

CLIMATE AND EVOLUTION

Anthropologists are not satisfied by learning how humans evolved. The persistent question remains: Why did humans evolve? One answer that has received considerable backing from assorted physical evidence involves the effect of climate change, which is believed to have played an important role in causing ape-like primates to transform themselves into human-like primates. After its initiation, the process may have been self-reinforcing, leading almost inexorably to modern *Homo sapiens*.

Case for Climate Causing Human Evolution

One prominent paleontologist and evolutionist who has put forward a specific climate-based theory of human evolution is Elizabeth Vrba of Yale University. Her theory has been largely supported by evidence collected in the early 1990s that shows the climate in Africa varying on a periodic basis ever since humans emerged on that continent.

About 5 to 6 million years ago, a general cooling of Earth's climate set the stage for evolution of a biped to exploit the newly developed African savanna. Before this time, Africa was warmer and moister, resulting in a forest ecosystem instead of the partly treeless savanna. At that time, the first australopithecines or their ancestors developed (*see* "New Root for Human Family Tree," p 32), although the first known fossil australopithecines known come from a warmer period about 4 million years ago. During that period, the woodland environment returned.

About 2.8 million years ago, the woodlands disappeared again as the climate turned much colder. Faced with this environmental challenge, the australopithecines took two different pathways. One branch evolved into a larger, tougher australopithecine, which lived largely on roots and other vegetation. The other stayed fairly small, developing a varied diet that included more meat which was probably obtained by scavenging. This second group, with its smaller teeth adapted for the life of an omnivore, became the first humans, the species we know as *Homo habilis*.

About 1 million years ago, ice sheets covered northern Europe and much of North America, cooling the whole Northern Hemisphere. Africa became cooler and drier also, even though it was not covered with ice. The savanna environment became more difficult. As the australopithecine line died out completely, the human line evolved into *Homo erectus*, hunters who would eventually develop fire, as well as improved stone tools.

This scenario is based not only on fossils of humans and other life forms from Africa, but also on specific layers of volcanic ash. This ash can be used to pin down the dates of specific climate changes to within a period of 10,000–20,000 years. Furthermore, the overall dryness of the African climate has been estimated by measuring the amount of African dust from different time periods found as mud on the ocean bottom. Because a dry climate produces more dust, a thick layer of dust between two markers of volcanic ash means that the climate was dry. The analysis of the ash and dust is largely the work of Peter deMenocal and coworkers at the Lamont-Doherty Earth Observatory at Columbia University.

The climate theory is also invoked by members of the Hominid Corridor Research Project (Christian G. Betzler of Universität Frankfurt in Germany, Timothy G. Bromage of Hunter College in New York City, Yusuf M. Juwayeyi of the Department of Antiquities in Malawi, Uwe Ring of Universität Mainz in Germany, and Friedemann Schrenk of Hessisches Landesmuseum in Darmstadt, Germany). They use this theory to describe not only the origin but also the distribution of early humans. In the October 28, 1993 issue of *Nature*, the Hominid Corridor team reports that a jawbone found in Malawi is between 1.8 and 2.4 million years old, and is a fossil of an early human, classified by the team as *Homo rudolfensis*. Furthermore, they argue that all human species about this time originated in tropical East Africa, dispersing south after the climate cooled. The part of the western branch of Africa's Rift system in Malawi connects fossil-rich areas of East Africa and South Africa, which suggests that the Rift was a corridor between the regions when the climate was suitable. Among the evidence cited for this theory are the fossil records of the evolution and dispersal of large mammals other than hominids during the same period. The assumption is that the same climate forces that affect gazelles and giraffes would also have affected humans.

Case Against Climate as a Cause

John D. Kingston and Andrew Hill of Yale University, along with Bruno D. Marino of Harvard University, examined layers of soil from sites in the Tugen Hills, which are within the Rift Valley, south of Lake Turkana in Kenya. The Lake Turkana region, the Awash River Valley in Ethiopia just to the north of Lake Turkana, and Olduvai Gorge to the south of the Tugen Hills are prime sites for fossils of the earliest hominids, as are parts of the Rift Valley. The scientists' motive was to find *local* evidence of climate change—nearly all the data on which the climate-change theory of human evolution is based comes from measurements of global change. The investigators chose the Tugen Hills because of its location amid the basic hominid territory, and because the region has had its soil layers of the past 15 million years exposed. Furthermore, the

---

Background

As early as 1975, Yves Coppens of the College of France in Paris postulated climate as a possible cause for the development of human traits (*see also* "The State of Paleoanthropology," p 21). His first hypothesis was based upon excavations of deposits from 4 million years ago to 1 million years ago along the Omo River in Ethiopia, where he observed evidence of vegetation changing in response to a drying of the climate during that period. In 1982, Coppens expanded his original idea to include all of Africa east of the Rift Valley, which he claimed had become drier after formation of the Rift and its associated mountain ranges about 8 million years ago. This theory was given wider prominence by a detailed article on the subject published in the May 1994 *Scientific American*.

period of greatest interest, the late Miocene and Pliocene epochs, when the first hominids arose, was already securely dated.

Climate is often inferred by locating plant spores or seeds or other fossils. However, not only are these clues sometimes absent, but they can also fluctuate wildly with short-term climate variations (decades to thousands of years). Soils are thought to reflect long-term climate by another type of "fossil" entirely—the ratio of two different stable isotopes of carbon that the soil contains. This ratio is a marker for the

TWO PATHS FOR CARBON

Green, or photosynthesizing, plants have evolved two different pathways for carbon utilization, known as C3 and C4. The C4 pathway is used by woody plants, such as trees and shrubs, along with forbs (annual plants other than grasses) and certain grasses that prefer wet, cool growing seasons. The C3 pathway is the more common one for grasses, many of which are found in drier or warmer conditions—think of the American Great Plains, where the native prairies form in a continental climate (hot summers, cold winters) with very limited rainfall. Sedges, which are very similar to grasses but have solid instead of hollow stems, also use the C3 pathway.

The designations C4 and C3 refer to the number of carbon atoms in the acid produced as the first stable stage in photosynthesis. The type of acid produced by C4 is called malic acid, while the C3 acid is phosphoglyceric acid.

The differences between C3 and C4 plants go beyond their direct response to climate. The two types of plants are also different in their response to carbon dioxide in the air, with C4 plants better able to utilize low levels of carbon dioxide than C3 plants. Thus, recent experiments have been conducted to determine what the effect will be of human activities that increase the amount of carbon dioxide in the air. At high levels of carbon dioxide, the C3 plants become more efficient than the C4 plants. Thus, global warming caused by carbon dioxide's greenhouse effect might change the types of plants suitable for agriculture (see "The Future of Climate," p 525).

It is thought that green plants originally appeared at a time in Earth's history when the climate was warm and humid due to increased carbon dioxide in the air and the resulting greenhouse effect. Thus, the first green plants used the C3 pathway. Later, when the climate changed and there was also less carbon dioxide in the air (see "Climate Clues to the Past," p 515), C4 plants evolved to deal with changing conditions.

Normal carbon dioxide comes in two forms: one in which the carbon is the very common carbon 12, and another in which the carbon is the slightly heavier carbon 13 (ignoring for the moment the possible kinds of oxygen that can be in a molecule of carbon dioxide). The ratio of carbon 13 to carbon 12 that is stabilized by photosynthesis is slightly higher for a plant using the C4 pathway than for one using C3. Thus, carbon remains from plants can be used to identify the plant that took the carbon from the air and put it to use.

mix of plants preferring a particular climate. At Tugen Hills, the ratio is the same in Miocene soils as it is in present-day soils, which suggests that the vegetation—and therefore the climate—is the same today as it was then. Today, the vegetation of the Tugen Hills consists of grassy bushland set about with acacia trees, but forested in regions along the rivers. Furthermore, the samples from different ages suggest that this mixed dry-climate ecosystem has not changed significantly in the past 15 million years. Kingston notes, "we didn't see any evidence for a savanna." Thus, the proposed abrupt change from tropical rain forest to either pure savanna or pure woodland did not happen in this part of the Rift Valley.

(Periodical References and Additional Reading: *Nature* 10-28-93, p 833; *New York Times* 12-14-93, p C1; *Science* 5-13-94, p 955; *Scientific American* 5-94, p 88; *New York Times* 5-17-94, p C1; *Discover* 1-95, p 38)

OUT OF AFRICA, BUT WHEN?

Very few today doubt Darwin's theory that the human line originated in Africa—even Jeffrey H. Schwartz, who believes humans are more closely related to orangutans than to African great apes, would put the ancestors of humans in Africa. Despite the agreement on an African origin, almost every anthropologist seems to have a different notion of when the transition out of Africa to the rest of the world took place. Schwartz thinks that the ancestors moved out as soon as the African tectonic plate drifted into the Eurasian one—that is, as soon as was geographically possible. This event was millions of years before the first hominids appeared, setting the stage for humans to evolve in Asia, and hominid great apes to evolve in Africa.

A more conventional view has been that humans evolved in Africa and were confined to that continent until about 1 million years ago. While in Africa, various hominids arose and became extinct, leaving only a single species, *Homo erectus*, by about that million-year-ago date. *H. erectus* prospered and spread into Asia and Europe. A few hundred thousand years later—no one is really sure how many—some population on one of the three Old World continents evolved from *H. erectus* into *H. sapiens*, which then displaced the remaining *H. erectus* populations.

The broad understanding of the origin of humans was complicated in the late 1980s and early 1990s by a succession of arguments about human migrations based on DNA analysis. The 1987 analysis of mitochondrial DNA that came to be known as the "African Eve " hypothesis claims that every human alive today is descended from one woman who lived in Africa about 200,000 years ago. After the research supporting the hypothesis was published, the first paper advocating this hypothesis was shown to be flawed, and a revised version was developed in September 1991 by members of the original team. This analysis reached essentially the same conclusion, but in February 1992 the revised version was shown also to be flawed. Even very general treatments of the use of DNA to locate an ancestor in time and space were shown to be inadequate. In April 1994, however, Sarah Tishkoff and Ken Kidd of Yale University in New Haven, Connecticut, used something completely different in DNA studies of the problem, avoiding the pitfalls of the previous approaches. Their new method is based on nuclear DNA instead of mitochondrial DNA, and is closely related to techniques used for determining the presence of a "genetic bottleneck" in a population. Tishkoff found great variation in a DNA segment in sub-Saharan Africans, much less in Northeastern Africans, and little variation in people from anywhere else. Kidd interprets this as meaning that a small population of Africans became isolated in the Northeast. Members of that Northeastern group eventually set out from there and became the ancestors of all other living human beings. This concept is at least consistent with the original African Eve hypothesis. Thus, although considerable evidence points toward the idea that all present-day humans are descended from a recent African ancestor, this hypothesis is not, in the eyes of most anthropologists, proved. Even without the confusion and controversy engendered by DNA analysis, plenty of controversy abounds based on various interpretations of the fossil record itself.

Early Humans in Asia

Yuri A. Mochanov, a respected Russian anthropologist, claims to have evidence that the first humans evolved in Siberia about 3 million years ago, but he has no support for this view among other anthropologists, even in Russia. Eugène Dubois was the first individual to search in Asia for the earliest human ancestors, succeeding in discovering the first known generally agreed-upon member of our genus other than *Homo sapiens*. The fossils he found in 1891 in Java are now recognized as *Homo erectus*. In the 1920s, Roy Chapman Andrews led well-publicized expeditions to the Gobi Desert in central Asia, the main purpose of which was to search for fossils of human ancestors, although Andrews did not find any. In the 1930s, however, excellent fossils of *H. erectus* were found in China. It was not until recent decades that *H. erectus* fossils were found to some degree in Africa.

Some anthropologists began to reason as follows: Although some of the specific fossils establishing the travels are in dispute, *Homo erectus* evidently traveled far. There is some evidence that *H. erectus* was present on all the Old World continents a half million years ago or more. Since the various "races" of humans seem to be associated to some degree with the different continents, perhaps *Homo erectus* evolved into *Homo sapiens* at least three times, and possibly more often. The proponents of this idea purport to see typical Mongolian features in *H. erectus* specimens from Asia, and Caucasian features in those from Europe.

Other anthropologists hold to the Darwinian notion that humans evolved once in Africa, and then migrated. In the African-origin theory, *Homo erectus* spread about the world, and then at a later date was displaced by *Homo sapiens*, which also arose in Africa (although not necessarily as early as in the African Eve hypothesis).

In 1992, Li Tianyuan of the Hubei Institute of Archaeology and Dennis A. Etler of the University of California at Berkeley entered the controversy. They reported on two largely crushed skulls discovered by Li and coworkers in 1989-90 in Hubei Province. When they claimed that the skulls, dated at 350,000 years before present, were archaic *Homo sapiens*, Li and Etler were supporting the multiregional side of the controversy. At that early date, any *H. sapiens* found in Asia would have to have evolved there from the *H. erectus* that had come to Asia from Africa more than half a million years earlier. But many anthropologists felt that the skulls' condition was too poor, and therefore the dating too imprecise to provide grounds for definite conclusions.

Another issue involves the timing of the arrival of *H. erectus* in Asia. In 1992 and 1993, Garniss Curtis and Carl C. Swisher III of the Berkeley Geochronology Center used new argon-isotope techniques to redate *erectus* skulls from Java, concluding that they were much older than previously believed. The new dates were published in the February 25, 1994 *Science* and announced by Swisher the day before. Curtis and Swisher concluded from potassium-argon dating that pumice samples from two sites where fossils had been found were spewed from volcanoes 1.7 and 1.8 million years ago. Thus, the fossils are probably also that old. Not everyone agrees with the dating, mainly because it is not completely clear that the pumice is associated directly with the fossils, but most anthropologists think Curtis and Swisher have established the new dates.

The Curtis and Swisher dates indicated that *erectus* reached Java just a short time after evolving in Africa, if indeed the first protohumans in Asia were *H. erectus* (*see* "The Species Problem," p 35). Swisher suggested that at least two species had been lumped together as *H. erectus*. Indeed, the new dates open up the possibility that some earlier ancestor could have traveled from Africa to Asia, although there is currently no fossil evidence to support that idea.

Meanwhile, although there is still considerable doubt about the 3-million-years-ago date that Mochanov favors for Siberian occupation, new dating techniques suggest that the stone tools he has been finding for the past dozen years at Diring-Yuiakh are about 500,000 years old, which would put them in *H. erectus* territory. This provides yet another piece of evidence that *erectus* could get about, even in the cold, far north.

Apparently one of the *erectus* homelands in Asia was what is now northern Vietnam. Vietnamese scientists have found large numbers of teeth from assorted species of primates in a limestone cave in the region, some of which are identified as *H. erectus* in a paper describing the finds (published in November 1994 by the American Museum of Natural History in New York City). Teeth last longer than bones and, because they are smaller than many bones, are also more mobile. The teeth are thought to have washed into the cave, leaving any bones behind to disintegrate with age. Teeth found thus far include some from orangutans and a close orangutan relative, three from *Gigantopithecus blacki* (thought to be a giant orangutan relative), and nine thought to be *erectus*. Some of the hominoid teeth are not identified with any known species. Among the scientists who have studied the site and the fossils is Jeffrey H. Schwartz, who hopes to find evidence to support his hypothesis that humans are closely related to orangutans.

In any case, neither the Hubei skulls, the Siberian tools, the Java dates, or the Vietnamese teeth will settle the question of human evolution in Asia. For example, not all anthropologists see in the Hubei skulls the modern features that Li and Etler find. Better dates for Asian and African *H. erectus* fossils are required to achieve a consensus. The discovery of some new bones would be helpful, as well.

Early Humans in Europe

After restudying evidence of the earliest Europeans, Wil Roebroeks and Thijs van Kolfschoten of Leiden University in the Netherlands wrote their conclusion in the September 1994 *Antiquity* stating that wherever else humans roamed millions of years ago, it was not Europe. In their view, the first evidence they can find of hominids on the European continent appears in the record only half a million years ago. Roebroeks and van Kolfschoten believe that apparent stone tools found in sites dated earlier than 500,000 years ago are either intrusions from a later time or natural rock formations mistaken for primitive tools. The two had put forth the same point of view in November 1993 at a workshop on the earliest Europeans. Further support for their conclusions was delivered at that workshop. For example, a well-known site in central Italy at Isernia was redated from 730,000 years ago to 500,000 years ago.

However, the earliest skeleton found in Europe has also been dated at 500,000 years before present. If the date is right (and it could be 100,000 years later), this would mean that its owner was among the early immigrants. The intact skeleton, found by cavers near Altamura, Italy, on October 7, 1993, appears to be archaic *Homo sapiens*, not *H. erectus*. In either case, the skeleton appears to approach typical Neandertal features.

The Georgia Jawbone   On the other hand, a lower jawbone with 16 teeth still in place was found by archaeologists near Tbilisi, Georgia, and appears to most experts to belong to *H. erectus*. The archaeological team was led by Vachtang Dzarparidee of the Georgian Academy of Sciences and Gerhard Bosinski of the University of Cologne.

The Georgian mandible, first described in December, 1991, cannot be dated precisely with methods now available, but it was found on top of lava that can be identified by paleomagnetic reversals as belonging either to a period 1.6 million years ago or one 0.9 million years ago. Furthermore, the assemblage of animal fossils found with the jawbone (saber-tooth cats and other animals) is more in line with the older date. A date of 1.6 million years ago matches the newly established time for *erectus* fossils in Java. When the Georgia jawbone and the Solo river dates are looked at together, they suggest that *H. erectus* left Africa for the East (although Georgia is technically in Europe, it is east of all of Africa except the horn of Somalia) about 200,000 years after the species first appeared.

Boxgrove Man, an Early Englishman   Late in 1993, a shinbone (tibia) found at the Boxgrove quarry near Chichester in Sussex, England, was identified as *H. heidelbergensis*. Furthermore, it was dated as between 524,000 and 478,000 years old, so Geoffrey Wainwright of English Heritage—a government group that financed the dig—called Boxgrove Man "our earliest European." The May 26, 1994 issue of *Nature* that carried the discovery of Boxgrove Man asked this question as the main headline on its cover: "The Earliest European?"

The discovery of Boxgrove Man was made in December 1993 by amateur archaeologist Roger Pederson, working with professionals Mark B. Roberts and Simon A. Parfitt of University College London, and Christopher B. Stringer , a paleontologist from the Natural History Museum in London. Boxgrove Man, whose tibia resembles that of a 180 lb (80 kg), 6 ft (3 m) tall male, was located in geologic time by examining the skeletons of a species of water vole (*Arvicola terrestris*) found in the same deposit. The water vole jaws found near Boxgrove Man's tibia contained a molar of a type known to have evolved in voles about half a million years ago.

An important advantage of Boxgrove Man over other early European fossils, such as the Mauer Mandible from Europe, is that the bone was found amid ample evidence of the hunting and butchering life of the time. Animals such as horses were evidently killed at the site, and stone butchering tools—hand axes, as these tools are often called—were made at the site, in order to cut the animals into usable pieces. The site contains many bones showing signs of expert dismemberment, flakes from construction of tools, and the leftover tools themselves. Apparently Boxgrove Man and his cohorts did not carry tools from place to place, as would modern hunters.

Early Spaniards   The cave of Sima de los Huyesos ("pit of bones") in the Sierra de Atapuerca near Burgos, Spain, was inhabited by cave bears and used by humans some 300,000 years ago. The cave bears seem to have been the primary occupants, since the bones of 250 bears have been found from that period. Since the mid-1970s, human fossil fragments have been discovered amid the cave-bear bones. The bones of humans seem to have been tossed into a pit in the cave, perhaps as part of a very early "cult of the cave bear." The popular novel about a cave-bear cult is set in Europe more than 250,000 years later, however. The Sima de los Huyesos people were very early humans.

Although some Neandertals date from this early time, the Spanish bones may not be Neandertal remains. Opinions differ. Some say that the bones are the remains of *Homo erectus*, *H. ergaster*, or *H. heidelbergensis*, the species (by one or more of those names) that preceded modern humans. By most accounts, however, the Spaniards were *H. sapiens*, although of the race anthropologists term "archaic." Three well-preserved skulls provide the best evidence for this. One of the skulls has a brain capacity of 1,390 cubic centimeters, well within the 1,200-1,700 cc range of modern humans. On the other hand, the Neandertals also had large brains, and the heavy brow ridges noted on the Spanish skulls are similar to those of the Neandertals.

However, two other fragmentary skulls found in the Sierra de Atapuerca are believed to be *H. erectus*, and to date from half a million years ago. These were found in a cave called Gran Dolina, which previously had yielded a rock fragment believed to be a crude stone tool from 700,000 years ago. Like Boxgrove Man, these fragments were dated by the remains of rodent fossils found with the hominid teeth and bone fragments. The hominid remains were identified as being from a child and a teenager.

The lack of tools or human garbage found in the cave seems to support the idea that the cave was not a residence. Instead it appears that the bones (both the older and the more recent specimens) were placed in a pit in the cave.

The research at Sierra de Atapuerca continues under the leadership of Juan-Luis Arsuaga of the Complutense University in Madrid, Spain.

And Yet Another Idea

The argument between proponents of an African genesis and people who believe in parallel evolution may be moot if an entirely different point of view is true. Suppose, as this third alternative states, both sides are partly right. Say *Homo sapiens* developed first in Africa, where all other hominid species evolved, at some yet-to-be-determined time greater than 100,000 years ago, and suppose that after small populations of *H. sapiens* established themselves in Asia and Europe, something else happened. Perhaps 80,000, 70,000, or 40,000 years ago, events caused those small populations in Africa, Asia, or Europe to proliferate wildly. Because each proliferation came from a different subpopulation, the ancestors would be distinctly different in their genetic makeup. The result would be similar to the parallel-evolution hypothesis, but it would be achieved without the need for three or more different groups of *H. erectus* (or other pre-*sapiens* hominids) to evolve convergently but separately into modern *H. sapiens*.

In the fall of 1993, the alternative theory and evidence supporting it was put forth in *Current Anthropology* by Henry Harpending, Mark Stoneking, and Stephen Sherry of Pennsylvania State University, Alan Rogers of the University of Utah, and coworkers. Harpending recognized that graphs of data on mitochondrial DNA gathered by researchers pursuing the African Eve hypothesis fit a different hypothesis better. He and Rogers had been using mitochondrial DNA to determine how large populations had been in the past, as well as their speed of growth from a founding group to the present. When viewed in this context, the DNA samples collected from human populations around the world formed a pattern consistent with a past that includes periods of small founder populations followed by periods of very rapid growth. Rogers used the DNA data to calculate that different populations of humans had suddenly expanded after various population bottlenecks—times with very few breeding individuals—in the past. In Africa, there was the imprint of an expansion about 80,000 years ago. The DNA record is consistent with a European expansion about 40,000 years ago, which matches the sudden displacement of Neandertals by Cro Magnons about that time. Several different populations appear to have passed through the bottleneck and begun to grow rapidly about 50,000 years ago. Among the explanations for rapid population growth are changes in climate—perhaps following a bottleneck caused by a giant volcanic eruption that lowered temperatures in an already cold global climate—and development of new cultural improvements, such as language or better tools.

(Periodical References and Additional Reading: *New York Times* 2-3-92, p B7; *New York Times* 5-19-92, p C1; *Nature* 6-4-92, p 404; *New York Times* 6-4-92, p A8; *Science* 6-12-92, p 1521; *Science* 8-14-92, p 873; *Discover* 1-93, p 24; *Nature* 4-8-93, pp 501 & 534; *New York Times* 4-20-93, p C8; *Science* 10-1-93, p 27; *Science* 11-12-93, p 991; *Discover* 1-94, p 40; *New York Times* 1-11-94, p C1; *Science* 2-4-94, p 61; *New York Times* 2-24-94, p A1; *Science* 2-25-94, pp 1087 & 1118; *Science* 4-15-94, p 350; *Nature* 5-25-94, pp 275 & 311; *New York Times* 5-25-94, p A8; *Science* 5-27-94, p 1248; *Archaeology* 5/6-94, p 16; *Science* 8-5-94, p 735; *Discover* 9-94, p 80; *Science News*, 10-8-94, p 235; *Archaeology* 11/12-94, p 251; *Science* 12-4-94, p 1726)

TWO NEW CAVES

It is remarkable that any paintings or engravings from the latest Ice Age have survived a passage of time in excess of 10,000 years. Even tombstones from just a hundred years ago bear messages that have already faded badly, and those from several hundred years ago are often totally obliterated. Outdoor paint must be renewed every few years on houses, and even museum paintings several hundred years old, although kept in a sheltered environment, need special protection and restoration to be visible. Thus, the most ancient paintings and engravings known tend to be from special environments—within sealed tombs, or deep in caves with a constant environment. Rarely do works exposed to the open survive by some series of happy accidents. Even then, they can succumb to less happy circumstances before our eyes. Cave paintings at Lascaux and Altamira have been injured by the breath of visitors. Few are allowed to visit any more.

Despite the difficulties associated with preservation, more than 200 painted or engraved caves from the Ice Age have been found in southeastern Europe, most containing just a few rude sketches. In the early 1990s, however, two new sets of spectacular paintings from the very distant past were found after the paintings had spent thousands of years in environments even more closed than most. Meanwhile, outdoor engravings that have survived up to now have been threatened with flooding and lost as a result of governmental action.

Art Before the Flood

Water is a particular enemy of ancient art, especially acidic water which can dissolve some common rock minerals. In one of the big stories of Ice Age art in the early 1990s, however, water preserved a portion of the art in a cave, while destroying the rest of it.

Sea level has varied considerably even in fairly recent times, rising dramatically when the glaciers melted at the end of the most recent cold phase of the Ice Age. The level of the oceans and connected seas, such as the Mediterranean, then rose about 360 feet (110 meters). Consequently, we know little about human settlements along the sea from the last glacial phase, except for a few places where there is a steep vertical descent into the ocean. In those places, sites that were rather far above the sea during the time of ice are closer to it today, but there may still be evidence of hunting for shellfish or other food associated with the seacoast.

In 1985, however, diver Henri Cosquer made a discovery that attaches his name to one of the most remarkable sites from human prehistory. At a depth of 121 ft (37 m), he found the underwater entrance to a large cave under Cap Morgiou on the French Mediterranean Coast, 7.5 mi (12 km) from Marseilles. A 450 ft (140 m) tunnel filled with water leads to two large limestone caverns that are otherwise unconnected to the outside world. Cosquer continued to explore his cave, and in July 1991 Cosquer encountered and photographed the first evidence that the cave had been visited by earlier humans, who had used the cave as a place to leave graffiti and artwork. Specifically, he found hand stencils painted on cavern walls. Returning to the cave, Cosquer also found drawings and engravings of animals, as well as geometric signs and other artwork.

Expeditions to Cosquer Cave (as it is now known) by the French Department of Sub-Marine Archaeological Research, which began in 1991 when Cosquer brought French underwater archaeologist Jean Courtin to see the artwork, have established that humans were using the cave as early as 26,000 B.C. At that time, the cave entrance was on dry land and open.

The first art created by human visitors are hand stencils, similar to those that occur throughout recent human history (found as petroglyphs in the southwestern United States and still made by Australian aborigines). A hand stencil is usually made by placing the hand palm down against a wall and blowing paint on it. Removing the hand leaves its outline as the unpainted part of the surface. Because the paint used in early Cosquer hand stencils came from organic sources, it was possible to use carbon 14 to date the stencils. One date was 25,000 B.C. Other radiocarbon dates from pieces of charcoal found in the Cosquer cave are as early as 25,870 B.C.

The hundred or so animal paintings and engravings came almost 10,000 years after the earliest hand stencils, also according to carbon-14 dating. The paintings are almost all done in charcoal or with a black oxide of manganese. The paintings are Solutrean, in the conventional system for naming this period in early French prehistory. These paintings are much earlier than the famous cave paintings at Lascaux (circa 15,000 B.C.) or Altamira (the red-and-black bison there were thought to be from about 12,000 B.C. until a 1992 carbon-14 dating of 14,000 B.C.), although close in age to the earliest animal engravings at Altamira. The common species shown at Cosquer cave are horses, chamois, and ibex, common game animals of the time. There are also several bison, "Irish elk," and red deer, as well as the head of some sort of cat. There were formerly many more paintings, but rising water flooded the lower part of Cosquer

WHEN IS A PENGUIN NOT A PENGUIN?

The answer to the riddle is either "when it is French" or "when it is an auk." At one time, a very common flightless bird, *Pinguinus impennis*, was found both in northern waters, and as far south as the Mediterranean during the Ice Age. In English, this bird was called the auk or the great auk. (Auks are related to various similar-looking black-and-white birds, such as the murre and the puffin). Sailors were quite familiar with auks because ships often docked at islands where the great auk was plentiful. The islands were chosen to provide the crew an opportunity to take birds and eggs to replenish the ship's stores. Taking the flightless birds and their eggs, both concentrated in giant rookeries, was easily accomplished. The result of this practice was that the last great auks were killed on June 4, 1844, and the bird has been extinct since that time.

The French word for the great auk was and is *pingouin*, pronounced "PEE-gwaaa." The great auk is white on the front and black on the back, flightless, and walks on two short legs. When sailors on Vasco de Gama's and Magellan's voyages encountered similar birds below the equator, they no doubt thought that the birds were a species of auk. From the French name, which has stuck, the "southern auk" was called the *pingouin* in French, which the English and Spanish heard as penguin and *pingüino*. As a result, the English distinguish between the northern auks and the unrelated southern penguins when speaking and writing. Similarly, Spanish speakers know the difference between *alca* and *pingüino*. But to this day, the French call the black-and-white birds of both north and south by the same name.

cave long ago, eroding any paintings at those levels. Some paintings today simply stop at the water line.

The location of Cosquer cave near the sea may account for some unusual animals not found in other early cave paintings, notably three depictions of birds that were first announced in the popular press as "penguins," although no penguins have ever lived in the Mediterranean. In reality, the birds are clearly auks of some sort, but because the cave is being investigated by the French, who use the same word for penguin and auk, mistranslation is easy. The other sea animals that definitely can be identified are eight seals. Some viewers also find sea motifs in some of the more abstract figures.

Further study of the cave may take years. In the meantime, the French government has sealed the underwater entrance to the cave to prevent damage by unauthorized persons—three swimmers from Grenoble drowned on September 1991 in their effort to reach the paintings. There is apparently no entrance from dry land at this time, and there is some fear that opening one would adversely affect the condition of the paintings. Any change in humidity is dangerous for cave paintings, but the paintings in Cosquer cave, where humidity is always very high, are also easily obliterated by just the slightest touch.

Outdoor Art Before the Flood

In another newly discovered Ice Age gallery, human-induced flooding almost resulted in water encroaching upon and obliterating ancient engravings instead of protecting them.

Although archaeologists believe that art was placed in the inaccessible depths of caves for ritual reasons, there is no reason why similar art could not also have been painted or engraved in places where it would have been more easily seen. But most exposed art from Ice Age times would long ago have been erased by weathering and erosion, so there is little of it around today. In desert conditions, such as the southwestern United States, parts of Australia, and northern Africa, exposed rock art is well known, although more often thought to be hundreds, not thousands, of years old. Such art is often in less accessible canyons. Local people have always known that the art was there; science has learned of it more recently.

In 1981, a young archaeology student reported to his professor that near his home village in the region near Vila Nova de Fozcôa, Portugal, were engravings of animals that appeared to be from Stone Age times. Investigation showed that the site, known as Mazouco after the local village, did contain a few engravings from the late Ice Age. The professor published the description of the work. Apparently the student, Nelson Rebanda, determined that he would get the credit for his next discovery himself.

Eleven years later, Rebanda learned that local people knew of mysterious animal engravings along a part of the Côa River known locally as Hell's Canyon, probably because local tradition had it that the engravings were magic and dangerous. But the lower part of Hell's Canyon had already been flooded by a hydroelectric project built in 1982. Furthermore, the rest of the canyon was scheduled for flooding by yet another hydroelectric dam. Rebanda located the engravings and set to work copying the art. He also notified his principal employer, the Portuguese Institute for Architectural and Archaeological Patrimony, and the government-owned electric utility that the new dam would cover important archaeological remains. But his message failed at first to sufficiently impress either the institute (which has more to do with architecture than with archaeology) or the utility. Although Electricidade de Portugal paid Rebanda to copy the art, the utility refused to halt work or to lower temporarily the level of the existing reservoir so that already flooded images could be copied. Construction of the new dam started in September of 1994. Finally, on November 7, 1994, Rebanda realized that he could not save the images by himself. Furthermore, he realized that few would believe he was working on anything of significance if they had only his word, his photographs, and his tracings to go by. Panicked, he hastily summoned experts in rock art from Lisbon, phoning Mila Simoes in the middle of the night. Simoes and her husband, archaeologist Ludwig Jaffe, confirmed the antiquity of the art—she believes that it dates back about 20,000 years—and demanded that the discovery be made public. Rebanda was aghast at the suggestion, because he wanted credit for first publication; he was writing a book about the engravings. Simoes went public with the story anyway. On December 13, 1994, the Portuguese government made Hell's Canyon a national monument and finally stopped construction on the dam.

More than 60 images have been discovered—typical Ice Age animals such as aurochs (the ancestor of cattle), horses, ibexes, and deer, all of which inhabited the somewhat inhospitable interior of the Iberian peninsula near the end of the recent Ice Age. Many are pictured in ways common to cave art of about the same time, but some are more freely depicted than in other known engravings. The images, hammered or scratched into the rock, range in size from about 6 in (15 cm) to 6 ft (2 m). Rebanda thinks that about 30 more engravings may be in the previously flooded part of the canyon.

Various experts have now visited the site. Dating of engravings is more difficult than dating of paintings. Style is important, as well as such details as matching the patina of the incised rock to the rock around it (recent forgeries would not match). The choice of animals and the style of drawing have convinced experts that the engravings are from about 18,000 B.C. Archaeologists are now searching the region for tools or other evidence that could be used to more firmly establish this date.

Chauvet Cave, near Vallon Pont-d'Arc, France

The main known sites for Ice Age art have been southern France and northern Spain. At the very end of 1994, three explorers in southern France, led by Jean-Marie Chauvet, found what appears to be the greatest group of cave paintings of them all. The paintings are on the limestone walls of a series of caverns near the Ardèche River and the town of Vallon Pont-d'Arc. A landslide had sealed off the entrance for thousands of years. On December 18, 1994, the three explorers became perhaps the first humans to enter the cave since the paintings had been made. Mindful of past accidental destruction of important archaeological evidence, the explorers put plastic down where they walked in the caverns. As a result, even the footprints of cave bears are preserved. Archaeologists have found, in addition to more than 300 large paintings, various other undisturbed evidence of the passage of Ice Age humans through

ICE AGE ANIMALS

The various mammals and birds depicted in Cosquer and Chauvet caves include most of the larger creatures that could be seen in southern Europe some 20,000 years ago. Many of these have since become extinct, often through the actions of humans, although a hotly debated issue concerns whether or not humans hunted animals to extinction some 10,000 years ago. Among the large animals of the Ice Age that have since become extinct, or nearly so, are the following:

Aurochs,   or *Bos taurus.* The auroch is classed as the same species as modern domestic cattle, but the "breed" was a giant one—with a shoulder height up to 6 ft (1.85 m) in males as compared to 3.5 ft (1.1 m) in modern cattle. The auroch survived the Ice Age in isolated pockets, but the last known aurochs had died by 1627. There is a disputed report of a herd surviving after this time in northern Iraq.

Bison (European, or Wisent),   or *Bison bonasus.* When this close relative of the American buffalo became extinct in the wild during and immediately following World War II, it was luckier than the aurochs: a few animals had survived in zoos. When things quieted down after the war, zoo-bred bison were reintroduced to forests along the border between Poland and Belarus (then part of the Soviet Union), and in the Caucasus mountains, where a few hundred continue to survive in the wild today.

Cave bear,   or *Ursus spelaeus.* About the size of a large grizzly bear of today, the cave bear was primarily a plant eater. It became extinct about 10,000 years ago.

Great Auk   *See* "When is a Penguin Not a Penguin," above.

"Irish elk" (Great Elk),   or *Megaloceros.* The common name for this large deer comes because huge antlers and even entire skeletons have been found in Irish peat bogs. However, *Megaloceros* was actually known all over Europe until it became extinct near the end of the Ice Age. Although its antlers exceed those of all other deer, sometimes measuring 11.5 ft (3.5 m) across, its body was somewhat smaller than that of the largest moose.

Woolly Mammoth,   or *Mammuthus primigenius.* This species, which roamed over much of the Northern Hemisphere during the Ice Age, became extinct about 10,000 years ago. Remains of about 40 mammoth, preserved in permafrost, are known to science. Furthermore, a pygmy species of mammoth is now known to have survived until about 2000 B.C. on Wrangell's Island off the north coast of Siberia. It is thought that the woolly mammoth was one of the main targets of Ice Age hunters. In some cultures, houses were constructed of mammoth bones.

Woolly Rhinoceros,   or *Coelodonta antiquitatis.* Although most modern rhinos are known for their nearly hairless skins, the nearly extinct Sumatran rhinoceros, a relative of the now extinct woolly rhino, is covered in hair. Although Ice Age humans may have hunted the woolly rhino for its meat, they were never the staple that the mammoths were in some cultures.

the caves, including bones, footprints, tools, and hearths. The caverns have been christened Chauvet Cave, after the leader of the discovery party.

The paintings in Chauvet Cave range from simple but inscrutable geometric designs and hand stencils to great groups of painted animal figures. New animals to the Ice Age bestiary include a panther and several owls never before found in Ice Age art, as well as a rare depiction of a hyena. Preliminary estimates indicate that the paintings were made about 18,000 B.C. This was at a peak of the extension of ice into Europe, and many of the animals illustrated are cold-adapted species that have long since vanished—woolly-haired rhinoceroses and mammoths , along with cave bears and cave lions. Unlike other caves, which rarely depict the rhinos or bears, these two species are prominent. There are more than 50 woolly rhinos at Chauvet. Furthermore, a cave bear skull is set before a painting of a group of bears, perhaps as part of a religious rite. Other species include the European bison, the aurochs (ancestor to modern cattle), the goat, and the horse.

The paintings are produced using charcoal and the minerals limonite (also known as yellow ocher), black manganese oxides, black hematite, and red hematite (also known as red ocher). According to prehistorian Michel Lorblanchet of the National Center of Scientific Research in France, cave paintings such as those in Chauvet were not made using brushes. Instead, a pigment was chewed until it was well mixed with saliva, and the mixture was spit onto the cave wall. Sometimes it is suggested that a tube, like a blowgun, was used to direct the color. The most obvious example of this technique are the many ancient hand stencils, clearly made by placing a hand against the wall, and blowing or spitting paint on its back to leave the outline.

(Periodical References and Additional Reading: *New York Times* 10-20-92, p C1; *Archaeology* 1/ 2-93, p 27; *Archaeology* 5/6-93, p 37; *New York Times* 12-27-94, p C1; *New York Times* 1-19-95, p A1; *New York Times* 1-22-95, p IV-6; *New York Times* 1-24-95, p C10)

THE STATE OF ARCHAEOLOGY

Archaeology is still a relative new science when compared to astronomy, biology, chemistry, or physics, and is often omitted from the natural sciences altogether, being considered more closely allied to the social sciences. Indeed, archaeology is a close cousin to history. In this book, archaeology is treated as a science largely because its results are regularly reported in the science pages of newspapers and popular magazines, as well as in journals. Furthermore, the activities of archaeologists fit within the parameters described by the new sociologists of science, who define science as a kind of activity practiced by scientists, and not as a body of knowledge. That is, societies of archaeologists are organized and publish in the same general way as, for example, societies of astronomers. Modern archaeologists are also increasingly dependent upon such scientific tools as radiocarbon or other physical means of dating, as well as on computers and electromagnetic methods of finding objects. Recently, earth-orbiting satellites have also been employed for site detection and mapping. Finally, archaeology is closely allied to studies that are more often considered science, such as physical anthropology or paleoanthropology.

Archaeology grew out of two main interests in Europe, Africa, and Asia—the rediscovery of the classical world of antiquity and fascination with the clearly observable distant past of Egypt. From the beginning, the Christian and Jewish desire to locate history associated with the Bible was also a factor. This Old World archaeology then moved on from the Egyptian desert to the deserts of the Near East, where the fabled cities of Ninevah and Babylon—famous in the West from Biblical and Greek accounts—were found to be part of a much larger chunk of the history of Western civilization. More recently, Old World archaeology has moved on to the Far East.

When Europeans came into regular contact with the Americas, they were astonished to find not only entirely new and different civilizations extant in Mexico, Central America, and Peru, but also people living in all sorts of societies across both the Americas. It was not until the rise of Old World archaeology in the nineteenth century, however, that scientists and historians began to investigate the antecedents of the societies of the Americas, finding a rich group of previously unknown civilizations and a set of mysteries that have fascinated the archaeologically minded ever since.

Beginnings of Boating

Human beings certainly crossed ocean waters as early as the settlement of Australia, which was probably about 60,000 years ago. Some archaeologists also think that early humans crossed even longer stretches of either the Pacific or Atlantic to settle the Americas, perhaps as early as 40,000 years ago. Thus, there must have been boats, since the distances involved are far too great to swim across. But until recent times, evidence of early boats of any kind was slim. New discoveries in the early 1990s greatly improved our knowledge of boats from early times, although nothing is known yet about what must have been the very earliest boats of all (see also "The Glory that was Greece," p 90).

It has long been known that the Egyptians were sailors. The importance of boats in the Egyptian culture is so great that ships were often buried in tombs or at the base of pyramids. In 1992, archaeologists from the University of Pennsylvania and Yale University, working at Abydos in southern Egypt, found a dozen ships from the third millennium B.C. buried side by side. Although the ships were several hundred years earlier than the earliest pyramids, they may, like those found at pyramids, have been intended to convey a pharaoh to the next world.

Some very early dugout canoes were found in Bercy, France, a suburb of Paris, by workmen digging the foundation of a new building. The canoes are the oldest wooden boats ever found. They had been preserved by being buried below the present bottom of the Seine, so groundwater aided in preservation of the oak. In all, eleven boats were found—two completely intact, one nearly so, and eight in fragments. The largest canoe was about 16 ft (5 m) long. All the canoes are thought to date from between 4300 B.C. and 3700 B.C.

Among the earliest ocean-going boats to be preserved and found is a 54 ft (16.5 m) long oak planked boat, held together with yew bindings, that was probably used in trade across the English Channel about 1400 B.C. It was discovered during road building in Dover, England.

Other Archaeological News

Among the miscellaneous discoveries of the early 1990s, not discussed elsewhere in this book, are the following:

- In 1992, Mark Patton uncovered an enormous cairn made of stone blocks beneath the surface of a hill known to contain a tomb from about 4000 to 3000 B.C. Because it had been hidden in a mound of earth known as La Hougue Bie, this cairn on the isle of Jersey (largest of the Channel Islands) is the best preserved of a number of such structures along the Atlantic coast of Europe.
- A building from about 3900 B.C. was found buried in Lake Constance between Switzerland and Germany. It had been decorated with wall sculptures of female human breasts, some painted with polka dots.
- The earliest known chariots were recognized in 1994, when David W. Anthony of Hatwick College in Oneonta, New York, asked experts from the University of Arizona at Tucson to apply mass spectrometer radiocarbon dating to bone samples from a grave in northern Kazakhstan. The grave was part of a complex of tombs belonging to people known as the Sintashta-Petrovka culture, who are considered a likely ancestor group of the Aryans who later invaded India. Imprints of spoked wheels in the tombs show that the Sintashta-Petrovka were the first, or among the first, to build lightweight wheels. The new dates, ranging around 2000 B.C., are a hundred or more years earlier than clay seals from Anatolia which bore the previously identified first evidence of spokes.
- In the summer of 1992, Canadian archaeologist Grace Rajnovich discovered a 10 in (25 cm) high sandstone head of a man in the Gotteschall Rockshelter of southwestern Wisconsin. The rockshelter, known for its wall paintings, may have been used by Native Americans for religious purposes around 1000 B.C.

- Itil, the lost capital of Khazaria, was found 100 ft (30.5 m) below the surface of the Caspian sea. Russian archaeologist Muran Magomedov of the State University of Dagestan used Arab sources along with aerial photography and underwater archaeology to locate the city that ruled much of southern Russia and Ukraine, including the Caucasus and Crimea, in the fourth to tenth centuries A.D.

(Periodical References and Additional Reading: *Archaeology*, 7/8-92, p 17; *Archaeology*, 9/10-92, p 24; *Archaeology*, 3/4-93, p 22, 23 & 24; *Archaeology*, 5/6-93, p 27; *Science* 10-8-93, p 179)

TIMETABLE OF ARCHAEOLOGY TO 1992

1401 B.C.   The first restoration of the Sphinx is conducted in Egypt. During ancient times two more efforts to restore the giant statue will take place.

1350 B.C.   A freighter loaded with copper, tin, resins used in making perfume, and scrap gold and silver, as well as miscellaneous treasures such as hippopotamus teeth, is wrecked on the southwest coast of Asia Minor (Turkey) at what we now call Ulu Burun. It will be excavated starting in 1982.

1200 B.C.   A small freighter loaded with copper and bronze is wrecked on the southwest coast of Asia Minor (Turkey) at what we now call Cape Gelidonya. It will be excavated starting in 1960.

About this time, Khaemwase, high priest of Ptah and favorite son of Rameses II, becomes the first known Egyptologist, exploring the tombs of his ancestors.

150 B.C.   A large decorated silver bowl 27 in (70 cm) in diameter, now known as the Gundestrup Cauldron, was made about this time. It was probably hidden in grasses in northern Jutland (now Denmark) about 100 B.C., later buried in peat, and found again in 1891.

36 B.C.   A stone monument in Chiapa de Corzo, Mexico, erected at this time contains the earliest known Mesoamerican date that appears to correspond to a real event.

1586   A 327 ton Egyptian obelisk, brought by the Romans from Egypt in ancient times, is raised to a vertical position by a team led by Domenico Fontana.

1650   William Dougdale, in his History of Warwickshire, notes that stone points found in England are weapons from before the time of smelting, although the common belief of the day is that they were weapons used by fairies that fell from the sky. Dougdale's view was echoed in 1699 by Edward Lhwyd.

Irish Bishop James Ussher sets the date of Creation at 4004 B.C., and the date of Noah's flood as 2349 B.C., based on accounts in the Bible.

1679   Sailor Robert Knox escapes captivity in Ceylon, reporting to the English that the ruins of a great city, Anuradhapura, are present on the island. Later investigators date the founding of Anuradhapura to the fourth century B.C., the time of Alexander the Great.

1722   On Easter Sunday, Dutch admiral Jakob Roggeveen discovers Easter Island, with its mysterious giant stone statues.

1762   The first of the four-volume *Antiquities of Athens* by James Stuart and Nicholas Revett is published. The work will be completed in 1816.

1763   The ruins of Pompeii, already under random excavation, are identified for the first time from inscriptions found at the site; that the town had been covered by volcanic ash in 79 A.D. was well known from Roman writings contemporary with the disaster.

1780   About this time Thomas Jefferson [American: 1743-1826], while governor of Virginia, digs a trench through a mound in the Rivanna river valley near his home, Monticello. He establishes that the mound had been used for burial by earlier Native Americans, although few accept his conclusions at the time, thinking that the numerous large mounds found throughout eastern North America, and especially in the Mississippi river valley, could not have been built by "Indians." Popular explanations of the origins of these mounds include Spanish explorers, displaced Canaanites, or the lost tribes of Israel.

1790   Archaeologist John Frere [English: 1740-1807] discovers flint tools that he believes were made by prehistoric humans. Few agree with him, at the time.

1798   On July 1, a French Army invades Egypt under the command of Napoleon Bonaparte. Included in the army are over 150 scientists, who undertake a thorough analysis of Egyptian life and land. One result is the beginning of archaeology as a science.

1799   Soldiers in Napoleon's army, digging near the Rosetta branch of the Nile, uncover a stone engraved in three different scripts. Translation begins in 1822, and ultimately the Rosetta stone becomes the key to unlocking Egyptian hieroglyphics.

1802   On a bet with drinking companions, Georg Friedrich Grotefend becomes the first to translate a cuneiform text, a Persian inscription from Persepolis. Credit for such translations is usually given to Henry Creswicke Rawlinson, whose translation of old Persian cuneiform is published in 1846.

1803   After 3 years of imprisonment on Malta as a result of the Napoleonic wars, the seventh Lord Elgin (family name Bruce) brings the first of a large number of Greek monuments and statues he has purchased in Athens to England, a task completed in 1812. The Elgin marbles reside today in the British Museum in London.

1809   Between this year and 1828, the monumental *La description de l'Egypte* ("Description of Egypt") is published. Based on drawings, maps, memoirs, and commentary by scientists in Napoleon's expeditionary force, the 20-volume set describes ancient and modern Egypt, using 837 large copper engravings, 47 map sheets, and about 7,000 pages of text.

1815   *Memoir on the Ruin of Babylon*, by Claudius James Rich [English: 1787-1821], is the first factual account in the West that correctly locates Babylon (in Iraq), the ancient capitol of Chaldea, famed from Biblical accounts.

1816   Christian Jorgensen Thomsen [Danish: 1788-1865] classifies objects for a museum collection into those from humanity's Stone Age, Bronze Age, and Iron Age. This is the first time this classification scheme is used.

1817   Giovanni Battista Belzoni [Italian: 1778-1823] opens and enters the great temple at Abu Simbel in Egypt, as well as a number of royal tombs.

1818   Giovanni Belzoni is the first modern European to enter the second largest of the great pyramids of Egypt, the tomb of Pharaoh Khafre (Chephren).

1820   Claudius James Rich learns from local scholars the location of the site of Ninevah (in Iraq), eighth century B.C. capital of Assyria, founded by Sargon II.

1822   Jean-François Champollion translates the hieroglyphic panel of the Rosetta stone, the first translation of Egyptian hieroglyphics.

English antiquarian William Bullock explores Teotihuacán in the Valley of Mexico. He removes objects and casts of works of art for display in England. The exhibit, in 1824, creates the first interest among the English in Mesoamerican civilizations.

1824   *The Wonders of Elora*, by John Seely, describes the great Kailasa Temple in India which was carved from a mountain; this account alerts Westerners to the ancient ruins of the East.

1836   The museum catalog of Christian Thomsen is published, giving wider usage to his classification scheme of a Stone Age, a Bronze Age, and an Iron Age.

1837   Henry Creswicke Rawlinson [English: 1810-1895] succeeds in translating two paragraphs of Persian cuneiform text from an inscription hundreds of feet above the ground that had been placed there by Darius the Great.

1839   John Lloyd Stephens [American: 1805-1852] and artist Frederick Catherwood [English: 1799-1854] embark on an expedition to Central America, with the goal of locating the ruins of Copán and investigating them.

1841   Stephens' and Catherwood's *Incidents of Travel in Central America, Chiapas, and Yucatán* reports their discovery of ruins of the Maya civilization. Its great popularity introduces Mayan studies in the United States and Europe.

1843   Paul-Emile Botta excavates Assyrian sculptures from a mound near Khorsabad (Iraq), the site of the ancient Assyrian city of Dur Sharrukin.

1845   Austen Henry Layard [English: 1817-1894] begins excavation of the ruins of Ninevah.

1846   Ephraim George Squier [American: 1821-1888] and Edwin Hamilton Davis [American: 1811-1888] explore the mounds built by Native Americans in the Mississippi River valley. Among other great mounds, they locate the Serpent Mound in Highland County Ohio.

Jacques Boucher de Crévecour de Perthes [French: 1778-1868] convinces other scientists that prehistoric flint objects are actually tools made by early humans.

1848   Guatemalan administrators Modesto Méndez and Ambrosio Tut are the first outsiders to call attention to the Maya ruins at Tikal, known in previous centuries only to the local Maya descendants.

Lieutenant J. H. Simpson observes the cliff houses of Mesa Verde in Colorado from a distance, but does not investigate.

1850   Auguste-Ferdinand-François Mariette [French: 1821-1881] arrives in Egypt and begins a program of surveying and excavating that leads him to locate the burial places of the sacred Apis bulls in Saqqara, the necropolis of Memphis, an early capital of Egypt.

E. George Squier reports on his discovery of mysterious statues on islands in Lake Nicaragua in Central America.

1862   Edwin Smith, an American Egyptologist, purchases a papyrus that, when translated in 1930, is recognized as a summary of the surgical practices of ancient Egypt of around 2500 B.C. (although the papyrus itself is a copy made around 1550 B.C.).

1864	French botanist Henri Mouhot's posthumous account of the great temples at Angkor in Cambodia is published.
1869	Antonio Zannoni undertakes the first systematic excavation of Etruscan remains, probing a cemetery near Bologna, Italy.
1871	German adventurer Karl Mauch becomes the first European to explore the ruins of Great Zimbawae, a huge ceremonial center built by native Africans in the fourteenth and fifteenth centuries.
1872	The discovery by George Smith in the British Museum of clay-tablet fragments containing a Sumerian account of a world flood, similar in many respects to the Biblical story of Noah's Flood, stimulates interest in Mesopotamia. The following year, Smith heads an expedition that recovers missing parts of the narrative from the tablets in Ninevah.
1873	Heinrich Schliemann [German: 1822–1890] announces on June 15 that he has found in Turkey the site of Homer's Troy.
	Georg M. Ebers, a German Egyptologist, purchases a papyrus scroll that, when translated, is found to be a summary of Egyptian medical practices. The papyrus had been copied about 1600 B.C. from a much earlier source.
1876	Heinrich Schliemann finds tombs of the pre-Hellenic Mycenaean civilization in Greece, which he incorrectly identifies as the tomb of Agamemnon.
1877	E. George Squier reports that several civilizations preceded the Inca in South America, including those now called the Chimú and Moche.
	French archaeologist Ernest de Sarzec begins an 11-season investigation in the south of what is now Iraq. This uncovers the first monuments and representations of the people of Sumer, as well as libraries of clay tablets.
	Edward S. Morse, in excavations at Omori, Japan, discovers the remains of the Neolithic Jomon culture.
1878	Baron Louis Gerhard de Geer establishes the science of using varves (layers of sediment) to determine actual dates. He extends a series of varves in Scandinavia back to the end of the most recent Ice Age, about 10,000 B.C.
1879	A. H. Sauce is the first to recognize the Hittite civilization, although Sauce was preceded in 1874 by William Wright, who made a brief observation that a previously unidentified script was that of the Biblical Hittim, who we know today as the Hittites.
	Maria Sautuola and her father discover the Cro-Magnon cave paintings at Altamira, Spain.
1880	Augustus Henry Pitt-Rivers [English: 1827–1900] introduces scientific methods of excavation to England as he explores the Rushmore estate he has inherited from his great uncle Lord Rivers. Rushmore's fields had once been a Roman battleground and had not been changed much since Roman times in England, when several Roman villages occupied the site.
	Gaston Maspero discovers five buried pyramids which include 7,000 lines of hieroglyphics inscribed on the walls of the tombs, the first such discovery.
1881	Emile Brugsch becomes the first archeologist to enter the graves near Luxor, Egypt, where members of the royal family had been reburied after depredations by grave robbers at the original sites.
1883	Louis Julian, a French chemist, discovers the first of the so-called Venus figurines at the Grimaldi site, a cliff in Italy near Monaco overlooking the

Mediterranean. The 15 figurines he finds are of women (some pregnant) with exaggerated sexual features. The statuettes have been dated from 16,000 to 23,000 B.C.

The pyramids and temples of Gizeh by William Matthew Flinders Petrie [English: 1853–1942] reveals that the Great Pyramid is aligned only a bit more than one-twelfth of a degree away from a perfect north-south line.

1887   Flinders Petrie begins to excavate the pyramid of Senurset II.

The Tell el-Amarna letters, diplomatic correspondence of Pharaohs Amenophis III and Akhenaten, are unearthed.

1888   Richard Wetherill discovers Cliff Palace, the largest of the cliff dwellings at Mesa Verde in Colorado.

1890   Swedish archaeologist Gustav Oscar Montelius correlates various local chronologies of artifacts from different parts of Europe, the beginning of an understanding of European prehistory.

1891   On May 28, peat cutters working near Gundestrup in Northern Jutland (continental Denmark) find a large decorated silver bowl 27 in (70 cm) in diameter, now known as the Gundestrup Cauldron. Variously claimed as Germanic, Celtic, and Thracian, and dated from some time in the century prior to 150 B.C., it depicts such enigmatic figures as a man with antlers wearing a torc (a neck ring made of braided wire), elephants, and a woman flanked by birds. It may have been made from melted Persian coins by five craftspeople of a separate ethnic group, similar in some ways to present-day Gypsies.

1893   Richard Wetherill discovers mummies of the Basket Makers in southern Utah. They represent a culture earlier than that of Mesa Verde.

1896   Max Uhle begins the excavation of Pachacamac in Peru. It is the first careful stratigraphic excavation in either Mesoamerica or South America.

1899   In March, Robert Koldewey and Walter Andrae undertake the first large-scale scientific excavation of the ruins of Babylon, a project which continues until interrupted by war in 1917.

1900   Arthur John Evans [English: 1851–1941] discovers on Crete the palace at Knossos, the central site of the Minoan civilization. Luigi Pernier discovers a second great Minoan palace at Phaistos.

Greek sponge divers accidentally discover the wreck of a first century B.C. freighter sunk off the island of Antikythera, off the southern coast of Greece. The vessel sank while carrying works of art to Italy. Eventually, fine bronze statues, glass bowls, and even a now-famous bronze orrery (an astronomical instrument used for calculating positions of planets) are recovered.

1902   A French expedition discovers at Susa (Shûsh, Iran), the ancient capital of Elam, tablets engraved with the first known collection of laws, the code of Hammurabi [c. 1955 B.C. to 1933 B.C.].

1904   American archaeologist Edward Herbert Thompson begins a long investigation of materials preserved in the sacred cenote, a natural sinkhole at Chichén Itzá in Yucatan, Mexico. He continues the investigation for most of the rest of his life, publishing in 1932 a popular account of the many relics and skeletons found.

1906   Hugo Winckler begins excavation of Hattusas, the capital of the Hittite civilization in the seventeenth century B.C., at Boghazköy, Turkey.

1907   After Greek sponge divers observe a row of columns on the sea floor (which they mistake for cannons), Alfred Merlin directs a group of divers in what turns out to be a seven-year effort to recover the contents of an early first century B.C. freighter sunk off Mahdia, Tunisia. The wreck includes not only the columns and other architectural elements, but many sculptures as well.

1908   Black cowboy George McJunkin discovers the first known Folsom points—flint arrowheads associated with fossil bison bones—in New Mexico.

1910   Gustave Oscar Montelius establishes the first absolute dates for European prehistory.

1911   Hiram Bingham discovers Machu Picchu, a well-preserved Inca city high in the Andes, that had previously been unknown to science.

    Manuel Gamio undertakes stratigraphic excavations in the Valley of Mexico that identify a sequence of three separate pottery styles—Aztec, the earlier Teotihuacán, and an even earlier style now known as preclassic or formative.

    Workers from Harvard University's Peabody Museum make a major site map of the Guatemalan Maya ruins at Tikal, improving on maps made in the 1880s by British explorer Alfred P. Maudslay.

1921   The ancient Indus River city of Harappa (Pakistan) is discovered. The city existed from about 3000 B.C. to 1500 B.C., when it was destroyed either by floods or by invasion of the Aryans.

1922   On November 4, after six years of excavation, Howard Carter uncovers the first of the steps leading to the tomb of Pharaoh Tutankhamen [fl. c. 1358 B.C.].

    Mohenjo-Daro (Pakistan), sister city to Harappa, although 400 mi (600 km) from Harappa on the Indus River, is discovered.

1923   Charles Leonard Woolley [English: 1880–1960] begins excavations at Ur (Iraq), a city that was already a metropolis by 4000 B.C.

1924   Karel Absolon, who excavates Stone Age sites at Dolni Vestonice, Pavlov, and Predmosti in Moravia (Czechoslovakia) throughout the 1920s, discovers the statuette known as the Venus of Dolni Vestonice, the oldest known fired ceramic ever found, dating from about 24,000 B.C.. Other fired ceramics of various kinds, including many parts of animals, are also found.

1926   A French team of engineers uncovers the body of the Sphinx from the desert sand as part of a restoration effort. Ironically, this speeds deterioration.

1930   Andrew Ellicott Douglass determines the date of a Native American site by using tree rings, establishing the science of dendrochronology. His work in developing this project had been started as early as 1901.

    Two Roman ships, long known to have been on the bottom of Lake Nemi in Italy, are recovered after the failure of multiple salvage efforts dating from 1446. The Italian government obtains the ships, which are well preserved, by draining the lake over a period of four years. In 1944, however, the ships are burned by the retreating German army.

1940   R. D. Evans suggests the method of potassium-argon dating, useful for dates more than half a million years old.

    In September, four young men discover the cave paintings of Lascaux in France, which date from 15,000 B.C.

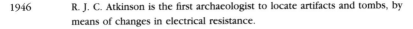

1946   R. J. C. Atkinson is the first archaeologist to locate artifacts and tombs, by means of changes in electrical resistance.

1947   Two young shepherds discover the Dead Sea Scrolls, Jewish religious documents from around the first century B.C., in a cave near Qumran in Palestine (occupied by Israel in 1967).

Willard Frank Libby [American: 1908-1980] proposes the use of carbon-14 for dating, which is suitable for dating events that are thousands, or tens of thousands, of years old.

1949   A diver spots amphoras (Greek storage jars) on the sea bottom off the tiny island of Grand Congloué, a few kilometers outside the harbor at Marseilles. In 1952, this location is determined to contain one or two shipwrecks from classical times.

1950   A 40-acre field in eastern England at Ken Hill yields a Celtic hoard of Iron Age torcs. The hoard includes the great gold Snettisham torc, considered one of Britain's finest archaeological treasures.

1951   John Papadimitrio and George Mylonas discover 24 early Mycenaean graves containing gold and silver face masks, diadems, bronze swords and daggers, and other objects dating from the seventeenth century B.C.

1952   Michael Ventris deciphers Linear B, one of the ancient languages of Crete, which he finds is a script for writing an archaic form of Greek. Linear A remains undeciphered.

Kathleen Kenyon begins a seven-year scientific investigation of Jericho (Jordan), showing it to be among the earliest urban sites ever located, having been occupied as early as 7000 B.C.

Oscar T. Broneer [Swedish: 1895-1992] discovers the site of the Temple of Poseidon, a fourth-century B.C. Corinthian shrine that, along with Olympia, Delphi, and Nemea, was one of the main temples at which all Greeks worshiped at that time. The Temple of Poseidon was famed as the site of the Isthmian games, second only to the Olympic games.

Jacques-Yves Cousteau begins to use his self-contained underwater breathing apparatus (scuba) on the first expedition of his research vessel *Calypso*, mainly investigating apparent classical shipwrecks off the coast of France.

Russian scholar Yuri Knorozov is the first to determine the correct approach to deciphering Mayan hieroglyphs, but few believe his results at this time.

1954   Kamal el Malakh and other archaeologists find the remains of a 142 ft (43 m) cedar boat in a chamber at the base of the Great Pyramid in Egypt.

1960   George Bass heads a team of underwater archaeologists who determine that a wreck on the sea bottom near Cape Gelidonya, Turkey, is that of the oldest ship found at that time: a small freighter loaded with copper and bronze from about 1200 B.C.

Helge and Anne Stine Ingstad discover the site of the only known Norse or Viking settlement in North America (excluding Greenland) at L'Anse-aux-Meadows, Newfoundland, Canada. Proof of their discovery comes with excavations of the eight large houses at the site over the period 1961-68.

Vernon L. Scarborough and Gary V. Gallopin, studying site maps of the Tikal ruins in Guatemala, recognize that most of the construction is arranged to divert water to a series of connected reservoirs in and around the site.

1961   The raising of the Swedish warship *Vasa*—sunk in 110 ft (24 m) of water outside Stockholm harbor on her maiden voyage in 1628—and preservation of the waterlogged vessel with chemical sprays marks the beginning of modern underwater archaeology.

1965   In *Stonehenge Decoded*, Gerald Stanley Hawkins [English: 1928-] argues that the 4,000-year-old site was originally an astronomical observatory.

1967   J. R. Cann, J. E. Dixon, and Colin Renfrew demonstrate that there were complex trade routes in the Mediterranean region as early as the period between 8000 B.C. and 5000 B.C.

Angelos Galanopoulos suggests that a great volcanic explosion destroyed most of the Mediterranean island of Thera (also known as Santorini), giving rise to the legend of Atlantis.

1970   In June, J. M. Adovasio and students begin excavating the Meadowcroft site near Pittsburgh, Pennsylvania. By 1978, it seems apparent that the site was inhabited by humans as early as 17,000 B.C., making it much older than other known American sites.

1974   Chinese archaeologists find a terra cotta army guarding the tomb of the first Ch'in emperor near the city of Xian in China.

1977   Manolis Andronicos [Turkish: 1919-1992] discovers a tomb in Macedonia that internal evidence and chemical dating suggests is that of King Philip II [382-336 B.C.], father of Alexander the Great. Although lacking definite proof, most authorities today believe that it is Philip's tomb.

1978   A team of anthropologists and archaeologists discover the earliest known religious sanctuary, dating from about 12,000 B.C., in Spain's El Juyo cave.

1979   Lothar Haselberger discovers the complete plans for the Temple of Apollo at Didyma, Turkey, which started construction in 334 B.C.

1980   Israeli divers in the Mediterranean off Athlit, Israel, just south of Haifa, recover the ram from a quinquereme or perhaps a quadrireme (naval vessels with five or four banks of rowers) that had been sunk around 150 B.C. It is the first physical evidence of the main naval weapon of classical times.

Evidence from airplane and satellite observations shows that Mayan cities were agricultural centers surrounded by extensive canals.

1982   George Bass leads a team of archaeologists who excavate the oldest shipwreck found so far—a freighter wrecked at Ulu Burun on the southwest coast of Turkey around 1350 B.C. The *Mary Rose*, a sixteenth-century English warship, is raised in Portsmouth, England, and found to be filled with Tudor artifacts.

1983   Peatcutters in the Lindow Moss in Cheshire, England, find what appears to be a woman's skull. Although a local man confesses that it is the skull of his murdered wife, it is later dated to be at least 1,700 years old.

The method of dating obsidian from chemical changes is introduced to archaeology.

1984   Peatcutters in the Lindow Moss in Cheshire, England, find the head and torso of a young bearded man preserved in the peat. The now-famous Lindow Man, who dates from the later Iron Age or early Roman occupation, now rests freeze-dried in the British Museum.

Michael G .L. Baillie and coworkers establish an unbroken chronology, based on Irish oak trees, that extends back 7,272 years.

1986   Harold L. Dibble invents a surveying system that records the exact three-dimensional position of artifacts as they are found on site at a "dig."

1987   In May, workers digging a ditch near Wenatchee, Washington find the first undisturbed Clovis site known. The six spear points encountered date from about 9500 B.C.

Susan Pevonak and Dale Croes find a microlith (a small blade tool) attached to a wooden handle at a site near Sekiu, Washington. This is the first microlith-based tool from any site around the world that is intact (i.e. the microlith is attached to the wooden handle) when uncovered.

Most of an adult male body is uncovered in the Lindow Moss in Cheshire, England. The body, which is badly damaged by the peat-cutting process that led to its discovery, may belong with the skull discovered in 1983, which had previously thought to belong to a woman. Examination of skin from one shoulder suggests that the body may have been painted. Early Britons were described by Roman writers as painting their bodies.

1988   In June, Andrea Caradini and coworkers discover a wall on the Palatine Hill in Rome that dates from the seventh century B.C., tending to confirm the legendary date for the founding of Rome.

Charcoal dates of artifacts found at the Monte Verde site in southern Chile indicate that humans have been in South America for at least 30,000 years.

In June, peatcutters in the Lindow Moss in Cheshire, England, discover a left knee and a large piece of skin from the lower torso of a human. In September, a piece of thigh is found. It is thought that these fragments might belong to the famous Lindow Man whose upper torso and head were found in the same bog in 1984.

1990   Cecil Hodder, who in 1989 began using a metal detector in the 40 acre field where the Snettisham torc, a braided gold neck ring, was found in 1950, discovers a bronze vessel containing fragments of about 50 torcs, 70 rings or bracelets, coins, and other bits of metal. Further investigation by the British Museum uncovers another five hoards, including 38 well-preserved gold and silver torcs.

The tomb of Caiaphas, high priest who sought the death of Jesus, is found in Jerusalem.

1991   The mummified body of a man who lived about 3300 B.C. is found in a melting glacier in the Tyrolean Alps on the border of Austria and Italy. The body wears leather clothes and shoes, and carries weapons and a box of charcoal wrapped in maple leaves. The weapons include a copper ax, a bow and bone-tipped arrows, and a flint dagger with a wooden handle.

The French diver Henri Cosquer discovers a submerged prehistoric cave near Marseilles, containing a variety of cave paintings. It is believed that people lived there more than 21,000 years ago, and that the cave's entrance has been covered by the sea for 20,000 years.

A series of scholars building on each others' work completes development of a way to decipher Mayan hieroglyphs.

OLD WORLD DIGS AND DISCOVERIES

Iceman UPDATE

On September 19, 1991, a pair of German hikers (Helmut and Erika Simon) on the Smilaun Glacier in the Ötz valley of the Tyrolean Alps, discovered the frozen and mummified remains of a body that had apparently been trapped in the ice for more than 5,000 years. The age of the body was not clear at first, and the mummy was hacked out of the ice with ski poles and ice picks by mountain climbers and people searching for evidence of foul play. The elevation was approximately 10,530 ft (3,210 m). Newspapers in the United States soon dubbed the remains the "Iceman," although he is known as Smilaun man, or even "Ötzi," in Europe. Evidence of killed game (ibex bones with cut marks) and picked berries were found nearby, thought to have been a last meal.

Since the time of the discovery, the site, mummy, and associated artifacts have been under intense international scrutiny and squabbling. The body was ultimately determined to have been found in Italy. By the time this was established, the mummy had been flown by helicopter to Innsbruck, Austria, where it remains frozen, after having been wrapped for a time in a sheet soaked in carbolic acid. Researchers working with the body must spend less than 30 minutes at a time with it so that it will stay frozen. Artifacts found with the body were taken to the Roman-Germanic Museum in Mainz, Germany, although finds subsequent to the original discovery have been transported to institutions in Italy. The main project has been directed by Konrad Spindler of the Institute of Archaeology at the University of Innsbruck.

The discovery of the Iceman was not the only 1990s surprise found high in the mountains. In the United States, in White River National Forest in the Colorado Rockies, a skeleton dated at 6000 B.C. — nearly 3,000 years before the Iceman — was found at about the same high altitude (more than 10,000 ft, or about 3,200 m). The skeleton was of a robust man who died in a cave for unknown reasons. Unlike the Iceman, however, the man in the cave was not mummified, and his tools and utensils were not preserved.

In 1994, an analysis of the Iceman's mitochondrial DNA was performed by a team led by Svante Pääbo of the University of Munich in Germany. This study confirmed that he was similar in heredity to the people who live north of the Alps today, which was not a surprise. The DNA contained a sequence that has never been found in sub-Saharan Africans, nor in Native Americans. That information convinced doubters that Ötzi was not a mummy plucked from Egypt or Peru and planted in the ice as a scientific hoax. Pääbo next plans to examine the Iceman's tissues for ancient viruses, and to use DNA to trace Ötzi's exact relationship to specific groups living today.

Another study by William Murphy of the University of Texas, conducted with x rays of the Iceman's skull, showed that his brain had dehydrated after death, either before the ice covered him or perhaps during a period when his head was briefly thawed later. The x-ray studies were also used to create an exact replica of Ötzi's skull, which will be used in comparison with skulls of modern populations.

Gradually, throughout the early 1990s, a picture emerged of the Iceman's life and death, and of one part of the culture of central Europe in the brief Copper Age that preceded the Bronze Age.

Current Reconstruction of Events

Between 3300 and 3100 b.c. the Iceman, who probably was a herder or shepherd (most likely a cattle herder, since there is no sign of wool or sheep leather), went on an expedition to replace a broken bow. He became trapped in the snow and froze to death . He appears to have been doomed by an unexpectedly early autumn storm. The berries found with him are the type that ripen in the fall, fixing the time of year to some degree.

Ötzi was between 25 and 40 years old at the time of the accident. CT scans reveal that at some earlier stage in his life, the Iceman's nose had been broken but had healed. Eight ribs, however, were cracked, three of them possibly shortly before his death, and may have contributed to his demise. He had other medical problems that most likely caused chronic back, knee, and ankle pain. Medically, his diagnosis today would be fairly severe osteoarthritis of the neck, lower back, and hip. Tattoos found on his skin in the same areas as the bone degeneration are thought to have been intended as medical relief—sort of a permanent acupuncture. Calcium deposits in the blood vessels of the cheek, pelvis, and neck would be diagnosed as arteriosclerosis. The teeth were quite worn down from chewing something tough, either food or fibers. He was about 5 ft 2 in (1.6 m) tall.

Surprisingly, the body was naked, and his clothes and other objects were found at a slight remove. Sometimes people freezing to death remove all their clothes, which may have happened in this case. There were seven articles of clothing preserved, which have been analyzed mainly by archaeologist Markus Egg of the Roman-Germanic Central Museum in Mainz, Germany. Grain found in Ötzi's clothes suggests that he came from an agricultural community. His "foundation garment" was a leather belt with a fanny pack. The belt held up a leather loincloth and separate leather leggings. Leather sewn together with sinews was used for clothing, including his jacket (made from alternating strips of tanned deerskin) and his shoes (size 6), which were stuffed with grass for insulation. Repairs in the leather were also made with grass, and grass was also used to bind some objects together, apparently as a temporary measure. This is the oldest known leather clothing to have survived, although remnants of even older woven clothing have been found at other sites, such as in Egypt. He had an outer cape of woven grasses or reeds—a type of cape that was still worn in the remote parts of the Alps until early in the twentieth century. He wore a conical fur cap, with the fur on the inside, which fastened to his head with a chin strap. Among the animals whose skins were used for all the leather were certainly goat, calf, red deer, and brown bear; and probably roe deer, ibex, and chamois as well.

The artifacts found with the mummy have revealed much about material use in such a society at this time. The Iceman carried an ax whose flanged head was of copper, although most of his other tools were imported Italian flint. These consisted of a small flint dagger with a wooden handle, the oldest known; a sharpened flint that appears to have been for cutting grass or sharpening arrowheads; and another piece of flint that may have been a drill. Tools made from other materials were an antler tool similar to an awl; two birch-bark canisters possibly for carrying embers; and a grass-net bag. These tools were carried in a rucksack with a wooden frame. Some articles were made from wood, including an unfinished bow of yew wood, arrow shafts of viburnum and dogwood, and the rucksack frame. It is interesting to note that wood continues to be the material of choice for such objects today. Bowstrings may have been the Achilles tendons from large animals. The Iceman carried in his quiver two finished arrows, with feathers to cause spin, and 12 unfinished arrows. One of the finished arrows was a type called a composite arrow that has a short shaft attached to a longer one, designed so the longer part can be broken off by a wounded animal, while the short part remains. Also in the quiver were sinews that probably were bowstrings, a line made from tree fiber, bone points, and a curved point made from antler. Further studies will eventually be conducted on such potential sources of information as the contents of the Iceman's stomach and his DNA, along with a search for any surviving microbes or parasites. These and other efforts have been frustrated to some degree by debates over who will control the research.

Melting of the glacier is taken to indicate that the climate in the Tyrol is warmer now than it has been in the past 5,000 years. Hikers are keeping their eyes peeled for the emergence of other ancient bodies.

(Periodical References and Additional Reading: *New York Times* 2-22-92, p 5; *New York Times* 7-21-92, p C1; *Nature* 3-4-93, p 11; *New York Times* 10-3-93, p I-23; *New York Times* 6-21-94, p C1; *Science* 6-17-94, pp 1669 & 1775; *Archaeology* 7/8-94, p 21; *New York Times* 11-2-94, p A18; *Discover* 1-95, p 57)

ANATOLIA AS CRADLE OF CIVILIZATION

Because Anatolia was either the origin of the farming, Indo-European cultural group, or among the first places to be occupied by that group, it is also the site of many of the earliest known examples of the technology that has been basic to Western civilization.

Very Early Pig Farmers

The classical concept of the Agricultural Revolution states that people in the Levant who had come to subsist on gathering and storing wild grains began to insure a more bountiful and accessible supply by developing grain farming. Around the same time, other people in the hills and mountains to the east managed to domesticate sheep and goats, while in far-off Mexico, members of the squash family were also being tamed. In 1994, archaeologists could also report that they had located another center of domestication. Far up on the Batman River—a tributary of the Tigris—in Anatolia, the pig was the first domesticate, according to evidence gathered at the ancient village of Hallan Ceni and announced early in 1994. The date of domestication of swine, according to radiocarbon analysis, was about 8400 B.C. to 8000 B.C. Depending upon whom one consults, this would either make pigs the first animals domesticated for food use, or tied for first with sheep and goats. Domestication of the dog for companionship and hunting preceded all other known domestic animals.

Pigs are a logical choice for domestication. Young pigs are easily tamed. There is little need to give pigs special care. They eat the same foods as humans, but will devour human garbage and also forage for acorns or other foods not favored by humans. And they produce a lot of meat for a small amount of food input. However, they cannot be allowed to forage among plants that humans are growing for food. Indeed, at Hallan Ceni, the place of pigs in the diet declined sharply after farming techniques and plant farming were introduced from the Levant.

The dig at Hallan Ceni, inspired by construction of a dam on the Batman River that will flood the site, was directed by Michael Rosenberg of the University of Delaware. Hallan Ceni was a small village of stone houses and stone statues that contained about 150 permanent inhabitants. In addition to the porkers, the people of Hallan Ceni were fond of freshwater clams. They also traded with other villages throughout the region, even obtaining shells from the Mediterranean.

RECOGNIZING FARMING

Domestication appears in the archaeological record because domesticated animals and plants have different characteristics from their wild ancestors. Domesticated animals are bred to be less dangerous than wild ones. Many domesticates are smaller, for example, than their wild counterparts (dogs and cattle come to mind). Pigs have their tusks bred away. Smaller molars in the first domestic pigs are the first sign to appear in the archaeological record. Frequently, the breeder aims for neonaty, the evolution of the appearance and behavior of a newborn that persists into adulthood. In other words, dogs are bred to be more like puppies, cats to be more like kittens, and pigs more like piglets. Neonaty often results in a foreshortening of the face that is observable in the archaeological record. If preserved, the presence of traits such as increased woolliness in sheep is also a good sign of domestication. Archaeologists also look at sex ratios and butchered body parts found in the record. Breeders eat most of the males and keep the females for further breeding or for milk. At Hallan Ceni, the majority of the pig bones were those of young males, mostly under a year old. Also, farmers consume all of a pig butchered in the village or at home, while hunters only bring home the best bacon from one killed far off in the forest. An abundance of pig bones in the village of Hallan Ceni was another clue pointing toward early domestication.

From this kind of evidence, and also from drawings and engravings showing people tending animals or plants, archaeologists can trace the step-by-step process of domestication, a process that continues today, as we find new uses for plants or even use genetic engineering to produce new forms of domesticated plants and animals.

The First Known Cloth

One of the discoveries in the early 1990s was the oldest fragment of woven cloth ever found. The fossilized fragments were found at Çayönü (chi-OH-noo), near the headwaters of the Tigris River, and have been radiocarbon dated at 7000 B.C., making this cloth some 500 years older than the previously oldest known fabric. The white cloth had been wrapped around the handle of a tool made from a deer antler, and perhaps was part of a bag that held the tool.

Woven cloth was known to have existed in Turkey at this time, about 9,000 years ago, through impressions of the weave preserved in clay. Pottery figures from as early as 18,000 B.C. show figures dressed in clothing—probably animals skins or simple skirts of cord or twine. No clothing made of actual cloth is shown as early as this, however.

The cloth from Çayönü is most likely linen, since flaxseed has been found at the site. Flax was probably domesticated about this time for its oil-bearing seeds (linseed). Wool might also have been woven into cloth at this time, but would not have lasted as long as linen does. The cloth was not made from thread, but from fibers twisted as they are in twine. The method of weaving was more similar to basket-making than to the way more recent cloth is made: two weft threads were apparently twined around each warp thread, making for a loose, or open, weave.

Some archaeologists believe that textile-making preceded agriculture and pottery, although plants had already been domesticated for about a thousand years when the Çayönü cloth was woven. Çayönü was an agricultural village with permanent houses and partly domesticated sheep and goats.

Excavations at the Çayönü site have continued for about 30 years under the direction of Robert Braidwood of the University of Chicago and Halet Cambel of Istanbul University. Actual identification of the cloth fragments, which were found in 1988, came in 1993 when Gillian Vogelsang-Eastwood of the National Museum of Ethnology at Leiden in the Netherlands examined the fragments under an electron microscope, revealing the woven structure.

A Source for Tin

Aslihan Yener of the Oriental Institute of the University of Chicago announced at the beginning of 1994 that she had found a major source of tin in the Taurus Mountains of Turkey. Not only had she found an extensively worked mine at a mountain called Kestel, but she had also discovered the smelting center in a nearby site called Goltepe.

From the mine site alone, it is difficult to prove that tin ore was the product. The mountain is riddled with small tunnels through which whatever was being mined was extracted. If that material was tin ore, it is now all gone, removed by miners in the second millennium B.C. Nevertheless, the smelting operation at nearby Goltepe was definitely directed toward tin, and that tin certainly had to come from somewhere. Slag, high in tin residues, comprises part of the evidence for tin smelting at Goltepe.

Tin mining and smelting can be traced back as early as 2870 B.C. at Kestel and Goltepe. Goltepe was occupied until about 1840 B.C., which may be when the tin ore ran out at Kestel.

The entire process of tin production is illustrated by the evidence at Kestel and

Background

When humans first started making metal tools, they used copper, which can sometimes be obtained without any smelting and which is easy to produce at low temperatures from abundant ores. Bronze is the name given to copper that has been alloyed with tin, or sometimes with arsenic. This produces a harder metal, which is easier to cast than copper. Shortly after people began to prefer metal to stone tools for some purposes, they discovered bronze and quickly abandoned copper. Thus, the earliest classification system for toolmaking referred to a Stone Age, followed by a Bronze Age, followed by the Iron Age, and omitting the very brief Copper Age. In the Middle East, the Bronze Age ran from about 3000 B.C. to about 1100 B.C. It ended when people learned how to make good iron from plentiful ores, releasing metallurgists from dependence on a limited supply of tin.

The very earliest bronze was made with copper that had been alloyed with arsenic by nature. It is not known how copper-tin bronze, which is stronger and less toxic than copper-arsenic bronze, was discovered. Sources of tin are not common and appear to have been absent altogether in the civilized regions of Mesopotamia and Egypt. An early source was in Afghanistan. Another source in classical times was in Cornwall, England. Both of these mining regions were far from the centers of civilization.

Goltepe. First the ore was taken from narrow shafts, most likely by teenagers (the mine shafts are too narrow for most adults, and skeletons believed to have been left due to mine accidents were of young people between the ages of 12 and 15). Fire was used to soften the ore enough to be removed with stone tools. The ore was further separated from other rock, using a washing process similar to panning for gold. The ore was layered with powdered charcoal and heated while air was blown at it through reeds. The tin formed inside lumps of slag and could be removed by breaking open the slag with stone tools. This method was quite "low-tech" when compared to large-scale copper smelting at the time, but since good bronze requires only 1 part tin to 10 to 20 parts copper, copper was needed in much greater amounts. Copper smelting also had a longer history.

What's New At Troy

The place of the Greek city-states in the history of Western civilization is well known, but the most the average person recalls about Troy is that unfortunate business about the return of Helen. Classicists also recognize Troy as the homeland of Aeneas and his followers. According to one set of legends, the influence of Aeneas on Western civilization included the founding of Rome. Despite its notoriety and fame, Troy is generally dismissed as a backwater city of little actual importance outside of poetry and legend.

In 1992 and 1993, however, archaeologists assembled new evidence that shows that ancient Troy was an important late Bronze Age city and important again in classical times. The first find came from a magnetometer reading that indicated a large underground feature, which excavation revealed to be a fortification ditch cut directly into the bedrock. Because of the location of the ditch, Troy of that period—the time of the *Iliad*—is now known to have been much larger than previously thought. The survey was conducted by Helmut Becker and Jörg Fassbinder, both of the German State Office for the Preservation of Historical Monuments in Munich.

Because of Homer's *Iliad*, Troy was a famous site in Hellenic and Roman times, even though the Homeric Troy had been abandoned long before. The site of the *Iliad* was known then (although later lost until the nineteenth-century rediscovery), and a new city, called Ilion, was built there. Ilion was primarily a tourist attraction, rather like Colonial Williamsburg. Much of Ilion was destroyed in 85 B.C. by the Roman legate Fimbia, but parts of the city were rebuilt under Augustus. Excavations in the early 1990s discovered a monumental building that had not been restored under Augustus and which may have been intentionally left as a war memorial.

In 1993, a Roman odeum—an auditorium for concerts, sometimes called an odeion—was partly restored, repaired, and reconstructed at the Ilion site. The building had originally been excavated some 50 years previously, and the ruins had been used for theatrical or musical performances. The new reconstruction not only made the audience's seats safer, but also uncovered a statue of the emperor Hadrian, coins, and other artifacts, as well as layers of destruction reflecting the sack of Ilion in A.D. 267 and a major earthquake in A.D. 475.

(Periodical References and Additional Reading: *New York Times* 2-23-93, p C1; *New York Times* 7-13-93, p C1; *New York Times* 11-9-93, p C1; *Archaeology* 11/12-93, p 20; *New York Times* 5-31-94, p C1; *New York Times* 1-4-94, p C1; *Archaeology* 1/2-94, p 18; *Science* 6-3-94, p 1398)

Mesopotamia UPDATE

Among the earliest sites to be investigated by archaeologists were the well-known cities of Babylon, Ninevah, and Ur. Although less well-known perhaps to the general public than the ruins and tombs of Egypt, the revelations of the great cities of Mesopotamia, along with millions of cuneiform baked clay tablets, most of them readable by scholars, have been central to the development of archaeology as a science.

Mesopotamian Empires

From a time before Sargon the Great through Nebuchadnezzar to the Persian rulers Cyrus and Darius, the land between the Tigris and Euphrates rivers has been subject to autocratic, top-down rule. The extent of the earliest empires, or even of their influence, has never been entirely clear. Modern archaeology, however, is beginning to explore the edges of power and to gain new insights as a result.

In 1992, the first archeological excavations began at Hacinebi Tepe, a site on the fringe of Mesopotamia, 700 mi (1,100 km) up the Euphrates from the centers of civilization in what is now a part of Turkey. The investigation continues to be directed by Gil J. Stein of Northwestern University in Evanston, Illinois. Hacinebi was a village settled by people from Anatolia (*see* "Anatolia as Cradle of Civilization," p 72) long before it came under the influence of the Mesopotamian culture. The Mesopotamian influence began to be felt about 3100 B.C., when the city-state of Uruk was dominant in the Fertile Crescent. After about 175 years, the decline of Uruk returned Hacinebi to its former Anatolian culture. After that, Sumer became dominant throughout the region.

The clearest evidence of Sumerian influence in the north was also found in the early 1990s. In 1992, Patricia Wattenmaker of the University of Virginia noticed a promising site in southern Turkey on a tributary of the Euphrates called the Balikh River. She investigated a large mound near an irrigation canal and found potsherds littering the site, some apparently as much as 7,000 years old. Later, a worker found a baked clay tablet covered with the type of early cuneiform in use about 3000 B.C. The site, known as Kazane Hoyuck, has since been found to have been occupied as early as 5000 B.C. and by about 2600 B.C. had become a large city, possible the one known as Urshu. The city declined and was abandoned some 600 years later.

In 1993, another, although smaller, city of that period was found on a different tributary of the Euphrates in Syria, almost to the Turkish border. This site, known as Tell Beidar, has also yielded clay tablets, as well as a temple and administrative buildings. The heyday of Tell Beidar was evidently later than that of Kazane Hoyuck and about a hundred years before the start of the Akkadian empire. Both cities show clearly that Sumerian influence was great in the north as well as in the south.

The first actual empire in Mesopotamia was that of Sargon the Great, an Akkadian, who consolidated his power about 2300 B.C. Akkad, the original center of Sargon's empire, is thought to have been about a third of the way up the land between the rivers from Uruk to Hacinebi Tepe, near Baghdad. But Akkad has not been definitely located yet. The Akkadian Empire assembled by Sargon stretched all the way from the Persian Gulf to the headwaters of the Euphrates, approximately up to Hacinebi Tepe. The Empire was about 800 mi (1,300 km) in length.

In 1993, Harvey Weiss of Yale University marshaled evidence suggesting that the empire established by Sargon foundered as a result of a major drought that struck the region of Subir in northern Mesopotamia about a hundred years after Sargon the Great. This drought seems to have lasted for 300 years. Subir is a fertile plain along the Habur River and its tributaries which together feed the upper Euphrates. Weiss based his conclusions on soil tests at three different sites in Subir (now in Syria). The drought appears to have begun with major volcanic eruptions in Anatolia that left a half-inch layer of volcanic ash over the rooftops of Tell Leilan, one of the sites far up a Habur tributary. Tell Leilan was a large Subiran city and a political capital of the region in Akkadian days, but was abandoned after deposition of the volcanic ash.

Volcanoes can affect climate in various ways, although there is no known historical instance of a volcano changing climate for longer than two or three years worldwide. Local climate could have been affected by haze from gases and dust of a continuously erupting volcano, although the volcano or volcanoes involved in the eruption around 2200 B.C. were quite some distance away from Tell Leilan (the most probably culprits, Nemrut and Karacilag, are a more than 100 miles away from Tell Leilan). However, the soil for 2 ft (100 cm) above the layer of volcanic ash is clearly dust from a very dry climate—quite different from the soil just below the ash layer, according to an analysis by Marie-Agnes Courty of the Centre National de Recherche Scientifique in Paris, France. Soil above that layer was evidently well-watered by seasonal rainfall.

TIMETABLE OF MESOPOTAMIA

8000 B.C.	Stone houses begin to appear at Tell Mureybit in what is now Syria.
5000 B.C.	Farmers are living in reed houses in the delta of the Euphrates River.
4000 B.C.	Uruk, perhaps the first great city, is founded on the banks of the Euphrates.
3500 B.C.	Sumer, the world's first civilization, formed from the city-states of Uruk, Ur, Lagesh, and others, develops along the banks of the Euphrates and the neighboring Tigris rivers. The Sumerians are the first to use the wheel, to develop writing and arithmetic, and to build monumental architecture— towers we know as ziggurats.
3000 B.C.	Semitic-speaking people, perhaps originally from Arabia, begin to appear in Northern Mesopotamia. These become known as the Akkadians.
2316 B.C.	End of the rein of Lugalzagesi, last of the kings of Sumeria. He is defeated by Sargon of Akkad, who founds the Akkadian Empire.
2193 B.C.	End of the reign of the last Akkadian emperor, Shar-kali-sharri, followed by a period of anarchy.
2113 B.C.	Ur-Nammu becomes the King of Ur and begins a stable dynasty in that city, although constantly besieged by potential invaders known as Amorites and Elamites.
1894 B.C.	About this time, a stable Amorite dynasty, ruling the empire from Babylon, develops in the wake of see-saw leadership. Before the Amorite empire, various Elamites and Amorites, as well as kings of Ur, capture and lose control of Mesopotamia in rapid succession.
1792 B.C.	Hammurabi the lawgiver, a descendant of the Amorites, becomes the first well-known king of Babylonia.
1595 B.C.	The Hittites, Indo-European speakers from Anatolia, sack Babylon.
1530 B.C.	The Kassite, nomads from what is now Iran, invade Babylon.
1350 B.C.	About this time, the Assyrians come to power in Mesopotamia.
1124 B.C.	Nebuchadnezzar, an Amorite, comes to power in Babylon.
1100 B.C.	Tiglath-pileser rules all of Mesopotamia.
612 B.C.	The Medes, from what is now Iran, destroy Ninevah and replace the Assyrians in power there, although not in Babylon as yet.
600 B.C.	Under Chaldean king Nebuchadnezzar II, Babylon becomes the largest city on Earth about this time.
540 B.C.	Persia, under Cyrus the Great, gains control of Mesopotamia. His allies conquer Babylon on October 13, 539.

Beer and Wine in History

Beer has a long record in human history. There are written references to Egyptian beer as well as to beers from other ancient civilizations. Indeed, some archaeologists have maintained that grains were originally domesticated to provide a steady source of raw material for beer making, and that the invention of bread occurred as a spinoff from brewing. Wine also has a long pedigree, perhaps originating as a way of preserving fruit juices. On the other hand, ancient people may simply have made beer and wine for the purpose of getting drunk.

In the 1990s, the earliest actual traces of wine and beer storage were found in dregs at the bottom of jars in a storage room at Godin Tepe, a trading post on the eastern frontier of Sumer in what is now Iran. Wine was recognized first, in 1990, by the chemist Rudolph H. Michael at the University Museum of Archeology and Anthropology at the University of Pennsylvania, using residues called to his attention by archaeologists Patrick E. McGovern from the same institution and Virginia R. Badler from the University of Toronto. Two years later, the same team found beer residues in other jars from the same storeroom. Although the chemical residues could be produced by other plants (notably spinach or rhubarb), the circumstances of the find, the type of jars used, and the prevalence of barley (the grain used for making beer) at the site all seem to indicate that the jars once held beer. Late in 1994, Badler told the American Anthropological Association that the storeroom appears to have held a supply of beer and wine and "ammunition" for an army post at Godin Tepe. The ammunition consisted of about 2,000 clay balls that were hurled with slings at the opposing army. Windows in the storeroom may have been used to pass out rations of booze and balls to the troops, in preparation for battle.

Beer, not wine, was the favorite drink of the Sumerians. This is well known from written references and from illustrations of Sumerians drinking beer through straws.

Background

Mesopotamia is a broad term that comes from the Greek for "between the rivers." Indeed, most of civilized Mesopotamia is between or on the banks of the Tigris and Euphrates rivers in what are now Syria and Iraq. The two rivers also extend into what is now southern Turkey. The land to the east of the Tigris (now Iran) was politically involved with Mesopotamia, and is often considered along with Mesopotamia proper. The mouth of the Tigris-Euphrates system is the Persian Gulf, so any civilization in the Gulf also interacted with Mesopotamia. The Arabian Peninsula to the west of the Persian Gulf connects directly to Mesopotamia, and many of the peoples of Mesopotamia (including the Akkadians) are thought to have come from Arabia. Nations on the Mediterranean shores west of Mesopotamia, such as Lebanon and Israel, were often dominated by Mesopotamian empires when not ruled by the Egyptians.

In many ways, Mesopotamia antedates Greece as a fountain of Western culture. Among the important contributions are the first cities, the first written language (probably), the wheel and attendant vehicles, and the alphabet. In 1992, it was discovered that there were domesticated horses in Mesopotamia at the time of the Akkadian Empire, although the horse was domesticated originally in what is now the Ukraine or in Russia. Familiar holdovers from Mesopotamia are legends (such as Creation and Flood accounts similar to those in the Bible); hours, minutes, and degrees (both for time and for angle measure); the zodiac and astrology; and the Pythagorean theorem.

The sign for "beer" is among the most commonly used symbols on Sumerian cuneiform tablets.

Islands in the Gulf

Mesopotamian texts from as early as the third century B.C. refer to the holy land of Dilmun, which was also an important commercial center. Scholars have identified Dilmun as a group of islands in the Persian Gulf close to the Arabian peninsula, the largest of which is Bahrain. Four archaeological expeditions have investigated sites on Bahrain, long known for its 170,000 burial mounds. The most recent expedition, in the early 1990s (the London-Bahrain Archaeological Expedition), found the remains of a large town. Working with the Bahraini Department of Antiquities, the expedition has uncovered about 40 stone buildings, including a temple, all preserved extremely well. Apparently the town was abandoned around 1750 B.C. when its harbor filled with silt.

Dilmun was a supplier of pearls as well as various kinds of raw materials. In addition to four pearls that did not get shipped out, the archaeologists have located pottery and weights from the Indus River civilization, and stone bowls and copper implements from Oman, confirming the role of Dilmun as a major trade center. The independent emirate of Bahrain continues to be a trade center, but today is primarily a port for Saudi Arabia (to which it is connected by a bridge) and other Persian gulf states.

(Periodical References and Additional Reading: *Nature*, 11-5-92, p 24; *New York Times*, 11-5-92, p A16; *New York Times*, 1-3-93, p I-10; *New York Times*, 5-25-93, p C1; *Science* 8-20-93, pp 985 & 995; *New York Times*, 8-24-93, p C1; *Archaeology* 7/8-94, p 20; *New York Times*, 11-9-93, p C1; *Science News* 12-10-94, p 390)

EGYPTIAN ARCHAEOLOGY NEWS

The fascination with Egyptian tombs and temples, which goes back to Herodotus and before, is one of the enduring parts of archaeological studies. But there is more to Egyptian prehistory and history than monuments. Much of technology still in use today originated in Egypt, including farm technology.

Egyptian Farming

Egypt was an African entity long before the rise of the kingdoms that made it a part of the cultural heritage of the West. In 1992, agronomist Jack R. Harlan of the University of Illinois, Fred Wendorf of Southern Methodist University in Dallas, Texas, and a team of Polish and American archaeologists proposed that their evidence showed that the two characteristic grains of Africa, sorghum and millet, were domesticated in lower (southern) Egypt about 6000 B.C., some 1,600 years before that part of Egypt knew of the domestication of wheat in the Near East (although about 2,000 years after the actual domestication of wheat). Thus, the Egyptian feat occurred independent of the domestication of wheat.

The evidence at the Egyptian site, 60 mi (100 km) west of Abu Simbel, does not show actual domestication. Instead, it demonstrates extensive use of sorghum and millet for food at the site, along with about 40 different other crops, including legumes, crucifers (mustard), fruits, nuts, and tubers. The sorghum seems well on its way to domestication, however, with a chemical composition closer to that of modern domesticated sorghum than to any of the wild species that grow in that part of Egypt.

Sorghum and millets grow wild on the African sub-Saharan savanna, and domestication was thought previously to have taken place at some unknown site there.

Later Egyptians favored bread made from barley. At Nazlet al-Simaan, the burial ground for the workers who built the pyramids, which was discovered in 1989, excavations by the Oriental Institute of the University of Chicago have revealed a vast bakery in which large 22 lb (10 kg) loaves of dense barley bread were made in pots 33 ft (10 m) high. It is not known why barley bread, which is harder to make and less versatile than wheat bread, was the choice for pyramid workers. Other investigators have suggested that the invention of bread baking developed as an outgrowth of beer brewing. In early beers, barley was baked into loaves and then fermented to make beer.

It would be more reasonable to expect that Egyptians would be growing wheat at about the same time, far to the north, in the Nile Delta. Not only is the delta the best farmland in Egypt, consisting of fertile soil carried there by the river, but it is also relatively close to the original centers of domestication in the Near East. But in 1993, archaeological research by Daniel Stanley and Andrew Warned of the Smithsonian Institution demonstrated that there was no Nile Delta as early as 6000 B.C. Because of rapidly changing sea levels in the Mediterranean, the present delta did not start forming until about 8,000 years ago. After another thousand years—around 5000 B.C.—enough land had been deposited by the river to make farming worthwhile, and the first Egyptian farms on the Delta were started.

SORGHUM AND MILLET

Sorghum and millet are two of the staple grains of humans around the world, but in the United States are primarily constituents of animal feed. Both are grasses, as are wheat, rice, oats, rye, and maize. Sorghum looks something like a cross between a giant wheat plant and an oddly earless form of maize, while millet at first glance looks something like a cattail. Many Americans would only recognize sorghum because one variety was commonly used as broom straw (before the introduction of synthetic straws). Sorghum molasses is a sweetener made from certain other varieties of sorghum.

Millet and sorghum are used like other grains for human food in Africa, and millet is also a popular grain in Asia. Both plants are suitable for hot, dry growing conditions. Millet can also be grown in a very short growing season.

Other Technological Advances

In other recent news, Carol Meyer of the University of Chicago and coworkers in the early 1990s discovered a sprawling gold-mining settlement in the Egyptian desert. The mining town, known as Bir Umm Fawakhir, operated during the Byzantine period of Egyptian history, the fifth and sixth centuries A.D.

In 1993, James A. Harrell of the University of Toledo in Ohio and Thomas Brown of the U.S. Geological Survey located and mapped the earliest known paved road, which is in Egypt, according to their 1994 report to the Geological Society of America. It had been observed by archaeologists as early as 1905 but not investigated or surveyed. The Egyptian road displaces the previous record holder, a flagstone-paved road in Crete from about 2000 B.C.

The Egyptian road was built between 2200 and 2600 B.C. to carry rock from a quarry near Giza to a water route that led to the construction site of the Pyramids and other tombs. About 7.5 mi (12 km) long and 6.5 feet (2 m) wide, the road was paved mostly with slabs of sandstone and limestone, but a few of the paving stones were logs of petrified wood. Although the wheel was already in use in Mesopotamia well before the construction of this road, it is believed that wheels were not yet used in Egypt, which suggests that the road was intended to make sledges move more easily. It is not clear exactly how stones were transported over the roads, since no grooves of the type that would be caused by sledges are evident on the paving stones.

By way of contrast, another paved road in the same region became a source of controversy in 1994. A secret government plan to build a modern eight-lane highway 1 mi (2 km) south of the pyramids became known to archaeologists. The cultural heritage division of Unesco, which includes the pyramids and related structures at Giza on its World Heritage list of places, called the road "illegal and in clear violation" of the 1972 World Cultural and Natural Heritage Convention. Archaeologists also claim that the road violates Egyptian laws intended to protect the site as well. Construction, which had begun already, was halted while the Egyptian government tried to find a compromise between its highway builders and its archaeologists.

Lost and Found Archaeology

Sometimes an archaeological discovery is made in a museum instead of at a dig. This is often the case in Egyptian archaeology, since many objects have been removed from tombs or other locations without proper identification of the site, or details of the original state of the artifact. Without this information, conclusions are suspect.

Usually the problem with gaining information from a museum collection stems from lack of care by the grave robbers or pot collectors who originally extracted the material. Sometimes Egyptian material was excavated by early archaeologists who did not follow methods considered acceptable today.

A large collection housed at the University of Chicago and at the Cairo Museum has lost its anchor in the earth for another reason entirely. Although meticulously excavated by a team of German archaeologists and James H. Breasted of the University of Chicago between 1927 and 1933 with all finds carefully annotated, the notes vanished as a result of World War II. When the war began, the ten notebooks of data were being prepared for publication at the Bode Museum in Berlin, Germany. Rudolf Anthes, the Egyptologist who was working on the book, fled Berlin. The notebooks vanished. After the war, it was presumed that the notebooks had been lost in the bombing, or perhaps seized by the Russian troops who liberated Berlin. In any case, no one seemed to know if they still existed, and most who once knew of the situation assumed long ago that the notebooks were gone forever.

In 1991, however, a year after the collapse of communism in Germany, officials of the Bode Museum contacted the University of Chicago to say that they had some notebooks that belonged to the University. Upon the return of the notebooks in 1993, it was clear that all ten volumes of the field notes had survived. They were immediately put to use in interpreting the museum's previously useless collection from the Breasted investigation. Furthermore, news of the recovery led directly to a reunion of the field notes of Uvo Hölscher, the German director of the dig, with the other notebooks. The handwritten daily log of the excavations (4 volumes), lists of all the artifacts, site plans, and architectural drawings (7 volumes), and about a thousand photographs were recovered from the Hölschers' attic in Hanover, Germany, by his grandson, who had read of the return of the first batch of notebooks in the *New York Times*.

The artifacts involved were all excavated from a site called Medinet Habu, near Luxor, Egypt (ancient Thebes). They consisted of statues, jewelry, erotic pieces of various kinds, and other objects thought to have religious significance. Without the notes, it had been impossible to tell whether the objects came from temples, tombs, or homes. With the notes, scholars have already discerned that much of the erotica was found in private homes, suggesting to some that they may have been connected to a fertility cult.

(Periodical References and Additional Reading: *New York Times* 2-9-92, p I-34; *New York Times* 2-25-92, p C8; *New York Times* 3-10-92, p C4; *Nature* 10-22-92, p 721; *New York Times* 10-27-92, p C2; *New York Times* 9-21-93, p C1; *Archaeology* 11/12-93, p 19; *Discover* 1-94, p 81; *New York Times* 5-10-94, p C13; *New York Times* 12-10-94, p 4; *Archaeology* 1/2-95, p 17)

NEW FINDS IN THE HOLY LANDS

The Near East has many important associations for millions of people. The region, which is now largely in the states of Israel, Jordan, Syria, Iraq, Egypt, and Saudi Arabia, includes nearly all the sites connected with the origins and early history of the religions of Judaism, Christianity, and Islam. Commonly, Christians call the valley of the Jordan River, including the Sea of Galilee and the Dead Sea, the Holy Land. But the entire region is a series of Holy Lands. Understanding the history of these Holy Lands through archaeology has long been one of the main motivations of the science.

In the early 1990s, a continuing improvement in political conditions in Israel and Jordan aided archaeologists as they sought the origins of Judaism and Christianity.

Around the Time of Abraham

You may recall from the Bible that when the Hebrew God destroyed the sinful city of Sodom, the good man Lot escaped but his wife was turned into a pillar of salt. The story of what happened after that is less well known, although it is in the Bible a verse or two later. Lot and his two daughters hid in a cave in a place called Zoar. After a time, the daughters, thinking perhaps that they were alone with the last man on Earth after all that brimstone and fire, got Lot so drunk that he did not know what he was doing, and then caused him to impregnate them. This, according to the Hebrews, was the origin of their enemies, the Moabites and the Ammonites. As proof of this story, everyone around those parts (now western Jordan near the Dead Sea) could point out the very cave where Lot's incest took place. In the early 1990s, an expedition studying an ancient stone reservoir in the area discovered a monastery built in A.D. 691 including a church building that protected the ancient cave. Further proof of authenticity was found inside the cave: drinking cups and a pot dating from the period of the Bronze Age when Lot is thought to have lived.

Around the Time of King David

Skeptics after the Enlightenment began to view most parts of the Bible as myths, with no connection to actual events. Archaeology in this century, however, has supported many of the events and major rulers mentioned in the Bible. Physical evidence for the chief actors, however, such as Abraham, Moses, David, and Jesus, has been scarce or nonexistent. But recent discoveries have included some familiar Biblical names.

Since prophets go without honor in their own countries, they also tend to go through history without leaving behind documented records. It is perhaps more surprising that kings, such as David and Solomon, have remained as undocumented as the prophets. But it is typical not to know specific information about the rulers of small kingdoms from as long ago as David—thought to be around 1000 B.C. In 1993, however, Avraham Biran of Hebrew Union College and the Jewish Institute of Religion in Jerusalem found specific evidence of a Biblical king from a short time after David—about 850 B.C. Furthermore, the inscription referring to the king uses the specific phrase "House of David," making it the first known archaeological reference to David from about his own time.

The inscription, on a broken monument found at a northern Israeli site called Tel Dan, is not complete, but provides enough clues to allow identification of the involved ruler: Asa, king of Judah, whose story is told in the Bible in chapters 14-16 of Second Chronicles. A broken commemorative pillar apparently refers to the wars and related events described in chapter 16. Identification is not complete, however, as the specific name Asa is not mentioned.

It was David who took Jerusalem from the Canaanites. Although the exact translation of the Hebrew text has been disputed, most authorities today believe that the Biblical account has the Israeli warriors entering Jerusalem through a water shaft. In 1991, Dan Gill of the Geological Survey of Israel showed that the Gihon Spring (which provided the stable water supply of ancient Jerusalem, even in time of drought or siege) and the tunnels that connect the spring to locations in the city are embedded in the kind of limestone formation that produces caves throughout the world—what geologists call a karst landscape. Further study showed Gill, as he reported in 1991 and elaborated in 1994, that the tunnels and shafts were all originally formed by nature, although some, such as Warren's shaft and Hezekiah's tunnel, were then altered to make the passages more uniform and wider. This explains how the tunnels could have existed when David's army attacked, even though the Canaanites were not at that time technologically capable of making long tunnels through many feet of rock.

Of course, David himself was famous for his battle against the giant Philistine, Goliath. Considering the intense warfare between the Philistines and the Israelites described in the Bible, it seems surprising that until the 1990s very little was known about who exactly the Philistines were. Today, archaeology and some interesting theories suggest a surprising origin for the Philistines. In 1992, an expedition from Harvard led by Lawrence E. Stager uncovered the remains of destruction of the Philistine city of Ashkelon by Babylonia's Nebuchadnezzar (*see also* "Mesopotamia UPDATE," p 76) in 604 B.C. Like other digs at Philistine sites, the Harvard reconstructions showed that the Philistines were an advanced culture, very likely more sophisticated about technology and art than the Israelites.

Ashkelon was the main port and a principal center of the Philistine culture, although it had been a Canaanite city even earlier. Apparently, the Philistines arrived from somewhere else around 1175 B.C. and, like the Hebrews, moved in on the fertile land of Canaan. But there is no written record that has been found to tell where the Philistines came from.

Lawrence Stager and the Israeli team of Trude and Moshe Dothan of Hebrew University in Jerusalem think that they have solved the riddle. In 1991 (Stager) and 1992 (the Dothans) wrote that the pottery, loom weights, brick ovens, and other details found in Philistine sites, especially when considered along with clues from the literature of other nations—the Hebrew Bible, the *Odyssey*, and Egyptian inscriptions—show that the Philistines were actually Mycenaean Greeks who had colonized Canaan. One reference from the Bible, for example, describes Goliath as wearing the type of armor used by the Mycenaeans, the same type described by Homer. Furthermore, other evidence, well known in the archaeological community, connects both the Mycenaeans and the Philistines with the People of the Sea, mysterious invaders mentioned in Egyptian sources. If the Dothans are correct, should any Philistine writing ever be found, it would be written in Linear B, the script of the Mycenaeans.

Around the Time of Jesus of Nazareth

Christians are all familiar with the names of the principal officials involved in the events described in the New Testament—Herod, Caesar Augustus, Pontius Pilate, Caiaphas (the High Priest). There is a rich historical record regarding Augustus, who was the first Roman Emperor [27 B.C. to A.D. 14]. Two Herods are also well known: Herod the Great, who died in 4 B.C. and was king at the time of the birth of Jesus; and Herod Antipas, the Herod who had John the Baptist killed, and was king at the time of the death of Jesus. However, it was not until the 1960s that any archaeological or historical evidence outside the Bible was found to contain a specific mention of any of the other familiar figures from the New Testament. At that time, the name "Pontius

Pilate, Prefect of Judea" was unearthed as part of a fragmentary Latin inscription on a building. The purpose of the inscription mentioning Pilate was apparently to identify the Roman administrator who had commissioned that building in Caesarea, on the Mediterranean Sea.

In 1990, a family tomb on the outskirts of Jerusalem was found with the name of Joseph, son of Caiaphas–although it took more than a year to be certain of the meaning of the Aramaic handwriting on the side of the ossuary where the bones are kept. This would appear to be the family tomb of the High Priest to whom Jesus was taken for judgement, and who presided over the council that condemned Jesus to death. The cave where the tomb is located was accidentally discovered by workmen in Jerusalem's Peace Forest.

Although neither Jesus nor John the Baptist is specifically mentioned in the ancient texts known as the Dead Sea Scrolls, there has been persistent speculation that some of the figures mentioned in those documents could have been disciples, or even the Baptist himself.

According to the Bible, the "wise men from the East" offered the infant Jesus gold, frankincense, and myrrh. While there is no historical record of the wise men, it is known that frankincense does come from the small, scraggly trees that grow in the Qara Mountains of Oman, hundreds of miles to the southeast of Jerusalem. The Koran relates that the wicked city of Iram, thought to be another name for Ubar, was once the main source of frankincense but was buried by a sandstorm because of its wickedness. Later historians remember Ubar as an "imitation of Paradise."

Until 1992, that was all that was remembered about Ubar. Documentary filmmaker Nicholas Clapp used old maps, however, to search out a possible location of Ubar, finding a candidate on a map from the Hellenic astronomer Ptolemy that had been drawn about A.D. 200. On that map, Ubar was known by its classical name of Omanum Emporium. Clapp asked Ronald Blom of the Jet Propulsion Laboratory to use satellite images to look for evidence of Ubar in Oman. Blom arranged for an already-planned space shuttle experiment to include the Arabian desert. Blom and coworkers at the laboratory then used the shuttle radar images along with images from two satellites to locate ancient camel trails across the desert—trails that converged on a location covered with giant sand dunes near a well that is known today as Ash Shisar. Archaeologist Juris Zarins of Southwest Missouri State University in Springfield, also working with Clapp, led an expedition across the desert to the site, where he found the ruins of Ubar in a dig that started in November 1991. News of the find emerged early in 1992 in a series of interviews with various participants.

According to Zarins, Ubar, or its predecessors in that region, had existed for perhaps 5,000 years, most likely by producing and trading frankincense. When the end came for Ubar, it was even more dramatic than burial in a sandstorm. The evidence suggests that an earthquake caused the city to fall into a giant sinkhole in the limestone underlying the region. The sand dunes that now cover it came later.

The following year, the same team discovered the remains of another lost city once known for its role in the frankincense trade. This city was known in classical times as Saffara Metropolis and now called Ain Humran. While Ubar was an inland trade center, Saffara Metropolis was near the Arabian Sea, but in the mountains where frankincense is harvested. The seaport connected with this frankincense trade,

Moscha, was discovered and excavated in the 1950s. It is on the Arabian Sea, a short distance from Saffara Metropolis.

Fighting over the Dead Sea Scrolls

In the early 1990s, the editing of the Dead Sea Scrolls became a matter of controversy. The team of editors authorized to work on the documents was very slow to publish any of the material, so in 1991, the Biblical Archeology Society published a collection of photographs of the scrolls with an appendix reconstructing one of the fragments. The authorized scholars sued for copyright violation. On March 30, 1993, an Israeli court sided with the authorized editors, and fined the Biblical Archeology Society, but various appeals were still pending.

In 1991, a new set of carbon-14 dates showed that the Scrolls date from the second century B.C. to the first century A.D.

In 1994, a new set of scrolls, this time from the city of Petra in what is now southwestern Jordan, were found in the fire-ravaged ruins of a church that burned in A.D. 521, and was later knocked to pieces by an earthquake. While these fifty-odd scrolls from about A.D. 500 were carbonized, they are still readable and provide the first significant information about the early Christian community in that part of the Byzantine Empire.

THE DEAD SEA SCROLLS

In 1947, a chance discovery by goatherds eventually led to the discovery of some 800 separate documents dating from around the time of Jesus, about the second century B.C. to A.D. 68. All the documents—as many as 600 of them in fragments that needed reassembly to be read—were found in dry caves near an early settlement on the Dead Sea (between Israel and Jordan) known as Qumran (in the Israeli-occupied West Bank regions). The texts have come to be known as the Dead Sea Scrolls. The documents range from copies of material already incorporated into the Talmud or Old Testament and speculations about the coming of the messiah to legal codes and commentaries.

Between the time of their discovery and 1991, access to the fragmentary Scrolls was restricted to a small group of scholars working in Israel, although all the full-length documents were published soon after discovery. Most of the fragmentary material is in the form of prayers, commentaries, prophecies, and rules of worship or conduct. In 1991, the Huntington Library in San Marino, California, permitted anyone to look through its store of photographs of the fragments. This was followed by several 1992 publications which assembled the fragments for the interested reader. The original band of scholars often disputed details of the published work. Nothing in the fragments changed the realization that came with the publication of the full-length documents: that early Christianity was more Jewish than had been understood before the discovery of the Scrolls.

(Periodical References and Additional Reading: *Science* 12-6-91, p 1467; *New York Times*, 2-5-92, p A1; *New York Times*, 4-19-92, p I-22; *New York Times*, 4-21-92, p C1; *New York Times*, 8-16-92, p IV-2; *New York Times*, 9-29-92, p C1; *Discover* 1-93, p 56; *New York Times*, 3-31-93, p A12; *New York Times*, 8-6-93, p A1; *Archaeology* 1/2-94, p 20; *New York Times*, 8-9-94, p C1; *Archaeology* 11/12-94, p 31; *Discover* 1-95, p 57)

CHINA AND THE SILK ROAD

Since the 1975 discovery of the buried terra cotta army of China's first Emperor near Sian, the archaeology of China has become noticed by scientists from the West. While much of China's past, a history nearly as long as that of Babylon and Egypt, has been known only in legend, Chinese and international investigators are beginning to uncover the hard evidence.

Over the next 10 years, however, archaeologists will have their last chance at one of the cradles of Chinese civilization—the Three Gorges region of the Yangtze River. The world's largest dam is being constructed along the river, and large parts of the state of Hubei and Sichuan will be inundated. Chinese archaeologists have already located 200 ancient tombs and sites, some dated back 7,000 years. In addition to excavating important sites, the archaeologists hope to disassemble the most important stone temples and move them to high ground, as the Abu Simbel temple of Rameses II was moved out of the way of Lake Aswan in Egypt.

Although the Three Gorges region has long been known as an early Chinese site, the other traditional "cradle of Chinese civilization," the Hwang Ho (Yellow River), which also dates back to about 5000 B.C., is considered the more important. But both river civilizations apparently were preceded by a hundred years or so by people living in the Mongolian region of China, also known as Inner Mongolia. A village from that time, now called Xinglongwa, has yielded pottery and jade ornaments that are the oldest so far found in China. Like the other early centers, Xinglongwa was also on a river, but not one so grand as the Yangtze or Hwang Ho. Instead, Xinglongwa was on a small hill overlooking the Mangniu River, a tributary of the Dalinghe River.

The Silk Road

Although people usually think of China and the West as isolated from each other until the time of Marco Polo [Italian: 1254–1324], scholars have long known that there was overland trade between China and the Roman Empire. The route between the two empires—known as "the Silk Road," in recognition of its most important commodity—is traditionally dated as beginning about the second century B.C. In 1991, however, the mummy of a woman who had been buried at Thebes, Egypt, in about 1000 B.C. was found to have strands of silk woven into her hair, suggesting that trade between China and Egypt greatly predates the known opening of the Silk Road. Further study is needed to reveal more definitively the chemical signature of Chinese silk, however. Otherwise, there is also the possibility that the strands are unprocessed silk from Egyptian caterpillars. Literary evidence has also suggested the early presence in Egypt of Chinese silk, notably the description of Cleopatra's "white breasts resplendent through [fabric] wrought in close texture through the skill of the Seres [people of eastern Asia, thought now to be the Chinese]." Cleopatra lived from 69 B.C. to 30 B.C.

Another new discovery in 1991 consisted of the ruins of one of the main stops along the traditional Silk Road. The city of Ubar on the Arabian peninsula was located by satellite-borne radar that penetrated beneath the surface of the sand. This led to study of the Silk Road using sand-penetrating radar on several U.S. space shuttle missions of the early 1990s. Derrold W. Holcomb of the Earth Resources Data Analysis Systems used data from a *Columbia* flight to locate ancient rivers and lakes in the Taklamakan Desert of northwestern China, for example, suggesting places for ground-based missions to look for archaeological sites.

The Silk Road may even be far older than the 3,000 years suggested by the remains of silk strands in the mummy's hair. Anthropologists who believe that humans evolved in Africa and then traveled to Asia (*see* "Out of Africa, But When?" p 47) believe that the Silk Road was a likely path for *Homo erectus* to get from Ethiopia to China. Russian geologists claim that they have found confirming evidence for this theory—remain of humans from 800,000 years ago in the Pamir Mountains, along the southern route of the Silk Road, from Samarkand (Uzbekistan) to Kashgar (Kashi, China).

Tattooed Lady of Scythia

Groucho Marx sang of "Lydia, the Tattooed Lady." In 1993, a group of Russian archaeologists found the earliest known tattooed lady in a grave near the point where Siberia, China, Mongolia, and Kazakhstan all meet, a few hundred kilometers north of Tashkent on the Silk Road. Unlike Lydia, whose tattoos were maps, the mummy known as "Lady" was engraved with swirling blue animals along her arms and hands, including carnivores and a flying horse. Lady is presumed to be a Scythian noble who died near the age of 25 around 500 B.C. She was buried with various grave goods, including a small stash of marijuana, and wore gold earrings, a white silk blouse, a long red-striped skirt, white stockings, and a felt-and-wood plume used as a hair grip. In addition to being embalmed, with her internal organs replaced with moss and peat, Lady was further preserved by Siberian permafrost.

Mongols in the News

For a variety of reasons, archaeologists have been rethinking the role of the Mongol Empire in history. In 1994, the American Museum of Natural History showed an exhibit of Mongol and pre-Mongol artifacts, "Empires Beyond the Great Wall," which contributed to the revival of interest.

On August 18, 1994, it was announced that the amateur archaeologist Maury Kravitz of Chicago, Illinois, had signed an agreement with the Mongolian Academy of Arts and Sciences to search for the tomb of Ghengis Khan, the thirteenth-century Mongol emperor. Kravitz noted that he believed that the tomb would contain the "greatest unretrieved treasure trove in the history of the world." Certainly if Kravitz were to find the tomb, it would rank among the great achievements of archaeology.

Legend says that the 2,000 servants who built the tomb and buried the emperor were then killed to prevent the location from being revealed, after which the soldiers who killed the servants were slain as well.

(Periodical References and Additional Reading: *New York Times* 3-16-93, p C1; *Archaeology* 3/4-93, p 21; *Archaeology* 7/8-93, p 22; *New York Times* 7-13-94, p A4; *Archaeology* 9/10-94, p 27; *New York Times* 9-10-94, p 2)

THE GLORY THAT WAS GREECE

People picture archaeologists as working in remote settings where they live in tents and endure desert or tropical rigors, but many of the most important sites are major cities that have been occupied by humans for millennia: Rome, London, Marseilles, Beijing, Mexico City. Notable among the occupied sites are the cities of Greece: Sparta, Corinth, Samos, Andros, and other classical cities of the Greek mainland or islands are still bustling communities, while Athens both was and remains a metropolis. Archaeologists working in Athens are often frustrated by the city that is in the way of their work. Sometimes, however, the needs of the city aid archaeology. The Athens Metro (a subway scheduled to be completed in 1998) tunnels below archaeological levels, but will have five stations in the ancient part of the city. Work on the stations was halted for as long as 17 months in the early 1990s, while archaeologists located and rescued or recorded fifth century B.C. graves, a large, previously unknown Roman bath, and Byzantine and Turkish reservoirs. Among the finds at modern-day Athens' Syntagma (Constitution) Square were the bed of the Eridanos River, a fifth century B.C. aqueduct, and a factory for molding bronze statues that also dates from the fifth century B.C. The Athens Metro company plans to include mini-museums in the Syntagma Square station, and perhaps in other stations as well.

The removal and recording of archeological treasures is delaying the construction of the subway at a cost of about $600,000 per day. After study, however, the architectural finds that cannot be removed will be destroyed to make way for the trains.

Meanwhile, above ground, work continues on restoring the temples atop the Acropolis. Since 1990, Greek architects have been in charge of a new effort to bring restore the hilltop shrine. The Propylaia, the gateway building, is being re-restored with the most recent noticeable step being removal of the main beams and cross beams of the ceiling. The ceiling itself had been removed in 1990. In the 1994 portion of the restoration, titanium was used to replace rusting iron reinforcements that ultimately caused more problems than they solved.

Other buildings are also changing. In 1993, the remaining frieze blocks of the Parthenon were shipped to the Acropolis Museum for conservation and cleaning of acid rain damage. The whole frieze may be replaced with copies of the original blocks. On another part of the hill, the Temple of Athena Nike will be dismantled and rebuilt for the third time since its original construction in 427-424 B.C.

Ancient Ships

Greeks are sailors and ship operators today, just as they were in classical times. Until recently little was known of ancient Greek ships, apart from a few written descriptions. Throughout the twentieth century, however, an increasing number of sites of ancient Greek shipwrecks have been found. In rare instances, the sea has preserved parts of the wooden ships, although more often the only evidence is ceramic or metal material from the cargo. The hull of a 47 ft (14 m) single-masted ship from the fourth century B.C. was located and raised off the northern coast of Cyprus. Another wreck from that century found near Sicily was also about that size, although only a few timbers were preserved.

About 1983, a Greek fisherman found evidence of an ancient shipwreck in the Aegean Sea off the island of Alonisos. Not until August 1992, however, did archaeologists begin a full-scale study of the wreck. When they did, they found that the ship was apparently about twice the size of the other two known. The Alonisos ship was probably built between 400 B.C. and 380 B.C.

Restoration of the Alonisos ship is proceeding cautiously, layer by layer. The top two layers consist of a cargo of wine in amphoras, the large jars that were used in classical times the way wooden barrels were used in northern Europe (where there were forests). The Alonisos ship may have carried as many as 3,000 amphoras, corresponding to a modern 150-ton cargo ship. Underneath two layers of amphoras, fine ceramic and metal utensils used by the officers were found, indicating that the captain and mates lived well.

The biggest surprise of the initial investigation was the size of the ship. From the pattern of amphoras, the ship was about 85 ft (26 m) long and 35 ft (11 m) wide. Although conditions suggest that the hull may be preserved, the cautious excavators had not reached far enough to be certain of that by the last available reports.

The Hellenic civilization is the extension of Greek language and culture throughout the Mediterranean as a result of colonization and conquest. Many Hellenic cities are still important today, and their archaeological remains are also buried beneath modern structures. One of the Hellenic city-states farthest from the Greek homeland, and one that has thrived since its founding about 600 B.C., was Massalia (today Marseilles, France). After six centuries, the Hellenic city-state at the mouth of the Rhone River was captured by the Rome in 49 B.C. and was already a Roman province when the Roman Empire began a couple of decades later. Today it is the second largest city in France.

Even more than Athens does, Marseilles hides its archeological treasures beneath the modern city. But when a large underground parking lot was built next to City Hall in the early 1990s, archaeologists (led by French archaeologist Antoinette Hesnard) had an opportunity to investigate previously hidden sites. The location was over part of the ancient shoreline, so the main discoveries were ship hulls, docks, and warehouses—including two ships from near the founding of Massalia by Greek sailors, and three from later Roman times. The hulls have been preserved by the wet conditions in the clay below the city, and will be dried out for further preservation. Both of the Greek ships, thought to date from about 525 B.C. to 480 B.C., were put together without nails. Ropes were used as bindings for both vessels, while one also employed pegs, mortises, and tenons.

Because of the preservative properties of the wet clay, baskets and objects of leather and wood were also found in the warehouse sites. Additionally, much ancient pottery was recovered.

The Sporting Life

Central to Greek and Hellenic experience were sports, especially the Olympic Games and, later, the Panhellenic Games. Since 1974, Stephen G. Miller of the University of California at Berkeley and coworkers have been investigating a large stadium and associated facilities at Neméa, south of Corinth. They have located such important adjuncts as the first known locker room and evidence of cheering sections in the stands, as well as such typically Greek components as a temple to Zeus and a sacred grove of cypresses.

In 1993, Miller described how various pieces of evidence were put together to ascertain the nature of the hysplex, which was the system Greeks used for starting a foot race. Instead of firing a gun, as we do today, they used an ingenious string-and-pole device (the hysplex itself) that had the same function as the starting gate does for race horses. The hysplex held the runners with a cord across their waist and another across their knees. To start the race, an official shouted *apite* (take off) and let go of a rope attached to the hysplex. Both sets of cords fell to the ground along with the poles holding up the waist-high cords. In 1993 Miller and Pano Valavanis of the University of Athens, following clues from drawings on a vase as well as from their excavations at the site, built their own hysplex and used it to start runners. It worked fine, starting the runners with an equality greater than a pistol shot provides, but not tangling the runners' feet in any way.

The hysplex, which was probably developed about 340 B.C., appears to derive from a mechanism invented about 10 years earlier to launch several catapults at once, showering opponents with rocks. Thus, it represents an early example of technological fallout from a military application.

Lead Face Powder

It is a commonly held view among the environmentally aware that lead in water pipes and lead dinnerware caused the decline of the Roman Empire. In this view, the known toxic effects of lead weakened the aristocracy (who could afford piped water and good dinnerware), lowered the aristocrats' reproductive rate, and made them stupid or insane. While historians think that this is an oversimplification, it remains clear that the unrestricted use of lead in classical times posed hazards for many.

In 1994, a team of archaeologists reported on an earlier lead hazard in Mycenaean Greece. The white powder they found in a Mycenaean tomb of about 1000 B.C. was most likely used as a cosmetic. This type of face powder is well-known from later Greek times and from the Roman Empire, when even generals used lead-based face powder on formal occasions. Greeks women from the sixth century B.C. also used lead-based creams to lighten their skin. The Mycenaean powder was four-fifths calcium carbonate (lime) and one fifth lead sulfate hydrate—a typical composition for face powder in classical times. Use of the powder could trigger various symptoms of lead poisoning, ranging from headache or vomiting to personality changes, according to modern analyses.

(Periodical References and Additional Reading: *New York Times* 4-13-93, p C1; *Archaeology* 7/8-93, p 21; *New York Times* 12-7-93, p C14; *New York Times* 5-13-94, p A4; *Archaeology* 5/6-94, p 15; *Science* 9-16-94, p 1655; *Archaeology* 9/10-94, p 19; *Archaeology* 1/2-95, p 44)

AND THE EMPIRE THAT WAS ROME

Although Rome itself is another modern city built on an archaeological site, the two most interesting new discoveries from the time of the Roman empire (in addition to discoveries at Troy and Marseilles that are discussed elsewhere) come from another site almost in Italy and one in a corner of Great Britain.

The Cemetery of Statues

In late July 1992, Major Luigi Robusto, while diving for sport a half mile off the coast of Italy, made what is among the most remarkable finds in underwater archaeology—the so-called "cemetery of statues" in the Adriatic Sea off Brindisi. Major Robusto observed a large foot sticking out of the sand in 50 ft (15 m) of water. Further investigation, largely by STAS, the Technical Service of Underwater Archaeology of the Italian Ministry of Culture, located more than a thousand pieces of Roman sculpture, mostly bronzes. Surprisingly, the statues, life-size or larger, are from various periods in Roman history. Apparently, they landed in the "cemetery" when a boat transporting them from one or more sites was shipwrecked on the way to Brindisi, sometime between the third century A.D. and the Middle Ages.

One possible explanation of the distribution of statues is that they had been collected to be melted down for their bronze. Another is that all of the statues were once in a city that was plundered, and the statues were booty. Many of the statues have parts missing, although it is not clear whether or not the missing heads or feet were lost as a result of the presumed wreck or at some earlier point.

The absence of clear evidence of the cargo boat presumed to have been shipwrecked makes it even more difficult to determine the reason for the collection to be there on the sea bottom. However, they got there, the find is the largest collection of large bronzes from antiquity ever discovered.

The Hoard at Hoxne

A farmer in Hoxne, Suffolk, northeast of London, lost his hammer in one of his fields in 1992. In November, the farmer, Peter Whatling, asked his neighbor to use a metal detector to look for the lost hammer. Instead of the errant hammer, Eric Lawes began finding very old coins. Following the trail, Lawes and Whatling soon located many Roman gold coins buried together in the field. With the help of the Suffolk Archaeological Unit, further investigation located a large hoard, including an astonishing 14,780 gold and silver coins, some 200 pieces of gold and silver jewelry, and five carefully packed silver bowls. There is reason to think that more treasure was also buried by the unknown family, apparently rich Romans who feared pillage. A matching set of silver spoons bearing the name Aurelius Ursicinus is the only clue to the identity of the family. Previously, there had been no records or evidence of Roman occupation in that region of England.

British law calls objects that were hidden but never recovered by owners long gone a "treasure trove," and turns it over to the Crown. The government then compensates the persons who located such a treasure trove. The Hoxne Treasure Trove, the largest ever found in Britain, is now at the British Museum in London.

Oddly, the treasure trove seems to be missing parts of the silver service. Since the part that was found is still in good condition, there is a good chance that the remainder of the service and possibly other treasures were buried elsewhere and remain to be found. In the eighteenth century, a much smaller hoard of gold coins was found in the neighborhood. Perhaps this part of Suffolk abounds in buried treasure.

(Periodical References and Additional Reading: *Archaeology* 11/12-92, p 46; *Archaeology* 5/6-93, p 27; *Archaeology* 1/2-94, p 22)

NEW WORLD MONUMENTS AND MYSTERIES

THE EARLIEST AMERICANS

The date of the earliest arrivals in the Western hemisphere has long been debated. By far the most popular view among twentieth-century archaeologists has been that hunting people from Siberia crossed over to the Alaskan peninsula near the end of the most recent Ice Age, making their way down an ice-free corridor east of the Canadian Rockies to what is now the United States over a few hundred years. Great quantities of easily hunted game then led to a population explosion that filled both North and South America in another few hundred years. This process began around 9500 B.C. This dominant point of view was challenged by a number of pieces of evidence in the early 1990s.

- Bones now called the Midland Woman found in Midland, Texas, in 1953, were dated in 1992 by a technique using radioactive uranium. The calculated date of 9,600 B.C. makes the bones of Midland Woman the oldest found in the Americas.
- In 1991 and 1992, 11 dates for the Mesa site in the Brooks Range of Alaska were calculated using a mass spectrometer and radioactive carbon-14. These dates all fell between 7,730 B.C. and 9,660 B.C. John Hoffecker of Argonne National Laboratory and Roger Powers and Ted Goebel of the University of Alaska in Fairbanks used these dates, combined with information from other sites in Alaska, known glaciation and ocean levels in the past, and data about past vegetation, to arrive at 12,000 B.C. to 10,000 B.C. as the most likely date for the first human settlers in Alaska. This is 500–2,500 years earlier than previously believed.
- In 1993, three post holes and two pits found in 1990 at the Paleo Crossing Site in Medina, Ohio, were radiocarbon dated at 10,200 B.C. The remains, along with Clovis points, were found at a construction site and later dated by the University of Arizona.
- In 1994, the remains of a mastodon showing marks of human butchering were found in the sediments laid down by the Aucilla River near Tallahassee, Florida. In addition to cutting away flesh, the hunters, who killed the mastodon in the spring (according to growth rings on the tusk), also removed the tusk from the jawbone, with the apparent intention of using the ivory. Ivory and chert tools were also found at the site. Radiocarbon dating of gourd seeds found with the bones, as reported by David Webb of the University of Florida in Gainesville, shows that the mastodon bones were buried about 10,200 B.C., 700 years too early for humans to be on the scene anywhere in what is now the United States. Furthermore, the Florida location suggests an unconventional path, with early Americans crossing eastward to the Great Lakes while still in Canada and then spreading into the United States.
- In 1992, a caver found remains of grizzly and black bears on Prince Edward Island. Carbon-14 dating the following year placed the remains at 10,300 B.C. Conventional theory has always proposed that Prince Edward Island was covered with a glacier at

that stage of the Ice Age, but the presence of bears suggests that it was ice free. This possibility led archaeologist James Dixon of the Denver Museum of Natural History to theorize that humans reached Prince Edward Island by boat about 12,000 B.C. Dixon has long believed that Asians could have reached the West Coast of North America that way, an idea first put forth in the 1970s by Knut Fladmark of Simon Fraser University in Vancouver, British Columbia.

- Ronald Dorn of Arizona State University and David Whitney of the University of California used carbon-14 dating on stone tools from the Mojave Desert in California and on petroglyphs (rock paintings) in Arizona's Petrified Forests. The petroglyph dates ranged from 16,000 B.C. to 14,000 B.C., while the stone tools yielded dates from 24,000 B.C. to 10,000 B.C.

- In 1994, a team led by Antonio Torroni of Emory University in Druid Hills, Georgia, reported on a study of mitochondrial DNA in 25 Native American groups from various parts of North and South America. They found that the most likely date for the arrival in the Americas of the ancestors of these groups was between 20,000 B.C. and 27,000 B.C.,with the earlier date considered more likely by the team. The research was reported in the February 1, 1994 *Proceedings of the National Academy of Sciences.*

- Johanna Nichols of the University of California at Berkeley analyzed some 200 language families worldwide to determine when one language split from another. Her analysis, published in 1994, suggests that humans reached the Americas from Asia between 33,000 B.C. and 38,000 B.C.

- The doctoral dissertation of Fabio Parenti, which received the highest possible grade from the jury at the École des Hautes Étude Sciences Sociales in Paris, reports on Parenti's conclusion after eight years of research that the Pedra Furada rock shelter in northeast Brazil was in use by humans as early as 48,000 B.C.

(Periodical References and Additional Reading: *New York Times* 3-17-92, p C5; *Science* 1-1-93, p 46; *Archaeology* 1/2-93, p 28; *New York Times* 2-1-94, p C10; *Science* 2-11-94, p 753; *New York Times* 3-23-93, p A1; *Science* 4-2-93, p 22; *Archaeology* 5/6-93, p 27; *Science* 2-4-94, p 660; *Science* 2-25-94, p 1088; *Science* 4-15-94, p 347; *New York Times* 6-8-94, p A18; *New York Times* 6-12-94, p IV-2; *Archaeology* 7/8-93, p 36; *Archaeology* 7/8-94, p 17; *Archaeology* 9/10-94, p 28; *Archaeology* 11/12-94, p 22)

CRADLE OF MESOAMERICA

Because of the obvious relationships among the Mesoamerican civilizations of southern North America (Mexico and perhaps the southern United States) and Central America, many archaeologists have assumed that the culture of these societies emerged from a common ancestor civilization. The Olmec civilization of the Mexican Gulf Coast, centered somewhat south of Veracruz, has the best credentials for being the originator, showing from as early as 1200 B.C. such cultural icons as the sacred ballgame, temples built on pyramids, ritual bloodletting from the penis, and human sacrifice.

But Olmec history, and the pre- and post-history of the Olmec, are poorly understood. For example, prior to the 1950s most scholars thought that the Olmec followed the Maya; carbon-14 dating showed that the reverse is true. Olmec writing is scarce and primitive and has not been deciphered, although its hieroglyphic forms appear to be ancestral to other writing in Mesoamerica. Therefore, it was considered to be a breakthrough when anthropologist John S. Justeson of the State University of New York at Albany and linguist Terrence Kaufman of the University of Pittsburgh announced in the March 9, 1992 *Science* that they had deciphered 80% of the 21 columns of glyphs on a monument from Veracruz that dates from A.D. 159. The date is the earliest one known for a specific object from the Americas (older sites exist, but not with readable dates).

The monument is not written in "Olmec," but in a glyph script used in the region after Olmec influence waned. Perhaps the people were descendants of the Olmec, but perhaps not. In any case, the language is an ancestor of the language called Zoquean, which is the native language of many of the people who still live near Veracruz.

Like many achievements in archeology, deciphering this ancient writing took many years of patient effort. As a graduate student in the 1960's, Kaufman began a project of recreating the ancestral language of the Mixe-Zoquean speakers of the Gulf Coast of Mexico. By 1976, Kaufman could claim that Olmec was a Mixe-Zoquean tongue. The monument that was to be the test of his theories was an engraved column, or stele, found in 1986 in the Acula River near the village of La Mojarra, 25 miles from the Gulf. It was notable both for the length of the text and its preservation. Kaufman now had a text to which he could apply his reconstruction. The text on the La Mojarra stele had to be compared with other, shorter texts in the same script, which was dubbed epi-Olmec ("on the Olmec model"). It also became more apparent that Mayan script, only recently deciphered itself, was closely related to epi-Olmec writing. Justeson and Kaufman were then able to put together a sensible tale from the mysterious glyphs.

The text, as deciphered, tells of a ruler called Harvest Mountain Lord, his battles, ritual bloodletting from his penis, and his eventual triumph over his main enemy, thought to be his brother-in-law. At the end, Harvest Mountain Lord distributes ritual objects to his supporters as rewards for their help.

Other ancient systems of writing around the world continue to present puzzles for archaeologists. In addition to Olmec itself, among the unread are the language of Harappa and Mohenjo-Daro, civilized cities that flourished along the Indus River in Asia around 2300 B.C.; a Cretan form known as Linear A from the Mediterranean, written from about 1650 B.C.; and the written language of Easter Island in the remotest South Pacific from about A.D. 1500, although the Easter Island hieroglyphics may not be an actual writing system.

Along the Nautla River

The discovery of a new major site north of Veracruz, known as El Pital, also fills some gaps in Olmec history.

El Pital, 9 mi (15 km) inland from the Gulf of Mexico and the head of navigation on the Nautla River, along with several other sites along the river, has been known to archaeology since the 1930s, but the Nautla sites were thought to be minor and had not been scientifically investigated. In the 1930s, banana farmers began to clear much

Background

Although orthodox theory says that humans have been present in the New World for only a few more than 10,000 years, a number of ancient civilizations developed. Those most commonly considered are the Aztec, Maya, and Inca—but that perception occurs because these societies were the only ones in evidence at the time of extensive European contact. As scientists have learned more about the predecessors of these three civilizations, archaeologists have come to recognize that the Aztec-Maya-Inca classification system needs to be replaced by a broader understanding.

In South America, the Moche and Chavín (along with a number of other different civilizations) preceded and led to the Inca Empire that existed at the time of the European conquest. Similarly, various different societies existed or had existed in Mexico and Central America. These societies shared many cultural features, although the culture of those in South America was noticeably different from those to the north, despite a certain amount of intercontinental trade. The civilizations of Mexico and Central America—and those of the southern United States to a lesser degree—were all manifestations of a larger cultural entity. These civilizations have gradually come to be known by the generic name "Mesoamerican."

more land in the region, exposing the true and great extent of El Pital. However, thinking the temple mounds were simply natural hills, they planted over them.

Archaeologist S. Jeffrey K. Wilkerson of the Institute for Cultural Ecology of the Tropics in Veracruz surveyed the area and recognized that the hills were the remains of a large city. He located more than a hundred structures as well as satellite communities all occupying a region of about 40 sq mi (100 sq km). Wilkerson's report appeared in February 1994 in the *National Geographic and Research* quarterly. Some of the structures are as tall as 130 ft (40 m). Ball courts and plazas indicate a typical Mesoamerican culture. From its strategic position, El Pital could control much Mesoamerican trade along the Gulf Coast and inland. The community, which may have numbered 20,000 inhabitants at its height, flourished from about A.D. 100 to 600. The ancient city is now known as El Pital, after the farming village that occupies the site today.

Cuajilotes, another site further up the Nautla River, has also captured the imagination of archaeologists and others. Cuajilotes was first reported by a rafting party in 1990, but was not visited by archaeologists until 1992. Soon after, on August 5, 1992, the governments of Mexico and Veracruz State declared that the region around Cuajilotes would be listed as a protected natural area, and that a massive reconstruction of the site would begin immediately. Cuajilotes is not Olmec, but flourished sometime before A.D. 1000.

It is becoming increasingly clear that the Nautla was one of the main corridors from Teotihuacán in the Valley of Mexico to the Caribbean Sea.

(Periodical References and Additional Reading: *Science* 3-19-94, pp 1700 & 1703; *New York Times* 3-23-93, p C1; *Discover* 1-94, p 82; *New York Times*, 2-4-94, p A10; *New York Times* 8-7-94, p A1; *Archaeology*, 5/6-94, p 16; *Discover* 1-95, p 50)

MEET THE MAYA

The Maya are the indigenous population of the Yucatán peninsula and the immediately adjacent regions of Mexico and Central America. Although not the oldest civilization in the Americas, the Maya have captured the imagination of many archaeologists and mystics, partly because the Mayan civilization had already declined when the Spanish arrived; thus, the air of mystery surrounding the Maya exceeds that of the two civilizations at their peak during the time of the Conquest—the Aztec of central Mexico and the Inca of the Andes. Furthermore, unlike the Aztec (who wrote only numbers and dates) and the people of the Inca civilization (who used knots in ropes to record data) the Maya had a well-developed system for writing. However, the meaning of Mayan hieroglyphics remained obscured until the past few years. The recently developed ability to translate their texts has lifted the veil from the mysterious Maya.

Royal Tombs

Perhaps nothing is as exciting in archaeology as locating and excavating the tomb of a king or queen that has never before been opened. Robert Sharer and coworkers from the University of Pennsylvania enjoyed that particular thrill when they located the tomb of a king of Copán, a Mayan city in what is now Honduras. The tomb had been closed for about 1,500 years. Further north on the Yucatán peninsula, a 1993 team led by David Freidel of Southern Methodist University in Dallas, Texas, succeeded in locating two royal tombs that had managed to escape grave robbers for an even longer period.

The Copán tomb contained not only the skeleton of the king and attendant jewelry, but an even more valuable horde of two dozen decorated vases, bowls, and plates, painted with colored stucco in vibrant colors that somehow survived in the closed environment of the tomb. The information in and on the various vessels is the real treasure. Sharer and his staff moved the ceramics to a controlled environment similar to that within the tomb to protect both the contents and the fragile stucco coatings. The temperature and humidity there will gradually be adjusted to that of the ambient air, then the glyphs on the vessels can be read and analyzed.

In Yucatán, the Freidel dig was at a site called Yaxuna and known primarily as two mounds of rubble in a flat plain 10 mi (16 km) south of Chichén Itzá. Upon excavation, one mound was discovered to be an enormous pyramid that contained within it a vaulted chamber, the tomb of a ruler of Yaxuna in the period from about A.D. 300 to 350. The former king occupied the tomb alone, his skeleton adorned with the jeweled regalia of Mayan royalty. Nearby, the other mound also contained a pyramid, smaller than the first, but containing a mass grave of a dozen or more men, women, and children. One of the three women was in the late stages of pregnancy, and one of the three children held a doll. Another child wore a royal pendant indicating that he would have been a future king. Other jewelry associated with royalty was also found in the mass grave.

Freidel has reconstructed the events leading up to this pair of tombs. He believes that the small pyramid was built from the remains of an earlier temple, which may have been leveled by an outside invader. The invader dedicated the new temple atop the pyramid by killing the ruling family and burying them with all their royal trappings. He or his descendants then built a new and larger pyramid and temple in which the single monarch—perhaps the conqueror himself or a powerful later ruler—was buried alone.

Aside from the drama of the royal tombs, the excavations at Yaxuna reveal that the city was more powerful than expected, and may have held political sway over the region ruled by Chichén Itzá a half a millennium later. But even during the earlier dominance of Yaxuna, the cultural influence of Teotihuacán in the Valley of Mexico had begun to take hold. Goods and ideas from Teotihuacán are prevalent during much of the early classic period of the Maya.

Two other royal Mayan tombs were also opened for the first time in the early 1990s—see "Class Structure and Society," below.

The Two-Empire Theory

Late in 1994, a paper proposing a new theory of the history of the classic stage of Mayan civilization began to circulate among Maya specialists. The paper, by Simon Martin of University College, London, and Nikolai Grube of the University of Bonn in Germany, suggests that during the classic period, the Yucatán peninsula was dominated by two political organizations which can be compared to alliances somewhat like the North Atlantic Treaty Organization and the Warsaw Pact during the Cold War. Perhaps the two groups of interlocked city-states were even more closely bound, as in the known Aztec "empire" that existed at the time of the Spanish invasion. That sort of empire leaves much of the governing to the individual states within it, operating largely on the basis of tribute and loyalty in war, more like the feudal system in the middle ages in Europe than the Roman Empire.

In the Martin-Grube reconstruction of the Maya past, the two dominant powers were the well-known city—or city-state—Tikal in Guatemala and the much lesser known, but larger, city of Calakmul, in Campeche state, Mexico, about 60 mi (100 km) to the northwest of Tikal. The primary evidence Martin and Grube find for their theory are glyphs that identify rulers of smaller cities as having gained power "by the doing of" the king of either Tikal or Calakmul. Alternatively, they find glyphs that name the ruler of one of the two large cities as the "lord" of the local king. Tracing these statements of allegiance to the main cites in the base of the Yucatán peninsula during the classic period indicates, to Martin and Grube at least, that the rulers of Tikal and Calakmul had amassed small "empires" by the end of the seventh century. Calakmul had the most cities that owed it sovereignty—Caracol, Dos Pilas, El Peru, and Naranjo—and they occupied a good strategic position, encircling Tikal. Tikal, however, held sway over some cities outside that perimeter, including Palenque, far to the west. Despite the initial dominance of Calakmul, by the middle of the eighth century Tikal had taken over the lead. But by the time that Tikal had defeated Calakmul in a series of wars, the classic period was coming to an end, with a spectacular collapse of the whole lowland Mayan civilization. The Mayan civilization of the postclassic period would be centered further out on the Yucatán peninsula and involve entirely different city-states.

The theory is controversial, in part because it is based almost entirely on translating the still somewhat enigmatic glyphs. Certain specialists have suspended judgement while others have chosen to totally reject the theory unless it can be confirmed by direct evidence of tribute flowing into Calakmul and Tikal from their supposed vassal states.

The Maya of Belize

Most of the famous Mayan sites are on the Yucatán peninsula, which is often thought of as part of Mexico. Much of the peninsula is not Mexican, however. Guatemala protrudes into the base of the peninsula, while the small state of Belize is entirely on the peninsula, where it forms the lower part of the eastern side, facing the Caribbean Sea. The Maya, of course, knew nothing of these modern political divisions, and occupied the entire peninsula, along with adjacent regions that include part of what are now Honduras, El Salvador, and non-Yucatecan Mexico. In the early 1990s, Belize was the site of many new discoveries made about the Maya.

Most of the previously known Mayan cities of Belize were in the northern portion which, like most of the rest of the Yucatán peninsula, consists of lowlands. The southern part of the nation contains a mountain range called the Maya Mountains, throughout which Mayan sites were discovered by Peter S. Sunham of Cleveland State University in Ohio. The Maya Mountains are largely uninhabited today; neither the surviving Maya descendants nor anyone else lives there. During the classical period of the Maya, however, there were mining towns scattered in arable valleys amid the mountains. In 1993, four such sites were discovered in the Maya Mountains, each featuring stone monuments or buildings and reservoirs that still contain water. The largest site, on an island in the Monkey River, includes a plaza and a pyramid. The other sites also include a ball court, a raised causeway, and a number of stone monuments. Most likely the inhabitants of these sites, whose total population at their height was about 7,000, traded minerals or other goods with the more populous cities in the lowlands of the peninsula.

The largest single deposit of Mayan jade artifacts, not counting the various pieces of jade recovered from the Sacred Cenote at Chichén Itzá in Yucatán, was discovered by Thomas Guderjan and coworkers from St. Mary's University in San Antonio, Texas. The cache was discovered at a site known as Blue Creek in northwestern Belize. Blue Creek, which includes about 80 buildings that have survived, flourished between A.D. 600 and 750. In addition to the jade, which included several pendants, the pieces found included ceramics, obsidian, and shells, along with fragmentary human remains. All were deposited in a stone-lined vertical shaft which was capped with an uncarved upright stone slab (known as a stele when carved, which is usually the case). Archaeologists think that the jade and other articles were intended as offerings to mark the dedication of the 20 ft (6 m) high building in which the objects were found.

Large Mayan centers persisted in Belize well after the collapse of the classic Mayan civilization in the decades before A.D. 900. In 1993, it was shown that the city of Xunantunich in west-central Belize prospered until some unknown date after A.D. 1000, about 150 to 200 years longer than most of the better-known classic Mayan cities. Xunantunich is not the only Mayan city in current-day Belize to have lasted past the collapse, however. In the 1970s, the northern Belize site of Lamanai was shown to have flourished in the period immediately following the classic period. One theory to account for the survival of Lamanai and Xunantunich, despite the fall of so many other cities, is that the cities of Belize were on major waterways leading to the Caribbean. With water access, they could trade with other cities in what is now Mexico. Because of a combination of geological and meteorological factors, there are no major rivers nearer the tip of the peninsula than the two in Belize and a couple of rivers on the other side of the base of the peninsula.

Class Structure and Society

On January 4, 1993, Arlen F. Chase and Diane Z. Chase, both from the University of Central Florida in Orlando, announced that excavations in 1992 at the Caracol site in Belize—not to be confused with the building El Caracol at Chichén Itzá in Mexico—demonstrated that the Maya had a large middle class interposed between the small group of ruling kings and priests and the peasant farmers previously thought to constitute the bulk of Mayan society. In the period between A.D. 500 and 900, the Caracol middle class included artisans, soldiers, bureaucrats, and professionals such as bone-setters. Caracol occupied a large site, covering 55 sq mi (140 sq km). The city rose to power after its conquest of the better-known Tikal in 562.

The evidence cited for the existence of a middle class includes the remains of large workshops for producing jade and shell jewelry. The output of these workshops would have been too great for the needs of the nobility or priesthood, yet peasants could not have afforded fine jewelry. Hence, there must have been middle-class buyers.

Another site, Dos Pilas in Guatemala, also shows signs of the Mayan middle class, according to its excavation director, Arthur Demarest of Vanderbilt University in Nashville, Tennessee.

Although the Chases announced their conclusions at a press conference in 1993, they had already published their discoveries about Mayan society in a 1992 book. The occasion of the press conference was to describe two royal tombs located in Caracol during the 1992 digging season, both discovered intact. One tomb, that of a ruler and another individual, dates from about 400, while the second, identified by a painted date as from 686, contains remains of four members of the royal family.

Maya Astronomy

In the early 1990s, astronomers began to develop a theory that the main religious beliefs of the Maya, especially their concept of creation, were a direct reflection of what the Maya priests saw in the sky. According to Maya translator Linda Schele, as the constellations appear and disappear at the zenith, they depict the events described in the Popul Vuh (the basic Mayan creation story as transcribed after the conquest). The Milky Way is a prominent part of the story, as are the planet Venus and the constellations of the Pleiades and Orion.

Background

Visitors to such Mayan sites as Chichén Itzá may be informed by their guide that a particular building was a Mayan observatory—and some of the buildings so designated, such as El Caracol, even look like modern astronomical observatories. The traditional view, reflected by the modern guides, has been that Mayan priests were great astronomers who used their knowledge both to refine the calendar and for astrological purposes.

Detailed measurements of some monuments confirm Mayan knowledge of some basic astronomical concepts. A plaza in Uaxactún is arranged so that on the annual solstices and equinoxes the sun rises over significant architectural points. El Caracol in Chichén Itzá has windows containing sight lines to the northern and southern extremes of the setting of the planet Venus, as well as windows aligned with the equinoxes; however, modern archaeologists no longer think of El Caracol as an observatory, but rather as a temple built with astronomical alignments. The best evidence that the Maya knew astronomy is illustrated by their calendar system, which was too sophisticated to have been constructed without a good knowledge of the exact length of the year.

Astronomer Jay M. Pasachoff and Samuel Y. Edgerton, both of Williams College, claimed in the early 1990s that a figure of a Mayan god smoking a cigar is actually a personification of Comet Halley. The date on the Mayan calendar that is inscribed on the panel showing the cigar-smoker is January 17, A.D. 684 in the Julian calendar, which is a half-year earlier than a known visit from Comet Halley. Furthermore, Pasachoff and Edgerton claim that the Maya of that time were already aware of the periodic nature of visits by Comet Halley and had employed the comet's period of about 76 years in calculating the birth dates of various local gods (the gods' birth dates are separated by periods that are multiples of 76 years).

TIMETABLE OF THE MAYA

1500 B.C.	Village life is established in the Yucatán peninsula. This is considered the start of the pre-classic period of Mayan history.
1000 B.C.	Villagers at Cuello (northern Belize) are making pottery around this time.
500 B.C.	Yaxuna in central Yucatán peninsula is founded.
100 B.C.	First large buildings are being constructed at Yaxuna.
250	The date arbitrarily chosen as the border in time between the pre-classic and the early classic periods of the Maya, although the traditional date is 292.
292	On July 8, the first dated stele, or inscribed column, is erected in Tikal, marking the traditional beginning of the classic period.
426	The first king of Copán (Guatemala), Yax-Kuk-Mo' or Blue-Quetzal-Macaw, comes to power.
600	The date chosen as the border in time between the early and the late classic period. By this time, Yaxuna and the northern Maya enter a period of decline.

615   Lord Pacal ascends to the throne in Palenque (Chiapas state, Mexico), beginning a period of great prosperity in western Mayan lands.

628   Smoke-Jaguar ascends to the throne in Copán (Guatemala) and a similar period of economic success is initiated there.

682   The accession of Ah Cacaw to the throne in Tikal initiates that city-state's period of prosperity as well.

            By this time, according to the Martin and Grube theory of Maya history, all the city-states in the Mayan lowlands have been absorbed into the "empires" of either Tikal or Calakmul.

750   By this time, Tikal has overpowered Calakmul to take control of the lowland Mayan region, according to the Martin and Grube theory. From here on, however, Tikal and Copán begin to decline, enabling the more northern cities, such as Yaxuna, to prosper once more.

761   Dos Pilas (Guatemala) is destroyed by war.

780   The first dated steles at Cobá (Quintana Roo state, Mexico) are erected.

822   Copán collapses in political turmoil.

869   The last known dated stele at Tikal is erected.

900   The date chosen as the border in time between the last classic period of Maya history and the post-classic period. Uxmal (Yucatán state, Mexico) is at its height around this time.

1000   About this time, the city of Chichén Itzá (Yucatán state, Mexico) arises. Cobá goes into a decline. Yaxuna is defeated in battle, and effectively ceases to exist for the next 200 years.

1200   About this time, Chichén Itzá collapses.

1441   The Mayan city of Mayapán (Yucatán state, Mexico) is destroyed, and what is left of Mayan civilization collapses along with it.

1502   Christopher Columbus encounters Mayan traders at sea.

1517   The first encounter between the Spanish and the Maya on the mainland takes place.

1547   The Spanish complete their conquest of Yucatán, although a few pockets of Mayan resistance continue for another 150 years.

The Mysterious Neighbors

As more history is uncovered, the Maya have become less of a subject for speculation. However, spelunkers and archaeologists have uncovered new mysteries just to the southeast of Maya country. At about the same time as the pre-classic or very early classic Maya, people living in what is now central Honduras made distinctive pottery and other wares, creating a culture not closely related to the Maya. The discovery of this culture took place in April 1994, when two Honduran and two American spelunkers were exploring the Calve de Río Talgua, a large cave system near Catacamas, Honduras. One of the explorers, Jorge Yañez, became intrigued with a small opening high above the cave floor. He returned to Catacamas for a ladder, and became the first to enter what has been dramatically christened "The Cave of the Glowing Skulls." Within the almost inaccessible cavern, Yañez found piles of human bones, including skulls. The "glow" refers to a coating of crystallized calcium carbonate that reflects light shone upon the skulls. All in all, the tropical setting and the mysterious bones are strikingly reminiscent of the opening sequence of *Raiders of the Lost Ark* in which archaeologist Indiana Jones escapes several booby traps to collect a precious artifact.

The cavern of the bones is a typical limestone room, similar to those in many caves that may be visited in the United States and elsewhere, complete with stalactites, stalagmites, calcite curtains, and other formations formed by water dripping through a limestone formation. The cavern is approximately 100 ft (30 m) long, 25 ft (8 m) high, and 12 ft (4 m) wide. It is a side branch of a cave through which the Talgua river flows for about 2 mi (3 km).

In addition to the bones, the cavern contained about 20 ceramic bowls and two marble vases, none of which are in the style of either the Maya or the known cultures in Honduras. A mile or so from the cave, the buried remains of a settlement has also produced pottery fragments (shards of pottery which were actually found lying on the surface) that match the style within the cave. The site, in a cow pasture, has not been excavated.

In the cave itself, the bones are separated and stacked, suggesting that the bodies had been left exposed until the flesh had rotted, after which the bones were brought to the cave for a second burial. The bones were also painted with red ocher in places.

The date of the cave burials is far from certain. Different authorities think that it could be anywhere from 800 B.C. to A.D. 500. It is hoped that there will be enough protein left in some of the bones to perform radiocarbon dating.

James E. Brady of George Washington University is heading the investigation of the cave, an investigation partly being financed by the making of a movie about the site.

(Periodical References and Additional Reading: *Discover* 1-93, p 58; *New York Times*, 1-5-93, p C1; *Archaeology*, 7/8-93, p 26 & 33; *New York Times*, 8-3-93, p C7; *New York Times*, 10-5-93, p C1; *Archaeology*, 11/12-93, p 24; *New York Times*, 1-18-94, p C1; *New York Times*, 10-4-94, p C1; *Science* 11-4-94, p 733; *Archaeology*, 11/12-94, p 18; *Science News* 12-17-94, p 415)

Native American UPDATE

Although Columbus was clearly mistaken when he identified the people he found in the Caribbean as Indians (some were Arawak, while others were known as Carib by their enemies and by various local names among themselves), the incorrect name stuck and was even applied to all the other peoples who lived in most of North or South America in pre-Columbian times. "Eskimos" were an exception to the general rule, but they also were not called by their own name, which, for the most familiar group, was Inuit. Properly, one should refer to the pre-Columbian people of what is now the United States by their own names, such as Mahican, Lakota, or Hopi, but it is convenient to have a collective name for all these groups. The politically correct currently use "Native American," recognizing that this designation is also flawed, since it would seem to refer to any person of any people who happened to be born, or born and reared, in the Americas.

Ancient Cannibals?

The Inuit and similar people of the far north do not like to be called Eskimo; that is a derogatory nickname that is usually translated as "raw fish eater." The "Carib," who are nearly extinct today, suffered an even more distasteful nickname, since their commonly used designation means "cannibal." Indeed, the word cannibal is commonly traced to the Spanish version of the name Carib, which in turn is what the Arawak, another people from the same region, called their main enemies. Although Eskimos sometimes are known to eat raw fish, it is far less clear whether or not the Carib ever ate fellow humans. Cannibalism is easy to make as an accusation, but hard to prove as a fact.

In 1973, archaeologist Larry Nordby excavated a small pueblo in Mancos Canyon, Colorado. This site had been occupied by the Old Ones, or Anasazi, about A.D. 1600. Underneath that pueblo he found yet another from a half century earlier which had been occupied by two to six families. He also found several beds of human bones that appeared to have been broken into pieces prior to being dumped in the beds. The bones could be evidence of early Native American cannibalism. Tim D. White and coworkers undertook a five-year-long study of the bones, which numbered 2,106 fragments, in an effort to establish whether cannibalism could account for the damage

---

THE CANNIBAL QUESTION

Some anthropologists believe that humans never cannibalize each other as part of rituals or in search of a more varied diet. Others, notably Marvin Harris, find rampant cannibalism around the world.

The anthropologists who find no evidence for regular cannibalism have to discount not only historical accounts but also tales of travelers and even reports of some cultural anthropologists. Those holding cannibalism to be a myth claim that the sources of such stories are simply biased against the society described as cannibalistic. Cannibalism is always a practice of the people on the next island, or of the tribe in the next valley, or of ancient peoples who came before. Observers never report that their own society is cannibalistic. Furthermore, personal encounters with modern cannibalism by unbiased observers are hard to come by (perhaps for a good reason, say those with the opposite view).

The Marvin Harris school finds cannibalism a natural way for a society living in an impoverished ecosystem to obtain protein. They note that the Aztecs, for example, lived in a region where there were no large game animals nor domestic animals, so they became cannibals, consuming the flesh from the prisoners of their almost constant state of war. The theory continues by claiming that the practice of cannibalism was ultimately invested with ritual and given a spiritual meaning in order to allow acceptance of the practice.

Opinions on this matter do not seem to be settled by the discovery of new evidence. Instead, new evidence seems to promote new arguments for both points of view.

---

observed. Slowly, they pieced together the fragments, revealing that they came from the skeletons of 17 young adults and 12 children. White and his team noted a common characteristic of the longer bones, such as limb or rib bones: a rounding and abrasion at the end of the bone. A similar pattern was observed in longer bones of large mammals from the same site. This kind of pattern could have resulted from use of bones as crude tools, but White thought otherwise.

White performed an experiment in which he broke up some mule-deer limb bones and ribs with stone hammers and then boiled the bones in an earthenware pot for three hours. The cooked bones, when compared with similar ones that had not been cooked, displayed the same rounded and abraded ends, suggesting that the human bones had also been boiled in rough pots. Essentially, the aim of such cooking would be to release fat that could be skimmed from the cooking pot, to cook marrow for direct consumption, or to produce the liquid soup base known to cooks as stock. White also observed that bones having little nutritional value, such as human facial bones, were not broken up. His conclusion, expressed in a 1992 report, is that the Anasazi practiced cannibalism for food, at least on some occasions. Famine is difficult to prove from bones, but the ecosystem where the Anasazi lived is fragile, and famine is therefore a ready suspect.

Even if White is correct in his analysis of the Mancos site, it is far from clear whether Anasazi cannibalism was an important ritual, as it is supposed to be in some societies, or a desperate response similar to that of the Donner Party when faced with starvation.

Native American Skeleton

Maintaining security at archeological sites has become a problem around the world, but it is particularly difficult at sites on public lands in remote places. For this reason, when an 8,000-year-old skeleton was found in a high mountain cave in the White River National Forest in Colorado in 1988, the discovery was not announced until 1994. Because of its location—above 10,000 feet in the Rocky Mountains—and its age, the skeleton has been compared to the mummified Ice Man of the Alps, which is about 2,700 years younger. But the White River skeleton, while a complete set of bones, was found without tools or clothing, and had been disarticulated by rodents.

After the bones had been examined utilizing a CT scanner, they were molded in casts and sampled for DNA (which was well preserved). They were then turned over to the local Southern Ute Tribe of Native Americans for appropriate reburial.

Great Serpents

The Great Serpent Mound in Ohio is one of the best-known Native American sites. People are less likely to have heard of the giant snakes that were carved into the ground by Plains Indians at an unknown date. These Kansas carvings are the most dramatic part of the remains of vast ceremonial centers. The Plains serpent sites are hard to date, but the one near Lyons, Kansas, appears to have been carved by Wichita sometime prior to A.D. 1541. A similar site is near Waconda Lake, Kansas.

One of these ceremonial centers was discovered by accident in Kansas in the early 1990s. A large field had been left uncultivated in an effort to preserve native prairie. After the field was sold, the new owner burned off the native grasses and forbs to produce pasture. With the vegetation gone, the field revealed the outline of a 90 ft (27.5 m) snake, a turtle, and the embankments, causeways, and trenches characteristic of a ceremonial serpent site. Investigations are continuing by Donald Blakeslee and coworker David Hughes of Wichita State University. Blakeslee and Hughes think that the causeways and animal figures at the site have astronomical alignments.

(Periodical References and Additional Reading: *Archaeology* 1/2-94, p 11; *Archaeology* 7/8-94, p 16; *Archaeology* 9/10-94, p 22)

REMOTE ISLAND LIFE AND DEATH

The archaeology of the Pacific Islands provides a set of puzzles and proposed solutions like no other on Earth. Although there are questions regarding the origins of specific groups of people on continents, it is easy to picture large or small groups easily moving from place to place on land. Some of the Pacific Islands, however, are the most remote places known to be occupied by humans, and it is not always clear how they could have come to be populated in the first place. Furthermore, one island will be populated by a people whose language and mien do not resemble those of people on other islands, with little in the way of a discernable pattern. Finally, Europeans reached these islands late, so scientific study of these issues is also recent.

A standard explanation of these patterns in the Pacific has evolved. The following timetable includes this standard explanation, some anchor dates to help understand the large-scale history of the region, and some newly developed dates that tell the fate of populations on islands that are too small to support humans without outside trade and genetic inputs.

TIMETABLE OF THE PACIFIC (STANDARD VERSION)

1.8 million B.C.	According to the latest dates, *Homo erectus* is in Asia, or at least in the land that is now Java, which is sometimes connected to Asia by dry land when the sea level is low.
60,000 B.C.	Carbon-14 dating suggests that *H. sapiens* is in Australia, having probably colonized New Guinea along the way. Other, more definitive evidence definitely puts humans in Australia and New Guinea by 35,000 B.C.
8000 B.C.	Pigs, native to Southeast Asia, are imported to New Guinea by boat, where they become an important part of a farming economy.
5000 B.C.	About this time, give or take a thousand years or so, the original Austronesian language develops, most likely in Taiwan. All inhabitants of remote Pacific islands, as well as inhabitants of Indonesia, the Philippines, parts of New Guinea, and even Madagascar, speak Austronesian languages.

4000 B.C.	Neolithic cultures using pottery along with shell and stone tools develop in Taiwan and southern China. Slightly modified traditions from the region will reappear in Polynesia.
1600 B.C.	Archaeological evidence shows that a culture now called Lapita that appears to stem from China and Taiwan has established itself on the islands to the east of Australia—Vanuatu, New Caledonia, Fiji, and adjacent islands. This culture is based on a mixture of hunting and gathering (especially of shellfish), as well as farming with pigs and chickens. There is no evidence of Melanesian culture at this time on those islands. Later, however, these islands will all be Melanesian.
200 B.C.	About this time, Polynesians are found in the Marquesas Islands, according to carbon-14 dates. Some artifacts from this time appear to have been made in Fiji, 3,000 mi (5,000 km) to the west, suggesting that the Marquesas were settled by people coming from Fiji, not by people from Tahiti. Others believe that Tahiti is the homeland of all Polynesians.
A.D. 400	Earliest date by which Hawaii is thought to be settled by Polynesians, although it could be as much as 500 years later.
900	Maoris are the first humans to colonize New Zealand.
1000	Polynesians, probably from Mangareva Island, colonize the remote coral atoll of Henderson Island and the nearby extinct volcano known as Pitcairn Island, among the most remote tiny islands in the Pacific. Mangareva is 250 mi (400 km) to the west of Henderson.
1450	The small populations of Polynesians on Henderson Island and on Pitcairn become detached from humanity and from each other when tribal warfare disrupts civilization on Mangareva, which has remained their chief trading partner.
1521	Ferdinand Magellan reaches the Mariana Islands, the first European to visit the islands of the remote Pacific.
1600	About this time, the small isolated populations on Henderson and Pitcairn become extinct.
1642	Abel Tasman becomes the first European to visit New Zealand.
1767	Samuel Wallis becomes the first European to visit Tahiti.
1789	Nine members of the crew of the H.M.S. *Bounty* settle on Pitcairn Island and import women from Tahiti to found a new population. Descendants of this settlement remain to this day.

The timetable proposed above conflicts with some evidence obtained in 1993. Cambridge and Oxford biologists Erika Hagelberg and John B. Clegg were following a trail of DNA studies that began in the 1980s with Rebecca Cann, Mark Stoneking, Koji Lum, and Bryan Sykes. In October 1993, Hagelberg and Clegg reported on a DNA analysis that they have conducted on older remains of Polynesians. They found that the very oldest in their sample, from about 700 B.C., appeared to have DNA of the type associated with Melanesians. Hagelberg, Clegg, and other researchers involved also found DNA evidence of some Melanesian ancestry in what are thought to be purebred

BLACK, SMALL, AND MANY ISLANDS

There are several separate cultures that have long existed in the South Pacific. They are physically different and exist on different islands, although most of their languages are related. These languages form the Austronesian family, perhaps the most diverse of the great language families. The combining form − nesia, from the Greek for "islands," is a common clue. Indonesia contains various cultures; its main islands were once known as the Indies. The three other groups with the − nesia suffix are the "black" islands of Melanesia, the small islands of Micronesia, and the many islands of Polynesia. The culture and language of the Melanesians is clearly different from that of the Polynesians. The Micronesians appear to be a mixture of the Polynesian with the Australian or the Melanesian.

Polynesians of today. Hagelberg and Clegg theorize that the islands where the Lapita culture took hold were first settled by Melanesians. Thereafter, Southeast Asians brought the Lapita culture to the islands, intermarried somewhat with the local Melanesians, and moved on. Others suggest that the Melanesian genes could have come later, which would be more in accordance with the standard theory of Pacific colonization.

Thor Heyerdahl has conceived a famous alternative theory explaining how the most remote Pacific Islands were colonized. Heyerdahl hypothesized that the Polynesian Islands were settled by sailors from South America. In favor of Heyerdahl's concept have been two main facts:

- The sweet potato, which was cultivated by Polynesian populations when the first Europeans encountered them, is native to South America, although all other domestic plants and animals cultivated by Polynesians are Asian in origin.
- Heyerdahl's raft *Kon Tiki,* when set adrift off the west coast of South America, did reach an atoll near Tahiti after about 100 days.

Easter Island, which is comparatively close to South America, has often been cited as the source of more evidence in favor of a South American origin for at least some Polynesians. Oral history suggests that Polynesian settlers arrived from the east, not the west. Furthermore, at one time there was a dominant population on Easter Island that stretched their ear lobes, a trait best known from the ruling class of the Inca culture.

DNA studies have been brought to play in this case as well. In 1993, Hagelberg reported that DNA from the bones of living and dead Easter Islanders matches that of other Polynesians and fails to match that of native populations from the west coast of South America. In this case, Hagelberg's work supports the more traditional theory that Easter Island, as well as the other Polynesian islands, were settled by people whose ancestors came from Asia.

(Periodical References and Additional Reading: *Science* 10-29-93, p 656; *Science* 1-7-94, p 32; *Nature* 8-4-94, p 331; *Natural History* 4-95, p 4)

RETHINKING THE EUROPEAN ARRIVAL

Although 1992 marked the 500th anniversary of the arrival of Christopher Columbus in the West Indies—the beginning of the permanent European occupation of the Americas—the event was not celebrated as much as had been expected. By 1992, even the descendants of the Europeans realized that the European arrival was hardly a blessing for either the Native American population or for the land. Furthermore, it was clear that even though the Columbus voyages in the fifteenth century were the start of the current phase of European occupation, they were certainly not the first encounters with Native Americans by Europeans.

Viking Visits

Historians have long known that Scandinavians settling Iceland and Greenland also claimed to have colonized a region that was apparently in the Americas. Yet there existed no definitive archaeological confirmation of a Viking presence on the continent until the discovery by Norwegian archaeologists Helge and Anne Stine Ingstad of a Norse site in Newfoundland in 1961. Since the discovery of the site, known as L'Anse aux Meadows, various other evidence has been collected, although no other settlement has been found. A conference in 1992 established the North American Bicultural Organization to collect and analyze information relating to the Norse period of American history. Among the topics discussed were:

- A Norse penny from about 1065 that was found at Blue Hill Bay near Brooklin, Maine. The coin, discovered amid evidence of an old Indian camp, was found with no other Viking artifacts, suggesting that it had reached Maine via trade routes. It does not establish a Viking presence that far south.
- Various Norse artifacts at a Native American site in Canada's Northwest Territories on Ellesmere Island, which is to the west of northwest Greenland.
- A statuette found further south, on Baffin Island. The statuette appears to depict a cloaked Christian bishop.

The following year, an analysis was conducted of a Newport, Rhode Island building, which people locally had long thought to have been built by Vikings. Bicultural Organization scientists demonstrated that construction of the tower was too late for the Norse period in North America. On the other hand, the scientists felt that their results established the tower as having been built before the occupation of Rhode Island by English Colonists, which took place in 1634. The main evidence was radiocarbon dating of mortar used in construction, which showed that the construction was too late for the Vikings. Archaeologists had previously searched in vain for Norse remains at the site. In addition, computer analysis of the building showed that the unit of measure used in constructing it was not the version of the English foot used in the oldest securely dated house in Newport, which dates from between 1676 and 1698. Thus, they think the tower was probably built by explorers, traders, or fishers other than the British, probably between 1492 and 1634. If so, that might account for the local belief, evidenced in Newport's Viking Hotel and other local place names, that the structure was Norse, since the tower would have been built before the colonists arrived. Finding a building already there in 1634 might have been interpreted as proof that the unknown builders were Vikings.

Oldest Site Permanently Occupied by Europeans

While L'Anse aux Meadows is the earliest site in North America known for sure to have been occupied by Europeans, it was abandoned after a maximum of two or three years. Furthermore, Newfoundland, the site of L'Anse aux Meadows, is an island — the mainland part of the province is Labrador. The oldest site on the North American continent known to have been occupied by Europeans was almost certainly in what is now the state of Florida (the name Florida was used originally by the Spanish to mean the entire East Coast of the continent, but here refers only to the territory occupied by the present state).

Spanish explorer Ponce de Leon tried to start a colony in Florida in 1521, but he was killed that same year and the project abandoned. Another Spanish colony began in 1559, at Pensacola Bay, and lasted two years before being wiped out by a hurricane. A French Protestant outpost, Fort Caroline (near Jacksonville), was started in 1564 but eliminated by Spanish forces led by Pedro Menéndez de Avilés, who hanged most of the settlers there. Thus, the oldest permanently occupied site is St. Augustine, Florida, which was founded by Menéndez in 1565, primarily in response to the French efforts at Fort Caroline.

Archaeological investigation of St. Augustine over the past 20 years has uncovered various remains of what appears to be the original settlement, which was about a mile north of present downtown St. Augustine. Finally, in July 1993, Kathleen Deagan of the Florida Museum of Natural History in Gainesville announced that she found remains of the original fort on the grounds of the shrine of Nuestra Señora de la Leche. Luckily, the remains are beneath the popular Roman Catholic shrine and next to the Fountain of Youth Park, a tourist attraction so designated in the mistaken belief that Ponce de Leon had sought the Fountain of Youth at that site. The shrine and tourist park serve to protect the archaeological site from otherwise ubiquitous beach development.

The chief evidence that this is the original fort includes a moat and palisade posts that show evidence of having been burned. This combination matches only one of several early European forts in St. Augustine — the earliest. Apparently, the first fort was built by erecting an earthen wall, a wooden palisade, and a moat filled with prickly plants around a thatched longhouse obtained from the local Native Americans. When the natives and the Spanish eventually fought, the Native Americans simply set fire to the fort by loosing burning arrows at the thatch, thus bypassing the Spanish defenses and destroying the fort. Undaunted, the Spanish moved across the Bay, built another fort, and stayed on.

(Periodical References and Additional Reading: *New York Times* 7-7-92, p C1; *New York Times* 7-27-93, p C1; *New York Times* 9-28-93, p C6; *Discover* 1-94, p 80)

ARCHAEOLOGY AND ANTHROPOLOGY ISSUES

RETURN OF THE HERITAGE

In the sixteenth through nineteenth centuries, native peoples were often subjugated by colonialism, primarily European. Although overt colonialism has mainly been eradicated, its effects continue in a number of ways. For example, the preservation of artifacts and remains by museums is viewed by some native peoples as an insult to their ancestors and traditions.

Late in the twentieth century, various injured parties began to demand that museums and private collections return both sacred objects and the bones of their ancestors. Remarkably, many such battles are being won by those seeking the return of their people's artifacts. Victory has required neither a mummy's curse nor the evil consequences that legend tells come from stealing the jeweled stone from an idol's forehead. For the most part, peaceful protests and vigils (although occasionally intermingled with more dramatic scenes) have gradually raised the consciousness of museum directors and officials, who have come to understand that the tribal treasures they display are most often stolen goods in one way or another.

Furthermore, similar considerations apply to new acquisitions. From an archaeologist's point of view, the situation is most problematic for artifacts taken from long-sealed graves. Archaeologists have typically assumed that they could unseal any tomb or excavate any graveyard. This is still largely the case when the original burial occurred sufficiently far in the past. Although nations such as Greece or Egypt insist that their own museums have the right to display ancient pots and mummies, no one expects the Mycenaean warriors or New Kingdom pharaohs—or even their presumed descendants—to come forth and demand reburial. Problems arise, however, in the case of more recently buried materials. Skeletons of native Americans killed by the United States Army late in the nineteenth century are a different matter from bones of *Australopithecus africanus* killed by leopards some two million years ago, although in a strict sense the difference is only a matter of degree.

After winning several recent battles over the contents of nineteenth- or even twentieth-century burials and tribal relics, Native Americans increasingly consider objects reflecting the more distant past of their heritage as worthy of rescue also. The relics of the Anasazi, over a thousand years old, are gaining protection by present-day Native Americans, for example. Even earlier remains may have to be handed over to modern people who may or may not be actual descendants of those who lived in the same region several thousand years ago. Museums are under great pressure to justify their collections, and archaeologists must keep in mind the living descendants who may hold views on a dig.

Native American Claims

The U.S. Congress passed the Native American Graves Protection and Repatriation Act on November 16, 1990. The main purpose of the law is to protect burial sites located on federal lands or on lands held by Native Americans, including Inuit and Hawaiians. Although the regulations affect pot hunter and serious archaeologist alike, there is little controversy over the need for site protection in the future. Another provision of the law, however, attempts to right past practices, which is more difficult. If Native Americans ask for the return of materials taken from graves, including skeletons, for which a cultural affiliation can be proved, the present holder of the materials must relinquish them. Museums acted under a deadline: they were required to notify the appropriate Native American groups by May 1996 regarding what they have in their collections.

More than a dozen museums had already been affected by the law in the early 1990s, well before the deadline. A typical example is the small collection of Sioux tribal clothing and ornaments that has been displayed since 1892 in a room of the Barre, Massachusetts, library. Exhibited are more than a hundred items that were apparently taken from the dead or from the field of battle at the massacre at Wounded Knee, South Dakota, on December 29, 1890. These items had all been purchased by a Barre resident and then given to the library for display. More than 200 Sioux were killed during the battle, including many women and children. The Barre display is the largest collection of Sioux-related materials from that event. It includes items that range from a doll and doll's cradle, to clothing, to hair identified as scalps of both Native Americans and whites.

A sociology professor from nearby Worcester, Massachusetts, knew of the Barre exhibit and informed the Oglala Sioux. Many Sioux, including some who traveled to Massachusetts just to view the exhibit, believe strongly that the materials should be returned to the tribe. Museum officials are concerned that important documentation may be lost if the collection is buried along with the bodies of the native Americans, who were tossed into mass graves after the battle. The Wounded Knee Survivors Association (which owes its name to the "second battle of Wounded Knee," a 1973 protest by the American Indian Movement that led to two native American deaths and many wounded) has proposed a plan in which the actual collection would be returned to the Sioux, in return for which Sioux artists could be hired to recreate clothing and similar everyday items for the library. Not all the Oglala Sioux think such a plan is necessary, since they feel that the material is rightfully theirs.

Another result of the 1990 act was the 1993 return of four Inuit skeletons. In 1897, Robert Peary brought four living Inuit from Greenland to the United States. These living "exhibits" were placed on display in the American Museum of Natural History, where archaeologists studied them for a brief time. Shortly after their arrival, however, they all contracted tuberculosis and died. Until their return to Greenland, the remains were kept in numbered boxes in the museum.

The disposition of relics and bones gathered without respect for native traditions has generated less controversy than have new excavations at very old sites. Some 2,000 years ago, some Native Americans buried some of their dead near great mounds in what are now eastern Ohio and West Virginia. We now know this Native American culture as the Adena Mound Builders, since they did not leave written records that might indicate what name they called themselves.

When a highway was to be built near an Adena mound in West Virginia in the early 1990s, Native American activists from around the United States and their allies tried to protect the mound and its associated burials. In 1991, they persuaded the West Virginia highway department to agree to rebury not only any bones they encountered in construction but also all soil or waste from the area near the bones. The agreement stated that no samples could be retained after pollen analysis or any other scientific study. Human remains, following an alleged ancient custom, would be covered with red flannel at all times and could not be touched by menstruating women. Native Americans would have the right to censor the publication of any photographs of the excavated materials. Some archaeologists cite this agreement as an example of the new reburial movement's negative effect on scientific inquiry, especially since the individuals on the Native American side of the agreement could not actually prove any direct connection with the Adena Mound Builders.

Organizations of archaeologists in the United States disagree on the fundamental principles involved in these matters. These disagreements stem from the fact that physical anthropologists need bones to study, while cultural anthropologists do not. This has led members of the American Committee for the Preservation of Archaeological Collections to state their position that archaeological science is the best way for Native Americans to learn about their own past, while opposing members of the World Archaeological Congress have suggested that oral tradition may be superior to scientific archaeology, as far as Native Americans are concerned.

Outside the United States

Canada is home to many of the same Native American tribes as the United States. At one site in Saskatchewan, the Cree have worked with Canadian officials and local archaeologists to produce an educational park in a region used as a special hunting ground since at least A.D. 410 and containing sites still held sacred by modern Cree. In 1986, the Cree elders declared that every detail of the park had to meet with their approval. The result has been park buildings, signs, and even traffic patterns that reflect Native American beliefs and customs. The 150,000 visitors to Wanuskewin, as the archeological park is known, are the beneficiaries of this policy, which most agree has produced an unusually moving and effective exhibit building in addition to preserving the sites themselves.

In regard to recently buried items, a situation similar to that of the Native Americans in the United States and Canada exists with Australian Aborigines. Aborigines have been reasonably successful in protecting the bones and artifacts of their recent dead. However, the human occupation of Australia goes back 40,000–60,000 years. By explaining what their work entails and what they hope to learn, archaeologists can often obtain the cooperation of Aboriginal groups living near a site.

(Periodical References and Additional Reading: *New York Times* 2-19-93, p A11; *Archaeology* 5/6-94, pp 18 & 37; *Archaeology* 11/12-94, pp 64 & 65)

BANDITS AND THIEVES

When it comes to the "oldest professions," grave robbing has to be among the top contenders for the number three spot. At least since Neandertal times, it has been the custom of many human cultures to bury goods of various kinds along with the deceased, and it has also been the custom for someone else to come along and remove those goods. When archaeology developed as a science, humans finally developed a plausible excuse, other than greed, for grave robbing.

In Peru, where grave-robbing has flourished due to both generally poor economic conditions as well as recent terrorist domination of large parts of the country, thieves have also taken to robbing museums of their collections of grave goods. Although trade in illegally acquired ancient artifacts dates back to the conquistadors, the present wave of looting began in 1987.

1987	Grave robbers in Sipán, some 380 mi (610 km) north of Lima, discover the richest burial ground ever found in the Americas: a graveyard of Moche priests and nobles who were invariably buried with gold ornaments and other gold pieces. Hundreds of villagers, learning of the gold, loot about 100 pieces before the government takes over the site. Armed guards are stationed at the site, and are still necessary today.
February 1992	A 2,000 year old Paracas mantle is stolen from the Museum of Archaeology, Anthropology and History in Lima, although it is recovered two months later. Most authorities consider Paracas textile manufacture to be the finest in history, and good examples are virtually priceless.
February 1993	Thieves steal 58 gold and silver pieces from the archaeological museum of San Antonio Abad University in Cuzco, melting down the necklaces, pins, figures, and so forth for bullion before they are caught.
July 1993	Officials discover that three more ancient textiles from Paracas are missing from the Museum of Archaeology, Anthropology, and History. In Peru, the legal status of ancient objects removed from the ground is not very clear. While in the earth, artifacts are treated legally the same way as minerals and are considered government property. However, both minerals and Paracas textiles are considered private property after excavation.

In the United States, antiquities from both pre-Columbian times and later Native American sites have been federally protected only since 1979. Until 1993, the law was considered to apply only to antiquities stolen from federal land or reservations set aside for Native Americans. However, in 1993, a federal appeals court agreed that relics taken from private land protected by state law (which is usually the case) cannot legally be traded across state boundaries. For practical purposes, the ruling extends federal protection to virtually all sites in the United States.

(Periodical References and Additional Reading: *New York Times* 7-25-93, p I-27; *New York Times* 8-25-93, p A3)

ARCHAEOLOGY RECORD HOLDERS

	Time	
Oldest cord or rope	17,300 B.C.	Found in the Ohalo II site in Israel, which was uncovered from 1989 through 1991 when the Sea of Galilee went through a period of low water levels. The purpose of the cord/rope is not known, although it may have been used in making nets.
Oldest underground entrance to a house	13,000 B.C.	To a mammoth-bone house found at Mezhirich, Ukraine.
Oldest evidence of a structure in North America	10,200 B.C.	Post holes at the Paleo Crossing Site near Medina, Ohio, where Clovis points have also been found.
Oldest pottery found in Americas	6000 to 5000 B.C.	Discovered in Santarem, Brazil, situated on the flood plain of the Amazon.
Oldest surviving wooden structure	5300 B.C.	A well found at Kuckhoven, Germany, preserved in underground deposits.
Oldest wooden boats	4300 to 3700 B.C.	Eleven oak dugout canoes found in Bercy, France.
Oldest paved road	2600 to 2200 B.C.	Egyptian road from a quarry to the shore of a lake, apparently used to transport stone to build tombs at Giza, site of the Pyramids.
Oldest load-bearing true arch	1900 B.C.	A Canaanite mud-brick arch at Askelon, a port on the Mediterranean.

(Periodical References and Additional Reading: *New York Times* 3-17-92, p C5; *Archaeology* 7/8-92, p 17; *New York Times* 9-29-92, p C1; *Archaeology* 3/4-93, p 24; *New York Times* 5-8-94, p I-13; *Science News* 10-9-94, p 235)

Astronomy and Space

THE STATE OF ASTRONOMY

Astronomy is the oldest science. As far back as we can go in the records of rocks and monuments, we can discern astronomy being practiced. Throughout time, recognition of the cycles of the sky and their influence on events on Earth has led to better farming as well as to the development of astrology. Early Greeks and Hellenic scholars were able to measure the distances to the Moon (fairly accurately) and the Sun (not so accurately) and to predict the positions of planets in the sky (with great success). The following pages will detail current progress toward modern goals in astronomy, such as measuring the distances to galaxies, measuring the distance to the edge of the universe, and uncovering and understanding more about the universe's past.

Tools of the Trade

From Stonehenge to the *Hubble Space Telescope*, the tools of the astronomer have been so specialized that they become the starting point for any discussion of astronomy. Astronomers' tools are primarily devices for gathering information from space and for measuring various aspects of that information. For example, most telescopes collect photons, the particles of electromagnetic radiation, and analyze their location or origin and their wave characteristics. This is true whether the telescope consists of a simple pair of refracting lenses or a network of giant radio telescopes. Since 1912, astronomers and other scientists have also collected other subatomic particles from space (including cosmic rays, neutrinos, and rarely such exotic particles as gravitons and a magnetic monopole) and searched for these particles' origin in space or their wave characteristics.

The changing nature of astronomy comes largely from changes in the tools, although sometimes great strides are made by physicists working far from telescopes. In fact, observational astronomers are doing increasingly more work far from their telescopes. Information from light-gathering devices is collected as digital data, rather than by using the traditional analog processes of the human eye or photographic plates. Once information exists as digital pulses of electricity or magnetism, it can be transmitted anywhere on Earth with essentially no loss of data. Thus, the astronomer collecting data from a mountaintop telescope operating in freezing cold and darkness may herself be in a comfortable, well-lighted office hundreds or thousands of miles away. Most dramatically, the telescope may be orbiting the Earth, as are the *Hubble Space Telescope* and the Cosmic Background Explorer.

Despite this new flexibility, funding for such observation remains a problem for astronomers. Even though many new giant telescopes of various kinds are being built in the 1990s (*see* "Giant Telescopes, Optical and Otherwise," p 130 and "Major Telescopes," p 247), money to operate these telescopes is sometimes in short supply. In the United States, the National Science Foundation (NSF) felt it necessary to review funding for all ground-based optical and infrared astronomy projects in an effort to use available funding in a way that will most benefit the progress of astronomy. Part of the difficulty seems to be the common dilemma of Big vs. Small Science (*see* "The End of Big Science in the U.S.? " p 13). If new money is to go to such projects as the very large Gemini telescopes, and if there is no additional money, then existing telescope funding will have to be reduced.

Astronomy and Space Science

The NSF review concerned only ground-based optical and infrared telescopes, which rely on whatever photons can get past Earth's atmosphere, as do radio telescopes. A few ground-based telescopes use other subatomic particles, such as neutrinos (*see* "Neutrino Astronomy Comes of Age," p 136) or even cosmic rays and gravitons (*see* "Giant Telescopes: Optical and Otherwise," p 130).

Photons and other subatomic particles are almost the only things known to reach Earth from space on a regular basis, although there is some evidence that cosmic dust also regularly drops in on Earth. On a less regular basis, larger objects such as meteors arrive on Earth, often with a bang. The most famous and devastating of these dropped onto Earth around 65 million years ago, causing a mass extinction, but many small meteoroids strike the upper atmosphere daily.

The larger objects that reach Earth are all thought to be from our immediate astronomical neighborhood, the solar system. Although much of the solar system has been known since ancient times, astronomers learned little detail about the planets and other solar-system bodies, and even less about meteoroids and asteroids, based on the ordinary collection of photons at the surface of Earth, even after the invention of the astronomical telescope in 1609. Photons that pass through Earth's atmosphere, however, are not the only evidence available in the space age. The photons that have been most useful in reevaluating the solar system have been those sent to Earth from space probes and satellites. Space missions to all the planets except Pluto, to asteroids, and to comets have given Earth-based astronomers a close look at much of the solar system. Such missions have continued into the early 1990s and continue to provide surprising new information.

Although space probes from Earth are now slowly making their way out of the solar system, for the most part astronomers must return to reliance on photons and other subatomic particles for information about the universe outside the solar system. Thus, astronomers do not have the close-up pictures of stars and galaxies that space probes have been providing for study of the solar system. Nevertheless, the evidence that has been collected from space-based telescopes near Earth has clarified data from the far reaches of the universe.

One observation reveals that the universe is noticeably lumpy. Individual objects and small groupings of objects are not scattered randomly or evenly across space-time. Instead, there seem to be a couple of hundred thousand groups here, and another couple of hundred thousand there, with here and there separated by distances vaster even than the enormous lengths across the groupings. For those of us on Earth, "here" refers to our own giant grouping, the Milky Way galaxy, while "there" consists of most of the universe, distributed in lumps, clusters of lumps, and great voids between the clusters. Explaining the origin of this lumpiness has become one of the great tasks in astronomy in the 1990s.

Astronomy has a track record for the twentieth century of having at least one totally unexpected observational or theoretical discovery each decade: 1900s, interstellar matter; 1910s, cosmic rays and distance estimates from Cepheid variables; 1920s, the expanding universe; 1930s, radio astronomy; 1940s, the big bang theory and radio galaxies; 1950s, the Oort cloud and superclusters of galaxies; 1960s, quasars and pulsars; 1970s, rings around Uranus and Jupiter, and the mysterious Chiron; 1980s, dust clouds around stars. No doubt the 1990s will also bring further unexpected discoveries and ideas.

(Periodical References and Additional Reading: *Science* 10-21-94, p 356)

TIMETABLE OF ASTRONOMY AND SPACE TO 1992

2296 B.C.   Chinese astronomers begin recording the appearance of "hairy stars" (comets).

585 B.C.   The scientist Thales [Greek: 624 B.C.–546 B.C.] predicts the solar eclipse that appears this year in the Near East.

480 B.C.   Astronomer Oenopides of Chios [Greek: fl. c. 480 B.C.] discovers that Earth is tilted with respect to the Sun.

410 B.C.   The first horoscopes become available in Mesopotamia (Iraq, Syria, and Turkey).

340 B.C.   Astronomer Kiddinu [Babylonian: fl. c. 379 B.C.] discovers the precession of the equinoxes—the apparent change in the position of the stars caused by Earth's wobbling on its orbit.

300 B.C.   Chinese astronomers compile accurate star maps.

240 B.C.   Chinese astronomers observe Halley's comet.

Astronomer Eratosthenes [Greek: c. 276 B.C.–c. 196 B.C.] correctly calculates the size of the Earth.

165 B.C.   Chinese astronomers are the first to notice sunspots.

130 B.C.   Astronomer Hipparchus of Nicea [Greek: c. 190 B.C.–c. 120 B.C.] correctly determines the distance to the Moon.

140   The *Almagest* of Ptolemy [Alexandria: c. 100–c. 170] develops the astronomy of the solar system in a model that has the Sun and planets rotating about Earth.

1200   The Chinese build an observatory which allows them to calculate with high precision the length of the year by measuring shadows projected on the ground.

1543	*De Revolutionibus* by Nicolaus Copernicus [Polish: 1473–1543] presents convincing arguments that Earth and the other planets orbit the Sun.
1577	Tycho Brahe [Danish: 1546–1601] proves that comets are visitors from space, not weather phenomena, as previously believed.
1592	David Fabricius [German: 1564–1617] discovers a star, later named Mira, that gradually disappears. When it is studied by Phocyclides Holawarda in 1638, Holawarda recognizes that it appears and reappears on a regular basis. It is the first known variable star.
1609	Galileo [Italian: 1564–1642] makes the first astronomical telescope.
	Johannes Kepler [German: 1571–1630] discovers that the planets move in elliptical orbits.
1610	Galileo observes the moons of Jupiter, the phases of Venus, and (although he does not recognize what they are) the rings of Saturn.
1611	Around the same time, several astronomers in Europe are the first in the West to discover sunspots.
1630	Galileo determines that the star Vega has an angular diameter of less than 2 arc seconds. He arrives at this determination by hanging a thin cord between himself and the star, and observing that the star can be hidden by the cord.
1633	Galileo is forced by the Roman Catholic Church to recant his support of Copernicus's theory that Earth revolves about the Sun.
1668	Isaac Newton [English: 1642–1727] invents the reflecting telescope.
1671	Giovanni Domenico Cassini [Italian-French: 1625–1712] correctly determines the distances of the planets from the Sun.
1682	Edmond Halley [English: 1656–1742] describes the comet now known as Halley's comet. In 1705, he correctly predicts that it will return in 1758.
1718	Halley discovers that stars move with respect to each other.
1755	Immanuel Kant [German: 1724–1804] proposes that many nebulas are actually composed of millions of stars, and that the solar system formed when a giant cloud of dust condensed.
1758	Halley's comet returns as predicted by Halley.
1773	William Herschel [German-English: 1738–1822] shows that the solar system is moving toward the constellation Hercules.
1781	Herschel discovers the planet Uranus.
1784	John Goodricke [English: 1764–1786] observes the star Delta Cephei and recognizes that its brightness varies regularly. This is the first notice of what are now called the Cepheid variable stars, which have been important in determining galactic distances.
1785	Herschel demonstrates that the Milky Way is a disk- or lens-shaped group of many stars, one of which is the Sun.
1801	Giuseppe Piazzi [Italian: 1746–1826] discovers Ceres, the first known asteroid.
1838	Friedrich Bessel [German: 1784–1846] is the first to determine the distance to a star other than the Sun.
1846	Johann Galle [German: 1812–1910] discovers the planet Neptune using predictions of Urbain-Jean-Joseph Leverrier [French: 1811–1877] and John

Couch Adams [English: 1819–1892].

1868   Armand-Hippolyte-Louis Fizeau [French: 1819–1896] proposes the use of an interferometric method for avoiding the effect of atmospheric turbulence. In his design, he puts a mask with two holes on the telescope objective, and observes the superimposed images in the focal plane.

1872   Henry Draper [American: 1837–1882] is the first astronomer to photograph the spectrum of a star.

1877   Asaph Hall [American: 1829–1907] discovers the two moons of Mars.

1912   Victor Franz Hess [Austrian-American: 1883–1964] discovers cosmic rays.

1924   Edwin Powell Hubble [American: 1889–1953] shows that galaxies are "island universes"—giant aggregations of stars as large as the Milky Way.

1929   Hubble establishes that the universe is expanding.

1930   Clyde William Tombaugh [American: 1906–] discovers the planet Pluto.

1931   Karl Guthe Jansky [American: 1905–1950] discovers that radio waves are coming from space, leading to the founding of radio astronomy.

1935   A. M. Skellet [American: 1901–] observes that radio waves reflected from the ionosphere appear to have been caused by meteors passing through that atmospheric layer.

1939   Sir Edward Victor Appleton [English: 1892–1965] confirms Skellet's observation of 1935: radio waves are reflected ionized particles caused as meteors pass through upper atmospheric layers.

1945   Robert Henry Dicke [American: 1916–] discovers thermal radiation at radio frequencies emitted by the Moon.

1948   George Gamow [Russian-American: 1904–1968], Ralph Asher Alpher [American: 1921–], and Robert Herman [American: 1914–] develop the big bang theory of the origin of the universe.

1957   The Soviet Union (Russia) launches *Sputnik I*, the first artificial satellite.

1959   The Soviet Union (Russia) launches a rocket that hits the Moon.

1961   Cosmonaut Yuri Gagarin [Russian: 1934–1968] becomes the first human to orbit Earth.

1962   The United States space probe *Mariner 2* is the first artificial object to reach the vicinity of another planet, Venus.

1963   Maarten Schmidt [Dutch: 1929–] is the first astronomer to recognize a quasar.

1965   Arno Allan Penzias [German: 1933–] and Robert Woodrow Wilson [American: 1936–] find radio waves caused by the big bang, proving to most astronomers that the big bang actually occurred.

1966   A space probe launched by the Soviet Union is the first object produced by humans to actually land on the Moon (as opposed to the probe which the Soviet Union crashed onto the moon in 1959). Another Soviet probe becomes the first to orbit the Moon.

1967   Susan Jocelyn Bell Burnell [Irish: 1943–] discovers the first known pulsar while working for Antony Hewish [English: 1924–]; Hewish later gets the Nobel Prize for the discovery.

At a meeting of the Institute for Space Studies in New York, John Archibald

Wheeler [American: 1911-] uses the term "black hole" to describe a gravitationally collapsed star that exhibits a Schwarzchild singularity. The term continues to be used for this phenomenon.

1969   The United States lands two people on the moon: Neil Alden Armstrong [American: 1930-] and Edwin Eugene "Buzz" Aldrin [American: 1930-].

1971   A U.S. spacecraft is the first to orbit another planet, Mars.

1975   A Soviet (Russian) space probe transmits pictures from the surface of Venus.

1976   The U.S. *Viking* space probes begin transmitting pictures of the surface of Mars; they do not detect life on the planet.

The rings of Uranus are discovered.

1978   A U.S. surveillance satellite detects an unexplained huge explosion over the South Atlantic Ocean. Although thought at the time to be a secret nuclear weapons test, possibly by South Africa and/or Israel, experts later decide that the explosion results from the impact of a meteor from space, such as a small asteroid. The energy released is calculated to be equivalent to 100 kilotons of TNT.

1979   The U.S. space probe *Voyager 1* discovers that Jupiter also has rings.

1980   Alan Guth [American: 1947-] develops the theory of the inflationary universe, an explanation of what happened after the big bang.

1981   The United States introduces the space shuttle, a reusable spacecraft.

Robert Kirshner, August Oemler, and Paul Schechter perform a sky survey concentrated in the direction of the constellation Boötes, and discover that there is a giant void measuring about 300,000,000 light-years across and containing no observable matter. At the time of discovery, it is the largest known entity in the universe.

1982   A thorough investigation of the 30,000 pictures taken by *Voyager 2* during its August, 1981 flight past Saturn allows astronomers to confirm the existence of an 18th moon of Saturn.

1986   The space shuttle *Challenger* blows up, killing all seven aboard.

1987   During a news conference, Alan Dressler uses the term "Great Attractor" to describe an unusual concentration of mass that causes galaxies to move toward it. This term continues to be used to identify this phenomenon.

1988   On August 29, an error in a computer command causes *Phobos 1*, a Soviet (Russian) space probe to Mars, to tumble, resulting in a loss of all communication with the probe.

1989   On March 27, just as the Soviet (Russian) space probe *Phobos 2* is reaching its destination—the Martian moon Phobos, where it is to deploy two surface landers—contact with the spacecraft is suddenly lost. Seventeen minutes of garbled but decipherable data are returned before the shutdown.

On August 24, the U.S. space probe *Voyager 2* flies by Neptune, which at the time is the planet farthest from the Sun (most of the time Pluto is the farthest). The probe passes by Neptune, its rings, and its moons, revealing that Neptune's rings appear "clumpy" but complete. It is not clear how many rings there are: some scientists count three, while others see five. Finding the magnetic poles gives scientists the first accurate measurement of Neptune's rate of rotation, which is computed to be about 16 hours, 3 minutes (give or take 4 minutes).

In September, Martha P. Haynes [American: 1951-] of Cornell University and Riccardo Giovanelli [Italian: 1946-] announce their observation of a large cloud of gas that they believe is a galaxy in formation.

Duane Muhleman uses the 210-foot antenna of the Deep-Space Network at Goldstone to bounce radar signals off Titan, Saturn's largest moon. The returning signals are picked up by the 27 dishes at the Very Large Array in New Mexico, and reveal that part of Titan is covered by a liquid.

1990   The space probe *Magellan*, launched in 1989, reaches Venus and produces a detailed map of its surface using radar and a 12 ft (3.66 m) diameter antenna. These radar images of Venus' surface possess a resolution 10 to 100 times better than previous images.

The Cosmic Background Explorer (COBE), launched by NASA in 1989, precisely measures the cosmic background radiation, confirming that it perfectly matches to a black-body radiation spectrum of a body at a temperature of $-455°F$ (2.735K), and that it has the same intensity in all directions, thus supporting the big bang theory.

The New Technology Telescope, built by the European Southern Observatory at La Silla, Chile, contains a 140 in (355 cm) mirror, the shape of which is continuously corrected by 75 computer-controlled actuators.

Mark Showalter searches through 30,000 *Voyager* images, and discovers a tiny moon (12 mi or 19.3 km across) circling Saturn. This moon, later named Pan, is responsible for creating the Encke Gap in Saturn's ring system.

1991   The 33 ft (10 m) Keck telescope at Mauna Kea in Hawaii is completed. The main mirror consists of 36 individual segments whose form and position are controlled by computer.

On April 10, Joss Bland-Hawthorn, Andrew Wilson, and Brent Tully announce their discovery of an object which possesses a mass several billion times that of the Sun, and which is possibly the largest black hole known. It is detected by observing the gravitational attraction it exerts on a large rotating disk of gas in the galaxy NC 6240.

METHODS OF OBSERVATION

HUBBLE REPAIRED AND MAKING DISCOVERIES

The problems with the satellite-borne *Hubble Space Telescope* (*HST*) were well known to most Americans—*HST* became a symbol for poor vision and expensive technological mistakes. However, the repair of *HST* ultimately proved the worth of space flight incorporating human pilots and crew when, in December 1993, astronauts used the space shuttle to travel to the vicinity of *HST*, brought it aboard the shuttle, successfully corrected the telescopes's problems, and relaunched it.

A Successful Mission

On January 13, 1994, a month after the *HST* repair, the United States National Aeronautics and Space Administration (NASA) called a press conference to show how well the new parts worked. The main camera, known as the Wide Field and Planetary Camera, had been completely replaced with a version containing built-in correcting lenses, rather like giving a nearsighted child glasses. New pictures taken with the "Correcting Camera" showed individual stars in the M100 galaxy where none had been seen before in uncorrected *HST* images. Before 1994 ended, astronomers were to use the newly visible stars to make one of the most dramatic astronomical findings of the year (*see* October 27, 1994, below). Successful images from the corrected Faint Object Camera were also displayed. Because of improved detectors in the Wide Field and Planetary Camera, new computer power, improved solar arrays, and better equipment for aiming the cameras, *HST* is now even better than it was originally intended to be.

The first correction was made in the Wide Field and Planetary Camera, which was replaced on December 7 with an improved camera that also contained four small corrective lenses. The Ball Corporation of Broomfield, Colorado, developed the main set of corrective lenses, named the Corrective Optics Space Telescope Axial Replacement, or Costar, which cost $50 million to develop and build. Costar includes 10 corrective lenses that work together to correct the light going to the Faint Object Camera and two spectrographs. Later on December 7, Costar was installed in place of the high-speed photometer, the most expendable part of *HST*.

Other problems had also been discovered before the space mission as *HST* continued to operate. For one thing, the gyroscopes that were supposed to ensure stability began to fail, one at a time. Gyroscopes are used to keep airplanes, ships, and spacecraft on course, by virtue of the fact that a gyroscope's spinning body resists any change in orientation. *HST* started off its trip with six gyroscopes, twice as many as it needed. By 1993 only three of the gyroscopes still worked. The gyroscopes were the first repair job for the shuttle crew, who installed a set of six new gyroscopes on December 5, 1993, even before starting repair of the telescope itself.

From the start of *HST*'s mission, the solar panels (packs of solar cells that power the telescope) caused a wobbling motion of the telescope, due to uneven heating as the craft moved in and out of Earth's shadow. New solar panels were installed on December 6, 1993, complete with modifications to reduce the wobble.

Another problem concerned one of two spectrographs aboard *HST*: the Goddard High-Resolution Spectrograph, which had to be shut down in September 1991. Engineers think that a defective solder joint caused the problem. Although the spectrograph still could be used in a limited way, it seemed more prudent to shut the spectrograph down completely, thus allocating its observing time to other instruments. Correcting the problem in space during the 1993 shuttle mission to the space telescope meant adding another day to the mission.

Two of three computer memory units in *HST* had also failed. The December repair mission used the extra day needed for the spectrograph repair to install a coprocessor to supply the missing memory.

Today's *HST* can see with clarity objects which are 10 times as far away as those which can be seen with similar clarity from ground-based telescopes. The Faint Object Spectrometer can see images that are six times fainter than it could capture before the repair. Before repair, only 12% of the light from a star could be concentrated in the central disk that is the main image of a star—some light inevitably misses the disk and appears in rings around that disk. After repair, more than 70% of the light could be concentrated in the central disk. As a result, *HST* has the resolving power needed to separate the lights of two fireflies 10 ft (3 m) apart at a distance as great as that from Washington, D.C., to Tokyo, Japan.

The annual budget for the *HST* currently runs $250 million, making it "big science" (*see* "The End of Big Science in the U.S.? " p 13). *HST* is a noteworthy example of big science producing results that would be difficult, if not completely impossible, to obtain any other way.

Checklist of Findings

Of course, the *Hubble Space Telescope* does not actually discover anything, any more than a balance scale or galvanometer makes discoveries. *HST* is just a tool that scientists use. It is common, however, to personify expensive tools, and to attribute the discoveries of scientists using those tools to the machinery that made the discovery possible. With that convention in mind, here is a brief checklist of discoveries that have been attributed to *HST*. The more important of these are discussed elsewhere in more detail, as indicated by cross references.

• April 20, 1992: Announcement of the discovery of the hottest known star and of photographs of the auroras of Jupiter and intergalactic clouds of hydrogen.
• June 3, 1992: Photographs confirming that galaxies collide with each other, changing galactic shapes and inducing star formation.
• June 29, 1992: Measurements based on type Ia supernovas in far distant galaxies show that the age of the universe is at least 15 billion years, a finding contradicted two years later by another set of measurements (*see* below and also "Cosmology UPDATE," p 236).

- May 1993: A unique view of the expansion of the gas shell around Nova Cygni, which would have passed from view before ground-based telescopes would have been able to resolve the faint image.
- July 17–18, 1993: Dramatic auroras on Jupiter.
- December 1993: Photographs of a region in the Orion Nebula reveal many disks of gas that appear to be stars and planets forming.
- January 15, 1994: First announced discovery based on the newly corrected optics: a Faint Object Camera photograph of white dwarf stars in the globular cluster 47 Tucanae, 16,000 light-years away.
- January 1994: Striking photograph of a violent explosion on the star Eta Carinae revealing hitherto hidden details and completely revising astronomers' views of the event.
- June 1994. A thin atmosphere of molecular oxygen found on icy Europa, the smallest of the four large satellites of Jupiter.
- July 7, 1994. Location of helium in the very early universe of about 11 billion years ago is announced in *Nature*.
- July 1994: Photographs of Neptune revealing that its Dark Spot has vanished and been replaced by wispy white clouds (*see* "Gas Giants and Fellow Travelers," p 168).
- August 14, 1994: First good near-Earth-based photographs of Uranus, its rings, and its moons. At least 11 concentric rings are seen. Images of two new Uranian clouds are also captured.
- September 1994: The inner structure of the shells of gas formed as a star dies appear in photographs of NGC 6543, known as the Cat's Eye planetary nebula.
- October 27, 1994: Measurements of distance to the M100 galaxy using Cepheid variable stars imply that the universe may be only 8 billion years old (*see* "Cosmology UPDATE," p 236).
- October 31, 1994: New images of Neptune are collected by Heidi B. Hammel of the University of Massachusetts and Wes Lockwood of Lowell Observatory, taken between October 17 and October 31. These reveal that Neptune had mysteriously brightened to its highest level in 22 years. The brightness is caused by clouds clearly visible in the Hubble images, but no one knows what causes the clouds.
- November 1, 1994: A sprinkling of very young stars (4 million years old) are found scattered among the 50-million-year-old stars that make up most of the open star cluster NGC 1850 in the Large Magellanic Cloud.
- November 1, 1994: No sign of the famous missing mass (*see* "Dark Matters," p 226).
- November 1994: First map of landforms and surface features of Saturn's largest moon, Titan (*see* "Gas Giants and Fellow Travelers," p 168).
- November 22, 1994: A nearby quasar is discovered hidden in the dust of the Cygnus A, the second strongest radio source observable from Earth (*see* "Quasars," p 233). Astronomers did not photograph the actual quasar, only its ultraviolet spectrum.
- December 6, 1994: Photographs of the "long-sought population of primeval galaxies," those formed when the universe was a tenth of its present age (*see* "Shape and Evolution of Galaxies," p 216).
- March 1995: Preliminary results from the first map of surface features on Pluto, being made by Alan Stern of the Southwest Research Institute in Boulder, Colorado,

were announced. Similar planetary studies announced at the March 21 press conference included reports on the weather on Mars, the atmosphere of Jupiter's satellite Europa, and atmospheric results of volcanic action on Venus.

- March 1995: Allan Sandage announced that his group's new measurements with *HST* showed that the universe is about 20 billion years old.
- April 17, 1995: Anita L. Cochran of the University of Texas at Austin announced the sighting about 30 comets in the Kuiper belt beyond Pluto and Neptune.
- July 1995: Astronomers announced the discovery of two new moons of Saturn observed by *HST* on May 22, 1995.
- July–August 1995: Various astronomers reported that the *HST* had found a new class of irregular galaxies from early in the history of the universe.
- September 7, 1995: A team from the University of Cambridge in the U.K. led by Nial R. Tanvir announced that their *HST* observations, reported in *Nature*, showed that the universe is 9.5 billion years old.
- October 1995: Astronomers showed *HST* photographers of volcanic activity on Jupiter's moon Io.
- October 1995: Dramatic *Hubble* views of pillars of gas in the Eagle Nebula are shown.

While many more discoveries are expected from *HST* in its present configuration, at some point in the future *HST* will be serviced again and become the "new, improved *Hubble*" all over again. Among the upgrades planned is a new infrared camera that

Background

The *Hubble Space Telescope*, the first space-based optical telescope, was designed to carry out 467 projects over 15 years, providing data for use in determining the correct age of the universe, observing the absorption spectra of quasars, surveying the universe for faint objects, finding accurate distances to stars, making observations in the ultraviolet range, providing good views of the planets, studying black holes, and investigating the origins and composition of comets. It was launched from the space shuttle *Discovery* on April 25, 1990, after seven years of delays and a cost of $2.1 billion.

After a successful launch, however, images taken with the spacecraft's main telescope were found to be blurred, no matter how carefully the telescope was focused. The blurring was of a type called spherical aberration, which gives each image a fuzzy halo. Spherical aberration results when rays from a telescope's central part and edges do not come to the same focus point. "The outside part of the mirror is lower than the rest," James Westphal of Caltech explained. The aberration resulted from an incorrect shaping of the 94.5-in (2.4-m) primary parabolic mirror. The problem was suspected as early as May 20, 1990, but announced to the public on June 27 of that year.

A crucial testing device used in building the main mirror had been constructed with a built-in error of 1/20 in (1 mm). This basic problem limited the telescope's usefulness, until December 1993, when NASA fitted *HST* with a corrective lens. A 1993 space shuttle mission had already been scheduled to provide maintenance and installation of improved equipment.

will, according to theory, permit *HST* to see some of the objects formed just after the big bang.

(Periodical References and Additional Reading: *New York Times*, 4-21-92, p C5; *New York Times*, 6-30-92, p C2; *New York Times*, 12-5-93, p A1; *New York Times*, 12-6-93, p A1; *New York Times*, 12-8-93, p A1; *New York Times*, 12-9-93, p A1; *Astronomy* 1-94, p 18; *New York Times*, 1-13-94, p A15; *New York Times*, 1-14-94, p A1; *New York Times*, 1-15-94, p 10; *Science* 1-21-94, p 323; *Astronomy* 4-94, p 44; *Science* 6-24-94, p 1850; *New York Times*, 10-27-94, p A1; *Science News* 10-29-94, p 278; *Astronomy* 11-94, pp 24 & 41; *Science News* 11-12-94, p 309; *New York Times*, 12-7-94, p A20; *Science* 12-9-94, p 1675; *Science* 12-16-94, p 1806; *New York Times*, 12-25-94, p IV-5; *Astronomy* 1-95, p 20; *Astronomy* 2-95, pp 22, 24, 44; *Astronomy* 3-95, pp 22 & 24; *New York Times*, 2-23-95, p B7; *Astronomy* 4-95, pp 22; *Science News* 4-1-95, p 204; *Astronomy* 5-95, pp 24)

GIANT TELESCOPES, OPTICAL AND OTHERWISE

In the history of telescopes, size has always impressed. Herschel's "giant telescope" with a mirror 48 in (122 cm) in diameter mounted in a cannon-like tube was considered a wonder in 1789. Some two hundred years later, the Keck's "giant mirror" was more than 8 times the diameter of Herschel's telescope and its light-gathering power was almost 70 times as great, with even larger optical telescopes just around the corner.

The larger a telescope mirror is, the more light it can collect, which is the bottom line for any telescope, although a good location, free from light pollution, clouds, and other distractions in the atmosphere, is almost as important. Although the *Hubble Space Telescope* has the best viewing location of any telescope, its main mirror is only 94.5 in (2.4 m) in diameter, one-fourth the size of the first Keck Telescope on Mauna Kea in Hawaii, the largest ground-based optical telescope operating in the 1990s.

Meanwhile, radio telescopes have begun to use interferometry to simulate telescopes far greater than any using a mirror. The largest in the 1990s was strung over 5,000 mi (8,000 km). There has been talk of building an Earth-Moon telescope that would effectively be 240,000 mi (380,000 km) wide.

Gravity Waves

One of the most expensive, if not the largest, telescopes ever built is designed to detect waves that everyone feels, but that no one has previously captured: gravity waves. Probably no unobserved entity has the full faith and credit of the astrophysics community as much as do gravity waves. Gravity waves are caused when anything with mass moves or interacts gravitationally with any other mass. The catch is that both masses must be huge before the waves are detectable. A good gravity wave detector, or telescope if you will, would open a new window on such events as black holes, supernovas, and binary neutron stars. There is even a chance that there exists a gravity wave background left over from the big bang. Like the now familiar cosmic microwave background of electromagnetism, a cosmic gravity-wave background could be used to determine very early events in the life of the universe.

Four gravity wave detectors are planned for the late 1990s: one in Italy near Pisa, one in Australia or Japan, and two in the United States. The U.S. effort, the Laser Interferometer Gravitational Wave Observatory (LIGO), will use both American detectors (one in Hanford, Washington, and one in Livingston Parish, Louisiana), so that each can verify the other.

The way to observe a gravity wave is to notice the ripple in space-time as the wave passes. This can be observed in several ways, but LIGO plans to measure the change in two free-hanging perpendicular masses at each end of a pair of perpendicular 2.5 mi (4 km) evacuated tubes. In theory, one tube will lengthen and the other will contract when a gravity wave passes. A laser beam will be used to measure the distances between the masses at the ends of the tubes, and thus the lengths of the vacuum tubes, with interferometry. If the weights are shifted by as little as 3.9–6.3 in (10–16 cm), or about a hundred-millionth of the diameter of a hydrogen atom, the interference patterns should show the effect, thereby indicating the passing of an energetic gravity wave. Such waves might be caused by supernova explosions or by collisions between neutron stars anywhere in the universe.

A prototype of LIGO has been operating with 130 ft (40 m) tubes, but it has not definitely detected gravity waves. The full-scale version will be a hundred times more sensitive, but it will cost $297 million to build and $68 million to operate for four years, an amount the U.S. Congress has been reluctant to fund, particularly since this 1994 estimate of costs is almost half again the size of the estimates made in 1990 when the project was originally proposed. In 1996, LIGO was projected to require seven years to build, three years longer than the original plan. Slippage continued. The original completion date had been 1994, but now the facilities are expected to be complete in 1998 with full operation starting in 2000.

Found in the Fly's Eye

In its own way, the Fly's Eye telescope is as unusual as LIGO, although the phenomena it collects—cosmic rays—have long been observed. Cosmic rays are particles, essentially nuclei of elements (including the protons that are hydrogen nuclei), that come from space. Their existence has been known since 1912. Some cosmic rays are caused by events on the Sun, while others are associated with supernovas. However, many cosmic rays of great energy could not be caused by any known mechanism. These are the objects that the Fly's Eye studies.

The Fly's Eye, like many modern telescopes, does not look at all like an optical telescope, although its basic operation uses light to detect cosmic rays. Picture a pair of hillsides dotted with oil-storage tanks. Each "tank" is actually a separate detector containing photomultiplier tubes. These detectors react to fluorescent bluish light from ionized nitrogen in the night sky as it arrives from the direction in which the detector is oriented. Nitrogen is, of course, the most abundant element in Earth's atmosphere, so entering cosmic rays tend to strike and ionize nitrogen molecules more often than any others. The Fly's Eye is run by physicists at the University of Utah who correlate information supplied by the various detectors in an effort to determine the height of the fluorescence and its intensity. The assumption is that heavier nuclei encounter nitrogen molecules at higher altitudes than do less massive nuclei, due to the fact that a more massive particle has a greater cross section. Although there are fewer nitrogen molecules at higher altitudes, this is a know factor. The unknown factor, the size of the incoming particle, can be approximated by how far it travels in the atmosphere before hitting a nitrogen molecule. Furthermore, the intensity of the light indicates the amount of energy released by the impact, which implies the amount in the original cosmic ray. This method is doubly indirect, in that most of the fluorescence is not caused by the original cosmic ray, but by the shower of subatomic particles the original ray tears out of the nucleus of the first particle it runs into.

The Fly's Eye does not so much collect particles or waves into one location, which is a common function of telescopes, but instead studies individual particles that arrive at wide intervals. Thus it collects particles over time, rather than over space. Because the cosmic ray events it observes are so rare, the telescope had to run for ten years before it collected any useful information. The main result obtained thus far is that medium-energy cosmic rays seem to be heavier nuclei such as the nuclei of iron atoms, and very high energy cosmic rays are all lighter hydrogen nuclei (protons). One interpretation of this is that the heavy nuclei are from within the Milky Way, since they reach Earth at moderate energies, while the very high energy cosmic rays come from outside the Milky Way. The explanation for this paradoxical conclusion is that cosmic rays, as charged particles, interact with the magnetic field of the galaxy. The moderately energetic cosmic rays are kept in the Milky Way by these fields, but high-energy cosmic rays are less affected by the fields. Thus, a high-energy cosmic ray from within the galaxy would quickly depart and not be detected on Earth. However, the much larger number of cosmic rays from outside the galaxy (since most of the universe is outside the galaxy) would ultimately be detected on Earth.

Yet very energetic cosmic rays could not travel far, as astronomers measure things, at high energies without interacting with the cosmic microwave background, causing the cosmic ray to lose energy. Thus, a very high-energy cosmic ray detected by the Fly's Eye on October 15, 1991, probably originated outside our galaxy but within the local supercluster of galaxies (see "Clusters and Walls of Galaxies," p 222).

Another approach to detecting high-energy information is Milagro, scheduled to be installed on a hill in the Jemez Mountains of New Mexico. It will consist of a 5 million gal (19 million L) reservoir of water containing 700 photomultipliers that can detect Cerenkov radiation (see "Neutrino Astronomy Comes of Age," p 136) caused by high-energy photons, or gamma rays. The photons are not detected directly; instead, when very high-energy gamma rays collide with atoms in the air, particles are produced which are hurled forward by the impact. Those particles entering Milagro in excess of the speed of light in water produce the detectable Cerenkov radiation. Unlike the Fly's Eye, which seeks cosmic rays, such rays and their associated particles have to be screened out from Milagro, either by the covering over the reservoir or by computer programs designed to pick out characteristic signals from the gamma rays.

The One-Week Telescope

For six days, the second-largest space telescope ever flown, the *Orbital Retrievable Far and Extreme Ultraviolet Spectrometer* (*ORFEUS*), floated through space as part of a special German scientific platform designed for launch and retrieval by the U.S. space shuttle. The excursion took place in September, 1994.

ORFEUS is designed to provide more detail than previous orbiting ultraviolet telescopes, which need to operate above Earth's atmosphere; that famous ozone layer which protects the surface from damage by ultraviolet radiation also serves to keep out information that such radiation might contain. *ORFEUS* was trained on more than thirty targets, most of them stars. Previous ultraviolet telescopes, such as the Extreme Ultraviolet Explorer launched in June 1992 and still in operation in 1995, have found that active stars and hot white dwarf stars produce strong signals in the ultraviolet. This information has helped provide new knowledge about the production of energy by stars.

Because *ORFEUS* is recoverable, it can be flown repeatedly. *ORFEUS* is already scheduled for future missions, extending its observing time well beyond the first week.

Keck Telescope UPDATE

For more than 40 years beginning in 1948, the 200 in (5 m) Hale Telescope on Mt. Palomar in California reigned as the leading optical astronomical instrument available. In the early 1990s its preeminence was challenged by a host of new devices, ranging from the *Hubble Space Telescope* to the New Technology Telescope (NTT). But the most clear-cut challenge in light-gathering comes from the new record-holder in size, the W. M. Keck Telescope on Mauna Kea on the "big island" of Hawaii. The $94-million Keck is twice the size of the Hale in diameter, measuring 394 in (10 m). Therefore the Keck has four times the Hale's light-gathering power, although it does not have the resolving power of some smaller telescopes (*see* "Optical Telescope UPDATE," p 140). Light-gathering power is important for a number of different reasons, one of which is because such power can speed up operations by factors of six or more, meaning more data for more astronomers.

The Keck can achieve such a large size without sagging under gravity because it is built in 36 segments, each measuring 6 ft (1.8 m) across and weighing 880 lb (400 kg). Each of the segments is separately supported by three movable pistons that continually adjust the segment to align it properly with the others. Furthermore, the Keck's location on Mauna Kea ensures much better viewing, both because Mauna Kea is higher than Mount Palomar (Keck is at 13,600 ft, or 4,145 m) and because Keck does not have the distraction of lights from a nearby major city.

Keck was supported by a grant of $70 million to Caltech and grew out of an ongoing large telescope project of the University of California, which funds Keck's operating costs. The University of Hawaii is also a partner in the enterprise. Keck's main instruments include a near-infrared camera that will use wavelengths ranging from 1 to 5 microns, a long-wavelength infrared camera for wavelengths between 8 and 20 microns, and both low- and high-resolution spectrographs. Its specialties are the near infrared wavelengths, where the youngest and most distant quasars and galaxies lurk, and the spectroscopic analysis of very faint objects.

In November 1990, nine of its mirror segments were in place, which created a light-gathering power equivalent to that of the Hale Telescope. Astronomers were able to obtain the first images (known as first light) from the Keck. The actual object imaged was galaxy NGC 1232, thought to be about 65 million light-years away.

Encouraged by the promising results of the first tests, a group consisting of the Keck Foundation, Caltech, the University of California, and NASA started planning in March 1991 for Keck II, which will be an identical twin 275 ft (85 m) away from what has already been renamed Keck I. Its object: optical interferometry. When used as an interferometer, the Kecks will be able to resolve the headlights of an oncoming car 16,000 mi (25,750 m) away. Both Kecks can also be used independently for twice the observing time. With a grant of $74.6 million from the Keck Foundation covering 80% of the cost, Keck II is scheduled for completion in 1996.

Keck II mirror segments were actually already in use two years earlier. After the mirrors for Keck I were complete, Kodak developed a method called ion-figuring which gives the mirrors a more precise polish. Instead of hand-polishing the mirror, a beam of ions is used to bump atoms off the mirror's surface. This technique produced superior mirror segments, which were originally intended for use in Keck II. However, astronomers instead transferred the Keck II segments into the Keck I reflector as they arrived. Then they shipped the Keck I segments to Kodak for ion-figuring. After that, the improved former Keck I mirror segments will be installed in Keck II.

The Very Large Telescope (VLT)

The Keck era may not last as long as the Hale era did, however. If various disputes and problems can be solved, the world's largest telescope in the first part of the twenty-first century will be called simply the Very Large Telescope, or VLT. Planned by the European Southern Observatory or ESO (a consortium of eight nations), the VLT is designed to be a set of four 27 ft (8.2 m) mirrors based on the NTT design, to be used either separately or in unison as an interferometer. The new telescope is to be located on the summit of Cerro Paranal in the Atacama Desert of northern Chile. It is expected to begin operation in 2000, making the Keck era only nine years long.

The four telescopes of the VLT will each have a mirror slightly smaller than that of a Keck, just 27 ft (8.2 m) in diameter. However, the mirrors are designed to be used together. In combination, they will produce light-gathering ability equivalent to a single telescope 54 ft (16.4 m) in diameter, with more than two-and-a-half times the light-gathering power of a single Keck, or about 25% more than the Kecks combined. Used as an interferometer, the VLT will have the resolution of a 394 ft (120 m) mirror. It is not clear, however, how soon the interferometer capability will be in place. Although the tunnels and other physical facilities will be built for the interferometer, money has been lacking for completion of the optics and electronics needed. The interferometer idea was an add-on to the original 1987 design of the VLT and was not funded along with the rest of the project.

Spin-Casting and Fusing Large Telescopes

A number of the new large telescopes rely on advanced technology developed by Roger Angel and his staff at the Steward Observatory Mirror Laboratory in Tucson,

Arizona. Angel's telescopes use a honeycomb backing to reduce the weight of the mirror, and a novel method of spin casting to create the blank for the mirror. Every reflecting telescope begins with a thick segment of glass called the blank. In traditional mirror-making, the blank is a disc, flat on both sides, rather like a thin, glass hockey puck. Angel's group, however, produces a blank that is concave on one side. They accomplish this by putting the molten glass for the blank on a turntable. As the turntable spins, the outer edges of the blank rise higher that the middle. As predicted by the laws of motion, the surface of the spinning glass forms a section of a parabola while it is being cast. The spin casting technique for mirror blanks greatly improves the whole manufacturing process.

So far the laboratory has built three 138 in (3.5 m) mirrors that were installed in the early 1990s. The first of these mirrors was placed at the Astrophysical Research Consortium Telescope at Apache Point Observatory in Sunspot, New Mexico, and received first light in 1991. In April 1992, the laboratory spun-cast a 256 in (6.5 m) mirror to replace the six small ones at the Multiple Mirror Telescope (MMT) on Mount Hopkins in Arizona. Polishing and mounting take years, however, so first light is not expected until early 1997. The name MMT will be retained, although it will no longer stand for multiple mirrors. The existing smaller mirrors will be used to make an interferometer.

Angel's laboratory then moved on to spin-casting another mirror the same size as that in the new MMT. This mirror is for the Magellan telescope, which will be part of the Carnegie Institution of Washington's observatory at Las Campana in Chile.

Angel's laboratory is also planning a series of five new 27.5 ft (8.4 m) mirrors for

Background

One technique for obtaining more precise information from anything that can be treated as a wave is called interferometry. Interferometry can be used in either a traditional form or in a modified version called aperture synthesis.

The basic idea of interferometry is that the same wave from two slightly displaced sources (in time or in position) will either interfere with or reinforce itself, making a pattern that is much larger and more observable than the size of the wave. Interferometry with light waves became famous when Albert A. Michelson (German/American: 1852–1931) and Edward W. Morley (American: 1838–1923) in 1887 used the technique to measure the speed of light in 1887. Recently, the main use of the technique has been in radio astronomy, where it is utilized to locate the precise origin of radio waves. This precision grants combinations of widely spaced radio telescopes a resolving power greater than that of any single optical telescope.

In the meantime, optical images of stars have already been produced by a variation of interferometry called aperture synthesis. In aperture synthesis, two different sources of light emanating from two sides of the same star are collected at two different locations. The two different locations for the collectors make the distance the light travels from each source different, thereby creating an interference pattern. A computer is used to convert the interference pattern back into an image. When this technique is used in radio telescopes, two or more telescopes are placed; the farther apart the placement, the sharper the final image.

projects that will probably achieve first light during the latter half of the 1990s. The first will be used as one of two eyes in the Large Binocular Telescope planned for Mount Graham in Arizona. The other eye will be used later in the series. While none of these will be as large as the Keck twins (*see* above), they individually and collectively will provide major new capabilities for the world's astronomers. Roger Angel and coworkers estimate that the 1990s will see four times the total worldwide light-collecting ability from the world's optical telescopes than in previous years.

But Angel is not alone in doing spin-casting. Schott Glaswerke also spin-casts large mirrors, although its technique is slightly different from Angel's. The VLT is being cast by Schott Glaswerke using a method which produces mirrors measuring only 12 in (30 cm) in thickness, although they are 27 ft (8.2 m) across.

Another new technique is used by Corning, which first casts the mirror in segments, then partly melts the edges to fuse them into a giant mirror. Among the telescopes that will use this type of mirror are the Japanese National Telescope, known as Subaru (first light expected around 1999), which will be 27.5 ft (8.3 m) in diameter and only 8 in (20 cm) thick, as well as both of the 26 ft (8 m) Gemini project mirrors (that on Mauna Kea in Hawaii is scheduled for 1998, and its twin at Cerro Pachón in Chile for 2000). Corning used a furnace 31 ft (9.4 m) in diameter for the final fusing and shaping prior to grinding and polishing. This takes about three years per mirror.

(Periodical References and Additional Reading: *New York Times* 2-4-92, p C1; *Astronomy* 6-93, p 20; *Astronomy* 8-93, p 48; *Science* 12-10-93, p 1649; *Science* 1-14-94, p 176; *Science* 3-11-94, p 1370; *Nature* 11-24-94, p 311; *Science* 11-25-94, p 1314)

NEUTRINO ASTRONOMY COMES OF AGE

Humans have organs that can detect certain subatomic particles, collecting vast numbers of them and extracting from them a great deal of information. The particles are photons, and we call the process seeing. Some photons are either too energetic or not energetic enough for human detection; these particles are usually called x-rays or radio waves. Of course, we are used to thinking of light, x rays, and radio in terms of waves, not photons. Even when we explain how an electron microscope works, mentioning the matter wave associated with an electron can clarify the concept better than an explanation involving the more common particle interpretation of an electron.

Collecting other kinds of subatomic particles and interpreting their information, then, is a process analogous to that which we humans use in order to see. Thus, a device for collecting particles that have traveled great distances is called a telescope, whether the particles are photons or something else. Yet a telescope designed to detect elusive particles differs from the process used in human vision because so few of these elusive particles are captured that the common wave interpretation is not especially helpful. To understand this different situation, it is important to remember that even the unaided dark-adapted human eye can detect a single photon at a time and derive information from it. The collection of individual particles is one form of seeing. A neutrino telescope is one of the best examples of a telescope for counting the number of individual particles captured, noting primarily their direction and energy, to obtain information from distant events.

Solar and Supernova Neutrinos

Neutrino astronomers have been kept busy attempting to determine why neutrino telescopes detect fewer-than-anticipated neutrinos caused by nuclear fusion in the Sun. As of the end of 1995, the Homestake neutrino telescope had observed only a third of the expected number of neutrinos. Newer telescopes, designed in part to detect missed by the Homestake device, have found, at best, only two-thirds of what theorists would predict, although margins of error are sufficiently great to encompass the low end of theoretical predictions. This problem remains unresolved, although a number of different ideas have been proposed by theorists and experiments have tested various possible solutions. Theory alone cannot accomplish much, and astronomers await new telescopes or adjuncts to present ones, known by such odd names as BOREXINO, HERON, HELLAX, and the LiF Cryogenic Detector. These will introduce new methods of neutrino detection which are hoped to provide sufficient information to support a theory of solar neutrinos.

The most dramatic success of neutrino astronomy, however, had nothing to do with the problematic solar neutrinos. This success was the observation of neutrinos from supernova 1987A, which confirmed the major theory about how supernovas explode. Many theorists believe that the missing mass in the universe (see "Dark Matters," p 226) is the cumulative effect of a very small mass possessed by neutrinos. If so, detection of the distribution of neutrinos will be recognized as important in understanding the past and future of the universe. New and unconventional neutrino telescopes are being designed for capture of high-energy galactic, black-hole, and supernova neutrinos, rather than simply for counting the low-energy neutrino flux from the Sun.

Natural Neutrino Telescopes

"Conventional" neutrino telescopes are based on putting pure or nearly pure elements or compounds far below Earth's surface, where they can be monitored for changes caused by neutrinos. A new generation of neutrino telescopes utilizes existing features of the Earth as the collector. Thus far, the main requirement for this approach has been that the feature be both transparent and available in great quantity.

One place that meets these requirements is the continuing ice cap at the South Pole, which supplies transparent ice as its medium. The AMANDA (Antarctic Muon and Neutrino Detector Array) is designed to have ten strings of sensors, each string 656 ft (200 m) long and containing 20 evenly spaced sensors enclosed in protective glass spheres. Half-mile (0.8 km) deep holes in the ice are melted out with boiling water, allowing placement of the strings at the bottom of the holes. High-energy neutrinos that have passed through Earth from pole to pole are the main object of AMANDA. The use of 200 sensors enables astronomers to pinpoint the direction from which the neutrino arrives, which may indicate the object or event that produced the neutrinos.

A similar project in the Pacific, the Deep Underwater Muon and Neutrino Detector (DUMAND), lowered its first string of detectors into the ocean off Hawaii on December 14, 1993. The ocean floor where the 1,431 ft (436 m) long string rests is 15,750 ft (4,800 m) below the surface. Three more strings were placed in 1994. Eventually, DUMAND will consist of an array of nine such strings. Because each string has 24 widely spaced photon detectors, DUMAND will also be able to determine the direction of neutrinos reaching it more precisely than can smaller telescopes.

Both AMANDA and DUMAND have been in development for many years because of the tremendous technical problems presented by working at the South Pole and at the bottom of the ocean. AMANDA was just 11 days behind DUMAND in lowering its first string into place, although DUMAND had been in the planning stage for far longer than AMANDA. Furthermore, DUMAND's first string only worked for 10 hours before its communication system went bad and had to be hauled out for repair.

DUMAND was first conceived in the early 1980s by John Learned of the University of Hawaii and coworkers, and its first grant proposal was submitted in 1988 and approved in 1989. Learned was also among the scientists who conceived of AMANDA in 1988. AMANDA's grant proposal was approved in 1992, but there are somewhat fewer technical difficulties in working with ice than occur when operating at the

Background

There are three families of the light particles known as leptons (see also "Subatomic Particles," p 666): the electron, muon, and tauon families. Ordinary matter is represented by the light electron and its associated neutrino. The heavier muon and much heavier tauon are not constituents of atoms. They result from various subatomic reactions and subsequently decay into lighter particles. The muon and tauon each have their own associated neutrinos. Thus, neutrinos come in three flavors: electron, muon, and tauon. Like all particles, they also have antiparticles, although it is possible that neutrinos, like the photon, are their own antiparticles.

Since neutrinos have little or no mass and interact very little with matter, they are extremely difficult to detect. All known ways of detecting a neutrino depend on the weak interaction, a force that acts only on particles with a left-handed spin. If neutrinos with a right-handed spin exist, instead of interacting with matter slightly, by the weak interaction, they fail to interact at all. Thus, right-handed neutrinos, if they exist, are undetectable. But the known left-handed neutrinos, of which 10,000,000,000 pass through each 0.155 sq in (1 sq cm) of Earth each second, can occasionally be detected. Indeed, one of them is observed every few days in one or another of the existing neutrino telescopes.

The reason behind the small number of observed neutrinos is not at all well understood. All the neutrino telescopes detect far too few neutrinos in comparison with the quantity postulated by theorists. One possible explanation for this discrepancy involves a possible tendency for neutrinos associated with electrons to sometimes change into muon or tauon neutrinos. Such transformed neutrinos would then escape detection. This popular idea has not been experimentally verified. One of the main reasons for the interest in neutrino astronomy is the desire to bring theory and observation into better agreement.

The original "neutrino telescope" was a 100,000 gal (375,000 L) tank of cleaning fluid (perchloroethylene) at the bottom of the Homestake Gold Mine in Lead, South Dakota. The chlorine in the cleaning fluid at the Homestake Mine is the sensor for the neutrino telescope. A neutrino is sometimes captured by one of the chlorine atoms, which causes it to decay into an electron and an atom of radioactive argon-37. The argon-37 has a characteristic energy profile which can be detected by radiation counters.

Two other detectors, developed by the United States with Russia and by Italy, use the element gallium instead of chlorine. Even a low-energy neutrino can convert gallium-71 to radioactive germanium-71. Measurement of radioactivity can be used to determine the number of neutrinos that have caused such conversions.

An entirely different method for neutrino detection is used by the Japanese Kamiokande II experiment. This technique looks for flashes of light in pure water, created when a neutrino slams into an atomic nucleus, jolting the particle into briefly moving faster than the speed of light in water. The resulting flash, called Cerenkov radiation, cannot be mistaken for anything else. This radiation is produced only by high-energy neutrinos.

Cerenkov radiation results from the shock wave produced by a particle exceeding the speed of light in any material (although, of course, nothing can go faster than the speed of light in a vacuum). This process for producing photons was discovered in 1934 by the Russian physicist Pavel Alekseyevich Cerenkov. Because the electromagnetic radiation produced varies in wavelength, brightness, and the angle between the particle path and the radiation, with the mass, charge, and velocity of the particle producing the effect, a Cerenkov counter (detector) can be used to analyze fast moving particles of all types.

All of the detectors described so far can only observe electron neutrinos. The new Sudbury Neutrino Observatory heavy water in an experiment that should be able to observe muon and tau neutrinos as well as electron neutrinos. DUMAND uses even more water, although it is salty, not heavy. DUMAND is suspended 3 mi (5 km) deep in the Pacific Ocean and also looks for flashes of Cerenkov radiation. AMANDA at the South Pole looks for flashes in water, but these are in water frozen into ice.

bottom of the ocean. On the other hand, DUMAND is better situated for the elimination of background radiation and for repair and upkeep.

Similar detectors are planned for the Aegean Sea and Lake Baikal in Russia. The Lake Baikal project combines ideas from DUMAND and AMANDA. Lake Baikal contains fresh water (like Antarctic ice) and, to make the resemblance to AMANDA even greater, the strings are actually laid while the lake is frozen. The Aegean project, named NESTOR, is running about a year later than the Lake Baikal telescope. NESTOR is technologically more sophisticated, which probably means more problems in getting it to work in the first place.

According to theory, however, all four of the new detectors are too small to accomplish much. A really effective neutrino telescope might have to be 10-100 times the size of DUMAND, which itself when finished will cover about 5 acres (20,000 sq m).

(Periodical References and Additional Reading: *New York Times* 2-18-92, p C6 *Science* 6-12-92, p 1512; *New York Times* 6-13-92, p 1; *New York Times* 12-21-93, p C1; *Science* 1-7-94, p 28; *Astronomy* 4-94, p 22; *Science* 11-18-94, p 1157; *Science* 1-6-95, p 45)

Optical Telescope UPDATE

Galileo built the first optical astronomical telescope in 1609. Since Galileo's day there have been very many improvements and changes, although much public recognition has simply gone to the largest telescopes, or more recently to space-based instruments. Optical astronomers today still cope with numerous problems, including distorted collections of photons, not enough money available for chasing photons, and the collection of too many photons from all the wrong sources. One ongoing problem is a fight between forces trying to build new telescopes atop Mount Graham in Arizona and forces trying to protect a subspecies of red squirrel already occupying the mountain. Recently, however, astronomers have come up with clever new solutions for some of these problems, although others continue to gnaw away at optical astronomy.

New Techniques Yield Better Images

Many of the improvements in optical telescopes have not concerned lenses or mirrors at all. Instead, better observing power has come from new telescope mountings; clocks, computers, and other devices to keep the telescope pointed at the right part of the sky; light recording devices; methods of observing changes in images; and climate control and other methods of reducing distortion. The New Technology Telescope (NTT) run by the European Southern Observatory at Cerro Tololo, Chile, for example, automatically corrects distortions caused by shifting mass as the telescope moves, as well as distortions caused by the wind at its location on a high peak 7,000 ft (2,130 m) above the dry Chilean desert. Thanks to the low humidity of its desert mountaintop and to the new technology, astronomers using NTT brag that it produces images six times as sharp as other telescopes of the same size.

Adaptive optics has been called by *Science* magazine "untwinkling the stars," since its main purpose is to remove the twinkling caused by distortion of light as it passes through the atmosphere. If Earth's atmosphere were a perfectly homogeneous gas, untwinkling would not be necessary. However, because air is filled with dust, and because different parts of the atmosphere have different densities due to variations in heat or humidity, starlight is bounced about in an essentially unpredictable way during its passage from space to Earth. At optical wavelengths, this turbulence means that the 33 ft (10 m) Keck telescopes have no better resolution than a 9.85 in (25 cm) reflector, although they do have other advantages. One extremely expensive solution to this problem is to put a telescope in space, while somewhat less expensive efforts to minimize turbulence have involved putting telescopes on high mountain peaks.

If one knew how the light was changed by turbulence at a given instant, however, it would be possible to compensate for the twinkling and produce a steady image. This is a difficult feat, made more difficult because the change itself varies from instant to instant. Since the 1950s, the astronomy community has understood that these difficult problems might be resolved by adaptive optics, but the earliest known attempts to apply these techniques were tentative steps taken in the 1980s.

On May 27, 1991, the astronomy community learned that the U.S. military's Strategic Defense Initiative (SDI, or "Star Wars") program had developed a system of adaptive optics better than any previously employed by astronomers. Developed to help correct the paths of laser weapons, the system was declassified after the SDI laser weapon program was terminated. The basic method used—creation of an artificial guide star with a laser, and adjustment of a mirror to eliminate distortion—is no different from methods previously proposed by astronomers, but never implemented due to lack of money. With much greater sums at their disposal, however, the SDI team developed working equipment. The laser is reflected by sodium ions that are abundant in a layer about 60 mi (95 km) high in the atmosphere. The result is a yellow artificial star that can be monitored for changes in its image caused by atmospheric conditions. A computer can make corresponding corrections in the image of the real target. These corrections can be effected as frequently as 150 times per second.

An important feature of adaptive optics is that existing telescopes can be retrofitted with the necessary modifications at a reasonable cost. Another virtue lies in its ability to improve even further the results obtained with optical or infrared interferometry. The only drawback of adaptive optics is that it works only for a small viewing area; it cannot correct for the image of an extended nebula, for instance.

Early in 1992, one of the instruments was loaned by the Defense Department to a team headed by Edward Kibblewhite of the University of Chicago. According to Kibblewhite, "If it lives up to its tremendous progress, [adaptive optics] will revolutionize astronomy." At longer wavelengths, adaptive optics should produce better results than the *Hubble Space Telescope*. The system, installed on the ARC Telescope at Apache Point, New Mexico for a five-year trial, was designed originally to be used in SDI experiments aboard the space shuttle. Some other notable systems using adaptive optics include operations at the La Silla telescope of France and the European Southern Observatory consortium; the Canada-France-Hawaii Telescope on Mauna Kea in Hawaii; and a medium-sized telescope operated by the Mount Wilson Institute. The military telescope at the Philips Laboratory has been made available to a limited number of astronomers. The Lawrence Livermore National Laboratory ran tests for 9 months in 1992-93 that caused considerable local excitement, as the beam producing the artificial star was clearly visible. Furthermore, even amateur astronomers found they could use the Livermore artificial star to correct their own observations. The Livermore experiments were then applied to a telescope on Mount Hamilton in California. A version of the same technology is being developed for the Keck telescopes on Mauna Kea.

A major new adaptive system will be installed on the MMT (formerly the Multiple Mirror Telescope) at Mount Hopkins, Arizona. This system is based on tests with the earlier MMT that took place early in 1994.

Appreciation of improved observation thanks to adaptive optics is so great, in fact, that nearly all the new large telescopes being built (*see also* "Major Telescopes," p 247) are expected to incorporate adaptive systems.

Another new technique involves application of the concept behind the interferometer (*see* "Giant Telescopes, Optical and Otherwise," p 130), so successful with radio telescopes, to optical telescopes. One of the first two-telescope optical interferometers is the Mark III on Mount Wilson in California. Its resolution is about 250 times better than normal ground-based telescopes, allowing it to measure directly the diameters of nearby stars. This success has led astronomers to plan arrays of telescopes that will be capable of revealing actual images of stars (until recently, the most one could see of any star other than the Sun was a point of light). The Mark III is not quite able to do all that astronomers plan for these future telescopic arrays, but it can measure the diameters of stars, observe changes in brightness across the face of a star, and resolve very close binary stars.

So far, astronomers have not been able to obtain the exact distance between two different mirrors with the precision needed for light waves, which are much smaller than radio waves. Instead, they have cleverly masked individual large mirrors so that the light is collected separately on two sides of the same mirror and then combined in the interferometer. This procedure is called aperture synthesis. The first such experiment was accomplished by a group headed by John Baldwin of Cambridge University, who used the 165 in (419 cm) William Herschel Telescope in the Canary Islands to produce detailed images of Betelgeuse. Another team, led by Shrinivas Kulkarni of Caltech, applied aperture synthesis to the Hale Telescope at Mount Palomar, improving its resolution by about 30 times.

The Cambridge group is also proceeding to the next logical step—construction of independent mirrors about 328 ft (100 m) apart to improve the technique. As with radio astronomy interferometry, the further apart the collectors of waves, the more precise the images.

The Very Cheap Telescope and Another Cheap Idea

Most recent professional astronomical telescopes involve "big science" (*see* "The End of Big Science in the U.S.? " p 13), but when the U.S. National Research Council looked at goals for astronomy in the 1990s, the panel recognized that many purposes and many more astronomers could be served by "small science" devices. Accordingly, the U.S. National Science Foundation (NSF) put forth a small amount of money (by big science standards) to fund a special-purpose telescope for Georgia State University. With about $53,000 from the NSF and half as much from other sources, the Multi-Telescope Telescope (MTT) was built at Hard Labor Creek, Georgia. Its main purpose is to collect data on stars forming double systems with other stars. In some cases, one of the pair will be a neutron star.

In studying stars, the main object is to gather photons in order to measure their energy (and variations in that energy) over time. This kind of information can be used to reconstruct the orbit of one star about the mutual center of gravity of a double-star system. From the duration and path of the orbit (and an assumption or two), the masses of the stars in the system can be determined using the laws of gravity. Furthermore, this knowledge can lead to approximations of other physical characteristics of the stars.

A picture of an individual star, or even a close system of two stars, was just a point of light until 1996—only recently has the first disk been seen and the first double system resolved into two separate points. Without advanced techniques, the more photons gathered, the brighter the point, but it still remains only a point. But, using several small mirrors to gather the photons and then combining the photons collected is as good as using one large mirror as long as you don't need to produce a picture.

The MTT at Hard Labor Creek is built from nine off-the-shelf mirrors ordered from an ad in *Sky and Telescope* magazine. The total cost of the mirrors was only $2,160, a drop in the bucket even for "small science." The rest of the $80,000 was spent on mounting the telescopes in a frame (made in the University's machine shop) that can be aimed at the desired point in the sky; a computerized aiming system to direct the frame; fiber-optic cables to carry the photons from the mirrors to a nearby shed; and a spectrograph to analyze the photons. Although the spectrograph is in the shed, the MTT is out in the open, without even a dome. A dome would only add to the expense (although a small commercial dome can be ordered for as little as $1,950). It is interesting to note that despite the 13.1 in (33 cm) diameter of each of its mirrors, the MTT has the light-gathering capacity of an expensive 40 in (100 cm) telescope.

Other astronomers are also looking for solutions to the high cost of seeing stars. The original plan for a successor to the *Hubble Space Telescope* involved a larger orbiting telescope that would cost $4 billion. This does not seem economically feasible, so Holland Ford of Johns Hopkins University and Pierre Belay of the European Space Agency have proposed an effective cheap alternative. Their solution is to mount seven mirrors on a tethered balloon to be flown in a cold climate, such as that of northern Alaska. This cheaper alternative (to be called the Polar Stratospheric Telescope or POST) would be nearly as good as a space-based telescope with a 200 ft (6 m) mirror at infrared wavelengths. Furthermore, according to Ford and Belay, POST would cost "only" about $60 million to build, about 1.5% of the price of the space-based alternative.

Light Pollution

One of the greatest observational astronomers of all time was Tycho Brahe, who died eight years before the first astronomical telescope was constructed by Galileo. Even Brahe needed a dark island to collect much of his data. Today, the astronomer with a telescope still needs such darkness.

When the Hale Telescope was launched at Mount Palomar just after World War II, the sky over the southern fringe of California was quite dark. The pre-war population of San Diego, the nearest large city, had been about 100,000. As San Diego (some 50 mi or 80 km away) and surrounding communities grew, light from the towns began to interfere with the operation of the telescope. Today, San Diego's population is one million greater than it was when the site was chosen for the telescope. Similar problems involving the growth of Los Angeles and light pollution from its suburbs caused the 1985 closing of Mount Wilson observatory, with its 100 in (254 cm) telescope.

In 1984, when new street lights were installed in San Diego, both the city and the telescope got a break. The lights chosen were low-pressure sodium lights, which

IMAGES INSTEAD OF PHOTOGRAPHS

The science of astronomy was given an enormous advance when the first astronomical photographs began to be used in the mid-1850s. For example, in 1856 photography pioneer George Phillips Bond (American: 1825–1865) was the first to observe that photographs of stars revealed their magnitudes. As photography advanced, it was clear that photographs revealed much more than that, since film could capture hours of light from a single object, while the human eye was limited to an instant.

Although photography continues to improve, it has in many ways run into technological walls. Photographs are limited by the chemical nature of the reactions used in making them. A photon of light has to have sufficient energy to change a molecule from one state to another. Furthermore, the resolution of the image is ultimately limited by the size of the molecules that make up the physical image.

Astronomers have therefore turned to new ways of capturing the light gathered by their telescopes. Instead of the coated film of photography, they utilize the silicon wafers of modern electronics. In particular, electronic light detectors called charge-coupled devices can count individual photons of light as they strike the mirror, recording the data in digital form. The resulting images often make all objects appear square to the uninitiated, yet their efficiency in capturing information far surpasses that of photographic film.

A new telescope planned in 1995 is using the largest array ever of such charge-coupled devices to map the entire universe in three dimensions. Such a map will be a major tool in cosmology. The 30 charged-coupled devices in the digital sky survey telescope, supported by the University of Chicago, Princeton, and the Institute for Advanced Study, is able to record as many as 120 million points of light at a time from a unique wide-field 100 in (2.5 m) telescope. It was installed in the telescope complex at Apache Point in the Sacramento Mountains of New Mexico on October 10, 1995. The Apache Point complex is run by a group that also includes scientists from New Mexico State University, Washington State, and the University of Washington.

---

produce definite wavelengths of light that astronomers can exclude or ignore, and which cost the city about $3 million a year less to operate than broad-spectrum white lights. The decision to use the low-sodium lights was widely hailed by astronomers.

But in 1993, the San Diego city council wanted to renege on its commitments to astronomy and saving money. Tourists were unhappy with the way low-sodium light makes skin appear pallid, and some also felt that the lights were not bright enough to deter crime. Plans were instituted to switch high-crime areas and the downtown area patronized by tourists back to full-spectrum white lights. Although only a quarter of the lights would be switched, the change might well mark the beginning of the end for Palomar and the Hale telescope, which was the world's greatest for over 40 years.

(Periodical References and Additional Reading: *New York Times* 2-4-92, p C1; *New York Times* 2-18-92, p C1; *New York Times* 3-17-92, p C8; *Science News* 7-23-94, p 56; *Science* 9-2-94, p 1356; *Astronomy* 10-93, p 22; *New York Times* 10-10-93, p I-26; *Science* 1-14-94, p 167; *Science* 10-21-94, p 356; *Astronomy* 12-93, p 20; *Physics Today* 12-94, p 24; *Astronomy* 1-94, p 20)

WHAT'S NEW IN THE SOLAR SYSTEM

STORIES ON THE SOLAR WIND

Although we on Earth often forget it, the true story of the solar system is the story of the Sun. The Roman Catholic Church on October 31, 1992, admitted that Galileo had been right in 1632 in saying that Earth travels around the Sun, rather than the Sun around Earth. Virtually everyone in the Church had long since admitted that Galileo had been unfairly made to recant, but it took a 13-year investigation and a statement from Pope John Paul II to make acceptance of the heliocentric theory official.

The Sun not only reigns as the leading player in the solar system by virtue of its size and central role in the creation of the whole system, but also it bathes the system in three forms of radiation: electromagnetic waves; the stream of charged particles known as the solar wind; and neutral neutrinos. These forms of energy, along with the gravitational energy that holds the whole system together, are the basis of much that happens throughout the solar system.

Particles from the Sun

The Sun is an oblate spheroid that propels particles in all directions through space, rather as the mirror ball at a disco sends moving spots of light around the room. The outward force behind these particles is largely a result of fusion in the core, although electromagnetic fields are involved to some degree as well.

The Sun and other stars are ultimately powered by gravitational forces that compress hydrogen. The hydrogen molecules come so close together that they begin to fuse first into helium and then more massive atoms. About 93% of the particles released from the Sun result from the fusion of hydrogen into helium, with the rest coming mostly from the production of beryllium and boron. Over billions of years (or less, depending on the size of the star), the history of a star is controlled by the uneven balance between expansive forces, such as fusion and repulsion between charged particles, on the one hand, and the relentless contraction of gravity opposing fusion and electromagnetism. Eventually, slow but steady gravity creates a stable core of some kind, but along the way expansive forces may make sudden flurries that take complete control, notably in supernovas. Boxing fans would recognize this interplay as a rather good fight in which stellar explosions are knockdowns, but gravity usually wins anyway.

Not all punches that connect are knockdowns, of course. Solar flares are brief excursions in which outer-directed forces, spurred primarily by magnetism in these cases, overcome inner-directed gravity. The result is that during a flare the number of particles of nearly all types stream away in much greater numbers than usual from a small region of the Sun's surface. Such solar flares often have dramatic effects on the flux of particles reaching Earth as well as on other regions in space. The cause of solar flares is thought to be streams of particles that have been first accelerated by the Sun's magnetic field and then directed by magnetic lines of force back to the surface of the Sun, where they strike like a hail of bullets, spewing more particles from the surface into space.

Many different types of particles from the Sun form what is called the solar wind. However, astronomers usually separate the solar outflow into two components, which might be termed radiation and particles. A physicist would recognize that radiation—including visible, infrared, and ultraviolet light along with radio signals and x rays—is really streams of particles as much as it is waves of different frequencies. Similarly, electrons and atomic nuclei are waves with very short frequencies as much as they are particles. Nevertheless, astronomers who refer to the solar wind mean particles, especially atomic nuclei, not waves. In the list below, which takes a point of view more like that of a physicist, all forms of radiation are counted as particles.

Photons   For human beings the most obvious set of particles that stream from the Sun are the photons of visible light and discernable warmth. But the processes of solar fusion and the electromagnetic field of the Sun produce photons in all parts of the electromagnetic spectrum, ranging from low energy particles (long radio waves) to very high energy particles (short gamma rays). The visible light is produced largely by a region considered the surface of the Sun, and known as the photosphere. In an eclipse, the hotter outer reaches of the Sun's atmosphere become visible.

An unusual source of gamma rays was detected for the first time in 1991 and recognized a year later. The Compton Gamma Ray Observer recorded a glow of gamma rays that followed large solar flares of June 11 and June 15, 1991, the last days of the Great Solar Flare of 1991. James M. Ryan of the University of New Hampshire in Durham reported the gamma-ray afterglow on July 15, 1992, and proposed that it was caused by collisions of particles from the solar flare with parts of the Sun's atmosphere.

Neutrinos   Some of the energy released by fusion is in the form of neutrinos (*see* "Neutrino Astronomy Comes of Age," p 136, and "Subatomic Particles," p 666), which are one of the by-products of several different fusion steps, notably those involving the element beryllium. The neutrinos, which are produced in the Sun's core, interact so little with other matter that most of them pass through the Sun's overlying mass and out into space. Even though each neutrino takes very little energy from the Sun and probably no mass in a form like matter (although some theories and experiments suggest a very small mass for the neutrino, which had been thought previously to be, like the photon, massless), the total amount of energy pouring forth in this form is quite great. Nevertheless, neutrinos are not usually considered a part of the solar wind.

A continuing problem throughout the 1990s, although first noted 20 years earlier, has been the contradiction between the theory of fusion, which predicts a basic production of solar neutrinos at a rate which would be measured as 132 solar neutrino units, and the findings of neutrino astronomers, whose best telescopes record about 83 solar neutrino units. Overall, the record is even worse than that: all the neutrino telescopes on Earth should register about one solar neutrino a day, but instead they average about one every three days.

There are theories about what might have happened to the missing neutrinos, but proof of these theories is lacking. Some of the newer neutrino telescopes continue to reduce the gap between theory and observation, but the long record of the oldest telescope offers a big clue to the mystery: the Homestake neutrino count varies somewhat with the sunspot cycle and therefore with the solar wind.

Electrons   A study announced on August 15, 1994, by Richard Belian and Tom Cayton of the Los Alamos National Laboratory showed that electrons from the Sun reach Earth in rates that vary with the solar sunspot cycle of 22 years. When sunspot activity is at a low, more electrons reach Earth, where most become trapped in Earth's magnetic field. The electrons are accelerated by Earth's magnetism and are dangerous to spacecraft, although harmless to anything on the surface of the Earth. Some communication satellites have been put out of commission by the high-energy electrons, notably a Canadian satellite in 1994.

Protons   Much of the solar wind consists of protons that are the nuclei of hydrogen atoms that have been stripped of their electrons. These protons are the cause of the auroras on Earth and form the main component of solar flares.

Neutrons   Although one does not usually think of free neutrons as part of the solar wind, the Compton gamma ray telescope observed about 250 neutrons as a result of the Great Solar Flare of 1991, as Stephen P. Maran of the Goddard Space Flight Center in Greenbelt, Maryland, reported on July 15, 1992.

Heavier nuclei   The Sun is composed mostly of light elements, but nearly every element has been detected in the Sun. Stripped of their electrons by electromagnetic forces, the nuclei of these atoms also form part of the solar wind. Helium-4 nuclei, also called alpha particles, are the most abundant.

All the particles that stream from the Sun are implicated in one of the great unresolved paradoxes of astronomy: unlike other heat-producing bodies, the Sun produces much higher temperatures at distances millions of miles from its surface than it does at the surface itself. The hot outer atmosphere of the Sun, called the corona, is millions of degrees Fahrenheit above the temperature of the surface. In the mid-1990s three different theories of how the corona could be heated competed with each other. One suggests that magnetic lines of force that are thought to produce giant flares might produce smaller flares on a regular basis, and that such tiny flares, called nanoflares, might then heat the corona. Another theory, proposed by Jack Scudder of the University of Iowa, is that less energetic particles of the solar wind are kept close to the Sun by gravity, but high-energy particles travel farther from the Sun and heat the corona. The third theory is that electric currents produced by the magnetic field accomplish the heating. This last theory has long been the most popular, and it has been to some degree supported by satellite studies of x rays in the early 1990s.

The corona is not generally visible from Earth, but can be seen with a coronograph or during an eclipse, such as the November 3, 1994, eclipse as observed in Putre, Chile, by the High Altitude Observatory eclipse expedition. In the eclipse, great streamers of bright light appeared where electric currents separated streams of charged solar wind particles. The orientation of streamers seen in this eclipse also tend to support the theory of heating of the corona by electrical currents.

The End of Influence

How far out does the solar wind stream? Photons and neutrinos are bosons that, in principle, travel in groups forever without interfering with themselves or each other, although encounters with dust or gas molecules eventually extinguish photons from the Sun and even slightly reduce the stream of neutrinos. Electrons and protons are fermions, interfering with each other and themselves, as well as being absorbed by dust or gas. The solar wind, made from particles that are fermions, eventually stops itself. The distance at which the solar wind dies down, however, is difficult to determine from Earth, since it is beyond the limits of the known part of the solar system.

Space probes, however, are traveling to the edge of the solar system and beyond, providing scientists with the first hard information on where the boundary is. The solar system extends well beyond the orbits of Neptune and Pluto to include the Oort cloud of comets (*see* "Comet, Meteoroid, Asteroid UPDATE," p 171) and the farthest reach of the solar wind. The edge of the solar system as defined by the solar wind is called the heliopause. No spacecraft have yet reached the heliopause, but information from two of the farthest space probes revealed in 1993 a general range in which the heliopause might be found. In May and June of that year, the Sun produced several of its characteristic giant solar flares. The flares increased the solar wind for a time, so that when the wind gusts reached the edge of the solar system—or the edge of the rest of the universe, depending on one's point of view—radio signals were generated. Although these signals could not be directly detected on Earth, information about them was relayed back to Earth by the far-out probes *Voyager 1* and *Voyager 2*. Study of the radio signals indicates that the edge of the solar system is between 82 and 130 astronomical units from the Sun. (An astronomical unit is the average distance of the Earth from the Sun—93 million mi or 150 million km.)

Voyager 1 in the mid-1990s is about 52 astronomical units from the Sun on one side of the plane in which the planets rotate, while *Voyager 2* is 40 astronomical units from the Sun on the other side of the plane. *Voyager 1*, traveling faster, is expected to reach the heliopause first, sometime around the year 2010. The *Voyager* scientists, who think that the probes may still be functioning and returning information well beyond that time, include Don Gurnett of the University of Iowa, Ralph McNutt of Johns Hopkins University, and Edward C. Stone of the Jet Propulsion Laboratory in Pasadena, California. McNutt made the estimate of the distance to the heliopause, which was announced at a meeting of the American Geophysical Union in Baltimore, Maryland, on December 16, 1993.

The Total End of Influence

In 1994, I. Juliana Sackman of the California Institute of Technology and Arnold I. Boothroyd of the University of Toronto, using calculations from James F. Kasting and Kenneth Caldeira of Pennsylvania State University, produced new descriptions of the last days and end of the Sun, publishing their results in *The Astrophysical Journal*. They conclude that after about 5.7 billion years of its lifespan, of which about 4.6 billion years have already been used up, the Sun will expand enough and become hot enough to destroy all life on Earth, although not the planet itself. By that time, about 40% of the original mass of the Sun will have been carried off to space by the solar wind.

One result of the loss of mass will be the expansion of orbits of the planets, since less mass means less gravitational force holding the planets in their orbit. Earth might drift from its present orbit at about 93 million mi (150 million km) to an orbit that averages about 172 million mi (277 million km).

As the Sun dies, there will be a period in which the energy balance between fusion heat and gravitational force will favor fusion. As a result, the Sun will expand from its present size to become a red giant star. This period will begin with solar storms that raise the Sun's temperature enough to evaporate Earth's oceans. If the Earth were not farther away as a result of outward drifting, these great storms would do even more damage to Earth. The expansion of the Sun to a red giant was predicted dozens of years ago, but both the final size of the red giant and its effect on Earth has been less certain. Some have predicted that Earth would be vaporized. Others have said that even though Earth would be inside the outer boundary of the red giant, its gases would be too tenuous to destroy the planet. The prediction by Sackman and Boothroyd is that after 11 billion years of its existence, the Sun will expand beyond even the enlarged orbit of Mercury, destroying the innermost planet, but the rest of the planets will continue to orbit the red giant as burnt crisps.

After about 12.4 billion years of life, the Sun is expected to form a planetary nebula, an expanding shell of gases, centered around a remnant of its former self. The remnant will not maintain enough fusion energy to counteract gravity—which always wins in the long run—and the remnant will collapse into a white dwarf at first and eventually into a black dwarf when fusion ceases entirely. The planetary cinders will continue to orbit the black dwarf, cinders around cinders for eternity.

(Periodical References and Additional Reading: *New York Times* 6-13-92, p 1; *New York Times* 7-16-92, p A21; *New York Times* 10-31-92, p 1; *New York Times* 12-27-93, p A1; *Science* 2-11-94, p 756; *New York Times* 8-16-94, p C11; *New York Times* 9-20-94, p C1; *EOS* 1-3-95, p 1)

MAGELLAN AT VENUS

The *Magellan* space probe, launched by the United States on May 4, 1989, orbited Venus for more than four years, using radar to map 99% of the planet. Scientists also used *Magellan*'s own mass and exact measurements of its position in space to locate changes in gravity caused by differences within the body of the planet.

Magellan's main goal was a complete map of Venus, and it was remarkably successful. According to Joe Boyce, NASA program scientist for *Magellan*, the probe has produced a better map of Venus than any we have of Earth. Scientists still do not have maps of the ocean basins on Earth—more than half the planet—as good as those they now have for four-fifths of the surface of Venus.

In all, *Magellan* sent about a trillion bytes of data back to Earth during more than 15,000 orbits of Venus. The radar maps increased the number of Venusian landforms significant enough to be named from about 300 to about 1,000. Eventually, features as small as 350 ft (100 m) in diameter were recorded. There is so much data that scientists will spend decades analyzing it.

Volcanoes of Venus and Other Geology

Active volcanoes are rare in the solar system. Aside from Earth, the only places where volcanic eruptions have definitely been seen are on Jupiter's moon Io. Something that might be a volcanic eruption, but probably isn't, has been seen on Titan. Most astronomers believe that there are active volcanoes on Venus as well, but none has ever been observed directly. *Magellan* did find at least 175 Venusian volcanoes at least 31 mi (50 km) in diameter. Large volcanoes constitute at least 5% of the surface area of the planet.

Although *Magellan* failed to glimpse a volcano in actual eruption, it saw much evidence of the effects of volcanism. Large lava channels were observed, longer and more regular than flows on Earth. Domes, some as much as 1 mi (1.6 km) high and 95 mi (60 km) in diameter, also larger than similar protrusions on Earth, were observed and are thought to be volcanic features. When the domes and lava channels combine with circular and oval landforms around the domes, the result is a characteristic Venusian feature scientists have named arachnoids.

Opinions as to what the *Magellan* data mean for Venusian geology continually changed during the mapping. For example, early on its second pass at mapping the planet, *Magellan* found what appeared to be evidence of a recent giant landslide. On August 29, 1991, Jeffrey Plaut of the Jet Propulsion Laboratory in Pasadena, California, observed the change attributed to the landslide in images taken eight months apart. If confirmed by the third pass, the landslide would be in the same class as the largest known on Earth. By October, however, many scientists had come to believe that the apparent evidence was an artifact of the radar imaging process. For similar reasons, Venusian rock was once thought to be weak, but now is believed to be strong.

There is little erosion on Venus, which lacks both water and strong winds, although a 1 mph (1.6 kph) wind on Venus has the effect of a 15 mph (25 kph) wind on Earth because of the dense atmosphere. The small amount of erosion observed in *Magellan* images suggests to some scientists that the surface is on average less than 400 million years old, with no part of the crust older than about a billion years. Probable explanations focus on the volcanic eruptions that have resurfaced the entire planet.

In addition to other surface features, *Magellan* found impact craters, but all are at least 4 mi (6.5 km) in diameter. No smaller ones exist because any objects that could cause them burn up in the thick atmosphere before reaching the surface of the planet. The total number of craters located was more than 1,000—more than are known on Earth, where erosion removes signs of craters relatively quickly, but far fewer per unit of total surface area than on Mercury, the Moon, or Mars. All large craters on Venus have their interiors partially flooded with lava. The impact craters that dot the surface of Venus are randomly placed over the entire surface with one significant exception. This means that geologically speaking all of the surface must be about the same age; otherwise, older parts would have more impact craters than newer ones. The significant exception is that there are fewer impact craters inside the large volcanoes. The common age for impact craters is estimated to be well over 500 million years, roughly during the period when life on Earth was moving from the oceans to occupy the land as well. The volcanoes, however, must have been active during the past 500 million years.

Of course, the recent age of the crust, geologically speaking, does not mean that Venus is a young planet. All the planets formed about the same time, as far as we now know, but they have aged at different rates. In the view of planetary geophysicists such as Roger Phillips of Washington University in St. Louis, Missouri, and William Kaula of the University of California, Los Angeles, Mars is the oldest-appearing planet, then Venus, then Earth, which has probably been kept young in part by life. Phillips thinks Venus looks like the Mars of a billion years ago. Since then, Mars has shut down because most of its internal heat has been lost to space. Now that process seems to be underway for Venus.

Plate tectonics as it is known on Earth appears to be absent on Venus. The crust is thick and strong. There seems to be nothing like Earth's worldwide network of rift valleys and ridges or the deep trenches into which plates plunge. One theory in this regard is that the lack of surface water on Venus makes the difference, since there is considerable evidence that Earth's surface water percolates deep into the crust and mantle, causing some of the observed effects of plate tectonics. On Venus, the water may have evaporated from the surface faster simply because the planet is closer to the Sun. With less tectonic activity, what water remains may be buried in deeper layers. Experiments with Earth rocks that have been dried to match levels of water found on Venus show that the rocks become 5 to 10 times stronger. The dry, thick crust has locked the geological processes in place. The end of change is the beginning of old age.

All of the data from *Magellan* has been put onto a CD-ROM that is available from NASA. Some images are available on the Internet World Wide Web at http://stardust.jpl.nasa.gov.

The Death of *Magellan*

Even when it was so crippled by old age and failing solar panels that it could no longer perform the mapping duties for which it was designed, *Magellan* proved to be an invaluable scientific tool. Scientists commanded the craft to commit suicide by lowering itself into the top levels of the Venusian atmosphere, enabling tests of atmospheric braking and measurements of density. In October 1994, *Magellan* began its irreversible descent. With its solar panels set like a windmill's blades, a rotational force was created by the atmosphere; countering that force with thrusters to keep the craft stable was then used to measure the density of the atmosphere. The same experiment had been previously conducted at a higher altitude. Comparison of the two forces suggested that the upper atmosphere was only about half as dense as had been expected at about 100-90 mi (160-150 km) above the surface, but seemed to rise dramatically to about 1.5 times the expected density by 86 mi (138 km) above the surface, where contact was lost.

As it entered the atmosphere, *Magellan*'s fragile solar panels sheared off, its antennas collapsed, and it broke into pieces. The two largest pieces finally crashed on the high slopes of Maxwell Montes.

TIMETABLE OF A SPACE MISSION

May 4, 1989   The *Magellan* space probe is launched from the space shuttle *Atlantis,* aimed at Venus.

August 10, 1990   *Magellan* enters its first orbit about Venus.

September 1, 1990   Mapping, scheduled to begin on this day, is delayed because of two mysterious losses of radio contact of 13 and 21 hours in late August.

September 15, 1990 *Magellan* begins its first attempt at using radar to map the surface of Venus.

November 15, 1990 Another radio interruption occurs, but lasts only 40 minutes, as software is now in place to correct malfunctions in transmissions.

May 1991   *Magellan* completes its first Venusian year (243 Earth days) and its first round of mapping. It has transmitted to Earth images of any feature as large or larger than a football field for about 84% of the planet's surface.

June 1991   Technicians working with a model of *Magellan* on Earth accidentally discover the cause of the mysterious interruptions in radio signals that dogged the early part of the mission; a better program for handling the difficulty is instituted.

January 15, 1992   *Magellan* completes its second round of mapping, covering 95% of the surface of Venus. NASA announces that it is considering cutting costs by ending the *Magellan* mission early, in May 1993.

July 7, 1992   A second transmitter on *Magellan* begins overheating and ceases to send reliable data. This condition lasts for over a week before engineers are able to solve the problem.

October 9, 1992   *Pioneer 12*, which had been circling Venus for 14 years (two of them with *Magellan* as company) studying the Venusian atmosphere and making the first radar and gravity maps of Venus, having run out of fuel on October 6, glides into the atmosphere, breaks up, and burns. *Pioneer* had continued sending useful data to Earth right up to the last.

August 1993   *Magellan* performs a series of complicated maneuvers to change its orbit from an elongated ellipse, traveling as close to the planet as 182 mi (294 km) and as far away as 530 mi (8,543 km), into a more circular ellipse, so that the craft roams from a near point of 110 mi (180 km) to a far point of 600 km (370 mi). With this more circular orbit *Magellan* is able to collect data on gravity variations on Venus.

September 10, 1994 About 113 mi (182 km) from the surface, *Magellan* is turned and its solar panels twisted to make it rotate like a pinwheel. Measurement of the twisting force produced is used to determine the density of the atmosphere.

October 11, 1994   At 7:21 a.m. California time, researchers at the Jet Propulsion Laboratory of NASA in Pasadena cause *Magellan* to lower itself into the atmosphere of Venus, a maneuver that will cause the craft to crash on the surface of the planet. Engineers plan to use *Magellan*'s flight data to study the dynamics of traveling through a dense atmosphere.

Background

Venus and Earth—next to each other in proximity to the Sun, and manifestly the same size—were long thought to be sister planets. They are alike in many ways, an exception being the fact that Earth has a Moon while Venus does not. Both have perceptible atmospheres filled with clouds, although Venus has more of the latter than Earth does; this has led to the idea that Venus might have more water than Earth and might be largely ocean below those inscrutable clouds.

But when the first space probes, as well as radar contact from Earth in 1979, were used to explore Venus, more differences were found than similarities. Rather than there being more water on Venus than on Earth, water is notably absent; the clouds are sulfuric acid. Furthermore, the atmosphere of Venus is not only extremely dense, but has trapped the Sun's energy well enough to raise Venusian surface temperature to a level that would melt lead.

Further exploration by space probes also demonstrated that Venus is not covered with floating continent-sized plates, now known to be the main geological force on Earth. Among the differences between Earth and Venus from a tectonic point of view is the way that highlands and lowlands are borne. Gravity anomalies mapped by *Magellan* suggested at first that Venusian highlands—corresponding to continents on Earth—are pushed up by energy from within, rather as hot spots are created on Earth. A few highlands on Earth are also the result of internal forces (for example, the Black Hills of South Dakota), but continents and mountain ranges all float on a partly molten layer where Earth's crust meets the mantle. But later studies of gravitational anomalies suggested that the hot spots on Venus were not hot any longer. Instead, the crust of the planet might be as much as 60 mi (100 km) thick, supporting the highlands and volcanoes by strength rather than buoyancy.

Volcanic activity appears to be dominant on Venus. All of its rocks seem to be volcanic in origin. About 90% of its surface resembles the recent lava fields of Hawaii or Iceland rather than the typical topography of the rest of today's Earth. On the whole, Venus is thought to be what Earth was like during the Archaean era, the period from about 4.5 to 2.5 billion years ago.

Theories abound as to why Venus and Earth are different in the way crust is built and recycled. One of the main theories proposes that the runaway greenhouse effect caused by the thick atmosphere, rich in carbon dioxide, accounts for the tectonic effects. A very hot lithosphere has different properties from stone and minerals at what we Earthlings call "room temperature." Another possibility is that Earthlike plate tectonics is possible only in the presence of water at fault lines, and water is lacking on Venus.

October 12, 1994   At 3:02 a.m. California time, the last piece of data is received from *Magellan*. Shortly afterward it crashes, probably on October 13.

(Periodical References and Additional Reading: *New York Times* 7-15-92, p A16; *New York Times* 10-10-92, p 6; *Science* 2-11-94, p 759; *Science* 5-6-94, p 798; *Science* 6-17-94, p 1666; *Science* 8-12-94, p 929; *New York Times* 8-23-94, p C10; *Science* 9-2-94, p 1360; *New York Times* 9-11-94, p I-18; *New York Times* 10-11-94, p C5; *New York Times* 10-12-94, p A15; *New York Times* 10-13-94, p A24; *Science News* 10-22-94, p 262; *Astronomy* 2-95, pp 28 & 32)

MORE ON THE MOON

Although the Sun and the Moon are the most recognized heavenly bodies, easily seen and located with the naked eye and no special knowledge, they are far from completely known or understood. Even the Moon, upon which humans have walked and hopped, continues to be the source of new discoveries. The same is true, of course, for the planet Earth, upon which humans have lived for hundreds of thousands of years, so it should be no surprise that the Moon is not fully understood.

But new techniques of mapping are making both Earth and the Moon better known. Looking at a region using wavelengths other than those of visible light enables a satellite or space probe to gather information about the molecules that reflect those wavelengths preferentially. Longer wavelengths, such as those used in radar, can be chosen so that dry sand or water on Earth or the "soil" on the Moon will be nearly transparent, allowing scientists to see what lurks below. For example, scientists using Earth-based telescopes have been able to see into the interior of the large Copernicus impact crater on the Moon, revealing the subsurface geology of that part of the Moon.

Another form of mapping that was pioneered on Earth and is now paying added dividends on the Moon uses a satellite or space probe to measure gravity anomalies by the deviations in its flight path caused by variations in gravity on the planet or satellite as the craft passes over it. The same concept has also been useful in studying Venus (*see* "*Magellan* at Venus," p 150). Finally, laser mapping can be used to pinpoint the exact location of physical features more closely than any previous means.

GALILEO AT THE MOON, PASS 2

In 1990 and 1992 the *Galileo* space probe passed near the Moon on its way to Jupiter (*see* "*Galileo* UPDATE," p 278). Astronomers did not waste these opportunities for a closer look, especially since the 1992 pass included views at various wavelengths of the seldom-seen far side. Images of features as small as 0.7 mi (1.1 km) were observed as single pixels in the digital imaging system. The second pass, known among *Galileo* scientists as EM-2 (Earth-Moon 2), also mapped the western edge of the part of the Moon visible from Earth, known as the western limb.

Galileo observed the Moon at wavelengths that appear as a different "color" for lava from volcanoes. One of the questions that apparently was resolved by the *Galileo* images is whether the northern light plains, which are made from less dark rock than the great maria ("seas" without water), are caused by volcanism, which is thought to be the origin of the maria, or by impacts, known to be the origin of large craters. The verdict was for impacts. But many smaller regions for which the origin had been somewhat in doubt were settled in favor of past volcanism. Ancient impacts and volcanoes combined to make the surface of the Moon as we know it today.

Instruments aboard the *Galileo* spacecraft were also able to analyze rock for the presence of several important minerals, including titanium and iron. Together EM-1 and EM-2 mapped about three-quarters of the total surface of the Moon. Some of this region had been well known, because it is easily visible from Earth and in some cases had been visited by astronauts in 1969–72 or by remote-controlled vehicles in 1976.

These well-studied places allowed scientists to adjust their instruments, making the results from previously little known regions more certain.

The Lunar Atmosphere

Although it is commonly believed that the Moon has no atmosphere, scientists have long known that a thin layer of gases clings to the "airless" surface of the Moon. These gases are thought to be generated primarily by the impact of photons from the Sun and by the solar wind (see "Stories on the Solar Wind," p 146) on the lunar surface.

In 1993, Michael Mendillo of Boston University and graduate student Brian Flynn became the first to measure the extent and thinness of the atmosphere of the Moon. They used a telescope equipped with a light detector designed to detect sodium atoms. The sodium atoms are not thought to be one of the main components of the atmosphere, but they can be used as a tracer for the other gases. About one of every 200,000 atoms of the Moon's atmosphere is thought to be sodium. As with other spectral images, choice of the right frequencies reveals the sodium, from which one can work backwards to determine the density of all other atoms.

Mendillo and Flynn observed that the atmosphere extends at least 5,000 mi (8,000 km) from the surface of the Moon. This seems at first thought to make the extent of the lunar atmosphere greater than that of Earth, but if one were to measure Earth's atmosphere at very low numbers of molecules of gas, Earth's atmosphere would be much more extensive than the comparatively thick air—colloquially known as "thin air"—known close to Earth's surface. By way of comparison, Earth's atmosphere is a trillion times as dense as that of the Moon near the surface, although the Moon's atmosphere includes as many as 160 million atoms per cubic inch (10 million atoms per cubic centimeter), which sounds more substantial than it is. For all practical purposes, an atmosphere that thin can be treated as if it were not there at all, just as we have long thought about the Moon.

Clementine Visits the Moon

When the space probe *Clementine* was launched on January 25, 1994, under the joint auspices of the U.S. military and the National Aeronautics and Space Administration, there was a limited popular expectation of success, but more hope from its creators. *Clementine* was intended primarily to be the first of a generation of cheap, all-purpose satellites. In addition to a study of the Moon, *Clementine* was scheduled to visit the asteroid Geographos. All the scientific goals were add-ons to the original military purpose, which was simply to test some hardware space scientists had developed. Many scientists thought that the experiments seemed like afterthoughts.

The mission to Geographos had to be scrubbed because of computer errors that caused *Clementine* to spin wildly for a time. The data from the Moon, however, were so good that no one deemed *Clementine* a scientific failure. Unfortunately, the probe's physical problems after four months in space suggest that the military goals were not met. Furthermore, the design did not seem to be a very good model for a new generation to follow.

Clementine's big success, however, consisted of its 71 days surveying the moon, which produced the first-ever map to include about 99.9% of the surface of the Moon, both near and far sides, photographing the Moon at wavelengths that ranged from infrared to ultraviolet.

Among the other observations from *Clementine* that reveal new details about the Moon is a map of gravity anomalies. When combined with the surface topography, the gravity map reveals that the Moon's crust is far thicker than that of Earth, with a lunar average thickness of about 66 mi (106 km) compared to the terrestrial average of 22 mi (35 km). The gravity map also reveals why the Moon's center of mass is away from the Earth, which accounts for the Moon's having a perpetual far side. The center of mass is not the center of the Moon, largely because the crust is thickest in one region on the far side. Despite this, the thinnest crust is found in another region of the far side. This thin crust is a direct result of impacts. Finally, the same methods were used to show that part of the interior of the Moon is still molten, since the shape changes too much as a result of reverse tidal forces caused by Earth's gravity for the Moon to be completely solid.

Clementine also used a laser some half-million times to precisely measure its distance from the Moon. This experiment had been derided for using military laser methods intended to measure the distances to enemy equipment, but the same methods proved to be extremely successful for measuring the distances to crater rims and floors. The results of the laser survey provide new details about the overall shape of the Moon. They also demonstrate that the relief (the distance from low valleys to high mountain tops) on the Moon is about twice what had been previously suggested, primarily because a feature on the far side known as the South Pole-Aitken basin is a staggering 7 mi (12 km) deep. It is therefore the deepest known crater in the whole solar system. Other basins are also deeper than expected.

Keeping Earth Stable, Not Crazy

Although the Moon is popularly associated with periodic psychosis, there is one sense at least in which the Moon keeps the Earth stable. Computer simulations demonstrate that the Moon provides the force that keeps Earth from bouncing around like a top that is about to end its spin.

Spinning bodies are in some ways all alike. When a top is well thrown, it begins to spin with its axis almost straight up. For a planet this would be called an *obliquity* of 0°; measurements are taken from the vertical, often called the *normal*. As the spinning top slows down, its axis begins to shift. The "north pole" of the top makes a small circle at first, but a larger circle as time goes on. This motion of the axis is called *precession*. As the top begins to precess, it also is leaning away from the normal. The obliquity is increasing. Soon after the obliquity increases to about 30° or a bit more, the top may spin out of control. Often the top will fall and roll across the floor. Now its axis is perpendicular to the normal, so the obliquity is 90°.

Earth is spinning like a top at the stage where the axis slowly precesses and the obliquity is about 23.5°, not too far from the normal. The evidence, which includes the geological record for millions of years, is that Earth has not varied far from either this amount of precession or this degree of obliquity. The reason, according to Jacques Laskar and coworkers at the Bureau of Longitudes in Paris, France, is Earth's

Moon, which is the largest satellite relative to its planet in the solar system. The giant Moon acts as a brake and as a reservoir of energy, stabilizing Earth into that obliquity and that precession. Take away the Moon, and Earth would vary wildly from an obliquity of 0° to one of 85° and all angles in between.

Were that to happen, seasons on Earth would no longer make sense. Climate would vary unpredictably. It is difficult to see how life could have evolved in such chaos. Neither of Earth's immediate neighbors has a similarly large Moon. Although they both appear stable from our short, 400-year-history with the telescope, computer stimulations of Mars show that its obliquity has varied from 0° to 60° in the past; some geological evidence from space probes tends to confirm this kind of crazy climate change. Mars has two moons, but both are very small in relation to their planet. Venus has no moon at all and a current obliquity that is very close to the normal. Venus rotates backward very slowly, however, with a "day" that is longer than its "year." Thus, Venus is barely a spinning body at all.

A problem with relying on the Moon for stability is that the Moon is moving slowly out of the neighborhood. In about a million years it will be too far away from Earth to have any effect on obliquity; then Earth's top can spin where it pleases.

(Periodical References and Additional Reading: *Discover* 1-94, p 72; *Science* 5-20-94, p 1059; *Science* 6-17-94, p 1666; *Nature* 7-14-94, p 100; *New York Times* 8-10-93, p C10; *EOS*, 11-15-94, p 537)

EARTH THREATENED BY ASTEROIDS; ASTEROIDS THREATENED BY EARTH

A lot of people were startled when newspapers reported that a U.S. citizen is six times as likely to be killed by an asteroid impact as by traveling on a U.S. airline—a prediction attributed to Clark R. Chapman of the Planetary Science Institute and David Morrison of the National Aeronautics and Space Administration (NASA) Ames Research Center. As Chapman and Morrison reported their actual calculation in the January 6, 1994, *Nature,* it would be more accurate to say that the risk of dying as an airline passenger and the risk of dying as a result of a massive impact by an asteroid or comet are about the same. They explain that an impact by a large asteroid or a comet is a low-frequency event that kills many people, so the possibility of a given person being killed by one is as high or higher than the possibility of being killed by a more common event that kills only a few people at a time. Thus, using various estimates from different sources, they developed the following table:

Cause of death	Chances
Motor vehicle accident	1 in 100
Murder	1 in 300
Fire	1 in 800
Firearms accident	1 in 2,500
Asteroid/comet impact (lower limit)	1 in 3,000

Cause of death	Chances
Electrocution	1 in 5,000
Asteroid/comet impact	1 in 20,000
Passenger aircraft crash	1 in 20,000
Flood	1 in 30,000
Tornado	1 in 60,000
Venomous bite or sting	1 in 100,000
Asteroid/comet impact (upper limit)	1 in 250,000
Fireworks accident	1 in 1,000,000
Food poisoning by botulism	1 in 3,000,000
Drinking water with EPA limit of trichloroethylene	1 in 10,000,000

Their argument was based on statistics, not on calculations of the paths of asteroids. Also, so far as is known, no human being has ever been killed by an asteroid or comet impact, but many have been killed in motor vehicle accidents, during floods, or by botulism.

The Nearness of Them

Some asteroids, however, come uncomfortably close to Earth; they cross its orbit. These are called the Apollo asteroids (after the asteroid Apollo, the first of its class to be discovered, by Karl Reinmuth in 1932). So far the Palomar Planet-Crossing Asteroid Survey (PPCAS), a project headed by Eleanor Helin, has found more than 40 Apollo asteroids, which together with those found by other groups make a total of more than 60 so far. But there are other threatening asteroids as well. With the Apollos, they form a group of more than 160 known asteroids collectively called Earth-approaching asteroids, although Chapman and Morrison used the scarier expression "Earth-crossing asteroid." The Earth approachers, or crossers, include the Atens, inside Earth's orbit, and the Amos, just outside, as well as the Apollo orbit-crossers.

How close do they come? Since the PPCAS project started, the recognized approaches keep getting nearer and nearer to us. One of the first close approaches to be observed in the early 1990s came on July 10, 1990. An asteroid named 1990 MF whizzed by only 3 million mi (4.83 million km) from Earth at a speed of 22,000 mph (35,000 kph). Since it is only 300 ft (100 m) by 1000 ft (300 m) or so, 1990 MF is not a very big asteroid. At that speed, however, it could do a lot of damage if it hit Earth. Six months later, in January 1991, asteroid 1991 BA passed only 105,000 mi (170,000 km) from Earth, about half the distance between Earth and the Moon. But 1991 BA is even smaller than 1990 MF, only about 33 ft (10 m) in diameter. While Toutatis on December 8, 1993, came no closer than 2.2 million mi (3.5 million km), publicized radar images of it caused the fearful some alarm. In March 1994 a new record was set when 1994 ES_1 passed less than 96,000 mi (150,000 km) away. But the closest recognized approach came about noon EST on December 9, 1994, when 1994 XM_1 was just 63,000 mi (101,000 km) away. Like the other recognized close-encounter asteroids, 1994 XM_1 is small, only about 43 ft (16 m) in diameter.

Earth Takes a Few Hits

These approaches were all misses, of course. Meteorites often strike Earth, including a notable example on October 9, 1992, that hit a 1980 red Chevrolet Malibu in Peekskill, New York. After passing through the car, the remaining object was a rock about the size of a football.

By the standards used today, astronomers could call the "meteor" that sliced through the top of the atmosphere in 1972 a medium-sized asteroid. It was estimated to be about 260 ft (80 m) in diameter.

Some think that the mysterious event on July 30, 1908, near the Tunguska River in Siberia was a collision between Earth and an asteroid that caused widespread devastation even though the asteroid exploded high in the atmosphere. According to 1993 calculations by White House fellow Christopher Chyba, the Tunguska event can be explained by the explosion of a stony meteorite about 100 yd (100 m) in diameter striking the atmosphere at a speed of mach 45 and exploding in a cloud of superhot gas about 5 mi (8 km) about Earth 's surface.

There is, however, considerable evidence of many collisions in the past in which an asteroid or comet reached Earth's crust. Of these, the one 65,000,000 years ago, thought to have caused mass extinctions at that time, was probably more than 6 mi (10 km) in diameter, leaving the 185 mi (180 km) Chicxulub crater on the Yucatán peninsula of Mexico and releasing as much energy as 200,000,000 hydrogen bombs, according to measurements of the crater made in September, 1993.

By early 1994 virtually all scientists had come to accept the reality of the Chicxulub impact, as a major conference on catastrophes at that time revealed. The scientists for the most part also accepted that an impact of that size could cause the sudden extinction of many species all at once. So-called "blind tests" found even skeptics identifying rock strata separated by short intervals of time as having come before and after extinction of common species, such as the forams common in all seas then and now. Specifically, at least a few forams of species thought by doubters to have become extinct gradually for millions of years before the impact were found in samples that immediately preceded impact. Not all doubters were convinced, but the evidence was seen by most as showing that this impact at least was the proximate cause of species extinction, not just a coincidental impact at a time that a lot of extinctions were taking place for unrelated reasons.

As the breakup and crash of comet Shoemaker-Levy 9 into Jupiter revealed (*see* "Comet Hits Jupiter," p 165), gravity from a nearby planet can cause more than a single impact from what was originally one comet or asteroid. Robert Rocchia and coworkers from the French Atomic Energy Commission in Gif-sur-Yvette, France, reported evidence for a multiple impact caused by the Chicxulub body. The second fragment, smaller than the main body but still well over 0.6 mi (1 km) in diameter, appears to have hit what is now the floor of the Pacific Ocean. The evidence consists of small grains of melted rock that contain a high amount of the element iridium, a known feature of the mass extinction era associated with the Chicxulub impact. Earlier, astronomers had also found evidence that asteroid 951 Gaspra and Apollo asteroid 3103 1982 BB each had been part of a collision or had broken apart while in orbit; pieces of 3103 1982 BB have apparently come to earth as a series of recognizable meteorites called enstatite achondrites.

Another crater caused by a large impact, as reported in the August 1994 *Geology* by C. Wylie Poag and coworkers at the U.S. Geological Survey, lies at the mouth of what is now Chesapeake Bay. The object struck some 35 million years ago. The impact caused a crater at least 50 mi (85 km) in diameter in the Atlantic Ocean, since the site was at that time about 60 mi (100 km) offshore.

The largest suspected impact, however, was some 4 billion years ago. At that time, a collision with an object about the size of the planet Mars occurred, according to Luke Dones of NASA's Ames Research Center in California and Scott Tremaine of the Canadian Institute for Theoretical Astrophysics. Their evidence is something anyone can observe: the rotation of the Earth and the existence of the Moon, both thought to have been produced by the collision. Although there are various theories to account for planetary rotations, the idea that they might be caused by massive collisions fits particularly well with the evidence. Venus, for example, rotates hardly at all today, and the rotation it has is opposite that of most of the planets. Mars has a day about the same length as that of Earth, but giant Jupiter spins around every 10 hours or so.

Based on craters on the Moon and Earth, one estimate is that an asteroid strikes our planet or the Moon once on the average of every 65,000-70,000 years. Other estimates range from one every 250,000 years to one every million years for a sizable asteroid more than 0.5 mi (1 km) in diameter. There are thought to be about a thousand asteroids larger than a kilometer that cross Earth's orbit. The largest known Earth-crossing asteroid is 1627 Ivar, with a diameter of about 5 mi (8 km).

A team of astronomers led by Paolo Farinella of the University of Pisa investigated the eventual fate of 47 Earth-crossing asteroids. They found that nearly a third of them will crash into the Sun in the next 2.5 million years and about a twelfth of them will shift their orbits so that the asteroids fly out of the solar system. The astronomers did not determine whether or not their models showed any crashing into Earth. Although the orbits of Earth-crossing asteroids are extremely variable, the results for the 47 studied suggest that many of the asteroids will fall into the Sun over time. Thus, there must be a continuing source of such asteroids or by now most of them would have left their Earth-crossing path. Their results were published in the September 22, 1994 *Nature*.

Not Just Asteroids

The details of the differences between asteroids and comets are becoming clearer (*see* "Comet, Meteoroid, Asteroid UPDATE," p 171), but if one lands on your house, these details may not matter much.

The majority of comets are thought to spend most of their existence in the Oort Cloud, a shell of debris surrounding the solar system at a distance that is perhaps as great as two light-years. When evidence of periodic mass extinctions appeared in the late 1980s, one of the potential culprits was comets from the Oort Cloud whose orbits had been disrupted by some regularly passing far-out planet (*see* "Newly Found Planets," p 178), star, or other force. This concept was revived in 1994 by Robert Matthews of the University of Aston in England. He calculated that the Alpha Centauri binary star system, already the closest to the solar system, will approach even closer, to within 3.12 light-years from Earth. This is close enough to affect the outer fringes of the Oort Cloud, sending perhaps 100,000 comets toward the Sun. Some of these might be expected to strike Earth along the way. Other nearby stars over the next few years, as astronomers count years, will do the same. But astronomical time is lengthy, like geological time. The six stars or star systems that will come close enough to have an effect (within 4.22 light-years according to Matthews) will pass by at various times over the next 45,000 years. The first comets from these sources might reach Earth's orbit about 20,000,000 years from now. No need to get into your comet shelter just yet.

However, another series of attacks comes from objects that are generally considered to be meteoroids, rather than asteroids or comets. Because a large group of these were discovered in 1993 by the Spacewatch team of the University of Arizona, the group is often called the "Spacewatch objects." These small bodies are so numerous and so located that some of them must strike Earth from time to time, but they are too small to cause more than local damage. After their presence was announced, the U.S. Defense Department released previously secret data showing that between 1975 and 1992 military satellites had detected 136 explosions of the type expected from collisions with meteoroids slightly smaller than the Spacewatch objects. The brightest recorded by the Defense Department was on April 15, 1988, above Indonesia. Another bright explosion, with about a fifth the energy of the Indonesian blast, was analyzed in detail and determined to be consistent with the explosion of a stony meteoroid of about 100 tons. This one took place over the Pacific on October 1, 1990.

Earth Fights Back

One presumes that the dinosaurs had no way of detecting the Chicxulub object before it struck—nor any way of defending against it had they known. Humans, however, have become increasingly capable of detecting bodies in space that are headed our way. Furthermore, as several different visits to asteroids and comets by space probes have shown, we can—given enough time—intercept a body if need be. It is not proven that we have the power to destroy a large body, but there is reason to believe that a well-placed hydrogen bomb, or several of them, would alter the course of most bodies of the size likely to cause trouble for Earth.

Is government worried about asteroid impacts? Officially, yes. In 1992 NASA, in response to a request from Congress, produced the report *Spaceguard Survey* on collision warning, although no specific avoidance plans emerged. The report called for spending $50 million on new telescopes and $10 million a year to pay for operating them. No action on this plan was taken, however. Also in 1992 NASA held a conference on how to deal with asteroid and comet threats. Plans that were suggested at the conference included putting the world's nuclear arsenal on 1,200 specially constructed missiles and firing them at the threat. Edward Teller's notion included building a thermonuclear weapon too big to be tested on Earth.

A later conference that year focused on the specific problem of deflecting Comet Swift-Tuttle away from a path intersecting with Earth on August 14, 2126, which seemed to be a remote possibility at the time of the conference. The most likely scenario would be to divert it with a thermonuclear blast equivalent to about 100,000,000 tons of high explosives—almost twice as great as the largest hydrogen bomb ever tested on Earth—while the comet was still beyond the orbit of Saturn. Failing that, an explosion when the comet came around the Sun for its final pass at Earth would have to be two orders of magnitude, or 100 times, greater to sufficiently divert it from its path. Later calculations, however, showed that Swift-Tuttle was not on a collision course with Earth after all.

On July 20, 1994, as Comet Shoemaker-Levy 9 plowed into Jupiter, the U.S. House Science, Space and Technology Committee asked NASA to develop a new 10-year plan that would enable it to identify and catalog the paths of all large asteroids, and comets, too. NASA responded with a new feasibility study, the Near Earth-Object Search Committee, headed by Eugene Shoemaker, intended to lead to development of just such a system, provided funds were forthcoming. Funding is likely to be the most difficult part of the problem, since NASA's budget has been severely reduced in the mid-1990s.

The U.S. Air Force Space Command, which for reasons of its own already has five telescopes watching the skies for unexpected artificial satellites, is undertaking modifications of at least one of these for watching for asteroids. The Air Force thinks that it will be able to spot an asteroid with a diameter of 0.6 mi (1 km) at a distance of 60 million mi (100 million km). It is not clear, however, what the Air Force plans to do if they discover one of these heading our way.

Comet expert Clark Chapman notes that comets that are aimed at Earth need to be identified at least a year in advance to have even "a fighting chance" of deflecting them. Furthermore, if they are piles of rubble, blasting them apart would not make much of a difference. Instead, they would have to be nudged away by a series of nearby explosions. The same might even be true of more compact and connected asteroids. A blast that turned a solid asteroid or small asteroid system into a pile of rubble in space might not change its path. Finally, although the Strategic Defense Initiative spent billions determining how to knock out a relatively small, slow-moving intercontinental ballistic missile, there is not very good evidence that it ever succeeded in doing so.

(Periodical References and Additional Reading: *New York Times* 3-25-92, p A23; *New York Times* 4-1-92, p A16; *New York Times* 4-5-92, p IV-5; *New York Times* 4-7-92, p C1; *New York Times* 11-1-92, p C1; *Discover* 1-93, p 34; *New York Times* 1-4-93, p B12; *New York Times* 6-24-93, p A1; *Discover* 1-94, p 32; *Nature* 1-6-94, p 33;*New York Times* 1-25-94, p C1; *Science* 3-11-94, p 1371; *Astronomy* 5-94, p 18; *Astronomy* 7-94, p 21; *Science* 7-29-94, p 595; *New York Times* 8-1-94, p A1; *EOS* 8-30-94, p 402; *Astronomy* 1-95, p 22; *Astronomy* 4-95, p 22)

Mars UPDATE

A new picture: in 1995 Mars was photographed with the *Hubble Space Telescope* while in line with the Earth and Sun—a configuration known as opposition. The resulting views showed features on a scale for which each pixel (picture element) was only 15 mi (25 km) across, strikingly good views to get from Earth. If the project proceeds on schedule, the next best views of Mars will come in 1998, the project when the first of several space probes, planned for launch in 1996 and later, will reach the vicinity of the planet (*see* "Mars: Loss and New Plans," p 292).

An old picture: old pictures sometimes contain new news, or at least information that was not noted before. In March 1992, Baerbel Lucchitta of the United States Geological Service in Flagstaff, Arizona, reported that she found evidence of a landslide on Mars in 1978. Two pictures taken by the *Viking 1* orbiter 2 minutes and 23 seconds apart show what appears to be a Martian landslide along the edge of a plateau. A dark cloud of dust appears in the second photograph, along with the shadow of the cloud.

Clouds of dust are hardly rare on Mars. Dust storms caused by the thin Martian winds can sweep through the whole planet, as was the case on November 13, 1971, when the U.S. space probe *Mariner 9* arrived to map Mars from orbit—fortunately, the storms died down over a period of three months, and *Mariner 9* was able to map Mars during the next year as it orbited the planet. Other planetary dust storms were observed from Earth in 1909, 1911, and 1956. Although there were planet-wide dust storms in 1992 and 1994, as seen from Earth, the storms seem to have decreased in frequency in recent years. Dust storms are recognized today by taking the temperature of Mars with a radio telescope. When temperatures suddenly rise about 100° F (50°C) in a short period of time, it signals the start of a storm. The temperature rises because the small particles of dust in the lower atmosphere absorb sunlight better than the clear atmosphere does.

Water on Mars

It is clear that Mars has plenty of dust. But when it comes to life and its problems, dust doesn't matter much. As everyone has known from Gunga Din to the Sons of the Pioneers, the more important factor is water.

Isotope ratios are used by planetary scientists who study all the terrestrial planets (Mercury, Venus, Earth, and Mars) in determining the history and state of both the atmosphere and water on the planet. For Mars, the ratio of deuterium (heavy hydrogen) to ordinary hydrogen in the atmosphere has been found to be five times that of the ratio in Earth's atmosphere. The standard explanation of this difference is that over planetary lifetimes, the smaller gravitational force exerted by Mars lets more lightweight hydrogen escape into space than does the gravitational force of Earth. That explanation would have as a corollary that the two planets began with similar ratios of deuterium to hydrogen. If so, given the lack of recycling of materials by plate tectonics on Mars, water from the interior of Mars would have an Earthlike ratio.

One factor involved in whether or not there was ever rain and snow from water on Mars is the suspected high carbon dioxide level of the early Martian atmosphere. It had been thought that this might have produced enough greenhouse warming to make water precipitation feasible; but more recent calculations show that the carbon dioxide itself would have frozen initially. Thus, instead of a gas blanket, transparent to incoming sunlight and opaque to heat from the surface, there would have been clouds of dry ice particles. These would have reflected sunlight, making the planet even colder than it was before the carbon dioxide entered the atmosphere.

It might seem impossible to measure water from the interior of Mars at this time, but most scientists believe that there are 10 samples of Mars rocks on Earth. These samples are known as the SNC (shergottite-nakhlite-chassignite) meteorites. SNC meteorites are thought to have been blasted into space by meteorites that collided with Mars, with some of the blasted-off material becoming meteorites that landed on Earth. The team of Laurie Lachine Watson, Ian D. Hutcheon, Samuel Epstein, and Edward M. Stopler of the California Institute of Technology in Pasadena studied water in the SNC meteorites, however, and found that the very small amount of water contained in them has a deuterium-to-ordinary-hydrogen ratio that is like that of the Martian atmosphere. They interpret this result as meaning that there has been interchange of hydrogen between the atmosphere and near-surface supplies of water, but the implications are more significant. The SNC meteorites date from ancient events, as much as 1.3 billion years in the past; thus, if ordinary hydrogen has escaped, it must have done so before the SNC meteorites were blasted out of the crust.

There is also other evidence of liquid water flowing on the surface of Mars, primarily small river valleys visible in regions that have not been much affected by later meteorite bombardment. While these valleys are generally thought to have developed as creeks and rivers, many astronomers doubt that rain ever fell to feed the streams. Instead, they believe, the water was all geothermal, coming forth from the interior as springs powered by the internal heat of the planet. SNC meteorites appear to have formed at high temperatures, evidence that Mars has a hot interior.

(Periodical References and Additional Reading: *New York Times* 3-24-92, p C7; *EOS* 4-15-94, p 538; *Science* 7-1-94, p 86; *Science* 8-5-94, p 744; *Science* 12-16-94, p 1799; *Astronomy* 6-95, p 20)

COMET HITS JUPITER

Eugene and Carolyn Shoemaker and David H. Levy had already discovered a number of comets, but Comet Shoemaker-Levy 9 (1993e), first sighted by Carolyn on May 22, 1993, was the one that would make their names household words in much of the world. The reason: as astronomer Brian G. Marsden of the Smithsonian Astrophysical Observatory in Cambridge, Massachusetts, had predicted in May 1993, Comet Shoemaker-Levy struck the planet Jupiter. Marsden also calculated that in 1992 the comet had passed within 10,000 mi (16,000 km) of Jupiter, where the force of the giant planet's gravity had torn the comet into the 21 pieces that were observed a year later. By 1993 all 21 were in orbit around Jupiter on a collision course with the planet. The impact would be in July, 1994, when the "string of pearls" would have become 186 million mi (300 million km) long.

Each day for nearly a week from July 16-22, 1994, the comet fragments pounded the giant planet, striking at a speed of 40 mi (60 km) per second, or 130,000 mph (210,000 kph). The fragments, now broken into 23 pieces, had become known by the letters of the alphabet in the order of their journey around the Sun. Fragment A struck a little after 4 p.m. EDT on July 16.

Seen from Outer Space

Although the comet fragments came from the side away from Earth and therefore struck the far side of the planet, they were observed by the *Galileo* space craft, which was coasting some 150 million mi (240 million km) beyond Jupiter on a roundabout path to the planet. This was less than a third as far away from the big event as Earth was at that time. Other space-based observations included images from the Earth-orbiting *Hubble Space Telescope* and data of various kinds from the solar satellite *Ulysses* and the Earth satellites *International Ultraviolet Explorer* and *Extreme Ultraviolet Explorer*.

Because of an antenna problem, *Galileo* could not observe, store, and transmit information on all of the impacts. Among the observations reported to Earth were the following:

• *Galileo* observed fragments B, G, H, L, and Q with an instrument called a photopolarimeter-radiometer, which measures brightness of light. The brightest of these fragments, G, H, and L, produced flashes that increased the brightness of Jupiter by about 3-5% for as long as 15 seconds.

• The G and R impacts were also measured by a near-infrared mapping spectrometer and by an ultraviolet spectrometer.

• A charge-coupled device camera took images of the K and N impacts. Each impact had a pattern of about 5 seconds of an initial bright flash that then dimmed slowly for another 10 seconds, at which point it brightened again and then faded, disappearing finally from the camera image after about 30 seconds. Data from the impacts took about six months to transmit because of *Galileo*'s faulty antenna. On December 7, 1995, *Galileo*'s atmospheric probe will sample Jupiter's atmosphere directly, which may help clarify some of the mysteries that remain, of which there are many.

Surprise, Surprise

When the first of Shoemaker-Levy's 23 fragments struck the far side of Jupiter, most knowledgeable observers expected to see nothing from Earth; some scientists doubted that there would be any observable effect from the whole encounter. Perhaps small white clouds might appear. Thus, the large dark region that appeared on Jupiter's horizon a short time after the impact was somewhat of a surprise, although media hype before the event had encouraged the public to expect spectacular results. This time the media were right and the scientists were wrong. The giant blotches from each event swung into view one after another, clearly visible from Earth in good telescopes. Some of the impact effects produced disturbances as large in diameter as the entire Earth, more prominent because of their color than even the Great Red Spot. The larger spots were still visible as much as seven weeks after the impact.

Among the other unexpected results:

• The blotches remained dark. If the comet fragments entered the atmosphere, the fireballs produced ought to have condensed into white clouds. But no white was observed.

- The blotches kept their shape, despite well-known fierce winds on Jupiter.
- There were no reactions observable in the Jovian atmosphere as a whole, such as might have been caused by giant shock waves. Astronomers expected to find both immediate wave patterns like those surrounding a rock tossed into a still pool as well as later effects called inertia-gravity waves. The ripples and waves never came.
- Strange dark rings faintly appeared from time to time around the smudges caused by the impacts. So far, no one has any idea what might have caused the rings, which appear in some images and not in others.
- Dust from the impact was expected to collect and hold electrons that are normally present in Jupiter's equivalent of Earth's Van Allen belts; instead, the dust produced additional electrons that spiraled around Jupiter 's magnetic field and increased radio emissions—the increase in the amplitude at high radio frequencies tipped off Earthbound astronomers to the extra electrons.
- The Hubble measurements showed that all fragments, large or small, produced plumes that rose to the same height of about 2,000 mi (3,300 km). Most of the flashes, except for those of the smallest fragments, were also about the same size. Despite this, the black spots varied in size with the different impacts.

The pattern of unexpected results clearly points to one major interpretation. The fragments all exploded high in the atmosphere with very little penetration. There were no white clouds from atmospheric fireballs because not even the largest fragment traveled far enough into the atmosphere to make such a fireball. There were no shock waves for the same reason. Furthermore, the implication of early and high explosions is that the fragments were loosely connected ices. Some astronomers had predicted that because comet Shoemaker-Levy 9 had broken into so many pieces as a result of tidal forces, it could not be much more than an orbiting pile of rubble. Although this description of the comet may have been correct, the inference that these astronomers made from this was that there would be little or no observable effect from the impact, which was far from true.

Galileo took infrared spectrograms that could be used to determine the temperature of the fireballs caused by the impacts, which reached as high as 13,000°F (7,200°C) for the G impact, higher than the temperature of the surface of the Sun. The measurements could also be used to determine the height of the top of the fireball. This latter result comes from measuring the amount of absorption by methane and molecular hydrogen of light from the fireball, which indicates the depth of the two gases above the source of the radiation. Since there was little radiation absorbed at the characteristic wavelengths of the two gases, the amount of the gases above the fireball was small.

The infrared spectra suggest that the fragments were smaller than expected, with the largest less than 0.3 mi (0.5 km) in diameter, but everyone interprets the data differently. Estimates of size for the largest fragments go to as high as 1.9 mi (3 km) in diameter, but opinions have swung toward smaller sizes as more evidence has surfaced.

Some astronomers questioned whether Shoemaker-Levy 9 was truly a comet; it could have been an asteroid. More water would have been expected from a comet than was actually observed. An asteroid would have been less likely to have broken into so many pieces. Although it probably will never be known for sure, the dust that streamed from the objects would tend to imply that they were parts of a comet after all.

New Knowledge about the Planet

Aside from information about the comet and crash itself, what did the event reveal about the planet? Not as much as one might hope, due largely to the high altitude of the explosions. Drifting debris did reveal new facets to the wind patterns on Jupiter, including better estimates of wind speed on high.

(Periodical References and Additional Reading: *Science News* 7-9-94, p 26; *New York Times* 7-17-94, p I-12; *Science News* 7-23-94, p 55; *New York Times* 7-26-94, p C1; *Nature* 7-28-94, p 245; *Science* 7-29-94, p 6-1; *Science News* 7-30-94, p 68; *Science News* 8-27-94, p 133; *Science News* 9-24-94, p 196; *Science* 10-7-94, p 31; *Science News* 10-8-94, p 229; *Astronomy* 11-94, p 35; *Science* 11-11-94, p 975; *Planetary Report* 11/12-94, p 8; *Science News* 12-17-94, p 412; *Discover* 1-95, p 28; *Astronomy* 4-95, p 30; *Science News* 3-18-95, p 167; *Astronomy* 5-95, p 49)

GAS GIANTS AND FELLOW TRAVELERS

The planets beyond Mars are not anything like the planets encountered by the starship *Enterprise* in its various treks. Planets Jupiter, Saturn, Uranus, and Neptune are known as gas giants because they are mostly hydrogen and other gases. Because they are cold from being far from the Sun and because the large planets have high gravitational forces, most of the gases are liquid or even slushy solids. The gas giants have thick atmospheres that probably gradually shade into the slush, so that there is no actual surface to the planets. Each planet may also sport some heavier elements forming solid core under all that slush. The gas giants are so large that the cores are about the same size and probably about the same composition as the whole planet Earth. Pluto is not a gas giant, and might be more like a tiny Star Trek planet, but it is so far away that no one can be sure. It may just be large comet, not a planet at all.

Jupiter underwent a major event during the early 1990s, so it deserves an article of its own (*see* "Comet Hits Jupiter," p 165).

Weather on Saturn

In some ways, climate and weather on Saturn are like climate and weather on Earth— there are seasons, winds, and storms, for instance. There are seasons because Saturn's axis tilts 26.7°, just as seasons on Earth are caused by an obliquity of 23.27°. When it is late summer for Saturn's northern hemisphere, there are apt to be big storms. But because Saturn's year is 29.5 Earth-years long, a Saturn summer from our point of view comes around infrequently and lasts more than 7 years at a time. Big, late summer storms in Saturn's northern hemisphere were observed during the past four Saturnian summers, in 1876, 1903, 1933, and 1960, as well as during 1990 and 1994.

In other ways, the climate and weather on Saturn are very different from that on Earth. Winds whip around Saturn from east to west at speeds averaging about 1,000 mph (1,600 kph), faster than the planet rotates. No one knows what causes the winds to blow in this direction or to be so very fast. Like Jupiter and Neptune, which also have high-speed winds, Saturn has an internal heat source that no doubt helps power the winds. Despite there being a Saturnian summer, temperatures during it are not noticeably different from temperatures during the rest of Saturn's year. Saturn is far from the Sun, and most of the atmospheric heat comes from inside the planet.

Also, when storms form on a giant gas planet, they are much larger and more persistent than storms on Earth. Our largest late-summer storms, the hurricanes, may cover regions a few hundred miles in diameter and last for three weeks at most. The Great Red Spot on Jupiter is 20,000 mi (32,000 km) long and 8,000 mi (13,000 km) wide and has persisted for hundreds of years.

On September 24, 1990, two amateur astronomers, Alberto Montalvo in Los Angeles and Stuart Wilber at Las Cruces, New Mexico, observed the start of Saturn's summer storm of 1990, quickly dubbed (in imitation of Jupiter's storm) the Great White Spot. Wilber was the first to report his observation to the Smithsonian Astrophysical Observatory in Cambridge, Massachusetts, so the storm is sometimes called Wilber's Spot. It was the biggest such storm since 1933, about 13,000 mi (21,000 km) long and covering about a sixth of the visible surface of Saturn. For unknown reasons, much larger storms occur every other summer, or about 60 years apart.

Like most very large storms on Saturn, the Great White Spot appeared just north of the equator about three years into summer. Within weeks, Saturn's swift winds had spread it from a spot to a band, stretching all the way around the planet by October 26, about a month after it was first observed. Shortly after that, Saturn moved behind the Sun, so observations from Earth ceased. The *Hubble Space Telescope* obtained some detailed photographs of the spot before Saturn was lost from view.

After Saturn's return to our field of vision, a larger arrowhead-shaped storm arose near the equator in September 1994. It was about the length of Earth's diameter when first observed. Like the Great White Spot, the arrowhead storm also appeared white against the reddish-brown of Saturn itself.

Like hurricanes on Earth, Saturnian storms are powered by heat from the Sun, little though that heat might be. The whiteness of the Spot is probably a result of crystals of new ammonia that form when warm gas in the storm cools (just as water vapor in a hurricane cools to liquid water). Older ammonia crystals are yellow, contributing to the predominant color of Saturn. Ground-based observations by Agustin Sanches-Lavega and coworkers from the University of Pais Vasco in Bilbao, Spain, support this explanation, as do observations from the Hubble Space Telescope.

According to theory and computer simulations, even larger spots should form 30° south of Saturn's equator, but the view from Earth of any such spots is blocked by the rings. The only way to check this prediction would entail using a space probe to observe from a different angle.

Saturn's Satellites and Rings

In 1994 Peter Smith of the University of Arizona and coworkers used the *Hubble Space Telescope*'s Wide-Field/Planetary Camera 2 to make 50 images of Saturn's largest moon, Titan, during one 16-day orbit of the planet. Processing of the images, which were taken in the near infrared to keep Titan's smoggy atmosphere as transparent as possible, showed that Titan's surface includes at least one very bright feature, which is about the size of Australia, and one large dark region. The images also confirmed that Titan always keeps one face toward Saturn, just as our own Moon keeps one face toward Earth.

Titan is so far away that it is impossible to be certain what the two large features are. Speculation centered on the bright feature. If, as is thought, much of the surface of Titan is covered with a dark goo similar to asphalt or crude oil, then the bright region could be an elevated "continent" that rises out of the goo. The *Cassini* space probe is scheduled to reach the vicinity of Saturn in 2004; at that time, a probe to Titan may answer many of the unanswered questions about the giant satellite, which is larger than the planets Pluto or Mercury and nearly as large as Mars.

The rings of Saturn are its most observable feature from Earth. In recent years, as a result primarily of better ground-based and airplane-based observations, astronomers have learned that all of the gas giants have some sort of ring system, but Saturn remains the winner for extent, complexity, and beauty of its rings.

The rings of Saturn are affected by a number of features of the planet and its environment. The planet, a fast-spinning semisolid for the most part, has a predictable mid-section bulge. Also predictable from Newtonian theory is the effect of the bulge on objects orbiting the planet, whether they be giant Titan or tiny ring particles; it causes their elliptical orbits to precess. Saturn's magnetic field does not affect large satellites, but many of the tiny particles in the rings become charged, which causes their orbits to regress; to regress is to move in the opposite direction from the precession caused by the bulge. For the E ring, which was studied in detail by Douglas P. Hamilton of the Max Planck Institut für Kernphysik in Heidelberg, Germany, and Joseph A. Burns of Cornell University in Ithaca, New York, in 1994, the amount of precession and regression almost exactly cancel, keeping the ring in one place. Very small particles in the ring, however, can be pushed around by the solar wind (*see* "Stories on the Solar Wind," p 146), which is still powerful at Saturn. The study of the E ring suggests that it is well named, being caused by particles from the satellite Enceladus, which are forced into forming the ring by the factors just mentioned.

Enceladus is actually *in* the E ring, and earlier calculations showed that it should have swept up all the particles in the ring after a brief 20 years. This contradiction between theory and observation prompted Hamilton and Burns to look for a new theory. They concluded that particles from the E ring that struck Enceladus knocked off more particles in a kind of chain reaction. The surface of Enceladus appears to be a layer of such particles. Some have suggested that Enceladus is just particles, held together in a sphere by gravity. But this process of particles being knocked from Enceladus by itself would produce, as chain reactions do, an exponential growth in particles in the ring, provided there was not some mechanism that would cause particles to go away. Hamilton and Burns, writing in *Science*, propose that particles striking each other grind each other down into dust, which can then be swept into nearby rings by the solar wind. A computer stimulation based on this theory did produce the computer equivalent of the observed E ring.

Neptune and its Satellites

Although Neptune is often thought of as the twin of Uranus, rather as Venus has been considered a twin planet with Earth, the truth is far more complicated. Among the characteristics that separate Neptune from its neighbor is weather. Neptune has the swiftest winds of any planet—about 900 mi (1,400 km) per hour. But superfast winds make Neptune more like Jupiter or Saturn than like its supposed "twin," Uranus, which does not have winds in that category.

Swift winds bring great storms. In 1989 when *Voyager* took a close look at Neptune, it found a spot on the bluish uniformity of the planet. In imitation of the well-known Great Red Spot of Jupiter and the previously discovered Great White Spot of Saturn, Neptune's feature was christened the Great Dark Spot. Furthermore, Neptune also had a smaller spot, simply called Dark Spot 2. It was thought that the Great Dark Spot and perhaps even Dark Spot 2 would be semipermanent features of Neptune, just as Jupiter's Great Red Spot has lingered for at least 300 years. But when the *Hubble Space Telescope* looked toward Neptune in 1994 it found that the Great Dark Spot was already gone, replaced by a few streaks of cloud.

(Periodical References and Additional Reading: *Science* 4-22-94, p 550; *Science* 7-22-94, p 470; *Science* 9-2-94, p 1360; *New York Times* 9-27-94, p C11; *New York Times* 11-8-94, p C10; *Astronomy* 2-95, p 44; *Astronomy* 4-95, p 32; *Astronomy* 5-95, pp 24)

Comet, Meteoroid, Asteroid UPDATE

The line separating comets and asteroids or asteroids and meteoroids is quite thin (*see* below); thus, these small bodies, which may also be called planetesimals in another context, are discussed here collectively. Our knowledge of what these bodies are like, where they come from, and what effects they have had or may have on Earth has increased greatly in the past half-century. The 1990s have been a time of greatly increased interest in these issues and progress in such understanding.

Far-Out Comets

During the late 1980s and early 1990s, astronomers identified an enigmatic body named Chiron as the largest known comet. Since then it has remained an object of considerable interest.

In November 1994 astronomers Karen Meech of the Institute for Astronomy in Hawaii, Marc Buie of Lowell Observatory in Arizona, and Michael Belton of the National Optical Astronomy Observatories in Arizona reported on their analysis of images of comet 2060 Chiron taken by the Planetary Camera on the *Hubble Space Telescope* in 1993. They identified the comet's still newly forming coma as a cloud of dust that was pushed away from the comet by gases that are evaporating in the Sun's heat as Chiron comes closer. Up to about 750 mi (1,200 km) from Chiron, the dust is thicker because it is still in orbit about the giant comet. This halo of thicker dust is surrounded by a less-dense halo of dust that is escaping from the comet, although Chiron is not close enough to the Sun to have formed a visible tail as yet.

Estimates based on a guess of what the size of the dust particles might be suggest that Chiron has a density that is less than half the density of water. Thus, it is mostly frozen light gases, not rock and metal like an asteroid.

Until 1994, one mystery concerning comets at long distances from the Sun was that some of them, including the famous Comet Halley, showed activity similar to that caused by the Sun's heat when a comet approaches the Sun. Although the Sun can heat dark objects at great distances, these comets develop comas (clouds of gas and dust) at distances much too far away for hot water, which causes coma close to the Sun, to occur. In 1994, Matthew Senay and David Jewitt of the University of Hawaii in Manoa demonstrated that, for some comets at least, the far-away comas are caused by melting or boiling carbon monoxide, with perhaps some other gases with similar melting and boiling points as well, such as nitrogen.

COMET OR ASTEROID?

Sometimes it is not easy to tell whether a body in space is an asteroid or a comet. It was once thought that the asteroids were all in orbit between Mars and Jupiter, so any largish object with an orbit beyond Jupiter might be expected to be a comet. Now we know that asteroids can be found in various parts of the solar system, although more are found between Mars and Jupiter; so if an object is found far beyond Jupiter, further evidence is needed to identify it.

In 1977 Charles T. Kowal of Lowell Observatory discovered 2060 Chiron, a large (but not planet-sized) body orbiting the Sun between Saturn and Neptune. This was the first such body ever observed beyond the orbit of Saturn. From the beginning no one was quite sure what it was—it was an odd place for an asteroid to be, but not an impossible place; and it was too big to be a comet. Mark Sykes of the University of Arizona and Russell Walker of Jameieson Science and Engineering calculated Chiron's size in 1991, which they determined to be 230 mi (372 km) in diameter—vastly bigger than any other comet known. Halley's Comet, for example, is only about 10 mi (16 km) long.

If Chiron was an asteroid, it was the most distant from the Sun then known; its orbit of 50.7 years would take it outside the orbit of Uranus at its farthest point. When discovered, however, it was heading for its perihelion, the closest point to the Sun, 2,053 million mi (1,257 million km) away, just inside the orbit of Saturn. It reached perihelion in 1996.

On February 20, 1988, Chiron was found to be getting brighter. Comets brighten as they approach the Sun, before they look fuzzy and develop a tail. Checking earlier records showed that Chiron had started getting brighter—much brighter than expected from just being closer to the Sun—as early as mid-1987. Thus, Chiron is apparently a very large comet and not an asteroid. Following the rule for comet naming, it is now officially designated Comet 2060 Chiron.

What is the farthest away an asteroid has ever been found? Since September of 1992 astronomers have been discovering objects beyond the orbits of Neptune and Pluto. In October 1994, Oliver Hinaut of the European Southern Observatory discovered asteroid or comet 1994 TG_2 about 42 AU (astronomical units—1 AU is the average distance of Earth from the Sun) from the Sun. This brings to 22 the number of asteroid-comets found by ground-based telescopes beyond the orbit of Neptune, in what might be called Hades, since it is the domain of Pluto. But the record distance is even farther from the Sun than 1994 TG_2 since two of these bodies, 1994 ES_2 and 1994 EV_3, travel out to as much as 45 AU, 7% farther from the Sun than 1994 TG_2. In 1995 the *Hubble Space Telescope* in one day found what appear to be 30 more smaller bodies in Hades.

Moons of the Asteroids

When Galileo turned his new telescope to Saturn in 1610, the image was not distinct enough for him to recognize the now famous rings, but he could see that the image had an odd shape. The same situation confronted Eleanor Helin in 1989 when she and coworkers looked at the asteroid she had discovered (1989 PB, subsequently named 4769 Castalia). Although the asteroid came within 2.5 million mi (4.05 million km) from Earth on August 25, 1989, the image obtained using radar with giant radio telescopes was too fuzzy to be sure what it was; but it looked more like a peanut than the spheroid expected of an asteroid. Three years later, in December 1992, the radio telescopes got a better shot at asteroid 4179 Toutatis which was even closer—2.15 million mi (3.5 million km)—and this time it seemed more likely that the peanut shape was actually two lumps traveling together. Could most asteroids be double?

In 1993, the space probe *Galileo* got close to its second asteroid. Although 951 Gaspra, an asteroid imaged by *Galileo* on October 29, 1991, had looked threatening, it was just a single misshapen piece of rock. But the new quarry of August 28, 1993, asteroid 243 Ida, was not alone in space. It had a tiny satellite orbiting at a distance of about 60 mi (100 km). Although the satellite, announced on March 3, 1994, and since named Dactyl, was not part of a double asteroid system—which, in retrospect, astronomers now think is the situation with Castalia and Toutatis—it was traveling along with the much larger Ida. Ida is like a 34 mi (55 km) long potato, while Dactyl is a sphere about 0.9 mi (1.4 km) in diameter. Dactyl is the name of a demon said to roam Mount Ida, near ancient Troy; it has nothing to do with the combining form for *finger* or the verse foot named after finger joints.

Both contact binaries, in which two fragments touch each other or are almost touching, and orbiting minisystems, like Ida and Dactyl, probably formed in the same way—as fragments from a collision between two larger asteroids. Although Dactyl is a slightly different color from Ida, it is still thought that they are different parts of a broken parent body. Some evidence suggests that Dactyl, which appears to be rounder than other asteroids, may be a collection itself. It could be a pile of impact gravel and rocks that gravity has formed into a sphere in space, not a solid body at all. Another theory is that Dactyl is covered with space gravel that it has swept up as it orbits Ida.

In 1994, R. Scott Hudson of Washington State University in Pullman re-analyzed the radio signals from Castalia with a technique that revealed details about the double shape. He found two lobes, each about 0.6 mi (1 km) in diameter, with a crevice between them that is 300-500 ft (100-150 m) deep.

Many impact craters on Earth and on its Moon appear to have formed in pairs—about 5-10% of all of them, depending on what exactly is counted as a pair. Thus, for near-Earth asteroids at least, as many as 1 out of 10 may be double. There is another possible mechanism for forming double craters, however. Tidal forces could break up a single asteroid as it approached its end. This is what happened to Comet Shoemaker-Levy before it struck Jupiter, although it was broken into a couple dozen pieces when it passed close, and then the pieces struck the giant planet on the next orbit. On August 13, 1994, shortly after the Shoemaker-Levy impacts, amateur astronomer Don Machholz discovered Comet Machholz 2, which was soon seen to be in at least five pieces, demonstrating that broken comets, like broken asteroids, are not so rare either. Indeed, Jun Chen and David Jewitt reported in the April 1994 *Icarus* on their study of 49 comets that had been imaged with charge-coupled devices—light recorders that are much more sensitive than photographic film. Chen and Jewitt found that 3 of the 49 had been caught in the act of breaking up, a much higher percentage (6% of the total studied) than anyone expected. Most comets, however, could not break into large pieces very often, or they would not last for very long. Instead, most fragmentation is thought to consist of smaller pieces breaking off. The small pieces could then become the meteor showers associated with specific comets and other small bodies in space that have been observed.

More about Ida

Asteroid 243 Ida turned out to be more interesting by far than astronomers expected. It had been chosen for the *Galileo* rendezvous mainly because it was not far out of the way for the Jupiter-bound space probe—thus, it was chosen nearly at random. But astronomers should always expect the unexpected when they have a chance of observing any heavenly body from a new vantage point.

Ida was already known to be twice as large as Gaspra, the other asteroid imaged by *Galileo*, and to spin twice as fast. But some of the conclusions reached from Earth observation failed to hold up with a closer look.

Ida was thought to belong to the same asteroid family as Gaspra; thus, not much new information was anticipated. But the first surprising discovery was that Ida is not part of the Koronis family, all of which are thought to have originated in the breakup of the same astronomical body. But since Ida was found to be perhaps as much as 10 times as old as Gaspra, the family idea was out the window from the start. Indeed, astronomers now think Ida may be as much as 2 billion years old. The evidence for this age is Ida's cratered and pockmarked appearance, more like the Moon than Gaspra. Dactyl also has at least a dozen craters, which is remarkable for so small a body. Some astronomers argue that Ida is even more primitive than 2 million years in age—that the asteroid is twice as old, made from the first kind of matter to condense when the solar system formed.

An important part of the evidence for Ida is its density. While *Galileo* was not expected to be able to provide information on Ida's density, the unexpected discov-

Background

Throughout the solar system there are small bodies that orbit the Sun on their own. If they are small enough, they are called **meteoroids**. If they impact on Earth, the recovered bodies of stone, iron, or carbon are called **meteorites**.

For a long time scientists did not believe meteorites existed, although a fall of about 200 in L'Aigle, France, in 1803 convinced most of them. Then scientists came to believe that meteorites are remnants of smashed planets or, later, part of the cloud of dust and gas that formed the planets. In 1983 researchers learned that at least some meteorites come from the Moon or from Mars. Five years later Edward Anders of the University of Chicago proposed that specks of dust embedded in some meteorites come from the stars.

Bigger bodies orbiting the Sun by themselves are called **asteroids** or **comets** (or planets, if they are sufficiently big). About 5,000 asteroids have had their precise orbits determined; this allows them to be designated with a number signifying their order of entry into this group, such as 1 Ceres (discovered first) or 3200 Phaëthon, although the number is usually dropped after the first reference. Some use only the number in parentheses to refer to the asteroid, using (1) for Ceres or (4789) for Sprattia, one of the most recent asteroids to have been assigned a number.

Another 13,000 asteroids have been found, but their orbits are yet to be calculated. They get temporary names consisting of the year found followed by a letter code that indicates the date they were first observed, such as 1991 BA or 1989 PB. Theory suggests that there are about a million more asteroids with a diameter greater than 0.5 mi (1 km) that have yet to be discovered.

Comets are recognized today as dusty frozen gases. Aristotle believed that comets were made from fire and air, atmospheric phenomena like lightning and rainbows. In 1577, Tycho Brahe showed that a great comet seen that year was at least four times farther away than the moon, which ruled out Aristotle's theory; then no one had any idea what comets are. Edmond Halley showed that the behavior of comets is consistent with Newton's theory of gravity, and for the next hundred years comets were thought to be large bodies in space, similar to the planets or at least to the moons of planets. But in 1770 Anders Jean Lexell showed that a comet he was studying had a very small mass, so comets were really not much like planets or moons. Finally, in 1949 and 1950 astronomers developed a theory that explained almost everything known about comets. They described comets as mixtures of dust and frozen gases in what has come to be called the "dirty snowball" theory of comet composition. With some minor modifications, this theory has held up as the main way to interpret comets even today.

Asteroids are mostly rocky or metallic. Seen through a telescope, asteroids are pointlike, like stars (hence their name), while comets are fuzzy. If an asteroid or a comet strikes Earth, it is sometimes called a **bolide.** Some astronomers do not make distinctions between medium-sized bodies in space. They call both comets and asteroids either **planetoids** or **planetesimals.**

A meteoroid that burns up in Earth's atmosphere is a **meteor.** Until recently, it was assumed that all meteors are made of rock or iron, like the meteoroids that do not burn up and reach the surface to become meteorites; there is good reason, however, to suspect that some if not most meteors are really little comets.

ery of Dactyl did just that. Astronomers have long known how to compute densities from orbital motion. In this case, Ida's density can be determined with great certainty to be between 2 and 3 grams per cubic centimeter. This is the same as the average density of rocks that make up Earth's crust, such as granite or limestone. Thus, Ida is more like a chunk of rock than a small planet, which typically would have a metal core.

Another unexpected result was that Ida has its own magnetic field; but this should not have been so surprising, since Gaspra also has a magnetic field. But the main surprise, of course, was Dactyl, Ida's own private satellite.

Meteor Showers

Each year on certain dates meteors begin to appear in the sky. These meteors seem to come from particular constellations, but the constellation actually has nothing to do with the shower of meteors; it merely appears in the portion of the sky from which the meteors come. The meteors are caused when Earth passes through the debris left by a comet that crosses Earth's orbit. Each year Earth passes through the debris on or about the same day, since a specific day corresponds to a given part of the orbit. The line of meteors is especially dense when the stream of debris is young, and it gets wider and less dense as the stream ages—rather like the contrail of a jet airplane. A particularly dense shower is called a storm. At the height of a meteor storm from the Leonid group in 1966, about 72,000 meteors an hour could be seen, but the storm only lasted about an hour.

The connection between comets and meteor showers has been observed since 1863, when it was first postulated by American astronomer Hubert Newton. Three years later further observations by Giovanni Schiaparelli clinched the connection. If a comet crosses Earth's orbit twice, it can leave behind two meteor showers that come at different times of the year. In some cases, the associated comet has never been observed.

Peak Date	Duration	Constellation	Meteor Shower	Associated Comet
January 4	2 days	Boötes, Draco, and Hercules	Quarantids	none
April 22	5 days	Lyra	Lyrids	Thatcher 1861 I
May 5	6 days	Aquarius	Eta Aquarids	Halley
July 28	14 days	Aquarius	Delta Aquarids	none
August 12	9 days	Perseus	Perseids	Swift-Tuttle 1862 III
October 21	4 days	Orion	Orionids	Halley
November 8	30 days	Taurus	Taurids	Enke 1786 I
November 17–18		Leo	Leonids	Tempel-Tuttle 1866 I
December 13	5 days	Gemini	Geminids	none

(Periodical References and Additional Reading: *New York Times* 9-16-92, p A22; *Nature* 8-12-93, p 574; *New York Times* 3-24-94, p A20; *Astronomy* 4-94, p 29; *Science* 4-1-94, p 35; *Science* 5-13-94, p 907; *Astronomy,* 6-94, p 24; *Science* 6-17-94, p 1667; *Astronomy* 7-94, pp 18 & 21; *Science* 7-22-94, p 470; *Science News* 8-6-94, p 93; *Science* 9-9-94, p 1543; *Science* 9-16-94, p 1658; *Science* 11-25-94, p 1322; *Astronomy* 12-94, p 26; *Astronomy* 1-95, pp 24 & 30; *Discover* 1-95, p 32; *Astronomy* 3-95, p 28; *Science News* 3-25-95, p 191; *Astronomy* 4-95, p 30)

STARS AND OTHER INTERESTING BODIES

NEWLY FOUND PLANETS

The ancient followers of Pythagoras had numerological reasons for believing that there must be exactly ten planets. The Pythagorean astronomer Philolaus [Greek: fl. fifth century B.C.] could only count nine, however. In addition to Mercury, Venus, Earth, Mars, Jupiter, and Saturn, the nine included the Sun and the Moon, all revolving about a Central Fire that was counted as number nine. So Philolaus tossed in a tenth planet that revolved about the Central Fire with the same period as Earth, but which always stayed on the far side of the fire where it could not be observed. (A period is the length of time in one orbit. The period of Earth is one year.) More recently the recognition of Uranus and the discoveries of Neptune and Pluto, have once again produced a set of nine known planets.

For reasons that are probably more psychological than numerological, scientists have also looked for a tenth planet, both in and out of the solar system. After many false discoveries over the past century or so (see box), astronomers in 1992 finally found number ten—and numbers eleven, twelve, and possibly more than that. These were confirmed more convincingly in 1994.

The Pulsar Planets

The first newly found planets, discovered by Alexander Wolszczan of Penn State University and Dale Frail of the National Radio Astronomy Observatory, orbit a millisecond pulsar about 1,500 light years from Earth. The pulsar can be observed in the constellation Virgo, and has the code designation PSR B1257+5 (see also "Pulsars," p 195).

Pulsars have properties that make it unusually easy to detect planets in orbit about them. Each pulsar produces pulses of electromagnetic radiation that are an exact reflection of the rapid rotation of the pulsar in space. A millisecond pulsar is so called because its rotation period is so rapid that it is measured in thousands of a second. In the absence of an orbiting planet, the pulses slow down gradually, but do not change in periodicity.

Although an orbiting planet would have little effect on the rotation period, the presence of one or more planets produces a noticeable change to the pulse. Here's why: the pulsar and the planet each orbit the system's center of mass. The smaller does not orbit the larger as people commonly suppose. From the point of view of an observer not gravitationally bound to the system, the more massive pulsar appears to approach the observer when the less massive planet is on the near side, and to recede when the less massive is behind. Because the planet is invisible from Earth, the pulsar itself seems to move back and forth for no reason. The amount of motion is relatively slight, but even the tiniest changes in the regular pulses can be greatly magnified by various wave-interference methods.

Slight changes in wavelength that result from an emitting body moving away from or toward an observer can be detected by optical interferometry. These changes are called Doppler shifts, and are similar to the changes in pitch noticeable when a siren on a police car approaches a listener and then recedes. Changes in the distance that the waves travel between the pulsar and the observer are reflected in pulses from further away arriving later than those from nearer positions. When every other factor is accounted for, especially any motion by the observer, changes in distance to the pulsar appear as changes in frequency of the pulse.

Wolszczan began using the 1,000 ft (305 m) fixed-dish radio telescope at Arecibo in Puerto Rico to study pulsars in 1990, when he was able to obtain observing time because regular operations were shut down for instrument repairs. PSR B1257 + 5 in Virgo, a rapidly rotating neutron star with a diameter of 12 mi (20 km) and a mass about 1.4 times that of the Sun, was found to have periodic variations in the rate at which it pulsed, which was approximately 160 times per second. Calculations showed that the variations in frequency could be caused by at least two planets orbiting the pulsar, which Wolszczan announced early in 1992. Although no one could suggest any other explanation for the data, there was some suspicion that the conclusion might be false, because a previously identified "pulsar planet" had been found to be spurious (see below). In April 1994, however, Wolszczan was able to calculate minute variations caused by the planets' affecting each other, an effect unlikely to be an artifact of any other process. For most astronomers, this last piece of data clinched the case in favor of existence of the planets.

The specific changes in the pulses can best be explained by two planets orbiting PSR B1257 + 5, with fairly good evidence for a third planet. Earlier hints of a fourth planet were not confirmed by the data on planetary interactions. The innermost planet is small, about the size of the Moon, while each of the other two is larger than Earth (seemingly three times Earth's mass), but not so large as Jupiter. The small planet maintains an orbit very close to the pulsar, and the two large planets are also closer to the pulsar than Earth is to the Sun (the distance designated as one astronomical unit, or A.U.). The first planet is about a fifth of an A.U. away from the pulsar, with a period of 25.3 days; planet number two is about a third an A.U. away from PSR B1257 + with a period of 66.6 days; and suspected planet number three is 0.5 A.U. from the pulsar with a period of 98.2 days.

It is easy to speculate that planets so close to a pulsar would have no atmosphere, since the solar wind from a pulsar is thought to be a strong gale. Beyond that, it is difficult to know much more about these planets, other than their masses and orbits.

Disks of Dust

The Sun did not always have planets around it. According to the most widely accepted theories of the last 200 years, the planets formed out of a cloud of dust that once surrounded the star we now call the Sun. In late 1983 and early 1984, the *Infrared Astronomical Satellite* (*IRAS*) detected just such clouds around several nearby stars— Vega, Formalhaut, HL Tau 500, and R Mon. *IRAS* also located about 70 other possible candidates for these dust clouds. Eventually, more than a hundred such dust clouds surrounding stars were located by various means.

At the end of 1984, Bradford A. Smith of the Jet Propulsion Laboratory and Richard J. Terrile of the University of Arizona made an even more spectacular discovery while following up on the *IRAS* candidates. They were able to photograph a disk of gas and dust surrounding Beta Pictoris, a bright star about twice the mass of the Sun that is visible from the Southern Hemisphere. Beta Pictoris is 51 light years away—close, as star distances go. The disk extends some 100 billion mi (160 billion km) from the star, about 30 times the average distance between the Sun and Pluto. The photograph Smith and Terrile produced appears to be dramatic evidence of planet formation in a very early state. In 1992, astronomers reported that the chemicals found in the disk around Beta Pictoris are essentially the same as those thought to have been in a disk around the Sun early in the formation of the solar system.

Shortly after it began operation, the *Hubble Space Telescope* (*HST*) was able to observe Beta Pictoris using its Goddard High Resolution Spectrograph. Observing in the ultraviolet region, where the flaw in the *HST* mirror was not much of a problem, the spectrograph found what appear to be large clumps of matter in the disk, clumps that spiral in toward the star. On May 17, 1991, Al Boggess of Goddard Space Flight Center announced the discovery at a press conference at the Space Telescope Science Institute in Baltimore, Maryland. He would not say what it meant, but the phenomenon engendered speculation that *HST* was observing something very like the processes hypothesized to occur in the early stages of planet formation.

In 1993, Pierre-Olivier Lagage and E. Pantin of the Service d'Astrophysique at Saclay in Gif-sur-Yvette, France, made a careful study of Beta Pictoris using infrared radiation at a frequency where the dust is most visible. They theorized that any nascent planets would have swept up parts of the disk, decreasing the quantity of dust in the regions where planets were forming. Their infrared photograph, using a new instrument at the European Southern Observatory in La Silla, Chile, shows two effects that suggest the presence of planets. One is the aforementioned depletion of dust in the inner 40 A.U. of the 1,000 A.U. diameter disk. The other is a noticeable lack of symmetry for the entire disk, which could be caused by planets with elliptical orbits, as in our solar system. It is worth noting that the orbit of Pluto is also about 40 A.U. from the Sun.

In addition to Beta Pictoris, the similar stars Vega (Alpha Lyrae), Fomalhaut (Alpha Piscis Austrini), and Epsilon Eridani have disks that can be detected with ground-based telescopes. These are all A-type stars about the same age as the Sun, although somewhat larger. A disk has also been found around HL Tauri and some other similar stars, called T Tauri stars. In these cases, the disk consists of gas rather than dust. HL Tauri, however, is a young star, more like what the Sun must have been like when the solar system formed. In searches by space-based telescopes, hundreds of other stars have been found with disks of dust. In some cases, the disks have observable gaps that suggest that planets are sweeping dust from their paths around the stars.

In 1991, Robert O'Dell and Zheng Wen also used *HST* to look for disks of dust around stars, this time using the flawed Wide Field-Planetary Camera. Because of the poor *HST* optics, however, dusty disks surrounding 15 young stars in the Orion nebula were not sufficiently clear to be certain of their true existence. The stars were thought to be less than 300,000 years old.

When the *HST* optics were corrected, the same region in the Orion nebula was photographed through the Wide Field-Planetary Camera as a test, which it passed with flying colors. The new, sharp photographs were available to O'Dell and Wen, who looked for dust disks. For 56 of the 110 moderately bright stars in the Orion photographs, O'Dell and Wen were able to discern dust disks. The disks are typically about 10 times as large around the star as the orbit of Pluto is about the Sun, which is what would be expected for a disk that is on its way toward collapsing into a group of planetoids that will then form planets. The mass of the outer rims of the disks was calculated to be several times that of Earth. Earlier studies of other young star populations also located dusk disks surrounding about half of the stars.

Condensation of a star and its associated planets from a dust disk may be seen in cartoon form during the opening credits of any episode of "Star Trek: The Next Generation." Even astronomers think this is a fairly accurate depiction.

Planets Around Ordinary Stars

Planets smaller than Jupiter cannot be observed about most ordinary stars—not because the planets are not there (as far as we know they *should* be there) but because such small planets and their effects are below the limits of detection. The light from the Sun, for example, is 100 times brighter than Jupiter, so searching for planets is like looking for a candle placed in front of a searchlight. The *Hubble Space Telescope*, however, has a chance of detecting a large planet in orbit around a normal, sufficiently close star. The method would not involve trying to resolve the two separate sources of light; instead, wobbles observed in the star's motion would imply planets in orbit, just as for the pulsar planets described above.

So far, *HST* has been used to look for Jupiter-sized planets around two of the stars nearest Earth, where any effects would be easier to detect. However, *HST* found no planet-indicating wobbles for Proxima Centauri after a two-year study. The other star, Barnard's Star, has also yielded nothing, but it is too early to rule out eventual discovery of a wobble in the data. Also, Proxima Centauri might well have planets smaller than Jupiter. Geoffrey Marcy of San Francisco State University and Gordon Walker of the University of British Columbia each conducted separate studies of nearby stars using data other than that provided by *HST*. Marcy and Walker independently reached the same conclusion. Of the 30 nearby stars they studied, there was no sign of a wobble caused by a Jupiter-sized planet, although they could not rule out smaller planets.

Finally, in 1995, Michel Mayor and Didier Queloz of Geneva Observatory reported a wobble caused by a planet around the star 51 Pegasi. Their result was quickly confirmed and soon other astronomers began to detect wobbles with Earth-based spectroscopes. The first planets found this way were Jupiter-sized or larger.

Alan Stern calculates that the Keck twins (*see* "Giant Telescopes, Optical and Otherwise," p 130) would have the light-gathering and resolving power to observe planets early in their formation history. According to our best current understanding, a planet would be red hot for about a thousand years after participating in one of the giant impacts that are thought to commonly occur early in the formation of a planetary system. Thus, it would emit more light than simply the reflected light of its star, making the planet observable by a sufficiently powerful telescope. Even Keck I by itself, according to Stern, could see a large planet if it were caught just after the impact.

Others involved in the search have also found the power of the Kecks intriguing. The U.S. National Aeronautics and Space Administration (NASA) is behind a program called Astronomical Studies of Extrasolar Planetary Systems (ASEPS) that, if sufficient funding is granted, will purchase up to a sixth of the Kecks' observing time to examine nearby stars for the presence of planets. Keck I and Keck II can combine the use of interferometry (see "Giant Telescopes, Optical and Otherwise," p 130) to catch wobbles with a precision far beyond any of the earlier studies. ASEPS is also funding adaptive optics experiments (see "Optical Telescope UPDATE," p 140) with Keck I that could produce direct images of a Jupiter-sized planet. Some astronomers would prefer such direct evidence to the indirect evidence of a wobble.

And On Those Planets, Could There Be Life?

One of the many programs killed or cut by the U.S. Congress in its effort to balance the budget was the High Resolution Microwave Survey, the NASA version of an attempt to contact life outside the solar system. The project involved sensitive radios that could listen to 14 million radio channels for signals from other planets. The effort began symbolically on Columbus Day, 1992, with a study by the Arecibo radio telescope that would target 800 to 1,000 nearby stars, beginning with GL615.1A, and a fairly random survey using the Gold stone radio telescope in the Mojave Desert.

Originally intended to be a 10-year search, Congress called it off after a year, during which the two radio telescopes had checked out 10,000 stars of the same general type as the Sun. About 25 of those stars appeared to be sufficiently interesting that they would need further monitoring. The second year was funded by $4.4 million in private donations, somewhat short of the $7.3 million needed that year for full operation. Continuation of the project has been dependent upon private contributions, which have continued to reach about $3 million each year.

Some of the additional funding went to scanning the other half of the universe, looking at the sky from the Southern Hemisphere using the Parkes radio telescope in Australia. Of particular interest were four stars that have unusual microwave signals for which no explanation is currently known. Most astronomers, however, believe that the microwave stars would be found to have some natural mechanism by which they themselves produced the signals.

PLANETS FOUND AND PLANETS LOST

In mathematics, proof of the existence of an object is often easier than proof that something fails to exist. Astronomy, however, is an observational science. Observing something proves that it exists, but failure to observe it could come from many possible causes. Surprisingly, both the prediction of new planets and the proof that others fail to exist usually comes from mathematical calculations.

Planet X   The discovery of Neptune in 1846, based on irregularities in the orbits of Uranus and Saturn, inspired astronomers to use the same method to predict other planets beyond Neptune. Although Pluto is beyond Neptune most of the time (its irregular orbit passes closer to the Sun than Neptune's orbit for about a quarter of a century every few hundred years—a situation that occurred during the 1990s), it is too small to account for observed irregularities in the orbits of Neptune and Uranus. Thus, reasoned astronomers, there must be a tenth planet, which was happily christened Planet X, invoking both the Roman numeral for ten and the algebraic symbol of the unknown. Searches failed to find it, and Clyde Tombaugh, who found Pluto but failed to find Planet X, calculated that it must be less than half as bright as Pluto or he would have seen it. If it is truly less than half as bright as Pluto, Planet X could be massive but dark. However, a 1983 study by the *Infrared Astronomy Satellite* (*IRAS*) covered 99% of the sky without finding a previously unknown warm body. Finally, a 1993 study by Myles Standish of the Jet Propulsion Laboratory showed that the supposed irregularities come from bad calculations based on incorrect masses for Neptune and Uranus, rather than from a previously undiscovered distant planet.

The planet Vulcan   For a long time, astronomers believed that deviations of Mercury's orbit from the precise predictions of Newtonian physics must be caused by another planet, which would have to be located nearer to the Sun than Mercury. The planet was named Vulcan, searched for, and several times thought to have been found, but never confirmed. Perhaps the astronomers were seeing sunspots when they observed what appeared to be the disk of a small planet orbiting the Sun. Finally, Einstein's theory of relativity accounted for the changes in Mercury's orbit without requiring the existence of another planet. Astronomers stopped looking for Vulcan.

The planet Nemesis   In 1983, J. John Sepkoski Jr. and David M. Raup calculated that mass extinctions occur every 26 million years, a regularity that cried out for explanation. One popular idea was that a planet (or perhaps a star) orbiting the Sun far from Earth followed an elongated path that every 26 million years took it through the Oort Cloud of comets, knocking hordes of the comets toward Earth, where they could rain destructively on the planet. Richard A. Muller, Marc Davis, and Piet Hut, thought a companion star might do the trick, and named their postulated star Nemesis. When no such star could be found, the name was passed along to the hypothetical planet. Eventually, calculations showed that a planet with the required characteristics could not exist either.

First pulsar planet   In 1990, Andrew G. Lyne, Matthew Bailes, and Setnam Shemar at the University of Manchester in England thought they found a planet revolving about the pulsar PSR 1829–10. Lyne, Bailes, and Shemar observed regular changes in the pulsar's position, which they postulated were the result of gravitational attraction from an orbiting planet. The team reported on their calculations in 1991, causing quite a stir. But in January 1992, they realized that the apparent regular wobble of the pulsar, from which they had inferred the presence of a planet orbiting the pulsar, was actually evidence of a planet orbiting the Sun. The apparent regular deviations in the position of the pulsar were subsequently revealed to be an artifact of Earth's motion. The astronomers had left Earth out of their calculations.

(Periodical References and Additional Reading: *New York Times* 6-12-92, p A1; *New York Times*, 10-6-92, p C1; *New York Times*, 10-13-92, p C1; *New York Times*, 12-17-92, p B18; *Physics Today* 4-93, p 22; *Astronomy* 10-93, p 18; *New York Times*, 10-7-93, p B12; *New York Times*, 3-1-94, p C13; *Science* 4-22-94, pp 506 & 538; *New York Times*, 4-24-95, p A1; *New York Times*, 6-14-94, p C1; *Nature* 6-23-94, pp 607, 610, & 628; *New York Times*, 7-5-94, p C8; *Physics Today* 8-94, p 20; *Science* 9-9-94, p 1527; *New York Times*, 9-20-94, p C11; *Astronomy* 11-94, p 24; *Discover* 1-95, p 34)

STARS: THEIR LIVES AND HARD TIMES

It is amazing that astronomers can learn anything about stars beyond their position and brightness. The discovery in 1859 of spectroscopy—the connection between lines in diffracted light and specific kinds of atoms—made it possible to learn the elements in stars. Similarly, the light from a star could also be used to determine its temperature. When astronomers realized that the vast population of stars represented similar objects at different stages of development, they were further able to characterize stars.

Even though many of these stages take millions of years, there are at any one instant large numbers of stars available for viewing at every conceivable stage. The problem in organizing this data is similar to that of determining the life cycle of a butterfly when given various butterfly-stage photographs presented at random. You need some additional information to determine that the small white or colored dot is the earliest stage, that the caterpillar (which may itself go through different stages that are not much alike) is the next stage, followed by the chrysalis, and finally the adult winged form. Similarly, astronomers observed giant red stars, small white ones, supernovas, medium-sized yellow stars, large blue stars, and clouds of gas. The life cycle of a star was there amidst thousands of photographs presented at random.

In 1910, two astronomers working independently found that plotting each star on a diagram using intrinsic brightness as one axis, and temperature as the other axis, allowed them to sort out the photographs of stars into patterns from which the life cycle of an individual star became apparent. Astronomers were given another tool when the principles of nuclear physics were developed—these could be used to explain how stars evolve.

Stars begin as condensation in clouds of gas, heated by gravitational energy and nuclear fusion. Different fusion reactions take place as the star evolves. Initially, the young star is not very bright or hot; in this stage, it fuses hydrogen to helium. Later, the star brightens and cools as it fuses helium into carbon. After that, a star's fate depends on how massive it is.

A star more-or-less the size of the Sun grows in diameter, becoming a red giant. It may then become smaller, perhaps becoming a variable star, waxing and waning. It then swells to a red giant again and blows off its outer layer. At first this shedding takes place slowly, but later there is a sudden increase in the velocity of material. This causes the gas around the star to form a planetary nebula, a spherical cloud of glowing gas that has a disk appearance when seen from Earth, similar to a planet. The planetary nebula is centered around what is left of the star. Half or more of the mass of a star is lost in the process. A star retaining less than 1.4 times the mass of the sun is known as a white dwarf. According to theory, after billions of years a white dwarf will burn out to become a black dwarf. The time involved for this transformation is so great that it may never have happened as yet. If the star is larger than 1.4 solar masses, it may vacillate between states, explode into a supernova, or, if larger than 3.2 solar masses, collapse into a black hole.

The *Hubble Space Telescope* (*HST*) has located tens of thousands of new stars forming in galaxies such as Arp 220, where two galaxies are colliding. Although several other mechanisms for star formation have been proposed (including energy from exploding supernovas and from collapsing black holes), the concept of galactic collisions has become in the 1990s a fashionable explanation for producing stars. The theory was last in favor in the 1940s.

HST detected a new stage in some stars' histories in 1993 when it located "naked stars." These are former red giants that have lost their outer cloak of hydrogen as a result of crowded conditions in their part of the sky—the outer layer was pulled away by another passing star. In normal evolution, red giants usually blow away their outer layers slowly as a stellar wind. The difference between those stars that evolve normally and the naked stars is that the naked stars are much hotter—as much as 10 times as hot as the stars that lose their outer layer more slowly.

Brown Dwarfs

Finding a new star is no surprise; the number of stars has always seemed to be infinite. Finding a new planet is major news. Celestial bodies that occupy a classification in between planets and stars are hard to explain, but exciting for astronomers, if not a big story for the general public.

If it is bigger than a planet, yet not quite a star, an astronomical body is classed as a brown dwarf, a name coined in 1975 by Jill Tarter. It is "brown" because it does not have enough mass to ignite the fusion process that lights up a star. Instead, gravitational contraction supplies enough energy for the body to radiate at a low level of luminosity. Thus, although a dwarf when compared to a star, the body has to be much larger than a planet so that gravitational contraction will provide enough energy to radiate in the visible region at all. (Early in life a budding brown dwarf may temporarily fuse heavy hydrogen, but the supply of heavy hydrogen runs out quickly, and fusion halts.)

These considerations put limits on what can be called a brown dwarf. The maximum size is about 8% the mass of the Sun, or about 80 times the mass of Jupiter; any larger, and hydrogen would fuse. It is less clear what the minimum size might be, but most astronomers would say the body has to be at least 10 times Jupiter's mass to qualify. Size, however, is not the only criterion, because certainly a body 10 times Jupiter's mass or less could be a large planet. The current thinking is that a planet forms when small bodies—planetesimals—crash into each other, resulting in a planet growing from a small body to a large one. A brown dwarf, on the other hand, starts out as a big gas cloud and just collapses. Therefore, distinguishing a big planet from a small brown dwarf requires knowledge of the process by which that body was formed.

In recent years a small number of astronomical bodies have been touted as brown dwarfs. Most of these are companion bodies in binary systems, observed or inferred. None have ever been firmly identified. In 1994, *HST* was pointed at a couple of suspected brown dwarfs and found two near misses. One is Gl 623B, the companion to the red dwarf Gl 623A in the constellation Lacerta. The two small stars are each clearly visible in the *Hubble* image. The pair are separated by 2 times the distance of Earth from the Sun (about the distance of the asteroid belt between Mars and Jupiter) and the suspected brown dwarf has only about 10% as much mass as the Sun, about a quarter more mass than the largest brown dwarf. In the other near miss, the situation is much the same. Gl 752B, more commonly known as Van Biesbroeck 10, contains about 9% of the Sun's mass.

Another suspect was a small object in a dust cloud near Rho Ophiuchi that seems to have an energy output less than 1% that of the Sun, which might indicate a brown dwarf. Theoretical considerations suggest a mass only about 3% of the Sun, which would definitely qualify, but the data is not good enough yet to be certain. The object was found using infrared light by George Rieke and Douglas Williams of the University of Arizona, along with Fernando Comeron of the University of Barcelona.

Late in 1995 a team led by Shinivas Kulkarni of CalTech in Pasadena announced that they had obtained spectra that established the previously unknown companion to Gl 229 as a brown dwarf. The object, known as Gl 229B, was then photographed with the *Hubble Space Telescope.*

White Dwarfs

The original discovery of white dwarf stars during World War I would have astounded the public then if its attention had not been elsewhere. The discovery came with Sirius B, the dim companion of Sirius A, which is the brightest star in apparent magnitude. By World War I, Sirius B had been known to exist for nearly three-quarters of a century. Since Sirius A and B formed a binary system, astronomers could calculate their masses from Newton's laws. The surprise came when Walter Sydney Adams measured the temperature of Sirius B. The tiny star had a temperature substantially higher than that of the Sun; the nearly invisible star was found to be white-hot instead of yellow-hot. Because Sirius B was emitting so much visible-light energy per unit of surface area and yet remained dim, the only possibility was that it had a small diameter. Soon someone calculated that a single teaspoon of material from a white dwarf would weigh about 5 tons, and the small, degenerate stars were on their way to the fame they achieved in popular science books of the 1920s and 1930s. More recently, with new knowledge regarding the even denser neutron stars, and the dense-beyond-measuring black holes, white dwarf stars are not as famous as they once were. Nevertheless, they remain extremely important to the theory of the life cycle of a star.

Until 1994, a difficulty with the star life-cycle theory was that two stages had never been observed. Both come at the very beginning of the white-dwarf phase, and represent the two different ways that a white dwarf can be born.

Becoming a Dwarf: First Way There are large numbers of stars like the Sun that are expected to retire to white dwarfdom late in their stellar lives. There are also sufficient numbers of white dwarfs of the type the Sun is expected to become. What was missing from the continuum was a Sun-like star caught in the act of retirement. Finally, in the November 1, 1994, issue of *Monthly Notices of the Royal Astronomical Society*, Martin Barstow of the University of Leicester in England and coworkers reported that the star RE 1738 + 665 fits into the missing stage, confirming the existence of this part of the star cycle.

The key piece of information was a measurement of the temperature of RE 1738 + 665 made by the *ROSAT* orbiting x-ray telescope. With a temperature of about 88,000K and an atmosphere of pure hydrogen, the star fit the profile of the core of a hot star that had recently (in astronomical terms) cast off its outer layers to become a white dwarf. The next business for RE 1738 + 665 will be to cool down to temperatures similar to other white dwarfs with pure hydrogen atmospheres. The study of RE 1738 + 665 was also important because the star failed to fit a different model of stellar evolution, known as the single-channel evolution of white dwarfs. Although the single-channel model may be helpful in other ways, the discovery of RE 1738 + 665 means that it is not a complete theory for all white dwarfs, and probably does not even describe the vast majority of them.

Becoming a Dwarf: Second Way The other way a white dwarf begins was also observed in 1994. According to the theory, at some point a star less than 1.4 Suns in mass can throw off the material that becomes a planetary nebula. The white dwarf forms out of the material left behind by the nebula. This stage takes place very quickly when compared to other parts of a star's life cycle, being completed in a few dozen millennia. George Jacoby of the U.S. National Optical Astronomy Observatories and Griet C. Van de Steene of the Kapteyn Institute in the Netherlands located a new white dwarf surrounded with a planetary nebula. In 1992, the *Hubble Space Telescope* also was able to pick out the white dwarf illuminating the planetary nebula NGC 2440 at a somewhat later stage of development.

Variations on a Pulse

Among the most famous types of stars are the Cepheid variables, stars whose intrinsic brightness is directly related to their period of variation in brightness. This relationship, first observed by Henrietta Leavitt in 1912 and used soon after by Harlow Shapley to determine the size of the Milky Way, became the basis of Edwin Hubble's proof of the expansion of the Universe (*see also* "Cosmology UPDATE," p 236). Although it is easy to think of these variable stars as exotic and uncommon, some of the best known and familiar stars are Cepheids. Among these is, or has been, Polaris, also known as "the north star" after its apparent position above Earth's north pole. Polaris may be a Cepheid no longer, although this change is neither known to be definite nor permanent.

Chief among the distinguishing characteristics of a Cepheid variable is the steadiness of its period, a property first recognized as early as 1784. Cepheid periods range from 1-100 days. For Polaris, the period has been known since 1913 to be about 4 days. Careful measurements showed a period of 3.96809 days up until the 1940s, when the period began to lengthen slowly. By 1945, it was two minutes longer than it had been early in the century. By 1983, the period was about five minutes longer than when first observed. Furthermore, the change in brightness over each period was decreasing at the same time. The flattening out of the brightness occurred at a much faster rate than the lengthening of the period. By 1994, the period seemed to disappear completely, perhaps because the brightness did not change enough to be measured. Later measurements of higher precision, however, continued to show a feeble, but stable, pulse.

Will Polaris cease its Cepheid behavior? It is not certain. Perhaps the variation will begin to increase, as did the less-well-known RU Camelopardalis, which stopped pulsating in the 1960s but has since begun again.

Clearly variation in stars occurs often, despite a stellar reputation for constancy (with the exception of "falling stars"). Another surprising change came from the star Zeta Orinis, the leftmost star in Orion's belt. In this case, the change involved the emission of x rays. For reasons that are poorly understood, Zeta Orinis and similar stars normally produce a steady stream of x rays along with their visible light. In September 1992, however, the *ROSAT* orbiting x-ray telescope was trained on Zeta Orinis when its output of energetic x rays suddenly rose 30% over a two-day period. *ROSAT* then turned to other business, so it is not known what happened next. It was not until 1993 that the data was examined and the sudden burst was discovered by Thomas Berghofer and Jürgen Schmitt of the Max Planck Institute for Extraterrestrial Physics in Munich, Germany. In February and September of 1993, *ROSAT* got a second and third look at Zeta Orinis as a part of its regular review of space. At second glance, the emission of high-energy x rays was normal again, but the output of low-energy x rays was increased. However, by the third look, everything was back to normal. Theorists were split on what might have caused the burst. Berghofer and Schmitt thought that the cause was a one-two emission of a gas cloud followed by a burst of stellar wind, with the x rays produced when the wind hit the cloud. Others thought that Zeta Orinis might be part of a binary system that somehow produced x rays on occasion.

A slower evolution in brightness was detected by Mart de Groot at Armagh Observatory in Northern Ireland. He used records dating back to 1600 to show that the star P Cygni has brightened steadily. The observations of William Herschel were particularly careful and clinched the case. According to de Groot, P Cygni has brightened by about 45% since 1712, going from magnitude 5.23 then to magnitude 4.82 now. The reason for the brightening, according to standard theories of star evolution, is that P Cygni was a very hot star undergoing cooling. As it cools, the energy it emits as electromagnetic radiation is lowered, with the peak energy moving from the ultraviolet (which is not observable from Earth) to visible light. Although there is less total radiation, more of it is in a form that we can detect. Although P Cygni is cooling in a generally predictable way, records show that the cooling process for this star is at least twice as fast as was predicted.

Blue Straggler UPDATE

Even before the *Hubble Space Telescope* was repaired, it contributed greatly to new knowledge about an important class of stars called blue stragglers. These hot blue stars are called stragglers because they are found in places where all other stars are cool and red. Most often blue stragglers are found in the 150-odd, roughly spherical swarms of 10,000 to a million stars called globular clusters.

Globular clusters are thought to have formed early in the history of the Milky Way. Today, they are free of dust and gas between the stars, which has presumably been used up long ago. Most of the stars in a cluster appear to be the same general size and color, typically red or reddish yellow. The stars in such a cluster contain only small amounts of elements heavier than helium. Such heavier elements were not present early in the history of the universe. According to theories of stellar evolution, all the stars in a cluster formed around the same time. In general, red stars are old, and blue stars are young; but in these clusters, the blue stragglers *must* be old.

Several theories have been suggested to explain the existence of the blue stragglers. Stars in globular clusters move slowly with respect to one another. One theory postulates that blue stragglers form when two stars in a globular cluster gently bump into or graze each other. In the process, one of the stars acquires a lot of hydrogen from the other, rekindling its fires and making the cannibal star young. Alternatively, the two stars could fuse into one large star, which would also cause the fused star to burn with the brightness of youth. Another idea suggests that tidal forces between the close stars cause both to burn hydrogen more efficiently. This last hypothesis was first proposed in 1964, independently by Fred Hoyle [English: 1915–] and William H. McCrea.

The *Hubble Space Telescope* can observe globular clusters at ultraviolet wavelengths, an optical region in which the blue stragglers stand out against the reddish background. Ultraviolet light is absorbed by the ozone layer of the atmosphere, so it cannot be used by ground-based telescopes. Studies using the *Hubble Space Telescope* in 1991 showed that the Hoyle-McCrea scenario, or something like it, is very likely. Using its Faint Object Camera, *HST* examined a crowded field of more than 600 stars in the very dense globular cluster 47. The camera picked out 21 blue stragglers in the globular cluster, a number too high to make the collision or cannibalism theories the only possible mechanisms for their formation. Francesco Paresce and coworkers at the Space Telescope Science Institute in Baltimore, who reported on the blue stragglers in 47 Tucanae in *Nature*, suggested that most of the blue stragglers are double stars whose greater mass has caused them to drift toward the center of the cluster. More recent studies of globular cluster M15 also found many blue stragglers near the core of the cluster.

Computer models show that the core of a globular cluster becomes very dense with stars. The average distance between stars in the center is about one five-hundredth of the average distance between stars in our own neighborhood. Such close quarters would certainly allow for many interactions to occur between stars; the high potential for such interactions is thought to account for the blue stragglers. In 1993, Paresce and coworker Guido De Marchi found 15 apparent white dwarf stars in the heart of globular cluster M15. De Marchi and Paresce, however, think that these are not true white dwarfs. Instead, they believe that this cluster represents the cores of stars which have lost their outer layers in near misses with other stars. The other stars would, according to the theories above, have gone on to be blue stragglers. The naked stellar cores, although not the same as white dwarfs, would eventually evolve into white dwarfs by the non-explosive route (*see* above).

Also in 1993, *HST* was used to identify something entirely unexpected in M15. Not only are there blue stragglers, but there are also yellow stragglers. These are stars similar to the old basic issue globular-cluster reddish star in color, but as bright as a young star.

(Periodical References and Additional Reading: *New York Times*, 6-3-92, p A15; *Discover* 1-93, p 30; *New York Times*, 6-10-93, p A20; *Discover* 1-94, p 77; *Science* 1-7-94, p.44; *Astronomy* 5-94, p 24; *Science* 9-16-94, pp 1660 & 1689; *New York Times*, 9-18-94, p I-41; *Nature* 11-24-94, p 317; *Astronomy* 3-95, p 45; *Astronomy* 4-95, p 32; *Astronomy* 6-95, p 14)

SUPERNOVAS: THEN AND NOW

When a "new star," or nova, appears in the sky, it is not only not new, it is actually very old. A nova or a supernova is a flare-up (for a nova) or an explosion (supernova) of a preexisting and often observable star. A supernova is old because it comes at the end of the life of a large star which dies in a large explosion. Furthermore, the explosion itself is old news by the time it is first seen on Earth, due to the explosion's distance from Earth. Most recent supernovas have been observed in other galaxies. Even Supernova 1987A, which is comparatively close, is about 170,000 light-years away, so that when we say it exploded in February of 1987 we really mean that the light from the explosion arrived on Earth in February of 1987. The actual explosion took place 170,000 years earlier, when most people on Earth were Neandertals (*see* "The Species Problem," p 35).

Setting Off a Supernova

A supernova is the expected end of certain types of large stars, although the reason behind this phenomenon is not totally clear. In January 1995, Willy Benz of the University of Arizona and Marc Herant of Los Alamos National Laboratory in New Mexico presented to the American Astronomical Society their three-dimensional computer model of how a massive star becomes a Type II supernova. Previous one-dimensional computer models failed to produce supernovas, predicting black holes instead. At the same meeting in Tucson, Arizona, Adam Burrows and John Hayes of the University of Arizona and Bruce Fryxell also presented a computer model, this one in two dimensions.

What the new computer models show is that the explosion proceeds from the core of the star. The core collapses as the elements in the core fuse to iron, which causes the core to lose its support from electron exclusion. Electron exclusion, as explained in the principle formulated by Wolfgang Pauli, prevents two "things" from occupying the exact same location at the exact same time. In the case of a supernova, the electrons collapse into the atomic nuclei, where they fuse with protons to become neutrons. When both iron and electron-proton fusion occurs, the core, which weighs about half again as much as the Sun for these large stars, shrinks in less than a tenth of a second to a sphere only 60 mi (100 km) across.

The collapse produces a giant burst of neutrinos. Although neutrinos fail to interact with most forms of matter, there are so many released in this instance that they heat up the remainder of the star, and even blow it away.

The core, which after the collapse consists almost entirely of neutrons, is often shot out from the rest of the explosion. This occurs because the collapse does not happen symmetrically, although theorists are not sure what produces such asymmetry. The expanding gases and particle cloud of the explosion is propelled one way, while the old collapsed core is propelled the other way. After a time, the former core is a neutron star located light-years away from the main part of the explosion. The neutron star can be propelled at such a high velocity that it is fired completely out of the galaxy where the explosion first took place.

Supernova 1987A UPDATE

At its peak, Supernova 1987A in the Large Magellanic Cloud was the first supernova visible to the naked eye since a supernova studied by Kepler in 1604 (although the supernova in the Andromeda galaxy in 1885 was near naked-eye visibility). By late spring of 1987, supernova 1987A was visible from the Southern Hemisphere on Earth as a third-magnitude star (between the brightness of the stars in the handle of the Little Dipper and the brightness of Polaris and the brighter stars of the Big Dipper). Supernova 1987A then began to fade, soon becoming invisible to the naked eye, as it had been prior to exploding, when it was the star Sk –69°202.

Surprisingly, although this phenomenon was predicted by theory, Supernova 1987A became brighter in 1993, although not as bright as it had been in 1987. Even before it became a supernova, Sk –69°202 had developed a sphere of gas around it. This sphere started when Sk –69°202 was a red supergiant, which had been its state for millions of years before its explosion. During the red period, its stellar wind—a stellar wind is just another star's version of the Sun's solar wind of protons and atomic nuclei—was slow, since the cool surface of the star provided little kick for the wind. However, for about 100,000 years before the supernova, the red giant heated up and became a blue giant.

A second phenomenon, thought to be closely related to the spheres of expanding gas, was observed as early as 1989. During its blue period, Sk –69°202 was hot enough to have a fast-moving stellar wind. As the fast particles caught up to the slow ones, they produced a sphere of greater density about the blue giant star. Now the second sphere is illuminated by the supernova, so when viewed by telescopes from Earth, it appears as a bright oval ring. It is that ring which started brightening in 1993, when the debris from the explosion began to catch up to and collide with the sphere. Such collisions give off visible light. It is thought that the visible ring around the remains will become even brighter during the last half of the 1990s.

Some astronomers question this explanation of the brightening. Richard McCray of the University of Colorado at Boulder and Douglas N.C. Lin of the University of

California at Santa Cruz proposed in 1994 that the ring actually comes from the disk of gas that collapsed to form the star, not from the stellar wind generated after the star's formation. These competing theories will be spectacularly tested in 1999 when the debris from the explosion, which is moving fast but much slower than light, finally reaches the ring. Not only will the ring brighten by much more at that time, but it should also show features in the shell that are not otherwise visible from Earth.

The viability of all these explanations are somewhat endangered by virtue of the fact that the ring is actually a ring and not a sphere. Although a thin sphere might look like a ring under certain circumstances, careful study by the *Hubble Space Telescope* in 1990 established that the bright glow is a nearly circular ring that appears as an oval because it is inclined about 43° to our line of sight. Thus, whether it is debris from the red giant phase or dust from the young star phase, it must be distributed in some pattern other than a sphere about the supernova. For example, if the interior cavity of the debris or dust is egg shaped, radiation from the supernova would reach the middle (where the diameter is smaller) more quickly than it would reach the ends of the longer circumference. In that case, the bright glow would appear ring shaped for a time, but later on would be ovoid. Observations in the future may help clear up this mystery as well.

In February 1994, the *Hubble Space Telescope* imaged Supernova 1987A and detected two more rings. They were each about the same size, about 2.5 light-years in diameter—much larger than the bright ring described above. Neither of the two large faint rings is centered on the supernova itself; one is nearer Earth than the supernova, while the other is farther away. Even a line through their centers would not intersect the supernova.

So what are these two faint rings? No one is sure exactly, although they must have been initiated by processes similar to those that created the bright ring that is centered on the supernova. No one theory of formation has much support beyond the individual astrophysicist proposing it.

TIMETABLE OF SUPERNOVAS

352 or 393 B.C.	Chinese observers report a supernova, the earliest known for which there is a record.
A.D. 185	Chinese observers report a supernova in the constellation Centaurus; it remains visible for 20 months.
369	The Chinese report a star now thought likely to have been a supernova; it remains visible for 5 months.
386	The Chinese report a star that remains visible for 3 months.
1006	A supernova in the constellation Lupus that remains visible for several years appears, according to records from China, Japan, Italy, Switzerland, and the Near East. This is the first of the five supernovas definitely known to have occurred in the Milky Way galaxy.
1054	One of the best-known supernovas of antiquity flares up on July 4, and is recorded in the constellation Taurus in China, Japan, and Arab lands, and possibly by Native Americans in what is now the southwestern U.S. In China, it was visible in daylight for about three weeks. Today, the

remnants of this supernova form the Crab Nebula, and the core of the original star is a pulsar in the nebula. This is the second supernova definitely known to have occurred in the Milky Way. It is thought to have reached an apparent magnitude of about − 5, somewhat brighter than Venus at the planet's brightest.

1181   A supernova in the constellation Cassiopeia is reported in China and Japan; it is visible for 183 days. This is the third of the known supernovas from the Milky Way galaxy.

1572   The supernova of Tycho Brahe is observed by him in the constellation Cassiopeia on November 11 (Julian calendar), although it was originally spotted five days earlier by some observers. It shines as bright as the planet Venus, which can reach a magnitude of − 4. Chinese astronomers also report this supernova, which fades to magnitude 0 within 100 days, but remains visible for 15 months, until March 1574. This is the fourth known supernova from the Milky Way.

1604   The supernova of Johannes Kepler is observed by him in October in the constellation Ophiuchus. It is not quite as bright as Venus, reaching a magnitude of about − 2.5, and lasts for about 12 months into the winter of 1605. Chinese and Korean astronomers also report the supernova. This is the last known supernova from our Milky Way galaxy.

1885   A supernova in M31, or the Andromeda Galaxy, reaches the fringe of naked-eye visibility, almost at the sixth magnitude.

1937   Fritz Zwicky discovers three supernovas visible in other galaxies using a Schmidt astronomical telescope. This is the beginning of modern observation of supernovas on a regular basis.

1968   A pulsar is found at the heart of the Crab nebula.

1972   A supernova in the galaxy NGC 5252 is nearly as bright as the rest of the galaxy.

1987   Supernova 1987A is discovered by Ian Shelton. In the Large Magellanic Cloud, just beyond the Milky Way, it is the nearest supernova to Earth since Kepler's supernova of 1604. It has an apparent magnitude of about + 3 at its height.

1992   A supernova appears in galaxy NGC 3690, although it is not noticed until its image is recognized in photographs in 1994.

A Type Ia supernova observed starting on April 28 is determined to be the farthest ever seen from Earth. Astronomers estimate the distance as 5 billion light-years.

1993   Supernova 1993j in the spiral galaxy M81 in Ursa Major becomes the first supernova whose presence is announced to astronomers on a new Internet supernova network. Its explosion is announced on March 29, a few hours after its discovery, allowing thousands of astronomers around the world to study its changes in brightness, and giving astronomers in the Northern Hemisphere their first chance since 1937 to view a bright type Ib supernova (although some observers think it is an odd version of type II).

A second supernova, against the odds, appears in NGC 3690, so that the two supernovas are about a year apart. Given that astronomers usually see only 20 to 30 supernovas a year in all the thousands of galaxies near enough to Earth for such sightings, the odds are greatly against two in a single galaxy. In this case, the galaxy is thought to be the result of a collision of two galaxies, which would have triggered a lot of star formation. Stars formed about the same time would tend to reach the end of their lives and explode about the same time also.

1994   A bright supernova (apparent magnitude 13), 1994I, appears in the Whirlpool galaxy M51 in Canes Venatici, with its light first being observed on Earth on April 1 and radio signals just two days later. The type Ic supernova is actually about 15 million years old since it is that many light-years from Earth, but it is still the second-closest supernova to Earth since the invention of the telescope (second to 1987A).

SUPERNOVA TYPES

Type I   Originally what is now called Type Ia, although sometimes still used to mean the same as Type Ia.

Type Ia   Astounding explosions in just an instant. No hydrogen lines are observed. Expansion velocities can reach 3% of the speed of light. These occur in galactic disks and other places where stars are numerous and close together, such as in elliptical galaxies. Type Ia supernovas are not well understood, but are thought to involve binary systems of which at least one member is a white dwarf. Interactions between the members of the pair can cause the white dwarf to collapse and annihilate itself completely, destroying the other member in the process.

Type II   Somewhat slower explosions that result in a characteristic rise and fall of energy. Hydrogen lines are observed. Speed of expansion is about half that of Type I. These occur wherever massive stars are found, including not only the galactic disks but also in the spiral arms of galaxies. Type II are thought to be the natural end of very massive stars via collapse into a neutron star, or pulsar. The core collapses deep inside the star and blows everything else away.

Type Ib   Similar to Type Ia in observable characteristics, but occurring in spiral arms of galaxies. Despite an appearance similar to Type Ia, the actual mechanism of destruction is more like that of Type II, but the star involved has blown away its outer hydrogen shell before the collapse of the core.

Type Ic   Similar to Type Ib, but the giant star, more than 40 solar masses, has completely shed not only its hydrogen but also its helium outer shell, leaving behind a carbon-oxygen star. In 1994, Ken Nomoto of the University of Tokyo and coworkers theorized that supernova 1994I, a type Ic, lost its outer envelopes to a binary companion before exploding.

(Periodical References and Additional Reading: *New York Times*, 11-15-92, p I-19; *New York Times*, 4-7-93, p D24; *New York Times*, 4-13-93, p C11; *Discover* 1-94, p 76; *Astronomy* 2-94, p 41; *Astronomy* 4-94, p 26; *New York Times*, 4-5-94, p C10 *Science* 4-15-94, pp 342 & 345; *New York Times*, 5-20-94, p A19; *Nature* 6-2-94, pp 354, 378, & 380; *Science* 6-3-94, p 1405; *Astronomy* 7-94, p 21; *Science News* 8-6-94, p 90; *Scientific American* 9-94, p 14; *Nature* 9-15-94, pp 199, 227; *Discover* 1-95, p 35; *Astronomy* 5-95, p 30)

PULSARS

The neutron stars that are the former cores of supernovas are collapsed parts of rotating bodies (*see* "Supernovas, Then and Now," p 190). This collapse speeds the rotation to many revolutions per second due to the law of conservation of angular momentum. The neutron stars produce beams of electromagnetic signals from their poles. These beams sweep the universe as the stars rotate. When observed from Earth, the beams sweeping past appear to be a series of pulses, sometimes occurring as frequently as thousandths of a second apart, all precisely the same interval from each other; hence, neutron stars are also called pulsars. Since the discovery of pulsars in 1967, neutron stars have become familiar entities to astronomers but remain exotic in behavior and origin. Astrophysicists tend to think that they understand pulsars fairly well, yet new observations continue to surprise. For example, pulsars are supposed to be born in massive explosions, but at least one has a planetary system (*see* "Newly Found Planets," p 178), and another has a companion (*see* "The Black Widow Pulsar," below).

The February Pulsar

Since pulsars are thought to originate in supernova explosions (*see* "Supernovas, Then and Now," p 190), astronomers examine recently observed supernovas for the pulsars they have produced. A year after pulsars were discovered, this approach paid off with the observation of a pulsar in the remnants of the great supernova that produced the Crab nebula. Similarly, the remains of supernova 1987A displayed a pulsar in 1989—except that it was quickly revealed to be a fake. What appeared as a pulsar in 1989 was discovered in 1990 to be an artifact caused by electrical variations from a video camera used in recording the signal. A team of astronomers led by John Middleditch of Los Alamos National Laboratory in New Mexico continued observations from posts in Chile and were rewarded with apparent pulsar signals in February 1992 and in February 1993. The signals are 2.14 milliseconds apart. This means that the neutron star is revolving about 450 times each second, fast but not faster than other millisecond pulsars (the spurious video camera signal had implied a pulsar spinning four times as fast as this, which would have been astonishing if true).

After their false alarm of 1989, the Middleditch team did not want to publish the news of their February signals, especially since the signals had disappeared between sightings. Although they thought it possible that other debris from the explosion could be blocking the pulsar signals most of the time, it was also possible that the signals were not caused by the pulsar at all, but by some other effect in the debris. There was also the chance that it could be another spurious signal. In September 1994, the team decided to informally let other astronomers know what was happening, in case someone else could confirm the pulsar, or had an idea of why the signals had only been detected for a short time in 1992 and 1993 but not in February 1994.

The Black Widow Pulsar

In 1988, a binary system was discovered to include a pulsar which seemed to be devouring the other star in the system by blowing away its hydrogen. Astronomers named the pulsar the Black Widow after the spider famous for devouring her mate, although the pulsar's mode of operation more closely resembles that of the Big Bad Wolf. The observational evidence for the Black Widow includes the fact that every nine hours the pulsar signal disappears for about 50 minutes. Such an effect might be caused by a type of eclipsing binary in which the companion comes between the observer and the signal generator, but scientists found a different explanation. It appeared that radiation from the pulsar was creating a stellar gale of particles such as protons from the companion star; every nine hours the gale passed between the pulsar and Earth, blanking out the signal for a time.

As if this were not complicated enough, relations between the Widow and her mate are the archetype of a dysfunctional, abusive marriage. The companion keeps changing his orbit. For a time he approached the Widow, then he fled, and now he seems to be returning. Thus far, the only sensible explanation for this behavior involves variations in a strong magnetic field, part of a theory propounded by James Applegate of Columbia University in New York City in the late 1980s. But the companion star seems to be too small to have much of a magnetic field. Applegate and coworker Jacob Shaham, also from Columbia, think that the Widow supplies the energy allowing the companion to shine, rotate, and produce a magnetic field. She does this with her powerful gravitational force, which sets up tides in the companion, and by producing the stellar gale that is destroying the companion. The companion's induced magnetic field interacts with the gale, causing the companion to keep twisting and turning so that he can continue to shine. In other words, he can't live with her and he can't live without her.

To prove that this theory is correct, astronomers need to observe the glow caused by tidal forces, which should be separate from another glow on the companion caused by radiant heating from the pulsar. Studies so far have been inconclusive.

Other Pulsars in the News

Geminga a Pulsar   Starting in 1972, astronomers were puzzled by a strong source of gamma rays that they could not match with any physical object. The apparently unobservable source was named Geminga, Italian for "it's not there." However, it was there, and was eventually it was located as a source of other radiation, including a little visual light. In 1992, Jules P. Halpern of Columbia University in New York City and Stephen S. Holt of the National Aeronautics and Space Administration found some wavelengths at the x-ray and the strong gamma-ray energy levels to be pulsing four times each second. Thus, Geminga is now known to be a pulsar, presumably the remnant of a supernova explosion in the distant past.

Background

Neutron stars were first proposed as theoretically possible by Walter Baade [German/American: 1893–1960] and Fritz Zwicky [Swiss: 1898–1974] in a 1934 paper on supernova explosions, just two years after the discovery of neutrons. It was not until 1968, however, that a neutron star was actually observed (in the debris of the explosion of the supernova of 1054, now known as the Crab nebula). The original concept of a neutron star involved the force of a supernova explosion fusing the protons and electrons of atoms together to form neutrons. This idea does not make sense in terms of modern concepts of physics, although it is often still given as an explanation of how a neutron star forms. Instead, it is more correct to say that gravitational forces squeeze the electrons out of the atoms, leaving just the protons and neutrons (collectively, nucleons) behind. The electrons do not really go away, however. Overcome by gravity, they blend into each other. The protons and neutrons cannot collapse further unless gravitational strength is much greater. A great enough gravitational force blends nucleons together resulting in a black hole instead of a neutron star.

A better name than neutron star would be "nucleon star" or "nucleus star," since the resulting object is essentially the same as an atomic nucleus, except that it is electrically neutral instead of positive. The object has essentially the same density as an atomic nucleus. Thus, even though a neutron star's diameter is thought to be on the order of 6 mi (10 km), its mass may be nearly three times as great as that of the Sun.

Closest Pulsar   PSR J0437-4715, known as pulsar 437 for short, is a typical pulsar. It is about 10 mi (16 km) in diameter, has a mass about the same as that of the Sun, and a period of rotation of about 175 times per second. Pulsar 437 is part of a binary system with a white dwarf. The main claim to fame for pulsar 437 is its location, which is virtually in our own neighborhood. Although literally very far away (at 400 light-years), it is very close for a pulsar. Located in the southern constellation Pictor, the pulsar is too far away to be seen from Earth, although its companion can be seen with a sufficiently large telescope.

Fastest Pulsar   PSR 2224 + 65 is smaller and denser than pulsar 437 and rotates a bit more slowly. It would be another typical pulsar except for its proper motion of 2,232,000 mph (3,593,000 kph), making it not only the fastest pulsar but also the fastest star of any kind ever to be observed. Its discovery was announced in the March 11, 1993, *Nature* by a team headed by James M. Cordes of Cornell University in Ithaca, New York. PSR 2224 + 65 gets its impetus from a supernova explosion. Its wake in the interstellar gas forms the visible Guitar nebula, about 300 light-years long.

The Pulsar Halo   Andrew G. Lyne and D. R. Lorimer of the University of Manchester were inspired by the discovery of the fastest pulsar to look at pulsar speeds in general. They determined that the average speed of a pulsar is about 280 mi (450 km) per second. This great speed implies that about half the pulsars formed in supernova explosions are propelled from their galaxies. Thus, galaxies must be surrounded by a faint halo of pulsars. Furthermore, Lyne and Lorimer think that the pulsars may be the source of the enigmatic gamma-ray bursts (*see* "Gamma Ray Burster Mysteries and Solutions," p 202).

(Periodical References and Additional Reading: *New York Times*, 5-24-92, p I-20; *New York Times*, 2-23-93, p C6; *Nature* 5-12-94, p 127; *Science* 8-12-94, p 869; *Scientific American* 9-94, p 14; *Science* 11-4-94, p 731)

BLACK HOLES SIGHTED

It is suspected that those astronomical bodies possessing the greatest mass do not shine. These are the black holes that have become such a familiar part of people's thinking in recent times. In fact, the term "black hole" is often used for anything that completely absorbs something—money or love, for example—and fails to give it back. Even casual acquaintance with the idea of a black hole suggests why it is difficult to actually observe one. The name comes from the black hole's property of gravitationally attracting light with such force that no light can be emitted from it.

Despite the difficulty of observing something that emits neither electromagnetic radiation nor any particles (other than the as-yet unobserved graviton), astronomers frequently announce that they have located a black hole. Since these discoveries are always on the basis of circumstantial evidence, no one is absolutely certain that any of the black hole "sightings" are real. Nevertheless, evidence in favor of black holes continues to accumulate.

Ordinary Black Holes: Dark Companions

Until recently, perhaps the best black hole candidates have been binary systems in which a black hole is paired with an ordinary star, such as V616 Monocerotis or V404 Cygni. These possible black-hole/ordinary binaries were first identified in 1989 and 1990. A binary system is a good place to look for a black hole. Even though the black hole may not be visible, the other member of the pair, if an ordinary star, will be. Thus, the ordinary star will appear to be orbiting a mutual center of gravity with something unseen. If the unseen something is large enough, it could be a black hole; otherwise it is likely to be a neutron star or a dim dwarf.

Another clue that a binary might include a black hole would be if it emitted energetic x rays. Ordinary stars do not emit much in the way of x rays, but binary pairs that include either a neutron star or a black hole pour them out. This occurs because the gravitational pull of the neutron star or black hole pulls matter from the ordinary star. The matter moves so quickly on its way to destruction that it emits x rays along the way. While astronomers were watching, V404 Cygni, or an unseen something from the same part of the sky, emitted a strong burst of x rays. Such a burst could have come from either a black hole or a neutron star (*see also* "Gamma Ray Burster Mysteries and Solutions," p 202).

By 1992, astronomers had determined that V404 Cygni was orbiting an unseen massive body every 6.5 days. The short period of the orbit could be used along with Newtonian gravitation theory to determine the approximate mass of the companion, which was found to be at least six times the mass of the Sun. This is far too great for a neutron star, which is constrained to have at most a mass about three times that of the Sun. Hence, the companion of V404 Cygni could only be a black hole. In the

USING SPECTRA TO RECOGNIZE BINARIES

Most stars are not isolated from each other, but appear in groups, of which the most common is the pairing of two stars called a binary system. A typical binary consists of two stars possessing somewhat different characteristics but orbiting a mutual center of gravity. It is sometimes possible to see the two stars with a good telescope, but more often one has to infer that there are two stars due to the detection of two different types of spectra. Ordinary stars can be classed by the types of spectra they show. The classes are assigned letters representing a range from hottest to coolest: O B A F G K M (memorized by the mnemonic "Oh, Be A Fine Girl/Gent, Kiss Me"). The presence of an O spectrum and a K spectrum in what appears to be a single star is a sure sign of a binary. Even if the two stars are of the same spectral type, the binary nature of the system can often be detected by the Doppler shifts in their spectra as they orbit the mutual center of gravity.

November 1, 1994 *Monthly Notices of the Royal Astronomical Society*, Phil Charles of the University of Oxford and coworkers reported that they had calculated the mass of the visible star using a different method, based on the spectrum of V404 Cygni. They found that V404 Cygni has a mass only 0.7 times that of the Sun. This mass, when used in calculations with the orbital information, would mean that the companion's mass is that of at least 10 Suns, making it very likely a black hole.

Giant Black Holes

Quasars are thought to be powered by giant black holes in the centers of galaxies. After the quasars die out, the black holes remain. A number of studies have found evidence for such galactic black holes. For example, jets of high-speed material streaming from the center of galaxy M106, discovered in 1994, are thought to be caused by a black hole that somehow focuses and accelerates the material out of the galaxy, where it streams for a distance of about 30,000 light-years. Another important clue has been the sighting of giant clouds of gas spinning too rapidly to be held together by anything but a black hole at their center.

An early report of such a rotating disk was published on April 10, 1991, in the *Astrophysical Journal Letters*. Using a new interferometer attached to the 88 in (224 cm) telescope the University of Hawaii runs on Mauna Kea, Joss Bland-Hawthorn of Rice University, Andrew Wilson of the University of Maryland, and Brent Tully of the University of Hawaii analyzed a strange object in the galaxy NGC 6240. With it, they were able to detect shifts of a spectral line of hydrogen gas that they thought were caused by a disk rotating about 650 mi (400 km) per second, behaving more like a solid disk than a cloud. This speed is consistent across the disk, which has a radius of about 4000 light-years. The disk is surrounded by more gas, which rotates at more reasonable rates.

The laws of gravity imply that the rotation speed of the disk is controlled by a massive concentration of matter—somewhere between 40 and 200 billion solar masses. This could be, say, 100 billion stars, but stars would emit light, and there is not much light coming from the disk. Another possibility would be even a greater number of brown dwarfs, all crowded together. Yet, given the difficulty in locating even one brown dwarf near Earth, this seems unlikely. The remaining possibility is a giant black hole, which could supply such mass and would not emit large amounts of light.

Over the next couple of years, astronomers discovered that other galaxies also seemed to have at their center massive black holes, which were thought to be the remnants of quasars (see "Quasars," p 233). In 1992, for example, John Kormendy of the University of Hawaii and Douglas O. Richstone of the University of Michigan used Doppler shifts to demonstrate that stars in galaxy NGC 3115 moved as if orbiting a black hole with a mass about a billion times that of the Sun. But such evidence is indirect—readings from an interferometer, while revealing, are not images.

Early observations with the *Hubble Space Telescope* (*HST*) could not find disks at the center of other galaxies because of the flaw in its main mirror—although, in some cases, what was observed appeared to be evidence for a black hole. A notable example was announced by Walter Jaffe of Leiden Observatory in the Netherlands and an international team on November 19, 1992. They photographed a disk in the heart of galaxy NGC 4261 in the Virgo Cluster which to some observers appeared to be a black hole, although the image was too fuzzy to be certain. The galaxy is one known for its energetic jets, which many also consider to be evidence of a black hole.

In 1992, Tod R. Lauer of Kitt Peak National Observatory and Sandra Faber used *HST* to look for a black hole in the heart of the elliptical galaxy M87 (see "Shape and Evolution of Galaxies," p 216) which, at 50 million light-years away, is moderately close. They found evidence strongly suggestive of something that was producing a peak in brightness near the center of the galaxy. Considered in combination with evidence from earlier studies of M87, a black hole seemed a good candidate for the cause of the peak.

In May 1994, Holland Ford of the Space Telescope Science Institute, Richard Harms of Applied Research Corporation, and eight coworkers were able to use the Wide Field-Planetary Camera of the newly corrected *HST* to examine the center of M87. The image from *HST* was immediately recognizable as a giant rotating disk of gas. The speed of rotation at 60 light-years from the center was clocked at 1.2 million mph (1.9 million kph). This was possible because we do not see the disk flat on, but at an angle of about 42°. Doppler effect measurements with *HST*'s Faint Object Spectrograph show a point about 60 light-years from the center to be approaching us, and another point about the same distance away receding. Working backward using the laws of geometry and gravity, the center of rotation must have a mass of about 2.4 billion Suns, not quite so great as that calculated in 1991, but still clearly a black hole candidate. Alternative explanations are great concentrations of dim stars or some form of exotic dark matter, because too little light is emitted by the region to be caused by any combination of ordinary bright stars.

In June 1994, astronomer John Biretta of the Space Telescope Science Institute obtained another piece of evidence in favor of a black hole in M87 by determining the speed of a jet of electrons being propelled from the heart of the rotating disk. Such jets, because of the angle at which they are observed, appear to be moving faster than

Background

The methods used by astronomers to observe distant objects of interest are often worthy of Sherlock Holmes. To locate what might be an extraordinarily large black hole, the actual light analyzed is that caused when an electron in a hydrogen atom drops from one possible orbit (designated 3) to another (designated 2). The amount of energy released when the electron drops is known precisely. As radiation, one way this energy manifests itself is as light of a specific wavelength and, therefore, color. Passed though a spectroscope, the light falls as a single line in a specific location.

Suppose that the source of the light is a hydrogen cloud, moving with respect to Earth. As a result of the Doppler effect, emitted light will have its wavelength lowered if the cloud is moving away from Earth and raised if it is moving toward Earth. Thus, observing the position of the hydrogen line with a spectrograph can provide information regarding the general movement of the cloud.

Once the overall motion of the cloud of hydrogen with respect to Earth is known, one can compensate for that motion, and look for another motion. If the cloud is rotating in any way except directly face on to Earth, part of it will be moving away from us at the same time that another part is moving toward us. While this motion may be much less than the overall movement of the cloud, it can be detected and its speed measured by an interferometer, which acts as a magnifying glass for differences in wavelengths.

light—in this case, 2.5 times as fast as the speed of light in a vacuum. In reality, the jets are merely traveling just short of the speed of light in a vacuum (in the case of the M87 jet, about 93% of that speed). Jets with this amount of energy are thought to be propelled by quasars, and quasars are now suspected of being black holes in a wrapper.

Indeed, after seeing the *HST* images, most astronomers have become believers. Although to the uninformed observer the bright disk in M87 looks like the Sun behind a translucent cloud, to astronomers it looks like a black hole.

(*See also* "Inside the Milky Way," p 206, for an account of similar but much smaller black holes found or suspected within our own galaxy.)

TIMETABLE OF BLACK HOLES

1687	Isaac Newton's theory of gravity is published.
1783	John Michell observes that if a body is both sufficiently massive and compact, light would be unable to escape due to the great force of gravity.
1795	Pierre-Simon Laplace calculates how light is affected by gravity, and notes that the largest luminous bodies in the universe in terms of mass might be invisible.
1915	Albert Einstein's theory of gravity (known as general relativity) is published.
1916	Karl Schwarzschild uses Einstein's theory of gravity to make the first calculations for a very massive luminous nonrotating body, and finds that relativity predicts that such a body will not permit light to escape.

1939	J. Robert Oppenheimer and Hartland Snyder show that a star greater than about 3.2 solar masses will collapse indefinitely after its fuel is exhausted, essentially concentrating all its mass in a point.
1958	David Finkelstein explains the event horizon, essentially the border of a sphere of space around a collapsed star. Within such a border, no light or matter can escape, but even outside the event horizon gravity affects time and space in unusual ways.
1963	Roy Kerr recalculates the gravitational field of a very massive body using general relativity and taking rotation into account for the first time. He confirms that light cannot escape.
1964	Edwin Salpeter and Yakov Zel'dovich demonstrate that even though a sufficiently massive body does not permit light to escape its surface, material that comes near the event horizon will radiate as it is being sucked below that horizon. Thus, even though a black hole itself cannot be seen, its presence is visible. Zel'dovich suggests that quasars may be powered by such massive bodies.
1967	John Wheeler invents the name "black hole" to describe bodies too massive to permit light or matter to escape from their surface.
1972	C. T. Bolton and coworkers identify the first known black hole candidate: the unseen x-ray-emitting companion object in the x-ray source Cygnus X-1.

Measurements show that the companion is at least 7 solar masses, too heavy by far to be a neutron star, the alternative explanation. |
| 1976 | Stephen W. Hawking calculates that black holes evaporate with time, taking information of their existence from the universe with them. |
| 1978 | Wallace Sargent and coworkers show that the velocity of stars rotating about the center of galaxy M87 could best be explained by assuming that a black hole of about 5 million solar masses is at the center, although other interpretations of the data are possible.

Martin J. Rees calculates that quasars must eventually evolve into black holes. |

(Periodical References and Additional Reading: *New York Times*, 7-14-92, p C2; *New York Times*, 11-20-92, p A1; *Nature* 5-12-94, p 102; *New York Times*, 5-26-94, p A1; *New York Times*, 5-29-93, p IV-2; *Nature* 6-2-94, p 345; *Science* 6-3-94, p 1405; *Physics Today*, 8-94, p 17; *Nature* 10-13-94, p 561; *Discover* 1-95, pp 32 & 33; *Astronomy* 3-95, p 30; *Astronomy* 5-95, p 26)

GAMMA-RAY BURSTER MYSTERIES AND SOLUTIONS

For about 20 years, astronomers have been investigating sudden bursts of gamma rays that seem unconnected with any visible feature of the sky. Many of these are very powerful and would most likely be caused by a cataclysmic event of some sort. These bursts seem to come at random from all parts of the universe, and like lightning, they have a reputation for not striking twice in the same place. Three that do recur, called "soft x-ray repeaters," are thought to be unrelated to the more energetic bursters, and are known to be associated with supernova remnants. Most bursts last only a few seconds, although a few last for minutes. According to some astronomers, the bursts cannot be linked to any known population of objects.

As part of the search for the cause of the bursts, scientists added the Burst and Transient Source Experiment (BATSE) to a spaceborne "telescope" for observing high-energy electromagnetic energy. The *Compton Gamma Ray Observatory* was launched on April 4, 1991. In its first three years of operation, the *Compton Observatory* recorded about a thousand bursts. It continues to record an average of almost exactly one burst each day. Despite all these observed bursts, astronomers have reached totally different conclusions about what the data mean. Jean Quashnock and Don Lamb of the University of Chicago conclude that the Compton record shows clumping of bursts, implying that the bursts are in the Milky Way galaxy. But a team led by Jay Norris of the Goddard Space Flight Center in Greenbelt, Maryland, used the same set of data to determine that the bursts are dilated by traveling great distances in an expanding universe, and that dimmer bursts are redder. These observations are taken to imply that the bursts are *not* in the Milky Way Galaxy, but are from distant galaxies. The theories behind these differing conclusions are discussed below.

On February 17, 1994, the *Compton Gamma Ray Observatory* encountered the longest and most energetic burst ever detected. The February event lasted more than an hour, instead of the more typical burst that gave the entities their name. Not only did the burst last a hundred times longer than any previously observed event, but during that period the burst consisted of gamma rays a dozen times as energetic as any previously observed in a burst.

Theories Come and Go, but the Bursts Go On

Since the discovery of the bursts, theories have abounded to explain them. As new evidence about the bursts has been gathered, either by specific searching or in random encounters, the old theories have been for the most part shown to be inadequate. Despite all efforts, the basic mystery remains: what causes the bursts? A summary discussion of the subject early in 1994 noted that gamma-ray bursts at that time were the only transient phenomena in astronomy for which there was no satisfactory explanation on which most astronomers could agree.

Here is an account of how astronomers have been groping for a theory:

- Bursts were first observed in 1968 by cold-war-inspired military satellites looking for secret nuclear blasts on Earth. By 1973, when news of the bursts was declassified, it was clear that they came from space. For many years thereafter, gamma-ray bursts were thought to be caused by neutron stars in the Milky Way. One of the main pieces of evidence leading to such a supposition was the great energy of the bursts. If the bursts were in far-off galaxies, there was no mechanism known that could produce the energy that was being measured in space near Earth. A few astronomers even put forth the idea that the bursters were something in the solar system or otherwise really close to the Sun, which would require the production of much less energy and therefore not require the exotic neutron stars. Among the arguments against such a local origin is that x rays would be detected if the bursters produce a typical electromagnetic spectrum and are close enough for the x rays to arrive without being absorbed by interstellar dust—but no associated x rays have been identified.

- In 1992, when the BATSE results showed that bursts are distributed equally in all directions, it became common to claim that they could not be caused by anything localized in the Milky Way, which is shaped like a disk and is far from being centered about Earth. Too much energy was being produced for a single neutron star to be the cause of the bursts; perhaps they were caused by two neutron stars colliding. This concept also provided an explanation for why bursts are, for the most part, not repeated from the same location. After such a collision, there would be nothing left to collide a second time. Calculations soon showed, however, that gamma rays would not be produced in such a collision—and gamma rays completely formed the observable bursts.

- In 1993, Martin Rees of Cambridge University and coworkers first proposed that gamma-ray bursts are a delayed energy release from the interaction of debris with material in the space surrounding a fireball caused by the collision of two neutron stars. Rees assumed that such collisions take place at a rate of about one per galaxy per 100,000 years, which would be enough to match observation, given the great number of galaxies in the universe.

- About the same time in 1993, Peter Mézáros of the Institute of Astrophysics in Paris, France, and coworkers suggested that the cause of the bursts might be interactions between a neutron star and a black hole that would direct a beam of gamma rays away from the site.

- Another solution to the problem was put forward in 1993 by P. A. Harrison, A. G. Lyne, and B. Anderson, who suggested that the Milky Way might be enveloped in a large sphere (known to astronomers as a "halo") of neutron stars that reached toward the nearby galaxies in the Local Cluster. The halo, if it exists, would preserve the original concept of bursts caused by single neutron stars by orienting the bursts close enough to Earth so that their original energy did not have to be incredibly great. This idea looked less and less viable, however, as BATSE continued to report bursts. For the lack of pattern to be maintained, the halo had to be postulated as larger and larger with each new batch of bursts included. Soon the halo was large enough to include nearby galaxies, which should possess halos of their own. The whole halo idea was falling to pieces.

- Also in 1993, R. S. White suggested that the bursts might be caused by comets colliding.

- Quashnock and Lamb argue that repeated bursts from about the same region in space mean that the bursts must be smaller than faraway ones that would have to consume the entire energy of a large star in a few seconds in order to produce the effects felt near Earth by the BATSE experiment.

- On January 15, 1994, Jay Norris of Goddard Flight Center in Greenbelt, Maryland, and coworkers reported that dim bursts regularly have longer durations than more intense ones. To Norris and his coworker Robert Nemiroff of George Mason University in Fairfax, Virginia, the negative correlation between burst intensity and length implies that the bursts are caused by fairly uniform violent events throughout the entire universe, events that all have about the same strength and duration. Those that are close seem brighter—that is, with greater amplitude, or stronger—

while those from far away are dimmer, as would be expected. But not so obviously, because, according to an interpretation of the general theory of relativity, electromagnetic radiation is stretched slightly as it travels through space. Thus, those events of the same duration that are far away seem to last longer to us than those that are close. Hence, if the statistics show that bright events are shorter than dim events, the bursts must be randomly distributed through space. This argument failed to sway many astronomers, who suggested that other explanations of this correlation were possible. For example, each event might use up the same amount of energy, but the more quickly it did so, the brighter it would appear. This would be similar to comparing the bright, short fires of pine with the less intense, longer lasting fires of oak. No relativity need apply.

- In the May 12, 1994 *Nature*, Andrew Lyne, this time with D. R. Lorimer, renewed the idea that halo neutron stars might cause the bursts, based upon their calculations that suggest many more neutron stars in the halo than previously thought (*see* "Pulsars," p 195).
- In the August 1, 1994 *Astrophysical Journal Letters*, Rees and Mézáros expanded upon their earlier idea. This time they proposed that giant bursts, like the one in February 1994, were caused by some cataclysmic event, such as the explosive collision of neutron stars with each other or with a black hole, or the explosion of a collapsed star. In an event of this type, the first explosion debris is not moving as fast as debris from later in the explosion. As the faster debris catches the slower debris, it produces an initial quick burst of soft gamma rays, followed by a long stream of hard rays as all the debris catches up with the virtually stationary interstellar matter surrounding the explosion.
- Four Russians from the P. N. Lebedev Institute in Moscow (Aleksander Gurevich, Gely Zharkov, Kirill Zybin, and Michail Ptitsyn) argue that evidence points to the bursts emanating from neutron stars in the galactic halo, a large sphere of matter that surrounds the Milky Way. They observe a slight increase in the number of bursts from the direction of the Andromeda Galaxy, the closest similar galaxy to the Milky Way. If the same processes produce fairly small bursts in the halos of each giant galaxy, the bursts would seem to be almost evenly distributed because most bursts would come from the Milky Way; but the stronger bursts from the Andromeda galaxy would also reach us.

As a further complication, the *Compton Gamma Ray Observatory* has also observed gamma-ray bursts that clearly come from Earth. No one thinks that these are previously unsuspected nuclear blasts. Instead, it seems most likely that the Earthly bursts are caused by bad weather (*see* "Lightning and Related Phenomena," p 506).

(Periodical References and Additional Reading: *New York Times*, 1-26-93, p C1; *New York Times*, 4-23-93, p A22; *Science* 1-7-94, p 47; *New York Times* 1-16-94, p I-21; *Science* 1-28-94, p 467; *New York Times* 1-16-94, p I-21; *New York Times*, 3-15-94, p C15; *Astronomy* 5-94, p 20; *Nature* 5-12-94, p 127; *Science* 5-27-94, p 1250 & 1313; *Nature* 7-7-94, p 26; *Science* 8-26-94, p 1187; *Astronomy* 11-94, p 24; *Nature* 12-15-94, p 652; *Science News* 12-17-94, p 404; *Astronomy* 1-95, p 24; *Astronomy* 2-95, p 26; *Astronomy* 3-95, p 30)

OUR GALAXY AND ITS NEIGHBORS

INSIDE THE MILKY WAY

Although astronomers have known for about 200 years that our larger environment is the Milky Way galaxy, we are still in the process of learning surprisingly basic information about the home galaxy. Estimates of the actual number of stars in the galaxy, for example, vary by a factor of two. Rotation and proper motion are reasonably well known (but *see* "Dark Matters," p 226), while new information in the 1990s provides the first estimate of our home's mass.

Weighing Our Galaxy

Douglas N. C. Lin of the University of California at Santa Cruz and coworkers Burton Jones and Arnold Klemola have used the nearest neighbors of the Milky Way to study both the neighbors and our galaxy itself. Their basic method was a comparison of the 1974 positions of certain bright objects in the neighboring Large Magellanic Cloud (LMC) with the positions of the objects in 1993, as measured against the background of faraway galaxies. From this data, they could compute the orbital speed of the LMC travelling about the Milky Way as a satellite galaxy. This speed they found to be 147 mi (237 km) per second, as measured with respect to the center of the Milky Way—or 530,00 mph (850,000 kph). The period of revolution for the LMC is a rather long 2.5 billion years. If the universe is as old as new measurements suggest (*see* "Cosmology UPDATE," p 236), the LMC has only gone through four or five such "years" in its entire history.

With this information in hand, the astronomers could use Newtonian mechanics to determine the total mass involved—in effect "weighing" the Milky Way. Of course, *weight* is a measurement of a force caused by gravitational attraction, so for an object in space considered with respect to itself, the proper term is *mass*. The astronomers calculated that the mass of the Milky Way is equal to that of about 600 billion stars, each the size of the Sun. This calculated mass is about 10 times the mass observable with electromagnetic radiation, which includes light, x rays, infrared radiation, radio waves, and so forth—the principal way that we observe the universe. Thus, the galactic mass measurements become yet another piece of evidence that dark matter makes up most of the universe (*see* "Dark Matters," p 226).

Previously, astronomers calculated the approximate diameter and thinness of the Milky Way. Our galaxy is about 30 kiloparsecs or 100,000 light-years across, and the spiral arms are only about 0.6 kiloparsecs or 2,000 light-years thick.

As the LMC orbits the Milky Way, the LMC is so close (160,000 light-years) that it interacts directly with the outer part of the Milky Way, known as the halo. These interactions are removing chunks of the LMC, and sticking them on the outside of our galaxy. Lin calls the process "galactic cannibalism," and predicts that in another 20 billion years the Milky Way will have completely consumed the LMC.

Background

Nicolaus Copernicus [Polish: 1473-1543] is justly famed for teaching humans that Earth and other planets revolve around the Sun. However, William Herschel [English: 1738-1822] in the period between 1784 to 1818 and Edwin Hubble [American: 1889-1953] in 1925 may have accomplished as much by demonstrating the general nature and shape of our galaxy, the Milky Way, and its relation to the rest of the universe.

Today we know that the Milky Way galaxy is a spiral assembly of over 200 billion stars floating amid an uncounted number of other such galaxies. Since the solar system is a part of the Milky Way, all we see of the galaxy is a band of stars across the night sky, too many to resolve with the naked eye. To the ancient Greeks, the band looked like milk (gala) spilled across the sky, which gives us the word galaxy.

The band of light does not look like milk to most people today. It is so faint that people living near cities cannot see it at all. To be appreciated properly, the Milky Way must be seen far from any source of light pollution.

What Is It at the Center?

As observed from Earth, the center of the Milky Way is in the constellation Sagittarius, a bit below the not-very-bright star (fifth magnitude) X Sagittarii. The center is about 28,000 light-years from Earth. A number of interesting phenomena have been observed near our Galactic center (*see*, for example, "The Sigma Telescope and the Great Annihilator" below), even though we can just barely image the center at visible wavelengths. Radio and infrared observations show gas clouds and stars near the center behaving as if they were orbiting some object that has several million times the mass of Earth's sun. This would be an excellent candidate for a supermassive black hole if it emitted some high-energy radiation. This, however, it fails to do.

Radio telescope studies, which are not affected by dust and gas, reveal that there is considerable activity at the center. Most of this is from Sagittarius A* (A* is pronounced "ay star"), the strongest and most compact radio wavelength source of energy in the region. Sagittarius A* is smaller in diameter than Earth's orbit—less than two astronomical units.

The question of classifying Sagittarius A* remains open. Sagittarius A* is a high energy emitter, which means that most of its radiation is toward shorter wavelengths. In optical terms, this means blue light. Thus, the "object" could be a compact cluster of very blue stars, the kind astronomers call type O supergiants. In 1992, German astronomers Andreas Eckart, Reinhard Genzel, and coworkers detected 340 individual stars within 1.3 light-years of Sagittarius A*, showing that there is a dense cluster in the region. The 340 stars detected are thought to be only the most massive of that dense cluster. The nearby object IRS 16, which was once thought to be a single massive object, was later shown to be a cluster of at least 25 blue supergiants.

Another common idea classifies Sagittarius A* as a black hole. A black hole cannot be seen directly; what is observed, if anything at all, is a ring of dust and gas, with pieces of stars swirling around the black hole as they are consumed by it. Such a ring is known as the accretion disk of a black hole. Accretion disks become very hot as they

THE SIGMA TELESCOPE AND THE GREAT ANNIHILATOR

In December, 1989, a Russian satellite called *GRANAT* was launched into a high, elliptical, four-day orbit from Kazakhstan. The French telescope aboard it, called the Sigma Telescope, is the product of a collaboration between Jacques Paul of Saclay in Gif-sur-Yvette, France, and Pierre Mandrou of the Centre d'Etude Spatiale des Rayonnements in Toulouse.

The instrument is based on an idea of Robert Dicke of Princeton, which uses an essentially random array of pinhole cameras in front of a gamma-ray detector. A computer program takes the pattern of images on the detector and computes the direction of the gamma rays. Unlike charged particles, which are buffeted about on their journey through space by stray magnetic fields, gamma rays mostly travel in straight lines, as does light.

As Dicke's idea is implemented in the Sigma Telescope, this arrangement has an excellent spatial resolution, but not such a good energy resolution. It can tell almost exactly where the gamma rays are coming from, but the amount of energy of the gamma rays is somewhat blurred. At lower wavelengths of light, this would be described as successfully detecting the origin of the light, but being unable to recognize its exact color. This energy blurring results from Sigma's sodium-iodide detector. Future plans are to combine a high-resolution germanium detector with a Dicke pinhole array, to obtain a gamma-ray telescope that can identify both the origin and energy of gamma rays.

Germanium detectors that can quite precisely determine energy levels but have a low spatial resolution have been observing the sky since 1977, when a balloon-based detector observed what appeared to be electron-positron annihilations near the galactic center. Subsequent balloon flights sometimes observed repeats of these annihilations and sometimes did not. It appeared to be a phenomenon that waxed and waned, although without a regular periodicity.

On October 13–14, 1990, the Sigma Telescope detected tremendous 20-hour burst of gamma rays from an otherwise ordinary x-ray source. During this period, the source became 50,000 times as energy-emitting as the Sun. Furthermore, the energy of the photons that comprised the gamma rays appeared to be exactly the amount that would be caused by the mutual annihilation of about 1,044 electrons and their antiparticles (positrons) every second. Thus, Marvin Leventhal of Bell Laboratories suggested that the x-ray source (known formally as 1E1740.7-2942 and informally as the Einstein source) be called the Great Annihilator. Its nickname, "Einstein source," derives from the discovery of this x-ray emitter (among others) by the Einstein satellite in 1979.

The astrophysicists' best explanation for the gamma-ray burst theorizes that the Great Annihilator is a black hole and that the burst came as a gas cloud or star fell into it, never to be seen in this universe again. However, careful measurements show that the Great Annihilator is not the suspected black hole at the center of the Milky Way, but is instead located 100 to 300 light-years away from the center. It seems odd that there could be a supermassive black hole at the center, and another modest black hole hanging around in the general neighborhood.

Albert Einstein's famous 1905 equation $E = mc^2$ (energy equals mass times the square of the speed of light) implies that mass and energy are two different ways to think about the same thing. It is convenient to measure the very small masses of subatomic particles in terms of their energy equivalents, which are much greater than their masses. Even so, a very small unit, the electron volt (abbreviated eV), and its multiples (such as the kilo-, mega-, or giga-electron volt) is needed to quantify the energy of the particles. For example, a kilo-electron volt is 1,000 times as great as an electron volt (a further discussion of this topic can be found in "Sub-atomic Particles," p 666).

The gamma rays found coming from the Great Annihilator can be recognized because most of the rays have an energy of 511 keV (kilo-electron volts). The energy of 511 keV is one of the great unexplained numbers of modern physics, because it is precisely the rest mass of every electron ever observed (physicists say "rest mass" because moving objects acquire additional mass from their motion, an effect that is more noticeable at speeds near the speed of light). The electron's antiparticle, the positron, also has a rest mass of 511 keV.

Particles and antiparticles that meet annihilate each other. To imagine this process, think of the antiparticle as a precisely shaped hole that the particle just fits. When particle and antiparticle meet, the particle falls into the hole, and both hole and particle completely disappear with a thud. The "thud" results from all the energy that both the particle and the hole were using to maintain their shapes. Even though the particle and the hole disappear, the energy cannot vanish. In the particular type of annihilation suspected for the Great Annihilator (in which the electron and positron first orbit each other in a combination known as positronium) the energy appears as one gamma ray for the particle's mass and another for the hole's mass. Each of these masses is 511 keV when measured in terms of energy, so the appearance in one second of 1,044 gamma rays possessing that energy suggests that half of 1,044 such positronium annihilations have occurred during that second, although some astrophysicists think that some related annihilation processes may also be involved.

---

are pulled together by the black hole, which is why they are expected to emit blue light as well as other radiation. The other radiation knocks electrons out of the gas farther from the black hole, producing radio waves.

As attractive as the idea of a black hole is, studies conducted by Andrea Goldwurm of the Centre d'Etudes de Saclay in Gif-sur-Yvette, France, and Russian and French coworkers between 1990 and 1994 using the Sigma Telescope appear to have ruled out this idea, at least for the object at Sagittarius A*. On the other hand, their report in the October 13, 1994 *Nature* found three good black hole candidates near Sagittarius A*. These candidates produced gamma rays with energy levels above 75 keV (*see* "The Sigma Telescope and the Great Annihilator" below) in significant amounts, similar to the black hole candidate at Cygnus X-1.

However, according to this evidence, the object at Sagittarius A* simply does not produce enough gamma rays to be a black hole of either the Cygnus X-1 variety nor of the galactic center type. Many astronomers, however, persist in thinking that Sagittarius A* really is a black hole, and that the evidence against this can be discounted. It is possible that Sagittarius A* radiates at a frequency higher than the Goldwurm team were able to observe, just as some dog whistles cannot be heard by humans.

Meanwhile, astronomers at Kitt Peak in Arizona were having better luck at lower frequencies. As noted above, infrared radiation can penetrate most kinds of dust and gas clouds. A team of astronomers, including Laird M. Close and Donald W. McCarthy Jr., used a new high-speed infrared camera and adaptive optics (see "Optical Telescope UPDATE," p 140) to snap 121 images of an object at the center of the Milky Way, which (in infrared) looked like a black hole.

Faster than Light and Close to Home

In March 1994, astronomers I. Felix Mirabel of the Center d'Etudes de Saclay and Luis F. Rodriguez of Mexico's National Autonomous University and National Radio Astronomy Observatory, while using the Very Large Array (VLA; see "Major Telescopes," p 247) to look for radio counterparts of gamma-ray bursters (see "Gamma-ray Burster Mysteries and Solutions, p 000), observed what they termed "a remarkable ejection event." On March 24, they turned the VLA instruments to GRACE + 105, which is intermittently one of the brightest radio sources in the sky. GRACE + 105 has been known only since 1992, so it has not been well studied. The telescope caught a very powerful (at x-ray wavelengths) object possessing two components. In six more observations over a period somewhat longer than a month, the astronomers detected two objects moving away from each other at an apparent speed faster than that of light. Such objects are called superluminal. For 20 years previously, such superluminal events had only been observed at great distances from the Milky Way galaxy, but this one, known as GRACE + 105, appeared to be relatively close at hand: within the Milky Way, in fact.

Proof that the object is so close derives directly from the equations for relativistic motion. Limitations on the distance can be calculated from the apparent motions of the two objects with respect to the background of stars. When data from GRACE + 105 is thus calculated, the distance must be less than 13.7 kiloparsecs (about 40,000 light-years), which places the burst well within the Milky Way in the direction of the Galactic plane. Furthermore, although quasars had previously been found only at great distances from the galaxy, GRACE + 105 has been identified as a quasar, albeit a very small one.

Mirabel and Rodriguez believe that the event at GRACE + 105 began on March 19. According to their theory, GRACE + 105 is a binary object composed of a fairly normal star orbiting a neutron star or a black hole. Some of the radio brightness comes from the dense neutron star or black hole pulling material away from its companion and producing a disk around the dense central object. Mirabel and Rodriguez call the double object a microquasar. On March 19, the dense star in the microquasar propelled outward from its poles two clouds of material that together may be about a third of the mass of the Moon. It is these two clouds that emit the strong radio waves. Although the mass of the clouds is small, their great speed gives them a kinetic energy about equivalent to 100 million Suns. The faster of the two clouds is moving at 92% of the speed of light in a vacuum.

FASTER THAN A SPEEDING PHOTON?

Since 1971, astronomers have known that there are objects registering on their instruments at speeds faster than light, speeds that had been deemed impossible according to standard relativity theory. Scientists explain that the objects in question—giant jets of subatomic particles streaming from the active cores of distant galaxies or quasars—are not moving faster than light, despite appearances. Instead, the effect is caused by material extending for great distances which is moving at a speed close to that of light, combined with the angle at which the jet is viewed. If the angle is close enough to 90°, the jet is long enough, and the speed of particles being ejected is great enough, then light from the close end of the jet will travel a shorter distance than light from the far end. The result will appear to be a jet of particles travelling faster than light. Astronomers have invented the adjective **superluminal** to describe this situation.

In one example, the jet emanating from the elliptical galaxy M87 is thought to be 16,000 light-years long. The part of the jet that is aimed more-or-less at Earth appears to be moving 2.5 times the speed of light in a vacuum. The jet is no doubt expanding very rapidly, but at a speed less than that of light. Measurements were taken by the Very Large Array at various times between 1982 and 1993 to obtain this information.

Emissions from the microquasar are located more in the x-ray part of the spectrum than in the more energetic gamma-ray portion. Because of the poor resolution of most x-ray telescopes, it is not clear whether GRACE + 105 is the x-ray source from that sector of the sky. The mechanism by which the two clouds are thought to have been produced is similar to that thought to cause gamma-ray bursts.

GRACE + 105 is in too dusty a region of the Milky Way to be observed at optical wavelengths, although there is some hope that it could be "seen" at infrared wavelengths. If its spectrum could be obtained, the combination of the redshift information with the distance derived from the apparent speed of the clouds could be used to determine a definite distance. Such a technique might become useful in measuring cosmic distances (*see* "Cosmology UPDATE," p 236).

An even closer and more powerful microquasar may be a better bet. In August 1994, such an object was found soon after the *Compton Gamma Ray Observatory* had located it as an x-ray source. Known as X-ray Nova Scorpii, the microquasar has the same features as GRACE + 105, including the superluminal jets. At 11,000 light-years, it is much closer than GRACE + 105. Furthermore, X-ray Nova Scorpii is also 10 times brighter. The second galactic mircroquasar was discovered by Robert Hjellming of the U.S. National Radio Astronomy Observatory, and its superluminal component recognized by Derek McKay and Michael Kesteven of the Australia Telescope National Facility.

(Periodical References and Additional Reading:; *New York Times* 1-6-93, p A17; *New York Times* 6-8-93, p C1; *Nature* 9-1-94, pp 18 & 46; *Science* 9-2-94, p 1362; *Nature* 10-13-94, pp 561 & 589; *Astronomy* 12-94, pp 18 & 20; *New York Times* 6-21-94, p C11; *New York Times* 9-1-94, p b12; *Discover* 1-95, p 33; *Astronomy* 3-95, p 24; *Astronomy* 5-95, p 32)

NEW NEIGHBORS: DWINGELOO 1 AND MORE

The Milky Way is not exactly isolated in space. There is a group of about 10 small galaxies that are either orbiting or passing very close to our own. Furthermore, the Milky Way and a couple of other large galaxies are bound together by gravitation. The other large galaxies also have small ones in orbit about them, or at least in their immediate neighborhood. The entire group that travels together is called the Local Group. Furthermore, at least three galaxies (Maffei 1 and 2 and IC 342) appear to be passing through the domain of the Local Group at speeds too fast to join us gravitationally. One theory, however, holds that Maffei 1 and IC 342 are former members of the Local Group that have been tossed out by an encounter with the gravity of M31, the Andromeda galaxy (often still called by its former name, the Andromeda Nebula).

If Maffei 1 and IC 342 did encounter M31, the event took place about 8 billion years ago. Astronomers have nicknamed such a theoretical encounter the "little bang" for the influence it has since had on the Local Group. Evidence in favor of the little bang includes the great speed of Maffei 1 and IC 342; the motion of M31 toward the Milky Way, caused by recoil from the encounter; the discovery by astronomers using the *Hubble Space Telescope* that M31 has two central regions of high density, instead of the single luminous center of other spiral galaxies; and a general disorder of Andromeda. In the little bang scenario, bits and pieces of M31 formed other galaxies in the Local Group, including the two Magellanic Clouds. The small galaxy Leo I, which is moving faster with respect to the Milky Way than any other galaxy, could also be debris. Other astronomers feel that the evidence is not compelling, or even that there is conflicting evidence precluding the little bang theory. The double nucleus of M31, whose discovery was announced on July 19, 1993, may not have been caused by a collision, but by some other process, or even be an illusion caused by a band of dust across the middle of a single elongated nucleus.

Beyond the Seven Dwarfs

In the 1930s, astronomer Harlow Shapley, the first to realize the approximate size of the Milky Way, discovered two much smaller galaxies that are adjacent (in astronomical terms) to the Milky Way. These are called dwarf galaxies (*see also* "Shape and Evolution of Galaxies," p 216). A dwarf elliptical galaxy may contain only a few million stars, as compared to the hundreds of million stars in a regular spiral galaxy. The two Shapley dwarfs discovered are part of the Local Group of galaxies. Prior to 1994, there were a total of seven or eight generally recognized dwarf galaxies in the Local Group. However, some astronomers include as dwarf galaxies all of the galaxies in the Local Group which are smaller than the three largest.

In 1994, the British astronomers Rodrigo A. Ibata, G. Gilmore, and Mike J. Irwin from Cambridge University announced the most recently discovered dwarf galaxy, which they also found to be the closest dwarf galaxy in the local group. They used data collected by a large telescope situated in Australia. Ibata was the first to notice that a group of stars was moving as a group, after which it took him and his coworkers two years to be certain that the stars were part of a previously undiscovered galaxy. Following one convention for naming members of the Local Group by the constellation in which they are found, the Cambridge astronomers have named the new dwarf Sagittarius. Although a dwarf, it is comparable in size to the previously known largest dwarf in the Local Group: Fornax, which was one of the two original dwarfs found by Shapley. These large dwarfs are about 10 light-years in diameter. Sagittarius is between 70,000 and 80,000 light-years from the solar system, and about 50,000 miles from the Milky Way.

Because of its large size and proximity to the Milky Way, Sagittarius should be torn apart by tidal forces. That is, the part closest to the Milky Way is affected by gravity so much more than is the farthest part that the Milky Way's gravity supersedes the internal gravity of Sagittarius. Indeed, Sagittarius appears elongated in the proper direction to suggest such a phenomenon. If so, Sagittarius will eventually become incorporated into (astronomers say "cannibalized by") the Milky Way. It should be absorbed completely over the next 100 million years. The same fate will eventually befall the even larger Large Magellanic Cloud, perhaps some 10 billion years from now.

Depending on which astronomer is asked, there are between 20 and 32 galaxies in the Local Group, including the newly discovered Sagittarius. Here is one list of commonly accepted members, with remarks about some of the better known members.

THE LOCAL GROUP

(*See also* "Shape and Evolution of Galaxies," p 216.)

Name	Remarks
M31 (The Andromeda "Nebula")	Spiral (Sb). About 1.5 times the mass of the Milky Way, and more than two million light-years away from it. In 1993, images from the *Hubble Space Telescope* demonstrated that M31 has two nuclei.
Milky Way	Spiral (Sbc). Our home galaxy.
M33 (in Triangulum—also known as NGC 598)	Spiral (Scd). About 0.15 times the mass of the Milky Way. Almost three million light-years away from us.
Large Magellanic Cloud (LMC)	Irregular (SBm). In orbit about the Milky Way some 160,000 light-years away. About 0.03 times the size of the Milky Way.
Small Magellanic Cloud (SMC)	Dwarf irregular (Im). In orbit about the Milky Way, almost 200,000 light-years away. About 0.005 times the size of the Milky Way.
Mini-Magellanic Cloud	Dwarf irregular.
Draco	Dwarf irregular or dwarf elliptical (dE). In orbit about Milky Way, at a distance of about 260,000 light-years.
Fornax	Dwarf spheroidal or elliptical (dE) galaxy. About 800,000 light-years from Milky Way.
Sculptor	Dwarf elliptical (dE). Discovered by Shapely in the 1930s along with Fornax. Seems to contain some stars that it has captured from other galaxies. In orbit around the Milky Way at a distance of 275,000 light-years.
Sextans A	Dwarf elliptical (dE) that contains mostly old stars.

Name	Remarks
Leo I	Dwarf elliptical (dE) in orbit around the Milky Way at a distance of about 750,000 light years, but speeding away from us at about 110 mi (177 km) per second.
Leo II	Dwarf elliptical (dE) in orbit about Milky Way at a distance of about 750,000 light years.
Leo A	Dwarf elliptical (dE).
NGC 6822	Dwarf irregular (Im) galaxy cloaked in dust. Vigorously producing new stars. About 1,700,000 light-years away in Sagittarius.
NGC 147	Dwarf elliptical (dE5) somewhat less than two million light-years from Milky Way. Gravitationally bound to M31.
NGC 185	Dwarf elliptical (dE3) somewhat less than two million light-years from Milky Way. Gravitationally bound to M31.
NGC 205	Elliptical (E5) galaxy in orbit about M31. Somewhat less than two million light-years from us.
M32 (NGC 221)	Elliptical (E2) galaxy in orbit about M31. Somewhat less than two million light-years away from us.
IC 1613	Dwarf irregular (Im) galaxy in Cetus. More than two million light-years away.
Ursa Minor	Dwarf elliptical (dE) less than 250,000 light-years from the Milky Way.
Carina	Dwarf elliptical (dE) more than 500,000 light-years from the Milky Way.
And II	Dwarf elliptical (dE) somewhat less than two million light-years from the Milky Way.
Tucana	Dwarf elliptical (dE).
Pegasus	Dwarf elliptical (dE).
Sagittarius	Dwarf spheroidal (dE) galaxy about 50,000 light-years from center of Milky Way, containing about a billion stars.

Dwingeloo 1

The August 4, 1994 discovery of galaxy Dwingeloo 1 was announced later that month at the General Assembly of the International Astronomical Union and detailed in the November 3, 1994 *Nature*. Dwingeloo 1 was located for the first time by an international group of astronomers working together as the Dwingeloo Obscured Galaxy Survey. They used a radio telescope to survey the "zone of avoidance," where light is absorbed by dust and gas but longer wavelength electromagnetic waves can penetrate. The head of the survey team was Harry Ferguson of the Space Telescope Science Institute in Baltimore, Maryland, and the lead author of the *Nature* article is Renee Kraan-Korteweg of the University of Gronningen in the Netherlands. Both the team and the galaxy are named for the Dutch 82 ft (25 m) radio telescope used to penetrate dust in the Milky Way and pick up the characteristic radiation of hydrogen, a wave 8.25 in (21 cm) in length. Immediately upon detecting the radio signal of a spiral galaxy, the Dwingeloo group asked for and obtained confirmation of the new galaxy from a large radio telescope and from the United Kingdom Infra-Red Telescope on

ZONE OF AVOIDANCE

Early in the twentieth century, astronomers tried to account for the observation that there were no galaxies in a large region of the sky. Theories were proposed to explain why galaxies stayed elsewhere in the universe, and the region, about 15% of the apparently observable universe, was named the zone of avoidance.

Astronomers still use the term "zone of avoidance" to describe this region today, although they have long known that the region actually contains as many galaxies as normal. Instead, large quantities of dust, gas, and hidden stars in that part of our own Milky Way galaxy prevent light from galaxies opposite the zone from penetrating it and reaching our location in space.

Mauna Kea in Hawaii. The Hawaiian telescope was able to image a faint spiral through the intervening dust using infrared wavelengths.

The newly found barred spiral galaxy is only 10 million light years away, making it a close neighbor by galactic standards, although not close enough to be a member of the Local Group. It is also moving in a separate direction from the galaxies in the Local Group. Dwingeloo 1 is smaller than the Milky Way, having only about a third of its mass.

Knowledge of the distribution of nearby galaxies provides important clues toward solving various cosmological problems, since it helps astronomers calculate the amount of mass in the universe, an important constant for several reasons (*see* "Cosmology UPDATE," p 236). For example, the concentration of mass that includes the Great Attractor has the zone of avoidance right in its center.

(Periodical References and Additional Reading: *New York Times* 7-29-93, p C10; *Astronomy* 11-93, pp 18, 28 & 94; *Astronomy* 2-94, p 26; *Science* 4-15-94, p 343; *New York Times* 7-26-94, p C6; *Nature* 7-21-94, p 194; *Science* 9-9-94, p 1526; *Nature* 11-3-94, pp 38 & 77; *Science* 11-11-94, p 974; *Discover* 1-95, p 36; *Astronomy* 3-95, p 22; *Science News* 6-24-95, p 392)

A C C O U N T I N G   F O R   T H E   U N I V E R S E

THE SHAPE AND EVOLUTION OF GALAXIES

There would seem to be no reason for galaxies to exist at all. In fact, it is easy to imagine a universe filled with randomly spaced stars: random spacing, for example, is the way molecules are arranged in a gas or liquid. Another possible arrangement for the universe would be that of a crystalline solid, with stars at regular intervals in a giant pattern in space-time. However, observation tells astronomers that these simple ideas are far from the truth. The universe exists mostly as apparently empty space, with large, shaped groups of stars called galaxies arranged in clusters of galaxies, and clusters of clusters.

One of the main unsolved questions in astronomy involves finding the mechanism that causes stars to clump and form galaxies in the first place. Another involves determining why galaxies clump, or form clusters and superclusters. There is a somewhat better understanding of why many galaxies form into shapes that are usually identified as spirals and ellipses, although the three-dimensional shapes in each case are more complicated than that, and a very large class of galaxies has no well-defined shape at all.

Galaxies are Born

The most probable way for the the big bang to have happened is smoothly in all directions, like an expanding sphere. The most popular version of the big bang, called the new inflationary universe theory, has two possible interpretations: the inflation could have proceeded smoothly, or the early universe could have developed small fluctuations in temperature or other properties. The model with fluctuations is generally preferred. Fluctuations seem to be necessary to produce the clumped universe astronomers observe. Furthermore, the uncertainty principle of quantum theory suggests that some small fluctuations should appear. For inflation with fluctuation, as the universe inflates and then explodes, the original small fluctuations expand along with the universe and result in the large observable disparities of today.

The organization of the universe is generally assumed to have resulted from gravitational attraction, although a loud minority supports a competing theory of organization occurring due to electromagnetic fields. If gravity is the cause of clumping, then whatever fluctuated in the very early universe must have had or produced mass. One possibility would be that fluctuations produced local differences in the number of baryons formed. (Baryons are the familiar protons and neutrons, along with less familiar and more massive versions of the same particles that soon decay into protons or neutrons; protons and neutrons are the main basis of mass in atoms, and the main constituent of atoms of hydrogen and helium.) The production of more baryons in some places would result in variations in the amounts of hydrogen or helium occurring from place to place, which is what we observe.

As simple as this idea may sound, baryon variations are not easily extrapolated from the equations used to describe the early universe. The problem is that baryons interact too strongly with electromagnetic radiation. The radiation present during the early universe would tend to smooth out differences in baryon density. As a result, most cosmologists (astronomers or physicists who study the universe as a whole) postulate some other kind of massive particle that reflects the fluctuation. This other particle is a member of the class of weakly interacting massive particles, or WIMPS, which have never actually been observed (*see also* "Dark Matters," p 226).

If fluctuations existed in the early universe, they should also be reflected in small variations in the radiation that was produced when matter precipitated out of the universe some time after the big bang. Specifically, regions that were denser then should appear as regions of the cosmic background radiation which are warmer now. Finding and measuring those warmer regions could help explain the evolution of the universe. Earth-based measurements have suggested very good agreement betweem observation and theory, but measurements from space are able to be much more precise. For one thing, not all of the background radiation is at frequencies that can penetrate the atmosphere. Checking the smoothness of the radiation was the main goal of the satellite based Cosmic Background Explorer (COBE) when it was launched in 1989 (*see also* "Cosmology UPDATE," p 236).

At first COBE found only smoothness, but as it accumulated data it was able to refine the measurements. By 1992, George Smoot of Lawrence Berkeley Laboratory in Berkeley, California, could announce that COBE had found fluctuations. The variations dated from a time when the universe was less than 300,000 years old, early in the big bang explosion. This result appeared to confirm the origin of galaxies, although later calculations showed that the ripples found by COBE are too small in value to provide complete confirmation. These fluctuations can account for some of the observed structure, but not all of it.

For one thing, the COBE wrinkles, although small in value, are too large in extent to cause individual galaxies. New studies are being conducted or planned to identify fluctuations that could result in galactic formation. Meanwhile, computer simulation has been an important tool in trying to understand how galaxies form. In these computer simulations, dwarf galaxies begin to form first, with these dwarfs gradually coming together to form larger galaxies.

Odd Shapes and Active Galaxies

Gregory Wirth and David Koo of Lick Observatory and Richard Kron of Fermilab compared several galaxy clusters in order to observe the shapes of individual galaxies in those clusters. Cl 0016 + 16, with a relatively high redshift of 0.55, is somewhat far away in galactic terms. For Cl 0016 + 16, new stars appear to be forming in what otherwise appear to be old galaxies. The galaxies closer in distance and time to Earth show new stars forming only in young galaxies. Using the *Hubble Space Telescope*'s Wide Field Camera, the researchers found that normal-appearing elliptical or lenticular galaxies (E or S0 types) did not show new star formation, but various odd-shaped galaxies accounted for new star formation. When the astronomers compared the odd-shaped and active galaxies from the different clusters, they also found marked variations by cluster, suggesting that different forces shape galaxies and stars in individual clusters.

One object, thought by those studying it to be a galaxy located some 13 billion light-years away (although it could be a quasar), seems to be forming stars at a rate about 10,000,000 times the rate of the Milky Way. This object could be a galaxy very early in its development. If so, it could contain clues regarding the general development of galaxies. However, it could also be a peculiar galaxy that is not typical of the early universe after all. Studies of 10214 + 4724 have been conducted by Philip Solomon of the State University of New York at Stony Brook working with Dennis Downes and Simon Radford of the Institute for Millimeter-Wave Radio Astronomy in Grenoble, France, as well as a team led by Paul A. van den Bout of the U.S. National Radio Astronomy Observatory at Kitt Peak in Arizona and a team headed by Erik Becklin of the University of California at Los Angeles. Becklin's group thinks that a quasar, not star formation, accounts for the radiation observed in 10214 + 4724.

Very odd shapes and other unusual galaxy properties were cataloged separately by Halton Arp, so that galaxy IC4553 has come to be better known from its Arp Atlas of Peculiar Galaxies catalog number of 220 than from its nineteenth-century Index Catalog (IC) number of 4,553. Arp 220 has been of particular interest since 1982, not only for its odd looped shape but also because it produces coherent microwave emissions. That is, it acts as a maser, the microwave analog to the laser. In 1993, as reported in the July 14, 1994 *Nature*, Colin J. Lonsdale of M.I.T. in Cambridge, Massachusetts, and coworkers found that the maser radiation originated in a very small region of the galaxy. One explanation of this would be a black hole hidden in the galaxy. However, because Arp 220 has two strong energy sources, it appears to have resulted from the merger of two galaxies that some time ago ran into each other.

NAMING THE SHAPES

Edwin Hubble was the first to recognize that the universe is filled with galaxies. Like scientists throughout time, he immediately began to classify them according to the information he had available: their appearance. Hubble's scheme recognized 16 types: 3 types of ordinary spirals with central spheres; 3 types of barred spirals, in which the central region is a bar instead of a sphere; 7 types of galaxies shaped more or less like ellipsoids; 2 types of galaxies too irregular to be grouped with the afore-mentioned; and 1 largely spherical galaxy that resembles the central region of an ordinary spiral galaxy but lacks those spiral portions called arms. Further study by astronomers Allan Sandage, Gerard de Vaucouleurs, and Sidney van den Bergh has led to improvements or corrections in the Hubble classifications, although the basic concepts remain. Primarily, the improved version includes additional types which are in between those originally recognized by Hubble.

Classification schemes are usually based in theories of galactic evolution, such as elliptical galaxies evolving into spirals. The following list, however, is arranged alphabetically according to the abbreviations astronomers use, and is comprehensive in the sense that there may be overlap from one classification scheme to another.

AA, Aba, Ab, ...	Anemic spiral galaxies, with some more anemic than others. These come between the lenticular galaxies (see below) and ordinary spirals.
dE	Dwarf elliptical galaxies, probably the most common type in the universe. They are known for their low luminosity and brightness.
E0	A sphere, designated as an elliptical galaxy with an ellipticity of 0. The concept of ellipticity is similar to, but still different from, eccentricity as used to measure ellipses. In the case of **E0**, the eccentricity and the ellipticity are both 0, since each depends on subtracting the short dimension from the long, and for a spherical galaxy the dimensions are equal.
E1	A slightly elliptical ellipsoid of revolution, although perceived in photographs as an ellipse. The formula for ellipticity is $10(a - b)/a$, where a is the longer dimension of the ellipse, and b is the shorter. Thus, since an **E1** galaxy has an ellipticity of 1, if b is considered to be a unit length, a must be a tenth of that unit longer.
E2-7	Ellipsoids of increasing ellipticity. The extreme case, **E7**, has a long dimension that is 70% greater than its short dimension. No galaxies with an ellipticity greater than 7 have been found.
Im	Sometimes used instead of **Irr Type I**.
Irr Type I	Galaxies with no developed spiral structure; also known as the Magellanic Irregulars (although the Large Magellanic Cloud is not classed in this group any more). Most of these are dwarf galaxies, showing no symmetry or nucleus, and consisting of stars and clouds.
Irr Type II	Galaxies that do not fit into any of the other classifications. Many of these are also given special names to describe their particular type, such as ring or active galaxies.
L	These are the same as the **S0** galaxies described below. Sometimes the **L** or the **S0** galaxies are subdivided by adding a minus or a plus, as in **L-** or **S0 +**.
Sa	Also known as an "early" spiral, this is the galaxy one pictures when the term galaxy is used, since it is a relatively tight spiral comprised of two pairs of arms and a well-defined central bulge.
Sb	This spiral is more open than **Sa** and in general contains more gas and dust in the spiral arms. The familiar Andromeda galaxy, or M31, is an example.
SBa	The capital **B** indicates a barred spiral. There is a bar instead of a sphere at the galactic center. In general, the barred spirals are classed by the same scheme as the ordinary spirals, except for the bar.
SBb	This barred spiral is more open than **SBa**, and in general contains more gas and dust in the spiral arms.
SBc	Also known as a "late" barred spiral, the spiral is very open and includes a great deal of dust and gas.
Sc	Also known as a "late" spiral, the spiral is very open and contains a great deal of dust and gas. The bulge at the middle is much less prominent than in an **Sa** spiral.
Sd	Between **Sc** and **Sm**, as the spiral structure gradually fades toward irregularity.

| Sm | Between **Sd** and **Irr Type 1** in irregularity. |
| **S0** | Lenticular (lens-shaped) galaxies, consisting of what appears to be the central bulge of a spiral, but lacking arms, and containing little or no gas and dust. In some cases, the **S0** galaxy may merely be a **Sa** galaxy viewed edge on. |

Evolution

The *Hubble Space Telescope* (*HST*) was used to follow up on an observation made from the ground in Chile in 1993 by Duccio Macchetto of the Space Telescope Science Institute in Baltimore, Maryland. Macchetto had noticed that a very distant quasar appears to have a galaxy lying just in front of it, and nearly the same distance away. When Macchetto observed it with *HST*, he was able to photograph the galaxy, which is clearly elliptical. Because of its distance, it must have formed only a million years or so after the big bang. Macchetto interpreted this to mean that elliptical galaxies are the first type to form, or at least form very early.

However, some theories propose that elliptical galaxies form more slowly than spiral galaxies. Other theories suppose that the inner cores of galaxies form first, becoming elliptical galaxies later, with spirals coming still later. Today, about 5% of all galaxies are spirals. In 1992, it was announced that one set of *HST* photographs suggested that 4 billion years ago, a third of all galaxies were spirals. However, this would imply that spirals evolve into ellipticals. In 1993, another *HST* photograph of galaxy NGC 7252, shows it (like Arp 220) to be a collision between two galaxies. François Schweizer and Brad Whitmore of the Space Telescope Science Institute interpreted the photo of NGC 7252 to mean that ellipticals are formed when two spirals merge.

The study of galaxy evolution is not an exact science.

Other *HST* photographs, notably those of Richard Ellis of the University of Cambridge in England and Richard Griffiths of the Space Telescope Science Institute, also demonstrate the evolution of galaxies by using distance to see the past. Similar views can reach even farther into the past at infrared wavelengths for which intergalactic dust is transparent, an approach followed by Lennox Cowie of the University of Hawaii. Another approach to finding ancient galaxies uses radio frequencies. George Miley of Leiden University in the Netherlands and coworkers used reports of radio sources to find what they believe to be the oldest galaxy ever discovered, which was then confirmed by optical examination using the Anglo-Dutch William Herschel Telescope at La Palma in the Canary Islands. The visible radiation from this galaxy originated as ultraviolet radiation, but because of a high redshift factor of 4.25 the radiation reaches us as ordinary light. It is this redshift factor (the highest ever recorded for a galaxy at the time of the announcement on November 15, 1994) which implies that the galaxy, known as 8C1435 + 635, is the farthest, and therefore the oldest.

The spectrum of light from 8C1435 + 635 is continuous, like that of starlight, but the researchers who have studied the galaxy think that this radiation does not come from stars. Instead, they believe that the light originates with a quasar in the heart of the galaxy. Energy from the quasar is absorbed by the gas and dust of the galaxy, which then re-radiates it in a polarized form that otherwise resembles starlight. Polarization is the key clue that such galaxies are not lighted by stars, because stars produce unpolarized light. In the case of 8C1435+635, there is too little light reaching us to ascertain whether or not it is polarized. However, similar old and distant galaxies do shine with polarized light, and therefore are thought not to have begun star formation yet.

In 1994, Esther Hu and Susan Ridgway of the University of Highway used the 7.2 ft (2.2 m) telescope on Mauna Kea to identify two other galaxies. These two galaxies are very red, with a distribution of wavelengths causing them to appear to be older, reddish elliptical glaxies made even redder by their cosmological distance. Two possible explanations for the appearance of these galaxies each pose problems for theory. Hu and Ridgway think that these are three-billion-year-old galaxies, located at a distance implying a time five-sixths of the way back to the big bang. In that case, however, the big bang was 18 billion years ago, a time frame not consistant with most other calculations. The competing theory states that these are younger elliptical galaxies, even closer to the origin of the universe, which also implies that the big bang must have taken place earlier than expected.

In the following timetable, the age of the universe is assumed to be 12 billion years old (*see* "Cosmology UPDATE," p 236). The three studies on which most of the timetable is based, however, reported their findings on December 6, 1994, using an estimated age of 14 billion years, so the dates reported in news accounts would be slightly different from those below.

Today	Spiral galaxies are large, bright, and well-formed, dominating the scene, although there exist 20 times as many elliptical galaxies. Many of the elliptical galaxies are smaller or less bright than the spirals.
2 billion years ago	Numerous collisions between galaxies in the Coma Cluster occurred, according to a study by Jack Burns of New Mexico State University in Las Cruces and coworkers. Such collisions are thought to have re-shaped both galaxies and clusters throughout time.
5 billion years ago	Spiral galaxies are more common than they are today, but appear ragged and not fully formed, according to an *HST*-based study led by Alan Dressler of the Carnegie Institution of Washington. Elliptical galaxies are the same as today.
6 billion years ago	Infrared studies show that there were three or four times as many dwarf galaxies as we see today. Since then, many of them have either faded away or merged with others to form normal galaxies.
8 billion years ago	Many galaxies appear to be ragged precursors of the spiral galaxies, according to an *HST*-based study led by Mark Dickinson of the Space Telescope Science Institute. Others, however, are elliptical galaxies that look just like the galaxies do today. Also present are fragments that appear to be forerunners of spiral galaxies.
10 billion years ago	The most ancient galaxy observed is an elliptical type that would appear perfectly at home today, according to Machetto's study. Gas clouds during this period have much more hydrogen than gas clouds do

THE GALACTIC TIME MACHINE

Time travel is not only a popular subject in science fiction, but also forms the basis of a kind of speculative science in which black holes, white holes, or so-called worm holes in space are used to bypass the rules of time and space. There is, however, a well-known effective method of time travel that has been recognized to exist since 1675, when Ole Römer established that light has a finite speed: everything we see is something that happened in the past. No event outside our own body occurs instantaneously with our experience of the event.

The main difference between time travel in astronomy and time travel in science fiction is the degree of this effect. Light from the Sun, for example, takes a bit more than eight minutes to reach Earth. But the Sun appears to move about two degrees across the sky in eight minutes, and since the image we see of the Sun is only about half a degree, we never see the Sun where it actually is. We see the Sun where it was eight minutes in the past. Similarly, light from the nearest known star takes about four years to reach Earth, and so our view of that star reflects the conditions existing four years previously. If the star suddenly exploded, we would continue to see the whole star for four years after the explosion.

The furthest observable parts of the universe may be very different today than they were some 10 billion years ago. We will never know what those conditions are today, but thanks to the galactic time machine, the images that reach us today reveal the conditions of 10 billion years ago.

today; it is thought that the extra hydrogen goes off to form galaxies. Esther Hu and Susan Ridgway of the University of Hawaii also located evolved elliptical galaxies this early in time (if one places the big bang at 12 billion years ago).

11 billion years ago   A dust cloud around a quasar (observed by Kate Isaak and Richard McMann of the University of Cambridge, and reported in April 1994) suggests that large stars have already formed in a galaxy surrounding the quasar.

The even older 8C1435 + 635 also consists of dust around a quasar, but star formation has probably not begun.

(Periodical References and Additional Reading: *New York Times* 4-7-92, p C6; *New York Times* 6-16-92, p C1; *New York Times* 11-3-92, p C8; *New York Times* 5-26-93, p A16; *Science* 5-13-94, p 906; *Nature* 8-25-94, p 629; *Astronomy* 11-94, pp 20 & 22; *Science* 11-11-94, p 974; *Astronomy* 12-94, p 44; *Science* 12-16-94, p 1806; *New York Times* 12-7-94, p A20; *Astronomy* 2-95, p 24; *Astronomy* 4-95, p 44)

CLUSTERS AND WALLS OF GALAXIES

The doctrine of uniformitarianism has been important in the history of astronomy. Uniformitarianism began in the eighteenth century, when geologists realized that the assimption that uniform forces had been operating over long periods of time could account for most features of the Earth that they were then studying. Subsequent knowledge has recast this assumption for geology, but its initial success caused the idea to be adapted to biology in the original Darwinian theory of evolution and later to cosmology in various forms. As with geology, the more we learn about biological evolution or about the universe, the less useful this simple idea becomes.

In cosmology, a persistent problem has been to use uniform forces to explain the distribution of matter in the universe. The previous article ("Shape and Evolution of Galaxies") considered how galaxies originate in the first place. The following articles ("Dark Matters" and "Cosmology UPDATE") explore new findings that can help account for the distribution of matter that is known to exist but has not been observed, and related unsolved problems. In this article, data is reported regarding the structure of the universe on a very large scale.

The Greater Attractor

A survey by Tod R. Lauer of the National Optical Astronomy Observatories in Tucson, Arizona, and Marc Postman of the Space Telescope Science Institute in Baltimore, Maryland, examined galaxy motions out to 500,000,000 light-years from Earth. This region contains about 30 times the volume of space covered in previous surveys of this kind. Our own Milky Way, the Local Group cluster of which it is a member, and other nearby clusters were found to have a proper motion (motion against the background of galaxies across the universe) in common. The nearby clusters of galaxies all drift toward the same spot in the cosmos, identified from here as in the direction of the constellations Hydra and Centaurus. The speed is about 1,500,000 mph (2,500,000 kph), not what one usually thinks of as drift. The results were published in the April 20, 1994 issue of *Astrophysical Journal*.

Postman comments that he and Lauer could only explain the drift as a result of a large mass, equal to about 100,000 Milky Way galaxies, situated about 300 million light-years away from us.

The same proper motion is revealed by the Cosmic Background Explorer (COBE), which observed a Doppler effect in the radiation left over from the big bang. Such an effect could be caused by an asymmetry in the expansion of the universe, but is much more likely the result of motion toward Hydra. COBE's estimate of the velocity is about 390 mi (630 km) per second.

However, this motion is not toward the "Great Attractor" (previously identified in a smaller survey of proper motions), because that region of the universe is included as a part of the universe that is drifting in the more extensive study by Lauer. Thus, there seems to be an even "Greater Attractor" out there somewhere.

The main conclusion to be drawn from the Greater Attractor is that the universe is even less uniform than previously thought. First of all, the clusters of galaxies that are all moving together constitute a great structure by themselves. Furthermore, whatever the Greater Attractor might be, its gravitational force proclaims it to be a more concentrated mass than was previously expected.

Other Clumpiness

Uniform motion can occur even if there is no interaction between galaxies themselves. However, galaxies are generally found in groups, not as individuals (*see*, for example, "New Neighbors: Dwingeloo 1 and More," p 212). The pattern of these groups is part of an overall tendency of the universe toward what is often called clumpiness—the disposition of matter to clump together in various ways, leaving open space or voids in between. Clumpiness seems to happen at every level. The solar system is localized in its part of the universe, with the distance to the outer planet being only about 1/300th of the distance to the nearest stars. In our local group, the clump consisting of the Milky Way and its satellites are about ten times as close to each other as they are to nearby M31. The nearest big cluster of galaxies is about 80 times as far from us as we are from M31 (although the further away we go, the less accurate the measurements are); and, it appears, so on ad infinitum.

Studies of clusters using electromagnetic radiation at x-ray wavelengths were impossible until the space age. Since then, x rays have proved to be an important adjunct to optical and radio wavelengths because they reveal the hot, diffuse gas that pervades the cluster unevenly. This is thought, in turn, to reveal the mass distribution of the cluster. Analyses of increasing sophistication have relied on the *Uhuru*, *Einstein* (*HEAO-2*), *Exosat*, and *ROSAT* x-ray satellites, including a 1994 *ROSAT* study by astronomers from the Max Planck Institute for Extraterrestrial Physics of the Virgo cluster. Because this cluster is considered to be the nearest "rich" cluster to the Local Group, it is important to cosmology as the basis for most computations of the Hubble Constant (*see* "Cosmology UPDATE," p 236).

Another study using data from the *ROSAT* satellite, by A. Katherine Romer of Northwestern University in Evanston, Illinois, and an international team of coworkers, recorded clusters of galaxies using x rays to identify clusters of clusters, or superclusters. The x-ray technique is intrinsically different from identification based upon optical guidelines, which suffers from problems connected largely with looking at clusters in two dimensions. *ROSAT* data was able to analyze clusters at distances three times as far as those distances known from optical surveys. An elongation of clusters along the line of sight, or a somewhat different type of elongation christened "the finger of God" effect, is found in optical surveys. It is believed that this stretching illusion results from the inclusion of galaxies occurring on both the near and the far side of the cluster itself, causing the cluster to appear longer radially than it really is. At x-ray wavelengths, one of the main differences is that the finger of God and other cluster elongations were absent, probably because distances were more separable. In other words, the x-ray survey found no stretching away of clusters in directions radial to Earth, which would require explanation.

There are, however, other kinds of structures that need explanation. The most prominent are called "walls" because the galaxies form great sheets, larger structures than even the superclusters. The first recognition of this structure was the Great Wall, 500 million light-years long, discovered in 1986 by Margaret Geller and John Huchra. This discovery was an unexpected result of a plan to map galaxies in space. Astronomers searching the skies from the Northern Hemisphere found that the Great Wall was accompanied by large voids where no galaxies existed. Some theorists explained

Background

When we look at the night sky, it is easy to see patterns of stars which form the constellations. However, any astronomy student is soon taught that these groupings are only apparent. Stars that appear close together due to their minimal angular separation may actually be very far apart. One may be a relatively dim star that is close, while the other is a very bright star that is distant. The same thing happens with galaxies; to know their true relation to each other, it is necessary to determine how far away each one is.

When astronomers began to make three-dimensional maps of galaxies, they found that groups of galaxies huddled together, with huge amounts of apparently empty space separating the galaxies in one group from neighboring groups. An individual group is known as a **cluster** of galaxies, not to be confused with either a galactic cluster (a fairly small group of young stars that probably formed together) or a globular cluster (a large, well-formed group of older stars). Perhaps the most familiar cluster of galaxies is the **Local Group**, which includes the Milky Way and Andromeda (M31). About two dozen galaxies are in known to be in the Local Group (but *see also* "New Neighbors: Dwingeloo 1 and More," p 212), which is about a megaparsec (somewhat more than 3,000 light-years) across. Another well-known cluster of galaxies, known as the Virgo Cluster after the constellation in which it is seen, is much larger, containing thousands of galaxies. Its size and distance from us are the objects of considerable scrutiny (*see* "Cosmology UPDATE," p 236).

In 1986 and 1987, astronomers discovered that galaxies not only form clusters, but that these clusters themselves clumped up into clusters of clusters, and even into great sheets of galaxies. There was no reason not to think that the superclusters also formed supersuperclusters, and so forth, right up to the size of the universe.

these voids and sheets in terms of bubbles, proposing that as the universe expanded, bubbles of matter (the observed sheets of galaxies) formed around expanding voids.

This seemed to be what Geller and Huchra found. If the Great Wall was an accidental alignment of galaxies, one would not expect to find a similar wall when looking at the other half of the universe as seen from Earth's Southern Hemisphere. Working with Luis da Costa of the Brazilian National Observatory, Geller and Huchra mapped galaxies in a region extending as far as 450 million light-years from Earth. They found a precise analog to the Great Wall of the north: the Southern Wall, 300 million light-years long and made from more than a thousand galaxies, almost a third of all the galaxies located in that great volume of space. Furthermore, the overall structure of southern space was the same as that in the north: huge voids surrounded by sheets of galaxies.

(Periodical References and Additional Reading: *New York Times* 3-21-94, p A13; *Astronomy* 4-94, p 20; *Nature* 4-28-94, p 828; *Astronomy* 7-94, p 18; *Nature* 11-3-94, p 75; *Discover* 1-95, p 34; *Astronomy* 4-95, p 26)

DARK MATTERS

The problem of the evolution and distribution of galaxies is intimately involved with two other famous puzzles in cosmology, which A. Conan Doyle might have called *The Mass that Was Not Seen* or *A Study in Darkness*. Cosmologists, however, speak and write of the problem of the missing mass, and—on the assumption that this mass actually exists but cannot be detected—the question of what this "dark matter" is. In the early and middle 1990s, most astronomers accepted the view that more mass exists than has been detected, so the focus of attention was on the nature of this mass.

The missing mass itself is often called dark matter simply because it does not emit light, as stars do. The situation is more complex than simple darkness, however, since the missing mass must not be reflecting or blocking light to any great extent, either. Furthermore, it is evidently not interacting with any other form of electromagnetic radiation, because the dark matter is not obvious at any wavelength, from radio to gamma rays.

Astronomers know that dark matter actually exists. The best evidence shows that galaxies would fly apart if there were not something more than the observable matter to produce gravitational forces. Vera Rubin first demonstrated this in the early 1980s. In the early 1990s, Magda Arnaboldi and Ken Freeman of the Mount Stromlo Observatory showed that at least one galaxy, NGC 1399 in Fornax, must have even more dark matter than previously calculated—by a factor of 10! They measured the rotational velocities of the galaxy using the revolutionary periods its planetary nebula, which enabled them to calculate the galactic mass. Similarly, Richard Mushotzky of NASA and David Burstein of Arizona State University measured the mass of an intergalactic cloud of hot gases, which gravitational theory shows requires 30 times as much dark matter as gas to stay together. Also in 1993, Douglas Lin of the University of California at Santa Cruz developed evidence that the halo of dark matter about the Milky Way galaxy is large enough to contain the Large Magellanic Cloud, which orbits the Milky Way at a distance of about 160,000 light years (*see also* "Inside the Milky Way," p 206). Similar measurements of 45 distant spiral galaxies by Dennis Zaritsky of the Carnegie Observatories, and coworkers showed that they also were embedded in dark matter that must extend for 10 times the width of the visible stars, dust, and gas.

Many of the proposed explanations regarding the dark matter's composition rely on theoretical possibilities that have not and, for the most part, cannot be proved or disproved. Thus, critics say that these ideas are not part of science at all. But some astronomers have conducted studies that, so to speak, illuminate the subject.

Dark Matter Found?

Penny D. Sackett of the Institute for Advanced Study in Princeton, New Jersey, Heather L. Morrison of the U.S. National Optical Observatories in Tucson, Arizona, Paul D. Harding of the University of Arizona in Tucson, and Todd A. Boroson of the U.S. Gemini Project Office reported in the August 11, 1994 issue of *Nature* that they could see dark matter. They were using a charge-coupled device (CCD), which can detect very weak light, to look for a faint but thick disk in the spiral galaxy NGC 5907 in Draco, observed edge-on from our place in the universe. Although they failed to find such a disk, very careful separation of the faint light from the dark background revealed a blue glow that surrounds NGC 5907. This blue glow, about 0.05 times the brightness of the dark sky background, is not distributed in the ordinary manner of stars in a galaxy, which typically are much less common the further away one looks from the galactic center. Instead, the blue glow decreases in the same way that gravitational studies show the density distribution of dark galactic matter changes with distance from the center. The blue haze could be light from MACHOs (*see* Background) in NGC 5907. Studies in 1992 and 1993 by various astronomers suggest that dark matter in the Milky Way galaxy may be MACHOS, although the specific MACHOS have not been located. From the weakness of the light, the NGC 5907 MACHOS would have to be low-mass dwarf stars of the M type.

Antoinette Songaila and Lennox L. Cowie of the University of Hawaii in Honolulu, together with Craig J. Hogan and Martin Rugers of the University of Washington in Seattle, used the giant Keck telescope to test the predictions of the big bang theory and to set limits for dark matter. They carefully measured the ratio of deuterium, or hydrogen-2, to ordinary hydrogen in a distant cloud of gas. At about 11.8 billion light years away, the cloud was thought to represent the early stages of the universe after the big bang. The scientists, using one set of assumptions, found only one deuterium atom to each 4,000 hydrogen atoms and, using slightly different assumptions, the similar ratio of one deuterium to about 5,300 hydrogen. According to theory, and if the measurement is correct (which it may not be because of interference from closer gas clouds), the matter visible in galaxies accounts for all the normal matter. Thus, if there is some other matter, which there certainly appears to be, it cannot be the ordinary stuff. Instead, it must be some exotic particle. This is a lot of inference to derive from one possibly flawed measurement, however. From another standpoint, the detection of any deuterium outside the galaxy was a major achievement, proving to many the worth of the Keck telescope. As the first major result based on Keck observations, the study may have received additional publicity.

About the same time as the Songaila group was at work, a different team led by R. Carswell used the 16.4 ft (5 m) Kitt Peak National Observatory Telescope to observe deuterium in the same cloud that was the focus of the Keck telescope. Carswell's team obtained results similar to those of the Songaila group.

The amount of deuterium created during the big bang can be used by theorists to calculate the total amount of matter in the universe (*see also* "Cosmology UPDATE," p 236). The problem is that the amount of deuterium in our part of the universe has been lowered by participation in nuclear fusion in stars. At least three-fourths of the local deuterium has been fused into one of two isotopes of helium, known as helium-3 and helium-4 (helium-4 is the common form of helium). Looking about, astronomers can find about the right amount of helium-4, but there is not much helium-3. The helium-3 deficit, if real, would suggest that the high amount of deuterium found by the Songaila and Carswell groups may be inaccurate.

Another University of Hawaii team led by Esther Hu used the 87 in (2.2 m) reflector on Mauna Kea to search for dark matter much closer to home, around the Milky Way. Earlier astronomers observed that some stars mysteriously brightened, and concluded that this was caused by lenses of intervening dark matter (such as an unseen planet or dark star) focusing the starlight. Hu's team found that there was no extra infrared radiation coming from that region of the sky, so the dark matter, if there, could not even be as bright as a dark star. Instead, the matter had to be either exotic or planetary. The upper limit of normal-matter starlike objects that could evade their study would be brown dwarfs, which are intermediate between planets and stars in size (*see* "Stars: Their Lives and Hard Times," p 184).

Three different studies in 1993, led by Charles Alcock of the Lawrence Livermore National Laboratory in California, Michel Spiro of the French National Center for Scientific Research in Paris, and Bohdan Psczynski of Princeton University in New Jersey, were thought to show the presence of a lens effect caused by the gravitational effect of MACHOs on light. Not everyone agreed. In 1994, Kailash C. Sahu of the Instituto de Astrofisica de Canarias in Tenerife was able to demonstrate that ordinary stars could account for all the observations in the three 1993 studies. His work did not, however, rule out the future discovery of unequivocal lensing caused by MACHOs, and Alcock's team, at least, disputed Sahu's interpretation.

Ron Allen of the Space Telescope Science Institute also looked for dark matter in the Milky Way using a radio telescope. He found that molecules of cold gases, including molecular hydrogen and carbon monoxide, could account for a third of the dark matter known from gravitational evidence to be present.

An x-ray telescope also detected what was hailed as dark matter. *ROSAT*, the German-British-U.S. *Roentgen Satellite*, produced data on a giant cloud of gas (alluded to earlier) amid the NGC 2300 minicluster of galaxies. Dark matter was invoked to hold the cloud, some 1.3 million light-years in diameter, together and to provide its motive force, since the three small galaxies in the NGC group could not contain enough mass to do the job. The research was announced at the beginning of 1993, and was conducted by Richard F. Mushotzky of the Goddard Space Flight Center in Greenbelt, Maryland, David Burstein of Arizona State University in Tempe, and Jack S. Mulchaey and David S. Davis of the Space Telescope Science Institute in Baltimore, Maryland.

Dark Matter Sought, But Not Found

One verifiable theory for the location of dark matter suggests that it consists of MACHOs, a general category for various massive forms of dark matter including some small stars that glow like dying embers. If there were enough of such red dwarf stars, they might account for the missing mass, although the mass of each individual star would be only 10% that of the Sun. Since many astronomers thought that they had grounds to believe that there were more red dwarfs than any other stars, it seemed plausible to expect that there existed enough of the red dwarfs to make up 90% of the mass of the universe. The reason for believing that the number of red dwarfs would be so great is simple: at all observable levels from Earth, there are more stars of each smaller class than there are of the next larger class. Thus, one would expect most stars to be very small.

From Earth, the red dwarfs are not really dark, just very dim. However, because they are so dim and red, it is almost impossible to separate them from small distant galaxies. The main impediment consists of imperfections in the atmosphere which cause images of both the red dwarfs and the distant galaxies to spread out slightly.

These smeared images are alike. Astronomers know that there are galaxies almost everywhere one points a telescope, because of the vast reaches of the universe. The red dwarfs in a galaxy, such as our own Milky Way, were thought to be similarly abundant. This could be neither proved nor disproved, because atmospheric diffusion makes dwarf stars look like faraway galaxies.

Enter the *Hubble Space Telescope* (*HST*), for which there is no atmospheric interference. In *HST* images, red dwarfs are points and distant galaxies cover small regions. Furthermore, *HST* can detect red dwarfs that are 100 times dimmer than can be seen from the ground.

A team consisting of John N. Bahcall, Chris Flynn, and Sofia Kirhakos of the Institute for Advanced Study in Princeton, New Jersey, along with Andrew Gould of Ohio State University, used *HST* to search for red dwarfs in the Milky Way. Searching patches of sky at random, the astronomers did not find as many small stars as had been expected. Indeed, in the galactic halo, where the missing mass problem is especially acute, no more than 6% of all the stars found are red dwarfs. While the percentage is higher in the main disk of the galaxy, it is still no more than 15%. Thus, in the Milky Way at least, red dwarfs are not the missing dark matter. This report appeared in the November 1, 1994 *Astrophysical Journal Letters*.

At roughly the same time, a group of astronomers from Italy used *HST* to look for red dwarfs in a different part of the galaxy. Francesco Paresce of the Space Telescope Science Institute in Baltimore, Maryland, noticed that it was possible to see through the globular cluster of stars known as NGC 6397. Small stars in the cluster should have made NGC 6397 opaque. With the help of Guido Marchi of the University of Firenze, Martino Romaniello of the University of Pisa, and the *HST*, Paresce looked more closely at NGC 6397 to see if the anticipated small stars could be found there. Their report appeared in the February 10, 1995 *Astrophysical Journal*.

Although there are many small stars in NGC 6397 measuring about twice the size of the red dwarfs, red dwarfs themselves are extremely rare. Furthermore, although Paresce's red dwarfs are observed have be about one-fifth the mass of the Sun, theory had predicted that they would be smaller. Thus, although there are not so many as expected, the smallest stars tend to be larger than expected and rather uniform in size. A greater number of less massive red dwarfs would contribute more to the missing matter, however, than the small number of red dwarfs in their larger size. Also, if red dwarfs (which in principle are observable) do not form between the size of one-fifth of a solar mass and the size of an almost unobservable brown dwarf, then it seems less likely that the universe is filled with unobserved brown dwarfs, either.

Paresce thinks that there is a mass limit below which stars do not form: about one-fifth the mass of the Sun may be a natural cutoff in star size. If this is the case, the rule is not explained by theory, which predicts that masses of matter about one-twelfth the mass of the Sun would make viable, if tiny, stars.

Another possibility in the case of NGC 6397 proposes that red dwarfs might form, but somehow be ejected by gravitational forces from the globular cluster. This last idea, however, is unlikely, since the Bahcall team would have found such stars in the Milky Way. Thus, the idea that the galaxy is filled with tiny stars has not been verified. Dark matter must be something else—perhaps dust.

Dust is recognized by the amount of starlight it absorbs and re-radiates at infrared

Background

As a result of theoretical work published in 1974 and later, astronomers and physicists worried about mass missing from the universe throughout the 1980s and into the 1990s. Put simply, the most widely accepted theories of universe formation and structure all predict far more matter than can be detected. If this matter exists, it is invisible, because it does not reveal itself by giving off electromagnetic waves at any wavelength. The missing mass is often called **dark matter** for that reason. The only force to which it is known to respond is gravity. Dark matter must be affected by gravity, because its gravitational force is what is apparent, even though dark matter itself is not observed in any other way.

Various notions of what the missing mass might be have been proposed and discounted, mostly because any dark matter theory has to postulate something that is undetectable by the means we ordinarily use. As a result, the hypothetical source of the missing mass cannot be found, so few believe that the invisible material proposed really exists (except for the proposer and a few close friends).

Astronomers, however, are very clever. If the only way that dark matter can be detected is through gravity, then they will find a way to use gravity to locate it— not the long-sought gravity waves, which are expected to come mostly from violent events, but the observable effects of gravity in the universe. In the early 1980s, Vera Rubin of the Carnegie Institution used the rotational velocities of different parts of galaxies to show that each galaxy must be embedded in an envelope of unseen matter. This accounted for about 10% of the presumed missing mass. It did not, however, offer any clues as to the identity of the dark matter.

There are a number of possibilities as to the identity of that missing mass. Astronomers have categorized missing-mass candidates into two large groups, cold matter and hot matter. When it comes to dark matter, the adjectives "hot" and "cold" refer to speed, rather than temperature as it is ordinarily envisioned. If one thinks of the kinetic theory of heat, however, this designation makes sense. Hot substances contain fast-moving molecules, while in cold substances the speed of the particles is very slow. Absolute zero, if obtainable, would be complete absence of motion for the particles.

Some Like It Cold   Cold dark matter could consist simply of moderately large bodies of ordinary matter that are too cool to radiate energy and too far away to see. Such bodies include large planets and small red or brown dwarfs. The dwarfs could be orbiting ordinary stars, orbiting each other, or wandering loose in galaxies. It is also possible that they are wandering between galaxies, although this is thought to be much less likely. All such versions of dark matter tend to be called Massive Compact Halo Objects (**MACHOs**, an acronym attributed to K. Griest), which originally were defined as neutron stars, and the long-lived low-mass stars known as red dwarfs in the galactic halo. The term MACHO, however, has struck a responsive chord, and is used often without regard to the halo that provides the H in the acronym.

Another possibility for cold dark matter is that it consists of clouds of cool gas that would be undetectable unless energized by radiation from other forms of matter. Yet another proposal, states that dark matter is as dark as anything could be: black holes that do not have any matter near enough to radiate energy from accretion disks.

Theorists have also postulated a host of slow-moving Weakly Interacting Massive Particles (**WIMPs**, an acronym attributed to M. Turner), although no WIMPs have ever been observed. Recently, many writers have used the expression "cold dark matter" primarily to refer to slow-moving WIMPs. WIMPs are thought to be "dissipationless," or unable to lose energy, as well as "collisionless," because they show no interactions with particles of ordinary matter. **Some Like It Hot**   Hot, dark matter could consist of subatomic particles different from the cold WIMPs. The "hot" hot candidate is the neutrino, if it has mass, or other possible WIMPs, this time fast-moving ones.

A lot of theorists also favor mixed dark matter: some hot with a dash of cold, or perhaps mostly hot somewhat slowed by cold. However, there is yet another idea under consideration.

A Lukewarm, but Strange Possibility   There has also been proposed an intermediate class of matter: grains of strange matter. Strange matter contains the strange quark, as well as the up and down quarks that along with electrons constitute ordinary matter. The idea was first proposed by Arnold R. Bodmer of the University of Illinois in 1971, although he suggested that the strange particles (consisting of more quarks than a proton or cosmic-ray particle, but fewer than a neutron star) would be found in the cores of massive stars. In 1984, Edward Witten of Princeton University in New Jersey suggested that grains of strange matter might account for all the dark matter in the universe. If strange grains (also called strangelets) exist, they would be larger than ordinary subatomic particles, but smaller than planets or gas clouds. Their speed would also be intermediate. Consequently, strange grains might be termed "lukewarm dark matter."

Calculations suggest that strangelets would have to be fairly substantial to live up to their role as lukewarm dark matter, and would consist of more than 1,023 quarks each. Experiments searching for evidence that such particles can exist are ongoing at Brookhaven National Laboratory in the United States and at CERN in Europe.

frequencies. If you know the absorbing capacity of the dust, the frequency of the infrared radiation implies the temperature and also the mass of the dust for a given volume of space. Dark dust absorbs more light per dust particle than light dust. Forsterite is thought to be one of the main components of interstellar dust in the Milky Way. Albert J. Sievers and N. I. Agladze of Cornell University and coworkers conducted a study of the mineral forsterite, which they reported in the November 17, 1994 *Nature*. Sievers and Agladze showed that the mineral (in a form similar to that believed to be most like intragalactic dust at the cold temperatures of space) is six times as dark as had been thought. This chain of reasoning and evidence implies a galaxy that is only one-sixth as dusty as previously thought.

Furthermore, it is standard practice for astronomers to estimate the amount of hydrogen gas present in a cloud on the basis of the dust in the cloud. Thus, less dust means that estimates of the cloud's mass are decreased by that reduced amount of hydrogen as well. If forsterite had turned out to be lighter, dust and gas could have accounted for some of the missing matter. As it is, the result suggests that even more matter might be missing.

Another mineral study has also been directed at finding dark matter, in this case

WIMPs (*see* Background, below). Daniel Snowden-Ifft and coworkers at the University of California, Berkeley, are looking at billion-year old samples of mica. They reason that WIMPs that have passed through the mica would have disturbed its crystalline structure. Even if WIMPs are quite rare, ancient samples would act like detectors left "on" for a billion years and would record the WIMPs that passed during that time. Since WIMP tracks are expected to be extremely small, they must be looked for with a microscope that can image individual atoms. The atomic force microscope (AFM) can do that trick, so the Berkeley team borrowed an AFM and started scanning. Although their sample was small—only about a ten-thousandth of a square inch (a thousandth of a square centimeter) of mica—and revealed no evidence of WIMPs, the study began to set limits on the mass and abundance of WIMPs. Further study, if it continues to reveal a lack of WIMP tracks in mica, will gradually reduce the possible mass and abundance of WIMPs, which would remove them from consideration as the missing dark matter in the universe.

Hot and Cold Contentions

In the early 1980s, some theorists proposed that the cold, or slow-moving, particles called WIMPs could cause the universe to clump together into galaxies. Various experiments have looked for changes in the universal background radiation caused by the big bang or for patterns in clusters of galaxies that would demonstrate that the cold dark matter theory was true. Instead, they have shown that cold dark matter is extremely unlikely (*see also* "Clusters and Walls of Galaxies," p 222). For example, Ben Moore of the University of California, Berkeley, compared actual behavior of dwarf galaxies with simulated behavior, finding that dark matter cannot mainly consist of cold WIMPs.

The other possibility, then, is that hot, or fast-moving, dark matter might be at work. If neutrinos have no mass, they move at the speed of light and fail to interact gravitationally. However, if neutrinos have even a tiny amount of mass, they slow down slightly and could cause the gravitational clumping observed. In part, this is because neutrinos are thought to be amazingly abundant throughout the entire universe (but see "Neutrino Astronomy Comes of Age," p 136). Observation of a galaxy cluster from early in the history of the universe, as was accomplished in 1994 by astronomers from the Cerro Tololo Inter-American and European Southern Observatories in Chile, favors a hot dark matter explanation, since clusters can form faster when attracted by fluctuations in hot matter than they can when attracted by fluctuations in cold matter.

What about mixed dark and cold matter? Several researchers, including Joel Primack of the University of California at Santa Cruz and Renyue Cen and Jeremiah Ostriker of Princeton University, used computer simulations based on this idea to see if anything like the observed universe could be evolved. The best results, from Princeton, came from a mixture of 64% cold dark matter, 30% hot dark matter, and 6% ordinary observable matter. However, the simulation caused galaxies to form late, move too fast, and cluster too much. Better results, however, could be achieved by decreasing the amount of dark matter and allowing the observable matter to compose more of the actual universe. These proportions resulted in much better galaxy formation and clumping, although the universe has so little matter in it that it keeps on expanding forever, an undesirable (but not impossible) situation. By 1996, however, such a universe began to seem more likely for other reasons as well. Another mixed dark matter simulation, by Greg Brian and Michael Norma of the University of Illinois and Anatoly Klypin of New Mexico State University, achieved better results, but also failed to exactly match reality.

(Periodical References and Additional Reading: *New York Times* 10-9-92, p A17; *New York Times* 5-26-92 p C1; *New York Times* 1-5-93, p A1; *New York Times* 9-21-93, p C7; *New York Times* 10-10-93, p I-27; *New York Times* 10-19-93, p C6; *Discover* 1-94, p 74; *Scientific American* 1-94, p 72; *Nature* 4-14-94, pp 584 & 599; *New York Times* 4-15-94, p A22; *Science* 4-15-94, p 346; *Astronomy* 5-94, p 24; *Astronomy* 6-94, p 20; *Science* 6-24-94, p 1845; *Nature* 7-28-94, p 248 & 275; *Science* 7-29-94, p 607; *Nature* 8-11-94, p 441; *New York Times* 8-16-94, p C5; *Nature* 8-25-94, p 629; *Astronomy* 11-94, p 26; *New York Times* 11-16-94, p B8; *Nature* 11-17-94, pp 225 & 243; *Science* 11-25-95, p 1319; *Science News* 11-26-94, p 357; *New York Times* 11-29-94, p C1; *Astronomy* 12-94, p 18; *Astronomy* 2-95, pp 26 & 27; *Astronomy* 4-95, pp 22 & 26)

QUASARS

Quasars continue to require explanation. Although scientists increasingly believe that most quasars are black holes at the centers of young galaxies, a lot of the evidence required to back up such a belief remains circumstantial. Nearly all the recent evidence consists of lines on a spectrogram.

For example, a clue that quasars are located in the centers of massive galaxies comes from a study of two distant quasars reported in the January 20, 1994 *Nature* by Richard Elston of the Cerro Tololo Inter-American Observatory in La Serena, Chile, Keith L. Thompson of the U.S. Naval Research Laboratory in Washington, D.C., and Gary J. Hill of the University of Texas at Austin. As is generally the case for studies of distant astronomical objects, the evidence is indirect and the reasoning chain fairly long.

Elston, Thompson, and Hill looked for electromagnetic radiation emitted by ionized iron in quasars Q0014 + 813 and Q0663 + 680. The element iron is important in understanding nuclear processes because it is the highest element produced during ordinary stellar evolution. If iron is present in the gases that feed a quasar and therefore produce its electromagnetic emissions, that iron must have been spewed forth in the explosions of supernovas. If there is a great deal of iron present, there must have been many supernovas. Such conditions could only occur in the center of a large galaxy. Elston, Thompson, and Hill found that the amount of iron present did imply such a galaxy. Furthermore, it meant that the quasars themselves had to be a billion or so years old, since it takes that long for stars to evolve into the type Ia supernovas that could produce the iron (*see also* "Cosmology UPDATE," p 236). Hence, these quasars are in evolved galaxies.

The Cloverleaf quasar (so called because a gravitational lens splits its image into four separate "leafs"; its real name is H1413 + 117) is the source of data suggesting that some quasars are embedded in primitive galaxies for which most of the mass is molecular gases, not stars. Richard Barvainis of MIT and an international team of coworkers were able to detect carbon monoxide in the Cloverleaf, reporting their discovery in the October 13, 1994 *Nature*. Emission from carbon monoxide also provide an accurate cosmological red shift of 2.5 for the quasar. The presumed galaxy in which the Cloverleaf is embedded could not be all gas, although it contains 100 times as much diatomic hydrogen as is contained within the Milky Way galaxy. Carbon and oxygen must have come from early star formation, so carbon monoxide lines in the spectrum of the Cloverleaf represent stars.

Until 1994, the nearest quasar known to astronomers was 3C273, the first quasar ever recognized (by Maarten Schmidt in 1963). Because it is "only" 2 billion light years away, the red shift indicating the distance to 3C273 was then seen as amazing for so bright an object. In 1994, Anne Kinney of the Space Telescope and Science Institute in Baltimore, Maryland, along with Robert Antonucci and Todd Hurt of the University of California at Santa Barbara, found a quasar that was more than three times as close, but which they could not see at all. The quasar they found in the bright radio galaxy Cyngus A is only 600 million light years from Earth, but the dark dust in Cygnus A absorbs electromagnetic waves at visible frequencies. What Kinney, Antonucci, and Hurt observed was a broad line in the ultraviolet spectrum of Cyngus A, which they interpreted as being caused by the unseen quasar.

Even closer to home are the "microquasars" discovered in 1994 in the Milky Way galaxy itself (*see* "Inside the Milky Way," p 206).

Dusty Quasars?

Known quasars abound by the thousands. Nearly all of those found thus far are similar in appearance to hot stars, which produce mostly blue light, although this light is often shifted to the red by their distance from Earth. In the late 1980s and early 1990s, however, some astronomers proposed that the main reason so many quasars are especially bright in the blue part of the spectrum is that astronomers on quasar hunts only look for blue starlike objects. If they looked for quasars in the red part of the spectrum, they would find even more quasars than they do now, according to this minority of quasar specialists.

The original reason for supposing that it would be a good idea to look for reddish quasars came in 1988 from Julie Heisler and Jeremiah Ostriker of Princeton University. They suggested that since quasars are generally very far away, the light from them must pass through a lot of intergalactic and interstellar dust. The effect of all this dust on quasar light would be the same as the effect of earthly dust on light from the setting or rising Sun: red light gets through but blue light is scattered, giving the Sun a red appearance. Obliging their theory, John Huchra of the Harvard-Smithsonian Center for Astrophysics found some very red (infrared in fact) quasars with the *Infrared Astronomy Satellite (IRAS)*. Huchra proposed that there might be as many red quasars as blue ones, providing that one knew where to look. After a time, Rachel Webster of the University of Melbourne in Australia and coworkers decided to test the hypothesis. Their plan was to locate quasars using radio frequencies, which would not bias the kind of visible light produced. Having found a population of quasars with a radio

Background

Quasars are still mysterious quasi-stellar objects, a description that was shortened to become their present name. Optically, quasars appear at first glance to be distant reddish stars—nothing that unusual.

A closer look at the light from quasars shows that they are moving away from us at such high speeds that their light waves are shifted toward the red by large factors, often a factor of three or more. The simplest explanation for this redshift is that it is caused by a combination of universe expansion and age, although clever astronomers have sometimes proposed other theories. If quasars are indeed very old, they are also very distant, which implies in turn that they release great amounts of energy to produce so much radiation that reaches Earth. Some of them appear even more energetic in radio telescope images than at optical wavelengths. Examination at x-ray wavelengths from space generally shows even more energy than at the longer wavelengths. Other evidence suggests that quasars must be less than a few light-years across.

A large body of evidence suggests that quasars are found in the cores of early galaxies. The galaxies involved are so far away, as are any very old features of an expanding universe, that they are invisible or nearly so with even the best telescopes. Their quasar cores, however, burn so brightly that they produce the vast amounts of electromagnetic radiation mentioned above. Furthermore, quasars vary with time scales that seem too fast unless the objects producing the energy are fairly small. The argument for a small size is that one side of a quasar would need to communicate with the other in order to announce that it is time to increase or decrease energy. Information can travel no faster than the speed of light.

Many astronomers believe that these characteristics mean that quasars are disks of material, probably primarily gas, that form around black holes. The evidence suggests to speculative astronomers that gravitational forces in a galaxy cause stars to be thickly congregated at the center of the galaxy. Some stars collide or combine to produce the first black hole. Perhaps other black holes also form in the region, but in time, they too coalesce. Soon there is one giant black hole that sucks all nearby gas, dust, and stars toward it. An accretion disk several light-years across forms and produces enormous amounts of energy. Variation occurs when a large object, such as a star or a thick cloud, is being sucked into the black hole. The accretion disk is the actual quasar, not the black hole. After a few billion years, however, the material closest to the quasar has disappeared beyond the event horizon of the black hole (*see* "Black Holes Sighted," p 198). The quasar dims and disappears. Thus, there are no quasars visible in nearby galaxies, because these galaxies are closer to the present time, and the intense radiation has died down, although the black holes are thought to continue to be present.

Local quasars (*see* "Inside the Milky Way," p 206) are probably smaller black holes that are still active.

telescope, Webster and her coworkers would then use an optical telescope to find how blue or red they were.

In August 1994 Webster was ready to report on the outcome of the experiment. By then the team had identified about 300 quasars using the Parkes radio telescope. When checked at visual wavelengths, more than half the radio quasars were brighter at red or infrared wavelengths than at blue wavelengths. It would appear that the Heisler-Ostriker-Huchra hypothesis was vindicated by experiment.

But perhaps not. First of all, only a small number of quasars are bright at radio wavelengths. Knowing that more than half the radio quasars sampled are red does not necessarily imply that more than half the radio-quiet quasars are red. Redness may be a factor that is directly associated with radio loudness. Thus, to be really certain, a different experiment will have to be performed. Indeed, in 1994, Webster initiated a different experiment during which the quasars are located by looking at infrared radiation. This procedure is thought to be more inclusive of all types of quasar.

In addition, the hypothesis that intervening dust causes the reddening implies that more distant quasars would tend to be redder than closer ones, because more dust would be likely to be between us and them. However, Webster and her coworkers have begun to analyze this criterion by checking the redshifts of quasars for which they have enough information. Although the number checked is still small, there is no apparent correlation between amount of red shift and redness.

Finally, there is some evidence suggesting that dust clouds around the quasars themselves can sometimes alter the wavelengths of electromagnetic radiation. None of this evidence demonstrates that such a dust cloud can cause reddening of light, but Richard McMahon of Cambridge University was able to use radio waves to locate dust clouds around two quasars. Other astronomers have found that some quasars cannot be detected at higher wavelengths of electromagnetic radiations, which might be blocked by a dust cloud. The Cloverleaf quasar has lines for a large amount of dust in its spectrum, and, as noted above, the probable dust cloud surrounding quasar Cygnus A seems to be so dense that it stops visible light completely.

(Periodical References and Additional Reading: *Nature* 1-20-94, p 250; *Astronomy* 7-94, p 22; *Nature* 9-22-94, p 313; *Science* 10-30-94, p 2012; *Nature* 10-13-94, pp 559 & 586; *Astronomy* 1-95, p 20; *Natural History* 2-95, p 72)

Cosmology UPDATE

Both the missing mass problem and the excessive smoothness of the big bang are main players in the relatively new science of cosmology. These are the hard problems of cosmology. The early and mid-1990s may have failed to solve either problem completely, but there was progress in solving various other cosmological puzzles.

Sources of Radiation

Electromagnetic radiation to some degree fills the universe, and has from the beginning. Radiation from the beginning is now known as the Cosmic Microwave Background (discussed further in relation to "The Smoothness Problem" below). When

Background

Cosmology's effective birth as a theoretical science came in 1915 with Einstein's theory of general relativity, and as an observational science in 1929 with Edwin Hubble's work demonstrating an expanding universe. Although Einstein's equations indicated an expanding universe, no one paid much attention to the possibility until the expansion was observed. The big bang, or origin of the universe in the currently accepted theory, is in a slightly different category. Early advocates of the theory had little more than the evidence of expansion and Einstein's theory to justify their ideas. Since 1965, however, the cosmic background radiation, explained below, has been accepted as good evidence for this theory.

The big bang really was a kind of explosion, complete with the release of a vast amount of energy into all the space that then existed. The easiest way to describe the amount of energy is in terms of temperature. The temperature of the big bang itself is not very meaningful, because not much in the way of matter or ordinary forms of energy existed at the very beginning, although some scientists will state that the temperature one second after the big bang was something like 10,000,000,000K (essentially the same for very high temperatures in Celsius, and a little less than half what a Fahrenheit estimate would be), and that 100 seconds later it would have dropped to only about 1,000,000,000K. By about 100,000 to 300,000 years after the big bang, however, matter as we know it had begun to condense, and theorists are able to calculate its temperature: 30,000K (54,000°F or 30,000°C). At that time, the electromagnetic radiation would have existed as gamma rays.

The universe was expanding then, as it is now. Just as lowering the pressure of a gas lowers its temperature, the expansion of the universe lowered this primordial temperature. The energy had nowhere to go, so it could not change, but because the amount of space it occupied was increasing, the energy became dispersed and diminished. When the universe doubles in size, its temperature drops to half of its previous measure. The gamma rays became longer and longer, moving down through the electromagnetic spectrum, until today they are in the microwave region, corresponding to a temperature of only 2.73K (−454.75°F or −270.42°C). We detect this temperature by the electromagnetic waves it produces, just as a hot stove warms in part by radiation. Low temperatures produce long waves, while high temperatures make short waves. The agreement of the exact measurement of this temperature with its postulated measurement convinces most astronomers that the big bang actually happened.

If the big bang was smooth, looking in any one direction we should find electromagnetic waves from that direction registering the same temperature as waves from any other direction. In practice, however, waves from one specific direction have a higher frequency than waves from the opposite direction. This is caused by the Doppler effect, which states that as we move toward waves, they reach us more frequently and as we move away from waves, they arrive less frequently. This effect can be, and is, used to tell us for the first time exactly in what direction and how fast we are moving through the universe.

electromagnetic radiation is arranged by wavelength or frequency or energy—all of which amount to the same concept in different guises—the Cosmic Microwave Background, which is a mixture of wavelengths as would be radiated from a perfectly black body kept at a temperature of 2.73K ($-54.76°$F or $-270.42°$C), has its highest brightness at a wavelength of about one millimeter. The part of the electromagnetic spectrum ranging from about 0.04 in (1 mm) up to an inch or so (a few centimeters) is the microwave domain, which accounts for the "Microwave" in the Cosmic Background. However, other submillimeter wavelengths also pervade the universe and have nothing to do with the Cosmic Microwave Background. Here are some, considered in order of increasing wavelength:

Infrared Radiation   The part of space in the Milky Way galaxy that apparently contains almost no stars, dust, or gas nevertheless radiates weakly in the infrared portion of the spectrum, which is just below the microwave region and extends down to about a micron. The radiation appears to come from some kind of invisible permeating matter that is warmed by ultraviolet light (high energy or short wavelength electromagnetic radiation of a few hundred angstroms in length), producing a range of infrared radiation that peaks at 7.7 microns. However, until 1993, no one knew any form of matter that would radiate at the particular infrared wavelength observed. That year, Adrian Webster of the Royal Observatory in Edinburgh, Scotland, demonstrated that a modification of buckminsterfullerene, in which a hydrogen atom is attached to each of 60 carbon atoms, would produce the appropriate form of infrared radiation when bathed with ultraviolet light (*see* "Ornamented and Filled Buckyballs," p 457).

Visible Light   Although it is apparent to the naked eye that most of the night sky is dark (away from cities at least), philosophers and astronomers have long thought that the sky should be filled with light all the time. Edmond Halley of comet fame was among the first to note that there is a contradiction between observation and theory. This problem of missing light is known as Olbers' paradox, since one famous version was formulated by German astronomer Heinrich Olbers in 1823. The basic idea states that every line of sight in an infinite or even a very large universe must eventually reach a star. Thus, none of the sky should appear dark.

Blocking light with gas, dust, or larger objects does not remove the paradox, since light from farther out would still be re-radiated, raising the temperature of the universe to noticeable heights above that of the Cosmic Microwave Background. The most accepted explanation of why the sky is dark at night states that the universe has a finite age and is, furthermore, expanding at a great rate. Light from the far reaches of the universe fails to reach us, so the sky is dark.

Ultraviolet Light   Although the entire sky is not filled with visible light, it is pervaded with radiation at higher wavelengths, called the diffuse high-energy astronomical background. This is not well understood. Studies using the facilities of the *Extreme Ultra-Violet Explorer* satellite between 1992 and 1994 revealed that the pressure caused by ultraviolet radiation, although very low by human standards (roughly 0.000000000000001, or 10^{-15}, bar), is more than 3 to almost 30 times as much as various theoretical predictions.

X Rays   The x-ray region of the spectrum is also present. This radiation was discovered by Riccardo Giacconi in 1962, a couple of years before the discovery of the Cosmic Microwave Radiation. From the beginning, Giacconi and others suspected that the cause was distant (equivalent to ancient) active galaxies and quasars, which are known to produce copious x rays. The problem was that the pattern of frequencies for active galaxies and quasars is different from that of the x-ray background.

In 1993, new observations of faraway x-ray sources were analyzed by Julian Krolick of Johns Hopkins University in Baltimore, Maryland, and Piotr Zycki of the Copernicus Astronomical Center in Warsaw, Poland. The observations came from three satellite-borne x-ray telescopes launched in the early 1990s. The new data showed that the patterns of frequencies matched after all, suggesting that the far-off galaxies and quasars do account for the background radiation. Somewhat less clear was what might account for the galaxies and quasars producing such large amounts of x rays, but most astronomers believe that massive black holes are involved (*see* "Black Holes Sighted," p 198). A separate look at the data from one of the satellites, the Japanese-U.S. collaboration known as *ASCA*, by Hajime Inoue of the Institute of Space and Astronautical Science in Japan resulted in an April 1994 announcement that concurred with the view that the cause was an overall dense pattern of ordinary galaxies, each producing small amounts of x rays, probably from black holes at their centers.

The Old Cosmic Microwave Background   In addition to the various shorter wavelengths that do or do not permeate the present universe, another wavelength of cosmological significance is the Cosmic Microwave Background of the early universe. Clearly, this would be a black-body radiation with more energy, corresponding to a higher temperature. Using light from a distant quasar to look at a gas cloud almost equally as distant as the quasar (and therefore about as old), a team led by Antoinette Songaila (*see* "Dark Matters," p 226) was able in 1994 to calculate the maximum possible temperature for the Cosmic Background Radiation at the time in the universe's history represented by the cloud, which has a redshift of about 2.9. The temperature limit is 13.5K ($-435.5°F$ or $-259.7°C$). This figure corresponds well with the theory that the temperature at that redshift should be about 10.7K. In 1994, this concept was checked with another use of distance to look back in time at an earlier temperature. The temperature in a cloud of dust that has a redshift of about 1.776 was measured as $7.4 \pm 0.8K$, which is quite close to theoretical predictions. A redshift of 1.776 is often interpreted to imply a time of about 11.8 billion years ago.

Age and the Hubble Parameter

Careful reports of the distance to quasars or faraway galaxies often do not indicate in light-years or other measures of length just how far away the objects are. Instead, they provide us with a number that reveals how much the spectrum has been shifted to the red by the expansion of the universe. One reason for maintaining a low profile on distance is that there is considerable disagreement regarding the age of the universe is.

Until recently, many astronomers have assumed that the universe is about 10-12 billion years old, or perhaps as much as 15 billion years old. Yet, some of the most respected cosmologists maintain that it is 18-20 billion years old. Measurements in 1994 made headlines by predicting ages as young as 8 billion years.

The various estimates of age result mostly from measuring how fast the universe is expanding. In the big bang theory, the rate of expansion today is a result of an initial expansion beginning with nothing. Working backward, the time the expansion began can be computed from the observed expansion rate. If the expansion rate were known exactly, then it would be a simple matter to calculate an age that everyone could agree on. In turn, this would make distances to faraway objects realistic in terms of light-years, parsecs, or other common units of measure.

In 1994, Wendy L. Freedman of the Carnegie Observatories in Pasadena, California, Barry F. Madore of the California Institute of Technology, and a dozen coworkers on the *HST* Extragalactic Distance Scale Project used the *Hubble Space Telescope (HST)* to take measurements of the distance to the galaxy M100 in the Virgo cluster. Their results suggest that the universe is young—from only 8-12 billion years old. By examining the brightness of 20 Cepheid variable s (*see* "Stars: Their Lives and Hard Times," p 184) in M100, they calculate that M100 is between 50-62 million light-years from Earth. This implies a Hubble constant of 80 ± 17 km/sec/megaparsec (kilometers per second per million parsecs). Depending on the mass one assumes for the universe as a whole, this measurement is consistent with an age estimate of 8-12 billion years. An earlier set of 1994 measurements used adaptive optics with the land-based telescope Canada-France-Hawaii on Mauna Kea in Hawaii to examine another galaxy in the Virgo cluster: NGC 4571. This measurement of three Cepheid variables implies that NGC 4571 is about 50 million light-years distant, a result that translates to essentially the same age estimates for the Virgo cluster. This group, led by Michael J. Pierce of Indiana University, published their results in the September 29, 1994 *Nature*. They concluded that the Hubble constant is 87 ± 7 km/sec/parsec.

A value for the Hubble constant based on the Virgo cluster could be affected by local proper motions of the cluster, just as our own Local Group is moving with respect to the background of the most distant galaxies (*see* "Clusters and Walls of Galaxies," p 222). A cluster much further away would provide a better indication of the speed of galactic recession. Although it is just barely possible to determine the distance to the Virgo cluster, various considerations indicate that the Coma cluster is about six times as far away as the Virgo cluster. At that distance, galactic recession is so fast that local proper motions, if any, should be swamped. Thus, the Virgo calculations are used as a stepping stone to Coma calculations, from which the actual value of the Hubble constant is derived.

Allan Sandage leads the group of astronomers who think that the Hubble constant is much lower, around 50, and that the universe is therefore much older. Sandage has more credentials than anyone alive, since he first measured the Hubble constant while working as an assistant to Edwin Hubble. Sandage's main argument against the Freeman and Pierce measurements is simple: the galaxies M100 and NGC 4571 are on the near side of the Virgo cluster, and therefore measurement to them is not the same as a measurement to most of the Virgo cluster. It is as if a person in Yonkers, New York, were to say that he lives only three blocks from New York City. He may live only three blocks from the northern part of the Bronx, but it would still take about a hour to get to Midtown or Downtown, which are 10-15 miles south of the northern border of the Bronx. Sandage wants to know how far it is to Midtown, not how far to the Bronx. Sandage thinks the Virgo cluster is about 80 million light-years away.

Sandages's case had been greatly encouraged by *HST* measurements of 27 Cepheids in 1992 suggesting a value of 45 for the Hubble constant, which translates to a universe 15-22 billion years old. But that was with the old, unrepaired *HST*. When sharper *HST* images became available after repair, the 1992 measurement, which had aroused skepticism at the time, was forgotten by most astronomers—but probably not by Sandage.

Sandage actually prefers a different type of distance indicator from the Cepheids, a type of supernova that is thought always to have the same brightness. Using these type Ia supernovas (*see* "Supernovas: Then and Now," p 000) in the two other galaxies, IC 4182 and NGC 5253, Sandage obtained a value of 52 ± 6 km/sec/megaparsec. Another measurement based on Type Ia supernovas by Abhijit Saha in 1994 produced $H = 54 \pm 8$, but some other astronomers in 1994 argued that correct interpretation of these measurements would produce a higher value of 67 ± 7 km/sec/megaparsec.

But another measurement, based on type II supernovas, gives a very different result from Sandage's value based on type Ia supernovas. Brian Schmidt of the Harvard-Smithsonian Center for Astrophysics in Cambridge, Massachusetts, and coworkers measured the expansion of the type II supernova 1992AM in the constellation Cetus to obtain its distance of 590 million light-years, leading to a value for H of 80 km/sec/megaparsec. Robert Kirshner of the Harvard-Smithsonian Center for Astrophysics in Cambridge, Massachusetts, has also been measuring the distances to Type II supernovas. By September of 1994 he had reported on 18 such measurements, producing an average value of 73 ± 13 km/sec/megaparsec for H.

Another study, conducted by Sidney van den Bergh of the Dominion Astrophysical Observatory in Canada, combines results from the type Ia supernovas with several other methods of estimating the approximate distance to faraway galaxies. However, van den Bergh thinks that the type Ia supernovas are not especially reliable, nor does he put much faith in the luminosity of particular galaxy types (another measure that has been used to produce low Hubble values). When he combines all these different measure with estimated sizes of planetary nebulas, studies of globular clusters of stars, and other indications employed to estimate galactic brightness, he thinks he can set a definite lower limit on the Hubble constant of 75 km/sec/megaparsec. This would rule out Sandage's 52, but agree with Freeman's 80 and Pierce's 87.

On the other side of the argument is a measurement of the iron content of quasars Q 0014+813 (redshift 3.4) and Q 0636+680 (redshift 3.2), which implies that each quasar has been in operation for about a billion years. Yet at those high redshifts, the quasars must have formed very early in the history of the universe. If the universe is only 8-10 billion years old, the age of the universe is too small for the quasars to have formed and gained that much iron. Thus, the universe must be older and the Hubble Constant must be less than 100 km/sec/megaparsec, or 30 km/sec/million light-years.

Finally, it has been known for some time that some stars in the Milky Way, close enough to study more directly, are at least 13 billion and maybe as much as 16 billion years old. As with other age measurements, however, there is lot of inference involved: astronomers are like sideshow operators looking at wrinkles and hair in an effort to guess the sucker's age. The stars in question are found in tight clusters of 10,000-1,000,000 large stars in the region above and below the spiral of the Milky Way galaxy. The region outside the plane of the spiral is known as the halo. The globular clusters of stars are thought to have formed at the same time, since there seems to be no other way to account for their concentration and for the similarity of the stars in the cluster.

Spectrograms of stars in globular clusters reveal a high content of hydrogen and helium, but not much of heavier elements. In the theory of stellar evolution (*see also* "Stars: Their Lives and Hard Times," p 184), the heavy elements are formed as stars age, but paradoxically this means that stars with few heavy elements are older than those with an abundance of them. The reason is that stars formed a long time ago were formed from the early gas of the universe, nearly all hydrogen and a bit of helium, while stars that formed more recently condensed from a gas that contains a mixture of elements heavier than helium. Thus, an older star may still consist largely of hydrogen and helium, although heavier elements are forming in its core, while a younger star contains some heavy elements from the beginning. Thus, the heavy-element-poor globular clusters are all old stars.

How old are they? Theory says that the largest stars age the fastest, so the age of a globular cluster should be the same as, or older than, the age of the largest surviving stars. This has been calculated as being at least 13 billion years. It could be greater than that: a figure of 14 billion years is commonly used, but at least one estimate has 18.5 billion years as its upper limit. No one has yet offered a satisfactory reconciliation between the age of the Milky Way galaxy, calculated as being greater than 13 billion years, and the age of the whole universe, which according to both the big bang theory and distant galaxy measurements is less than 12 billion years.

THE HUBBLE "CONSTANT"

The expansion rate of the universe is a number that has been named the Hubble parameter or Hubble Constant. The Hubble parameter is a ratio that is designated H, after Edwin Hubble, who in 1929 established that the universe is expanding. H is the ratio of the expansion velocity of any observed astronomical object to its distance from the observer. Technically, then, H is not a constant, since it has varied over the history of the universe, due to the slowing of the expansion rate. In some theories of cosmology, the expansion will reach a rate of 0 and stop. In other versions, it will pass through 0 and become negative, resulting in an imploding universe. These differences in the fate of the universe are directly related to its mass. The rate of slowing is also determined by the mass of the universe, a value that is not known for sure, although a specific mass is predicted in some theories.

Since H is technically not a constant, careful astronomers call H the Hubble *parameter* instead of the Hubble *constant*. The value of H that applies today (a specific number) is H_0, but that is not exactly constant either, since it will be different tomorrow. Ordinary usage, however, favors "constant," and many writers drop the subscript of 0.

H is usually measured in units based on an expansion velocity in kilometers per second and a distance in millions of parsecs (a parsec is 3.258 light-years; often a million parsecs is termed a megaparsec). A value of H that is near 100 km/sec/megaparsec implies that the big bang happened a brief 10 billion years ago. Sometimes H is given in miles per second and the distance in megaparsecs, in which case 100 km/sec/megaparsec would be converted to 60 mi/sec/megaparsec. Occasionally, H is measured in km/sec/million light-years, which is different enough from the use of megaparsecs to cause considerable confusion if one fails to notice the unit. Since one megaparsec = 3,258,000 light-years, one needs to divide a value of H using light-years by 3.258 to get the more familiar value. 100 km/sec/megaparsec, for example, equals about 30 km/sec/million light-years.

The lower H is, however, the older the universe. The age of the universe is $1/H$ hundred billion years. Here is a conversion table with higher ages based on a mass value for the universe that corresponds to the amount of mass that we are certain is present in the universe, which is not enough to slow down expansion. Enough dark matter to slow down and eventually stop the expansion would result in the lower age estimates.

H km/sec/ megaparsec	H mi/sec/ megaparsec	H mi/sec/ million light-years	Age of the universe
33	20	10	25–30 billion years
50	30	15	18–20 billion years
67	40	20	12–15 billion years
75	47	23	11–13 billion years
80	50	25	10–12.5 billion years
100	62	30	8–10 billion years
125	78	40	6–8 billion years

Another equation commonly used is that the age of the universe is two-thirds of the reciprocal of the Hubble constant times a trillion (10^{12}) years. This produces, to the nearest billion, estimates of 13 billion years for $H = 50$ km/sec/megaparsec and 8 billion for $H = 80$ km/sec/megaparsec, two popular values for H.

The Smoothness Problem

Before 1992, one of the main theoretical difficulties with the big bang theory was that radiation in the universe left over from the big bang appeared to be uniform (or "smooth," in the common nontechnical language of astronomers and mathematician, to whom "uniform" has a technical meaning). Smoothness was unexpected in part because the rest of the universe was lumpy, or clumped (see "Clusters and Walls of Galaxies," p 222). If the big bang was smooth, it should have resulted in a smooth universe, not the lumpy one observed.

In 1992, however, George F. Smoot of Lawrence Berkeley Laboratory in Berkeley, California, reported that the Cosmic Background Explorer (COBE) satellite experiment found statistical evidence that the big bang was not smooth after all (see also "The Shape and Evolution of Galaxies," p 216). While many people considered this expected result to be among the great discoveries of cosmology—it was the lead story in the news on April 24, 1992, and some of the phrases used to describe it were "unbelievably important," "holy grail of cosmology," and "opened up the dark ages in cosmology"—a few pointed out that the COBE results were only averages. No one had found a specific fluctuation of the type that could lead to the development of structure in the later universe. However, such a wrinkle in the background energy was finally detected in an observational experiment specifically designed to find one. The variations in temperature, reflecting differences in amounts of energy and therefore mass, were on the order of 6 parts per 1,000,000, taken from 210 million measurement samples of data believed to derive from 300,000 years after the big bang—the earliest time for which such an observation is even theoretically possible, as is explained below.

In 1993 and 1994, there was confirmation of the COBE observations in a further and more detailed report on the "cosmic ripples" from balloon experiments and from a ground-based source. The balloon experiments were conducted in 1993 by the National Scientific Balloon Facility in Texas.

In the most important extension, reported in 1994, a team consisting of S. Hancock, Rod D. Davies, G. M. Gutierrez de la Cruz, A. N. Lasenby, and R. A. Watson of the University of Manchester in England along with R. Rebolo and J. E. Beckman of the Instituto de Astrofisica de Canarias on Tenerife, Spain, set up three coordinated radio telescopes in the Canary Islands (Tenerife is the largest of the Canaries). They were able to observe specific features in the Cosmic Microwave Background that correspond to the fluctuations observed by COBE, ripples that existed when the universe was about 300,000 years old. At that time, the energy level of the universe had cooled enough that atoms of hydrogen and helium could form. Before that time, the universe had been a plasma of free particles that interacted directly with photons of electromagnetic radiation. Atoms interact less directly, and are transparent to a large extent. Thus, we can still "see" how many photons there were at different locations in the early universe.

What the Tenerife experiment revealed are giant fluctuations, larger even than the observed superclusters of galaxies. Because of expansion, a wrinkle 300,000 years after the big bang would be about one-tenth to one-fourth the size that the Tenerife experiment can detect. Some astronomers see in the giant fluctuations the possibility of directly measuring the curvature of the universe.

COBE scored its second success, reported early in 1993, when an experiment announced by John C. Mather of the U.S. National Aeronautics and Space Administration's Goddard Space Flight Center in Greenbelt, Maryland, established that the big bang was indeed a singular event. If it had occurred in stages, or if it had been followed by smaller events of the same type, there would have been variations in temperature at different frequencies, which COBE failed to detect. In the process, COBE data established a more specific temperature for the Cosmic Microwave Background, usually quoted as 2.7K or 2.73K. COBE results were 2.726K ($-454.763°F$ or $-270.424°C$).

Fixing Omega

The Greek alphabet ended with omega: Ω as a capital and ω in lower case. The symbolism of a last letter is attractive, so mathematicians use ω for one kind of infinity, while cosmologists let Ω symbolize the density of matter in the universe. More specifically, Ω is the ratio of the actual density of the matter in the universe to the amount of matter needed to produce a solution to a cosmological equation called the Friedmann-Lemaître equation (discovered by Alexander Friedmann in 1922 and independently by Georges Lemaître in 1927) for which the universe gradually slows down and finally stops expanding, but does not then collapse. The critical density and today's Hubble constant, and therefore Ω and H, are intimately linked. For example, if today's Hubble constant were 100 km/sec/megaparsec, the critical density would be 10^{-29} gram per cubic centimeter. Then, if the density of the universe were also measured to be 10^{-29} gram per cubic centimeter, the value of Ω would be 1. This value is known as the "flat" universe value, since it corresponds to a solution of Einstein's general relativity equations in which space is not curved. Curvature of space is difficult to imagine, but curvature of a two-dimensional sheet can be visualized instead. Such a sheet is called flat, and by analogy so is uncurved space.

The relationship between Ω and H works both ways. Calculations of H from the distance to far-off galaxy clusters have to assume a particular value for Ω. Commonly astronomers report results based on $\Omega = 1$ and also on $\Omega = 0$. $\Omega = 0$ corresponds to space with a positive curvature (like the surface of a sphere) and expansion without ceasing. Results are usually not calculated for Ω greater than 1, which would imply a collapsing universe, since that is not what is observed. Indeed, some theorists, such as David H. Schramm of the University of Chicago in Illinois, suggest that a value greater than 1 would have caused the early universe to have recollapsed immediately after the big bang. Schramm also calculates that a value less than 1 would cause the universe to fly apart too quickly for stars to form. So $\Omega = 1$ is the popular choice for Schramm and other astrophysicists.

However, measurements of Ω based on actual densities are usually close to 0, with 0.2 a typical result (*see also* "Dark Matters," p 226). This is unsatisfactory for most cosmologists, who favor $\Omega = 1$ for theoretical, and perhaps psychological, reasons. One way out of this dilemma would be to change the Friedmann-Lemaître equation by adding a constant to it. Einstein himself, when he first realized that general relativity would produce an expanding universe, proposed such a cosmological constant, but later decided that this was a blunder. The physical interpretation of the cosmological constant is a force in empty space that opposes gravity. A suitable cosmological constant could raise Ω to 1, but even raising Ω to 0.8 would solve both the age paradoxes and the missing-mass problem. If $\Omega = 0.2$ and the cosmological constant is 0.8, the result adds to 1, with the energy from the cosmological constant replacing the dark matter needed to supply the missing mass. Furthermore, by slowing the earlier expansion of the universe, it would raise the age higher, since most age estimates are based on the rate of expansion observed today.

Research based on a study of gravitational lenses by Chris Kiochanek of the Harvard-Smithsonian Center for Astrophysics, however, suggests that there is no way, even with a cosmological constant, to produce a valid Ω greater than 0.5. Thus, although theorists continue to hold out for $\Omega = 1$, there is still no evidence that this value can be produced either by observation or by jiggering with the equations.

(Periodical References and Additional Reading: *New York Times* 4-24-92, p A1; *New York Times* 5-12-92, p C1; *New York Times* 6-30-92, p C2; *New York Times* 1-8-93, p A1; *New York Times* 4-14-93, p A18; *New York Times* 9-14-93, p C5; *Science* 11-5-93, p 846; *Nature* 1-20-94, p 250; *Nature* 1-27-94, pp 316 & 333; *New York Times* 2-1-94, p C6; *Nature* 4-14-94, pp 584 & 599; *Science* 4-15-94, p 346; *Science* 4-29-94, p 663; *Astronomy* 7-94, pp 22 & 24; *Scientific American* 8-94, p 32; *Science* 8-5-94, p 740; *Nature* 9-1-94, p 43; *Nature* 9-29-94, pp 374 & 785; *New York Times* 10-4-94, p C5; *New York Times* 10-27-94, p A1; *Nature* 10-27-94, pp 741 & 757; *Science* 10-28-94, p 539; *Astronomy* 11-94, pp 20 & 22; *New York Times* 11-1-94, p C1; *New York Times* 11-29-94, p C1; *New York Times* 12-27-94, p C1; *Astronomy* 1-95, p 20; *Discover* 1-95, p 35; *Astronomy* 2-95, p 22; *Natural History*, 2-95, p 72; *Astronomy* 3-95, p 49; *Astronomy* 4-95, pp 24 & 30)

ASTRONOMICAL LISTS AND TABLES

MAJOR TELESCOPES

Telescopes were first discovered in Holland (the Netherlands) about 400 years ago. The first telescopes used lenses, familiar at the time from lenses in spectacles, to gather light and focus it. Later in the seventeenth century, scientists realized that curved mirrors could also gather and focus light. Since the light did not need to pass through the mirror (as light passes through a lens), mirrors proved to be more efficient than lenses for large telescopes.

OPTICAL TELESCOPES

Year	Type	Importance
1608	Lens	Hans Lippershey in Holland applies for the first patent on a telescope.
1609	Lens	Galileo builds the first optical telescopes, eventually reaching 30 power.
1611	Lens	Johannes Kepler introduces the convex lens, producing even greater power.
1663	Metal mirror	James Gregory is the first to propose that a reflecting telescope can be made.
1668	Metal mirror	Isaac Newton builds the first telescope to use a mirror, rather than a lens, to collect light.
1723	Metal mirror	John Hadley invents a reflecting telescope based on the parabola, which concentrates light at a point.
1789	Metal mirror	William Herschel builds a telescope with a 48 in (122 cm) mirror, the largest for many years.
1840	Composite-metal mirror	William Parsons (who will become Lord Rosse in 1841) introduces a method for making the mirror of two different metals and builds a successful 36 in (91 cm) reflector.
1845	Composite-metal mirror	Lord Rosse builds a 72 in (183 cm) reflector, but it is no improvement in observational power over his smaller mirror, in part because of its great weight.
1856	Silvered-glass mirror	Léon Foucault makes the silvered-glass mirror popular, although the technique was invented earlier in the 1850s by Justus von Liebig and also applied to astronomical telescopes by Carl August von Steinheil at the same time as Foucault.

Year	Type	Importance
1897	Lens	Alvan Clark builds what is still the world's largest telescope to use a lens instead of a mirror, the Yerkes Observatory.
1917	Silvered-glass mirror	The Hooker Telescope at Mt. Wilson is put into operation; it proves to be the world's largest for about 30 years.
1929	Combination	Bernard Schmidt's telescopes, which combine lenses and mirrors, are made. The Schmidt telescope becomes the workhorse of astronomy.
1948	Silvered-glass mirror	The 200 in (5 m) Hale Telescope, located on Mt. Palomar, becomes the largest and best on Earth.
1962	Silvered-glass mirror	The largest telescope devoted to observing the Sun is erected at Kitt Peak in Arizona.
1974	Silvered-glass mirror	The Kuiper Airborne Observatory is a 36 in (90 cm) telescope mounted in an airplane of the type used for military cargo. By flying at a height of 41,000 ft (12,500 m), the telescope is largely above Earth's atmosphere.
1976	Silvered-glass mirror	The Soviet Zelenchuksaya Telescope becomes the world's largest, but various problems limit its effectiveness.
1979	Silvered-glass mirrors	The Multiple Mirror Telescope (MMT) uses six mirrors to obtain the equivalent light-gathering power of a 177 in (4.5 m) reflector.
1989	Silvered-glass mirror	The 94 in (2.4 m) *Hubble Space Telescope (HST)* becomes the first optical telescope in space.
1991	Silvered-glass mirror	The Keck I Telescope on Mauna Kea in Hawaii uses the world's largest mirror, 400 in (10 m) in diameter with four times the light-gathering power of the Hale. Scheduled observing did not begin until November 1993, after two years of adjustment and testing.
1994	Liquid mirror	In January the world's first astronomical mirror using a spinning dish of liquid mercury as a reflector has its debut outside of Vancouver, British Columbia, where it is operated by astronomers from Laval University in Quebec. When rotating, the mirror is a perfect parabola 106 in (2.7 m) in diameter. Although costing only $200,000 to build, the telescope is one of the world's 20 most powerful light-gathering tools. The main disadvantage is that it can only see objects that are directly overhead; a computer digitally compensates for the motion of Earth to allow tracking of specific objects.
1994	Silvered-glass mirror	The WIYN telescope on Kitt Peak in Arizona, named after the sponsoring institutions—the University of Wisconsin, Indiana University, Yale University, and the National Optical Astronomy Observatories—is only 138 in (3.5 m), but special computer-controlled actuators in the spun-cast mirror make its images very sharp.

A number of new optical telescopes are planned that will surpass all the existing ones in size and light-gathering ability. Although the dates of completion remain problematical, the basic designs and considerable development are already complete.

Year	Type	Importance
1996	Silvered-glass mirror	The Carnegie Institution of Washington, D.C., together with the University of Arizona is building a telescope with a 21 ft (6.5 m) reflector called Magellan to be placed at the Las Campanas Observatory in Chile.
1997	Silvered-glass mirrors	The Very Large Telescope (VLT) planned by the European Southern Observatory for Cerro Paranal in the Atacama Desert of Chile, which has the second lowest humidity (after Antarctica) and thus some of the best viewing on Earth, will consist of four 27 ft (8.2 m) mirrors designed to work together, producing the light-gathering power of one mirror 54 ft (16.4 m) in diameter with a surface area of 2,153 sq ft (200 sq m).
1999	Silvered-glass mirror	The Japanese National Large Telescope, also known as the Subaru telescope, will have a single mirror that is 327 in (8.3 m) in diameter, just large enough to be the largest single-mirror reflector ever built; it will also be located in Hawaii on Mauna Kea. The mirror is a thin, or meniscus, type that will be mounted on actuators to adjust the curvature. Construction of the mirror started in 1992.
2000	Silvered-glass mirrors	The Large Binocular Telescope will eclipse the Japanese National Large Telescope twice—it is planned to use *two* mirrors, each 331 in (8.4 m) in diameter—if it ever gets built. The planned site is Mount Graham in Arizona, where environmental groups have used the endangered status of the Mount Graham red squirrel to fight against the construction of the telescope.
2000	Silvered-glass mirror	A consortium of American, Argentinean, Brazilian, British, Canadian, and Chilean astronomers have begun to build a telescope with a 315 in (8 m) mirror on Mauna Kea, to be followed by a twin at Cerro Pachón in Chile. Thus, this is called the Gemini project.

RADIO TELESCOPES

Before 1931, all telescopes were optical—that is, they gathered and focused electromagnetic radiation in the range people can sense with their eyes. However, stars, planets, and other objects in the universe also produce other wavelengths of radiation. A radio telescope gathers and focuses radiation at long wavelengths, the same kind of electromagnetic radiation used for transmission of radio signals.

Year	Type	Importance
1931	Ordinary antenna	Karl Jansky accidentally discovers that radio waves are coming from space as he tries to track down sources of static.
1937	Parabolic dish	In Wheaton, Illinois, Grote Reber builds the first radio dish telescope.
1957	Steerable dish	The 250 ft (75 m) parabolic dish at Jodrell Bank in England is the first major radio telescope.
1962	Steerable dish	The 300 ft (90 m) dish at Green Bank, West Virginia, is the first used to search for extraterrestrial life. It collapses mysteriously on November 15, 1988.
1963	Fixed dish	The largest fixed-dish radio telescope, 1,000 ft (305 m) across, is built into a valley at Arecibo, Puerto Rico.

Year	Type	Importance
1970	Steerable dish	The world's largest steerable dish, 328 ft (100 m) in diameter, is installed at Effelsberg, Germany.
1977	Several antennas	The first Very Long Baseline interferometry array begins operating at Caltech's Owens Valley Radio Observatory.
1980	27 antennas	The Very Long Array (VLA) is built in the shape of a 13 mi (21 km) Y near Socorro, New Mexico. Each of the 82 ft (25 m) antennas is mounted on railroad track so that they can be moved closer or farther apart, depending on the task.
1993	10 antennas spread from Hawaii to the Virgin Islands, a distance of 5,000 mi (8,000 km)	The Very Long Baseline Array (VBLA) uses interferometry techniques that produce a resolution more than 1,000 times better than the best optical telescopes. Each antenna is an 82 ft (25 m) parabolic dish.
1995	30 antennas	The Giant Meterwave Radio Telescope (GMR) north of Pune, India, contains 30 parabolic 148 ft (45 m) antennas arranged in a 15 mi (25 km) Y. Not only is the telescope larger than the Very Long Array, but also, as its name suggests, the wavelengths GMR are intended to capture are longer than the centimeter-length waves that are the focus of the VLA.

OTHER SPACE-BASED TELESCOPES

Since both short and long wavelengths coming from space had been studied by optical and radio telescopes, it seemed likely that other wavelengths also could be detected. The problem is that Earth's atmosphere, relatively transparent to optical and radio waves, is almost opaque to other wavelengths of electromagnetic radiation. The solution is to put telescopes in satellites traveling above Earth's atmosphere to detect other wavelengths.

Year	Type	Importance
1961	Gamma rays	The first telescope in space observes gamma rays that do not penetrate Earth's atmosphere.
1970	X rays	*Uburu* ("freedom" in Swahili), the first telescope to detect x rays, is launched into space.
1972	Ultraviolet radiation	The *Copernicus* spacecraft incorporates a telescope designed to collect ultraviolet radiation.
1978	X rays	The *Einstein Observatory*, which detects x rays from space, becomes one of the most productive satellite-based telescopes.
1983	Infrared radiation	The *Infrared Astronomy Satellite* (*IRAS*) becomes the most successful satellite-based telescope to this date, detecting possible new planetary systems and the formation of stars.
1987	Ginga	The first Japanese orbiting x-ray telescope, which remains in space until 1991.
1990	X rays	The German *Roentgen Satellite* (*ROSAT*) extends the number of known galactic and cosmic x-ray sources to 100,000.

Year	Type	Importance
1991	Gamma rays	The *Compton Gamma Ray Observatory* contains a more sensitive telescope with a wider range than previous satellite telescopes.
1992	Ultraviolet radiation	The *Extreme Ultraviolet Explorer* (*EUVE*) is a U.S. satellite observatory launched in June. The telescope has studied very short wavelength ultraviolet radiation since that time, detecting new sources and probing late-type stars, hot white dwarfs, and the interstellar medium.
1993	X rays	The Japanese *ASCA* satellite, launched in February, was preceded by the Ginga Japanese orbiting x-ray telescope in 1987.
1999?	X rays and visible light	The *X-ray Multi-Mirror Mission* (*XMM*) will not only contain three x-ray telescopes, but also one 1 ft (30 cm) optical reflector alighted with the x-ray devices. The hope is to be able to see what is causing the x rays. At present, by the time an optical telescope can be trained on an interesting x-ray event, the event is long past.

OTHER GROUND-BASED TELESCOPES

The concept of a telescope was expanded considerably when the first radio dish was built by Grote Reber in 1937. If a device that gathers electromagnetic radiation at wavelengths that are invisible to humans is a telescope, then any method of detecting any kind of radiation from space might also be called a telescope. Using counters or photographic film to detect the particles and waves known as cosmic rays began before the first radio telescope, which may account for this form of telescope not going by that name. But when an experiment designed to trap neutrinos from the Sun was started in 1968, people called it a telescope. Today there are a number of existing and planned "telescopes" used to capture radiation, especially neutrinos, from space.

Year	Type	Importance
1968	Cleaning fluid	A tank containing 100,000 gal (375,000 L) of perchloroethylene (a common cleaning fluid) at the bottom of the Homestake Gold Mine in Lead, South Dakota, was the first neutrino telescope, installed by Raymond Davis Jr. The telescope works by detecting argon, which is sometimes produced by a collision between a neutrino and the chlorine in perchloroethylene. The Homestake telescope led to many others because it found far fewer neutrinos coming from the Sun than had been expected.
1988	Very pure water	Kamiokande II in Japan watches for neutrinos in a 21,000 gal (80,000 L) pool of water by looking for flashes of radiation caused when a neutrino smashes into an electron and propels it to a speed greater than the speed of light in water, which produces visible light known as Cerenkov radiation.
1990	Liquid gallium	The U.S. and Russian experiment called SAGE looks for neutrino reactions in a tank containing 30 tons of gallium, a metal that is a silvery liquid like mercury at a temperature of 86°F (30°C) and above; later the amount was increased to 60 tons of gallium. The basic concept is that neutrinos sometimes can change gallium to another metal, germanium, which can then be detected.

Year	Type	Importance
1991	Gallium chloride in solution	Similar to the original SAGE in its use of gallium, this American-European effort called GALLEX is 101 tons of a solution installed in Italy off the Gran Sasso highway tunnel through the Abruzzi mountains northeast of Rome.
1993	Ocean water	The DUMAND (Deep Underwater Muon and Neutrino Detector) neutrino telescope was under development from at least 1981 until 1993, when it lowered a 436 m (1,431 ft) string of sensors to the floor of the Pacific 18 miles west of Hawaii in water some 4,800 m (15,750 ft) deep. The telescope is planned to have nine similar strings in the ocean when it is completed, which may not be for several years. The method uses Cerenkov radiation in sea water produced by muons that in turn result from rare neutrino interactions.
1994	Ice	The AMANDA (Antarctic Muon and Neutrino Detector Array) is based on the same general principles as the DUMAND telescope, but its strings of sensors are lowered into holes drilled in the Antarctic Ice Cap near the U.S. South Polar Station. Ten strings in all, each 656 ft (200 m) long, were to be in place by the end of 1994.
1995	Deuterium oxide (heavy water)	The Sudbury Neutrino Observatory (SNO) operates in a nickel mine about 300 mi (320 km) north of Toronto, Canada, and uses 1,000 tons of deuterium oxide (water in which ordinary hydrogen has been replaced with "heavy" hydrogen, or deuterium—hence, "heavy water") to look for Cerenkov radiation caused not only by neutrinos or muons, but also for break-up of deuterium into hydrogen and a neutron resulting from collisions with muon or tauon neutrinos.

(Periodical References and Additional Reading: *New York Times* 2-4-92, p C1; *Science* 6-12-92, p 1512; *New York Times* 6-13-92, p 1; *New York Times* 11-25-92, p D1; *Astronomy* 6-93, p 20; *Astronomy* 8-93, p 48; *New York Times* 8-17-93, p C1; *New York Times* 1-6-94, p A4; *Science* 1-1-94, pp 42 & 55; *Science* 1-14-94, p 176; *Astronomy* 4-94, p 22; *Science* 4-14-94, p 348; *Astronomy* 6-94, p 26; *New York Times* 8-28-94, p I-20; *Science* 10-21-94, p 356; *Astronomy* 11-94, p 46; *Astronomy* 2-95, p 24; *Astronomy* 3-95, p 32)

SOME FORTHCOMING ECLIPSES

There are four types of eclipses, depending on the exact relationship of the Sun, Moon, and Earth. In a total eclipse, the Moon as seen from Earth covers the Sun completely. An annular eclipse occurs when the Moon is somewhat farther from Earth, so that even if the shadow is observed in a direct line with the Moon and Sun, a ring of sunlight remains around the Moon. A partial eclipse occurs when only part of the Moon's shadow touches Earth, while the rest passes beyond one of the poles. An uncommon type of eclipse is the annular/total, which is annular at each end of the path across Earth, but total in the middle of the path. Partial eclipses are omitted in the following table, along with many whose entire path of totality is in the Eastern Hemisphere. Annular eclipses are labeled A and total eclipses T.

Date	Type	Approximate Path	Longest Duration
March 9, 1997	T	Siberia	2 min 50 sec
February 26, 1998	T	Mid-Pacific through extreme N. of S. America and Caribbean	4 min 9 sec
August 11, 1999	T	N. Atlantic	2 min 23 sec
June 21, 2001	T	S. Atlantic, Africa, Madagascar, Indian Ocean	4 min 56 sec
December 14, 2001	A	Across Pacific to Central America	3 min 53 sec
June 10, 2002	A	N. Pacific to Mexico	23 sec
December 4, 2002	T	S. Atlantic through Africa and Indian Ocean to Australia	2 min 4 sec
November 23, 2003	T	Antarctica	1 min 57 sec
April 8, 2005	A/T	S. Pacific to northern S. America	42 sec
March 29, 2006	T	Atlantic through N. Africa across Mediterranean into Russia	4 min 7 sec
August 1, 2008	T	N. of Hudson Bay region	2 min 27 sec
July 22, 2009	T	India into S. China into S. Pacific	6 min 39 sec
July 11, 2010	T	S. Pacific to Chile	5 min 20 sec
May 20, 2012	A	N. Pacific to U.S. Northwest and Midwest	5 min 46 sec
November 13, 2012	T	Australia through S. Pacific	4 min 2 sec
November 3, 2013	A/T	Atlantic through central Africa	1 min 39 sec
March 20, 2015	T		2 min 47 sec
March 9, 2016	T	Indonesia	4 min 9 sec
August 21, 2017	T	N. Pacific across U.S. into Atlantic	2 min 40 sec
July 2, 2019	T	S. Pacific into southern S. America	4 min 32 sec

BASIC FACTS ABOUT THE PLANETS

The planets in our solar system are of two main types, which are grouped separately below. The inner planets are somewhat like Earth and are called the terrestrial planets. The outer planets, except for Pluto, are gas giants, similar to Jupiter. Pluto is not very much like the planets in either group as far as we know, but it is so far away that its exact nature is not known.

The numbers used below are sometimes given in scientific notation because of their size. A large number, such as 1,840,000,000, might be written as 1.84×10^9. The exponent 9 signifies the number of places after the first digit. A small number such as 0.000000000000001 is written as 10^{-15}, where the exponent -15 tells the number of zeroes, counting the one before the decimal place, before the first nonzero digit.

Some of the words and phrases used in the table are briefly defined below:

- **Bar** is a measure of pressure that is slightly less than Earth's air pressure at sea level under normal conditions, or about 29.92 in (76 cm) of mercury, as the weather report on television would state it.
- **Eccentricity** is a positive number that measures the shape of certain curves; the smaller the number, the more an ellipse is like a circle, which has an eccentricity of 0.
- **Ecliptic** refers to the apparent path of the Sun through the stars as viewed from Earth, which is in a plane inclined 23.5° to Earth's equator.
- **Escape velocity** is the speed needed for an object to be propelled from the surface of a planet and not fall back.
- **Inclination of axis** is the angle that the line about which a planet rotates makes with the plane defined by its path around the Sun.
- **Inclination of orbit to the ecliptic** is the angle that the plane defined by a planet's path around the Sun makes with the plane defined by the apparent path of the Sun among the stars as seen from Earth.
- **Orbital velocity** is the speed of a planet in its path around the Sun.
- **Retrograde** means in the opposite direction of other planets.
- **Revolution** is the trip a planet makes about the Sun.
- **Rotation** is the turning of a planet about a line through its center.

TERRESTRIAL PLANETS

	Mercury	*Venus*	*Earth*	*Mars*
Average distance from the Sun in miles (in kilometers)	35,900,000 (57,900,000)	67,200,000 (108,200,000)	92,960,000 (149,600,000)	141,600,000 (227,900,000)
Rotation period in Earth days	59	− 243.01 (retrograde)	1	1.0004
Period of rotation in hours	1,407.6	− 5,832.2	23.9	24.6
Period of revolution in Earth days	224.7	365.26	687	88
Average orbital velocity in mi per second (in km per second)	29.8 (48.0)	21.75 (35.00)	18.46 (29.71)	14.98 (24.11)
Inclination of axis	2°	3°	23° 27′	23° 59′
Inclination of orbit to the ecliptic	7°	3.39°	0°	1.9°
Eccentricity of orbit	0.206	0.007	0.017	0.093
Equatorial diameter	3,031 mi (4,878 km)	7,520 mi (12,104 km)	7,926 mi (12,756 km)	4,222 mi (6,794 km)
Diameter relative to Earth	0.382 times	0.949 times	—	0.532 times

	Mercury	*Venus*	*Earth*	*Mars*
Mass	3.303 × 10^{23} kg	4.87 × 10^{24} kg	5.97 × 10^{24} kg	6.42 × 10^{23} kg
Mass relative to Earth	0.055 times	0.815 times	—	0.1074 times
Mass of Sun relative to planet mass (with atmosphere and satellites)	5,972,000	408,520	328,900	3,098,710
Average density	3.13 oz/in ^{3} (5.41 g/cm³)	3.06 oz/in ^{3} (5.3 g/cm³)	3.19 oz/in ^{3} (5.52 g/cm³)	2.27 oz/in ^{3} (3.93 g/cm³)
Gravity (at equator surface)	12.4 ft/sec² (377.95 cm/sec²)	28.2 ft/sec² (859.54 cm/sec²)	32.1 ft/sec² (978.41 cm/sec²)	12.2 ft/sec² (371.86 cm/sec²)
Gravity (relative to Earth)	0.38	0.88	1	0.38
Escape velocity at equator	2.7 mi/sec (4.3 km/sec)	6.40 mi/sec (10.3 km/sec)	6.96 mi/sec (11.2 km/sec)	3.1 mi/sec (5 km/sec)
Average surface temperature	332°F (167°C)	854°F (457°C)	59°F (15°C)	−67°F (−55°C)
Atmospheric pressure at surface	10^{-14} bars	90 bars	1.01325 bar	0.007 bar
Atmosphere (main components)	Virtually none (but traces of sodium, potassium, oxygen, helium, and hydrogen)	Carbon dioxide 96%; nitrogen 3.5%; argon 0.006%; oxygen 0.003%	Nitrogen 77%; oxygen 21%; water 1%; argon 0.93%; carbon dioxide 0.03%	Carbon dioxide 95%; nitrogen 2.7%; argon 1.6%; oxygen 0.15%
Planetary rings	None	None	None	None
Planetary satellites	None	None	1 moon	2 moons

OUTER PLANETS

	Jupiter	Saturn
Average distance from Sun	483,600,000 mi (778,300,000 km)	886,700,000 mi (1,472,000,000 km)
Rotation period	9 hr, 55 min, 30 sec	10 hr, 39 min, 20 sec
Period of revolution (Earth days)	4,332.6	10,759.2
Period of revolution (Earth years)	11.86	29.46
Average orbital velocity	8.1 mi/sec (13.06 km/sec)	5.99 mi/sec (9.64 km/sec)
Inclination of axis	3°7'	26°44'
Inclination of orbit to ecliptic	1.3°	2.5°
Eccentricity of orbit	0.048	0.056
Equatorial diameter	88,700 mi (142,800 km)	74,800 mi (120,400 km)
Diameter relative to Earth	11.21 times	9.41 times
Mass	1.899×10^{27} kg	5.686×10^{26} kg
Mass relative to Earth	317.9 times	95.2 times
Mass of Sun relative to planet mass (with atmosphere and satellites)	1,047	3,498
Average density	0.759 oz/in³ (1.31 g/cm³)	0.40 oz/in³ (0.69 g/cm³)
Gravity (at equator surface)	75.06 ft/sec² (2287.83 cm/sec²)	29.69 ft/sec² (813.51 cm/sec²)
Gravity (relative to Earth)	2.34	0.92
Escape velocity at equator	37.0 mi/sec (59.5 km/sec)	22.1 mi/sec (35.6 km/sec)
Average temperature in atmosphere	(at surface) −162°F (−108°C)	(at surface) −208°F (−133°C)
Atmospheric pressure at surface	3,000,000 bars	8,000,000 bars
Atmosphere (main components)	(Near cloud tops) Hydrogen 90%; helium c 10%	Hydrogen 94%; helium c 6%
Planetary ring	Yes	Yes, extensive
Planetary satellites	16	20

THE CONSTELLATIONS

Constellations are small groups of stars that, from our vantage point on Earth, seem to form some particular shape. In actuality, however, the stars that form a constellation are usually at vastly different distances from the solar system. They only appear to form a particular figure from Earth.

Long ago, probably well before recorded history of any kind, people began naming these groups. By Sumerian times there were already stories being told about how

Uranus	Neptune	Pluto
1,783,000,000 mi (2,869,600,000 km)	2,794,000,000 mi (4,496,600,000 km)	3,666,100,000 mi (5,900,100,000 km)
17 hr 12 min	16 hr 6 min	6 days, 9 hr, 18 min (retrograde)
30,685.4	60,268	90,950
84.01	165	249
4.2 mi/sec (6.8 km/sec)	3.35 mi/sec (5.4 km/sec)	2.9 mi/sec (4.7 km/sec)
97°54′	28°48′	60°
0.8°	1.8°	17.2°
0.047	0.009	0.25
32,600 mi (52,400 km)	31,300 mi (50,400 km)	1,423 mi (2,290 km)
4.1 times	3.88 times	0.18 times
8.66×10^{25} kg	1.030×10^{26} kg	c 1.3×10^{22} kg
14.6 times	17.23 times	c 0.0021 times
22,759	19,332	3,000,000
c 0.7 oz/in^3 (1.21 g/cm^3)	0.9 oz/in^3 (1.5 g/cm^3)	c 1.3 oz/in^3 (2.2 g/cm^3)
25.5 ft/sec^2 (777.24 cm/sec^2)	36 ft/sec^2 (1097.28 cm/sec^2)	c 14.1 ft/sec^2 (c 164.59 cm/sec^2)
0.79	1.12	0.16
13.2 mi/sec (21.2 km/sec)	14.66 mi/sec (23.6 km/sec)	0.8 mi/sec (1.3 km/sec)
(cloud tops) −344°F (−209°C)	(cloud tops) −365°F (−220°C)	(at surface) c −355°F (c −215°C)
NA	NA	0.1 millibar (?)
Hydrogen, helium, methane	Hydrogen, helium, methane	Tenuous; methane & possibly neon
Yes	Yes, interrupted	No
15	8	1

particular constellations were formed. Even today there are groups of three to six constellations that are parts of a single story. Most of our present knowledge of such stories comes from the Greeks, who reflected much of their mythology in the stars and planets. Thus we have the stories of Perseus or the ship called the *Argo*.

One group of constellations has exerted a special influence on human thought, at least since 1500 B.C. As the Sun, Moon, and planets move through the sky, they pass through a group of twelve constellations: the constellations of the zodiac. Chaldean astronomers believed that the presence of the Sun, a planet, or even the Moon in one of these constellations influences happenings on Earth. This influence is considered especially powerful at the time of a person's birth or at other significant times. We call this belief *astrology*. Because of the precession of the equinoxes, the traditional twelve constellations are no longer where they were 3,500 years ago. Modern astrologers have divided the year into twelve "houses" based on where the signs of the zodiac used to be. Thus, when an astronomer and an astrologer refer to the zodiac, they mean quite different things.

Today astronomers use constellations for their own purposes, especially to map the sky. Each part of the sky is named after a particular constellation. These constellations, especially in the Southern Hemisphere, may not be traditional ones, but rather are groups of stars that astronomers have named so that all of the sky is covered (such constellations are labeled "Of modern origin" in the following table). The International Astronomical Union has decreed that each constellation be bounded by straight north-south and east-west lines. As a result, many of the larger traditional constellations extend beyond the boundaries of the astronomical constellation. The part of the sky that contains many constellations connected to water by name or legend is sometimes called the "wet quarter" of the sky.

While astronomers often use the traditional names of stars, most of which come to us from Latin or Arabic sources, they also use other systems for naming objects in the sky (galaxies, radio sources, quasars, and so forth) that are based on constellations. Generally, the brightest star in a particular astronomical constellation is called alpha, the next brightest beta, and so on through several letters of the Greek alphabet. Thus Sirius is also known as alpha Canis Majoris, usually abbreviated to α CMa, which means it is the brightest star in the constellation Big Dog (Sirius has long been known as the dog star). Since Sirius is a binary star, the much brighter main star is officially α CMa A while the white-dwarf companion is α CMa B. Bright radio sources were once designated by the name of a constellation followed by a letter of the Roman alphabet, such as Cassiopeia A (Cassiopeia is a character from Greek mythology) or Cygnus A (the Swan). Unfortunately, this system has been largely abandoned, so the same radio source may have several different names depending on the astronomer. X-ray sources are still designated by the name of the constellation followed by an X hyphen number. The number 1 is the brightest x-ray source in a given constellation—so, for example, Scorpius X-1 is the brightest x-ray source in the constellation Scorpio.

All astronomer's constellations are named in Latin. When astronomers use a constellation to locate a star, the genitive case (meaning "of the thing") is used. Thus the constellation Big Dog is officially Canis Major, but Sirius is alpha Canis Majoris, or "alpha of Big Dog."

In the table, the 25 brightest stars as seen from Earth are listed according to the constellation in which they can be seen. The number preceding the common name indicates the rank. Thus, Sirius is listed under Canis Major as "Contains #1: Sirius (α CMa A)," which means it is the brightest star, while Canopus, the second brightest, is listed under the constellation Carina as #2: Canopus.

Name	Genitive	Abb.	Translation	Remarks
Andromeda	Andromedae	And	Andromeda	Character in Greek myth; part of Perseus group
Antlia	Antliae	Ant	Air pump	Of modern origin
Apus	Apodis	Aps	Swift, or Bird of Paradise	Of modern origin
Aquarius	Aquarii	Aqr	Water Bearer	In zodiac and wet quarter of sky
Aquila	Aquilae	Aql	Eagle	Contains #12: Altair (α Aql)
Ara	Arae	Ara	Altar	Part of the Centaurus group
Aries	Arietis	Ari	Ram	In zodiac and part of Argo group
Auriga	Aurigae	Aur	Charioteer	Contains #6: Capella (α Aur A & B)
Boötes	Boötis	Boo	Herdsman	In Ursa Major group; contains #4: Arcturus (α Boo)
Caelum	Caeli	Cae	Chisel, or graving tool	Of modern origin
Camelopardalis	Camelopardalis	Cam	Giraffe	Of modern origin
Cancer	Cancri	Cnc	Crab	In zodiac
Canes Venatic	Canum Venaticorum	CV	Hunting Dog	Of modern origin
Canis Major	Canis Majoris	CMa	Big Dog	Part of Orion group; contains #1: Sirius (α CMa A) and #22: Adhara ($\in$ CMa A)
Canis Minor	Canis Minoris	CMi	Little Dog	Part of Orion group; contains #7: Procyon (β CMi)
Capricornus	Capricorni	Cap	Goat	In zodiac and in wet quarter
Carina	Carinae	Car	Ship's Keel	Of modern origin, but part of the ancient constellation *Argo*; contains #2: Canopus (α Car)
Cassiopeia	Cassiopeiae	Cas	Cassiopeia	Character in Greek myth
Centaurus	Centauri	Cen	Centaur	Character in Greek myth, part of the Centaurus group; contains #3: Rigil Kentaurus (α Cen A & B) and #11: Hadar (β Cen)
Cepheus	Cephei	Cep	Cepheus	Character in Greek myth; part of Perseus group
Cetus	Ceti	Cet	Whale	Part of Perseus group
Chamaeleon	Chamaeleontis	Cha	Chameleon	Of modern origin
Circinus	Circini	Cir	Compass	Of modern origin
Columba	Columbae	Col	Dove	Of modern origin

Name	Genitive	Abb.	Translation	Remarks
Coma Berenices	Comae Berenices	Com	Berenice's Hair	Berenice was a Third Century B.C. Egyptian Queen; in Greek times, this was Ariadne's hair
Corona Australis	Coronae Australis	CrA	Southern Crown	Sometimes considered as Sagittarius' crown
Corona Borealis	Coronae Borealis	CrB	Northern Crown	Also Ariadne's crown
Corvus	Corvi	Crv	Crow	Companion of Orpheus
Crater	Crateris	Crt	Cup	
Crux	Crucis	Cru	Southern Cross	Of modern origin, originally part of Centaurus; contains #19: Beta Crucis (β Cru) and #21: Acrux (α Cru A); smallest constellation
Cygnus	Cygni	Cyg	Swan	Contains #18: Deneb (α Cyg)
Delphinus	Delphini	Del	Dolphin	In wet quarter
Dorado	Doradus	Dor	Goldfish	Of modern origin; contains Large Magellanic Cloud and Southern ecliptic pole
Draco	Draconis	Dra	Dragon	Contains north ecliptic pole
Equuleus	Equulei	Equ	Little Horse	
Eridanus	Eridani	Eri	River Eridanus	Contains #9: Achernar (α Eri)
Fornax	Fornacis	For	Furnace	Of modern origin
Gemini	Geminorum	Gem	Twins	In zodiac; contains #17: Pollux (β Gem)
Grus	Gruis	Gru	Crane	Of modern origin
Hercules	Herculis	Her	Hercules	Character from Greek myth, part of the Argo group
Horologium	Horologii	Hor	Clock	Of modern origin
Hydra	Hydrae	Hya	Hydra (water monster)	Monster from Greek myth, part of the Argo group; largest constellation
Hydrus	Hydri	Hyi	Sea serpent, or water snake	Of modern origin
Indus	Indi	Ind	Indian	Of modern origin
Lacerta	Lacertae	Lac	Lizard	Of modern origin
Leo	Leonis	Leo	Lion	In zodiac; contains #20: Regulus (α Leo A)
Leo Minor	Leonis Minoris	LMi	Little Lion	Of modern origin
Lepus	Leporis	Lep	Hare	Part of Orion group
Libra	Librae	Lib	Scales	In zodiac; originally the claws of Scorpius
Lupus	Lupi	Lup	Wolf	Part of Centaurus group
Lynx	Lyncis	Lyn	Lynx	Of modern origin

Name	Genitive	Abb.	Translation	Remarks
Lyra	Lyrae	Lyr	Harp	The harp of Orpheus; contains #5: Vega (α Lyr)
Mensa	Mensae	Men	Table (mountain)	Of modern origin
Microscopium	Microscopii	Mic	Microscope	Of modern origin
Monoceros	Monocerotis	Mon	Unicorn	Of modern origin
Musca	Muscae	Mus	Fly	Of modern origin; originally Musca Australis
Norma	Normae	Nor	Square (level)	Of modern origin
Octans	Octanis	Oct	Octant	Of modern origin
Ophiuchus	Ophiuchi	Oph	Ophiuchus (serpent bearer)	Character in Greek myth, identified with physician Aesculapius
Orion	Orionis	Ori	Orion	The hunter, character in Greek myth; contains #7: Rigel (β Ori A), #10: Betelgeuse (α Ori), and #24: Bellatrix (gamma Ori)
Pavo	Pavonis	Pav	Peacock	Of modern origin
Pegasus	Pegasi	Peg	Pegasus	Winged horse in Greek myth; part of Perseus group
Perseus	Persei	Per	Perseus	Character in Greek myth, lead figure in a story used to name six constellations
Phoenix	Phoenicis	Phe	Phoenix	Of modern origin
Pictor	Pictoris	Pic	Easel	Of modern origin
Pisces	Piscium	Psc	Fish	In zodiac
Piscis Austrinus	Piscis Austrini	PsA	Southern Fish	In wet quarter; contains #16: Fomalhaut (α PsA)
Puppis	Puppis	Pup	Ship's Stern	Modern subdivision of ancient *Argo*, the ship that carried the Argonauts
Pyxis	Pyxidis	Pyx	Ship's Compass	Modern subdivision of ancient *Argo*
Reticulum	Reticuli	Ret	Net	Of modern origin
Sagitta	Sagittae	Sge	Arrow	
Sagittarius	Sagittarii	Sgr	Archer	In zodiac; contains center of Milky Way galaxy
Scorpius	Scorpii	Sco	Scorpion	In zodiac, the Scorpio of astrology; contains #15: Antares (α Sco A) and #23: Shaula (lambda Sco)
Sculptor	Sculptoris	Scl	Sculptor	Of modern origin; originally l'Atelier du Sculpteur; contains south galactic pole
Scutum	Scuti	Sct	Shield	Of modern origin; named after shield of Polish hero John Sobieski

Name	Genitive	Abb.	Translation	Remarks
Serpens	Serpentis	Ser	Serpent	Other part of Ophiuchus group
Sextans	Sextantis	Sex	Sextant	Of modern origin
Taurus	Tauri	Tau	Bull	In zodiac; contains #13: Aldebaran (α Tau A) and #25: Elnath (β Tau)
Telescopium	Telescopii	Tel	Telescope	Of modern origin
Triangulum	Trianguli	Tri	Triangle	
Triangulum Australe	Trianguli Australis	TrA	Southern Triangle	Of modern origin
Tucana	Tucanae	Tuc	Toucan	Of modern origin
Ursa Major	Ursae Majoris	UMa	Big Bear	Big Dipper to most Americans; part of Big Bear group
Ursa Minor	Ursae Minoris	UMi	Little Bear	Little Dipper to most Americans; part of Big Bear group. Contains north celestial pole
Vela	Velorum	Vel	Ship's Sails	Modern subdivision of ancient *Argo*, the ship that carried the Argonauts
Virgo	Virginis	Vir	Virgin	In zodiac; contains #14: Spica (α Vir)
Volans	Volantis	Vol	Flying Fish	Of modern origin
Vulpecula	Vulpeculae	Vul	Little Fox	Of modern origin

ABUNDANCE OF ELEMENTS IN THE UNIVERSE

Astronomers and physicists believe that the big bang produced a universe that contained 80% hydrogen, 20% helium, and probably no other elements at all. When clouds of hydrogen began to collapse into small spaces as a result of gravitational forces, the nuclei of hydrogen atoms were pressed so close together that they fused into heavier hydrogen (deuterium and tritium), which in turned fused to form helium. The process released energy, and the balls of hydrogen and helium became stars. The energy released balanced the force of gravity, and the stars stabilized in size. This process provided the source of energy for the Sun. In larger stars, the helium and hydrogen continued to fuse, producing nitrogen, carbon, neon, and some oxygen, as well as more helium. Even larger stars went farther and were able to produce magnesium, silicon, and iron, which have more protons and neutrons in their nuclei, and are therefore heavier than the elements mentioned previously.

Iron is the end of the line for this process, because fusing iron nuclei takes more energy than the process produces. Gravitational energy, however, causes a star's core to contract with great speed when the fusion process begins to slacken. This contraction provides the necessary energy to fuse iron, but it provides so much energy that the star blows up, becoming a supernova. In the process of exploding, the elements heavier than iron are created. Furthermore, the explosion sends all the elements from the supernova into space, creating clouds that contain all elements. The solar system apparently formed from such a cloud, since silicon, oxygen, and iron are abundant, and elements as heavy as uranium occur in smaller amounts.

Since the big bang, some hydrogen and helium has remained as interstellar gases and some has formed smaller stars in which the fusion process does not go beyond the fusion of helium. As a result, hydrogen and helium remain the most abundant elements. Carbon, nitrogen, and oxygen are the main components of the medium-sized star's fusion cycle, so they are the next most abundant elements. After iron is produced, the amount of heavier elements present through supernova explosions goes down considerably, although nickel occurs in quantities near that of iron. With few exceptions, fusion produces elements with even atomic numbers more easily than those with odd atomic numbers. Therefore, when arranged by atomic number, the abundance of the elements tends to seesaw back and forth. This effect becomes more pronounced for elements heavier than carbon.

In the following table, hydrogen is assumed to have an abundance of 1,000,000,000,000 (one trillion) units, arbitrarily chosen as a large number so that the other elements will not all be very small; then the other elements can be assigned the following amounts, based on data from astrophysical theories, astronomical measurements, elements found on Earth, and elements found in meteorites.

Atomic No.	Element	Abundance
1	Hydrogen	1,000,000,000,000
2	Helium	162,000,000,000
3	Lithium	3,160
4	Beryllium	631
5	Boron	758
6	Carbon	398,000,000
7	Nitrogen	112,000,000
8	Oxygen	891,000,000
9	Fluorine	1,000,000
10	Neon	551,000,000
11	Sodium	2,000,000
12	Magnesium	25,100,000
13	Aluminum	1,560,000
14	Silicon	31,700,000
15	Phosphorus	251,000
16	Sulfur	22,400,000
17	Chlorine	355,000
18	Argon	4,880,000
19	Potassium	66,100
20	Calcium	1,550,000
21	Scandium	708
22	Titanium	77,400

Atomic No.	Element	Abundance
23	Vanadium	6,610
24	Chromium	240,000
25	Manganese	132,000
26	Iron	3,710,000
27	Cobalt	56,200
28	Nickel	891,000
29	Copper	10,000
30	Zinc	19,100
31	Gallium	282
32	Germanium	1,590
33	Arsenic	129
34	Selenium	2,140
35	Bromine	446
36	Krypton	1,530
37	Rubidium	224
38	Strontium	501
39	Yttrium	56.2
40	Zirconium	316
41	Niobium	31.6
42	Molybdenum	75.8
43	Technetium	trace
44	Ruthenium	27.5
45	Rhodium	6.31
46	Palladium	18.2
47	Silver	6.61
48	Cadmium	28.2
49	Indium	5.12
50	Tin	37.2
51	Antimony	8.92
52	Tellurium	112
53	Iodine	22.4
54	Xenon	115
55	Cesium	14.5
56	Barium	120
57	Lanthanum	12.6
58	Cerium	19.5
59	Praseodymium	4.57
60	Neodymium	22.9
61	Promethium	trace
62	Samarium	7.76
63	Europium	3.02

Atomic No.	Element	Abundance
64	Gadolinium	11.2
65	Terbium	1.74
66	Dysprosium	12.0
67	Holmium	2.45
68	Erbium	6.92
69	Thulium	0.12
70	Ytterbium	6.03
71	Lutetium	0.115
72	Hafnium	2.51
73	Tantalum	0.0562
74	Tungsten	3.98
75	Rhenium	7.94
76	Osmium	25.1
77	Iridium	15.8
78	Platinum	50.1
79	Gold	4.63
80	Mercury	5.62
81	Thallium	3.59
82	Lead	31.6
83	Bismuth	3.16
84	Polonium	0.0000000316
85	Astatine	trace
86	Radon	0.0000000001
87	Francium	trace
88	Radium	0.000126
89	Actinium	0.0000000631
90	Thorium	1.00
91	Protactinium	0.00001
92	Uranium	0.0501

ASTRONOMICAL RECORD HOLDERS

BY SIZE

Largest comet nucleus	Chiron	Diameter: 230 mi (372 km), but Ceres Pallas may displace Chiron by being comets

Largest asteroid	Ceres	Diameter: 588 mi (947 km); but if Ceres ia a comet, then Pallas or Vesta is largest. Pallas, like Ceres, may turn out to be a comet, so Vesta at 349 mi (582 km) may be the largest
Largest natural satellite	Ganymede (satellite of Jupiter)	Diameter: 3,276 mi (5,274 km)
Largest known planet	Jupiter	Size: 317.9 times the mass of Earth
Largest known object	Combination of the Lynx-Ursa Major supercluster with the Pisces-Perseus supercluster of galaxies	Size: About 700 million light-years long
Smallest known natural satellite	Leda (satellite of Jupiter)	Diameter: 9.3 mi (15 km)
Smallest known planet	Pluto	Size: 0.0017 as big as Earth Diameter: about 1,400 mi (2,250 km)

BY DISTANCE

Distance records that are based on red shifts depend on what one thinks the Hubble constant is. In the following, a value for the constant is used that makes the universe 12 billion years old. Many astronomers would prefer values that result in ages of as much 15 or even 20 billion years, while one set of measurements in 1994 suggest the true age to be 8 billion years (*see* "Cosmology UPDATE," p 236).

Most distant supernova	In the galaxy cluster AC118	Distance: About three billion light-years from Earth
Most distant galaxy	8C 1435 + 635	(Reported in the November 15, 1994 *Monthly Notices of the Royal Astronomical Society* by a group directed by Mark Lacy of the University of Oxford in England) Red shift: 4.25. If the universe is about 12 billion years old, this translates to about 10 billion light-years away.
Most distant object	Quasar PC 1247 + 3406 as reported on August 26 1991, found by Donald Schneider of the Institute for Advanced Studies, Maarten Schmidt of Caltech, and James Gunn of Princeton	Distance: About 11 billion light-years from Earth; it has a red shift of 4.897, the greatest ever measured
Most distant star in Milky Way galaxy	Unnamed star in Virgo	Distance: About 160,000 light-years from Earth (about as far as the Magellanic Clouds, but in a different direction)

Most distant star ever observed	Supernova observed April 28, 1992, through Isaac Newton Telescope at La Palma, Canary Islands	Distance: three billion light-years from Earth
Nearest pulsar	PSR J0437-4715	400 light years
Nearest star	Proxima Centauri	Distance: 4.22 light-years, or 24.8 trillion mi (40 trillion km)
Closest known approach by an asteroid	About noon EST on December 9, 1994, by 1994 XM_1, a small asteroid about 43 ft (16 m) in diameter	Distance: 63,000 mi (101,000 km), about one quarter of the distance to the Moon
Farthest asteroid from Sun	1991 DA, found by Australian astronomers on February 18, 1991	Distance at aphelion: 22 astronomical units, or 2,046 million mi (3.3 billion km); beyond orbit of Uranus

BY BRIGHTNESS

Most luminous object in universe (produces the most light)	Probable gas cloud in Ursa Major found by Michael Rowan-Robinson of Queen Mary and Westfield College in London	More than 300 trillion times as luminous as the Sun and more than 20,000 times as luminous as the Milky Way galaxy
Brightest-appearing star at optical wavelengths	Sirius A	Brightness: Apparent magnitude −1.47; absolute magnitude +1.45, about 26 times as bright as the Sun; other stars are actually brighter, but much farther away
Brightest star in Milky Way in reality at optical wavelengths	Cygnus OB2 #12	Absolute visual magnitude of −9.9; would be first magnitude visually from Earth if not for dust
Brightest star in Milky Way including all wavelengths	HD 93129A	Absolute bolometric magnitude (including all wavelengths of electromagnetic radiation) of −12.0, about 5 million times as luminous as the Sun
Brightest object in universe (produces the most intense light)	Quasar BR 1202-07	Brightness: Absolute magnitude of −33 is about 1 quadrillion times as bright as the Sun
Faintest object outside the solar system	Unnamed probable brown dwarf discovered in 1988 by Michael R.S. Hawkins	Brightness: about 1/4,000 that of the Sun

BY OTHER MEASURES

Oldest star in Milky Way	CS 22876.3	Age: Between 15 and 20 billion years old, as calculated by Timothy Beers and coworkers at Michigan State University in 1990
Youngest star	VLA 1623 in Rho Ophiuchi, according to a 1994 report by Derek Ward-Thompson of Cambridge University in England	Age: About 10,000 years old
Hottest star	The NGC 2440 nucleus in the Large Magellanic Cloud, first identified by the *Hubble Space Telescope* in August 1990	Temperature: 200,000 K (200,000°C or 360,000°F) at the observable surface
Most energetic cosmic ray	Detected by the Fly's Eye at the Dugway Proving Grounds near Salt Lake City on October 15, 1991	Energy: 3×10^{20} electron volts (about 0.03 foot-pound, or almost 0.05 BTU)
Oldest observatory	Tomb at Newgrange, Ireland, aligned with the Sun at the winter solstice	Age: About 5,150 years old
Fastest star	PSR 2224+65	Speed: 2,232,000 mph (3,616,000 kph), about 1,000 times as fast as a typical star, such as the Sun
Most humans in space at one time	Starting on December 2, 1990 and lasting for 9 days	Total: 12 (7 aboard US space shuttle *Columbia*; 3 on USSR *Soyuz TM-11*; and 2 on USSR space station *Mir*)

(Periodical Reference and Additional Reading: *Science* 9-6-91, p 1094; *Astronomy* 11-91, p 28; *New York Times* 4-21-92, p C5; *Discover* 1-93, p 30; *New York Times* 3-11-93, p B13; *New York Times* 7-23-93, p C6; *Science* 12-10-93, p 1649; *Science* 11-11-94, p 974; *Astronomy* 4-95, p 28)

THE STATE OF EARTH'S SPACE PROGRAMS

A chronicle of space programs from October 4, 1957, to today reveals that activity has been cyclical, influenced heavily both by serious space accidents and politics on Earth. Major accidents have always led to a period of re-evaluation followed by renewed efforts. The influence of politics is more pervasive. The initial burst of activity in space was spurred by the politics of the cold war; popular United States programs were reduced during the Vietnam war; the European Space Agency (ESA) has been a political tie on a continent that resists union; the collapse of communism pulled the rug out from what may have been the most successful space program of all; the need for Japan to demonstrate its power peacefully has led it into space; the financial needs of both the United States and Russia forced their new cooperation on the most expensive projects; and so forth.

Although the main problems facing space programs in the 1990s came from well-known international financial difficulties, other earthly problems also had unexpected influences. For example, an earthquake in the California desert on July 28, 1992, damaged the Goldstone radio telescope used to keep in touch with space probes and satellites from the United States and from the European Space Agency. Communications were not lost, however, because the Goldstone duties were assigned to smaller radio telescopes. Other problems were more typical: a new rocket from Japan exploded several times before becoming operational, and a number of satellites and probes were lost, some because of launch-vehicle failure and a smaller number because of breakdowns aboard the spacecraft.

New Uses of Space

In the beginning, scientific uses for satellites were the primary goals of space programs. Remember that the first earth-orbiters were a part of the first International Geophysical Year effort of 1957–58. Scientific research is still the purpose of two important groups of satellites. Telescopes of one kind or another are put aboard satellites either to improve resolution (*see* "Hubble Repaired and Making Discoveries," p 126) or to observe at wavelengths that fail to penetrate Earth's atmosphere (*see* "Giant Telescopes, Optical and Otherwise," p 130). Other Earth-orbiting satellites aim instruments at Earth to locate archaeological ruins (*see* "New Finds in the Holy Lands," p 83 and "China and the Silk Road," p 88) or places where anthropological fossils are likely.

When the first Earth-orbiting satellites were put in place in the 1950s, everyone except the science-fiction buff was amazed, although few had any idea what practical consequences might come from the new gadgets. Weather forecasting was the earliest practical application, starting less than three years into the "space age." Meteorology was followed within a few months by the first navigational and communications satellites. All three applications were introduced by the United States and remain the practical basis of the commercial space program worldwide. About the same time as the first communications satellite experiment, the United States introduced another application that remains a foundation of space applications: military spying. More recently, another military application has come to earth in civilian clothes. The Global Positioning System (GPS) of 24 satellites (completed in June 1993) is designed for weapons targeting. It is beginning to be used for all sorts of navigational purposes, ranging from locating lost campers to landing aircraft (*see* "The State of Technology," p 679).

Space applications are "transparent" to the average user. Although most people are aware of satellite involvement in global television, weather pictures from space, and GPS maps on automobile dashboards, the actual operation of the satellite is not obvious. Even a modern satellite-based home television system seems not much different from cable or even broadcast television to the user. Other similar transparent applications are on the way, notably satellite-based telephone systems, coming on line at the end of the 1990s. Telephones in the near future will not only be portable, but will use satellites as their main link with each other. Today, a very busy executive may carry a bulky cellular telephone with her, even when she goes out to dinner with friends; tomorrow, housewives, retirees, and older children may carry small telephones about the size of present-day beepers with them wherever they go. The satellite systems that service the new generation of portable telephones may have hundreds of satellites, not just the two dozen used for GPS (*see* "Reinventing Communications," p 703).

Internationalizing Space

The space race was for a long time a match race between the United States and the Soviet Union. Although the two nations are still the leaders in terms of number of missions annually and the lofting of the most complex satellites and probes, other entrants to the field have begun to make it more of a horse race, as the saying goes. The ESA has often found that it was left alone or with fewer partners when the United States dropped out of such programs as the missions to Halley's comet, another comet flyby, the study of the poles of the Sun, and others. Finding itself alone so often, ESA has moved to take the initiative, although the Europeans hope that others will join them, perhaps including the U.S. The most ambitious ESA concept is a plan to return to the Moon in a program designed to conclude with a permanently occupied base for astronomical studies.

Like the Europeans, the Japanese have developed a full space program which includes excellent launch vehicles, commercial and scientific projects of their own, and contracts to launch satellites for other entities. Of course, neither the Europeans nor the Japanese have developed their own space transportation system for humans, but their military personnel, scientists, and even newspaper reporters are spending an increasing amount of time in space on Russian or U.S. vehicles. Japanese scientific missions, such as the *ASCA* x-ray satellite launched in February of 1993, have contributed to new and important discoveries about the universe.

The forthcoming international space station, discussed below, is the most dramatic example of the internationalization of space to date, and it is unlikely to be the last.

Getting There

When people first thought of traveling into space, they thought of birds, since some birds were observed to fly to great heights. Cyrano de Bergerac hitched his imaginary chariot to swans. Even after it was clear that the atmosphere needed for such transport does not extend indefinitely, Hugh Lofting whimsically sent Dr. Doolittle to the Moon on the back of a giant luna moth.

More practical ideas appeared during the nineteenth century. Jules Verne (many of whose ideas have since become reality) wrote in 1865 of a giant cannon, perhaps inspired by a thought experiment proposed by Isaac Newton using a cannon ball to orbit the Earth. Verne's cannon, in principle a barely workable arrangement, propelled three fictional astronauts from Cape Canaveral, Florida, to a lunar orbit, followed by splashdown in the Pacific. Thirty years later, the more practical Konstantin Tsiolkovsky worked out the details of using a three-stage liquid-fueled rocket, which some 60 years later became the method used. Solid-fuel staged rockets have also been successful, as have two-stage rockets, such as Japan's powerful new H-II.

The following are some recent improvements on the art.

Back to the Gun   Scientists from Lawrence Livermore Laboratory in California have been working on developing a type of gun so powerful that it will be able to launch a rocket to the Moon. The difference between a gun and a rocket, of course, is that a rocket carries its explosive charge along with it and uses the expulsion of gases over a period of time, while the gun explodes all of its fuel as quickly as possible, so that the gases expel a projectile from a stationary chamber. The gun in Jules Verne's *From the Earth to the Moon* was powered by guncotton (nitrocellulose), an unstable material produced by reacting cellulose with nitric acid. When ignited or even jarred, guncotton changes almost instantly from a solid to a high-volume mixture of gases, which is the main point of an explosion. Explosives such as guncotton are able to react on their own, and don't require oxygen from the atmosphere.

The Lawrence Livermore device, known as the Super High Altitude Research Project (SHARP), is a gun, but the high pressure gases used are not produced by destabilizing an explosive. Instead, the gas (hydrogen) is heated and compressed first, then injected into the chamber of the gun.

SHARP has an interesting history. The first version of a gas-injection gun was developed in Nazi Germany, where a weapons program called Centipede was unsuccessful in attempts to shell London from France. In the 1960s, the U.S. Defense Department had a similar project called HARP (same as SHARP, but not super). HARP led to the still-standing distance record for a gun: on November 19, 1966, a HARP gun fired an 185 lb (84 kg) missile to a height of 111.8 mi (180 km). HARP, however, lost funding. In the 1980s, the government of Iraq hired HARP scientists to develop an operational supergun for them, but the components to be used in its manufacture were recognized and intercepted by Western nations in 1990. The chief scientist involved was mysteriously assassinated, probably by spies from a nation with reason to fear Iraq. With that kind of interest being shown by a known enemy of the United States, scientists from Lawrence Livermore were able to get SHARP funded without difficulty.

The plans are for a complex gas-injection system. Methane burning in oxygen will heat the hydrogen and provide the power to compress it. The tube used for the projectile will be evacuated, so that there will be no air resistance until the projectile leaves the barrel of the gun. Hydrogen gas will be injected at several places behind the projectile, insuring that the force pushing it will be continuous all the time the projectile is in the gun. If all goes well, the missile will reach a velocity of about 9,000 mi (15,000 km) an hour as it leaves the gun. Pointed straight up, the projectile would reach an altitude of about 270 mi (430 km). Improved versions might be able to launch a 10-ton payload into space as far as a lunar orbit.

One promising idea is to combine three different launch methods: the light-gas gun would push the vehicle to high speeds; a scramjet engine (a supersonic ramjet) would continue propelling the craft to the outer fringes of the atmosphere, reaching a height at which the oxygen from air is too thin to burn in the jet; and then a rocket, which needs no atmosphere to operate, would finish lifting the satellite into orbit. There is no definite schedule, however, for the achievement of such a triple-threat lifter.

One problem with a fast launch from a gun is that the highest speed comes in the densest part of the atmosphere, so that the outer layer of any launched object will be burned off, just as the nose cone of a rocket re-entering the atmosphere is burned off. Gas-gun-launched satellites will have to have ablating heat shields as part of their original payload.

One-Stage Rocket   Although the first U.S. satellites were launched with a powerful one-stage rocket, the Russians from the beginning used Tsiolkovsky's concept of stages: a powerful first rocket lifting the whole mechanism through the denser lower atmosphere, with one or more smaller rockets that could take over after the weight of the first rocket body had been jettisoned. The only reason that the first U.S. rockets were one-stage was that the military Redstone rocket used for early satellite launches had not been designed for space travel, and was only converted to that objective in response to Soviet advances. Rockets intended to lift satellites or launch space probes are all two or more stages, with three stages being the most common configuration. The U.S. space shuttle uses a modification of the stage concept. A combination of two solid-state and three liquid-fuel rockets is used as a first stage. Then the solid-state rockets are let go and the liquid-fuel rockets continue as a second stage. When the external liquid-fuel tank is exhausted, it drops off as well, leaving the orbiter (which most of us think of as the space shuttle) as a kind of third stage, although the thrust into orbit actually depends on the first two stages.

In 1993, the first test-firings of a planned one-stage space rocket were conducted. The idea is to make the one-stage Delta Clipper rocket so light that it can lift its own fuel and 10 tons of payload into orbit. Ten tons is only half the payload of the three-stage Titan, but the Delta Clipper would be reusable and therefore much cheaper. Furthermore, most commercially useful and many scientific satellites weigh less than 10 tons. Larger payloads or long-range probes would still require the more powerful three-stage vehicles. The first stage of a Titan 4 launch vehicle alone weighs about as much as the entire Delta Clipper with its payload. NASA estimates that a shuttle flight costs somewhat more than $400 million, about 40 times as much as engineers think the Delta Clipper will cost per launch.

When successfully tested for the first time on August 18, 1993, the experimental prototype of the Delta Clipper (one-third the scale of the planned vehicle) became the first rocket ever to land vertically on Earth, using its rocket engine as a brake. After rising 150 ft (45 m) from the launch pad and hovering for a moment, it moved to one side, extended four legs, and landed on its tail. This form of landing has been a staple of science fiction stories and movies for so long that most people do not realize it had not been accomplished prior to the 1993 test flight. Later test flights also used vertical landing. The operational version is expected to land the same way. A full-scale version of the Delta Clipper may be launched as early as 1998.

The main way that the operational Delta Clipper is expected to reduce its own weight to 52 tons while retaining the capacity to lift 588 tons of fuel is by use of advanced materials. The outer skin of the experimental version of the rocket is seven layers of a graphite-epoxy material developed by Dick Rutan, who first became known to the general public in 1986 when his lightweight plane *Voyager* flew nonstop around the world. Most other launch vehicles use a much denser aluminum skin. The planned operational rocket is expected to use a similar lightweight outer skin. There may be problems, however: the fifth test of the experimental rocket on June 27, 1994, was halted by an explosion that peeled away a 60 sq ft (5.5 sq m) section of the skin. Other parts of the craft are also lightweight, including the steering jets, fuel valves, and onboard computer.

Living There

It seems as if the Americans have been saying "Space is a nice place to visit, but I wouldn't care to live there," while the Russians have treated space like a desirable dacha outside of Moscow, ready to be occupied for long periods of time each year. The truth, however, is that no one lives in space—yet. Someday we will be recording the first marriage in space, and the first birth, and perhaps the first divorce; but before all that the rest of the world has to learn to live in space at least as well as the Russians already do.

Perhaps the greatest cause for optimism regarding the future of living in space has been the beginning of joint programs by the U.S. and Russia, the two nations with by far the greatest experience in space development. Knowledge from extensive Russian experimentation with space stations has become available for use in the international space station of the future. Officially, the agreement to merge U.S. and Russian efforts was reached on September 2, 1993, and officially signed in December 1993, but the first fruits did not really come until a year later. Although dates are revised from time to time, it now appears that a new permanent international space station will be completely assembled sometime between the start of the new millennium and 2010. In the United States, where plans for a space station had been experiencing funding trouble in Congress for years, only a week after the U.S.-Russian accord was announced, the international station easily sailed through the House with a vote of nearly two to one. The other partners in the international effort are the European Space Agency, Japan, and Canada. Both the United States and Russia have abandoned plans for their independent stations, the *Freedom* space station and *Mir 2*.

Plans for the new international station included 10 U.S. shuttle flights to *Mir*, the Russian space station, as a warmup. The first such flight was scheduled for May 1995 and actually took place in June and July. There are plans to loft the first module of the

SPACE TALK

Sometimes it is difficult to keep straight all the words used to describe the machines that travel in the regions outside Earth's atmosphere—the region that we refer to as **space**. Such machines are lifted above the atmosphere by **rockets**, which are also called **launch vehicles** (although technically the rocket is just the part of the launch vehicle that spews forth gases to propel the machine, while other parts of the launch vehicle are used in steering or for structural purposes). A generic name for all machines that travel in space is **spacecraft**, which can be either singular or plural.

In space, a spacecraft may take several possible paths. Some spacecraft follow a path that leads around and around Earth. These are the Earth-orbiting artificial satellites, known familiarly as **satellites**. For the most part, all of the satellites eventually fall back into Earth's atmosphere, although some may stay in the path around Earth, or **orbit**, for hundreds of years. If a satellite returns to Earth in such a way that it can be reused, it is called a **shuttle**. If a satellite is designed so that people can visit it from Earth and return, leaving the satellite in space, it is called a **space station**.

Other spacecraft travel on paths that lead away from Earth and will never return. The most common name for this group is **space probes**. This term, however, is not very satisfactory, as the name fails to differentiate a probe that becomes a satellite of another planet, as did *Magellan* (*see* "*Magellan* at Venus," p 150, and also "*Galileo* UPDATE," p 278), from a probe that orbits a star (*see* "*Ulysses* to the South Pole," p 276) or from a probe that travels to another planet or natural satellite of such a planet and continues in a path that eventually takes it out of the solar system (*see* "Major Accomplishments of Satellites and Space Probes," p 294).

There is some confusion in the popular mind about what happens to gravity in space. One occasionally still reads the incorrect description of space as a "gravity-free environment." More correctly, the media write or speak of **weightlessness** in space, which is at least technically correct, although perhaps misleading. Although weight appears to vanish in an orbiting satellite or a coasting space probe, mass remains. A technically more meaningful term is **free fall**, which explains as well as describes what happens to objects and people inside an orbiting satellite. All the parts and contents of the satellite are falling toward Earth at the same rate, so a brick let loose in the cabin will tend to stay where it is with regard to everything around it. Gravitational force is reduced in space because the distance between the center of Earth's gravity and the satellite in orbit is greater than it was when the satellite was on the surface of Earth. However, in low Earth orbit, the increase is only about 3% in terms of distance, which works out to a gravity reduction of less than 6%. Thus, weightlessness is a result of free fall, not of a "gravity-free environment." On a coasting space probe a long way from Earth, however, there is both a significant reduction in gravitational force from Earth, as well as free fall.

international station a month after the last of the *Mir* missions, in December 1997. It would not be surprising, however, to see that date drift into 1998, or even later. The core segment, known as Alpha Station, is only one part of the final permanently occupied station. Early plans were to have a Russian "space tug" for control, and a Russian service module that would power the station. To this, the U.S. would add docking nodes for the shuttle and other spacecraft; laboratory modules; additional power systems; and so forth. The U.S. shuttle would also lift the Canadian robotics and European and Japanese laboratory modules to the station. By 2002, the station is expected to be complete. The finished space station will be 361 ft (110 m) long and 290 ft (88 m) wide. It will house 6 persons at a time, and is expected to be an operating space laboratory for about 10 years after construction is completed.

Living in space is not just free fall and games. Physiological changes include altered blood pressure, disorientation and vertigo, and loss of muscle mass and other key cells. Perhaps the most difficult long-term change involves loss of bone mass. Upon return to Earth, a person who has spent a long time in free fall may be temporarily crippled by such osteoposrosis.

Another set of problems might be labeled traffic. NASA estimates that there is a one in five chance that a dead satellite, shattered rocket stage, or other piece of junk put into space from Earth will strike the space station either during construction or occupation. The chance that such a collision could cause the death of an occupant, or even the complete destruction of the international space station, is high: one in ten. Three reasons for the high risk derive directly from changes in the space station as a result of Russian participation: the size of the station and especially its living area will be larger, in part because Russian rockets can lift heavier loads; the orbit will be more north-south to enable Russian space vehicles to reach the station more easily; and, as a result of the different orbit, the speed of the station will be increased by about 20%. Additional shielding may succeed in cutting these risks in half.

Russian scientists, noting that in eight years *Mir* has not suffered any penetrating collisions, think that NASA may be unduly alarmed. NASA plans to study *Mir*'s exterior for dents to improve its estimates.

The situation for all space travel, not just the space station, is increasingly complicated by traffic and space junk. Several satellites that have suddenly disappeared, or simply stopped working, are thought to have been victims of an encounter with fragments of old launch vehicles, deliberately destroyed satellites, or simply garbage left by people in space. A spacecraft that blows up can by itself create hundreds of pieces big enough to be tracked by radar from Earth. In fact, the total being tracked today exceeds 7,000 pieces. In addition, there are probably enough smaller pieces to account for a hundred times as much mass in orbit. It was not until 1993 that NASA established a deliberate policy to avoid creating more space junk. A further concern involves large pieces running into each other, and potentially starting a chain reaction of smaller and smaller pieces of debris. Such a scenario could make it impossible to occupy certain orbits for fear of encounters with undetectable clouds of junk that could collide with a satellite or space station, perhaps destroying it.

(Periodical References and Additional Reading: *New York Times* 3-10-92, p C1; *New York Times* 7-15-92, p A16; *New York Times* 7-25-92, p 9; *New York Times* 9-29-92, p C1; *New York Times* 8-10-93, p C1; *New York Times* 8-19-93, p A20; *New York Times* 9-3-93, p A1; *New*

York Times 11-5-93, p A20; Discover 1-94, pp 56, 58, & 60; Science 1-28-94, p 462; New York Times 2-2-94, p A12; New York Times 2-15-94, p C8; New York Times 3-25-94, p A14; Science 4-29-94, p 663; New York Times 5-17-94, p C1; Science 6-3-94, p 1396; New York Times 6-24-94, p A10; New York Times 6-27-94, p A1; New York Times 6-28-94, p C12; New York Times 6-30-94, p A20)

ULYSSES TO THE SOUTH POLE

In June 1994, the U.S. space probe *Ulysses* reached the first true goal after a journey that began with an idea in 1959 and a launch 30 years later on October 6, 1990. The idea was to fly a spacecraft around the Sun, perpendicular to Earth's orbit, permitting observation of phenomena that can never be seen from Earth itself. Along the way, *Ulysses* would also be able to observe the planet Jupiter, the fifth U.S. spacecraft to do so.

Ulysses at Jupiter

Ulysses made its closest approach to Jupiter on February 8, 1992, traveling at a speed of 61,249 mi (98,611 km) per hour. *Ulysses* collected data on Jupiter's magnetic field while in the neighborhood, but its main purpose in being there was to steal some of Jupiter's gravity—that is, to derive momentum from Jupiter. *Ulysses* started the trip to Jupiter with a push of nearly 18 mi per sec (30 km per sec) from the motion of Earth, in addition to the thrust of its rockets. That motion, however, is not in the proper direction to put a satellite into a nearly solar-polar orbit. The gravitational energy stolen from Jupiter redirected *Ulysses'* path toward the Sun and provided additional speed. Jupiter itself was slightly slowed by the maneuver, but not enough to count, since Jupiter weighs 522,225,000,000,000 times as much as the 800 lb (360 kg) Ulysses.

The exchange of momentum in a gravity assist can also change the direction of motion. Even with the gravity push from Jupiter, however, *Ulysses* orbits at an inclination of only about 80°; an orbit of 90° would be perfectly polar.

Ulysses obtained some surprising new data while at Jupiter. It found a magnetic field that was almost twice as extensive as had been observed in 1980-81 by the *Voyager* space craft, but only slightly more than the size of the magnetic field as studied by the *Pioneer* probes in the 1973-74. Dust in the space around Jupiter was present in a different quantity also, which was attributed to the state of volcanic eruptions on Jupiter's satellite Io. Streams of dust particles discovered by *Ulysses* may, however, have a source in a thin auxiliary ring, and be collimated by the planet's magnetic field.

The South Pole of the Sun

The goal *Ulysses* reached in 1994 was the south pole, but not the Antarctic destination of polar explorers. This was the south pole of the Sun.

Ulysses did not visit the pole itself, but merely flew above it, and not even directly

Background

Ulysses was designed by NASA and the European Space Agency to report back on those areas of the Sun we cannot observe from Earth: the polar regions. Since Earth's orbit is roughly parallel to the solar equator, the poles can be seen from Earth obliquely at best. The *Ulysses* spacecraft first traveled to Jupiter, where careful planning caused the Jovian gravity to whip it into a polar orbit over the Sun, something difficult to do from Earth.

One way to picture such a gravity assist is to think of *Ulysses* gaining momentum as it is drawn toward Jupiter. The gain is matched by a loss of momentum for Jupiter, which causes only a tiny effect, because momentum is the product of mass and velocity. Jupiter slows down by a very small amount, while *Ulysses* is speeded up enormously. Of course, when *Ulysses* passes Jupiter, the gravitational interaction is reversed, so that Jupiter's gravitational force begins to slow the spacecraft. From the point of view of an observer on the planet Jupiter, the speeding up and slowing down is entirely symmetrical, so *Ulysses* seems to lose what it had gained. However, from the point of view of an observer somewhere else, such as on Earth or on the Sun, the motion of Jupiter around the Sun is partly imparted to *Ulysses* as it goes by. It is as if a person in a moving car tossed a ball in the direction of the car's motion. Relative to the ground, the ball begins with the speed of the car, and also gets the impetus of the toss. It is less obvious that the car is slowed down when the ball is tossed, but it is. You can observe such an interchange of momentum better for masses that are nearly the same size. If you throw an anchor forward from a canoe as it is coasting toward shore, the canoe will slow or even move backward, depending on the momentum given the anchor, and the mass of the occupied canoe.

Ulysses is not the only spacecraft to orbit the Sun. Several early space probes that missed such targets as the Moon or Venus went into solar orbits. The oldest working space probe, *Pioneer 6*, was deliberately launched into a solar orbit on December 16, 1965, and continues to return useful data.

---

above it. The probe's closest approach was a latitude of 80.2°S at an altitude of 2.4 astronomical units (A.U.), or about 225 million mi (360 million km), on September 13, 1994. The spacecraft then continued in orbit about the Sun, passing through the plane of Earth's orbit in early March of 1995, attaining its closest approach to the Sun at about 1.4 A.U. (139 million mi, or 210 million km) a few days later, and reaching its highest northerly solar latitude in late summer of 1995.

Ulysses is conducting 11 investigations, the most important of which involves the Sun, its magnetic field, its corona, and the solar wind. One unusual experiment will be an attempt by B. Bertotti to use any sudden deviations of *Ulysses* from orbit to detect gravity waves.

Results from the south pole of the Sun have confirmed some of the expectations about the magnetic lines of force and solar wind, but have also included some surprises. There has been less of a flow of cosmic rays than expected; it had been thought that the Sun's magnetic field would channel cosmic rays toward the poles. The expected higher level of cosmic radiation does occur very close to the south pole, however. Further from the pole, *Ulysses* has found waves of magnetism that flow outward from the Sun. These waves push cosmic rays, reducing the level away from the pole itself. Also flowing from the polar regions is a faster solar wind, about 460 mi (750 km) per second, which had been predicted by theory and Earth-based observations. This speed near the pole is about twice the speed of the solar wind near the equator.

Some newspaper and television reports of *Ulysses* findings described the result as "finding no evidence of a magnetic south pole at the south pole." The explanation for this result is that *Ulysses* was too far away to observe the south magnetic pole. Thus, although the probe did not detect the magnetic pole, there is no reason to think that the pole does not exist.

One of several mysteries about the Sun is why the corona is hotter than the Sun's surface, a question that *Ulysses* may be helpful in solving. About 4.6 A.U. (430 million mi, or 690 million km) from the Sun, *Ulysses* reported a temperature of about 3 billion K (6,000,000,000°F) in the thin ionized gases of that region, much hotter even than the Sun's corona. For reasons that are poorly understood, temperatures increase with distance from the Sun for a long way from the Sun, although this was the first indication that there are pockets of such high temperatures so far from the Sun.

Space scientists hope to receive additional data from *Ulysses* as it orbits the Sun until at least 2001, which will be a year of maximum magnetic activity. The next pass at the poles will begin in November 2000.

(Periodical References and Additional Reading: *New York Times* 2-9-92, p I-25; *New York Times* 2-12-92, p A18; *Discover* 1-93, p 32; *Science News* 8-6-94, p 93; *Nature* 8-19-95, p 695; *EOS* 9-6-94, p 411; *Science* 9-16-94, p 1660; *Astronomy* 11-94, p 27; *Science News* 11-19-94, p 326; *EOS* 11-22-94, p 546; *New York Times* 12-20-94, p C1)

Galileo UPDATE

The space probe *Galileo*, while it does not hold the long-distance space probe record (*Voyager 1*) nor the record for most planets visited (*Voyager 2*), has had the most complex and interesting journey of them all. Taking the long way—2.3 billion mi (3.7 billion km)—to Jupiter, *Galileo* visited Venus, checked out Earth's Moon and Earth itself on its two swings past home, and flew by two asteroids as it wended its way to multiple destinations in the vicinity of the largest planet. Upon arrival one part of *Galileo* plunged to its doom in the thick atmosphere of Jupiter itself, while the main section, if the last stages of its journey go as planned, will travel about the Jovian system, reporting on such fabulous bodies as Io, Europa, Ganymede, and Callisto. As every traveler knows, such a journey is not cheap. The bill comes to about $1.4 billion.

Scenes Along the Way

Just as a vacationer sometimes finds that there is more to see on the way to Serendip than one would expect, the convoluted path *Galileo* followed to Jupiter provided a great deal of interesting science. The flyby of Venus on February 10, 1990, was notable for its maps of the atmosphere below the clouds. As *Galileo* came closer to the planet, observatories all over Earth trained their instruments on Venus so that Earth-based observations could be combined with the *Galileo* data. One team of astronomers, primarily from the University of Hawaii, used the Infrared Telescope Facility on Mauna Kea to study the night side of Venus. The team observed phenomena that they attributed to regions of high water vapor density in the atmosphere, and corresponding regions of low water vapor density that resulted in "dry spots" and "cold spots" that could be caused by different amounts of water vapor at different levels in the atmosphere. Comparison of atmospheric hot spots located by *Galileo* with geologic features mapped by the *Magellan* orbiter may produce the first clues to active volcanoes on Venus. Another coup from *Galileo*'s fly-by was the detection of the strongest evidence yet for lightning on Venus.

Passing by Earth and its Moon on December 8, 1990, *Galileo* made some excellent snapshots of both. Of most scientific interest were measurements of the magnetosphere and of the composition of the Moon.

The asteroid 951 Gaspra was observed by the space probe *Galileo* during a fly-by on October 29, 1991. Gaspra is no more than 12 mi (20 km) long, 8 mi (12 km) across, and 7 mi (11 km) wide. About 150 photographs were collected at Gaspra as *Galileo* passed 1,000 mi (1,600 km) from the planetoid at 18,000 mph (29,000 kph). The asteroid was 205 million mi (330 million km) from the Sun in the asteroid belt between the orbits of Mars and Jupiter. A few photographs were returned to Earth for immediate processing and were widely published on November 15, 1991. These photographs were laboriously transmitted by the two small antennas, the only ones working. Most of the remaining photographs and data were transmitted later, when *Galileo* was closer to Earth and the data rate could be improved.

On its return to Earth's vicinity on December 8, 1992, *Galileo* observed regions of the Moon that have never been seen previously. The next visit for *Galileo* was another asteroid, Ida, on August 29, 1993 (see "Comet, Meteoroid, Asteroid UPDATE," p 171).

Life on the Third Planet

When the *Galileo* spacecraft was forced by schedule changes to travel to Jupiter by way of detours past Earth (twice) and Venus, scientists took advantage of the opportunities presented in various ways. One of the most interesting projects was an attempt to detect life on one of the planets along the new route; not on the second planet from the Sun, where chances of life seem slim, but on the third planet, which is known to have conditions thought suitable for the evolution of living organisms.

Specific observations intended to detect life on the third planet were made during the December 1990 pass by the planet, 600 mi (960 km) above the Caribbean Sea. Analysis of the results was not completed and published for several years.

The scientific team, led by Carl Sagan of Cornell University, was able to report that the evidence suggested strongly that there is life on the third planet, although it did not prove conclusively that life exists there. The best evidence comes from an analysis of the gases in the atmosphere. For example, there is far too much molecular oxygen. Oxygen is highly reactive and would be expected to combine with other elements and be lost to the atmosphere. Some free oxygen could be produced from water molecules split by ultraviolet light, but the large amount detected by *Galileo* must have some other cause. Production by living organisms seems likely. Similarly, although not quite so dramatically, there is too much methane.

Camera images of the third planet taken by *Galileo* and relayed to the scientists found no evidence of living organisms or structures that such organisms might build. An excess of radio waves with regular changes in amplitude might even indicate the presence of *intelligent* life on the third planet from the Sun.

On its second pass at the third planet, *Galileo* came within 190 mi (302 km) of the surface, picking up a second dose of energy and speed from the planet, which did not seem to miss it. New pictures were taken of the Southern Hemisphere, this time mostly as practice for the main event at Jupiter.

While *Galileo* was watching Earth, Earth was watching it. Photographs taken through a telescope on November 28, 1992, show *Galileo* while it is still 5,000,000 mi (8,100,000 km) from Earth, the farthest ever optical detection of an artificial object, so far as is known.

The Main Events

As *Galileo* finally approached to within 50 million mi (80 million km) of Jupiter, it released what would appear to be a bomb, in its general form and attributes. After the hard time Jupiter had with Comet Shoemaker-Levy (see "Comet Hits Jupiter," p 165), one would think it unkind of *Galileo* to bomb the planet of its intentions. However, the probe, though aimed at Jupiter, did not explode. Indeed, as it neared Jupiter, it imploded. The probe was simply a bomblike sensor that used Jupiter's gravity to help it find its enormous target. Indeed, Jupiter pulled it so hard that the probe was traveling 106,000 mph (170,00 kph) when it reached the planet. After blazing though the outer reaches of the Jovian atmosphere, the probe deployed a parachute and descended more slowly to whatever height was allowed by pressure from the dense atmosphere and probable high temperatures. The 57 minutes or so of signals radioed from the probe on December 7, 1995 to *Galileo* and then passed on to Earth provided the first information on conditions deep in the atmosphere of Jupiter. Although Comet Shoemaker-Levy also provided some clues, the comet fragments apparently exploded quite high above the planet. Not only did the probe fail to reach the surface of Jupiter, it is not even clear that Jupiter has a surface that we would recognize — that is, a sharp boundary between a fluid and a solid. Scientists expect to spend years analyzing data from the probe.

Meanwhile, the main part of the spacecraft began a two-year tour of Jupiter's satellites, taking a lingering look at the planet from close at hand.

THE ANTENNA PROBLEM

The most common problem facing scientific experiments in space in the early 1990s has been difficulties with antenna which have failed to point in the right direction or failed to open up properly. Over a period of 17 months in the early 1990s, six major spacecraft costing $5.3 billion had antenna malfunctions, degrading the flow of information back to Earth. Perhaps the most disappointing was the failure of the main antenna on the *Galileo* space probe to open, discovered in April 1991.

The *Galileo* antenna problem was probably caused by numerous delays that pushed back its launch dates from a 1982 liftoff to a launch in 1989. During and after the final delay, caused by the long shutdown of the U.S. space program after the *Challenger* disaster, five trips eventually affected the antenna.

The first four were the trips between Florida and California's Jet Propulsion Laboratory in Pasadena, where *Galileo* waited out post-*Challenger* rethinking and retooling of the shuttle program. During those journeys aboard trucks, the 16 ft (5 m) main antenna lay on its side and bounced about a bit. Technicians think that the jostling caused lubricant to rub off from three or four of the 18 antenna ribs, those on which the spacecraft lay. As a result, small pins that keep the molybdenum mesh antenna furled until it is ready to be deployed became unlubricated and stuck in place. On NASA communications satellites, such pins are regularly replaced and relubricated just before launch, but this step was not taken for *Galileo*.

The fifth troublesome trip was that of *Galileo* through space. Safety considerations inspired by the *Challenger* disaster resulted in reduced power for *Galileo*'s engine, making the direct 18-month trip to Jupiter no longer feasible. Scientists worked out a route that involved three gravity assists from planets, including one from Venus. *Galileo*, however, was not designed to travel that close to the Sun, so other changes had to be made to protect the craft. One such change consisted of keeping the main antenna furled like an umbrella until after the Venus flyby. In April, 1991, when the first attempt to open the antenna was made after the first Earth gravity assist, the antenna stuck, opening only part of the way. Technicians suspect that the passage close to the Sun further complicated the problem caused by the stuck pins.

Consumers are used to *broadcast* radio, but for long-distance transmission of a weak signal *focused* waves are needed. Properly open, the *Galileo* main antenna forms a paraboloid to direct radio signals toward a specific target; partially open, it is worthless.

Initial efforts to fix the antenna involved heating or cooling it by turning it toward or away from the Sun. Heating can cause remaining lubricant to make parts more slippery. Cooling can shrink some materials more than others, perhaps freeing a tightly stuck pin. A major maneuver to freeze the antenna by putting it in the shade for 50 hours was attempted, but it failed to get the antenna cold enough to unstick it. The last try at freeing the antenna through heating was implemented in December, 1992, when *Galileo* had returned to the vicinity of Earth. When that effort also failed, the spacecraft's handlers tried to use mechanical force to open the antenna in late December and early January 1993, without success. On January 20, the technicians working on the antenna problem declared defeat. The antenna would remain furled for the duration. One last-ditch attempt was made after that, but it was also unsuccessful.

Scientists were forced to rely on the lesser amount of data that could be transmitted by *Galileo*'s two small antennas. These transmit data at a rate of 10 bits per second, as compared with the planned 134,000 bits per second for the main antenna. Using new receiving hardware at the tracking stations, as well as various software fixes developed on Earth and transmitted to *Galileo* early in 1995, however, the smaller antennas will be able to achieve about 70% of the planned scientific objectives of the mission. Many of the most interesting results from other planetary encounters, however, have come unexpectedly from transmitted images. The antenna problem will reduce the total number of images transmitted from 50,000 to about 1,000. Furthermore, since data compression is at the heart of the 1995 software fixes, each image will contain less actual data than had originally been expected.

(Periodical References and Additional Reading: *New York Times* 12-8-92, p C1; *New York Times* 12-31-92, p A16; *New York Times* 1-21-93, p A18; *Science* 2-4-94, pp 625 & 627; *Astronomy* 4-94, p 26; *Discover* 1-95, p 94; *Science News* 1-14-95, p 25; *New York Times* 7-11-95, p C1)

OTHER U.S. EXPLORATION OF SPACE

At the beginning of the space age, the efforts of the United States and the Soviet Union were recognized by all as primarily politically motivated, attempts to convince people around the world of the superior technological capabilities of their differing economic systems. Yet from the very beginning, the United States demonstrated that capitalism, even if was state sponsored, was interested in practical results. Space was used for astronomy, weather forecasting, communications, navigation, military spying, cartography, resource location and evaluation, and environmental monitoring. All were pioneered by the United States, and all were amazingly successful. By the mid-1990s, when almost all the world had become capitalistic, there seemed to be no need to demonstrate further technological superiority. By this point, however, other factors had begun to influence the direction of the U.S. space program.

Fighting Over Money

In the United States, the Republican administrations and Democratic congresses of the 1980s watched over a gradually declining space effort, as budget concerns worked against new programs. By the mid-1990s, a Democratic administration and a Republican congress faced even greater budget problems. The U.S. National Aeronautics and Space Agency (NASA) continued to put forth new proposals, or to work toward completion of old projects, but there was a sense that no program was so well established that it was immune to funding cuts. When Daniel S. Goldin was appointed head of NASA in March of 1992, he was expected to reduce costs further and to eliminate some programs, as well as to shift the emphasis away from both piloted missions and government funding of all space endeavors. He proceeded to do as expected, with the motto "cheaper, faster, better" continually on his lips. Commercial interests, such as Motorola's planned $3 billion Iridium project (*see* "Reinventing Communications," p 703), planned without NASA help in launching, have come to rival the $4 billion NASA now spends on the space shuttle annually. The Congressional Budget Office, however, did not think that Goldin was going far enough. It suggested even greater de-emphasis of the shuttle and more use of robot missions, and perhaps even cancelling the shuttle and the planned U.S. space station. Scientists have not always agreed with the downsizing of NASA. The U.S. National Research Council (NRC), for example, representing the National Academy of Sciences, called for more shuttle missions devoted to science.

Era of Small, Cheap Probes?

It was not the NRC call, however, that put the space probe *Clementine* in the service of science. *Clementine* began as a part of the Strategic Defense Initiative (SDI, or "Star Wars") program, later known as the Ballistic Missile Defense Organization. *Clementine*'s military purpose was to test the feasibility of using new low-density materials and low-mass electronics for eventual use in Brilliant Pebbles, an SDI scheme for filling the region near Earth with small interceptor satellites that would destroy intercontinental ballistic missiles (ICBMs) in space. The Anti-Ballistic Missile Treaty limited the kind of targets that could be used in the tests, but the military got around the treaty by identifying the Moon as an enemy ICBM. Depending upon who is telling the story, either scientists interested in the Moon learned of this and asked to be let on board, or the SDI people wanted an excuse to aim at the Moon and approached the scientists. NASA became involved in one way or another, and *Clementine* was redesigned for scientific goals compatible with the original military ones. Observation of the Moon and tracking an asteroid would establish the feasibility of *Clementine*'s systems for the military, while at the same time accomplishing something that the scientists involved found useful.

Clementine was developed by the U.S. Naval Research Laboratory with cost control in the forefront. The number of people who could work on development was limited, although the team did reach about 55 persons. Controls and reviews were reduced in number, and even real estate used by the contractor was in the low-rent district. A basic schedule of two years was allowed for development, with only 22 months of it actually needed. Everything on the spacecraft was miniaturized as much as possible, a throwback to the earlier days of the U.S. space program, when miniaturization was pioneered because launch vehicles were not very powerful. Some of the six on-board cameras had a mass less than 4% that of similar cameras used in the 1980s. Control systems, such as gyroscopes, were also made from new lightweight composite materials. As a result, the final mass was about 500 lb (227 kg) in a craft less than 3 ft (1 m) in diameter and about 6 ft (2 m) long, and with a cost that was kept to less than $80 million—only $5 million more than was in the budget. Launching was handled by the Air Force using a Titan 2G rocket.

The *Clementine* space probe was launched on January 25, 1994, and reached a pole-to-pole orbit about the Moon on February 19, sending back to Earth pictures that revealed new details never before seen. These included the far side of the Moon and other regions that are difficult to observe from Earth, such as the depths of craters and the polar regions. *Clementine* was able to pass only about 260 mi (425 km) above the Moon at the low point of its orbit. Everything worked well, and the spacecraft returned images of the Moon at a rate of about one every four seconds for 71 days, producing about 1.5 million images. Since *Clementine* had separate cameras for ultraviolet, visible, and infrared (two of these), it was able to report far more information than simply what the Moon looks like. Indeed, the origin of its name—after the miner's daughter in the old song—comes from its ability to detect minerals on the Moon. Camera resolution ranged from 100 ft (30 m) for the best visual images to 1,560 ft (475 m) at the worst infrared viewing point. The average resolution was about 650 ft (200 m) per pixel. Another capability was laser ranging, enabling the craft to locate positions on the Moon with great accuracy. This device was originally intended for use by the Department of Defense in determining the distance in space to an unidentified "dark metallic object," which sounds suspiciously like an ICBM (*see also* "More on the Moon," p 155).

The *Clementine* visit was the first lunar mission for the United States in over 20 years, but not the only lunar study during that time. A Soviet robot landed on the Moon and returned samples in 1976; the *Galileo* space probe passed closed to the Moon twice, in 1990 and 1992; and a Japanese test in 1991 put a small camera in orbit around the Moon, from which it returned photographs to Earth. Nevertheless, *Clementine* seemed to many observers to be a first step in a return to the Moon—still the most logical location for a scientific base in space, in the view of many astronomers.

On May 3, *Clementine* fired its main thrusters and headed off to a planned rendezvous with asteroid 1620 Geographos. Its start was hampered by trouble on Earth and in space. On Earth, the U.S. Defense Department decided that they were no longer interested in the remainder of the mission and tried to cut off funding. NASA was not much help; they seemed embarrassed that *Clementine* was succeeding with a modest budget, when it appeared that ordinary NASA projects could not get by without giant budgets. In space, an onboard computer went batty and switched on the steering thrusters. The thrusters used up nearly all the onboard fuel supply before they were turned off. Furthermore, the firing left the craft spinning at 80 revolutions per minute—faster than an old-fashioned 78 rpm phonograph record.

The Earthly problem was resolved by the intervention of influential members of Congress. The problem in space was harder to solve. Ultimately, the mission to Geographos and an opportunity along the way to image Comet Shoemaker-Levy plunging into Jupiter both had to be scrubbed. Instead, *Clementine* was set wandering in solar orbit to see how much its miniature instruments could take. The computer that made the mistake that ended the mission was not a new experimental device; it was one of the few standard components on the spacecraft.

While *Clementine* was being hailed for its observations of the Moon, NASA's Goldin announced the development of two new "smallsats" for the late 1990s. Dubbed *Lewis* and *Clark* after the two explorers of the northwestern United States, the cheap satellites would, like *Clementine*, be used in part to test new miniaturization techniques and cost and schedule controls. At the same time, they would also provide new commercially useful studies of the environment, including both cities and forests. *Lewis* is to cost less than $60 million, including the launch and operations, while *Clark* is scheduled to cost $10 million less than that. For most Earth-orbiting satellites in the mid-1990s, the launch alone costs $60 million.

A somewhat less heralded launch occurred on July 4, 1992, and was also supposed to mark the beginning of an era of small, cheap satellites. *Sampex* (the *Solar, Anomalous, and Magnetospheric Particle Explorer*) was about two-thirds the size of *Clementine* and cost about the same to develop over a three-year period. *Sampex* was intended to be the first of a group of Earth satellites to be called Small Explorers.

Not everyone was entranced with the idea of smaller, cheaper missions. Joseph A. Burns, head of the U.S. National Research Council's committee on planetary and lunar exploration, pointed to the limited sensing abilities of *Clementine*, and called for continued funding of large-scale and necessarily expensive missions to planets. Others noted that for some astronomical instruments size is of the utmost importance, as for example in orbiting telescopes, which need to gather as many photons as possible from distant sources *Clementine*'s cameras, designed for close-up views, are hardly a model for telescopes that must observe distant galaxies. Finally, even though its failure was apparently not caused by the fact that *Clementine* was smaller or cheaper, lack of total performance did tend to undermine the concept.

Earth from Space

The *Galileo* space probe conducted a famous experiment in the early 1990s, seeking life on Earth (see "*Galileo* UPDATE," p 278), but many satellites provide a wealth of data about life on Earth—not only on living organisms, but also on mineral resources and tectonic activity. Unfortunately, much of this data has been collected by the military, and has not been made available to scientists in the past. However, with the end of the cold war on Earth, scientists have been able to access such satellite information as gravity maps of the ocean floor and ice thicknesses of parts of the polar icecap. Future military satellites are expected to be designed, with scientific access to some of the data in mind from the beginning. In some cases, all data collected may be released to qualified scientists. Other projects may continue to be classified, such as the detection from space of heat from nuclear reactors, and the chemical sensing of by-products of nuclear fission, although the technology itself may become available for scientific purposes.

One tool, developed for observation of planets from space, has been dramatically successful in mapping the second planet from the Sun (see "*Magellan* at Venus," p 150). This tool is radar. A version of radar that uses three different wavelengths at once was tested from Earth orbit at a height of about 120 mi (193 km) in an orbiting spacecraft in April and October of 1994. The radar system (developed by the space agencies of the United States, Germany, and Italy) easily detected volcanoes, cleared patches of rain forest, glaciers, and even objects as small as oceangoing ships. Because the radar can penetrate dry sand and snow, it can detect features that are not even observable on the ground (*see* "New Finds in the Holy Lands," p 83).

Observing Earth's current weather with an eye toward predicting future weather patterns was one of the earliest applications of satellite technology. Three separate systems came into use in the United States, two for civilian weather and one for military weather. Inasmuch as the current weather observed by the Defense Department system was the same as the current weather reported by the National Oceanic and Atmospheric Administration (NOAA) system, the military and NOAA agreed in

1994, under fiscal pressure from the Clinton administration, to combine their efforts for a new generation of weather satellites. Furthermore, the joint European meteorological agency, Eumetsat, which already has close ties with NOAA, was invited to join the project. Instead of having four satellites in low polar orbits, the new program is aimed at putting three more sophisticated satellites in geostationary orbit, starting around 2004. Meanwhile, one of the current generation of NOAA satellites was short-circuited by a single screw that extended too far, knocking it out of commission just two weeks after its launch on August 9, 1993.

The NOAA and military satellites are all different from the Geostationary Operational Environmental Satellite (GOES) devices that already observe U.S. weather from geostationary orbits. The GOES program has been in a permanent state of crisis since 1986, when a replacement satellite blew up on launch. Problems were compounded in 1989, when GOES-6 failed. GOES-7 took over its duties, and a satellite was borrowed from the European weather program Eumetsat to supplant the coverage. Finally, after many delays, GOES-8 was successfully launched on April 13, 1994. GOES-7 has been declared near failure from fuel loss since the beginning of the 1990s, but has been kept working by careful fuel management. GOES-8 is the first of five greatly improved geostationary weather satellites; the second will replace GOES-7.

NOAA has also begun to operate the main U.S. remote-sensing system in space: the Landsat satellites. These were previously under the control of NASA, or under the joint control of NASA and the U.S. Defense Departments. The program, under a budget cloud for several years and outclassed by French and Russian competition in the early 1990s, suffered a further setback in October 1993, when Landsat 6 was lost—literally—as a result of a bad launch. While NOAA will operate a new generation of more competitive satellites, beginning with Landsat 7, design and manufacture will be handled by NASA. The military also fared badly in Earth observation, losing three ocean-surveillance satellites in a launch failure on August 2, 1993.

Locating and Communicating

In 1994, commercially available versions of the Global Positioning System (GPS) became available. These could be used by ships in conjunction with a known land-based signal to find positions at sea to within 3 ft (1 m) in accuracy. GPS is also available for automobiles, and even for hikers, with only slightly less accuracy, since these versions rely solely on the GPS satellites without the aid of a fixed Earth-based transmitter.

A study released in 1993 by the International Technology Research Institute at Loyola College in Maryland reported that the United States was falling behind Japan and Europe in the sophistication of its communications satellites. There are more than 120 communications satellites in geostationary orbits. Included among U.S. difficulties were launch failures for several commercial and military communications satellites, resulting in the destruction of billions of dollars in equipment.

Commercial communications were the subject of the Loyola report. The U.S. military spends billions of dollars maintaining its own secure communications system, with each satellite in the Milstar system costing $1 billion, not counting launch or maintenance. Despite these high costs, critics such as John Pike, director of the

Federation of American Scientist's space policy project, say that the Milstar satellites have "astonishingly modest capabilities." The first attempt to launch a Milstar also cost the program mightily, as the Titan rocket blew up on the pad. The first successful launch was on February 7, 1994, but a month later the satellite suffered a crippling power failure, leaving it dependent on its backup system. A different kind of launch failure put a Navy communications satellite into a useless orbit in March 1993.

Tethers in Space

The early and mid-1990s produced a number of experiments involving the concept of using a cord or wire in space. One possible application of such a device would be to hold a small payload at a fixed distance from a larger satellite, such as the space shuttle. Thus, the general name for all such endeavors has become "tethers," the name for a cord, wire, or chain used to keep an animal in a limited region. The tether experiments are easily confused with each other.

TIMETABLE OF TETHERS

1960	Y. N. Artsutanov proposes running a tether from Earth's surface to a satellite in geostationary orbit, enabling the tether to lift other payloads into orbit.
September 1966	*Gemini 11* astronauts dock with an empty Agena rocket, attach a tether to the shell, and keep the station tethered to rocket shell by a 100 ft (30 m) dacron strap. The linked pair undergo wild gyrations, to the considerable distress of the astronauts.
1972	Mario D. Grossi of Raytheon Corporation and the Smithsonian Astrophysical Observatory in Cambridge, Massachusetts, proposes dangling a wire some 12-65 mi (20-100 km) long from a satellite for use as a radio antenna or for studying Earth's magnetic field.
1974	Giuseppe Colombo suggests that a long wire dangled from a satellite could generate electricity more efficiently than a solar sail.
July 31, 1992	The space shuttle *Atlantis* lifts off in the first experiment designed to test a space tether since the 1966 *Gemini* trial. The planned tether will hold a 1,000 lb (500 kg) satellite at the end of 12 mi (20 km) of copper wire. The wire itself has a mass of 350 lb (160 kg). The purposes of the experiment are to test the feasibility of tethers and to determine whether or not usable amounts of electricity could be generated by this method.
August 6, 1992	After four frustrating attempts in which the tether snarled or simply failed to unreel beyond 840 ft (256 m), *Atlantis* astronauts abandon their efforts and with considerable difficulty reel the tethered satellite back to the shuttle. Even deployed for only a few meters, however, the tether generates 36 volts of electricity. Later investigation shows that a bolt added to strengthen the reel mechanism had jammed it instead.
March 30, 1993	A Delta rocket on a military mission also carries a tethered box on its second stage. The 57 lb (26 kg) box unreels successfully for a full length of 12.5 mi (20 km) of braided polyethylene, resembling braided

dental floss, and orbits smoothly for two hours before a command from the ground cuts the tether, allowing the tether and box to burn up in the outer atmosphere.

June 26, 1993   A second Delta rocket on another mission also carries a tether into orbit on its second stage. This "tether" is actually a fairly short 0.3 mi (0.5 km) copper wire with a xenon-filled cathode tube at each end. The combination acts as an electrical ground (something that is otherwise hard to come by in space) by releasing ionized xenon from one of the cathode tubes. After the experiment is completed, the tether remains on the second stage as its orbit gradually decays, and the whole operation burns up.

March 9, 1994   A 12.5 mi (20 km) length of polypropylene weighing 15 lb (6.8 kg) and resembling string is orbited on a satellite approximately 215 mi (350 km) above Earth's surface, kept extended by a small weight like a plumb bob. Although the main purpose of the flight is to test the possibility of using tethers in launching spacecraft into low Earth orbit, the most amazing result is that the cord is highly visible in the night sky, much easier to see than the space shuttle or most satellites.

September 16, 1994 Two U.S. astronauts demonstrate for the first time in 10 years that they can fly around in space *without* being tethered to their craft. Most space walks have included tethers to keep astronauts from floating away into their own orbits, but new backpacks enable the astronauts to maneuver well enough to rescue themselves if a tether breaks.

February 25, 1996   After reeling out a thin cord to 12.25 mi (19.7 km) of its full 24.75 mi (20 km) length, *Columbia* shuttle astronauts are surprised when the tether mysteriously snaps and an Italian satellite at the end of the tether floats away.

(Periodical References and Additional Reading: *New York Times* 3-12-92, p A1; *New York Times* 7-5-92, p I-17; *New York Times* 7-28-92, p C1; *New York Times* 8-6-92, p A1; *New York Times* 8-29-92, p 6; *New York Times* 8-30-92, p I-32; *New York Times* 3-27-93, p 9; *New York Times* 3-31-93, p A16; *New York Times* 6-27-93, p I-24; *New York Times* 7-28-93, p A14; *New York Times* 8-3-93, p C10; *New York Times* 8-14-93, p A1; *New York Times* 8-10-93, p C9; *New York Times* 9-15-93, p D1; *New York Times* 10-10-93, p I-29; *New York Times* 1-16-94, p IV-7; *New York Times* 1-26-94, p A15; *Science* 2-4-94, p 622; *New York Times* 2-8-94, p C10; *New York Times* 3-5-94, p 9; *New York Times* 3-15-94, p C8; *New York Times* 3-25-94, p A14; *EOS* 4-5-94, pp 161 & 163; *New York Times* 4-12-94, p C1; *New York Times* 4-14-94, p B11; *New York Times* 4-26-94, p A18; *New York Times* 5-10-94, p C6; *New York Times* 5-11-94, p A21; *New York Times* 5-17-94, p A18; *Nature* 5-19-94, p 176; *Science News* 5-21-94, p 326; *Astronomy* 6-94, p 24; *Science News* 6-11-94, p 383; *Science* 6-17-94, p 1666; *Nature* 7-14-94, p 100; *New York Times* 7-29-94, p A2; *New York Times* 9-17-94, p 9; *New York Times* 9-21-94, p B8; *EOS* 11-22-94, p 586; *Science* 12-16-94, p 1835; *Discover* 1-95, p 92)

RUSSIA: SPACE PROGRAM AMID CHAOS

Although many institutions in the post-communist Russian society have collapsed or been reinvented, the hugely successful Soviet space enterprise has continued even when everything about it has fallen. It is true that little has been mentioned recently concerning *Burak*, the planned Soviet (now Russian) space shuttle. However, work with the present space station *Mir* continues and has assumed new importance as a training ground for the international space station that will combine U.S. and Russian programs. Until 1995, Russian space scientists continued to speak of building *Mir 2*, but that economically unfeasible project was then replaced by plans for the international space station. As described below, Russia also recognized that plans for a Mars mission with the United States in 1998 as well as a large scientific Earth satellite carrying many astrophysics experiments were too costly. The Mars mission was dropped, while the Spectrum X-Gamma Earth satellite project was retained.

Russian space scientists have stayed in their home country, unlike Russian scientists in many other disciplines who have moved to the West when their government financing was cut.

Space Station Cooperation

One of the biggest differences between East and West in space exploration since the 1970s is that while the United States abandoned experiments leading toward a space station after the successful U.S.-Soviet linkup in space in 1975, the Soviet Union continued to make development of a space station a priority. The United States turned to creation of a partly reusable space vehicle. It is still not clear which was the wiser policy. With the cooperative spirit that developed in the mid-1990s, the differing approaches may be seen in retrospect as a happy accident. The future international space station will benefit from both, while the economic reality is that there has been less duplication of effort than was the case with earlier phases of the space program.

The Soviet Union launched its first space station, *Salyut 1*, in 1971. Despite tragic deaths and many more near-tragedies, as well as power failures and other difficulties, the *Salyut* series was continued for 15 years and included seven space stations in all. The United States during that time launched only three space station prototypes, all in 1973. The second generation of Soviet space stations began with *Mir* in 1986, and its single member continues in orbit and in use today, having been continuously occupied except for a few months in the late 1980s.

Thus, the Russians have about a quarter of a century of experience in the space station business, while the Americans have almost none. With the collapse of the United Soviet Socialist Republic (U.S.S.R.), the Soviet space program came under Russian control. As early as February 21, 1992, Russian space scientist Yuri Semenov told a U.S. Senate subcommittee that the Russians were willing to share their experience and equipment with the United States in a series of joint projects. U.S. National Aeronautics and Space Administration (NASA) administrator Richard Truly was at that time skeptical of using *Mir*, which some senators wanted to buy outright from Russia. He thought that *Mir* was no match for the planned U.S. space station *Freedom*. This may have been true, although *Mir* was already in existence while *Freedom* continued to be postponed and redesigned—with each redesign simpler, cheaper, and less versatile than the one before.

The *Freedom* project continued to be controversial throughout the early 1990s, always in danger of losing funding. It, too, will be subsumed by the international space station. Whether or not funding problems will be as hotly debated for the international station as they were for *Freedom* is yet to be seen.

Funding problems were at the heart of the loss of another planned cooperative venture. Russia and the United States were to collaborate on a 1998 Mars exploration program to be called Mars Together, but Russia decided it did not have the funds to follow through. The deal was declared dead late in 1994. Russia and the United States continued to plan for somewhat less ambitious separate Mars programs for late 1996, however. Mars Together, or something like it, was rescheduled for 2001 or later.

Russian Space Entrepreneurs

A collapsed economy and a confused central government is a good environment for the development of creative businessmen, even businessmen who began as scientists. Russians eager to sell something, anything, to the West quickly recognized that they had a winner in their highly successful space program. Russia had developed and manufactured the world's biggest rockets and the only permanent space station. Individual cosmonauts had spent periods of time in space that vastly exceeded records from anywhere else on Earth. While Russian skills might be lacking in space-related miniaturization and computers, Russian scientists were far ahead in space biology.

One of the earliest endeavors dates back to 1987, before the complete collapse of the Soviet Union. This is the sale of spy-satellite photographs to commercial interests in the West. These had a resolution of about 5 meters—that is, they could resolve or show individual objects as small as 31 ft (5 m)—twice as good as the next best available space photos from France, and six times as good as U.S. *Landsat* photos. Russia upped the ante in 1992 when it began to offer their military's better spy-satellite photographs, which have a resolution of only 6 ft (2 m), a vast improvement over any other declassified space photographs. Even so, Russian sources said that the resolution had been purposely degraded for commercial use—the originals were much better! No official source is telling how these Russian snapshots compare with the best classified U.S. spy photos, which are rumored to be able to show objects the size of baseballs. The sharp Russian photographs were priced about the same as the best French *SPOT* and U.S. *Landsat* photos.

By 1994, the Russians had plied their space expertise into over a billion dollars worth of orders, ranging from payments for joint research programs with Western scientists to launch vehicles for a 67-nation telecommunications consortium and for the giant Motorola Iridium project (*see* "Reinventing Communications," p 703). NASA expects to spend about $100 million a year for Russian space products in the late 1990s.

Although the Khrunichev State Research and Production Space Center is owned by the Russian government, it has turned away military orders in preference for long-term commercial gains (and did so even before the breakup of the Soviet Union). Khrunichev is even one of the Iridium investors, owning one-twentieth of the enterprise.

A Russian Experiment

Dating back to Konstantin E.Tsiolkovsky in 1895, Russia's space program has deep roots, as well as a few odd branches. Russian willingness to try new ideas contributed to their being the first in space, the first to hit other bodies in space with space probes, the first to recover objects from space, and the first to develop space stations. Often these projects were accomplished simply by trying several times until they got it right, utilizing brute force more than cleverness.

In 1993, one of the odder branches was the first giant mirror in space, another science fiction concept that, like satellites and space probes, Russia was willing to test by doing. The idea was to use a 66 ft (20 m) diameter aluminum and mylar disk to reflect sunlight from space to Earth. The earliest known proposal of this sort came from German space and rocket pioneer Hermann Oberth in 1929. In some science-fiction depictions of the future, the dark side of Earth is kept illuminated by larger versions of orbiting mirrors, while in others parabolic versions of mirrors in space are used to concentrate sunlight for energy production. Lighting from space could be used by farmers to extend the number of daylight hours in the short planting and harvesting seasons of the North; could illuminate places suffering a natural disaster such as an earthquake or hurricane; could reduce the darkness of winter in regions near poles; or perhaps could simply be directed at the downtown area of a large city to keep crime down. If the test satellite, known as *Znamya* (Banner), worked, the Russians hoped to put 30-100 larger mirrors in space to handle such chores. A mirror 650 ft (200 m) in diameter, placed in a high orbit, could illuminate a spot about 40 mi (250 km) across.

Znamya was unfolded on February 3, 1993. Although hailed by the Russian scientists involved as a success, the light from the test skittered across Europe and the Atlantic Ocean at 17,000 mph (27,000 kph) per hour. This proved too fast for illumination, although it could be seen as a flash. Cosmonauts aboard *Mir* could see the spot of light as it raced across the Earth for about six minutes. The Russian scientists behind the enterprise hoped to use this experiment as an aid in raising the $10 million or more needed for a test of a larger mirror that could be guided so that its light could be kept on a part of Earth's surface for an appreciable amount of time.

Znamya was originally designed for an entirely different purpose, although one equally at home in science fiction. Plans had been underway for a competition between spacecraft using solar sails—ships propelled by the solar wind in exactly the same manner (although in three dimensions instead of two) as sailboats or iceboats on Earth. Russian scientists, however, were the only ones to take the idea of a space regatta seriously, so they were left with plans for a solar sail and no competitors. The sail was then converted into the space mirror project.

(Periodical References and Additional Reading: *New York Times* 2-23-92, p I-26; *New York Times* 10-4-92, p I-10; *New York Times* 1-12-93, p C1; *New York Times* 2-5-93, p A6; *New York Times* 3-22-93, p D3; *New York Times* 11-23-93, p D4; *Discover* 1-94, p 59; *New York Times* 5-17-94, p A1; *Science* 12-16-94, p 1799; *Planetary Report* 7/8-95, p 15)

MARS: LOSS AND NEW PLANS

One of the great disappointments of planetary space exploration since the triumphs of the three V programs (two *Vikings* that landed on Mars, several *Veneras* that reached Venus, and two *Voyagers* that passed close to all the outer planets but Pluto) has been the failure of attempts to return to Mars. In 1988, the Soviet-launched *Phobos 1* went into a tumble and was lost. In 1989, the sister ship *Phobos 2* mysteriously lost contact with Earth just as it was reaching the vicinity of the planet. In 1993, virtually the same thing happened to the U.S. space probe *Mars Observer* just three days before it was to have entered Mars orbit.

Mars Observer

After a 17-year gap, the United States attempted to return to the Mars exploration game with the first of its new generation of less expensive planetary missions. The plan was to employ a smaller probe that reused equipment designed for other missions, thereby keeping costs down. If the mission had succeeded, the *Mars Observer* would have surveyed Mars with cameras similar to those used on weather satellites for Earth. Indeed, it would have been something like a weather satellite for Mars combined with a *Landsat*. It would also have had devices for measuring the magnetic field and conditions in the atmosphere. Maps that resulted would have become the basis for Mars landing expeditions in the future.

Launched on September 25, 1992, the *Mars Observer* fell silent on August 21, 1993. All efforts to revive it failed, and the embedded device telling the craft to phone home if in trouble also failed to work.

A number of theories were proposed to explain the loss. A special commission found that the most likely cause was a leak of the oxidizer used in controlling the vehicle. Over the 11 months that *Mars Observer* was in orbit, the oxidizer accumulated; when it came time to use the chemical to oxidize the fuel, there was too much in the wrong place, causing an explosion instead of a controlled burn. It is not clear whether the explosion destroyed the craft or simply sent it into a tumble, making communication impossible.

The panel also criticized the use of hardware and software designed for low Earth orbit in a probe that would travel great distances through space. Another problem identified by the panel was that the builder of the *Mars Observer* had been put under financial pressure to cut corners. Because it seemed possible that a similar disaster might occur if a backup of the same design were assembled and launched, a second attempt in the Mars Observer program was ruled out. Instead, the United States will rely on several other missions to Mars.

Next Steps on Mars

The Mars Surveyor program will include two space probes, one each launched in 1996 and 1998, while a separate *Mars Pathfinder* will also be launched in 1996. Each of the Surveyors, intended to orbit Mars, will carry about half the experiments that were planned for the Observer program. *Mars Pathfinder* will land in a region called

Ares Vallis, which is thought to be where water once flowed, forming the same kind of geological feature that on Earth is a flood plain. The main reason that Areas Vallis was chosen is that it is expected to have plenty of sunlight—essential because the *Pathfinder* lander is solar powered.

Pathfinder is expected to parachute onto Mars on or about July 4, 1997. It will carry a very small robot called a rover that has a mass of 22 lb (10 kg). The rover, named Sojourner by 12-year-old Valerie Ambroise, can move about the rocky plain in a limited way on its six wheels, keeping within 100 ft (30 m) of the solar-powered lander. The lander will also be the relay station for reports from the rover to a second relay on *Surveyor*. The main mission of the rover will be to use an alpha-proton-x-ray spectrometer to determine the composition of the rocks it encounters. Although this phase of the mission is scheduled to last only a month, the experience with *Viking*, which continued functioning long after its scheduled demise, makes one hopeful for additional reports.

The Surveyor program is scheduled to go far beyond the *Observer* experiments, however. If the first flights prove the program works, a *Surveyor* will be launched at each window of opportunity—approximately one every 25 months through the first part of the twenty-first century. For the first three launch windows (1996, 1998, and 2001), each Surveyor would be accompanied by a lander of some kind, although possibly of a different type than *Pathfinder*. Starting in 2001 or perhaps 2003, the program will involve the Russians as well, in what is currently known as Mars Together, but plans for the division of duties are not yet clear.

Russia is also planning to take advantage of the 1996 launch window, with a program to be called simply Mars '96.

The U.S. National Aeronautics and Space Agency (NASA) hopes to use the Surveyor program to help in planning for a mission that would not only land on Mars but also send back samples of the Martian surface to Earth. This would be a robot mission, similar to an early Soviet mission to the Moon that returned samples.

(Periodical References and Additional Reading: *Discover* 1-93, p 33; *Discover* 1-94, p 57; *New York Times* 1-6-94, p D18; *Science* 1-14-94, p 167; *New York Times* 2-15-94, p C5; *Astronomy* 6-94, p 22; *New York Times* 9-11-94, p I-19; *Science* 12-16-94, p 1799; *New York Times* 7-16-95, p I-19; *Planetary Report* 7/8-95, p 15)

SPACE PROGRAM LISTS AND TABLES

MAJOR ACCOMPLISHMENTS OF SATELLITES AND SPACE PROBES

The space program began as a scientific endeavor, part of the highly successful International Geophysical Year (IGY) program of the late 1950s. While space flight has provided moments of high drama and has been a tool of international politics, the space program has also yielded considerable scientific dividends, including new understanding of the Earth and its immediate environment; a totally new view of the solar system; and a greatly revised and more detailed view of the rest of the universe. Not since the spate of inventions at the beginning of the seventeenth century (the microscope, telescope, and calculus) has a new technological tool contributed so much to so many different branches of science.

Technologically as well as scientifically, space exploration is a success. The space program has revolutionized both international and (more recently) local communications, made weather forecasts that are largely reliable for short periods, and put sophisticated navigational tools in the hands of pilots, with drivers and walkers soon able to receive the same technological aid in locating themselves on the globe.

The first scientific and technological uses of space were fueled by the cold war between the United States and the U.S.S.R., but other nations soon came aboard. France and India saw the uses of space early, although French space efforts today are part of the European Space Union (ESA). The U.S.S.R. became separated into a number of states at the beginning of 1992, with the Russian Federation (commonly called Russia) taking over the space capabilities. Soon after, Russia and the United States began cooperating in space. It appears likely that nationalistic exploration of space is withering away. One international communications union had 67 nation members in the early 1990s. Major projects, such as the international space station and the exploration of Mars, are expected to be led by the United States, Russia, the ESA, and Japan working together.

Articles describing the current contributions of the space program to science and technology are so pervasive in this book that they cannot be listed here.

Critics of programs that put people in space — the "manned" space program, as it is often called — could point to these pages as evidence in support of their case. Nearly all the scientific and technological triumphs that extend beyond the space program itself rely on robot spacecraft, usually classed as satellites or probes. The following list contains the major robot satellites or space probes up to 1995, including a few spectacular failures accomplishing little directly, although contributing experience that would be used later. The "manned" space program is given a separate listing (*see* "Spaceflights Carrying People," p 303). In addition, while the greatest number of recent Earth-orbiting artificial satellites have been commercial or military, these are not listed below unless they break new ground or are notable for some other reason. For example, there are over 100 communications satellites, and there have been dozens of spy satellites, but most of them were simply copies of or slight improvements on the originals. In the following list, greater emphasis is given to recent satellites and probes. A number of planned projects are listed at the end, but dates for these change often, and programs are frequently eliminated; occasionally, new launches are added.

Launch	Name of Vehicle	Remarks	Nation
10-4-57	*Sputnik 1*	Part of IGY. First satellite to orbit Earth; burns up on reentry 1-4-58.	U.S.S.R.
11-3-57	*Sputnik 2*	Carries Laika, first dog in space. Burns up in atmosphere on 4-14-58 with dog still aboard.	U.S.S.R
1-31-58	*Explorer 1*	Satellite launched in desperate attempt to keep up with Russians, but proves its worth with first major scientific accomplishment of space program: detection of the Van Allen radiation belts. Reenters atmosphere 3-31-70.	U.S.
3-17-58	*Vanguard 1*	Grapefruit-sized IGY satellite that had been expected by U.S. to be first in space. Excellent orbit enables scientists to study shape of Earth, finding a bulge at South Pole. Still in orbit.	U.S.
3-26-58	*Explorer 3*	After launch failure of *Explorer 2* on 3-5-58, becomes third successful American satellite; studies radiation and micrometeoroid hazards.	U.S.
5-15-58	*Sputnik 3*	First Soviet scientific satellite. Burns up 4-6-60.	U.S.S.R.
7-26-58	*Explorer 4*	Studies radiation in space.	U.S.
8-24-58	*Explorer 5*	Last numbered failure in Explorer series. After this one, numbers are not assigned to Explorer satellites unless launch is successful.	U.S.
1-2-59	*Mechta (Lunik 1)*	Aimed at the Moon, it becomes the first space probe to go into orbit around the Sun, since it misses the Moon by about 5,000 mi (8,000 km).	U.S.S.R.
3-3-59	*Pioneer IV*	First U.S. probe aimed at the Moon. Like the first Soviet effort, it also misses and goes into orbit around the Sun.	U.S.
8-7-59	*Explorer 6*	Provides first television pictures of Earth from space.	U.S.
9-12-59	*Lunik II*	First probe to reach the Moon, where it crash lands on 9-13-59.	U.S.S.R.
10-4-59	*Lunik III*	First probe to photograph the far side of the Moon and radio back the pictures.	U.S.S.R.
4-1-60	*Tiros 1*	First satellite with a practical purpose: observation of weather on Earth.	U.S.
5-15-60	*Sputnik 4*	First satellite intended to be recovered from space, but rockets intended to slow it down actually lift it into higher orbit.	U.S.S.R.

Launch	Name of Vehicle	Remarks	Nation
6-22-60	Transit I-B	First satellite developed as an aid to navigation.	U.S.
8-12-60	Echo 1	First experimental communications satellite— actually a large orbiting balloon from which radio signals are bounced.	U.S.
8-18-60	Corona (Discover-14)	First known military spy satellite, it takes photographs of the Soviet Union and later parachutes them back to Earth, where they are recovered in midair or, in the case of a miss, at sea.	U.S.
8-19-60	Sputnik 5	Carries dogs Belka and Strelka; after 18 orbits, successfully returns them to Earth.	U.S.S.R.
12-1-60	Sputnik 6	Intended to duplicate Sputnik 5, but fails on re-entry and burns, killing dogs aboard.	U.S.S.R
2-4-61	Sputnik 7	First attempt to reach Venus. It fails.	U.S.S.R.
2-12-61	Sputnik 8 & Venera 1	Sputnik 8 launches Venera 1 in second attempt to reach Venus. It also fails, although it demonstrates that one satellite can be used as a launch pad for another.	U.S.S.R.
3-9-61	Sputnik 9	Successfully recovered with dog Chernushka intact after single orbit.	U.S.S.R.
3-25-61	Sputnik 10	Successfully recovered with dog Zvezdochka intact after single orbit.	U.S.S.R.
1-26-62	Ranger III	Misses Moon by 22,862 mi (36,793 km) on 1-28-62.	
3-7-62	Orbiting Solar Observatory (OSO 1)	First major satellite to be devoted to astronomy.	U.S.
4-23-62	Ranger IV	First American probe to reach the Moon. Lands on far side on 4-26-62, but fails to report back.	U.S.
4-26-62	Cosmos 4	First known Soviet military spy satellite.	U.S.S.R.
7-10-62	Telstar	First active communications satellite; allows direct television between Europe and North America when in the right part of the sky.	U.S.
7-22-62	Mariner 1	Aimed toward Venus, but fails on launch due to computer programming error.	U.S.
8-27-62	Mariner 2	First probe to reach vicinity of another planet (Venus) and return scientifically useful information.	U.S.
10-18-62	Ranger V	Misses moon by 450 mi (724 km) on 10-21-62.	U.S.
10-31-62	Anna-I-B	First satellite specifically intended for use in measuring exact shape of Earth.	U.S.
11-1-62	Mars 1	First probe aimed at Mars. Contact lost about 66 million mi (106 million km) from Earth.	U.S.S.R.
6-26-63	Syncom II	First communications satellite to be placed into a geostationary (geosynchronous) orbit, 22,300 mi (35,900 km) high.	U.S.

Launch	Name of Vehicle	Remarks	Nation
7-28-64	*Ranger VII*	Returns close-up photographs of the Moon before crashing into it.	U.S.
8-28-64	*Nimbus I*	First weather satellite to be stabilized by rotation so that its cameras always point toward Earth.	U.S.
11-28-64	*Mariner 4*	Flies by Mars and transmits 21 pictures of the planet's surface to Earth. Closest approach is 6,118 mi (9,850 km).	U.S.
4-6-65	*Early Bird*	First commercial satellite, used for communications between U.S. and Europe. From geostationary orbit it could carry either 240 voice channels or 1 television channel.	U.S.
4-23-65	*Molniya 1*	First Soviet communications satellite.	U.S.S.R.
7-16-65	*Proton 1*	Heaviest Earth orbiter to this date, with a mass of 12,225 kg (weight of 26,956 lb on Earth).	U.S.S.R
11-16-65	*Venera 3*	First probe to make physical contact with another planet, crash landing on Venus on 3-1-66, although radio contact is lost en route.	U.S.S.R.
4-8-66	*OAO 1*	The first Orbiting Astronomical Observatory, intended to record ultraviolet radiation, x rays, and gamma rays. Fails after two days in orbit.	U.S.
11-6-65	*GEOS-1 (Explorer 29)*	The first Geodetic Earth-Orbiting Satellite (GEOS), carrying lights and laser reflectors used in measuring the size and shape of Earth.	U.S.
11-26-65	*A-1*	First satellite to be launched by a nation other than the U.S.S.R. or U.S.	France
12-16-65	*Pioneer 6*	Launched into solar orbit; still functions today.	U.S.
1-31-66	*Luna 9*	Although main vehicle crash-lands, ejected capsule lands safely on the Moon and transmits photographs to Earth.	U.S.S.R.
3-31-66	*Luna 10*	First space vehicle to go into orbit about the Moon.	U.S.S.R.
5-30-66	*Surveyor 1*	First soft landing of complete vehicle on the Moon.	U.S.
8-10-66	*Lunar Orbiter 1*	First American space vehicle to go into orbit around the Moon.	U.S.
6-12-67	*Venera 4*	Ejects instrument package into atmosphere of Venus. Package parachutes toward surface, but contact is lost before reaching surface.	U.S.S.R.
6-14-67	*Mariner 5*	Second American satellite to reach vicinity of Venus.	U.S.
7-19-67	*IMPE (Explorer 35)*	Orbits the Moon and measures Earth's magnetic "tail."	U.S.

Launch	Name of Vehicle	Remarks	Nation
7-4-68	RAE-A (Explorer 38)	Radio Astronomy Explorer monitors radio waves from space, including emissions from Earth and from the Sun, other planets, and stars.	U.S.
9-15-68	Zond 5	First Soviet satellite to return to Earth from vicinity of the Moon.	U.S.S.R.
12-7-68	OAO 2	Second Orbiting Astronomical Earth Satellite is first ultraviolet telescope in space.	U.S.
2-11-70	Ohsumi	First satellite to be launched by Japan.	Japan
4-24-70	Mao 1	First Chinese satellite. It broadcasts song "The East is Red" once a minute, pausing at the end for other signals.	China
8-17-70	Venera 7	First Venus probe to return signals from planet's surface.	U.S.S.R.
9-12-70	Luna 16	First space probe to land on the Moon without humans aboard, scoop up samples, and return them to Earth.	U.S.S.R.
11-10-70	Luna 17	Carries roving vehicle to Moon's surface. Vehicle roams for two weeks at a time (during daylight), then "sleeps"; as it wanders, it returns photos and other data to Earth.	U.S.S.R.
12-12-70	Uhuru (SAS-A) (Explorer 42)	Officially the Small Astronomy Satellite, Uhuru, which means Freedom in Swahili, is the first orbiting x-ray telescope. Launched from Kenya.	U.S.
5-28-71	Mars 3	First space probe to soft-land on Mars, although it quickly ceases functioning.	U.S.S.R.
5-30-71	Mariner 9	First space probe to orbit another planet (Mars). Returns 7,329 photographs of planet.	U.S.
3-2-72	Pioneer 10	First space probe to study Jupiter and, on June 13, 1983, first to leave solar system.	U.S.
7-23-72	Landsat I	First Earth resources satellite.	U.S.
8-21-72	Copernicus (OAO 3)	Actually the fourth Orbiting Astronomical Observatory, Copernicus has a 32 in (81 cm) reflecting telescope for ultraviolet radiation studies and separate x-ray detectors.	U.S.
3-6-73	Pioneer 11	First space probe to reach vicinity of Saturn.	U.S.
11-3-73	Mariner 10	First space probe to observe two planets (Venus and Mercury) and only probe ever to observe Mercury.	U.S.
12-10-74	Helios	First West German space probe.	West Germany
6-8-75	Venera 9	Returns first photographs from surface of Venus.	U.S.S.R.
8-20-75	Viking 1	First American space probe to soft-land on Mars. Continues to return data until May 1983.	U.S.
9-20-75	Viking 2	Successfully soft-lands on Mars.	U.S.

Launch	Name of Vehicle	Remarks	Nation
8-20-77	Voyager 2	After studying Jupiter and Saturn, becomes first space probe to reach vicinities of Uranus and Neptune.	U.S.
9-5-77	Voyager 1	After studying Jupiter, becomes first space probe to reach vicinity of Saturn.	U.S.
1-26-78	IUE	*International Ultraviolet Explorer* becomes the only astronomical satellite to be placed in geostationary orbit. It is still sending back data.	ESA/U.S./ Britain
5-20-78	Pioneer 12 (Venus 1)	First space probe to go into orbit around Venus. Stays in orbit until October, 1993.	U.S.
6-26-78	Seasat	Analyzes ocean currents and ice flow.	U.S.
8-12-78	ISEE-3	Originally the third International Sun-Earth Explorer, space probe is renamed *International Cometary Explorer (ICE)* when it is redirected to study tail of comet Giacobini-Zinner in 1983.	U.S.
12-13-78	HEAO-2	*High-Energy Astronomy Observatory* (also known as the Einstein Observatory) makes high-resolution x-ray images of the universe.	U.S.
2-24-79	P78-1	Studies solar radiation until purposely shot down by U.S. Air Force 9-13-85. Still working at the time of its destruction, satellite is deemed by many scientists to be too valuable to be used as target.	U.S.
2-14-80	Solar Max	Studies solar radiation. After failure in November 1980, it is repaired and relaunched from space shuttle in April 1984; finally pushed to its destruction by a massive solar flare on December 2, 1989.	U.S.
1-25-83	IRAS	*Infrared Astronomical Satellite* studies galactic and extragalactic infrared sources and discovers new stars forming as well as possible planet formation.	U.S.
12-15-84	Vega 1	First Soviet mission to study Halley's comet. Along the way it drops a balloon probe into atmosphere of Venus.	U.S.S.R.
12-21-84	Vega 2	Second Soviet mission to Halley's comet; it also releases a balloon probe at Venus.	U.S.S.R.
1-7-85	Sakigake	First Japanese mission to study Halley's comet (this one from far away).	Japan
7-2-85	Giotto	Joint European mission to Halley's comet. It passes closest to the comet (375 mi or 600 km)—and later is redirected to comet Grigg-Skjellerup, which it passes on 7-10-92 at a distance of 125 mi (200 km).	ESA
8-18-85	Suisei	Japanese mission to Halley's comet.	Japan
2-21-86	SPOT	French satellite designed to photograph surface details of Earth as small as 30 ft (10 m) across.	France
5-4-89	Magellan	American probe that orbited Venus for two years and mapped the planet in detail with radar.	U.S.

Launch	Name of Vehicle	Remarks	Nation
10-18-89	Galileo	After passing near Venus and Earth (twice), it will orbit Jupiter, report on Jovian moons, and drop probe into Jupiter's atmosphere.	U.S.
11-18-89	Cosmic Background Explorer (COBE)	Studies cosmic background radiation in hopes of learning cause of galaxy formation, successfully detecting small fluctuations in the Cosmic Microwave Background. Retired from service on December 23, 1993.	U.S.
4-24-90	Hubble Space Telescope	Flawed optical telescope placed in orbit about Earth by U.S.; successfully repaired 12-10-93.	U.S.
10-6-90	Ulysses	After visiting Jupiter, goes into solar orbit to study previously unobserved north and south poles of the Sun.	ESA
1-24-91	Hiten-Hagoromo	After traveling near the Moon, Hiten launches a small probe that goes into lunar orbit, from which it photographs Moon and measures density of micrometeoroids.	Japan
4-5-91	Compton Gamma Ray Observatory	A 17-ton telescope for observing the universe at very short wavelengths.	
6-7-92	Extreme Ultraviolet Explorer (EUVA)	Satellite designed by researchers from the University of California at Berkeley to study the high range of ultraviolet radiation in universe.	U.S.
7-5-92	Sampex	The Solar, Anomalous, and Magnetospheric Particle Explorer is the first of the Small Explorer Project satellites. NASA hopes to launch one of these inexpensive scientific satellites every year.	U.S.
7-24-92	Geotail	Japanese satellite launched by NASA to study Earth's magnetosphere.	Japan
10-6-92	Freja	A Swedish satellite launched by a Chinese rocket from the Gobi Desert, Freja carries experiments from the U.S., Sweden, Canada, France, and Germany. The U.S. experiment is measuring Earth's magnetosphere.	Sweden
2-2-93	Znamya (banner)	Russia's 65 ft (19.8 m) diameter Mylar mirror, which reflects sunlight to nighttime regions of Earth in a feasibility experiment.	Russia
2-9-93	Pegasus 3	Brazilian satellite to monitor environment in Amazonia, launched by Orbital Sciences Corp. of Fairfax, Virginia.	Brazil
2-20-93	ASCA	X-ray telescope with improved resolution, enabling it to locate sources more accurately, named Astro-D when first launched.	Japan/U.S.
10-5-93	Landsat 6 (failed)	Explodes seven minutes after takeoff.	U.S.
11-22-93	Gorizant (horizon)	Russian communications satellite launched into a geostationary orbit controlled by Tonga as a commercial venture by Rimsat Corp. of Fort Wayne, Indiana.	Russia/Tonga

Launch	Name of Vehicle	Remarks	Nation
1-25-94	*Clementine*	Joint U.S. military-science mission to map the Moon from lunar orbit and later visit the asteroid Geographos.	U.S.
2-9-94	*Shijan 4*	Chinese space research satellite.	China
3-9-94	Tether experiment	A 15 lb (7 kg), 12 mi (20 km) plastic string is unwound from a small U.S. satellite, becoming the first space object to be visible from Earth with the naked eye as something other than a point of light.	U.S.
4-13-94	GOES-8	The first of five U.S. Geostationary Operational Environmental Satellites using new technology to track weather in North America.	U.S.
9-8-94	*Telstar 402* (failed)	$200,000 communications satellite lost.	U.S.
11-1-94	*WIND*	To go into a figure-eight orbit around the Earth and Moon at first, studying the solar wind. In 1996, it will move to a point in Earth's orbit and orbit the Sun itself, staying in the same relation to the Earth as it revolves, about a million miles from Earth.	U.S.
11-15-94	*Express* (failed)	Satellite designed to test new ceramic materials for use in heat shields. Explodes before entering orbit.	Germany/ Japan
11-94	GOMS	The first Russian geostationary weather satellite. Because it fills a big gap in the coverage, *GOMS* is expected to improve forecasts worldwide.	Russia
12-30-94	NOAA-14	Weather satellite placed in polar orbit. Part of a system of NOAA satellites.	U.S.
4-95	SAC-B	The *Satélite de Aplicaciones Científicas-B* is the first spacecraft from Argentina. Launched by NASA, *SAC-B* carries several instruments to study solar and background x rays, as well as gamma-ray bursters.	Argentina
4-95	*High Energy Transient Experiment (HETE)*	Satellite intended to study gamma ray bursts.	U.S./ Japan/ France
5-23-95	GOES-9	The second of five new Geostationary Operational Environmental Satellites, the basic weather-satellite system from NASA.	U.S.
7-95	*Submillimeter Wave Astronomy Satellite (SWAS)*	Orbiting radio telescope designed to study composition of gas clouds in the Milky Way and nearby galaxies.	U.S.
8-95	FAST	The *Fast Aurora l Snapshot* satellite will monitor the upper reaches of the auroras.	U.S.
11-95	*Infrared Space Observatory (ISO)*	A satellite from the European Space Agency that will extend the work of the earlier IRAS.	ESA
11-95	*Cluster*	Four identical satellites that will measure the ionized gases in Earth's magnetic field.	ESA/U.S.

Launch	Name of Vehicle	Remarks	Nation
12-30-95	*X-Ray Timing Explorer (XTE)*	First U.S x-ray telescope since 1979 will study x-ray sources including stars and possible black holes.	U.S.
2-16-96	*NEAR*	*Near Earth Asteroid Rendezvous*, the first of the new, less-expensive Discovery series of space probes, will visit and orbit asteroid 433 Eros starting in December, 1998, approaching as close as 15 mi (24 km). On the way it will pass near asteroid Iliya in August, 1996.	U.S.
2-24-96	*Polar*	Like WIND, *Polar* will study the interaction of the solar wind and Earth's magnetic field, concentrating especially on the poles.	U.S.
1996	*Mars '96*	A Russian spacecraft to orbit Mars and drop robot stations and probes to its surface.	Russia/ France/ U.S.
Summer 1996	*Muses-B*	A radio observatory intended to add a space component to the VLBA of radio telescopes on Earth (*see* "Major Telescopes," p 247).	Japan
8-96	*ADEOS*	The *Advanced Earth Observation Satellite* will study the ozone layer, air-sea interactions, and ocean circulation patterns.	U.S./ France/ Japan
11-96	*Mars Global Surveyor*	A probe to Mars that will orbit the planet and survey it from space.	U.S.
12-96	*Mars Pathfinder*	The second Discovery space probe, it will parachute a rover to the surface of Mars.	U.S.
Summer 1997	*Lunar-A*	Spacecraft will orbit the Moon. It will loose three projectiles, each expected to penetrate the lunar surface to about 4 ft (1.2 m).	Japan
10-97	*Cassini-Huygens*	Large-scale mission to orbit Saturn and to study its satellites. *Cassini* is the main probe that will orbit Saturn, while *Huygens* is a probe into the Saturnian atmosphere.	U.S./ESA/ Italy
1997	*DVSAGE*	The *Discovery-Venera Surface-Atmosphere Geochemistry Experiments* will be a joint mission to study further aspects of the geology of Venus revealed by the *Magellan* mission of 1989-1994.	U.S. & Russia
1997	*TRMM*	The *Tropical Rainfall Measuring Mission* will also observe lightning and measure energy reflected by clouds.	Japan/ U.S.
10-98	*WIRE*	*Wide Field Infrared Explorer* satellite will observe galaxies undergoing bursts of star formation, continuing the work begun in 1983 by *IRAS*.	U.S.
1998	*Landsat 7*	Will replace partially disabled *Landsat 5* and failed *Landsat 6*.	U.S.
1998	*FUSE*	The *Far Ultraviolet Spectroscopic Explorer* will study emissions in the high ultraviolet range.	U.S.
1998	*Planet-B*	Spacecraft will orbit Mars and measure its magnetic field in detail.	Japan

Launch	Name of Vehicle	Remarks	Nation
1998	Mars Global Surveyor 2	The second year of a planned program in which Mars will be explored. The plan is to continue this program whenever Mars and Earth are in a suitable relation to each other, which happens about once every two years.	U.S.
1998	Mars Pathfinder 2	The second in a program in which a roving robot vehicle is parachuted to the surface of Mars.	U.S.
1999	Gravity Probe B	A satellite designed to test predictions of Einstein's theory of general relativity.	U.S.

SPACE FLIGHTS CARRYING PEOPLE

The space programs of the Soviet Union (now Russia) and the United States both have had dramatic flights by human pilots as an important component, although many scientists felt that most goals of the space program could be achieved without risking lives. So far, no other nation has developed launch capability for spacecraft carrying people, although many have sent people into space on U.S. or Soviet/Russian vehicles. With the apparent internationalization of space, the concern about where on Earth a space traveler was born or whether the traveler is called a cosmonaut (Soviet/Russian terminology) or astronaut (U.S. term) is becoming increasingly irrelevant.

The *Vostok*, *Voskhod*, and *Soyuz* missions are part of the Soviet (later Russian) program; all other flights in this list are part of the U.S. program.

PROVING THAT PEOPLE CAN VENTURE INTO SPACE

Craft	Date	Duration	Crew	Remarks
Vostok 1	4-12-61	1 hour 48 minutes	Yuri Gagarin	First space flight by a human; 1 orbit.
Mercury 3	5-5-61	15 minutes	Alan B. Shepard Jr.	Freedom 7 (suborbital).
Mercury 4	7-21-61	16 minutes	Virgil I. Grissom	Liberty Bell 7 (suborbital).
Vostok 2	8-6-61	25 hours 18 minutes	Gherman S. Titov	First multi-orbit flight; 17 orbits.
Mercury 6	2-20-62	4 hours 55 minutes 23 seconds	John H. Glenn Jr.	Friendship 7; first orbital flight by American; 3 orbits.
Mercury 7	5-24-62	4 hours 56 minutes	M. Scott Carpenter	Aurora 7; 3 orbits.
Vostok 3	8-11-62	94 hours 24 minutes	Andrian G. Nikolayev	Landing by parachute; 64 orbits.
Vostok 4	8-12-62	70 hours 57 minutes	Pavel R. Popovitch	Dual launch with Vostok 3; 48 orbits.

Craft	Date	Duration	Crew	Remarks
Mercury 8	10-3-62	9 hours 13 minutes	Walter M. Schirra Jr.	*Sigma 7*; 6 orbits.
Mercury 9	5-15-63	34 hours 20 minutes	L. Gordon Cooper	*Faith 7*; 22 orbits.
Vostok 5	6-14-63	119 hours 6 minutes	Valery F. Bikovsky	81 orbits.
Vostok 6	6-16-63	70 hours 50 minutes	Valentina Tereshkova	First woman cosmonaut; dual launch with *Vostok 5*; 48 orbits.
Voskhod 1	10-12-64	24 hours 17 minutes	Vladimir M. Komarov; Konstantin P. Feotistov; Boris B. Yegorov	First multi-person crew; 16 orbits.

PLANNING FOR OPERATIONS IN SPACE

Craft	Date	Duration	Crew	Remarks
Voskhod 2	3-18-65	26 hours	Aleksei A. Leonov; Pavel I. Belyayev	First extravehicular activity (EVA) by Leonov (20 minutes); 17 orbits.
Gemini 3	3-23-65	4 hours 53 minutes	Virgil I. Grissom; John W. Young	First American multi-person crew; 3 orbits.
Gemini 4	6-3-65	97 hours 56 minutes	James A. McDivitt; Edward H. White II	First American EVA; first use of personal propulsion unit; 62 orbits.
Gemini 5	8-21-65	190 hours 56 minutes	L. Gordon Cooper; Charles Conrad Jr.	Demonstrates feasibility of lunar mission; simulated rendezvous; 120 orbits.
Gemini 7	12-4-65	330 hours 35 minutes	Frank Borman; James A. Lovell Jr.	Extends testing and performance; target for first rendezvous; 206 orbits.
Gemini 6A	12-15-65	25 hours 51 minutes	Walter M. Schirra Jr.; Thomas P. Stafford	First rendezvous (with *Gemini 7*); 15 orbits.
Gemini 8	3-16-66	10 hours 42 minutes	Neil A. Armstrong; David R. Scott	First dual launch and docking; first Pacific landing; 6.5 orbits.
Gemini 9A	6-3-66	72 hours 21 minutes	Thomas P. Stafford; Eugene A. Cernan	Unable to dock with target vehicle; 2 hours 7 minutes of EVA; 44 orbits.

Craft	Date	Duration	Crew	Remarks
Gemini 10	7-18-66	70 hours 47 minutes	John W. Young; Michael Collins	First dual rendezvous; docked vehicle maneuvers; umbilical EVA; 43 orbits.
Gemini 11	9-12-66	71 hours 17 minutes	Charles Conrad Jr.; Richard F. Gordon Jr.	Rendezvous and docking; 44 orbits.
Gemini 12	11-11-66	94 hours 34 minutes	James A. Lovell Jr.; Edwin E. Aldrin Jr.	Final Gemini mission; 5 hours of EVA; 59 orbits.

TO THE MOON AND EXPERIMENTS IN SPACE

Craft	Date	Duration	Crew	Remarks
Soyuz 1	4-23-67	1 day 2 hours 48 minutes	Vladimir M. Komarov	Komarov is killed when parachute fails, first fatality of space program; 18 orbits.
Apollo 7	10-11-68	10 days 20 hours 8 minutes	Walter M. Schirra; Donn F. Eisele; R. Walter Cunningham	Eight service propulsion firings; 7 live TV sessions with crew; performs rendezvous with S-IVB stage.
Soyuz 3	10-26-68	3 days 22 hours 51 minutes	Georgi T. Beregovoi	Approaches the unpiloted Soyuz 2 to a distance of 650 ft (200 m); 64 orbits.
Apollo 8	12-21-68	6 days 3 hours	Frank Borman; James A. Lovell Jr.; William A. Anders	First Saturn V propelled flight; first lunar orbital mission; returns good lunar photography; 10 orbits.
Soyuz 4	1-14-69	2 days 23 hours 14 minutes	Vladimir A. Shatalov	Docks with Soyuz 5 in first linkup of two space vehicles that both carry people; 48 orbits.
Soyuz 5	1-15-69	3 days 46 minutes	Boris V. Volynov; Alexei S. Yeliseyev; Yevgeni V. Khrunov	Three cosmonauts perform EVA and are transferred to Soyuz 4 in rescue rehearsal.
Apollo 9	3-3-69	10 days 1 hour 1 minute	James McDivitt; David R. Scott; Russell Schweickart	First flight of all lunar hardware in Earth orbit, including lunar module (LM).

Craft	Date	Duration	Crew	Remarks
Apollo 10	5-18-69	8 days 3 minutes	Eugene A. Cernan; John W. Young; Thomas P. Stafford	Lunar mission development flight to evaluate LM performance in lunar environment; descent to within 50,000 ft (15,240 m) of Moon.
Apollo 11	7-16-69	6 days 21 hours 18 minutes	Neil A. Armstrong; Michael Collins; Edwin E. Aldrin Jr.	First lunar landing; limited inspection, photography, evaluation, and sampling of lunar soil; touchdown July 20.
Soyuz 6	10-11-69	5 days	Georgi S. Shonin; Valery N. Kubasov	First triple launch (with Soyuz 7 and 8).
Soyuz 7	10-12-69	5 days	Anatoly V. Filipchenko; Vladislav N. Volkov; Viktor V. Gorbatko	Conducts experiments in navigation and photography with Soyuz 7 and 8.
Soyuz 8	10-13-69	5 days	Vladimir A. Shatalov; Aleksey S. Yeliseyev	80 orbits.
Apollo 12	11-14-69	10 days 4 hours 36 minutes	Charles Conrad Jr.; Richard F. Gordon Jr.; Alan L. Bean	Second lunar landing; demonstrates point landing capability; samples more area; total EVA time: 15 hours 32 minutes.
Apollo 13	4-11-70	5 days 22 hours 55 minutes	James A. Lovell Jr.; Fred W. Haise Jr.; John L. Swigert Jr.	Third lunar landing attempt aborted due to loss of pressure in liquid oxygen in service module and fuel cell failure.
Soyuz 9	6-2-70	17 days 16 hours	Andrian G. Nikolayev; Vitaly I. Sevastyanov	Longest space flight to this date.
Apollo 14	1-31-71	9 days 42 minutes	Alan B. Shepard Jr.; Stuart A. Roosa; Edgar D. Mitchell	Third lunar landing; returns 98 lbs (44 kg) of material.
Soyuz 10	4-23-71	2 days	Vladimir A. Shatalov; Aleksey S. Yeliseyev; Nikolai N. Rukavishnikov	Docks with Salyut 1, the first space station.
Soyuz 11	6-6-71	24 days	Georgi Dobrovolsky; Viktor I. Patsayev; Vladislav N. Volkov	All three cosmonauts killed during reentry.

Craft	Date	Duration	Crew	Remarks
Apollo 15	7-26-71	12 days 7 hours 12 minutes	David R. Scott; Alfred M. Worden; James B. Irwin	Fourth lunar landing; first to carry Lunar Roving Vehicle (LRV); total EVA time: 18 hours 46 minutes; returns 173 lb (79 kg) of material.
Apollo 16	4-16-72	11 days 1 hour 51 minutes	John W. Young; Thomas Mattingly II; Charles M. Duke Jr.	Fifth lunar landing; second to carry LRV; total EVA time: 20 hours 14 minutes; returns 213 lb (97 kg) of material.
Apollo 17	12-7-72	12 days 13 hours 52 minutes	Eugene A. Cernan; Ronald E. Evans; Harrison Schmitt	Last manned lunar landing; third with LRV; total EVA time: 44 hours 8 minutes; returns 243 lb (110 kg) of material.

FIRST STATIONS IN SPACE

Craft	Date	Duration	Crew	Remarks
Skylab 2	5-25-73	28 days 49 minutes	Charles Conrad Jr.; Joseph P. Kerwin; Paul J. Weitz	First *Skylab* launch; establishes Skylab Orbital Assembly in Earth orbit; conducts medical and other experiments.
Skylab 3	7-29-73	59 days 11 hours	Alan L. Bean; Owen K. Garriott; Jack R. Lousma	Second *Skylab*; crew performs systems and operational tests, experiments, and thermal shield deployment.
Soyuz 12	9-27-73	2 days	Vasily G. Lazarev; Oleg G. Makarov	First Soviet spaceflight to carry humans since *Soyuz 11* tragedy.
Skylab 4	11-16-73	84 days 1 hour	Gerald P. Carr; Edward G. Gibson; William R. Pogue	Third *Skylab*; crew performs unmanned Saturn workshop operations; obtains medical data for extending spaceflights.
Soyuz 13	12-18-73	8 days	Pyotr I. Klimuk; Valentin Lebedev	Performs astrophysical and biological experiments.

Craft	Date	Duration	Crew	Remarks
Soyuz 14	7-3-74	16 days	Pavel R. Popovitch; Yuri P. Artyukhin	Crew occupies Salyut 3 space station; studies Earth resources.
Soyuz 15	8-26-74	2 days	Gennady Sarafanov; Lev Demin	Makes unsuccessful attempt to dock with Salyut 3.
Soyuz 16	12-2-74	6 days	Anatoly V. Filipchenko; Nikolai N. Rukavishnikov	Taken to check modifications in Salyut system.
Soyuz 17	1-10-75	30 days	Alexei A. Gubarev; Georgi M. Grechko	Docks withSalyut 4; sets Soviet endurance record to date.
Soyuz 18A	4-5-75	22 minutes	Vasily G. Lazarev; Oleg G. Makarov	Separation from booster fails and craft does not reach orbit, but crew lands successfully in western Siberia.
Soyuz 18B	5-24-75	63 days	Pyotr I. Klimuk; Vitaly I. Sevastyanov	Docks with Salyut 4.
ASTP	7-15-75	9 days 1 hour	Thomas P. Stafford; Vance D. Brand; Donald K. Slayton	Apollo-Soyuz Test Project, cooperative U.S.-U.S.S.R. mission.
Soyuz 19	7-15-75	5 days 23 hours	Alexei A. Leonov; Valery N. Kubasov	Docks with ASTP, the U.S. Apollo capsule.

THE SOVIETS STUDY HUMAN BIOLOGY IN SPACE

Craft	Date	Duration	Crew	Remarks
Soyuz 20	11-17-75	90 days	No crew	Biological mission; docks with Salyut 4.
Soyuz 21	7-6-76	49 days	Boris V. Volynov; Vitaly Zholobov	Docks with Salyut 5 and performs Earth resource work.
Soyuz 22	9-15-76	8 days	Valery F. Bykovsky; Vladimir Aksenov	Takes Earth resource photographs.
Soyuz 23	10-14-76	2 days	Vyacheslav Zudov; Valery Rozhdestvensky	Unsuccessfully attempts to dock with Salyut 5.
Soyuz 24	2-7-77	18 days	Viktor V. Gorbatko; Yuri N. Glazkov	Docks with Salyut 5 for 18 days of experiments.

Craft	Date	Duration	Crew	Remarks
Soyuz 25	10-9-77	2 days	Vladimir Kovalyonok; Valery Ryumin	Unsuccessfully attempts to dock with *Salyut 6*.
Soyuz 26	12-10-77	96 days	Yuri V. Romanenko; Georgi M. Grechko	Docks with *Salyut 6*; crew sets endurance record for this time.
Soyuz 27	1-10-78	6 days	Vladimir Dzhanibekov; Oleg G. Makarov	Carries second crew to dock with *Salyut 6* space station.
Soyuz 28	3-2-78	8 days	Vladimir Remek; Alexei A. Gubarev	Carries third crew to board *Salyut 6*; first non-Russian, non-American in space (the Czech Remek).
Soyuz 29	6-15-78	140 days	Vladimir Kovalyonok; Aleksander S. Ivanchenkov	Docks with *Salyut 6*; crew sets new space endurance record for this time.
Soyuz 30	6-27-78	8 days	Pyotr I. Klimuk; Miroslaw Hermaszewski	Carries second international crew to *Salyut 6*; first Polish cosmonaut (Hermaszewski).
Soyuz 31	8-25-78	8 days	Valery F. Bykovsky; Sigmund Jahn	Carries third international crew to *Salyut 6*; first East German in space (Jahn).
Soyuz 32	2-25-79	175 days	Vladimir Lyakhov; Valery Ryumin	Carries crew to *Salyut 6*; new endurance record set.
Soyuz 33	4-10-79	2 days	Nikolai N. Rukavishnikov; Georgi Ivanov	Engine failure prior to docking forces early termination; first Bulgarian in space (Ivanov).
Soyuz 34	6-6-79	74 days	No crew	Launched with no crew; returns with crew from *Salyut 6*.
Soyuz 35	4-9-80	185 days	Valery Ryumin; Leonid Popov	Carries two crew members to *Salyut 6*.
Soyuz 36	5-26-80	8 days	Valery N. Kubasov; Bertalan Farkas	Carries two crew members to *Salyut 6*; crew returns in *Soyuz 35*; first Hungarian (Farkas).
Soyuz T-2	6-5-80	4 days	Yuri Malyshev; Vladimir Aksenov	Test of modified *Soyuz* craft; docks with *Salyut 6*.

Craft	Date	Duration	Crew	Remarks
Soyuz 37	7-23-80	8 days	Viktor V. Gorbatko; Pham Tuan	Exchanges cosmonauts in Salyut 6; returns Soyuz 35 crew after 185 days in orbit.
Soyuz 38	9-18-80	8 days	Yuri V. Romanenko; Arnaldo Tamayo-Mendez	Ferry to Salyut 6; First Cuban in space (Tamayo-Mendez).
Soyuz T-3	11-27-80	13 days	Leonid Kizim; Oleg G. Makarov; Gennadi M. Strekalov	Ferry to Salyut 6; first three-person crew since Soyuz 11.
Soyuz T-4	3-12-81	75 days	Vladimir Kovalyonok; Viktor Savinykh	Mission to Salyut 6.
Soyuz 39	3-22-81	8 days	Vladimir Dzhanibekov; Jugderdemuduyn Gurragcha	Docks with Salyut 6; first Mongolian (Gurragcha).

THE SPACE SHUTTLE: THE U.S. REENTERS SPACE

Craft	Date	Duration	Crew	Remarks
Columbia	4-12-81	2 days 6 hours	John W. Young; Robert L. Crippen	First flight of reusable space shuttle proves concept; first landing of U.S. spacecraft on land.
Soyuz 40	5-14-81	8 days	Leonid I. Popov; Dumitru Prunariu	First Rumanian (Prunariu) in space.
Columbia	11-12-81	2 days 6 hours	Joe H. Engle; Richard H. Truly	First reuse of space shuttle; ends early due to loss of fuel cell.
Columbia	3-22-82	8 days	Jack R. Lousma; C. Gordon Fullerton	Third shuttle flight; payload includes space science experiments.
Soyuz T-5	5-13-82	211 days	Anatoly Berezovoy; Valentin Lebedev	First flight to Salyut 7; space station equipped to measure body functions.
Soyuz T-6	6-24-82	8 days	Vladimir Dzhanibekov; Jean-Loup Chrétien; Aleksandr Ivanchenkov	Mission to Salyut 7; Soviet-French team.
Columbia	6-27-82	7 days 1 hour	Thomas K. Mattingly II; Henry Hartsfield Jr.	Fourth shuttle mission; first landing on hard surface.

Craft	Date	Duration	Crew	Remarks
Soyuz T-7	8-16-82	8 days	Leonid I. Popov; Svetlana Savitskaya; Alexander Serebrov	Mission to *Salyut 7*; second Soviet woman in space (Savitskaya).
Columbia	11-11-82	5 days 2 hours	Vance D. Brand; Robert F. Overmyer; Joseph P. Allen; William B. Lenoir	First operational mission; first four-man crew; first deployment of satellites from shuttle.
Challenger	4-4-83	5 days	Paul J. Weitz; Karol J. Bobko; Donald H. Peterson; F. Story Musgrave	Second shuttle joins fleet; deploys TDRS tracking satellite; first shuttle EVA.
Soyuz T-8	4-20-83	2 days	Vladimir G. Titov; Gennadi M. Strekalov; Aleksandr A. Serebrov	Cosmonauts fail in planned rendezvous with *Salyut 7*.
Challenger	6-18-83	6 days 2 hours	Robert L. Crippen; Frederick H. Hauck; John M. Fabian; Sally K. Ride; Norman E. Thagard	First five-person crew; first American woman in space (Ride); first use of Remote Manipulator Structure ("Arm") to deploy and retrieve satellite.
Soyuz T-9	6-27-83	150 days	Vladimir Lyakhov; Aleksandr P. Alexandrov	Crew spends 149 days in *Salyut 7* after *Soyuz T-10* fails in relief mission.
Challenger	8-30-83	6 days	Richard H. Truly; Daniel Brandenstein; William Thornton; Guion S. Bluford Jr.; Dale Gardner	First night launch; first African-American (Bluford) in space; launches weather communications satellite for India.
Columbia	11-28-83	10 days	John Young; Brewster Shaw Jr.; Robert Parker; Owen Garriott; Byron Lichtenberg; Ulf Merbold	Carries Spacelab; six-man crew performs numerous experiments in astronomy and medicine.
Challenger	2-3-84	8 days	Vance Brand; Bruce McCandless II; Robert Stewart; Ronald McNair; Robert Gibson	Jet-propelled backpacks carry two astronauts on first untethered space walks; two satellites (Western Union and Indonesian) lost; first landing at Kennedy Space Center.

Craft	Date	Duration	Crew	Remarks
Soyuz T-10B	2-8-84	237 days	Leonid Kizim; Vladimir Solovyov; Oleg Atkov	Mission to *Salyut* 7 to repair propulsion system; sets new duration in space record for crew.
Soyuz T-11	4-2-84	8 days	Yuri Malyshev; Gennadi M. Strekalov; Rakesh Sharma	Docks with *Salyut* 7; first Indian cosmonaut (Sharma).
Challenger	4-7-84	8 days	Robert L. Crippen; Francis R. Scobee; Terry Hart; George Nelson; James van Hoften	Deploys Long Duration Exposure Facility for experiments in space durability; snares and repairs attitude control system of *Solar Max* satellite.
Soyuz T-12	7-18-84	12 days	Svetlana Savitskaya; Vladimir Dzhanibekov; Igor Volk	Savitskaya becomes the first woman to walk in space.
Discovery	8-30-84	6 days	Henry W. Hartsfield Jr; Michael L. Coates; Steven A. Hawley; Judith Resnik; Richard M. Mullane; Charles D. Walker	Third shuttle in fleet deploys three satellites and tests a solar sail.
Challenger	10-5-84	7 days	Robert L. Crippen; Jon A. McBride; Kathryn D. Sullivan; Sally K. Ride; Marc Gameau; David C. Leestma; Paul D. Scully-Power	Carries first Canadian astronaut (Gameau); deploys *Earth Radiation Budget Satellite* and monitors land formations, ocean currents, and wind patterns; uses Sir-B radar system to see beneath surface of sand.
Discovery	11-8-84	7 days	Frederick H. Hauck; David M. Walker; Anna L. Fisher; Joseph P. Allen; Dale A. Gardner	Salvages two inoperative satellites and returns them to Earth for repair.
Discovery	1-24-85	2 days	Thomas K. Mattingly II; Loren J. Schriver; James F. Buchli; Ellison S. Onizuka; Gary E. Payton	"Secret" military mission.
Discovery	4-12-85	6 days	Karol J. Bobko; Donald E. Williams; Jake Garn; Charles D. Walker; Jeffrey A. Hoffman; S. David Griggs; Margaret Rhea Seddon	First U.S. senator in space (Garn).

Craft	Date	Duration	Crew	Remarks
Challenger	4-29-85	7 days	Robert F. Overmyer; Frederick D. Gregory; Don L. Lind; Taylor G. Wang; Lodewijk van den Berg; Norman E. Thagard; William Thornton	Carries European Spacelab module to conduct 15 experiments in space.
Soyuz T-13	6-6-85	112 days	Vladimir Dzhanibekov; Viktor Savinykh	Successfully repairs damage to *Salyut* 7, which suffered power failure.
Discovery	6-17-85	6 days	John O. Creighton; Shannon W. Lucid; Steven R. Nagel; Daniel C. Brandenstein; John M. Fabian; Salman al-Saud; Patrick Baudry	First Arab in space (Prince Sultan Salman al-Saud); successfully launches 4 satellites.
Challenger	7-29-85	7 days	Roy D. Bridges Jr; Anthony W. England; Karl G. Henize; F. Story Musgrave; C. Gordon Fullerton; Loren W. Acton; John-David F. Bartoe	Carries Spacelab 2, a group of scientific experiments.
Discovery	8-27-85	7 days	John M. Lounge; James D. van Hoften; William F. Fisher; Joe H. Engle; Richard O. Covey	Repairs satellite *Syncom 3*.
Soyuz T-14	9-17-85	65 days	Vladimir Vasyutin; Alexandr A. Volkov; Georgi M. Grechko	Takes supplies to *Salyut* 7; terminates early to return Vasyutin to Earth because he is ill.
Atlantis	10-4-85	2 days	Karol J. Bobko; Ronald J. Grabe; David C. Hilmers; William A. Pailes; Robert C. Stewart	Fourth shuttle brings fleet up to planned size.
Challenger	10-30-85	7 days	Henry W. Hartsfield Jr.; Steven R. Nagel; Bonnie J. Dunbar; Guion S. Bluford Jr; Ernst Messerschmid; Reinhard Furrer; Wubbo J. Ockels	Carries Spacelab 1-D scientific experiments conducted by Germans.
Atlantis	11-26-85	7 days	Brewster H. Shaw Jr.; Bryan D. O'Conner; Charles Walker; Rudolfo Neri Vela; Jerry L. Ross; Sherwood C. Spring; Mary L. Cleave	Works on techniques for assembling structures in space; first Mexican astronaut (Vela).

Craft	Date	Duration	Crew	Remarks
Columbia	1-12-86	5 days	Robert L. Gibson; Charles F. Bolden Jr.; George D. Nelson; Franklin R. Chang-Diaz; Steven A. Hawley; Robert J. Cenker	First U.S. congressman in space (Nelson).
Challenger	1-28-86	73 sec	Francis R. Scobee; Michael J. Smith; Ronald E. McNair; Ellison S. Onizuka; Judith A. Resnik; Gregory B. Jarvis; Christa McAuliffe	When O-rings in the solid-fuel boosters wear through, the entire fuel supply explodes, killing all six regular astronauts and elementary-school teacher Christa McAuliffe.

AFTER THE *CHALLENGER* DISASTER; THE FIRST REAL SPACE STATION

Craft	Date	Duration	Crew	Remarks
Mir	2-20-86	Still in space	Variable	Soviet space station is launched without a crew.
Soyuz T-15	5-5-86	125 days	Vladimir Solovyov; Leonid Kizim	First cosmonauts board *Mir* space station.
Soyuz TM-2	2-7-87	326 days	Yuri Romanenko; Aleksandr Laveykin	Both cosmonauts begin long tours in space.
Soyuz TM-3	7-23-87	8 days	Aleksandr P. Alexandrov; Aleksandr Viktorenko; Muhammad Faris	First Syrian in space (Faris).
Soyuz TM-4	12-20-87	366 days	Vladimir Titov; Musa Manarov; Anatoly Levchenko	Cosmonauts set new record of year in space about the *Mir* space station: 366 days.
Soyuz TM-5	6-7-88	10 days	Aleksandr Alexandrov; Viktor P. Savinykh; Anatoly Y. Solovyev	Alexandrov is not the same Alexandrov as on *Soyuz TM-3* flight.
Soyuz TM-6	8-29-88	9 days	Vladimir Lyakhov; Valery Polyakov; Abdul Mohmand	First Afghan in space (Mohmand); on 9-6-88 Lyakhov and Mohmand are stranded 24 hours as they attempt to return in *Soyuz TM-5*; they land safely on 9-7-88.

Craft	*Date*	*Duration*	*Crew*	*Remarks*
Discovery	9-29-88	4 days	John M. Lounge; David Hilmers; Frederick H. Hauck; George D. Nelson; Richard O. Covey	Redesigned shuttle makes first flight since the *Challenger* disaster.
Soyuz TM-7	11-26-88	152 days	Alexandr A. Volkov; Sergei Krikalev; Jean-Loup Chrétien	*Mir* is temporarily abandoned for first time when cosmonauts return to Earth.
Atlantis	12-2-88	4 days	Robert L. Gibson; Jerry L. Ross; William M. Shepherd; Guy S. Gardner; Richard M. Mullane	"Secret" military mission deploys a radar spy satellite.
Discovery	3-13-89	4 days	Michael L. Coats; John E. Blaha; James F. Buchli; James P. Bagian; Robert C. Springer	Deploys NASA's third relay satellite and tests thermal control system for proposed U.S. space station.
Atlantis	5-4-89	4 days	David M. Walker; Ronald J. Grabe; Mary L. Cleave; Norman E. Thagard; Mark C. Lee	Launches space probe *Magellan* on its way to map Venus with radar.
Columbia	8-9-89	5 days	Brewster H. Shaw Jr.; David C. Leestma; James C. Adamson; Mark N. Brown	"Secret" military mission launches spy satellite.
Atlantis	10-18-89	5 days	Donald E. Williams; Michael J. McCulley; Shannon W. Lucid; Franklin R. Chang-Diaz; Ellen S. Baker	Launches *Galileo* space probe, which travels to Venus, returns twice to vicinity of Earth and then will take up long-term orbit about Jupiter.
Discovery	11-22-89	5 days	Frederick D. Gregory; John E. Blaha; F. Story Musgrave; Kathryn C. Thornton; Manley Lanier Carter Jr.	"Secret" military mission.
Columbia	1-9-90	10 days 21 hours	Daniel C. Brandenstein; Bonnie J. Dunbar; Marsha S. Ivins; G. David Low; James D. Wetherbee	Launches communications satellite *Syncom IV* and retrieves the Long Duration Exposure Facility, which has been in orbit since 4-7-84; longest shuttle flight to date; after return, problems found with fibers in working parts of shuttle.

Craft	Date	Duration	Crew	Remarks
Soyuz TM-9	2-11-90	6 days	Anatoly Solovyev; Aleksandr Balandin	To relieve Alexander S. Vitorenko and Alexander A. Serebrov; on July 18, Solovyev and Balandin are briefly trapped outside *Mir* by a faulty hatch.
Atlantis	2-28-90	6 days	John O. Creighton; John H. Casper; David C. Hilmers; Richard M. Mullane; Pierre J. Thuot	"Secret" military mission launches spy satellite that fails and soon burns up in atmosphere.
Discovery	4-24-90	5 days	Loren J. Shriver; Charles F. Bolden Jr.; Bruce McCandless II; Steven H. Hawley; Kathryn D. Sullivan	Launches *Hubble Space Telescope*.
Soyuz TM-10	5-8-90	6 days	Gennadi Manakov; Gennadi Strekalov	Mission to replace Solovyev and Balandin.
Discovery	10-6-90	4 days	Richard N. Richards; Robert D. Cabana; Bruce E. Melnick; William M. Shepherd; Thomas D. Akers	Launches *Ulysses* space probe into orbit about Sun.
Atlantis	11-15-90	6 days	Richard O. Covey; Frank L. Culbertson Jr.; Charles D. Gemar; Carl J. Meade; Robert C. Springer	"Secret" military mission proclaimed as the last that will be kept from public.
Columbia	12-2-90	9 days	Vance D. Brand; Guy S. Gardner; Jeffrey A. Hoffman; John M. (Mike) Lounge; Robert A. R. Parker; Samuel T. Durrance; Ronald A. Parise	After long delays caused by leaking hydrogen, shuttle carries a set of three ultraviolet telescopes and one x-ray telescope, an instrument called ASTRO.
Soyuz TM-11	12-2-90	9 days	Viktor Afanasyev; Musa Manarov; Toyohiro Akiyama	Japanese journalist Akiyama visits *Mir* on an outing sponsored by Japanese corporations and returns; other two replace Gennadi Manakov and Gennadi Strekalov in *Mir*.

Craft	Date	Duration	Crew	Remarks
Atlantis	4-5-91	6 days	Steven R. Nagel; Kenneth D. Cameron; Linda M. Godwin; Jerry L. Ross; Jay Apt	Launches 17-ton *Gamma Ray Observatory*; an unscheduled spacewalk is required to get the satellite's antenna to open properly.
Discovery	4-28-91	5 days	Michael L. Coats; L. Blaine Hammond Jr.; Gregory J. Harbaugh; Charles Lacy Veach; Guion S. Bluford Jr.; Richard J. Hieb; Donald R. McMonagle	First military mission under new nonsecret policy; tests detection devices developed for space use.
Soyuz TM-12	5-18-91	8 days	Sergei Krikalev; Anatoly Artebarsky; Helen Sharman	Political problems in the Soviet Union result in Krikalev's spending an unexpected 313 days in space; Sharman is first Briton in space; returns with Manarov and Afansev.
Columbia	6-5-91	10 days	Bryan D. O'Connor; Sidney M. Gutierrez; James P. Bagian; Margaret Rhea Seddon; Francis A. Gaffney; Millie Hughes-Fulford; Tamara E. Jernigan	Performs experiments to test human and animal adaptation to space.
Atlantis	8-3-91	8 days	John E. Blaha; Michael A. Baker; Shannon W. Lucid; G. David Low; James C. Adamson	Performs 22 experiments and launches communications satellite.
Discovery	9-12-91	5 days	John O. Creighton; Kenneth S. Reightler Jr.; Mark N. Brown; James F. Buchli; Charles D. Gemar	Launches *Upper Atmosphere Research* satellite.

Craft	Date	Duration	Crew	Remarks
Atlantis	11-24-91	7 days	Frederick D. Gregory; Terence T. Henricks; James S. Voss; Mario Runco Jr.; F. Story Musgrave; Thomas J. Hennen	Delayed five days, this is the first completely nonsecret military flight; studies how well military installations can be seen from space, which turns out to be very well indeed; deploys satellite as well; lands early as a result of failure of a navigation unit (second time ever that early landing is forced by equipment failure).
Discovery	1-22-92	8 days	Ronald J. Grabe; Stephen S. Oswald; Norman E. Thagard; William F. Readdy; David Hilmers; Roberta L. Bondar; Ulf D. Merbold	Performs experiments as the First International Microgravity Laboratory.
Soyuz TM-14	3-17-92	8 days	Klaus-Dietrich Flade; Aleksandr Viktorenko; Aleksandr Kaleri	On March 25, Flade, a German, returns with Sergei Krikalev and Aleksandr Volkov, who were already aboard *Mir*.
Atlantis	3-24-92	9 days	Charles F. Bolden Jr.; Brian Duffy; Kathryn D. Sullivan; David C. Leestma; C. Michael Foale; Byron K. Lichtenberg; Dirk D. Frimout	Inaugural flight of Mission to Planet Earth studies Earth's atmosphere and auroras despite problems with a blown fuse.
Endeavor	5-7-92	9 days	Daniel C. Brandenstein; Kevin P. Chilton; Thomas D. Akers; Richard J. Hieb; Bruce E. Melnick; Kathryn C. Thornton; Pierre J. Thout	The first flight of the replacement for *Challenger* features capture and relaunch of an erring communications satellite; after two efforts at rescue fail, the first ever three-person EVA and manual capture succeeds.

Craft	Date	Duration	Crew	Remarks
Columbia	6-25-92	14 days	Richard N. Richards; Kenneth D. Bowersox; Bonnie J. Dunbar; Lawrence J. DeLucas; Ellen S. Baker; Eugene H. Trinh; Carl J. Meade	Sets record for duration of shuttle mission; carries U.S. Microgravity Laboratory for experiments in low gravity.
Soyuz TM-15	7-27-92	15 days	Anatoly Y. Solovyev; Sergei Avdeyev; Michael Tognini	The third trip by a French astronaut, Tognini this time, to the Russian (formerly Soviet) space station, costing the French about $12 million.
Atlantis	7-31-92	8 days	Loren J. Schriver; Andrew M. Allen; Marsha S. Ivins; Jeffrey A. Hoffman; Franklin R. Chang-Diaz; Claude Nicollier; Franco Mallerba	Main experiment, unrolling of a tethered satellite, fails because of a jammed bolt, but *European Retrievable Carrier* (*Eureca*) satellite is launched successfully. Crew includes first Swiss astronaut (Nicollier) and first Italian (Mallerba).
Endeavor	9-12-92	8 days	Robert L. Gibson; Curtis L. Brown Jr; Jay Apt; N. Jan Davis; Mae Carol Jemison; Mark C. Lee; Mamoru Mohri	First professional Japanese astronaut in space (Mamoru Mohri); first married couple in space (Mark Lee and Jan Davis); and first black female astronaut (Mae Jemison). Japanese-sponsored mission to study biology in space.
Columbia	10-22-92	10 days	James D. Wetherbee; Michael A. Baker; Glenwood MacLean; Tamara E. Jernigan; William M. Shepherd; Charles Lacy Veach	Launches the Italian-made satellite *LAGEOS-2*, designed to provide information about Earth's gravitational field.

Craft	Date	Duration	Crew	Remarks
Discovery	12-2-92	7 days	David M. Walker; Robert D. Cabana; Guion S. Bluford Jr.; James S. Voss; Michael R. Clifford	Last scheduled military flight for the space shuttle; future military flights, except for emergencies, to use disposable rockets. Launches secret military satellite known as *DOD-1*.
Endeavor	1-13-93	6 days	John H. Casper; Gregory Harbaugh; Donald R. McMonagle; Susan J. Helms; Mario Runco Jr.	Astronauts observe x rays and launch a *TDRS (Tracking and Data Relay Satellite)*.
Soyuz TM-16	1-24-93	8 days	Alexander Polishchuk; Gennadi Manakov	Docks with *Mir* on Kristall module, using U.S.-compatible port first developed for the joint Soviet-U.S. mission in 1975; Anatoly Solovyov and Sergei Avdeyev return after six months aboard space station.
Discovery	4-8-93	9 days	Kenneth D. Cameron; Stephen S. Oswald; Kenneth D. Cockrell; C. Michael Foale; Ellen Ochoa	Launches and retrieves the *Spartan* sun probe, used to study the solar corona, and observes the ozone layer of Earth's atmosphere.
Columbia	4-26-93	7 days	Steven R. Nagel; Terence T. Henricks; Jerry L. Ross; Charles Precourt; Bernard A. Harris Jr.; Ulrich Walter; Hans Schlegel	The European Spacelab, with two German scientists (Walter and Schlegel) aboard, conducts experiments on weightlessness, mostly paid for by Germany.

Craft	Date	Duration	Crew	Remarks
Endeavor	6-21-93	8 days	Ronald J. Grabe; Brian Duffy; G. David Low; Peter J. K. Wisoff; Nancy Jane Sherlock; Janice E. Voss	The crew recovers Europe's *Eureca* satellite for possible reuse and conducts experiments using a pressurized laboratory called Spacehab, built by McDonnell-Douglas Aerospace for commercial use, although mostly used by NASA for its initial flight.
Soyuz TM-17	7-1-93	28 days	Vasily Tsiblyev; Aleksandr Serebrov; Jean-Pierre Haignere	Haignere is a French astronaut who returns with Aleksandr Polishchuk and Gennadi Manakov on July 29.
Discovery	9-12-93	10 days	Frank L. Culbertson Jr.; William F. Readdy; James H. Newman; Daniel W. Bursch; Carl E. Walz	The crew tries out equipment and procedures to be used in repairing *Hubble Space Telescope*; first night landing at Kennedy.
Columbia	10-18-93	14 days	John E. Blaha; Richard A. Searfoss; William S. McArthur Jr.; Shannon W. Lucid; Martin J. Fettman; Margaret Rhea Seddon; David Wolf	Conducts biological experiments on humans and rats, including first dissection of animal in space.
Endeavor	12-2-93	11 days	Richard O. Covey; Kenneth D. Bowersox; Claude Nicollier; F. Story Musgrave; Thomas D. Akers; Kathryn C. Thornton; Jeffrey A. Hoffman	Installs corrective lenses and other replacement parts on defective *Hubble Space Telescope*, solving its focusing and stability problems.
Soyuz TM-18	1-8-94	21 days	Viktor Afanasyev; Yuri Usachyov; Valery Polyakov	Trip for Polyakov planned to last for 14 months (and does last that long plus a couple of weeks more); previous crew aboard *Mir* returns to Earth.
Discovery	2-3-94	8 days	Charles F. Bolden; Sergei M. Krikalev; Kenneth S. Reightler Jr.; N. Jan Davis; Franklin R. Chang-Diaz; Ronald M. Sega	Krikalev becomes the first Russian to fly on a U.S. space shuttle mission; he previously took part in flights to *Mir*.

Craft	Date	Duration	Crew	Remarks
Columbia	3-4-94	14 days	John H. Casper; Andrew M. Allen; Charles D. "Sam" Gemar; Marsha S. Ivins; Pierre J. Thuot	Low-key flight to develop techniques for use in space-station construction; also conducts biological and materials experiments.
Endeavour	4-9-94	11 days	Sidney M. Gutierrez; Kevin P. Chilton; Michael R. Clifford; Linda M. Godwin; Jay Apt; Thomas D. Jones	Uses radar to map details of Earth in 3-D and took photographs to check space radar against ground truth.
Soyuz TM-19	7-3-94	16 days	Yuri Malenchenko; Talgat Musabayev	Malenchenko from Russia and Musabayev of Kazakhstan travel to *Mir* to relieve Viktor Afanasyev and Yuri Usachev.
Columbia	7-8-94	14 days 17 hours 55 minutes	Robert D. Cabana; Leroy Chiao; James D. Halsell Jr.; Richard J. Hieb; Chiaki Naito-Mukai; Donald A. Thomas; Carl E. Walz	Sets new record for length of time a U.S. space shuttle is in orbit, orbiting Earth 236 times as crew studies biology in space; first Japanese woman in space (heart surgeon Mukai).
Discovery	9-9-94	11 days	Richard N. Richards; L. Blaine Hammond Jr.; Susan J. Helms; Mark C. Lee; J. M. Lineger; Carl J. Meade	Conducts laser experiments to measure pollution in atmosphere and releases and recaptures satellite that collects data on solar wind; Meade and Lee perform untethered spacewalk to test new jet pack designed for use in building space station.
Endeavour	9-30-94	11 days	Michael A. Baker; Daniel W. Bursch; Thomas D. Jones; Steven L. Smith; Peter "Jeff" J. K. Wisoff; Terrence W. Wukcytt	The Space Radar Laboratory 2 mission is a followup of a nearly identical mission flown in the spring; the purpose is to detect seasonal changes.

Craft	Date	Duration	Crew	Remarks
Soyuz TM-20	10-3-94	1 month	Aleksander Viktorenko; Yelena Kondakova; Ulf Merbold	Merbold, from the European Space Agency, studies space biology aboard *Mir* and returns to Earth on 11-4-94 along with Yuri Malenchenko and Talgat Musabayev, leaving Viktorenko and Kondakova on *Mir*.
Atlantis	11-3-94	11 days	Curtis L. Brown; Donald R. McMonagle; Jean-François Clervoy; Scott E. Parazynski; Joseph R. Tanner	Third shuttle flight since 1992 to carry the Atlas lab (Atmospheric Laboratory for Applications and Science); Clervoy represents the Agency on the mission.
Discovery	2-3-95	8 days	James D. Wetherbee; Eileen M. Collins; C. Michael Foale; Bernard A. Harris Jr.; Vladimir G. Titov; Janice Voss	Carrying Russian cosmonaut Titov, the shuttle travels to between 37-44 ft (11-13 m) of the space station *Mir*.
Endeavour	3-2-95	16 days 15 hours 8 minutes	Stephen S. Oswald; Samuel T. Durrance; Tamara E. Jernigan; William G. Gregory; John M. Grunsfeld; Wendy B. Lawrence; Ronald A. Parise	Mission carries the lab Astro 2, which centers on three ultraviolet telescopes and is used, among other things, to study volcanic eruptions on Jupiter's moon Io. Sets new duration record in space for a shuttle, orbiting Earth 263 times.

Craft	Date	Duration	Crew	Remarks
Soyuz TM-21	3-14-95	8 days	Norman E. Thagard; Vladimir N. Dezhurov; Gennady M. Strekalov	Thagard becomes the first U.S. astronaut to live aboard the Russian space station *Mir,* where he studies the body's reaction to weightlessness; Valery Polyakov, who returns on 3-22-94 with Yelena Kondakova and Aleksander Viktorenko, sets new space endurance record of 14.5 months in space on one trip.
Atlantis	6-27-95	10 days	Robert L. Gibson; Charles J. Precourt; Ellen S. Baker; Gregory J. Harbaugh; Bonnie Dunbar; Anatoly Y. Solovyev; Nikolai M. Budarin	Docks with *Mir* for five days, giving the space station a temporary crew of 10 and a mass of about 225 tons; Spacelab on *Atlantis* and Spektr module of *Mir* used for biomedical research; leaves Solovyev and Budarin as crew of *Mir,* returning with previous *Mir* crew of American Thagard and Russians Dezhurov and Strekalov.
Discovery	7-13-95	9 days	Terence "Tom" Henricks; Kevin Kregel; Nancy Currie; Donald Thomas; Mary Ellen Weber	Mission is delayed for five weeks by woodpeckers pecking holes in the fuel tank insulation; launches *Tracking and Data Relay Satellite* (*TDRS*) to replace one lost when *Challenger* blew up, and conducts medical and military experiments.

Biology

THE STATE OF BIOLOGICAL SCIENCE

The life sciences consist of the various branches of biology, medicine, physical anthropology, and, in a broad sense, paleontology. In this volume, the various parts of the life sciences are handled in somewhat different ways.

The **Biology** chapter is concerned primarily with genetics and development, as well as studies of organisms and their behavior. Most of molecular biology or biochemistry is too technical for the scope of this book. With the exception of some aspects of medical technology, medicine is not covered here, but is treated in a separate volume, *The Handbook of Current Health and Medicine* (available from the same publisher). Physical anthropology and related topics in paleoanthropology are addressed in the **Anthropology and Archaeology** chapter. Although the advent of DNA studies has shifted the study of fossilized ancient life into the realm of a life science, paleontology is included in its traditional niche within the **Earth Science** chapter.

From Molecules to Behavior

Throughout the twentieth century, the field of biology has been severely divided by the whole-organism biologists and those scientists who work with organisms at the level of the cell, chemical, or molecule. Although this dichotomy is not going to go away completely, the 1980s and early 1990s have seen a number of bridges built. For example, geneticists trace the effects of a specific gene (one part of a large molecule) on a protein (a whole molecule) which interacts with other molecules in chemical reactions, producing effects at the cellular level affecting the whole organism (*see* "Genetics and Beyond," p 336; "Genome Project UPDATE," p 366; and "The Threat of Organisms in New Genes UPDATE," p 372). Major investigations of this type concern efforts to reconstruct the entire development history of some organisms (*see* "Development Developments," p 342, and "*C. elegans* UPDATE," p 362). The driving force for studies at all levels of organization continues to be evolution by natural selection, which supplies the link between organisms and molecules from the other direction; that is, from organism to DNA as opposed to from DNA to development of an organism (*see* "Evolution UPDATE," p 412).

Almost without exception, the organisms studied by biologists before the twentieth century were plants and animals on a scale similar to that of humans. Although the microscopic world of protists, monerans (bacteria and certain algae), and cellular fungi (such as yeasts) was known earlier, these tiny creatures were too small to dissect with early microscopes, and it was thought unlikely that they would have any behaviors or traits worth studying. Toward the end of the nineteenth century, of course, the relationship between bacteria and disease was established, causing a great burst in the field of microbiology. As the twentieth century progressed, improved microscope technology, as well as the accumulating weight of experience with microorganisms, led to a new recognition of the importance and diversity of the microscopic living world (*see* "Bacteria: Giant, Deep, and Social," p 382; "The Kingdom of the Archaea," p 378; and "Fungi Surprises of the 1990s," p 388).

That microorganisms can display complex behavior patterns and perhaps even learn from experience is one of the big surprises of studying the small. Behavior and learning in animals has long been recognized and continues to be an important part of whole-organism biology (*see* "Finding Their Way," p 407 and "Thinking, Remembering, Feeling," p 354). More surprising, however, has been the discovery of behavior, communication, and other previously unsuspected traits in plants (*see* "Plant Behavior," p 401).

Attack of the Animal Lovers Continue

Ethical debates over animal rights have been a major problem in recent biological and medical research. Although nearly everyone agrees that a line must be drawn between permissible subjects for scientific experiments and those subjects who are too sentient for investigations that might cause physical pain or mental stress, there are widely divergent views of where that line should be drawn. At certain times in the past, some groups of people have accepted harmful or potentially harmful experimentation on people who were already dying for some other reason, on mentally retarded or demented people, on prisoners or slaves, on very young children, or on members of what were thought to be lower classes. Such experiments would be unacceptable today. Similar experiments with animals as subjects might or might not be acceptable, depending upon the kind of animals involved. Deadly procedures suitable for rats and mice might be rejected for apes, cats, or dogs. Some people would restrict all potentially harmful investigations using mammals, but allow experiments involving dismemberment or death of other vertebrates. Some might draw the line at vertebrates, but not be bothered in the least by what scientists do to crustaceans or worms. Recent research on plants might suggest to certain rights activists that living plants should be protected from biological experimentation along with all animals.

In the 1990s, serious confrontations have occurred between those who draw the lines for acceptable experimentation at different places in the nonhuman animal kingdom. Various groups have been organized to promote one set of criteria or another. One of the newest such groups is the European Biomedical Research Association, formed on November 9, 1994, by scientists who are in favor of safe and responsible animal research where necessary, and the use of non-animal alternatives where possible. These scientists feel that they need such an organization to counter publicity garnered by an older organization, the European Coalition to End Animal Experiments. The new group also hopes to promote better practices among scientists from Eastern Europe, who have long operated without the kinds of restraints and rules common in Western nations. In the United States, the group People for the

Ethical Treatment of Animals (PETA) opposes all research using animal subjects, while the more traditional Humane Society objects only to certain forms of such research. The actual effect of research on animals is poorly known. In the Netherlands, where data is systematically collected, researchers found that about 25% of laboratory animals suffer severe discomfort. As defined in this survey, "severe discomfort" includes prolonged deprivation of food and water, deliberate production of cancer in laboratory animals, and testing toxic substances by the LD-50 method (in which all the animal subjects in a population are fed increasing amounts of a toxin until half the subjects in the group have died—LD-50 means 50% get a lethal dose). Another 25% of animal subjects in the Netherlands were judged to experience moderate discomfort, while about 50% experienced no identifiable discomfort. In the United States, about 40% of all animal research involves experiments which cause the animal subjects pain and distress not relieved by drugs. (This percentage is based upon self-categorization by some laboratories.)

On February 25, 1993, a federal district judge told the U.S. Department of Agriculture that it had to rewrite the rules that had been in use since implementing the Improved Standards for Laboratory Animals Act of 1985. Judge Charles R. Richey ruled that the 1985 standards were not sufficient because they relied on committees at research institutions to determine specifics; because the rules for primate research did not comply with the law's intent of protecting them from psychological damage; because in some cases research was permitted using smaller cages than the rules as written allow; and because the rules failed to provide adequate exercise for dogs. Biological researchers who had already spent large sums changing their investigations to comply with the existing rules were appalled to think that even stricter regulations might be written and enforced.

Since 1987, biological researchers have also faced with alarm an event known as the World Laboratory Animal Liberation Week, or WLALW. In 1993, the U.S. Departments of Agriculture and Justice reported jointly to Congress that animal activists' attacks on laboratory facilities between 1977 and 1993 had cost about $137 million in various kinds of damage. During WLALW in 1993, the activists chose to protest by vandalizing the homes of scientists who use laboratory animals in their research, mainly by spraying slogans on houses and cars. One of these scientists has also has had fake bombs left at his doorstep.

One response to these protests shows up in statistics which reveal that the number of common pet mammals used in research fell about 50% between 1968 and 1992, according to a study released in 1994 by the Tufts University Center for Public Policy. The Tufts report was based upon U.S. Department of Agriculture figures for use of dogs, cats, primates, rabbits, hamsters, and guinea pigs. If Tufts' estimates are correct, the same time period may have seen an even greater drop (about 60%) in the use of rats and mice, from perhaps 50 million to as low as 20 million. Actual data on laboratory use of rats and mice is hard to come by, so estimates were based on data released by such groups as the U.S. Defense Department and the pharmaceutical company Hoffmann-La Roche.

Human Genome Diversity Project Problems

For at least the past two centuries, biologists have used the latest technical advances to study how human populations vary. Looking back, some of the methods used seem misguided, while others still seem useful. On the negative side, measurements of skull shapes and volumes were used to denigrate certain ethnic groups. Another technology, introduced in 1900 (the same year that Mendel's laws of genetics were rediscovered), was used with virtually no political impact or agenda, however. Statistical variations in blood type between populations became a useful tool for tracing ancient population shifts, such as the migration of Asians from Siberia to become "Native Americans."

When Luca Cavalli-Sforza of Stanford University in Palo Alto, California, proposed in 1991 that a library of genes from small and isolated populations be collected, he no doubt expected that this project would be more like the collection of blood-type information than like earlier population comparisons thought to have had a racist component. With Ken Kidd of Yale University in New Haven, Connecticut, and Mary-Claire King, Charles Cantor, and Allan Wilson of the University of California, Berkeley, Cavalli-Sforza prepared a proposal for a Human Genome Diversity Project (HGDP) that would collect DNA and cell lines from hair, saliva, and blood of a select list of indigenous and mostly endangered populations. The idea was to get the data before these groups either became extinct or merged into the larger human population. Cavalli-Sforza was already well known for his work in relating genetic profiles to languages. Planning was started, along with efforts to obtain the $30 million needed to fund the project. Preliminary lists of populations to be studied were compiled.

Although these early stages were relatively private, word of the planning conference reached an organization called the Rural Advancement Foundation International, a group based in Canada and dedicated to protecting plant species from Western exploitation (there is, of course, a history of such exploitation). In the 1970s, there had been a plant program similar in some ways to the Human Genome Diversity Project that collected samples of 125,000 plant varieties, most of them grown by the same groups of indigenous peoples targeted by HGDP. Although the plant varieties were in the public domain, genes from plants grown in developing countries were used in commercial, patented hybrids that made a lot of money for the companies that owned the patents. None of the money went back to the indigenous people whose plants contributed the genes that made the hybrids valuable. The Rural Advancement Foundation International is devoted to preventing this kind of sequence of events, which they consider exploitation. In the foundation's view, the HGDP raised the possibility of exploitation of genes from indigenous peoples in developing countries. So the Foundation obtained the preliminary lists of populations whose genes might be collected and proceeded to notify their contacts in those populations about what was happening.

The first reactions were fierce opposition, especially by representatives of peoples from Australia and India. The Australians were already fighting a tough but largely successful battle with archaeologists and anthropologists who had been excavating sites and collecting artifacts and bones. Contemporary Australian aborigines demanded the return of excavated materials for reburial. Representatives of the government of India suggested that no DNA or cell samples collected among the many small populations of that country would receive export licenses. Self-designated representatives of other listed groups also protested, some claiming that it was against their culture to take blood, and that the scientists involved were no better than vampires. The project was denounced as racist, and as "genetic colonialism." A conference of the World Council on Indigenous Peoples denounced the whole plan.

None of the HGDP researchers had expected this reaction, nor any opposition at all. Their primary concerns had been how to obtain funding for the project, and how to best preserve the DNA they expected to gather.

After some regrouping, the HGDP or something very much like it is getting back in business. The new organization will allow much greater local control over how samples are gathered and what becomes of them. Some groups have been found who are eager to be a part of the project. Many small populations in Europe, such as the Basques, are excited about the possibility that the data will help to explain their own origins. Some Native American groups also want to have their genetic makeup on record. A Howard University project which proposes to trace the African genes in the African American population has a lot of support from the subject group, especially since possible medical benefits are claimed.

Upward and Onward with Biology

Humans first explored the biology near the surface of the land in Africa and later around the world. Gradually, the region being studied expanded in other ways as well—to the tops of the highest mountains and recently to the depths of the deepest seas. In the 1990s, considerable close attention was paid to a region previously observed largely from afar: the forest canopy. In a forest, the topmost branches of tall trees touch or nearly touch each other, although trunks are much more widely spaced. The nearly continuous covering at the top of the forest, which can be hundreds of feet above the ground, is known as the canopy. It is a productive environment like none other on Earth. Some animals that live in the canopy are never found at ground level at all, obtaining all the moisture they need from leaves and fruit.

The first major canopy studies were in the Willamette National Forest in Oregon in the 1970s, where scientists used techniques developed for climbing vertical rock faces to reach the canopy. These techniques were then transferred to tropical rain forests, where canopy life is even more isolated, forming its own ecosystem. In rain forests, abundant moisture allows both animal and plant species unique to the canopy to thrive. These include epiphytes—plants that live on other plants and have no need for soil. They get their water and energy from rain, sunlight, air, and perhaps a little dust now and then.

Climbing tall trees is a difficult feat, requiring special skills and athletic prowess, and does not allow for very good access. Some scientists built observation posts in trees and used various means to get to them. One of the first was in Massachusetts, where Margaret Lowman built a 70 ft (21 m) high walkway in the canopy of a typical New England hardwood forest. A similar walkway 200 ft (60 m) high was installed in 1993 in the Willamette Forest.

Perhaps the best solution utilized a giant crane. The first one was erected in a wildlife park in Panama in 1990, allowing observers to reach the canopy easily, providing a base from which long-term studies could be made, and even giving a limited amount of mobility from site to site in the canopy. A construction crane on Barro Colorado Island in Panama lifts workers in a steel cage as high as 125 ft (38 m) from the forest floor.

More recently, canopy studies have included temperate forests. In 1994, a 300 ft (90 m) crane was installed in the old-growth forest in the state of Washington. The crane includes a long boom, at the end of which is a gondola to hold the scientists. Although the crane itself stays in a permanent location in the Wind River Experimental Forest, the gondola, which can move along the 270 ft (82 m) boom, can be placed anywhere within a region covering 5 acres (2 hectares). Furthermore, it can be raised and lowered to provide a third dimension of observation.

Scientists in the Wind River project are particularly interested in the ecological effects of lichens and fungi in such a setting. Observation has shown that lichens that have fallen off trees are unusually high in nitrogen, suggesting that lichens may have the ability to fix (convert to a usable form) atmospheric nitrogen (*see also* "Fungi Surprises of the 1990s," p 388).

Researchers had originally hoped to have the Wind River crane in the temperate rain forest of the Olympic peninsula, but local forest workers feared federal interference with timber removal and kept the crane out. Scientists from the University of Washington then found the Wind River site in southern Washington, near Portland, Oregon, and Mount Hood.

More Life to Study

It will be no surprise if the canopy studies reveal that there are more species and a greater abundance of organisms of all kinds than previously identified. Despite what is commonly believed, the 1990s are revealing new species much faster than anyone can document the extinction of previously known species. This does not mean, however, that biodiversity is rapidly rising. In fact, biodiversity is probably decreasing, just as those alarmed over damage to the environment have claimed. However, there are more scientists than ever before, and consequently more discoveries are being made.

The pace of discovery of new species, which one might assume to be diminishing, has actually risen rapidly in the 1990s (*see,* for example, "Newly Found Mammals," p 393). The discovery of great numbers of bacteria deep in Earth's crust (*see* "Bacteria: Giant, Deep, and Social," p 382) and of unexpected one-celled Archaea in the oceans (*see* "The Kingdom of the Archaea," p 378) have not only increased our knowledge of Earth's biodiversity, but also of its biomass (the total amount of the Earth's mass that consists of living organisms).

Another realm in which both biodiversity and biomass are greater than anyone had supposed is the middle-depth of the oceans. Many of the life forms which exist between the ocean's surface and its bottom are poorly known, in part because attempts to net them produce mostly blobs of jelly, and in part because the region involved is so vast that nets can only provide random samples. The only way to explore this region has involved the use of submersibles, either human-carrying small submarines or undersea robot cameras. These are still a relatively new development in ocean research, and have been used mostly to examine the geology and biology of the bottom and lower depths of the sea. This use pattern, however, is beginning to change. One robot-submersible study of the Monterey Canyon off California's coast, by Bruce H. Robison and coworkers at the Monterey Bay Aquarium Research Institute, has found about a dozen new species each year since 1988.

Another underwater canyon off California, the La Jolla Canyon, has revealed something unexpected about the biomass in the ocean. Eric W. Vetter of the Scripps Institution of Oceanography at La Jolla, California, studies crustaceans, mostly tiny amphipods and leptostracans. Over a period of a year, from March 1992 to March 1993, he dove with scuba gear and counted the organisms found in about 3,920 sq yd (2,000 sq m), out of a mat that is about 196,000 sq yd (100,000 sq m) in all. He found as many as 3.24 million crustaceans occupying a single square 39 in (1 m) on a side. Vetter thinks this assemblage of organisms is significantly denser (by a factor of four) than any other patch of life previously analyzed.

Vetter notes, however, that it is not the single largest assemblage of like organisms in one place in the sea. One part of the Bering Sea, measuring something between 8,500–14,500 sq mi (22,000–37,500 sq km), produces between 1.5 billion–3.3 billion lb (6.6 million–1.5 billion kg) of amphipods each year according to figures published in 1990.

(Periodical References and Additional Reading: *New York Times* 2-26-93, p A12; *New York Times* 2-22-94, p C1; *New York Times* 3-3-94, p A18; *Science* 4-15-94, p 335; *Science* 6-24-94, p 1824; *New York Times* 7-26-94, p C1; *Science News* 9-10-94, p 170; *Nature* 11-3-94, p 47; *Science* 11-4-94, p 720; *New York Times* 11-6-94, p I-25; *Science* 11-25-94, p 1327)

TIMETABLE OF LIFE SCIENCE TO 1991

9000 B.C.	The agricultural revolution starts in the Near East with the domestication of sheep, goats, and wheat.
8000 B.C.	The agricultural revolution starts independently in what are now Latin America and Indochina.
350 B.C.	Aristotle [Greek: 384-322 B.C.] classifies the known animals using a system that will continue to be used for the next 2,000 years.
1609	Jan Baptista van Helmont [Flemish: 1580-1635 or 1644] plants a bush in a pot. After a few years he weighs the mass of the plant and the earth in the pot, proving that most of the mass of the plant does not come from the earth in the pot, but is absorbed from water and from the air.
1648	Jan Baptista van Helmont's experiment showing that plants do not obtain the materials for their growth from the soil is published posthumously.
1665	Robert Hooke [English: 1635-1703] describes and names the cell.
1668	Francesco Redi [Italian: 1626-1697] shows that maggots in meat do not arise spontaneously, as most people then believed, but are hatched from flies' eggs.
1669	Anton van Leeuwenhoek [Dutch: 1632-1723] discovers microorganisms, creatures too small to see with the naked eye, and recognizes that sperm are a part of reproduction.
1680	Giovanni Alfonso Borelli [Italian: 1608-1679] shows in his book *De motu*

animalium (Concerning animal motion) that human muscles are not strong enough in proportion to human weight for flight similar to that of birds.

1683   Van Leeuwenhoek is the first to observe bacteria.

1735   Carolus Linnaeus [Swedish: 1707-1778] introduces the system still used for classifying plants and animals.

1779   Jan Ingenhousz [Dutch: 1730-1799] discovers that plants release oxygen when exposed to sunlight and that they consume carbon dioxide. This is the beginning of the understanding of photosynthesis.

1827   John James Audubon [French-American: 1785-1851] starts publication of *Birds of America*.

1834   Sergei N. Winogradsky discovers the mechanism of nitrogen fixation in plants.

1839   Theodor Schwann [German: 1810-1882], building on the work of Matthias Schleiden [German: 1804-1881] in 1838, develops the cell theory of life (generally attributed to both Schleiden and Schwann).

1856   Louis Pasteur [French: 1822-1895] discovers that fermentation is caused by microorganisms.

1858   The theory of evolution by natural selection as independently developed by Charles Darwin [English: 1809-1882] and Alfred Wallace [English: 1823-1913] is announced to the Linnaean Society.

1859   Darwin's *On the Origin of Species* is published.

1862   Observing that the Madagascaran orchid *Angraecum sesquipedale* hides about an inch of nectar at the base of a 1 ft (0.3 m) long tube, Charles Darwin predicts that there must be an unknown species of moth with an 11 in (28 cm) tongue that can feed on the nectar, at the same time pollinating the orchid.

1864   About this time, a captive duck-billed platypus is observed laying two eggs, convincing scientists that there are mammals that can reproduce by laying eggs.

The mathematical physicist George Stokes discovers that hemoglobin can link itself to oxygen and then release it.

1865   Gregor Mendel's [Austrian: 1822-1884] theory of dominant and recessive genes is published in an obscure local journal.

1898   The first identification of a virus occurs when tobacco mosaic disease is recognized as being virally caused, although viruses cannot yet be seen and are known only from their effects.

1900   Three different biologists, Hugo Marie De Vries [Dutch: 1848-1935], Karl Franz Joseph Correns [German: 1864-1933], and Erich Tschermak von Seysenegg [Austrian: 1871-1962], rediscover the laws of heredity which were originally published by Gregor Mendel in an obscure local journal in 1865 and subsequently ignored.

1901   The okapi is discovered.

1903   In Madagascar, two etymologists discover a previously unknown moth with an 11 in (28 cm) tongue. In 1862, Charles Darwin had predicted its existence on the basis of an orchid that would require such a creature for pollination.

1905   Clarence McClung finds that female mammals have two X chromosomes and males have an X paired with a Y.

1907   Thomas Hunt Morgan [American: 1866–1945] starts experiments with the fruit fly *Drosophila melanogaster*, establishing that the units of heredity are located within cells on small bodies called *chromosomes*.

1909   Wilhelm Johannsen [Danish: 1857–1927] coins the word *gene* to describe the unit of heredity.

1910   Thomas Hunt Morgan discovers that some genes are linked to a particular sex. In *D. melanogaster*, a white-eyed mutation appears in males only.

1911   Morgan starts mapping gene locations on *Drosophila* chromosomes.

1918   The number of human chromosomes is counted (incorrectly) for the first time. Herbert M. Evans counts 48.

1919   Karl von Frisch [Austrian-German: 1886–1982] discovers that bees have a language used to communicate good sources of flower nectar.

1936   Andrei Nikolaevitch Belozersky [USSR: 1905–1973] isolates deoxyribonucleic acid (DNA) in the pure state for the first time.

1938   The first known live coelacanth is captured. Scientists had believed that the species had been extinct for 60 million years.

1941   George Wells Beadle [American: 1903–1989] and Edward Lawrie Tatum [American:1909–1975] theorize that each enzyme is controlled by a single gene.

1944   Oswald Theodore Avery [Canadian/American: 1877–1955], Colin MacLeod, and Maclyn McCarthy determine that DNA is the chromosomal repository of genes, not proteins, as had been generally believed.

1952   Eugene Aserinsky [American: 1921–] discovers that sleep with rapid eye movements (REM) is a specific stage of sleep, later found to be associated with dreams.

        Alan Turing shows that spots and stripes on animal skins are caused by two interactive chemicals diffusing at different rates.

1953   James Watson [American: 1928–] and Francis Crick [English: 1918–] determine the structure of DNA, the basis of heredity. DNA is made up of various combinations of four bases; its architecture consists of two paired strands arranged in a double helix.

1954   J. Lin Tjio and Albert Levan show that humans have 46 chromosomes (arranged in 23 pairs) rather than 48, as was previously believed.

1958   Dorothy Crowfoot-Hodgkin uses x rays to determine the structure of vitamin B-12.

        Alfred Gierer and Gerhard Schramm discover the infecting properties of nucleic acids in viruses.

1959   Scientists establish that the chromosome known as Y confers maleness in humans. A human with two X chromosomes will normally be female, although there can be exceptions when a male gene crosses over from the Y to an X chromosome. The condition of two X chromosomes and no Y is later found to happen about once in every 20,000 males.

Christian Boehmer Anfinsen determines the sequence of the 24 amino acids in the molecule of ribonuclease.

1960   John Cowdery Kendrew [English: 1917-] and Max Perutz locate all of the atoms in the organic molecule myoglobin, a close relative of hemoglobin.

1961   Marshall Nirenberg [American: 1927-] and J. H. Matthaei of the U.S. National Institutes of Health learn to read one of the units of the genetic code when they find that a combination of three uridylic acid bases in a row (UUU — called a *codon*) in ribonucleic acid (RNA) always codes for the amino acid phenylalanine.

1962   Linus Pauling [American: 1901-1994] and Emile Zuckerkandl propose that the evolution of molecules can be used to determine how long one species has been separate from a related species.

1967   Charles Yanofsky is the first to prove that the sequence of codons in a gene determines exactly the sequence of amino acids in a protein.

Roger Payne and Scott McVay discover that the male humpback whale sings an intricate song that follows rules of composition similar to human music.

1968   Werner Arber [Swiss: 1929-] discovers restriction enzymes, a class of proteins that will make genetic engineering possible.

David Zipser discovers the meaning of the last remaining undeciphered codon of the genetic code, the pattern uracil-guanine-adenine (UGA), ultimately found to indicate "stop making this protein."

1969   Jonathan Beckwith [American: 1935-] and coworkers are the first to isolate a single gene: the bacterial gene for a step in the metabolism of sugar.

Roger Donahue is the first to map a human gene to a chromosome other than the X and Y chromosomes: the gene for the "Duffy" blood type.

1970   Har Gobin Khorana [Indian/American: 1922-] and coworkers produce the first artificial gene.

Howard Temin [American: 1934-] and David Baltimore [American: 1938-] discover the enzyme that causes RNA to be transcribed to DNA, a key step in the development of genetic engineering.

1971   Roger Y. Stanier establishes that the one-celled organisms previously known as blue-green algae are, along with bacteria, part of a separate kingdom of life (now known as Monera), and are not closely related to true algae at all.

1973   Stanley N. Cohen [American: 1917-] and Herbert W. Boyer [American: 1936-] succeed in putting a specific gene into a bacterium, the first instance of true genetic engineering.

1975   César Milstein [British: 1927-] announces the discovery of how to produce monoclonal antibodies.

1977   Phillip A. Sharp and, independently, Richard J. Roberts and coworkers discover that DNA in organisms more complex than bacteria contains long stretches of meaningless material that is not part of any gene. These stretches of meaningless material are labeled *introns*.

1980   Martin Cline and coworkers transfer genes from one mouse to another, and succeed in having genes function in the new organism.

The U.S. Supreme Court rules that a microbe developed by General Electric for oil cleanup can be patented. This is the first patent for a microorganism engineered by humans.

Stanley N. Cohen of Stanford University and Herbert W. Boyer of the University of California at San Francisco receive the first patent issued for a method of genetic engineering. Royalties on the patent will go to their universities, which had applied for the patent in 1974 along with an application for a patent on the results of using the technique.

1981   The Chinese produce a genetic copy (clone) of a zebra fish.

At Ohio University in Athens, Ohio, scientists transfer genes from other organisms into mice for the first time.

1983   Walther J. Gehring and coworkers discover the homeobox gene, essential in the development of a wide variety of organisms ranging from yeasts to humans.

The DNA polymerase chain reaction is invented by Kary B. Mullis.

1984   Alex Jeffreys develops "genetic fingerprinting," a method of identifying a specific individual using any DNA-containing material from that person (for example, white blood cells, skin cells, or sperm).

1985   Charles G. Sibley and Jon E. Ahlquist use DNA studies to revise the evolutionary history of perching birds and songbirds from Australia and New Guinea.

1986   The first genetically altered virus (for herpes in swine) is marketed and the first field trials of a genetically altered plant begin.

1988   On April 12, the U.S. Patent Office issues patent no. 4,736,866 to Harvard Medical School for a mouse developed by Philip Leder and Timothy Y. Stewart via genetic engineering. It is the first U.S. patent issued for a vertebrate.

1989   Francis Collins, Lap-Chee Tsui, and coworkers find and make copies of the gene that causes most cystic fibrosis.

1990   On July 31, the Recombinant DNA Advisory Committee, a watchdog group for genetic engineering, approves the first serious attempt to insert genetically altered cells into human patients in an effort to correct an inborn genetic defect.

Robin Lovell-Badge, Peter Goodfellow, and coworkers discover the gene on the Y chromosome that produces maleness in humans, which they name *SRY*.

Anne E. Houde and John A. Endler demonstrate in guppy experiments that mate preference can produce rapid evolutionary changes.

Fernando Nottebohm shows that a male canary grows new neurons in its brain each autumn, allowing it to produce a different song from the previous year.

1991   In June, Craig Venter of the NIH files for patents on 350 sequences of complementary DNA (cDNA; effectively parts of genes) from the human brain. His research group has used an automated process to find the sequence of bases in these fragments of cDNA, which were obtained from commercial sources of brain clones of cDNA.

GENES AND DEVELOPMENT

GENETICS AND BEYOND

Although scientists in 1911 had predicted the existence of some entity on chromosomes corresponding to that which is now understood to be the gene, for the first 50 years of the twentieth century genes were only theoretical concepts. In the second half of the century, genes were recognized as a chemical code on the DNA molecule, and the code was eventually deciphered. Today, totally new phenomena continue to be discovered, observations long known are still being explained, and heredity itself is seen as going beyond DNA in some ways.

It All Began with RNA

Most scenarios describing how life started assume that the most important step occurred when a molecule called ribonucleic acid (RNA) began to appear on Earth for one reason or another. After that first step, which is difficult to explain theoretically, the second step seems easy to postulate. Because RNA contains the instructions used for building proteins, once RNA is present proteins can be built. The combination of RNA and proteins duplicates itself to become the first form of life, and as life evolves, fragile RNA is supplemented by more durable deoxyribonucleic acid, or DNA, to maintain stability.

The problem with these scenarios, however, is that RNA simply carries the code (*see* Background). All the work of building a single protein is done by other proteins, not by RNA. Only the blueprints are in the RNA. The situation is a classic "chicken-or-egg" problem.

One way out of this dilemma would be either to show that proteins can make proteins without RNA, or to show that RNA can make proteins without the help of other proteins. In 1992, Harry Noller and coworkers at the University of California at Santa Cruz came very close to showing that the second possibility provides the solution. Noller and coworkers removed as much of the protein as possible from a ribosome without destroying the RNA itself. Then they took the RNA, which still had a few wisps of protein clinging to it, and added it to an amino-acid soup. Not only was the RNA in the protein-poor ribosomes able to stitch the amino acids into proteins, but it did so almost as effectively as do intact ribosomes in the same situation. Even though the ribosome still retained some protein, it was 95% pure RNA, suggesting that, under suitable conditions, the RNA could make proteins without help.

That may explain the second step, although there was still some original protein involved in this experiment. Now back to the first step: the making of RNA itself. In cells, this is once again performed by a combination of proteins. But in 1982, Thomas R. Cech and coworkers at the University of Colorado found a form of RNA in a one-celled pond animal called *Tetrahymena thermophila*. This RNA could both break and form bonds by itself, with no help from proteins. The RNA could act like a protein, cutting RNA strands into pieces and then splicing them together. If the original RNA also contained nonsensical strings of nucleotides—strings that do not code for anything—the RNA could discard them and form meaningful RNA from the nonsense.

Knowing that RNA can both duplicate itself, given the proper materials to work with, and that it can make a protein from a different set of given materials, gives considerable plausibility to the scenarios of life's origins.

Sex and the Single Chromosome

When it comes to human chromosomes, sex is the exception that proves the rule. In this case, the rule states that chromosomes come in pairs, with each member containing similar versions of the same gene, called *alleles*. Unlike the other 22 pairs of human chromosomes, the rule-breaking chromosome pair that we label X and Y are not much like each other, and for the most part do not contain the same alleles. X is similar to any one of the other 44 chromosomes, but Y, its pair-mate, is a tiny thing that is barely a chromosome at all.

We call X and Y the sex chromosomes because the presence or absence of the Y chromosome largely determines which sex a person will be. Although hormones produced by genes located on chromosomes other than the X or Y chromosomes do most of the actual work of turning a fetus into a male or a female, a gene on the Y chromosome sets the whole process in motion for males. Without that gene on the Y chromosome, the fetus continues to develop as a female. In 1990, the genetic switch on the Y chromosome was found, a single gene called *SRY* that turns on all the other genes needed to build a male. Thus, for humans, the combination XX (two X chromosomes) in the sex-determining pair produces a girl, but XY is a boy. Once in a while, however, the *SRY* gene crosses over from the Y to the X chromosome. When that happens, an XX gene combination can result in a male.

The same chromosomal arrangement for determining sex (XX = female, XY = male) is used by all other mammals, but is far from universal among all animals. If we agree to designate the large, normal-appearing chromosome in the sex-determining pair "X," and the smaller, insignificant-appearing chromosome "Y," then for birds the combination XX is male and XY female (biologists often designate these chromosomes as Z and W because of various differences from the mammalian X and Y). Butterflies and moths, reptiles, and some amphibians and fish use the same system as birds. Although some flowering plants use an X and Y system like mammals, with two resultant sexes, many plants have no sex chromosomes at all. Insects vary quite widely: in the *Drosophila* fruit flies so beloved of geneticists, XX means female and XY male. In *Drosophila*, however, the mechanism differs from that in humans: the extra X is the determining factor, while the Y is just along for the ride.

Biologists seriously question what biological advantage has resulted in the existence of different chromosomal sexes. A related question asks why the genetic structure of the X and Y chromosomes are so different from each other in so many species. This difference is maintained from generation to generation because the vastly different X and Y chromosomes do not recombine during reproduction.

One might think that disparate X and Y chromosomes can not recombine precisely because they are so different from each other. In 1931, however, the famous statistician and geneticist Ronald Aylmer Fisher proposed a different explanation. He argued that certain characteristics, especially those involved in sexual selection, had evolved to remain separate, which would not be the case if the chromosomes mixed it up every time meiosis occurred. For example, a long tail on a peahen would be a disadvantage in life and no use in attracting a male, who hardly needs to be attracted in any case. Male peafowl, called peacocks, have a long tail that the peacock spreads to attract a female peafowl, called a peahen. It works. Peahens are genetically programmed to respond to the tail of the peacock. The long tail is determined by genes on the chromosomes. If these genes sometimes crossed over, as other genes do, some peahens would acquire useless long tails. Evolution has arranged that such sex-related genes do not cross over (according to Fisher). But most other genes—e.g. for

---

MITOSIS AND MEIOSIS

Among the many unhappy memories of an American high-school education is the necessity of learning the definitions and distinguishing characteristics of mitosis and meiosis. Walther Flemming, who discovered chromosomes, coined the word *mitosis* in 1882 from the Greek *mitos*, meaning "thread," and *-osis*, meaning "a condition or process." The word *meiosis* also comes from Greek, specifically from *meioo*, which means "to make less," and the common ending *-osis*. The memorization process is complicated by the fact that, in English, both the *i* in *mitosis* and the *ei* in *meiosis* are pronounced with the same long "eye" sound.

Mitosis is the process during which the two strands of DNA in the double helix unwind, each separating from its complementary strand. The cell then uses each of these single strands to build two chromosomes, resulting in two new double helices. Because the cell then has twice as many chromosomes, the entire cell divides itself in two, resulting in two identical cells (daughter cells), each with the appropriate number of chromosomes. If all goes well, each daughter cell is exactly like its parent insofar as genes are concerned. All cells undergo mitosis somewhere along the way. Every organism starts with a single cell (a zygote), and every subsequent cell is the daughter of that original zygote, usually many generations removed.

Meiosis is a different process that occurs only in cells that give rise to ova (eggs) and sperm. It is more complex than mitosis. As a result of meiosis, four *nonidentical* daughter cells are produced, each with half the number of chromosomes of the original cell. If all has gone well, each daughter cell contains one and only one member of each of the original pairs. If all has not gone well, serious genetic disease can result in the offspring. Because early meiosis can include an event called *recombination*, in which genes are exchanged between members of a pair of chromosomes, the resulting chromosome in the daughter cell may not exactly resemble either of the original parental chromosomes. Recombination occurs between all pairs of genes except for the very different X and Y members of the sex-determining pair.

A mnenomic device can help differentiate between mitosis and meiosis. Think of the cell that divides asexually as *it*, as in mITosis, while the sexual one, mEiosis, has an *e* as in sEx.

longer legs—do cross over or recombine during meiosis. So, Fisher suggested, natural selection soon halts recombination of sex-determining genes.

In 1992, William R. Rice of the University of California, Santa Cruz, demonstrated that Fisher's explanation was correct. Initially, only the sex-determining genes and those determining secondary sexual characteristics fail to recombine. However, as more generations are bred, genes located near the sex and secondary sex genes also stop recombining. It is as if Nature took no chances and extended the region around the genes that should not recombine. Eventually this process results in complete chromosomes that essentially never recombine. Rice did his work with *Drosophila* fruit flies, which have an XY system (see above for difference from human XY system).

Rice went on to study the related question of what happens when an artificially constructed ordinary-sized Y chromosome is kept from recombination in a fruit fly population. He showed that harmful genes accumulate faster if recombination is prevented, and that the harmful genes soon become inactive. After that, with no evolutionary pressure to maintain the inactive gene, it gets deleted. The Y chromosome gets smaller.

The logical end of this process is elimination of the tiny Y chromosome entirely. In some species this seems to have already occurred. This process has reached its natural end, or close to it, in some species of birds, although with special biochemical methods it is sometimes possible to locate a very tiny remnant of the Y chromosome. In fish, an older phylum, all possible combinations are found, ranging from species with full-sized Y chromosomes that still recombine to species with no Y chromosomes at all.

A New Nature vs. Nurture

Two plants rooted from stems of the same parent will look quite different if one is kept in sunlight, watered, and given fertilizer, while the other grows up in partial shade with inadequate water and soil nutrients. Seeing the full-grown plants, it would be easy to think that they were genetically very different. All biologists know that individual organisms, even those born with the same genes, differ as adults in small ways at least. The influence of the environment accounts for nearly all of the ways in which the resulting organism, called the *phenotype* by geneticists, differs from its recipe for inheritance, called the *genotype*.

It is often difficult to tell whether a particular trait in a phenotype is a product of the genotype, or the result of external influences the organism experiences during development. The difficulty in determining the root of a trait is captured in the alliterative phrase "nature vs. nurture," in which biologists have extended the meaning of *nurture* to include all environmental effects on development.

In 1993, Eva Jablonka of Tel Aviv University and Eytan Avital of David Yelin Teacher's College, both in Israel, proposed a modification of genetic theory that they call phenotype cloning. They argue that mothers nurture their children largely as they themselves were nurtured. In turn, this way of bringing up a child acts as a hereditary influence that has little to do with the genotype. While at first thought this idea seems no more than a refinement of the ordinarily expected environmental factor, Jablonka and Avital argue that their theory explains behavioral practices that otherwise are poorly understood. Among these is the way that mothers of many species actively discourage the fathers from participating in the upbringing of the children. This protects the maternal influence on the hereditary pattern of phenotype cloning, because the traits passed on will be those of the mother, not those of the father. Experiments with mothers bringing up very young offspring of another species have shown that some maternal traits are even passed across species lines. The theory also explains why many mammals and birds are willing to become "foster parents" to offspring not their own.

Earlier studies have shown that maternal influence can be exerted physically as well as through upbringing. Different phenotypes may result from identical genes, depending on whether they are of maternal or paternal origin. Also, experiments have shown that some traits are affected by the egg shape or by the nutrients provided through the placenta. Heredity, therefore, is influenced by more than just DNA.

Background

Most cells contain recognizable small pairs of bodies called **chromosomes**. Each chromosome is observed to double in normal cell division, while in the division that produces egg or sperm cells for plants or animals, the pairs separate instead of doubling. Each chromosome consists primarily of a large molecule of deoxyribonucleic acid (DNA), which has the structure of a ladder that has been twisted. When chromosomes double, the DNA untwists; each new chromosome will receive one of the uprights of the original ladder. Each of these two uprights is then used to build a new chromosome that, barring mistakes, exactly resembles the original before the doubling. By the 1940s, it was already apparent that this process transmits hereditary characteristics from one cell to its daughters, although the role of DNA and the mechanism involved were not understood until the 1950s. The unpairing process occurring in eggs and sperm was matched by the formation of new pairs due to fertilization, carrying hereditary characteristics from one generation to the next. Since such characteristics are expressed in the cells and ultimately in the whole organism, the DNA can be said to control cell development as well as heredity.

Each upright of the DNA twisted ladder (or **double helix**) carries its necessary information as strings of four different subunits called **bases**. These bases are adenine, thymine, guanine, and cytosine (usually abbreviated A, T, G, and C). Although each DNA molecule contains the same four bases, their total number and arrangement varies enormously. The exact order and arrangement of the bases is preserved when a chromosome doubles itself.

A group of highly variable complex polymers called **proteins** contributes to the structure of a cell and carries out the majority of work performed by a given cell. Each protein is constructed from some combination of subunits called amino acids. The information in the sequence of bases in DNA is a method of describing each protein. Nearly every combination of three of the four bases A, T, G, and C exactly identifies one amino acid.

DNA is such a large molecule that the average chromosome can describe tens of thousands of proteins. A sequence of bases describing exactly one protein is called a **gene**, so the 23 pairs of human chromosomes carry the information for a couple of hundred thousand genes. Not all of these genes are active at any one time, but mechanisms based ultimately in the DNA turn genes on and off as needed. In addition to the genes, DNA of plants and animals contains long sequences of bases that seem to do nothing, although evidence is beginning to accumulate suggesting that these "nonsense" stretches have specific functions (*see* "Genome Project UPDATE," p 366).

DNA by itself does not do anything. It is merely a recording, like a CD, cassette, or floppy disk. One needs a complex machine to turn tiny pits in a CD or magnetic patterns on a cassette tape into music or pictures. Most functions of living cells are accomplished by complex molecules that are the equivalent of sound and pictures, if DNA is the equivalent of a CD or cassette. For the most part, the complex molecules are proteins, polymers built by connecting about 20 different amino acids in various configurations. Involved every step of the way is RNA, a polymer made from sequences of four bases that are strung out along a backbone made from a simple sugar and phosphoric acid. The four bases are like those of DNA except that thymine (T) is replaced with the similar base uracil (U), so the four bases of RNA are cytosine (C), guanine (G), adenine (A), and uracil.

In a simplified view of how a protein is made, the main actors are various forms of RNA. The first step is **transcription**, in which a type of RNA called messenger RNA forms along a single strand of the DNA, matching it base for base—that is, each C on the DNA matches a G on the RNA and vice versa, while each A on the DNA matches a U on RNA, and each T on DNA matches an A on RNA. The transcribed RNA is just like its complementary strand of DNA (except for the substitution of U for T).

The transcribed messenger RNA then moves to a ribosome, where the next process, **translation**, takes place. The messenger RNA is pulled through the ribosome one codon at a time (remember that a codon is a sequence of three bases which codes for a specific single amino acid or the beginning or end of a gene). As each codon passes through the ribosome, it is matched with a small molecule of RNA called a transfer RNA. The transfer RNA has brought along an amino acid from among those floating around in the cell. Twenty-odd special ligase enzymes (proteins) match the appropriate transfer RNA and amino acid. The transfer RNAs drop off their amino acids in the ribosome, where they are added one at a time to the new protein chain that is forming there. When a stop codon is reached, the protein chain pops out and the messenger RNA leaves the ribosome.

Note that the above description is simplified. One simplification is the omission of the action of most of the pre-existing proteins that makes this process possible. Left to themselves, DNA does nothing and RNA, versatile as it is, does not do much. A group of proteins works to assemble the RNA on the DNA template to begin with, for example. This group is called a **transcription factor** or **transcription machine**. Its main component is a protein called RNA polymerase. Another group of about 80 proteins operates the ribosome.

Another simplification involves the mechanism needed to turn the gene on in the first place to allow the various steps toward protein synthesis to take place. This "switching-on" mechanism requires a different gene and its protein product. Similarly, a third gene and concomitant protein is needed to turn the whole process off.

Other simplifications include omitting mention of the process of methylization, in which methyl groups are attached to the messenger RNA to protect parts of it from enzymes that could cut it to pieces, and omitting the processing that removes "junk" parts of RNA before it reaches the ribosome (for organisms other than bacteria, there is generally more junk than functioning RNA).

The hypothesis that each gene makes a single protein (through several steps involving RNA) and that each protein stems from a single gene goes back to 1941, when very little was understood about the physical nature of genes. At that time proteins were only moderately well understood. By 1953, Frederick Sanger was able, in a heroic effort, to analyze completely for the first time the structure of a fairly small protein: insulin. That was the same year that James Watson and Francis Crick deduced the structure of DNA, the key step in unraveling the physical basis of genetics. Within a few years, the principal focus had shifted from the study of proteins to the study of genes. The genetic code was deciphered, and machines were developed that could find the DNA base sequences that make up the code.

One result of cracking the genetic code was to replace a big part of the difficult biochemical problem of describing the structures of proteins with the largely solved problem of deciphering a gene. Once the gene that accomplishes something is found, the sequence of amino acids in the protein follows automatically. This is not the complete description of a protein, since the shape of a protein determines a large part of its action, but sequence may be enough for many purposes. Thus, a biochemist may start with a result and proceed directly to trying to find the gene, leaving the question of the protein that actually accomplishes the result until later, since that is the easier problem. Of course, if for some reason a protein's sequence of amino acids is already known, you can back into knowing a lot about the gene that makes the protein.

(Periodical References and Additional Reading: *Science* 6-5-92, p 1436; *Discover* 1-93, p 69; *New York Times* 1-3-94, p B13; *Science* 1-14-94, pp 171 & 230)

DEVELOPMENT DEVELOPMENTS

Rudyard Kipling's *Just So Stories* of 1902 included fanciful tales of "How the Elephant Got His Trunk" or "How the Camel Got His Hump." The just-so stories of today are equally remarkable, and have the advantage of being true, as developmental biologists continue to make amazing progress toward learning exactly how organs originate. Today it is possible to use microscopic magnetic resonance imaging to watch a labeled cell or group of cells travel through an early-stage embryo to its destiny, which perhaps is an early death. Furthermore, this view appears on a television monitor, allowing scientists to watch images of the whole embryo as it develops over several days. Kept on videotape, the whole thing can be played backwards or forwards at slow or fast speeds or stopped on a particular image. Furthermore, the image is in 3-D form, allowing any desired anatomical slice to be viewed as needed. Kipling would have been amazed and delighted.

However, it should not be thought that biologists have completely unraveled the mysteries of development. Each step in the developmental sequence seems to require a combination of inducers and inhibitors. While some of these have been identified, many remain only suspected. Biologists can watch waves of cells originate, move to their intended place in the embryo, transform themselves into adult tissue, and, in many cases, die to get out of the way; but what mysterious perfume summons them to their places or effects transformation or death is still imperfectly known.

One of the basic modern just-so stories might be called "How the Worm Got Its Body," the cell-by-cell analysis of the development of the nematode *C. elegans* (*see* "*C. elegans* UPDATE," p 362). Here are some of the other new tales.

How the Agouti Got His Coat

Dyes have been important in biology since the 1880s, when Paul Ehrlich and Hans Gram pioneered the use of dyes to stain bacteria. Dyes that are most useful in biology are those that react with a particular tissue, making the stained structure visible, when it would otherwise be difficult or impossible to see. Ehrlich also recognized that a dye which reacts with a particular tissue could be used in the diagnosis or treatment of disease. More recently, scientists have discovered that an organism's color patterns, caused by differential effects of natural hormones, can reveal secrets of normal and abnormal growth and development.

Patterns can be formed by changing concentrations of two or more chemicals that affect color. In 1952, mathematician Alan Turing showed that it was theoretically possible to produce stripes or spots by diffusing two chemicals at slightly different rates. Patterns produced in this way are called Turing structures, and were first observed experimentally in 1990.

In the 1990s, an important field of experimental study involves fur coloration in mammals, including solid colors, patches, stripes, and spots. One common coloration is a subtle arrangement of stripes that derives from a basic pattern of striping on each hair, rather than from alternating bands of different colored hairs. This is called *agouti fur*, named after the South American rodent that displays the pattern.

In an agouti, each hair is black at its tip and base, but yellow-red in the midsection. As in a Turing structure, the agouti fur results from the interaction of two chemicals, although in agouti fur there is also a third factor involved. Animal coloration involves cells called melanocytes. In the case of the agouti, the chemicals react with the melanocytes and not with each other. Both chemicals in the coloring process for agouti fur are peptides, structures made from amino acids that are too small to be labeled proteins. One of these peptides is produced in the brain, while the other is made in the skin around the base of each hair. Melanocyte-stimulating hormone (MSH), the peptide from the brain, interacts with melanocytes in growing hair to cause black or brown coloring. The agouti peptide manufactured in the skin causes the hair melanocytes to be reddish yellow. Whenever the agouti peptide is present, it blocks the melanocyte-stimulating hormone. If the agouti peptide is present for only a short time during the growth of the fur, the end of each hair will be black while the middle will be the reddish-yellow produced in the absence of MSH. This is the pattern observed in the agouti.

Slight hereditary differences in peptides or in melanocytes lead to variations that appear quite different from agouti coloring. In golden retrievers, foxes, and chestnut horses, the melanocytes are unresponsive to MSH. The spots in leopards are thought to be Turing structures caused by diffusion of agouti peptide and MSH. When a mutation in the melanocyte makes it respond as though it were constantly being stimulated by MSH, the leopard with the all-black coat is called a black panther. Zebra stripes are Turing structures caused by two different mutations: one like that in the black panther and one like that in albino animals.

Studies of how genes affect fur coloration have contributed to an understanding of how other features of organisms develop, as well as an understanding of the causes of certain genetic diseases. Mechanisms similar to those resulting in stripes can also cause segments to form, producing, for example, the basic plan for an insect's body. In mice, the color of the coat results from interactions of about 50 different genes. Among these are genes called *steel* and *kit*. Studies of *steel* and *kit* mice show that far more than fur color is affected; mutations in these genes can cause changes in apparently unrelated organs, including sex glands, the structures responsible for hearing, and red blood cells. These effects are due to the fact that proteins produced by *steel* and *kit* control migration of cells to their expected destinations in the growing mouse embryo. For the *steel* and *kit* proteins, the migrating cells are precursors of the mouse melanocytes; the stem cells that produce red blood cells; precursors of the cells of the inner ear; and the cells that will become the sex organs. Studies in mice show that when something is wrong with *steel* or *kit*, the proteins produced may fail to lead the developing cells to their expected designation, resulting in improper functioning. Similar genes are found in humans, for whom mutations in the equivalent of *steel* and *kit* genes can lead to forms of piebaldism, manifested by odd patches of white skin, and disorders that are often marked by deafness or infertility.

The agouti gene does more than influence color, although its actions are not well understood as of yet. Defects in the agouti gene that lead to all-yellow animals also are likely to cause obesity, diabetes, and tumors in the liver or other organs. The agouti gene was discovered by Richard P. Woychick of Oak Ridge National Laboratory and reported in *Cell* in December 1992.

Defects in another colorizing gene, found in mice, humans, and many other animals, can cause albinism. Similarly, animal breeders have long known that a hereditary factor called extension will result in a yellow pattern that extends across the entire body of the animal that has it. Extension is due to a particular mutant form of the receptor protein in melanocytes. Roger D. Cone of Oregon Health Sciences University in Portland reported the discovery of the extension gene in *Cell* in March 1993.

BIOLOGICAL STYLE

Because life is so diverse, the discipline of biology is famous for the great variety of different organisms that biologists study. In a well-known anecdote, physicist Enrico Fermi, confronted with the discovery of many new subatomic particles in cosmic rays, commented that if he had to learn about all those particles, he might as well have become a biologist. Even more than astronomers, whose vast numbers of stars and galaxies tend to be more alike each other than are living creatures and their components, biologists are fussy about how their subjects are named and how those names are written.

Taxonomists have long used italics to separate genus and species names from other classifications, so it may have seemed natural to geneticists to use italics also. The official name of a gene is normally in italics, with lower-case letters for recessive genes and an initial cap for dominant ones. Thus, some genes known to be involved in cell development are *ras*, *Sry,* and *myc,* following the early tradition of using three letters to abbreviate longer gene names. Later, a gene found in humans that corresponds to a gene previously identified in another animal might be given in capital letters in order to distinguish it from the gene originally found: thus, *Sry* in mice corresponds to *SRY* in humans. Some biologists avoid the three-letter rule entirely, using the whole word or words, as in *noggin* and *Sonic hedgehog* or *bride of sevenless*. This practice has become current, so that a three-letter italic abbreviation usually means that the gene has been known for some years. For example, *Sonic hedgehog* is rarely spoken of by its abbreviation, which is *Shh*. More recently, some genes are given abbreviations of fewer or more than three letters: the gene *wingless* is abbreviated *wg*, while a gene variation that is an early gene in the Hox family is labeled *Hoxa*. Note that no italics are used for gene families such as Hox, following the more familiar taxonomy style, which also uses italics for individual entities, but not for families or larger groups.

A gene is often located before its protein product is known, in which case the expressed protein has had no name assigned to it. One rule has been to give the protein the same name as the gene, but in uppercase Roman letters instead of italic lowercase. Thus, the protein BMP is the expression of the gene *bmp.* In other cases, biologists may simply call the protein by the name of the gene that produces it, as for example "the *ras* protein" or "the ras protein"—although some biologists would be more formal and write RAS. This is probably the most important rule to remember, as it can be confusing when a gene such as *Shh* and its protein SHH are mentioned in the same sentence or paragraph. Also, biologists may use just an initial cap, as in "the Ras protein" or "the Hedgehog protein," usually with the word *protein* added to make it clear.

Sometimes the protein is found before the gene. In that case, the name given the protein can be turned into the name for the gene simply by italicizing it. Thus, the genes for the proteins netrin-1 and netrin-2 are known as *netrin-1* and *netrin-2*. Since the protein was named first, the capital letters are unnecessary. In other cases, a well-known protein retains its name while the gene may get a different name or three-letter abbreviation. For example, the protein known as $G\alpha_0$, or "gee alpha sub zero," was well known before the gene that makes it in *C. elegans* was found and christened *goa-1.*

Finally, in many publications, this whole system is abandoned as simply too complicated. Thus, whatever the biologist writes is often ignored, and italics are dropped, or capitals added by the editor and copyeditor according to the style of the publication, not according to the rather complex conventions of biology.

Although humans are thought to have at least a half dozen genes that work together to produce skin and hair color, the details are not well understood. Carrot-top red hair appears to be a mutation in a gene similar to the extension gene, while albinism in humans is caused by the same gene as that leading to albinism in other mammals.

How the Rat Learned to Feel Pain

Development of the nerves and the brain is of particular interest for several reasons. In adult mammals, new cells normally no longer appear, even to replace those damaged or destroyed. Celldeath, however, continues. An older mammal continually loses brain cells without replacement—although this does not seem to interfere with thought or memory. Nerve cells make many precise and complicated connections with each other and other organs. Understanding how this occurs could explain many other aspects of development. In any case, a major goal of developmental biology is to learn how the long threads, called axons, find their way. Axons protrude from each nerve cell and then produce a thousand branches, each of which connects with another nerve at a junction called a synapse. Often, axons have to travel great distances: some spinal-cord axons in humans reach all the way from the base of the spine to the brain. To connect properly, these long axons must travel 8-10 in (20-25 cm) in the developing embryo.

Considerable progress toward answering questions concerning axon growth was announced in *Cell* on August 12, 1994, when Marc Tessier-Lavigne, Tito Serafina, and colleagues at the University of California in San Francisco identified two chemicals used to direct the connections between the brain and spinal cord in animals that have spinal cords or similar organs. The chemicals, proteins as one might suspect, have been named *netrins*, from a Sanskrit word meaning "one who guides." Both netrins are produced by genes that are closely related to a previously discovered protein, known as Unc-6, which in *C. elegans* guides the axons of sensory nerves to specific destinations. The two proteins in vertebrates, called netrin-1 and netrin-2, also are directed toward sensory nerves, specifically those which convey information about pain and temperature to the brain.

As early as 1988, when Tessier-Lavigne was at Columbia University in New York City, he and coworkers there had established that some sort of signal from nerve cells in the brain in rat embryos served to summon axons from the developing back of the spinal cord to a small set of nerve cells in a structure called the floor plate. In work at the University of California, small amounts of the specific chemicals involved were first isolated from the floor plates of 25,000 chick embryos, which enabled the researchers to genetically engineer cells that produce the chemicals in any desired amount. The proof that these chemicals are the correct ones comes when a piece of embryonic rat spinal cord is placed on a petri dish along with cells engineered to produce netrins. The tips of the axons, called growth cones by biologists, then grow in the direction of the netrin-producing cells. In a conference in November 1994, Tessier-Lavigne announced that while netrins attract the axons of spinal cells, they repel axons of some other cells.

Other chemicals discovered in the late 1980s or early 1990s also influence the direction of axon growth and the formation of synapses. Proteins of another group, much investigated in the early 1990s, promote the growth of nerve cells or neurons.

Brain-derived neurotrophic factor (BDNF)   BDNF is a nerve-cell factor that might also be called neurotrophin-2. According to research in Zurich in 1994, however, BDNF fails to cause rat spinal cords to sprout axons when injected into living animals.

Collapsins   Discovered by Jonathan A. Raper of the University of Pennsylvania in Philadelphia, and announced in the October 22, 1993 *Cell,* collapsins repel the growth cones of some axons, but (like netrins) attract the growth cones of others. As the name suggests, the repelling role was the first to be discovered in this case.

Connectin   Connectin is an axon guidance chemical that sits on the cell surface instead of diffusing through the environment as other guidance chemicals do. Originally thought to attract axons, in 1994 it was discovered that connectin actually works by repelling axon tips. Probably both attraction and repulsion occur, depending on the particular axon involved.

Docking sites   Presumed receptor molecules on axon growth cones that respond to particular chemicals in the environment are called the docking sites of those chemicals.

Epidermal Growth Factor (EGF)   The first protein known to cause nerve cells to grow was EGF, discovered by Samuel Weiss and Brent A. Reynolds of the University of Calgary in Alberta and announced in the March 27, 1992 *Science.*

The *fas IV* Protein   After it was recognized that this grasshopper gene was a Semaphorin, the *fas IV* protein was renamed Sema I.

Nerve Growth Factor (NGF)   This molecule got is name because it is the first known protein that causes new nerve cells to grow, which makes NGF the equivalent of neurotrophin-1. By 1987, however, it became clear that NGF does not attract axons to their targets in living organisms, although it does so *in vitro,* an attraction that was discovered in 1979.

Netrins 1 and 2   Evidence indicates that netrin-1 is the attractive molecule released by the incipient top part of the spinal cord and the lower part of the brain (called the floor plate). Netrin-2 is expressed lower down in the developing spinal cord. The two probably work together to direct growth of the entire organ.

Neurotrophins-3 (NT-3), -4/5 (NT-4./5), and -6 (NT-6)   More recently discovered nerve-cell growth factors are called neurotrophins. NT-3, according to research in Zurich in 1994, causes rat spinal cords to sprout axons when injected into living animals. The discovery of NT-6 was announced in the November 17, 1994 *Nature.*

Nitric Oxide   A theory proposed by P. Read Montague of the University of Alabama with Joseph A. Gally and Gerald M. Edelman of the Neurosciences Institute in New York City suggests that the nearly ubiquitous gas nitric oxide contributes to the development of the brain in an embryo by enhancing connections at synapses and also by dilating blood vessels.

Semaphorins   The semaphorins are a groups of axon guidance chemicals first found in insects by Corey S. Goodman of the University of California, Berkeley, but now also known in humans. This name is also used for the family of chemicals that includes the collapsins and a related protein in fruit flies.

The *unc-6* Protein   This product of the *unc-6* gene in a nematode summons axons and also helps cells find their correct locations. The *unc-6* protein appears to be an analogue of the netrins.

The Trick That the Hedgehog Plays

It is said that the hedgehog, a small Old World insectivore covered with sharp spines, knows only one trick, but it is a good one. When frightened, the hedgehog rolls into a tight ball with only its spines exposed. All its vulnerable soft parts are tucked inside.

The gene *hedgehog* got its name because a mutation in fruit flies results in a spiny fly, looking something like a hedgehog. By studying the gene in both its normal and mutated forms, scientists learned that the gene *hedgehog* somehow directs the insect larva to form as a series of segments. In 1992 the *hedgehog* gene was cloned, making it possible to use the fruit fly gene to look for similar genes in other creatures. Soon, variants of *hedgehog* were found in other segmented creatures and even in vertebrates, which scientists believed to have evolved from segmented animals. In January of 1994, three different groups of researchers learned what *hedgehog* genes do in such vertebrates as mice, zebra fish, and chickens—and, by inference, in humans. In each case, the *hedgehog* gene begins producing its protein early in the development of the embryo, at a stage when the embryo is an elongated blob with barely a differentiated organ of any kind. The most dramatic effect of one of the *hedgehog* genes is to make limbs form. Within the body, the same *hedgehog* gene connects and arranges the nervous system, causing the spinal cord to extend itself into the backbone, for example.

The main *hedgehog* gene in vertebrates is one named *Sonic hedgehog*, to the distress of some biologists. At one point, the individual genes of the *hedgehog* type were to be named after the 14 different species of hedgehogs. The first three names assigned with this in mind, however, suggested that geneticists are not very good taxonomists. One of the genes was named *Moonrat hedgehog,* although the moon rat is not a hedgehog, but a closely related insectivore. Another was called *Desert hedgehog,* which is a family of hedgehogs, not a species. Only *Indian hedgehog* uses the common name of a species. The fourth and most significant gene, *Sonic hedgehog*, was named, without any regard to conventional biological taxonomy, after a character in a video game.

Sonic hedgehog produces a protein that diffuses through the surrounding cells. It is thought that the cell receptors combine with the *hedgehog* protein to initiate specific changes. Since the changes caused by the *hedgehog* protein are changes in the shape of the organism, the protein is called a morphogen, or "shape generator." Part of the evidence that the protein produced by *Sonic hedgehog* is a morphogen lies in the location of the gene's expression—it produces its characteristic protein at the places where either limbs develop or the nervous system begins to take shape. Much less is known about how this gene expression results in specific changes in the cell, and that remains the subject of intense study. Other unresolved issues include how the morphogen manages to diffuse from cell to cell (since not much space exists between cells), and the distribution pattern of the morphogen in the developing organism.

How the Fruit Fly Got Its Claw and Eye

Much of the original work on development has been done with the fruit fly species *Drosophila melanogaster*. Even before DNA was well understood, scientists used radiation and other means to produce developmental disorders in the fruit fly. They also grafted cells from one part of a developing embryo to another to determine what would happen. As a result of this research program, many fundamental genes are identified by names coming from characteristics of fruit flies in which the particular gene has been mutated. Thus we have *hedgehog*, as described above, and other genes such as *wingless, sevenless,* and *dpp*, the abbreviation for *decapentaplegic.* More importantly, the effects of gene expression on development are better understood in the fruit fly than is in almost any other organism, with the exception of the nematode *C. elegans.*

Various models have been proposed to explain how genes interact to produce the results observed in experiments with grafting and with mutations. The appendages, which include six legs, two wings, two antennas, and two balancing organs called halteres, are especially suitable for study. Appendages are both discrete and asymmetrical. The group of cells destined to become a wing or a leg, called a disc or a bud at different stages, can be lifted and rotated, for example. The resulting adult can then be examined to see whether the rotation resulted in a backwards appendage, which would suggest that the disc or bud already had preformed the appendage and the information was stored within the small group of cells. The ultimate appearance of a normal appendage after such a rotation, however, might mean that the directions for making the disc or bud into a wing or leg were stored somewhere else in the body, and then transmitted to the developing appendage. Similarly, a leg disk or bud could be interchanged with a wing disk or bud to learn at which stage the wingness or legness became fixed. When the genes for development were identified and methods found to manufacture their products, these kind of experiments could be greatly refined.

Late in 1994, it appeared that the whole sequence of experiments was beginning to produce definite conclusions. A series of important papers in 1993 and 1994 established that a gene called *engrailed*, or *en*, specifies which part of a disc that is destined to become an appendage will become the back and which will be the front. This information is enough to cause the cells at the back to begin expressing *hedgehog*— that is, the Hedgehog protein begins to be produced, but only by the back cells. Some of the Hedgehog protein spills over into front part of the disc. When it does, it causes the expression of different genes in the top part of the disc from those in the bottom part. Hedgehog protein diffuses only slightly into the front, so it activated genes in the border only. On the top border, the turned-on gene is *dpp*, while on the bottom border it is *wingless*, or *wg*. This results in a line across the disk with a midpoint where the two proteins DPP and WG meet. As the appendage grows into a bud and later perhaps a leg, the midpoint stretches out to become an axis from one end of the leg to the other. Furthermore, the small region where *dpp* and *wg* are both expressed, or where DPP and WG as well as the Hedgehog protein all occur, is stimulated to express the genes needed to develop the tip structure of the appendage—a claw in the case of a developing leg. Thus, three genes work together to start the process of claw development.

In 1983, H. Meinhardt had predicted the majority of this scenario using theoretical knowledge and a bit of ordinary geometry. Experiments in which one or more of the genes involved is selectively inactivated or encouraged to be expressed across the whole disk have been used to help establish the truth of Meinhardt's predictions. The resulting types of observable mutations fit his theory exactly.

The story of *wingless* is much more involved than its role in developing the claw. For one thing, the whole process begins a stage earlier, when *wg* is the gene that, working through two other genes, causes *engrailed* to become active. In another related role, *wg* is also involved in setting up the 14 segments that eventually become the main part of the fly's body.

In 1995, the team of Walter J. Gehring, Georg Halder, and Patrick Calladerts of the University of Basel in Switzerland found that a single gene, called *eyeless,* could cause the development of a complete *Drosophila* eye, even when *eyeless* is expressed in a cell on an antenna, wing, or leg. This is a different kind of development gene from *wingless,* one that is totally astonishing to most biologists, even those working in the development field. The eyes that appear as a result of this single gene are normal and complete, and include lenses and cells that respond to light.

Analogs to the *eyeless* gene have been found in animals ranging from flatworms that have rudimentary eyespots to squid, whose eyes are among the most complex of all animals. It is not known, however, if the gene has the same function in animals other than fruit flies. On structural grounds, scientists have long believed that eyes of insects, vertebrates, and mollusks evolved independently from each other, but *eyeless* could still be a common precursor.

Research with fruit flies is now much easier as a result of a new technique announced at the end of 1992. Fruit fly embryos, unlike those of mice and humans, are difficult to keep alive when frozen. The new method enables frozen fruit fly embryos to be successfully revived. The trick is to uses a series of chemical treatments which removes a waxy membrane that otherwise interferes with freezing. The membrane keeps water in the embryo, where it freezes to a glassy ice that interferes with life processes when thawing occurs. The two teams who developed the trick—one headed by Peter Mazur of Oak Ridge National Laboratory in Tennessee and the other by Peter L. Steponikus of Cornell University in Ithaca, New York—also learned that frozen fruit fly embryos revive better when somewhat older. Other embryos freeze and revive better at stages where only a few cells have developed.

Life of an Animal Cell

In the early 1990s, a number of the just-so stories of development reached a point at which the tales could be strung together to provide a developmental overview of cell birth, growth, and death. Most of this narrative was originally discovered from fruit fly mutations or in the study of the development of the tiny *C. elegans* nematode. The astonishing part of the story was reached when researchers found how few tools Nature uses in development. Closely analogous genes in all animals perform the same or similar functions as do the development genes in fruit flies and nematodes.

What Makes the Cell Divide?   A short answer to that question would be the *ras* pathway, a chain of interactions for which activation of a gene called *ras* (originally discovered in rat sarcomas) is the central player. From 1980 to 1993, biologists gradually worked out the details of the *ras* pathway, which seemed especially important because a simple defect in the *ras* gene is one of the common causes of cancer. The normal *ras* protein is found on the underside of cell membranes. Another cell protein, called a receptor molecule, straddles the cell membrane, with the working end of the molecule actually outside the cell. When the receptor picks up a suitable chemical from the cellular environment, its other end, which extends into the cytoplasm, changes to a shape that grabs and holds one of the proteins already inside the cell as that protein floats by. That protein is the SOS enzyme, named for its gene, which is *sos* for "*son of sevenless.*" The oddly named *sevenless* is a fruit fly gene that initiates development of one kind of photoreceptor. The combination of the receptor and the SOS enzyme soon comes into the vicinity of the *ras* protein, which also travels inside the cell membrane. Something about SOS causes the *ras* protein to replace a small inactive molecule that it normally holds with an active version of the same molecule. The active combination is a *kinase,* an enzyme that adds phosphate groups to proteins. The phosphate groups act as signals for a protein to do something. In this case, the *ras* protein uses a phosphate group to tell another kinase, called Raf, to get going. It acts in turn. Eventually, one protein that has had several phosphate groups added to it becomes a signal to the DNA molecule, telling it to make a copy of itself. This is the first step in making a new cell.

What Tells a Cell What to Become?   In all kinds of animals there is an important structure found in the DNA that some have suggested is the defining feature of animals as opposed to other organisms, just as hair and mammary glands are defining features of mammals. As with so many discoveries relating to development in animals, the discovery of this structure occurred first to scientists studying mutations of fruit flies. Some mutations caused a group of cells to develop into organs that did not belong where they appeared—an extra pair of wings behind the first or legs sprouting from the head. The new structures were similar to the normal wings and legs, so they were called *homeotic,* which means "similar." Later, as scientists were able to isolate the genes involved in homeotic mutations, they found that all of them contained a small region in common. The sequence of 183 bases and their partners in the other strand of the DNA was named the homeobox. It is now known that the homeobox is the part of the gene that specifies a segment of a protein. This segment just fits around DNA and switches on the genes that it embraces.

It is the homeobox that defines animals. All invertebrates, from worms to fruit flies, contain 8 genes that have the homeobox, while all vertebrates, from fish to humans, contain 38 of the genes. No plants, fungi, monerans, protists, or archaeans contain genes with a homeobox. All genes that contain the homeobox are called Hox genes and are usually labeled with letters, such as *Hoxa* or *Hoxb.*

Since Hox genes seem to be essential to animal life, the question of what function they have is vital to understanding how animals function. By selectively mutating the Hox genes of mice and chicks, biologists have determined that Hox genes, when active, tell a cell what it is to become, especially when dictating whether a given cell will become part of the front of the animal or the rear.

In a developing vertebrate limb, a number of genes work together, among them *Hoxa, Hoxb-8, Hoxd, Sonic hedgehog, bmp-2,* and *bmp-4.* The key players seem to be

Hoxb-8 and *Sonic hedgehog*. The *bmp* genes produce substances called bone morphogenic proteins that cause cartilage to turn into bone.

What Tells the Cell to Die?   Although cells can be killed by outside forces (a process called necrosis), most cells die by committing suicide, which is called apoptosis. Often apoptosis occurs as an important part of development, shaping the final organism in the same way that a sculptor shapes a block of marble into a statue by removing pieces of the block. An example is the cell that at first blocks the end of the vas deferens in *C. elegans*. The vas deferens in the nematode has the same function as in the male human; it is the tube through which sperm pass from the testes to the outside world. The blocking cell in *C. elegans* is also the cell that knows where the opening of the vas deferens should be, and it leads the tube there. Then it undergoes apoptosis, or commits suicide, providing the tube with an opening.

Apoptosis is sometimes called "programmed cell death," because it seems to

---

Background

In the seventeenth and eighteenth centuries, a popular idea regarding human development postulated that conception somehow produces a very tiny but complete human being who simply grows larger before and after birth until reaching adult size. It is surprising that anything resembling this incorrect idea was ever able to take hold. Aristotle, the father of biology, had accurately described in great detail the regular formation of the organs of a chick in an egg, one of the classic studies of science. William Harvey had clearly stated in 1651 that animals have a uniform pattern of development out of formlessness that begins when an egg is fertilized by a sperm. It was not until late in the nineteenth century that biologists began to understand the pattern that is now so clearly understood, that development is a complex process orchestrated by heredity.

We have long known that as the tadpole grows into a frog, organs such as the gills and tail are dissolved and replaced by lungs and legs. Similarly, the fetus of a mammal may go through stages during which what appear to be organs similar to gills and tails are grown and then absorbed. However, a review in 1981 of the history of the study of how an organism develops from a single cell could accurately end with this summary: the process by which details latent in the germ become translated to adult form remains very little understood. Since then, research has shown that development of an adult organism from a single cell involves common sets of genes that are shared across species, phyla, and, in the most basic instances, across kingdoms of life. Each cell is influenced not only by its genes, but also by neighboring cells and by chemicals in its environment. Cells arise from precursor cells, may migrate some distance through the developing organism, and then die and be absorbed after their role in development is over.

The recognition that development includes cell death did not strike home until 1972, when J. F. R. Kerr, Andrew H. Wyllie, and A. R. Currie described it and gave it the name *apoptosis*, which is an ancient Greek medical term. More often in popular writing the process of apoptosis is termed "programmed cell death" or "cell suicide," although neither of these seems quite accurate. Perhaps "developmental cell death" would more precisely describe the way some cells in development are currently believed to be born for the sole purpose of dying.

happen according to a long-range plan and because it follows a regular sequence of steps: the cell nucleus condenses and its DNA is broken into tiny fragments; the other contents of the cell also seem to shrink as well; the whole cell is broken into small pieces held together by bits of cell membrane; the small pieces can then be engulfed by neighboring cells or by passing phagocytes so that their contents can be recycled.

Much of the work on apoptosis has been conducted as a part of the development analysis of *C. elegans*, which contains two different genes that promote apoptosis and one that interferes with apoptosis, even when the other two are active. According to one theory, the death promoters are always active. Apoptosis occurs when the preventive gene lets it happen. The preventive gene for *C. elegans* is called *ced-9*— the abbreviation *ced* means "cell death defective"—and it has a close analog in vertebrates in a gene named *bcl-2*. In fact the genes are so similar that human Bcl-2, the protein produced by *bcl-2* in humans, will discourage cell death in *C. elegans.*

Another set of studies involves a fruit fly gene with the appropriate name *reaper.* In sharp contrast to the system used by *C. elegans* for apoptosis, expression of the gene *reaper* in fruit flies is enough to start a cell death. It is not clear how *reaper* might interact with those other genes known to be involved in cell death. One possibility is that the *reaper* protein might inhibit the fruit fly equivalent of *ced-9* or *bcl-2*. Another possibility, although somewhat less likely, is that fundamental differences exist between the way that fruit fly cells die and the way worm and mammal cells die. In the July 15, 1995 issue of *Genes and Development*, the same team that found *reaper,* headed by Hermann Steller of the Howard Hughes Medical Institute and MIT, found that a gene near *reaper* named *hid*—for *head-involution defective*—initiates apoptosis in exactly the same way as *reaper.* Thus, even if one of the genes is defective, cell death takes place as programmed. The presence of such a back-up system suggests how important apoptosis is to development of organisms. One 1993 study even found a possible relationship between rapid evolution and apoptosis (*see* "Evolution UPDATE," p 412).

(Periodical References and Additional Reading: *Science* 3-27-92, pp 1646 & 1707; *Science News* 4-4-92, p 213; *New York Times* 5-5-92, p C3; *Science News* 11-7-92, p 317; *Science News* 12-18-92, pp 1896 & 1932; *New York Times* 12-22-92, p C2; *New York Times* 1-5-93, p C1; *New York Times* 2-23-93, p C1; *New York Times* 5-4-93, p C1; *New York Times* 6-29-93, p C1; *Nature* 1-13-94, p 170; *New York Times* 1-11-94, p C1; *Science News* 1-15-94, p 44; *Scientific American* 2-94, p 28; *Science* 2-4-94, pp 610 & 681; *Scientific American* 4-94, p 20; *Nature* 4-14-94, pp 587 & 639; *Scientific American* 5-94, p 16; *Nature* 5-26-94, pp 279; *New York Times* 8-16-94, p C1; *Science News* 8-27-94, p 135; *Nature* 9-1-94, p 15; *Nature* 10-6-94, p 477; *Nature* 10-13-94, pp 560 & 609; *Science* 10-28-94, pp 561, 564, 566, 568, 571, & 575; *Nature* 11-10-94, pp 132 & 175; *Nature* 11-17-94, p 266; *Science News* 11-19-94, p 325; *Scientific American* 1-95, p 17; *Science* 3-10-95, p 1445; *Science* 3-24-95, p 1766; *Nature* 8-3-95, p 380; *Science* 8-11-95, p 753; *Nature* 8-32-95, pp 722, 723, 765, & 768; *Nature* 9-21-95, p 248; *The Sciences* 9/10-95, p 12)

THINKING, REMEMBERING, FEELING

Until recently, most studies of cognition were limited to humans, in part due to the widespread belief among biologists that the perceptions and abilities of other animals differed significantly from those of humans. Today, the biological community is split on the question of whether or not any nonhuman animals have language skills. Many biologists have found a middle ground on the issue, and no one really doubts that advanced mental abilities are present in animals other than humans. Indeed, memory has been effectively studied in mice for many decades, and more recently in sea slugs, flatworms, and roundworms.

Despite the usefulness of animal studies, we have a unique view of human thought processes and emotions. Thus, no matter how much one can learn from observing octopi or chimpanzees, the human brain remains the great fount of cognitive and emotive studies. Furthermore, whatever activities one can dimly discern in animals are often strikingly present or absent in individual humans, enabling a scientist to relate a specific brain structure to thinking, remembering, feeling, speaking, viewing, or understanding in a way that is virtually impossible for animal subjects, even when parts of the animals' brains have been destroyed in the course of scientific experiments (*see* "The State of Biological Science," p 325).

Among the traits that are virtually impossible to duplicate in nonhuman subjects are mental illnesses caused by brain abnormalities. Although it sometimes appears that easily observed animals (such as pet dogs) are mentally ill, this is difficult to authenticate. It remains difficult for experts to agree even on whether a particular human is schizophrenic or not, although new methods of brain scanning suggest that there may be objective physical means for identifying this particular mental illness.

Thinking?

Sophisticated people make fun of movies in which dogs sense that their human friend is in some kind of trouble and come to the rescue with a clever plan boldly carried out. Lassie has done it again! A critique of such a scene might include the assumption that dogs cannot think and solve complex problems the way humans can. Scientists studying animals that seem capable of solving such problems are trained to assume that something other than human-type cognition is at work—instinctive behavior or conditioned responses, but not *thought*. Even scientists who admit that animals can solve problems often deny that creatures other than humans have the ability to be aware of themselves—to have what is often called *consciousness*.

Some experiments that purport to show animal cognition are difficult to dismiss. One series of experiments with tame ravens, reported in 1993 by Bernd Heinrich of the University of Vermont in Burlington, involved a piece of meat on a string lowered from a perch. To obtain the food, the bird had to raise the meat to perch level. Since a bird lacks grasping hands, it would need to lift the string a few inches with its beak, step on the string with its foot, and then repeat the process. The string was long enough that it would take about 20 repetitions to bring the meat to perch level. Crows had previously been given the same problem, but they did not succeed in solving it, although they pecked at the string for a few moments before giving up. When presented to five tame ravens, however, the five spent about half a day just looking at the apparently inaccessible food. Then one of them flew over to the perch, and immediately began to haul the meat up using the beak for lifting and the foot for holding in place. It performed flawlessly on its first attempt to get the meat, and repeated the performance flawlessly thereafter. Furthermore, three of the remaining four ravens also developed similar—but not the same—methods of using their beak and foot to get the food. Since their movements were somewhat different from the first, it would appear that they solved the problem on their own, rather than by copying. Further experiments in which the string setup was changed in various ways also suggested that ravens could solve problems unlike any encountered in the wild, and that each raven did so separately.

When the experiment was tried with wild ravens, 1 of 27 tested solved the problem in just fourteen minutes, but 20 failed to solve it at all. All of these results suggest that ravens can conceptualize problems and solve them in much the same way humans do.

Other wild birds also have a surprising ability to think. Recall the velociraptors in *Jurassic Park*—intelligent dinosaurs that hunted cooperatively. The author may have been inspired by the feats of intelligence of those modern dinosaurs (*see* "Dinosaurs and Birds," p 492), the raptors: hawks, eagles, falcons, and their relatives. These birds of prey have consistently demonstrated cooperative hunting behavior when attacking larger and more dangerous, or more elusive animals, although they tend to hunt alone for small animals such as mice and songbirds. In a 1993 article in *BioScience,* David H. Ellis of the U.S. Fish and Wildlife Service and coworkers cataloged numerous examples of raptor behavior involving a cooperative pair or even a pair of packs. Generally one bird would flush the prey for a closely following second bird. Harris's hawks patrol for rabbits in two groups that move throughout the scrub leapfrogging each other. When one group sights a rabbit, it signals the other team, and the hawks converge on the rabbit from all directions. The Harris's hawks also team up to flush rabbits out of the brush or use tandem pursuit of a rabbit to insure that a fresh hawk is leading the charge at all times. Every few moments, the lead hawk drops out and rests a bit while the next hawk in line takes the lead. Eventually, the rabbit tires and all the hawks share the food.

Another experiment with a different animal caused biology professor David Sloan Wilson of the State University of New York to state that "we have to revise upwards our estimates of the cognitive abilities of nonhuman animals." Guppies appear to be smart enough to make themselves look good to a potential mate by hanging with guppy nerds. In the investigation, male guppies were permitted to observe how close a female guppy would permit another male to approach. For male guppies, being allowed to approach is the equivalent of an invitation to mate. What the observing guppies didn't know was that some males were prevented from approaching the female by an invisible barrier placed by the researchers. To an observer who could not see the barrier, it looked as if these males had no sex appeal whatsoever.

After staging the scene, Lee Alan Dugatin of the University of Missouri in Columbia and Robert Craig Sargent of the University of Kentucky in Lexington allowed the observing guppy to mingle with the other males he had seen. The observer immediately headed to the side of the unsuccessful male and lingered there. Controls who had not observed the little scene of the successful and unsuccessful males did not

SOME ANIMALS THAT WERE THOUGHT TO THINK

Scientists have long been skeptical of thought in animals other than humans, particularly since the early twentieth century episode in which the horse Clever Hans was shown to be clever only in the presence of his owner. The owner, unknown even to himself, was signaling the horse to stop when a correct answer was reached. There is, however, a long tradition of animals whose actions appear to be informed by cognition. Those who want to believe will believe, and those who do not want to believe will look for other explanations.

In 1871, Darwin himself reported that chimpanzees use stones as tools and that orangutans appear to know the principle of the lever; but these behaviors could be attributed to an advanced sort of instinct. As early as 1917, however, one of six hungry chimpanzees showed that it had a different instinct or insight than its fellows when it pushed a crate under otherwise out-of-reach bananas. The instinct of the other five chimps was to leap in vain for the bananas.

In 1948, E. C. Tolman demonstrated that rats can learn from a single experience and can apply that learning to a somewhat different situation.

In 1953, scientists were observing and feeding a colony of Japanese macaques in order to study their behavior. A now-famous Japanese macaque named Imo learned to wash sand off a sweet potato before eating it, and the practice gradually spread over nine years to most of the other macaques in the colony. During that nine-year period, Imo discovered or invented a way to improve the quality of another provision provided by the scientists, using water to winnow out the grain from the sand; that practice also spread slowly through the colony.

In the 1960s, R. J. Herrnstein claimed that his experiments showed that pigeons regularly use abstract concepts.

In 1992, Graziano Fiorito and Pietro Scotto of the University of Reggio Calabria in Catanzaro, Italy, showed that the common octopus of the Mediterranean (*Octopus vulgaris*) can not only learn to choose between a red and a white ball, but that octopi learn faster by watching the training of another octopus in an adjacent aquarium than they do from being trained themselves. Most biologists believe that octopi are the most intelligent of the invertebrates.

---

choose between the two. Thus, the observing guppy was able to recognize at some level that he might have a better chance of mating if he hung around with a guppy that did not appeal to females. Whether or not this should be labeled cognition, however, is debatable.

Remembering

Although there is considerable disagreement on how well animals can reason, there is no doubt that even the lowliest animals learn from experience. Controlled experiments have shown that animals can learn to work their way through mazes or to perform various simple or even complex tasks to obtain rewards. Observation of animals in the wild shows that learning is not simply a laboratory feat—anyone who has ever seen a squirrel teach itself how to raid a bird feeder can attest to this. Even if several days of squirrelly attempts fail, a successful method, perhaps found by chance, is ultimately discovered and repeated regularly. The squirrel has learned the solution to the problem, whether or not it has thought it through.

Recent animal experiments have largely been concerned with finding where in the brain learning occurs, what actually happens chemically when some event is remembered, and ways to alter (especially to improve) memory.

Rats   In the January 1994 *Proceeding of the National Academy of Sciences*, Ursula Staubli of New York University in New York City, Gary Lynch of the University of California, Irvine, and coworkers reported that a drug they had developed significantly improves the ability of rats to solve mazes, presumably because the drug enables the rats to remember the maze better. Rats injected with the drug (one of a class called Ampakines) perform twice as well on a second exposure to a specific kind of maze as do control rats, even hours after all the Ampakine has left the drugged rats' brains.

The main theory of memory for all animals, from worms to humans, is that experience produces changes in the chemical and electrical connections between brain cells. These changes are often called long-term potentiation. At a subsequent time, the brain can somehow locate these enhanced pathways to partially recreate the experience. This is a bit like the much-better understood way that a computer stores the memory of keystrokes by enhancing a certain arrangement of magnetic domains on a disk, which can then be reconstructed by retracing that part of the disk. However, for animal memory, if the chemical theory of long-term potentiation as the cause of memory proves true, a more powerful chemical response might allow animal memory to develop faster or last longer, while a more powerful magnet would not necessarily improve computer memory.

Ampakines are not the only drugs known to increase the strength of chemical or electrical signals between nerve cells. Caffeine and certain other known drugs do that, and also are known to enhance learning slightly. Because caffeine affects the connections between all the nerves, it cannot make a very great difference in enhancing selected parts of the system (for example, those that are involved in learning). Instead, after that first small boost from a cup or two of coffee, additional caffeine makes an animal jumpy and irritable, not more learning retentive. Scientists working with Ampakines note that there is no increase in either the jumpiness or alertness of rats given the new drugs, so they hope that Ampakines simply increase activity within those connections used specifically in learning and not connections throughout the brain. Positron emission tomography, however, shows that Ampakines are active throughout the brain. Thus, many scientists not directly involved in this research remain skeptical about whether these drugs would meet the ultimate goal of improving long-term memory in humans.

Rats are among the animals studied in which one kind of memory storage seems to originate largely in a brain structure called the hippocampus (*see* Background *and also* Birds, below). The hippocampus is also thought to be used in establishing the position of objects in space, including the recognition of a particular place. Studies since the 1970s have shown that putting a rat in a particular location actually leads to specific patterns of activity in the hippocampus. There is also some evidence that these place-patterns are stored in long-term memory by signals generated within the hippocampus. This storage capacity sometimes occurs during sleep after a day spent exploring new places. In the July 29, 1994 *Science*, Matthew A. Wilson and Bruce L. McNaughton of the University of Arizona in Tucson, Arizona, reported that three rats they studied did replay the hippocampal place-patterns of the day's events during one phase of sleep. Their research suggests that this kind of declarative memory is formed when the hippocampus signals the rest of the brain to form the memory. That memory is then no longer localized in the hippocampus. Upon return to the same place on another occasion, the memory pattern in the brain is activated, causing the original pattern of nerve cells to fire.

Mice   Eric Kandel of Columbia University's College of Physicians and Surgeons in New York City is the dean of memory researchers. He discovered that the protein called CREB (short for Cyclic-AMP Response Element Binding protein) is involved in long-term memory. Alcino Silver of Cold Spring Harbor Laboratories in New York reported in the October 7, 1994 *Cell* that mice who are genetically engineered to lack CREB are still capable of learning tasks, but have much greater difficulty in remembering what they have learned. One such task is locating a submerged platform in a pool of murky water. Normal mice remember for a week the unpleasant experience of being tossed in the pool; they find the platform much faster when subjected to the same experience a week later. Mice with no CREB must be tossed in the water a dozen times a day before they can remember where the platform is from one swim to the next.

Flies   In the April 7, 1994 *Cell*, Jerry C. P. Yin of Cold Spring Harbor reported that fruit flies genetically engineered to produce a particular configuration of CREB can remember tasks after just one trial. Wild-type fruit flies with the normally occurring CREB configuration take about 10 practice sessions before remembering similar tasks equally well. In October 1994, Yin and his coworkers found that yet another configuration of the CREB molecule prevents memory in fruit flies.

Birds   In the November 1994 *Proceedings of the National Academy of Sciences*, Fernando Nottebohm of Rockefeller University in New York City and Anat Barnea of Tel Aviv University in Israel reported that, each October, black-capped chickadees in the wild grow a large number of cells in the hippocampus. Caged chickadees, fed on a regular basis and protected from the exigencies of autumn weather, show only half the hippocampal growth of the wild ones. The researchers attribute the difference to the challenges of autumn weather, which encourages new growth to face the coming rigors of winter.

Songbird brains appear to be more plastic than mammal brains. Although bird brains are notoriously small, birds actually show a great range of learned behavior. It would be unacceptable to say that the ability to lose old cells and grow new ones on a regular basis evolved in order to enable small brains to remember effectively; on the other hand, no matter how it came about, the plasticity of the bird brain seems to be an effective adaptation. Think of the song of the mockingbird, the nest of the bowerbird, the speech-imitation of the parrot, or the ability of the crow to count. A lot goes on in these small brains, and it seems to help to clear away the dead wood from time to time and bring in a few new cells. Humans sometimes wish that they could do the same, but this ability apparently does not reside in the mammal brain. Mammals, especially humans, usually have many brain cells to spare for most tasks, cells that seem to be unused and therefore available.

Petri dishes   Experiments *in vitro* can also be used to learn about memory processes based upon the theory of long-term potentiation. One set of investigations announced in 1994 by Daniel V. Madison of Stanford University School of Medicine in Stanford and Erin M. Schuman at the California Institute of Technology in Pasadena established the effects of the ubiquitous chemical messenger nitric oxide on long-term potentiation of synapses. As has been hinted at in earlier work by Schuman and Madison and a host of other investigators, nitric oxide produced at one synapse diffuses through the body to nearby synapses, where it strengthens connections in synapses that are receiving nerve impulses. Thus, memories appear to be formed in a group of adjacent synapses, not in a pattern scattered through the brain. Nitric oxide can only affect nearby synapses, as it degrades rapidly. However, other research in 1994 also suggested that nitric oxide, by diffusing through a small patch of brain, causes chemical changes that make an entire small region behave as a single entity.

Feeling

Scientists once believed that the most basic emotions, such as fear, anger, or love, originated in chemical reactions (hormones), rather than originating in the brain. Scientists now recognize that a different group of chemicals, those produced by nerves in the brain and elsewhere, produce the emotions. The hormones are merely messengers. Before the nerves begin to produce their chemicals, however, they have to be stimulated. The knotty origin of emotions, whether in humans or other animals, is difficult to unravel. Learning, hormones, and nerve interactions are all involved in the process together.

In the October 20, 1994 *Nature*, a report by Larry Cahill, Michael Weber, and James L. McGaugh of the University of California, Irvine, and Bruce Prins of the Long Beach Veteran's Affairs Medical Center in Long Beach, California, focused on one aspect of the interaction of emotions and the brain: memory. McGaugh's rat studies throughout the 1980s and into the 1990s suggested that the hormones produced in the stress reaction (primarily epinephrine and norepinephrine) enhance memory. The emotional signals used by rats, however, may differ from those used by humans. The study in *Nature* reported that memory in humans is affected by the administration of a beta blocker (propranolol hydrochloride, tradenamed Inderal). In a double-blind test with controls, subjects were given either a dose of Inderal or a placebo, and then asked to view one of two different tapes—either an emotionally charged story in which a young boy suffers a terrible accident that nearly kills him, or a story with similar details in which there is no accident. The beta blocker suppressed hormone production but did not prevent the subjects from becoming emotionally involved with and upset by the story of the accident. However, the memory of details in the accident story was significantly greater a week later among those who took the placebo. In fact, it appeared that memory was somewhat decreased in those subjects whose hormone production was suppressed by the beta blocker. Memory of the unemotional story was not affected by either the drug or the placebo. In all, the strongest, most detailed memories were demonstrated by the subjects who took the placebo and viewed the more emotional story. This study established a mechanism for the long-suspected connection between strong emotion and clear recollection.

Human Brains

The key question, or course, is whether the human brain accomplishes its astonishing feats as a result of its greater size (or at least a higher ratio of size to body weight) or because of some other factor. If intelligence is dictated only by larger size in proportion to body weight, then it can be (and has been) argued that dolphins are more intelligent than humans because of their even greater brain-to-mass ratio.

Scientists who measure the influence of heredity on human personality traits believe that half of the variance in personality traits expressed by individuals stems directly from genetic makeup, while the remaining half of the variance is due to some other factor, probably experience. The effects that genes can have on cognitive ability have been best explored through study of genetic disorders that include some form of mental retardation as part of the syndrome. In 1993, for example, researchers at the University of Utah in Salt Lake City established that a deletion on chromosome 7 causes a selective mental retardation. The deletion halts expression of the known gene for the protein elastin, which is the main protein in elastic fibers in the body. A region of the chromosome located near this gene is also lost because of this deletion. While abilities to perform problem solving, most spatial tasks, and mathematics are greatly impaired with this deletions, language, music, and sociability seem to be enhanced. This rare combination of traits was recognized as early as 1961 and is known as Williams syndrome.

Autopsies on two persons with Williams syndrome suggest that no one specific region of the brain is involved. Instead, most of the brain is shrunken in size, yet packed with nerve cells. The frontal lobes and neocerebellum, thought to be involved with meaningful language, are the exceptions to this, along with that part of the brain involved in auditory discrimination. One theory holds that the excess of nerve cells

---

Background

The human brain is thought to contain 100 billion or so nerve cells (neurons), as well as 10 other brain cells for every neuron. Furthermore, each neuron produces a long cable, called an axon, that leads to perhaps 1,000 nerve fibers, each connecting to suitable similar fibers put forth by other nerves. These 100 trillion connections are called synapses, and it is thought that most of the important stuff the brain accomplishes is done at the synapse level. In particular, memory is thought to occur through the selective strengthening of certain patterns of synapses, while emotions are mediated by chemicals that increase or decrease particular types of signaling systems used by the synapses.

Signals between synapses occur in two main ways: chemical and electrical. A rapid change in the electric potential of a cell is called **firing.** A cell fires when a signaling chemical opens a gate, permitting electric charges to move from one side of the gate to the other. However, electrical signals between nerves are different from electric currents in house wiring or in a computer, which are carried by the movement of loose electrons. In connections between nerves, electrical signals are carried by charged atoms, in which electrons are bound to the atom, not loose. Thus, even electrical signals between nerve cells are actually chemically based.

The chemicals used in transmitting signals are generally called **neurotransmitters**, but the neurotransmitters consist of a large number of different kinds of chemicals. They utilize different mechanisms of action and often act on different populations of synapses. Among the neurotransmitters thought to be involved in cognition is GABA (gamma amino-butyric acid). Chemicals that may have a role in memory include acetylcholine, norepinephrine, glutamate, serotonin, ACTH (adrenocorticotropic hormone), vasopressin, and carbon monoxide. Emotions are known to be affected by almost all the neurotransmitters, but particularly by dopamine, serotonin, enkephalins, endorphins, and anandamide. Emotional memory enhancement is apparently mediated by epinephrine (adrenalin) as well as norepinephrine, as discussed above.

Memory is not a single type of brain activity. There are several different types of memory, often categorized on the basis of duration:

Short-term memory   Most events can be stored for a few seconds or minutes, but cannot be recalled consciously, or perhaps even unconsciously (this is still debatable), after a longer period of time. There may be more than a single type of short-term memory. The kind of memory that holds the beginning of a sentence in your mind until the end is reached may not necessarily be the same as the type of memory used in looking down at a footpath and then away from it. Still different may be the type of memory involved in deliberately remembering a phone number until it is dialed, or deliberately remembering someone's name during a dinner party.

Long-term memory   As the name implies, these are the memories that seem to be permanently etched in the mind, although they often fade with disuse. Your current telephone number usually appears in your mind whenever you need it, but your telephone number from years ago may or may not be available. It often seems that long-term memories never leave the brain, although the ability to fish them out from the files gradually fades.

Long-term memory can also be classified by the way in which the memory is acquired.

Declarative (episodic) memory   Single events (such as a particular encounter on a vacation day or a burn from a hot stove) can be remembered indefinitely, although most encounters and days fade into each other, and touching a cool stove produces no particular memory.

Procedural memory   Repeated activities can become skills, such as playing the piano or riding a bicycle. Even if such skills grow rusty, less repetition is needed to bring them back.

Damage to the brain can result in the loss of one type of memory, while other types of memory are left intact. This implies that the various categories of memory must utilize different mechanisms. Furthermore, certain types of brain damage can interfere with the subsequent formation of new memories of a particular type, while leaving intact memories of that same type which were formed prior to the damage.

occurring in the smaller brain results from loss of a gene product that tells brain cells to die (*see* "Development Developments," p 342).

In most people, the brain is shaped by production of excess nerve cells which are later pruned away. Research has shown that male canaries use a similar mechanism for developing a new song each mating season. The old brain cells in the region responsible for song die, and new ones grow each year (*see* above for Chickadees). In humans, however, new brain cells either do not grow after such a pruning or do so only under very special circumstances. The high number of brain cells crammed into a smaller-than-normal brain in a person with Williams syndrome suggests that all the cells formed early in life are still there, taking up space and energy. As a consequence, those cells usually marked for further growth cannot grow.

Learning in humans appears to operate on the same basis as learning and memory in other animals. For example, the same issue of *Science* that reported on declarative memory in rats moving into long-term memory during sleep also included an article reporting that procedural memory in humans is enhanced during sleep. Other studies have shown that students retain more information if they sleep after studying than if they cram all night. Sleep in mammals, at least, appears to be necessary for proper memory functioning.

(Periodical References and Additional Reading: *New York Times* 1-19-93, p C1; *Natural History* 10-93, p 51; *Science* 1-28-94, pp 466 & 532; *Science* 2-18-94, p 973; *Scientific American* 5-94, p 16; *New York Times* 7-29-94, p A11; *Science* 7-29-94, pp 603, 676, & 679; *New York Times* 8-2-94, p C1; *Science* 10-14-94, pp 218, 221, 223, & 294; *Science News* 10-15-94, p 264; *Nature* 10-20-94, p 702; *New York Times* 10-25-94, p C1; *New York Times* 11-8-94, p C1; *New York Times* 11-15-94, p C1; *Nature* 12-8-94, pp 498 & 519; *Science News* 4-22-95, p 253)

C. elegans UPDATE

Since the year that the Beatles had their first hit and John F. Kennedy was shot, dedicated workers have been exploring everything there is to know about the tiny worm *Caenorhabditis elegans*. During this time, they have learned virtually all the physical details of the worm's makeup, and are beginning to understand exactly how the worm behaves, having closely examined the chemical signals passing from cell to cell that cause motion, sensation, reproduction, and the rest of the worm's rather limited repertoire. While this is a great intellectual adventure, it has also had important implications for developmental biology in general. For we now know that as the worm turns, so does the human move a muscle. Nature has been revealed to use a limited number of devices to achieve the incredible complexity of life on Earth.

Here are some of the more recent revelations about *C. elegans*.

Aging and Death

The nematode *C. elegans* figures in two surprises connected with mortality. As long ago as the August 24, 1990 *Science*, Thomas E. Johnson of the University of Colorado in Boulder had reported a connection between nematode reproduction and length of life. Mutant *C. elegans* with a defect in the gene *age-1* develop a four-fold decrease in reproductive capacity, but enjoy a 65% increase in average lifespan, and a 110% increase in the maximum lifespan. The improvement in length of life derives from a decrease in the mortality rate for older worms, not from a lower death rate at all ages. Interestingly enough, the gene *age-1* had originally been given its name for other reasons, despite its unexpected aptness.

The life-lengthening effect of *age-1* appeared to be related to another discovered cause for an increase in *C. elegans* lifespan, which was announced by Wayne A. Van Voorhies of the University of Arizona in Tucson in the December 3, 1992 *Nature*. Van Voorhies found that sperm production in both hermaphroditic nematodes (the most common sex for the nematode) and in male nematodes decreases length of life, which suggests that perhaps the mutant *age-1* effect described by Johnson was also a result of decreased reproductive ability. The factor that Van Voorhies found was related to sperm production only; nematodes that had been genetically manipulated to produce no sperm lived about twice as long as male nematodes who regularly make sperm. The spermless nematodes were able to conduct sexual intercourse, and did so normally and frequently. The research began when Van Voorhies discovered that prevention of mating increased lifespan somewhat, but later work clearly implicated sperm production as the causative factor. Some scientists, learning of Van Voorhies' discovery, suggested that the longer length of life of human females might also be due to lack of sperm production, although that is a big leap.

Factors other than sperm production are also at work in the death rate for the nematode. In the December 2, 1993 *Nature*, a team from the University of California, San Francisco, reported that a specific mutation of a single *C. elegans* gene more than doubles lifespan. The average worm with the mutated gene lived 48 days, compared to 18 days for the wild type. This time the gene involved is not *age-1*, but a developmental gene named *daf-2*.

The normal role of *daf-2* is to instigate a special kind of larval stage that conserves energy during bad environmental conditions, thus permitting the nematode to live longer in hopes of better times. The mutated *daf-2* fails to produce that special larval stage, yet still produces the longevity. If a human could do the same, life expectancy would increase to about 150 or 160 years. While there is no obvious application of this result to humans, the discovery is a powerful example of how genes affect length of life.

Unlike the effect of *age-1*, *daf-2* not only extends lifespan, but also affects the aging process in a positive way. Mutant worms appear healthier and more muscular at any adult age than do the wild types.

How the Worm Turns

Since publication of the Worm Book in 1988, which contains the developmental fate and wiring diagram of the nematode *Caenorhabditis elegans*, several research teams have used a number of different methods to determine exactly how the 302 nerve cells in *C. elegans* function. Perhaps the most successful has been that of Steven L. McIntire of the University of California, San Francisco. He and his coworkers have been able to trace exactly which nerve cells are active in a number of the worm's basic behavioral activities.

The behavior of a tiny roundworm is fairly limited. Among other occupations, *C. elegans* can wiggle in ways that move its body forward or backward; it can move its head back and forth as it looks for food; and it can defecate. In all of these cases, 26 of the nerve cells that use the neurotransmitter GABA (*see* "Thinking, Remembering, Feeling," p 354) are a large part of the action, so much so that the behaviors cease in worms that do not produce active GABA. For example, when nerves on one side of the worm's body tell muscles to contract to produce locomotion, they also activate the neurons on the other side of the body that produce GABA. The GABA inhibits contracting nerves on that side, so the worm wiggles. If GABA is inactivated, the muscles on both sides contract at once, and the worm shortens instead of wiggling. A similar GABA system controls the head sweeps. Something different happens in defecation, since the GABA-producing nerve cells cause muscles to contract, instead of inhibiting the nerve signals to the muscles.

One technique McIntire and his team used in this work involved selectively inactivating nerve cells with lasers. The team also studied various mutations, which led them to identify the genes behind these instinctive behaviors of the nematode. The identified genes are *unc-25,* which makes the protein that makes GABA; *unc-46* and *unc-47* that are needed to release GABA; *unc-49* which makes a protein needed to cause muscles to respond to GABA; and *unc-30,* which apparently encodes the

---

Background

Biological convention dictates that one should refer to a species with its genus abbreviated to its first letter for second or later references, yet *Caenorhabditis elegans* has become so famous that more and more first references use the abbreviation *C. elegans*. Almost as often, this tiny nematode—only 0.04 in (1 mm) long—is simply called "the worm." The worm is suitable for many projects for three basic reasons: it has a life cycle of just three days; it is nearly transparent; and because of its tiny size, up to 100,000 individuals can be kept in a laboratory dish.

Originally, the worm became famous because it was chosen to be the first multicellular organism for which the developmental fate of every cell was traced. The developmental fate of a cell is the history of the cell, from its origin as a daughter through its travels to its proper place in the organism, and to its final cell death, if that is the intended fate. This intimate knowledge of the worm's 959 cells is unprecedented. Many hope that this feat will eventually be duplicated for other animals, and ultimately for humans.

A decade-long project is still under way in mapping the worm's genome. In addition to its purely scientific interest, the gene mapping has other practical applications. Because so many genes are conserved by evolution, mammals (including humans) use the same genes to manufacture the same proteins. Mutations in worm genes corresponding to the human disease hypertrophic cardiomyopathy and to brittle bone disease have already been found. In November 1990, Paul Sternberg of Caltech described a worm gene that is similar to the *ras* gene that is implicated in many human cancers.

development of the ability to produce GABA in the 19 nerve cells used in wiggling, as well as in some other nerve cells.

In 1995, several workers also found that the protein $G\alpha_0$, previously thought to be involved in development of the nervous system, is part of pathways that control egg laying and male mating, as well as being another part of the system controlling locomotion.

TIMETABLE OF THE WORM

1963   Sydney Brenner of the Medical Research Council's Laboratory of Molecular Biology (MRC) in Cambridge, England, conceives of a project aimed at learning the complete development of a multicellular organism. He chooses *C. elegans* because of its transparency and its small number of cells.

1983   John Sulston of MRC and coworkers on the Worm Project succeed in sequencing the development of each of the 959 cells that make up a single *C. elegans*, establishing the parentage of each cell and the path to its adult location.

1986   John White, Eileen Southgate, and Nichol Thomson of MRC establish the location of every nerve connection among the 302 neurons of *C. elegans*, a result known as the wiring diagram.

1988   The Worm Book, formally known as *The Nematode* Caenorhabditis Elegans, edited by William B. Woodand the Community of *C. elegans* Researchers, is published by Cold Spring Harbor Laboratory. It contains all the data collected on the Worm Project to that date.

1990   In April, Alan Coulson and John Sulston of MRC, working with Richard Wilson, Robert Waterson, and coworkers at Washington University in St. Louis, Missouri, make available sheets of filter paper containing 95% of the *C. elegans* genome broken down by gene.

1991   The Cambridge and Washington University laboratories have sequenced 200,000 bases out of the expected 100 million bases in the DNA of *C. elegans*.

Catherine H. Rankin of the University of British Columbia and Norman Kumar of the University of Toronto report that *C. elegans* can be used in learning studies because it can be conditioned.

1994   Workers led by Robert Waterson of Washington University succeed in sequencing 5,000,000 base pairs of the genes of *C. elegans*, including a continuous 2,200,000 base-pair sequence.

1995   Gene sequencing on the Worm Project reaches a rate of 10 million base-pairs per year.

2000   Worm Project workers are scheduled to complete the base-by-base map of the *C. elegans* genome.

(Periodical References and Additional Reading: Nature 12-3-92, p 458; *New York Times* 12-3-92, p A18; *Nature* 7-22-93, pp 282, 334, & 337; *Nature* 12-2-93, pp 404 & 461; *Nature* 12-22/29-94, pp 780; *Science* 1-13-95, p 172; *Science* 3-17-95, pp 1597, 1648, & 1652; *Science* 6-2-95, p 1270; *Science* 7-28-95, p 468)

Genome Project UPDATE

After a very slow start, progress on the Human Genome Project (HGP) sped up rapidly in the mid-1990s. The project is divided among several nations and, in the U.S., nine different national laboratories. From the founding of the project in 1989 to the end of 1994, the U.S. laboratories primarily conducted preliminary studies. In France, however, researchers plunged into the development of what is termed a physical map of the human genome, the sum total of the genetic information of the human species.

The Physical Map and the Genome Directory

In December 1993, Daniel Cohen, director of the Centre d'Etude du Polymorphisme Humain in Paris, France, released a map of more than 90% of the human genome (a later, corrected version proved to cover only 75% of the genome). Cohen and his coworkers use large sections of yeast DNA called megaYACs to organize the fragments of human DNA. This is possible because yeasts, although fungi, have many more genes in common with animals than do bacteria, whose genes have also been intensively studied. The map consists of landmarks that can be used to find a particular gene on a chromosome, although the landmarks were still too far apart for that purpose at the end of 1993. Work sped along, however, and the Cohen group soon had a map with a resolution of one landmark every 100,000 base pairs or so. By early 1995, two chromosomes, 16 and 19, had been sequenced so extensively that the distance between landmarks was 10 times closer. By early 1996, a successor group at Généthon led by Jean Weissenbach had developed a map with a resolution of about one in every 700,000 base pairs.

The main secret of the French success has been intensive use of automated mass production, a technique that other researchers began to utilize more extensively in 1994 and 1995. The initial impetus for automation came from a campaign to locate those genes responsible for muscular dystrophy. Money raised towards this effort supplied Cohen with an advanced and well-funded laboratory called Généthon.

In September 1995, the magazine *Nature* made an important and unusual contribution to the field by publishing what they called *The Genome Directory,* which included the latest version of the physical map along with some 55,000 Expressed Sequence Tags (or ESTs), each corresponding to a gene. This will enable scientists to put the ESTs and the physical map together, greatly simplifying the search for a particular gene on the genome. Knowing an EST does not reveal the function of a particular gene, however. Most of the ESTs were developed by people working with J. Craig Venter at The Institute for Genomic Research (TIGR) in Gaithersberg, Maryland, although a second group at Washington University in St. Louis, Missouri, with funding from Merck pharmaceuticals, has also been developing ESTs, which were included in *The Genome Directory.* TIGR is a nonprofit institution, but it is closely allied to the commercial company Human Genome Sciences.

Because of certain attributes of ESTs, they can reveal not only the gene but also the body tissue in which the gene is expressed—an important benefit for researchers. Although ESTs are rapidly moving toward identification of every human gene, know-

WHAT ARE GENE MAPS?

Note that the question is not "What is a Gene Map?" There are a number of different kinds of maps, some more useful than others.

Linkage Maps   The first forms of gene maps were linkage maps that localized a particular gene on a long stretch of a specific chromosome. These regions are identified by a marker that is a specific fragment of a DNA molecule that appears when the whole molecule is sliced up by a given enzyme. Since thousands of these linkages are known, linkage maps can be used to give a rough idea of the locations of specific genes. Linkage maps are available now.

Physical Maps   Making a physical map begins by using several different enzymes to cut human DNA into pieces. The pieces are then assembled on large stretches of yeast DNA to make what are called "yeast artificial chromosomes," or YACs. Very long stretches are called megaYACs. These megaYACs can be put together like a jigsaw puzzle by matching fragments at each end. When all the megaYACs have been joined, the result is a complete physical map, so called because it can be used to locate the position of any specific sequence on the genome. For example, a disease gene located on a linkage map can be tossed in a pot with various megaYACs; the specific megaYACs to which the disease gene clings reveals the physical location of the disease gene on the genome, at least to the length of the megaYAC. By using overlapping megaYACs, or by breaking apart the existing megaYACs, the physical map can be improved, or given higher resolution. A physical map of the human genome is expected to be complete early in 1997.

The Complete Map   The complete map consists of a full list of the three billion base-pairs in their exact order on each chromosome. Because humans are not all identical, in most cases there are several versions present of each gene or intron. The differences between functional alleles (different versions of a gene that produce some workable protein) are small, however. A truly complete map would include the different alleles of each gene. The full-sequence map is expected by 2005.

In addition to maps, there are other useful ways of getting to the genes. Just as a person mailing a letter or making a phone call does not need a map, requiring only an address or telephone number, so it is possible to retrieve a gene without having much of an idea where the gene is on the genome. An equivalent of the address or phone number is the Expressed Sequence Tag, or EST. An EST is a piece of artificial DNA comprised of bases which are complementary (matching) to those bases which make up the sought-after gene. RNA made by the gene during the process of protein synthesis can be used to create the EST (*see* "Genetics and Beyond," p 336). When the EST is mixed in with bits and pieces of the genome, it gloms onto the gene from which it was formed. The concept of the EST and its continued use is due mostly to J. Craig Venter.

ing all the genes will not complete the Human Genome Project. As discussed below, the nongene DNA is now thought to be important, and ESTs do not reveal anything about that part of the DNA molecules.

Success with Bacteria and Other Organisms

Early in 1994,it was dogma that even the best biological laboratories could sequence only about two million base pairs each year—far too slow for the HGP's original goal of completing the sequencing in 2006. In 1994, however, J. Craig Venter stated that he expected the majority of genes to be located by the end of 1995 using methods he had developed while at the National Institutes of Health. For many purposes, mapping the location of genes is even better than simply knowing the base-pair sequence, the avowed goal of the Human Genome Project (*but see* below). Initially, few thought that Venter could contribute significantly to base pair sequencing. Although it was clear that his methods would identify genes, they would not always delineate function. Thus, many were surprised when Venter's team from TIGR in Gaithersberg, Maryland, working with Hamilton Smith of Johns Hopkins University in Baltimore, became the first to completely sequence all the base pairs in a free-living organism, the bacteria *Haemophilus influenzae*. This accomplishment was announced on May 24, 1995, and printed in the July 28, 1995 *Science*. The number of base pairs in the *H. influenzae* genome (1,800,000) and the number of genes (1,743) are both considerably less than those numbers in a human. Still, the methods that TIGR used for *H. influenzae* can be adapted to the HGP and could speed up the whole process.

The *H. influenzae* task took 13 months at a cost of $0.50 per base. The price is about what was originally expected for sequencing the human genome, but at that speed the task of sequencing the human genome would take nearly 2000 years. However, the Venter team has already demonstrated how much faster it can go. TIGR took only 4 months to sequence its second bacterium, *Mycoplasma genitalium*. Meanwhile, the sequencing of *Escherichia coli,* a high priority of the HGP since 1990, became bogged down in old technology and has not been completed. Venter's team may try to complete the *E. coli* project using its own approach.

Sequencing of the genome for the yeast *Saccharomyces cerevisiae*, the common yeast used to make bread, has also been completed. Chromosome 3 on *S. cerevisiae* was the first full chromosome of any kind to be entirely sequenced. This was accomplished in May 1992 as a cooperative project by the European Community. By late 1995, three of the yeast chromosomes had been fully sequenced and the others were on their way. In March 1996, the complete gene sequence was announced.

Although the base-pairs have been recorded, the functions of nearly half of the genes in the yeast genome have yet to be defined. Something similar could be true of the human genome. Because genes in yeast and in other higher organisms are remarkably similar, by the time the human genome is sequenced it is likely that many of the now unknown genes will have been identified in yeast or in some other organism with a smaller genome. Furthermore, the ESTs of the human genome are helpful, although they do not tell what the gene does either.

Progress also continued throughout the early 1990s toward a linkage map of the mouse genome, being developed by Eric Lander and a group at the Whitehead

Institute of MIT in Cambridge, Massachusetts. The group announced in the March 14, 1996 *Nature* that they had, after five years of effort, located 7377 genetic markers, which together constitute a dense genetic map of the domesticated laboratory mouse.

Because agricultural scientists have their own community, their work on gene engineering or gene mapping is often unnoticed by other researchers. For example, while much of the biological community was involved in debates over whether it would be safe to conduct field trials of genetically engineered organisms, the U.S. Department of Agriculture (USDA) had already approved such trials, and they had taken place. Similarly, when USDA scientists announced on January 9, 1994, that they had completed physical gene maps for cattle and swine, little notice was taken by scientists in general. The maps were developed by the Meat Animal Research Center in Clay Center, Nebraska.

Beyond and Between the Genes

Base-pair sequencing reveals a lot more than just the genes, which account for only about 3% of the three billion base-pairs. Although the material between genes was originally thought to be "junk," many biologists now think that these stretches are important features of the genome. Some nongenes are now known to regulate the expression of a particular gene, and while these regulatory factors do not themselves make proteins, they are targets for proteins synthesized by genes. Other intervening DNA portions may have a structural function, keeping the chromosome in shape.

In August 1993, Theodore Krontriris of Tufts University School of Medicine in Boston, Massachusetts, and coworkers found that a bit of nongene exists in a particular mutated form in about 10% of certain cancers. The mutation may occur in a stretch of DNA that serves a regulatory function.

In higher organisms, a DNA region including a gene also includes parts that are discarded upon protein assembly. These particular nongenes are called introns. In 1994, Ben F. Koop of the University of British Columbia in Victoria and Leroy Hood of the University of Washington in Seattle demonstrated that for one important gene found in both mice and men, the introns were remarkably similar in both species. Yet these introns are deleted and tossed aside when the gene makes its protein product, in this case the T-cell receptor. The usual belief holds that unless a part of the body is useful—even a part as small as a sequence of base pairs on a DNA molecule—it will develop random mutations and in most cases will even be deleted over time. Thus, the introns would appear to serve some purpose, although there is no hard evidence as to what that purpose might be.

Finally, a physicist from Boston University, Eugene Stanley, along with some coworkers from both Boston University and Harvard Medical School, reported in the December 5, 1994, *Physical Review Letters* that he and his group had results indicating that nongene DNA means something. They put nongene DNA through tests that are normally applied to strings of letters to identify languages as opposed to random strings. Two tests, one based on Zipf's law and one from Claude Shannon's information theory, both indicated that nongene DNA exhibits some kind of language, not just random sequences.

Moving Rapidly Ahead

One biologist willing to tackle big tasks, John Sulston of the Sanger Centre in Cambridge, England (*see* "*C. elegans* UPDATE," p 362), along with Robert Waterson of Washington University in St. Louis, Missouri, proposed that the time and the cost of the Human Genome Project would drop dramatically if the project aimed for 99.90% accuracy rather than the planned 99.99%. Based on their experience with the genome of *C. elegans,* Sulston and Waterson predicted that the cost for the human genome could be lowered to 10 to 12 cents per base and that a large team could finish the task by the year 2001. This might be possible, although the rate of sequencing would have to be dramatically accelerated. This proposal, officially put forward on December 16, 1994, but broached months earlier, had the effect of shaking up the HGP community and turning its attention to the actual task at hand, instead of the peripheral projects that most laboratories had begun. The true task, of course, is doing real sequencing. By the spring of 1995, less than a year after the Sulston-Waterson proposal, separate teams were at work on directly sequencing at least a half-dozen of the 23 pairs of chromosomes.

The vastly increased use of databases on the Internet has become an unexpected boon to the Human Genome Project and related gene-sequencing efforts. Scientists all over the world routinely turn to their computers to find out the latest in data that match with their own part of these vast projects. At the same time, they add their own data to the mix. The result is not only faster communication, but also the provision of

---

Background

The human being is a fairly complex animal that scientists believe is almost completely described by its genes. The Human Genome Project (HGP) intends to uncover the complete sequence of about three billion bases in human DNA, to identify the genes encoded therein, and to recognize the immediate purpose for the genes. Because most of the DNA base sequence is tossed aside in the process of making proteins, it is expected that only 3% of the base sequence actually encodes human genes. There are thought to be about 50,000–100,000 human genes in all. Since 3% of 3 billion is 90 million, one can calculate that the average gene probably consists of about 900 to 1,800 bases, although genes vary considerably in length.

Work performed on a related project, sequencing the genome of the nematode *Caenorhabditis elegans* (see "*C. elegans* UPDATE," p 362), indicates that there are twice as many nematode genes as initially expected. This may be a hint of what is yet to be learned about the human genome.

In 1937, J. Bell and F. R. S. Haldane determined that color blindness and hemophilia genes are located on the X chromosome. This was the first time that human genes had been directly associated with any specific location in the cell nucleus. The genesis of the Human Genome Project occurred nearly 50 years after these discoveries. In 1984, Robert Sinsheimer, chancellor of the University of California at Santa Cruz, was inspired by the high cost of major telescopes to conceive of a truly expensive project in biology: he proposed to map the human genome. Sinsheimer's ultimate goal would be a string of the letters T, G, C, and A that would fill 510 volumes of a standard-sized encyclopedia. The next year, he arranged a workshop to present his idea.

Sinsheimer was not the only one looking for a big project. The U.S. Department of Energy (DOE) was supporting national laboratories that were losing their funding. Sinsheimer's idea seemed to be just what the DOE was looking for, even though genes had no obvious relationship to energy. The tenuous link claimed by DOE was that the Department had previously funded studies of radiation damage to genes. DOE sponsored a second conference on Sinsheimer's idea, which led to others. The Office of Technology Assessment, the National Research Council, and (somewhat belatedly) the U.S. National Institutes of Health (NIH) got into the act. Along the way, the idea evolved into the Human Genome Project.

The Human Genome Project started formally on January 3, 1989. In the summer of 1989, the NIH and DOE sponsored a joint conference to determine how to get the job done. They developed a plan that would start in 1991, run for 15 years, and cost about a buck per nucleotide, or $3 billion. They also developed a strategy for doing the work, and put it out for public comment. The DOE continued to keep its hand in, but the NIH finally established leadership. By the fall of 1989, the basic nature of the project had taken shape; it involved many small groups of genetics researchers around the world cooperating under the leadership of James D. Watson [American: 1928–] and the NIH. Scientists from the European Community enthusiastically joined the project and the Japanese reluctantly fell in line as well.

The project has been marked by a high degree of cooperation among research groups throughout the genetics community. Work has been parceled out so that geneticists are cooperating rather than competing in the way that the first Genome Project leader Watson described so vividly in his book *The Double Helix*. Each of the main groups has its own chromosome or set of chromosomes to study. Researchers also agreed on common languages and tools to make sharing data easier. The final financial goal calls for the project to be completed for half of the originally estimated cost, or only $0.50 a base.

The Human Genome Project has faced serious opposition from outside the genetics community, some of it well organized. Some socially conscious scientists and nonscientists fear that the knowledge gained could be used to discriminate against people with "bad genes."

---

immense amounts of useful raw data that would never have become available in ordinary print media.

(Periodical References and Additional Reading: *Discover* 1-93, pp 86 & 88; *New York Times* 12-16-93, p A1; *New York Times* 12-28-93, p C1; *New York Times* 1-12-94, p A18; *New York Times* 1-30-94, p I-1 *Science* 2-4-94, p 608; *Scientific American* 3-94, p 21; *Science* 6-3-94, p 1404; *Science* 9-30-94, pp 2031, 2033, & 2049; *Science* 11-25-94, p 1320; *Science* 1-13-95, p 172; *Science* 2-10-95, p 783; *Science* 5-26-95, p 1134; *Science* 6-2-95, pp 1270 & 1273; *Science* 6-16-95, p 1560; *New York Times* 6-28-94, p C1; *Science* 7-28-95, pp 468, 469, 496, & 538; *Nature* 8-10-95, p 459)

The Threat of Organisms in New Genes UPDATE

Although a few people had strongly opposed the use of genetically engineered plants and animals, the first widespread reaction against genetic engineering by a large section of the general public involved a product produced by genetically engineered bacteria—bovine somatotropin, or BST. When BST, a hormone that increases milk production in cows, became commercially available in 1994, it aroused protests among many consumers, and some merchants claimed they would not stock milk from BST-treated cattle or that such milk would be labeled as such. Despite many letters to newspaper editors and articles attacking the use of genetically engineered BST, the hormone was quickly adopted by much of the dairy industry. The first BST-influenced milk appeared in markets in February 1994, and in 110 incidents that year protesters dumped the milk from BST-treated cows outside the supermarkets selling it. Protests then gradually began to die down.

A second product appeared even more controversial in principle, since the food product itself is genetically engineered. A few stores in the western and midwestern United States began to carry the Calgene Flavr Savr tomato in May of 1994, after official U.S. Food and Drug Administration (FDA) approval on May 18, 1994. The Flavr Savr contains two genes not normally found in the fruit. One of the genes changes the way that the tomatoes ripen, allowing them to be harvested at a later and therefore more flavorful stage. This new ripening gene cancels out the tomato's natural gene that produces the enzyme polygalacturonase, which breaks down pectin. The other new gene is a marker gene for resistance to an antibiotic, which is used to make certain that the tomato plant actually contains the first gene. Protests were much quieter than those against BST, however, in part because of caution in marketing the tomato, which kept it out of many stores and made for a low profile nationally.

Products to Come?

Throughout the early 1990s, many ideas for new genetically engineered consumer products were floated. By the end of 1994, more than 2,000 field trials of plants and animals with new genes inserted had been conducted, including vegetables, fruits, grains, fish, and chickens, as well as genetically engineered bacteria and viruses designed to protect plants or animals. Many of the tested plants or animals were transgenic, meaning that the new gene inserted came from a different species. For the most part, these ideas and tests engendered few or no protests, in part because they proved less controversial than milk that is fed to babies or a fruit that is part of nearly every person's diet.

Most of these new organisms are intended to become commercial products, and it is difficult to get much in the way of detailed information out of the corporations that are involved. However, word has leaked out that companies are working on virus-resistant alfalfa, cotton, cantaloupes, cucumbers, potatoes, papayas, squash, and tomatoes; herbicide-resistant alfalfa, cotton, potatoes, soybeans, and tomatoes; insect-resistant apples, potatoes, rice, strawberries, and walnuts; a genetically-engineered hormone that will produce low-fat pigs; high-starch potatoes that absorb less fat and other potato varieties that contain unique proteins; higher protein sunflower seeds and higher starch rice; cotton whose oil will be lower in trans fatty acids; and transgenic corn, rapeseed, sugar snaps, and tobacco. The pace of development has accelerated rapidly. By 1994, the number of field tests for transgenic crops reached 486. Still, the only one to reach the U.S. market is the virus-resistant yellow crookneck squash, introduced without fanfare or labeling during 1995. There is contradictory information about whether genetically engineered tobacco is being used in Chinese cigarettes sold to the general public.

In addition to foods or other goods consumed by the general public, there are a number of transgenic animals being developed specifically to produce medical products such as human blood factors, hormones, and infection fighters. These target products are being extracted from the milk or blood of various kinds of domesticated livestock, including pigs, goats, sheep, and cattle.

Here are some specific new ideas that have emerged from research scientists, or that have involved official government actions.

TIMETABLE OF IDEAS FOR ORGANISMS WITH NEW GENES

1990	The U.S. Food and Drug Administration (FDA) approves the use of a gene-engineered form of the hormone renin in the production of cheese.
February 14, 1992	William Woodson of Purdue University and coworkers find a gene that causes carnations to wither. He foresees using the gene to make flowers stay fresh longer.
April 7, 1992	Bruce Hammock and coworkers at the University of California at Davis announce that they have incorporated information from a scorpion gene to engineer a virus that infects plant-eating caterpillars. The virus causes the caterpillar's cells to produce a deadly toxin, resulting in quick death.
April 24, 1992	Christopher Somerville of Michigan State University and coworkers demonstrate that plants can be genetically engineered to produce small amounts of plastics.
May 26, 1992	The FDA declares that genetically engineered food will be treated as safe unless there is some specific reason to suspect that it might cause harm.
June, 1992	In *Bio/Technology*, Indra K. Vasil of the University of Florida in Gainesville and coworkers announce the insertion of a gene into wheat, conferring resistance to the powerful herbicide Basta, which ordinarily kills any plant it touches.
July 3, 1992	In *Science*, Toni A. Voelker announces that Calgene Inc. in Davis, California, has developed rapeseed that expresses a gene for the tropical oil laurate, an important ingredient in soaps and shampoos.
January 14, 1993	German scientists develop tobacco that expresses a gene from grapes, resulting in resistance to a *Botrytis* fungus infection.
February, 1993	Ciba-Geigy announces in *Bio/Technology* that it has developed a strain of maize (known as corn the United States) that resists the European corn borer, a major pest.

August, 1994	In *Bio/Technology,* a team led by Maarten Chrispeels of the University of California at San Diego, T. J. Higgins of the Division of Plant Industry in Canberra, Australia, and Larry Murdock of Purdue University in West Lafayette, Indiana, announce a genetically engineered pea that resists weevils after harvest. This is the first time that a seed has been engineered to resist damage in storage. The gene in question is already known to promote weevil resistance in beans and cowpeas, so it is not expected to harm the pea's food value.
September 15, 1994	Fishery scientists from Canada and the United States announce in *Nature* that they have introduced a growth-hormone gene from sockeye salmon into coho salmon, causing the salmon to grow 11 times heavier than normal in one year's time—one giant was 37 times normal size. All the salmon were harvested upon reaching 10 lb (5 kg), so it was unclear whether or not they would have continued growing. The scientist's actual goal involves growing farm-raised salmon faster, not growing larger salmon.
September 16, 1994	Botanists from Michigan State University in East Lansing announce that they have cloned from maize (corn) the first plant gene used in controlling growth, leading to speculation concerning giant vegetables and trees. Its first use, however, is insertion in tobacco plants to produce shorter, bushier plants that farmers prefer over long, lanky varieties.
September 23, 1994	Three different teams of researchers announce the discovery and cloning of two plant genes that fight bacteria and viruses. The bacteria-fighting gene is almost the same as one found earlier in 1994 by an Australian group. Attempts are already under way to introduce the genes into tomato plants.
November 22, 1994	Ralph Brinster of the University of Pennsylvania announces discovery of a technique for introducing new genes into a sperm line, from which the genes naturally transfer to offspring.
December 20, 1994	In the *Proceedings of the National Academy of Sciences*, Christopher Somerville, now of the Carnegie Institution of Washington at Stanford University, and coworkers announce that they have genetically engineered mustard plants to produce a plastic that is a biodegradable analog of polypropylene.
May 1995	Aya Jakobovits and coworkers at Cell Genesys in Foster City, California, and rivals Nils Lonberg and coworkers at GenPharm International in Mountain View, California, each develop mice that express human genes in their immune systems, allowing them to produce human instead of mouse antibodies. Such antibodies would be more useful than ordinary monoclonal antibodies, which have too much mouse in them, often precipitating adverse reactions in humans.
September 1995	Imutran, a British biotechnology firm, announces that they have bred transgenic pigs (and have a herd of about 400 of them) expressing a gene which blocks the transplant-rejection mechanism. Monkeys who received hearts from these pigs had already demonstrated that they could survive for as long as two months. The goal, still some years away, would be to use pig organs in humans.

Potential Problems

The main potential for negative outcomes of the BST and Flavr Savr introductions consisted of possible increased udder disease in overworked cows, resulting in higher levels of antibiotics in milk, and the development of antibiotic resistance related to the marker gene in the tomatoes. Tests have failed to show either result.

Some scientists are concerned about the possibility of adverse effects from other genetic experiments. Studies using transgenic plants, conventional varieties, and related weeds have been conducted by scientists who fear that transgenic plants may escape cultivation or interbreed with weeds. The first field trial of this kind, by M. J. Crawley of Britain's Imperial College, reported in the June 17, 1993 *Nature,* showed that genetically engineered rapeseed or canola (programmed to resist an antibiotic or a specific herbicide) is not more invasive than other varieties of rapeseed. On the other hand, Johanna Schmitt of Brown University in Providence, Rhode Island, showed that a hybrid bred with one transgenic and one wild parent grows as well as a hybrid with a domesticated parent and a wild parent. Schmitt worked with tame rapeseed plants and closely related wild mustard. Subsequently, as reported in the February 1994 *Ecology Applications,* Terrie Klinger and Norman C. Ellstrand of the University of California, Riverside, showed that hybrid crosses of wild and tame radishes are more successful than their wild parents, producing about 15% more seeds. The intention of these experiments is to predict whether a plant genetically engineered to resist an herbicide or disease could potentially contribute its resistance gene to an undesirable weed, making the weed even more of a pest.

Richard F. Allison and Ann E. Greene, both of Michigan State University in East Lansing, Michigan, reported in the March 11, 1994 *Science* that genes from existing viruses can exchange with engineered genes to produce unexpected and potentially harmful new virus varieties. In particular, plants whose new genes are designed to resist a viral disease might confer the resistance factor to infectious viruses. About a fifth of all genetically engineered plants are intended to resist viral diseases.

Allison and Greene worked with cowpea plants that had been engineered to contain a large part, but not all, of the gene for a protein that forms the outer coat of a virus that infects cowpeas. A similar damaged gene exists in a mutant form of the virus. The researchers thought that the gene from the virus might combine with the gene engineered into the plant, a process of recombination known from other viruses of this type. Of 125 genetically altered plants exposed to the mutant virus, 4 showed symptoms of infection. In these instances the gene in the plant and the gene in the mutant virus, neither of which could cause the viral disease themselves, recombined in ways that produced an infective virus. The actual recombinations were different from each other, which suggests that there are multiple ways to get an infective virus through recombination.

The numbers of plants involved in this study were too small to convince many scientists, especially those working on ways to protect plants from viruses through genetic engineering. So far, their field trials have shown no instances of recombination producing an infective virus. One difference, of course, is that Allison and Greene deliberately set out to produce and find recombination, whereas other experimenters have entirely different goals in mind and therefore conduct their experiments quite differently.

Background

Genetic engineering had its tentative beginning in 1973. Almost immediately, people began to worry about it. In July, 1974, Paul Berg [American: 1926-] of Stanford chaired an 11-scientist Committee on Recombinant DNA Molecules that wrote to *Science* and *Nature* about possible dangers from genetically engineered bacteria. The Berg committee proposed a voluntary ban on specific experiments with genetic engineering until their "potential hazards" could be evaluated. The following February, a conference of 139 scientists from 16 countries met at Asilomar, California, and adopted a set of voluntary guidelines for genetic engineering.

Despite controls put on fellow scientists, which eventually were codified into U.S. administrative rules and regulations, development of organisms with new genes (or at least old genes from different species) proceeded rapidly.

In 1975, there were six laboratories in the United States and a few outside the United States that had begun to conduct genetic engineering. The following year, Genentech established itself as the first commercial company to use genetic engineering. In January 1980, another commercial group, Biogen, announced that one of their scientists had used genetic engineering to develop a strain of bacteria able to produce human interferon. By 1981, the human gene for interferon was installed in yeast as well as bacteria, and by 1985, interferon produced in this way had become commercially available. By 1986, it was possible to purchase an almost-living organism—a virus, that is—produced by genetic engineering.

Organisms had been granted patent protection much earlier. Initially, patents became available for plants or other organisms bred in ordinary ways with no genetic engineering. By 1987, there were 600 applications put forth for patenting creatures produced by genetic engineering. That year, a ruling allowing genetically engineered oysters to be patented broke the ice. The following year, the first patent for a genetically engineered vertebrate, a mouse, was granted. Also by 1987, field trials had begun with genetically altered plants in open fields. There were five such trials in the United States that year, a number which more than tripled the next year, doubled again the next, and so forth. By 1992, the number of American field trials of transgenic plants had reached 161.

People opposed to genetic engineering were increasingly alarmed, but the government was becoming reassured by the results of the carefully monitored tests. By 1991, the U.S. White House Council on Competitiveness was calling for an end to controls on genetic engineering and proposed that government agencies treat all products of genetic engineering just as any other new breed of plant, animal, bacterium, or fungus. At that point, about 700 different varieties of vegetables, fruits, and grains (some with traits quite different from the wild type) had been developed throughout the 1980s by nongenetic traditional means, and no environmental or health problems were known to have been caused. Government officials saw no reason why transgenic plants should be any more troublesome.

Until the early 1990s, debates over the handling of genetically engineered products involved scientists, companies involved in genetic engineering, and government officials. At that point, genetically engineered food products became available, causing considerable controversy among the general public.

Another group with somewhat different goals, and very different methods for conducting experiments with transgenic organisms, consists of Chinese scientists backed by an aggressive government effort to use genetic engineering to improve China's food supply. Unlike the carefully controlled field trials of the West, Chinese scientists are planting or releasing genetically engineered organisms in huge numbers over thousands of acres. Furthermore, the products, such as genetically engineered soybeans, are not separated from the rest of the harvest. Thus, they may already be in use by the general public, even while still theoretically undergoing field trials. In December 1993, the Chinese government issued the first set of rules intended to keep field trials from causing disasters due to unexpected results of the genetic engineering involved.

Another set of researchers is dealing with a different kind of problem completely. Since most insecticides and antibiotics are quickly countered by resistant strains of insects and bacteria, it seems possible or even likely that plants genetically engineered to resist insects or diseases may also lose their effectiveness in short order. Surveys suggest that the same portion of crops is lost to insects and disease at the end of the twentieth century as was lost at the beginning, despite a century of "breakthroughs" in plant protection. In 1992, mathematical studies of this problem showed that transgenic plants will just as quickly induce the evolution of insect or bacteria populations able to overcome resistance. Two teams have considered a number of strategies intended to slow down the development of such populations or strains. The teams found that planting a mixture of seeds of resistant strains with those of wild strains or developing plants that included some resistant parts and some susceptible parts actually sped up the loss of resistance to pests and disease. However, planting some regions with resistant plants and others with susceptible varieties does slow down loss of resistance in computer simulations. Wild populations from the regions with susceptible plantings would tend to slow the evolution of insects or disease organisms with genes for overcoming resistance.

(Periodical References and Additional Reading: *New York Times* 2-9-92, p III-9; *New York Times* 2-14-92, p A14; *New York Times* 4-7-92, p C6; *New York Times* 4-28-92, p C2; Discover 1-93, p 44; *Nature* 1-14-93, pp 153; *Nature* 2-11-93, pp 500; *Nature* 2-18-93, pp 593; *Nature* 6-17-93, pp 580 & 620; *New York Times* 2-4-94, p A1; *New York Times* 2-8-94, p A1; *New York Times* 2-20-94, p IV-14; *Science News* 3-5-94, p 151; *New York Times* 3-9-94, p A12; *New York Times* 3-11-94, p A16; *Science* 3-11-94, p 1423; *New York Times* 4-9-94, p 7; *Science News* 4-30-94, p 276; *New York Times* 5-18-94, p C1; *New York Times* 5-19-94, p A1; *New York Times* 5-22-94, p IV-2; *New York Times* 5-25-94, p A14; *Science* 8-5-94, p 729; *Nature* 9-15-94, p 209; *Science* 9-16-94, p 1699; *New York Times* 9-20-94, pp C6 & C7; *New York Times* 9-23-94, p A21; *New York Times* 10-16-94, p I-27; *Science* 11-11-94, p 966; *New York Times* 11-22-94, p A1; *Science* 12-2-94, p 1472; *Science News* 12-24/31-94, p 420; *Discover* 1-95, p 96; *Science News* 2-4-95, p 72; *Nature* 9-21-95, p 185)

FROM BACTERIA TO FUNGI

THE KINGDOM OF THE ARCHAEA

Classification is among the most important tasks of science, because until workable groupings are recognized it is impossible to make sense of the world. If each event and each experience were treated as completely new and unprecedented, unrelated to other situations, there would be no science. In biology, the science of classification is called either taxonomy or systematics.

Our Archaean Cousins

A recent breakthrough in taxonomy, resisted at first by traditional systematicists, has been the ability to use DNA to establish relationships. Whether used for finding the evolutionary distance between an ape and human or between two forms of bacteria, DNA analysis has become the most basic form of taxonomic data. Because the same information is coded in DNA, RNA, and the protein produced by the combination, any of the three can be employed in such studies. When such studies were conducted in the early 1990s on the Archaea and various bacteria, scientists found that the evolutionary distance between the Archaea and bacteria is as great as between any two creatures known, as great as the distance between the bacteria and, say, humans.

In 1990, two different German research groups who were studying Archaea DNA discovered that a sequence Archaea use to regulate gene expression, one of the most fundamental activities of life, is similar to the one used by eukaryotes to regulate gene expression. In eukaryotes, the sequence is known as the TATA box. Following up on this discovery, as reported in the May 10, 1994 *Proceedings of the National Academy of Sciences,* Gary Olsen and coworkers at the University of Illinois looked at the protein used to work with the TATA-like sequence in Archaea. In both Archaea and humans, the protein is similar enough that one name, the TATA-binding protein (TBP), suffices for both. Although TBP is by some standards quite different in Archaea and humans, with only 40% of the working part completely identical, by the standards of protein construction the two kinds of TBP are essentially the same. About the same time as Olsen's work, Stephen P. Jackson and coworkers at Cambridge University in England copied off the gene for TBP from Archaea, and found the DNA sequences of the archaean gene to be like those found in the eukaryotic TBP gene. Jackson's group reported in the May 27, 1994 *Science* that the TBP of the archaean *Pryococcus woesi* has the same form and function as that in humans. Indeed, archaean TBP will bind to human proteins and function like the human form of the protein.

Monerans do not use anything like the TATA box and TBP. Thus, the presence of the TBP system in Archaea indicates that the Archaea are more closely related to eukaryotes than are Monera. A complex system like TBP and the TATA box is not likely to have arisen twice in evolutionary history. Thus, the implication is that the Archaea emerged later than such Monerans as eubacteria and that they share a common origin with the eukaryotes. It is not thought, however, that the eukaryotes evolved from Archaea themselves. Instead, as one writer for *Science* put it, the tree of

Background

Bacteria are still poorly classified. Because of their small size there is no truly meaningful evolutionary taxonomic scheme for bacteria. Until recently, the best classification schemes have simply grouped bacteria by their shapes (spheres, rods, spirals), size, observed features of cell structure, behavior, and so forth. As always with taxonomy, there are several competing systems. Furthermore, since bacteria reproduce by cell division, the classic definition of a species in terms of sexual reproduction is meaningless.

Despite these obstacles, bacteria are given specific and generic names, and assigned to families and phyla. What we ordinarily mean by bacteria are often called **eubacteria**, which literally translates to "good bacteria." In biology, however, the prefix *eu-* generally signifies "complete" or "including the necessary part." These eubacteria include the familiar bacillus *Escherichia coli*, and those cocci known as "staph" and "strep." A group of spiral creatures, including those that cause syphilis and Lyme disease, are **spirochetes**, not to be confused with the spiral eubacteria, which are spirilla. Tiny creatures that look a bit like miniature rod-shaped bacilli are the **rickettsiae**.

It is difficult to define whether certain creatures are bacteria or fungi. The **actinomycetes** that are ubiquitous in soil are often classed as bacteria, but seem more like fungi in lifestyle. The **mycobacteria**(sometimes mycoplasma) are quite similar to the actinomycetes, but because they include the organisms that cause tuberculosis and leprosy, mycobacteria are nearly always thought of as bacteria, or at least as a branch of the actinomycetes. Another group that is separate in lifestyle are the bacteria that can take nitrogen from the air and convert it to a usable chemical. These **nitrogen-fixers** include some organisms that probably are not closely related to each other.

It has long been recognized that certain photosynthetic, one-celled organisms, once known as algae, more closely resemble bacteria than they resemble other algae. The main classification scheme for algae uses color words, although the colors themselves are not always apparent in every organism. While the golden, green, brown, and red algae are considered either protists or plants, the blue-green variety has long been recognized as a close kin to the bacteria, now called **cyanobacteria**. Together, the various kinds of bacteria and cyanobacteria form a large group known as **Monera**. Members of Monera are called **Monerans**.

Recently, there has been another major shift in classification. Throughout the 1980s and into the 1990s, evidence began to accumulate that certain organisms recognized largely by a lifestyle of chemical dependence and a fondness for heat include a group of organisms that differ significantly from the bacteria who also share these characteristics. This group is now known as the **Archaea**. As scientists learned more about the Archaea, the impression that these might be primitive ancestors to the eubacteria gave way to a surprising alternative conclusion.

life has three main branches: the prokaryotes, the eukaryotes, and the Archaea. Earlier studies separated the Archaea into two main lineages, the Euryarchaeota, who are primarily known for their ability to metabolize methane or various inorganic chemicals, and the Crenarchaeota, known primarily for their ability to live in extremely hot water.

A Sea of Archaea

Microscopic organisms are usually identified by growing them in cultures until there are enough in one place to display a large number of the same sorts of microbes on a slide made from the culture medium. Putting an uncultured drop of pond water on a slide will reveal a mixed host of microorganisms because they have been "cultured" by the rich growing conditions of the pond. The same experiment with a drop of sea water, taken from the vastness of the oceans, may either show nothing at all, or not enough of any one organism to make identification possible. Put the sea water in a culture at a pleasant temperature and, with plenty of food and light, there may be results. If sufficient care has been taken to eliminate contamination, the culture will consist of many individuals of a particular organism from the ocean water. This does not suggest, however, that the particular organism cultured out is necessarily the most common. In fact, another organism or organisms may have been present initially, but did not survive the culturing process. What is healthy for one microbe may be deadly or irrelevant to another. Thus, the method used for identifying microbes is only successful for some microbes, while completely overlooking others.

Analysis of DNA and RNA has had the unexpected side effect of providing a new way to identify microorganisms. Instead of growing microorganisms, one can simply perform the equivalent of DNA-fingerprinting on a sample. If the result matches the suspect, then you know that the suspect has been on the scene. In practice, scientists look for RNA, not DNA, but the principal is the same. Furthermore, RNA can be used to create matching DNA, which can then be "cultured" through a polymerase chain reaction to magnify amounts.

In 1992, this new technique was applied to water from the Pacific Ocean some 350 mi (550 km) off the coast of California, by Jed A. Fuhrman, Kirk McCallum, and Alison A. Davis of the University of Southern California, Los Angeles. The team was surprised to find that the RNA they identified included that from seven species of apparent Archaea. The Archaea were of the type known as thermophiles, or "heat lovers," previously found mainly in hot springs, geysers, and undersea geothermal vents. Surprisingly, the samples of seawater were taken far from any known sources of hot water. One question entertained by the Southern California team was whether the Archaea were actually gut-living, methane-digesting organisms that had been excreted by fish or some other sea creatures.

Scientists soon found similar indications of Archaea in other oceans. The biggest surprise came in studies conducted by Edward F. DeLong and coworkers from the University of California, Santa Barbara, of Archaea in both icy surface water and water directly under the pack ice in the ocean near Antarctica. Some 20-30% of the microbes found in these waters were Archaea. Not all ocean samples displayed this same composition. At one site in the North Pacific, almost all the microbes were Monerans, mixed in with a few Protists. With the exception of this site, lower depths (300–1,600 ft, or 100–500 m) at all sites tested demonstrated relatively higher proportions of Archaea, although none so great as in the Antarctic surface water.

Genetic studies comparing the DNA produced from the Archaean RNA detected in cold and deeper waters showed that some species appeared related to the Crenarchaeota, while others were Euryarchaeotes similar to *Thermoplasma acido-philum*. Thus, even though the Archaeans were found in cold ocean waters, their known relatives are most often found in hot springs and geysers.

In one sample, more than a third of all the microbes present were Archaea. These vast numbers of Archaea in the seas must exert a great effect on the ecology of the Earth, but at present not enough is known to define that effect.

Although the biggest surprise came from the number of new species of Archaea in the cold oceans, another 1994 study of the Archaea in a single hot spring in Yellowstone National Park (the Black Pool) more than tripled the number of species known among the Crenarchaeota. Clearly, there remains much to be discovered about our Archaean cousins.

ANIMAL, VEGETABLE, OR MINERAL? OR WHAT?

When Carolus Linnaeus divided up the world into Kingdoms in the eighteenth century, he used the scheme familiar today from the game "Twenty Questions." The three original Linnaean Kingdoms were Animal, Vegetable (or Plant), and Mineral. Bacteria were already known at that time, but not taken seriously. Anton van Leeuwenhoek had evidently observed bacteria with his single-lens microscope as early as 1681, although his first written description is dated September 17, 1683. Because the bacteria were moving, Leeuwenhoek referred to them as "Animals." In the eighteenth century, bacteria were sometimes viewed by other observers as transitional between the Animal and Plant kingdoms. As early as 1866, however, Ernst Haeckel proposed that all single-celled organisms, including bacteria, should have their own Kingdom, which he called Protista (note the taxonomic custom of capitalizing the names of higher divisions, such as Kingdoms).

Nearly a hundred years later, however, in the 1950s, the taxonomic status of bacteria was still not clear. A botany textbook from the 1950s states that "there is general agreement that bacteria belong in the Plant Kingdom." Yet there was just as good an argument in favor of bacteria being among the Animals, where Leeuwenhoek had automatically placed them. In 1957, the great evolutionist George Gaylord Simpson followed the Haeckel tradition of separating life into three Kingdoms, writing of Plants, Animals, and Protists. However, he clearly conceived of the various "Protists" as consisting of those Protists classed as almost-animals and those classed as almost-plants.

In 1969, R. H. Whittaker set up a different classification scheme altogether, giving Bacteria and the Fungi their own Kingdoms, along with the already recognized Kingdoms for Plants, Animals, and Protists. Whittaker's five-kingdom scheme, which was widely adopted, makes the greatest division between those organisms whose cells have nuclei (the eukaryotes), and those that do not (the prokaryotes). The bacteria are grouped with the only other prokaryotes, the blue-green algae, forming the Kingdom Monera. Eventually, the blue-green algae came to be known as blue-green bacteria or cyanobacteria (cyan- means blue-green).

About the same time that Whittaker proposed his five kingdoms, however, scientists began to find that "bacteria" inhabited some hot springs and geysers in geothermal fields. At first these were taken to be like other bacteria, simply heat-loving, but in 1977 Carl Woese of the University of Illinois recognized that the organisms he was studying differed from bacteria in more fundamental ways. He proposed the classification archaebacteria for the tiny, heat-loving single-celled organisms.

By 1981, Karl O. Stetter of Regensburg University in Germany was finding archaebacteria that lived at extremely high temperatures in hot springs in Iceland—temperatures higher than 194°F (90°C). In the early 1980s, also, tiny organisms that live by metabolizing sulfur at temperatures as high as 230°F (110°C) were found near undersea geothermal vents. At first, scientists continued to think of these archaebacteria as a special phylum of bacteria. Subsequently, most agreed that the organisms were sufficiently different from the known types of bacteria to be assigned their own Kingdom, the Archaea.

The original name of archaebacteria was proposed in the belief that Archaea were the first form of organized life, emerging before there was free oxygen, living on chemicals, aided by the great heat of a recently molten Earth. While the prefix *archae-*, meaning "from ancient times," has been retained, there is a fundamental problem with the above described scenario. The results of various genetic tests strongly suggest that the Archaea are more closely related to the eukaryotes than are the Monerans. Thus, it is unlikely that the Archaea existed first and resulted in bacteria or cyanobacteria. Instead, it appears likely that some Monerans gave rise to the Archaea, and some Archaea evolved into the first eukaryotes.

For now, the six-kingdom scheme satisfies most taxonomists. They realize, however, that the Protists are something of a grab-bag of misfit organisms, which is the main reason they are all classed together. Someday it is likely that another Kingdom or two will be split off. The two Empires, however (the prokaryotic Empire of the Kingdom Monera and the Empire of Archaea and the Eukaryotes), seem to accurately represent the main dichotomy of fundamental life forms, replacing the original two Kingdoms of Plants and Animals.

(Periodical References and Additional Reading: *Nature* 3-12-92, p 148; *Science* 5-27-94, pp 1251 & 1326; *Nature* 10-20-94, pp 657 & 695)

BACTERIA: GIANT, DEEP, AND SOCIAL

Although bacteria have been recognized for more than 300 years and are no doubt the most common form of life, we are really just getting to know these tiny creatures. For the first 200 years of human awareness of bacteria, almost nothing was known about them. Interest increased greatly when Pasteur and Koch showed that bacteria cause several common diseases. Once their attention was engaged, scientists found that bacteria are truly ubiquitous (*see* below for how much so), and play a vital and necessary role in the other kingdoms of life.

Scale remained a problem for scientists. Viewed under an optical microscope, bacteria are simply spheres, rods, and corkscrews with no internal features. The electron microscope, which enables scientists to see dim bacterial shapes more clearly, has only been generally available since 1935. Even with the electron micro-scope, however, the internal workings of bacteria are known primarily by inference, not observation. In the 1990s, scientists continue to be surprised by what bacteria are like, where they can be found, and how they behave.

The Giant Bacterium

Until 1993, the one sure thing that could be said about all bacteria, without exception, was that they were all microbes: living organisms so small that they can only be seen with a microscope. That year, using a technique similar to DNA fingerprinting, a mysterious creature more than half a millimeter (0.02 in) long was identified as a type of bacterium. A half millimeter, about the size of a printed hyphen (-), is large enough to be seen without a microscope. Because volume increases as the cube of dimen-sions, the volume of the newly discovered bacterium is a million times that of most ordinary bacteria. The previous record for length in a bacterium was held by a spirochete measuring only about a third the length of the newly found giant. While the spirochete is a slender thread, the behemoth is a fat rod, nearly cucumber-shaped.

The giant bacterium is *Epulopiscium fishelsoni*, discovered by L. Fishelson in 1985 but not clearly identified at that time. Fishelson and coworkers originally thought *E. fishelsoni* was some sort of protist, belonging to that large kingdom of one-celled and colonial creatures which ranges from Paramecia and Amoeba to giant algae and complex slime molds. Further study with the electron microscope showed that the hallmarks of a protist were missing: there was no apparent cell nucleus, and what had seemed to be the protist propulsive fibers called cilia were more similar to the kind of flagella used by some bacteria. Graduate student Esther R. Angert of Indiana Univer-sity, working with her advisor Norma R. Pace and with Kendall D. Clements of James Cook University in Australia, studied RNA sequences from samples collected by bursting *E. fishelsoni* cells and then amplifying the RNA with a polymerase chain reaction. Angert used the amplified RNA to produce DNA, essentially the same method used to find expressed sequence tags (ESTs—*see* "Genome Project UP-DATE," p 366). Comparison of the DNA sequences with other known sequences showed the closest match to be with *Clostridium* bacteria, the gram-positive bacterial family that is best known for producing botulism toxins in anaerobic situations, such as in canned vegetables.

Not only is *E. fishelsoni* larger than all other known bacteria, it is even larger than the smaller protists. One protist, *Nanochorum eukaryotum*, is only one or two microns across. Several hundred individuals of *N. eukaryotum* could be laid along the length of *E. fishelsoni*, and perhaps as many as three million could be packed into the space occupied by the body of *E. fishelsoni*.

Although there is always an interest in record-breaking creatures, the significance of *E. fishelsoni* goes beyond mere astonishment. For one thing, many scientists think that there are even larger bacteria waiting to be found, believing that *E. fishelsoni* is more likely the dawn of a new understanding rather than an isolated freak. Now that scientists know that bacteria can grow this large, more of giant size will be sought.

Even if *E. fishelsoni* emerges as a freak, it offers the potential to study bacterial internal processes that are unobservable in typical small bacteria; as Angert has said, "It's so huge that we could stick electrodes into it."

One problem confounding further study of *E. fishelsoni* is that no one knows how to keep the bacterium alive in a laboratory setting. In nature, it is found only in the gut of surgeonfish, common denizens of coral reefs in the Red Sea and off Australia. It appears that the bacterium has a symbiotic relationship with the fish, similar to many symbiotic arrangements between bacteria and other creatures (the bacteria that enable cows to digest grass and termites wood are well-known examples). The inability to culture *E. fishelsoni* was one of the main factors preventing its identification before Angert's work, which did not rely on living specimens.

Among the implications of the *E. fishelsoni* discovery is the need to re-evaluate fossilized one-celled creatures from the distant past. Until now, scientists have classed these fossil remains primarily by size. However, if some bacteria are large and some protists are small, other means will need to be used.

Bacteria of Underworld

Beginning with Jules Verne in 1864, early science-fiction writers suggested that there might be unknown life deep below the surface of the Earth. The concept lost popularity because few readers found such ideas plausible. Geologists had learned that the lower crust was very hot, under high pressure, and contained no oxygen, conditions that would seem to preclude much in the way of life.

By the early 1990s, about the only scientist of any repute to suggest that there might be life at the lower depths was astronomer Thomas Gold of Cornell University. Gold was already considered a crank by many, because of his early advocacy of the steady-state cosmology, and because of his belief that large amounts of abiotic hydrocarbons are found in deep rock. Gold's best success was long ago, in 1968, when he correctly predicted that pulsars would be found to be neutron stars. In the July, 1992 *Proceedings of the National Academy of Sciences*, Gold argued that there are "strong indications that microbial life is widespread at depth in the crust of the Earth." Furthermore, Gold argued that other planets and their satellites might also host bacteria deep under lifeless surfaces. On Earth, Gold calculated that bacteria were capable of living at least as far as 3 mi (5 km) below the surface. Gold proposed that bacteria would not only be found in such depths, but that they would be abundant. Gold's calculations implied that, if all the bacteria rose to the surface, they would form a layer of slime 5 ft (1.5 m) deep over the whole planet.

As with pulsars in 1968, Gold's predictions of bacteria existing at depths within the Earth was soon confirmed by observation. For reasons of its own, the U.S. Department of Energy (DOE) had, since 1985, been conducting studies of life in the lower depths. The first results of their test drills were not published until 1992, about the same time that Gold was writing. The DOE had to deal with deep groundwater pollution it had created, and was also concerned with the possibility of deep underground storage of nuclear wastes. Better knowledge of conditions at depths would be helpful for both endeavors. The DOE developed special techniques to be sure that samples from deep wells would not be contaminated by surface bacteria.

One DOE site studied in 1992 was in Triassic Period sediments in Virginia, where an oil company was drilling deep exploratory wells. DOE scientists were able to obtain samples from 8,700 ft (2,650 m) to 9,180 ft (2,800 m) below the surface. Double-blind tests were conducted using similar samples that had been baked at 1,100°F (600°C) for several hours to remove any traces of life from the rocks. The unbaked samples had emerged from sedimentary rock where the ambient temperatures ranged from 130-170°F (65-75°C). These samples produced bacteria, or the other tiny organisms most biologists today class as archaea or archaebacteria (*see* "The Kingdom of the Archaea," p 378), although it remains common to call the archaea "bacteria," even in scientific reports. About 100 different cultures were produced from organisms taken from the depths of the oil well.

Because the sediments had certainly contained bacteria and probably archaea when they were deposited some 200 million years ago, one theory suggests that the organisms found in the sediments today are descendants of the Triassic bacteria or archeabacteria, which have evolved separately from surface organisms since the Triassic Period. If that is the case, then the bacteria acquired the ability to live at high temperatures and pressures through evolutionary changes, perhaps evolving into archaea. Another possibility suggests that archaea already adapted to life in the crust gradually moved into the Triassic sediments when the rocks became hot enough for them.

The following year, evidence of deep archaebacteria emerged from another source—fluids released in a submarine volcanic eruption off the coast of Oregon. Samples of the vent fluids collected by scientists aboard the research submersible *Alvin* were rich with archaea of the type that live at very high temperatures, although the water from which the fluids were collected was not nearly hot enough for such archaebacteria. These findings implied that the archaea had been propelled from somewhere below the surface, where the high temperatures they would need to thrive could be found. A related discovery a year earlier, however, provided evidence that archaebacteria related to this group may be widespread in deeper ocean waters, in which case there may be another explanation for the *Alvin* findings.

In the October 21, 1993 *Nature*, Karl O. Stetter and coworkers from the petroleum industry reported that they had found populations of heat-loving, sulfur-eating bacteria and archaea in undersea oil reservoirs. It was not completely clear how the organisms got into the reservoirs, one located about 10,000 ft (3,000 m) beneath the floor of the North Sea, and three about the same distance beneath the surface of the North Slope in Alaska. In each case, the reservoirs had been injected with sterilized seawater in hopes of recovering additional petroleum from oil wells using the reservoirs. Stetter and his workers reported that the most likely possibility involved spores of the organisms in the sea water that were passed through the chemical sterilization process. If so, the organisms could have responded to the high-sulfur warm environment provided by the reservoirs, where temperatures range from 185-215°F (85-102°C), with a sudden population explosion. Such a scenario would also explain how communities of heat-loving bacteria or archaea develop at isolated geothermal vents. Given the other communities now known to exist beneath the surface of the Earth, however, the organisms could easily have reached the reservoirs directly from the rock.

More evidence piled up in 1994, all of it tending to confirm Gold's predictions and calculations. The DOE released more information about its quiet efforts, saying that in addition to the Triassic sediments in Virginia, it had found populations of heat-loving bacteria and archaea at five other sites. A group from Great Britain reported that they had found sulfur-eating bacteria or archaea in ocean sediments as deep as they could drill. Their deepest sample came from 1,700 ft (518 m) of sediments below 2,950 ft (900 m) of water in the Sea of Japan. Since there is no appreciable change in the environment of sediments, similar bacteria or archaea are likely to inhabit sediment depths twice as deep as those sampled, up to 3,300 ft (1,000 m) below the surface. All the bacteria found by the British researchers are anaerobic, and an analysis of their genes shows that many are distantly related to previously known bacteria and archaea.

In 1995, even more evidence surfaced, quite literally this time. Oil from a reservoir 5,500 ft (1,670 m) below ground—a reservoir that had never been pumped full of sea water and indeed is far from the sea—was examined and found to contain thriving colonies of thermophilic bacteria and hyperthermophilic archaea. The archaea, similar to those found at deep-sea geothermal vents, extended the known range of these creatures. All the bacteria and archaea thrived at the ambient temperature of the East Paris (France) Basin reservoir, which is between 150-160°F (65-70°C).

In addition to the surprising discovery that much of life on Earth is actually *in* Earth, there may be practical results of these discoveries as well. Previously discovered heat-loving bacteria have supplied the key enzymes that make many new biochemical processes practical, including the widely used DNA polymerase chain reaction.

Not all sulfur-metabolizing bacteria are at great depths. Wells and caves often host the same kind of bacteria. In 1986, for example, the Movile cave in Romania was discovered by accident during a construction project. The cave had been completely closed to the outside world for a long time, probably millions of years. Yet the Movile cave hosts a complex ecosystem, which included 31 invertebrates as well as the sulfur-metabolizing bacteria that are the base of the food web. The bacteria oxidize hydrogen sulfide using carbon dioxide, with the result being sulfuric acid. In other caves, and even at the bottom of some stratified pools, different bacteria use energy from organic sources to metabolize various sulfur compounds into hydrogen sulfide. Finally, some cave bacteria are able to oxidize hydrogen sulfide into pure sulfur, leaving great deposits behind. These and other results of recent investigations were shared at a conference in February 1994 at the University of Colorado at Colorado Springs, bringing together various specialists on cave life and formations to share information on how bacteria influence caves in regions where limestone is predominant.

Social Life of Bacteria

Scientists and poets alike have been fascinated by those social insects(bees, ants, and termites) who have developed colonies that resemble organisms in which the cells are replaced by individual insects, "organisms" whose tissues and organs are formed of many similar insects working together. To some, the social insects appear to be the future of life, and societies of humans should be heading the same way, in this view. However, it is only recently that people have begun to take seriously the similar organization of the social bacteria—myxobacteria similar to the better-known slime molds. Furthermore, scientists have observed apparent social divisions and activities in many common bacteria that previously were thought only to act independently of each other.

The behavior of myxobacteria, like that of slime molds, is social. Recent studies of myxobacteria have been carried out by Dale Kaiser of Stanford University. Myxobacteria are free-living independent creatures that, under certain conditions, come together to form an organism with parts and a single purpose. That purpose is usually reproduction, although aggregation may also be triggered by starvation. Between periods of aggregation, the myxobacteria work together in somewhat smaller groups to scavenge food. The myxobacteria are dependent upon excreted enzymes to digest their food partially before they absorb it. The enzymes are excreted into the region near the food source. However, if even a single bacterium attacks its target (usually the dead body of a bacterium of another species), these enzymes seep into the soil before achieving their purpose. Since the myxobacteria work together as a pack, they can all produce the enzymes at once. Additionally, by forming a sort of ball around the victim, the pack can keep its enzymes concentrated where they will do the most good.

The response to starvation is the most striking myxobacterial behavior. When the bacteria reach a certain state of food deprivation, they begin to release a chemical signal. This signal causes the bacteria to aggregate into huge groups. At some point, the bacteria sense that there are enough in one place to form a superorganism. When there are somewhere between 10,000 and 100,000 myxobacteria gathered, a peptide signal made from eight amino acids begins to be released. This peptide directs the bacteria to pile on top of each other. The result may be a cylinder, although near the top there is usually some spreading, resulting in a characteristic mushroom shape. When this shape has been achieved, the myxobacteria turn themselves into what bateriologists call "spores," although these differ from the spores of fungi, mosses, and ferns. Bacterial spores are cells in a state of suspended animation. Like the signal to form the shape, often called a "fruiting body" in analogy with fungal mushrooms, the signal to turn into spores is due to another small protein released by the myxobacteria. It appears that one myxobacterium needs to be told by its fellows how to behave and what to become.

Although the social behavior of myxobacteria has been long known, it is now recognized that all bacteria are capable of using chemical signals to communicate with each other. Given the right circumstances, common bacteria such as *Escherichia coli* and *Salmonella typhimurium* work together to achieve mutual goals. These bacteria possess a gene, discovered by Elene O. Budrene and Howard C. Berg of Harvard University, that causes the bacteria to aggregate into tight little balls in the face of unhealthy chemical conditions, such as exposure to hydrogen peroxide. Similarly, the disease-causing bacteria of the genera *Vibio, Neisseria,* and *Pseudomona* possess a gene, discovered by Gary Schoolnik of Stanford University and coworkers, that calls for the production of long fibers. These fibers link together so that a thousand or more bacteria can form a ball. This gene, also found in *E. coli,* appears to be triggered by some chemical in the host, since the bacteria only produce the fibers when they find a location suitable for feeding, such as mucus membranes.

(Periodical References and Additional Reading: *Nature* 3-12-92, p 148; *New York Times* 10-13-92, p C1; *Nature* 3-18-93, pp 207 & 239; *New York Times* 3-18-93, p D22; *Nature* 10-21-93, pp 694 & 743; *New York Times* 12-28-93, p C1; *Nature* 5-12-94, p 100; *Eos* 8-23-94, p 385; *Nature* 9-29-94, p 410; *New York Times* 10-4-94, p C1; *Discover* 1-95, p 66)

KILLER ALGAE

Algae have sometimes been considered plants, as have bacteria and fungi. In modern taxonomies, however, most algae are in one of several different phyla of the Protists, with the exception of the creatures once known as "blue-green algae," which have been classified as Monerans and renamed cyanobacteria. While the brown algae commonly called kelp are the largest protists known, most algae are microscopic, single-celled creatures.

Algae are famous for forming friendly combinations with other creatures. Perhaps the best known are the lichens (*see* "Fungi Surprises of the 1990s," below), but corals also get their color from algae that live in their flesh. Indeed, algae form mutually beneficial relationships with animals ranging from sponges to sea squirts (primitive chordates).

Algae, however, can also be deadly. In the July 30, 1992 *Nature*, J. M. Burkholder, Howard Glascow Jr., and coworkers at North Carolina State University in Raleigh wrote of their discovery of something they termed "phantom algae." These algae, a then unnamed species of microscopic dinoflagellates, lie resting on the bottom of the sea, waiting for a school of fish to begin feeding above them. When this occurs, the algae swarm upwards and release a harmful neurotoxin that stuns the fish, causing their skins to break out in sores. Particles of flesh come from the sores on the fish, and the algae puts forth special organs to allow themselves to ingest and digest the flesh particles from the fish. When the fish begin to die, the algae stop feeding, don a hard crust, and return to watchful waiting on the ocean floor. When a new group of striped bass, summer flounder, eel, or even lowly menhaden pass over, the algae shed their scaly coats and spring into action again.

The researchers reported that nearly a million Atlantic menhaden had been killed this way during one episode in the estuary of North Carolina's Pamlico River, one of the largest examples of algal attacks.

(Periodical References and Additional Reading: *New York Times,* 7-30-92, p A16; *Natural History* 9-95, p 12; *Nature* 9-21-95, p 223)

FUNGI SURPRISES OF THE 1990S

The fungi are a diverse group of both single and multicellular organisms that have in common a sedentary lifestyle (like plants) based upon obtaining nutrients from other living creatures, dead or alive (as animals do). Once thought of as defective plants, but now considered perhaps somewhat more closely related to animals, fungi have enough characteristics in common with each other to form a Kingdom of their own.

Fungi on the Tree of Life

Mushrooms that spring up in woods or on lawns appear to be odd plants, which may be why mycologists are to be found in the botany departments at universities. Yet mushrooms, the fruiting bodies of underground fungi, are not much like plants. None ever photosynthesize, flower, or produce leaves. Some other fungi are a bit like mosses in habit; that is, they are low-growing, spreading organisms. These include mildews, molds, blights, scabs, and rots. Truffles grow underground like tubers, but have no above-ground plant associated with them. Yeasts are more like protists than plants, being simple one-celled organisms that thrive in liquids if given a bit of a nutrient. Yet it is easy to see why most fungi were first thought of as plantlike, even after their significant differences from plants had become clear.

Drawings of an evolutionary "tree of life" from several decades ago either show the fungi as a lower branch on the main plant stem or as a fork with the plants as the other part.

In the April 16, 1993 *Science,* Patricia O. Wainwright of Rutgers University in Brunswick, New Jersey, along with Gregory Hinkle, Mitchell L. Sogin, and Shawn K. Stickel of Woods Hole Marine Biological Laboratory in Massachusetts, reported on a different tree of life based on RNA studies. In their tree, the fungi and the animals come off together as a separate branch. This branch joins with the plant branch to form a fork, with one side of the fork all plants, and the other side split into three branches: animals, fungi, and some protists that have long been thought to be one-celled versions of sponges. Sponges, of course, are generally considered animals, although just barely. In fact, the ancients classified sponges as odd marine plants.

In the 1980s, some biologists argued that fungi are more like animals than plants, basing their arguments primarily on the types of proteins produced, and the methods of production. The 1993 RNA study went further, suggesting specifically that fungi and animals had arisen from a common ancestor that was a flagellate protist. Not everyone agrees with this new study, although some scientists reported that the study explains such phenomena as the difficulty of treating fungal diseases in animals, a problem because both fungi and animals use the same biochemical pathways. Other evidence links yeasts, which are fungi, to animals. Yeasts have been a fruitful tool for study of animals genes, and can be genetically engineered to produce hormones that are more useful than the same hormones produced by bacteria.

Giant Fungi

Until 1992, the record size for the largest known organism on Earth went to the giant sequoia, with the blue whale considered the largest animal. The sequoia was a contender for age as well, although the longest-living organism was recognized to be an individual plant that had regenerated itself by vegetative reproduction for 11,700 years at least. This particular desert creosote plant was named King Clone by the biologists who established its age. In 1992, a couple of fungi surpassed both records, at least for a few months.

The first contender was an *Armillaria bulbosa* from a forest near Crystal Falls, Michigan. In 1988, scientists recognized that some samples they had taken of fruiting bodies in the forest were genetically parts of the same fungus. Over the next four years, the scientists studied the fungus, determined its dimensions, and estimated its age. When they announced in the April 4, 1992 *Nature* that the fungus was the oldest and largest organism on Earth, they got a lot of publicity. Even then, however, it was clear that the fungus probably did not deserve either title.

In size, the *Armillaria* covered 38 acres, but most of it, like almost all fungi, consisted of thin threads. Its total mass was perhaps as big as a large blue whale, but nowhere near the mass of a giant sequoia. Furthermore, the age could not be any greater than the last time that part of Michigan had been glaciated, which made the fungus slightly younger than King Clone in the desert.

A month later, inspired by the Crystal Falls fungus success, other scientists pointed out that a similar claim could be made for an *Armillaria ostoyae* fungus in southwestern Washington State, except that the Washington fungus covered about 1,500 acres, making it much larger than the one from Michigan. The Washington fungus was probably younger, however, as it was growing very fast, doubling in size every 20 years when conditions were suitable. Analysis of the Washington State fungus, however, had been conducted earlier and without DNA fingerprinting, so it was less clear that the giant fungus was actually a single individual.

Toward the end of 1992, a group of botanists laid claim to the "biggest" title with an aspen clone. While the 47,000 aspen trees from the same root system only cover 106 acres, trees are much heavier than fungi. Therefore, their combined mass was estimated at 13.2 million lb (600 m tons), 40 times the mass of the largest blue whale, but only slightly more than the mass of the greatest giant sequoias.

Origins of Lichens

Even when species are closely associated throughout their lives (for example, the monarch butterfly and the milkweed) we continue to think of them as two distinct species. Yet there is one large group of organisms, numbering in the thousands, so closely intertwined that we recognize the pair as a single organism. These are the lichens, symbiotic duos in which one partner is a fungus and the other is an alga or a cyanobacterium. Lichens vary considerably, however, in the degree of symbiosis exhibited. For example, in some cases the fungus and the alga reproduce separately, while in many other instances they produce the functional equivalent of a spore, a reproductive package containing cells of both the alga and the fungus. Often, the fungus is unable to exist without the alga, although generally the alga can survive on its own. Although various alga and cyanobacteria can be partners with fungi, about half of all types of lichen have the same species of green alga as one member. Thus, the type of partnering fungus generally determines the classification of the lichen.

The classic view suggests that lichens originally developed when a fungus regularly parasitized an alga. The alga probably fought back. Over evolutionary history, the fungus and the alga developed a tolerance to each other, so that the relationship changed from parasite and reluctant host to mutual partnership. However, recent research by Andrea Gargas and Paula DePriest of the Smithsonian Institution in Washington, D.C., and coworkers suggests that this picture of lichen evolution is not true.

Among the difficulties of studying the evolution of lichens is that the fungi, even when clearly different from one another, do not possess a lot of the kind of structure used in most comparative studies. Gargas and DePriest got around this problem by ignoring structure and going directly to the DNA. They were able to show that the fungal partner can be almost any type of fungus, implying that the lichen lifestyle has evolved many times. The fungal partners of different lichens are more closely related to non-lichen fungi than they are to each other. Furthermore, the types of fungi that have formed lichens normally have quite different lifestyles. While some fungi may have originally been destructive parasites of algae (following the common view), other fungi may have formed helpful partnerships with the algae all along. In either case, lichenization may have occurred recently or long ago in evolutionary terms. In most instances, the relationship between the fungus and alga can be expected to continue to evolve. A friend today may become a vicious ax murderer tomorrow, while an invasive parasite may become a partner.

Invasion of the Flower Snatchers

Fungi that have invaded or colonized alga are lichens, while fungi that invade or colonize plants are pests. The names "rust," "blight," "scale," and "smut" are among the cusswords of farmers and gardeners. While the name "powdery mildew" sounds more pleasant than blight or smut, the actual fungus is just as repulsive. Thus, it is a surprise in more ways than one to find that a fungus can turn into a flower, or at least something that looks quite like one.

B. A. Roy of the University of California, Davis, discovered that appendages appearing to be flowers on a wild mustard called *Arabis* spp. are something else entirely. The apparent flowers are good enough to fool botany students and, at a distance, even professional botanists. In a scenario familiar from science fictions movies, the *Arabis* have been taken over by an alien creature. The alien hides for a time, waiting for spring. In spring, the invader restructures the host. The mustard plant doubles its normal height and produces twice as many leaves, many of which form as a rosette at the top of a nearly naked stem. Leaves in the rosette turn bright yellow and become covered with a sweet syrup. The fake "flowers" "bloom" at the same time as the buttercups, which typically grow alongside *Arabis*. The inconspicuous flowers normally produced by a non-invaded mustard plant fail to appear.

There is a point to restructuring the *Arabis* in this way, of course. The invader has sex in mind. Flies, bees, and butterflies are attracted to the yellow "flowers" of the co-opted mustard, just as they are to the buttercups that grow in the same fields. When the insects come to feed on the syrup, they also pick up on their bodies the fungal equivalent of sperm from the invader, at the same time perhaps depositing sperm from other "flowers" of *Arabis* on the sexually receptive organs of the invader. After this happens, the work of the "flower" is over, so the invader stops wasting energy on the project. The "flower" is allowed to turn green so that the plant can devote more space to the main business of photosynthesis, producing energy to keep the invader well supplied with food.

The alien is a fungus—specifically a rust called *Puccinia monoica*. Like many rusts, it requires sex to reproduce, and pollination by insects is an effective way to achieve mating. The rust-induced "flower" is actually more attractive to insects than are the real flowers of species that grow in the same region as *Arabis,* high in the meadows of the Rocky Mountains. Insects spend more time on a pseudoflower than on a real flower, probably because the syrup is scattered over the whole leaf rather than being concentrated in a reservoir at the base of the flower.

Some other sexually reproducing rusts also produce sweet syrups on infected plants to attract insects. Some smuts use real flowers to attract insects that then disperse spores, although only dispersion and not mating is the object. The fungus that causes blueberry fruits to shrivel up like mummies also takes over leaves in the host, as *Puccinia* does with wild mustard, but the change caused by the mummy-berry fungus is not so dramatic. The infected blueberry leaves have their ultraviolet reflectivity changed so that they look rather like flowers to insects, many of whom can see at ultraviolet wavelengths, yet still look like leaves to us mammals. As with the smuts mentioned above, the mummy-berry is only interested in spore dispersion, not in sex. But the *Puccinia monoica* is the first known instance of a fungus causing a plant to create a structure solely for the purpose of fungal reproduction.

(Periodical References and Additional Reading: *New York Times* 4-2-92, p A1; *New York Times* 5-18-92, p A12; *Discover* 1-93, p 69; *Nature* 3-4-93, p 56; *New York Times* 4-16-93, p A18; *Science* 4-16-93, p 340; *Science* 6-9-95, pp 1437 & 1492)

PLANTS AND ANIMALS

NEWLY FOUND MAMMALS

During the early part of the twentieth century, scientists believed that the okapi(a creature resembling both a zebra and a giraffe), first identified in 1901, would be the last large mammal, and indeed the last large creature of any kind, to be become known to science. Few believed in legends of Himalayan Yeti, the Loch Ness Monster, the West Coast Sasquatch (Bigfoot), or Lost World dinosaurs on remote South American plateaus. Although no creatures as fabulous to the imagination as these have emerged from the real forest, the 1990s have provided a surprisingly rich array of new mammals, both large and small. When the kouprey (*Bos sauveli*) was named and described in 1937, there was less notice taken of it than one might expect, although in retrospect it is often cited as the last large mammal to be discovered before the 1990s.

If you count whales, sheep, pigs, and goats among the "large" mammals, there have been a baker's dozen of large ones discovered throughout the twentieth century. If you include small mammals, the number becomes greater. One count states that 746 mammals have been discovered since 1930 alone. And the pace of discovery, which one might expect to have fallen off, is actually accelerating. About 40 previously unknown mammal species were identified from the end of World War II until the beginning of the 1990s. About 50 new mammal species were identified in just the early 1990s, with (one presumes) more to be found in the late 1990s.

In addition to mammals, there are also a few other dramatic discoveries from time to time. While few persons become excited over the discovery of the ordinary new species of fish, a "living fossil" like the coelacanth, discovered by science in 1938, will make the newspapers as well as the scientific journals. One such discovery of the 1990s was park ranger David Nobel's recognition that some pine trees in Australia's Wollemi National Park in New South Wales were not like any he had seen before. Biologists were called in to investigate, and identified the trees as a type thought to have become extinct some 50 million years ago after having been widespread during the reign of the dinosaurs. A tree may not cause the public excitement of a previously undiscovered mammal, but the Australian pine trees, of which 39 individuals are known, are likely to be of more scientific value than finding another monkey or ox, since the trees are so radically different from other pines.

TIMETABLE OF NOTABLE MAMMALS IN THE EARLY 1990s

1990   Maria Lúcia Lorini, Vanessa Guerra Persson, and Dante Martins Texeira discover the black-faced lion tamarin, a monkey species previously unknown to science, on an island off the coast of Brazil.

1991   James Mead of the U.S. National Museum of Natural History recognizes that 11 examples of beaked whales, which washed up on various Pacific beaches in the late 1970s and 1980s, represent a previously unknown species, which he names *Mesoplodon* peruvianus. This is the third new beaked whale to be discovered since World War II.

1992   Marco Schwarz discovers a tiny new monkey in the Amazon rain forest of Brazil, naming it "the Maues marmoset" after the river near which it is found.

Jutta Schmid of the University of Tübingen in Germany thinks that she has found a new species of lemur in Madagascar. It is the world's smallest lemur, weighing just an ounce (a few grams). A later search through the literature, however, indicates that the lemur had previously been known to science, but somehow its scientific name, *Macrocebus myoxinus*, had been accidentally transferred to a different species in the same genus.

In May, Vietnamese scientists from the Ministry of Forestry led by Vu Van Dung, along with scientists from the World Wide Fund for Nature, are jointly surveying the Vu Quang Nature Reserve. In hunter's houses, they find three skulls of a goatlike bovine known to natives of the rain forest as the sa lao. Further studies, in which more bones and three complete skins are collected and analyzed, demonstrate that the sa lao is completely new to science.

1993   In the June 3 *Nature*, Vu Van Dung, James MacKinnon, and coworkers describe the sa lao as a new genus of bovid, which they name *Pseudoryx nghetinhensis*, with *Pseudoryx* meaning "similar to the oryx" (an animal that also has long curved horns), and *nghetinhensis* meaning "from Nghe tinh province" (the name of a former Vietnamese province that covers the range of the newly found mammal).

1994   A living sa lao, or *Pseudoryx nghetinhensis*, is liberated in June from a Vietnamese villager who had captured the calf. Later, a few other specimens are also captured and kept in captivity.

A new species of muntjac, or barking deer, is also found in the Vu Quang forest in Vietnam. It is distinguished from other muntjacs by a large size, larger antlers, and unusual curved canine teeth.

Tim Flannery of the Australian Museum in Sydney tracks down and captures a live black-and-white tree kangaroo in Indonesian New Guinea, which possesses the odd habit (for a tree kangaroo) of living on the ground. Its black-and-white color, unique in tree kangaroos, led to the discovery of this previously undiscovered species. Flannery had observed a photo showing the tree kangaroo some years earlier, but was unable to mount an expedition to the remote region high on the Mouke Mountain Range of Irian Jaya until 1994.

Phillip Morin of the University of California, Davis, Jane Goodall of the Combe Stream Research Centers, and coworkers use DNA taken from the hair of chimpanzees to show that a West African population of chimpanzees is different enough from other chimpanzees to constitute a new species, thus finding a new large mammal in a different way, as announced in the August 25, 1994 *Science*. Research by Maryellen Ruvolo of Harvard University, reported in the September 13, 1994 *Proceedings of the National Academy of Sciences*, suggests that the West African gorilla is also a separate species, different from the mountain gorilla.

David C. Oren of the Goeldi Natural History Museum in Belém, Brazil, launches an effort to find the giant ground sloth, thought to be extinct, or one of its previously unknown relatives. The search is based on descriptions from rubber tappers and Amazon natives of a 6 ft (2 m) tall creature encountered occasionally in the Amazon rain forest. No success is reported.

While not a new species, another development in the new-large-mammal field is the discovery by Michel Peissel in Tibet of an ancient, previously unknown pure breed of horse, unrelated to any modern purebred line.

Researchers offer convincing evidence that the most famous photograph of the Loch Ness monster, supposedly taken by Robert Wilson in 1934, is a fake.

And a New Mammal from the Recent Past

The Lost World idea includes island lost worlds, like the one where King Kong roamed with some leftover dinosaurs. In March of 1994, Andrei Sher of the Severtso Institute in Moscow and coworkers astonished a lot of people by reporting of a Lost World of sorts on Wrangel Island, about as far north of Siberia as Cuba is south of Florida. Although this Lost World was completely lost soon after the Great Pyramid and an early form of Stonehenge was built, it did contain a remnant of a giant animal believed to have become extinct much earlier.

The animal is the mammoth, thought to have become extinct about 8000 B.C. at the latest. The Russian scientists, however, found teeth and bone fragments from a pygmy mammoth on Wrangel, and were able to date these to as late as 1700 B.C.

During the last Ice Age, when sea levels were lower because so much water was tied up in ice, Wrangel Island was part of the mainland of Asia. Mammoths were creatures of the Ice Age, and evidently roamed the edges of ice sheets in search of grass. At the end of the Ice Age, Wrangel Island, because of hillier topography than neighboring Siberia, kept its grasses and also the mammoths. Then, when the ice melted and sea levels rose, the mammoths were trapped on what had become an island.

Large animals often develop into pygmy forms on islands. The pygmy rhinoceros of Sumatra is a prime example, although the rhino of nearby Java, a different species, is nearly as big as its mainland cousins. Conversely, the giant tortoise of the Galápagos has evolved in the other direction. In any case, more often than not, the island form of a large animal is smaller than the mainland form, although the reasons behind this dichotomy are far from clear. The pygmy mammoth of Wrangel Island is not a surprise in that way. The real shocker is that any species of mammoth survived until civilization was well under way, living on after the invention of writing and the agricultural revolution.

(Periodical References and Additional Reading: *New York Times* 10-20-92, p C7; *Nature* 6-3-93, pp 398 & 443; *New York Times* 6-8-93, p C1; *Discover* 1-94, pp 43 & 54; *New York Times* 2-8-94, p C1; *New York Times* 3-20-94, p IV-1; *Nature* 4-14-94, p 593; *New York Times* 4-26-94, p C6; *New York Times* 5-3-94, p C4; *New York Times* 5-24-94, p C4; *New York Times* 7-6-94, p C4; *New York Times* 8-26-94, p A20; *New York Times* 12-15-94, p A3; *Discover* 1-95, pp 60 & 61)

HORMONES AND SEX

It is fashionable to think that genes are destiny, and to disbelieve the old idea that events happening to a pregnant mother determine important features of the yet-to-be-born child. Life is seldom simple, however. Scientists today recognize that newborns, as well as adults, are products of their environment as well as of their genes. Furthermore, during the 1990s, researchers learned new ways in which a mother influences development after the embryo is conceived. While stopping short of finding that a craving for strawberries produces a child with a strawberry birthmark, scientists have been surprised to learn that sex is to some degree determined in the womb, not during conception.

The Case of the Hypermale Hyena

Among spotted hyenas (*Crocuta crocuta*), the normal complement of children per birth is twins. But what twins! Although hyenas are doglike in appearance, and their young are called "cubs" as if they were bears, the hyena is more closely related to the cat and the mongoose family, which would suggest that the young be called "kittens." Yet these offspring are hardly kittenlike. When the pair make their way into the world through their mother's clitoris (this odd mode of birth is explained below), their eyes are open, their muscles are working, and their teeth are ready to rip into something. Often that "something" is the other twin. Whether or not it succeeds in killing its twin may depend in part on sex. Among spotted hyenas, the female of the species is definitely deadlier and nearly always the boss, but most twin killings occur between cubs of the same sex.

The female hyena is clearly the more dominant, not only wearing the pants, but filling them with what at a casual glance appears to be a penis and testicles. This illusion is so sufficiently convincing that the first African big game hunters thought there was only a single sex among the spotted hyenas, a hermaphrodite who could impregnate, be impregnated, give birth to twins, and suckle the offspring. Looks, however, can be deceiving. In the female spotted hyena, what appears to be a penis (which can be limp or erect as the occasion demands) is actually a greatly enlarged clitoris. The apparent testicles are protruding, fat labia, fused together. These labia block the vagina. Another opening, small but extremely stretchable, exists at the end of the clitoris. Copulation and birth both take place through this clitoral opening into the vagina.

The source of this masculinization of the female hyena appears to be an unusually large amount of testosterone that is present in the womb. Both male and female cubs are born aggressive, but the masculinizing effect of the testosterone (which is responsible for the anatomic similarities of the male and female cubs) does not serve to make the male any more masculine than normal. Furthermore, scientists believe that the female spotted hyena also produces a second hormone for aggression that the male lacks. Part of the evidence for this is extrapolated from the fact that while the testosterone levels in the female decline as she ages, aggressiveness does not. Throughout life, the female spotted hyena remains, when viewed against a background of typical mammal sex behavior, a "hypermale," with the smallest female dominant over even the largest male.

The Case of the Gestating Gerbil

The effect of male hormones on a female Mongolian gerbil(*Meriones unguiculatus*) is much more subtle, and in some ways even odder. A gerbil mother gives birth to a litter of seven or eight offspring called pups. Since the 1970s, studies have shown that in various species of rodents, the effect of those hormones produced by a male fetus on its siblings can be observed in the siblings after birth. The most apparent differences would occur in a female pup who spent her gestation period developing between two males (a 2M female), compared to a female pup who developed in the womb between two sisters (2F). Social stress to the mother, which also changes the hormonal balance, can produce the same kinds of changes in daughters as does the 2M position. The differences, which include changes in the age of puberty, aggressiveness, size of home range, sexuality, and genital and brain anatomy, are particularly prominent among the gerbils.

A study of 2M gerbil females by Metrice M. Clark, Peter Karpiuk, and Bennett G. Galeef Jr. of McMaster University in Hamilton, Ontario, revealed what appears to be a form of Lamarckian evolution, or inheritance of an acquired characteristic. The Canadian psychologists found that 2M gerbils have litters with, on the average, more males than those of 2F gerbils. Thus, the hormonal influence of the 2M females' brothers includes the eventual production of litters in their sister that may be 60% male. In such a litter, since gerbil fetus positions are determined entirely on the basis of chance, there will be more 2M females than in the litter of a 2F female, who statistically has fewer male fetuses to females. Thus, the subsequent generation of gerbils should have even more 2M females born to the 2M mother, and even more 2F females born to the 2F mother. The actual probabilities, based upon analyzing 50 litters, are that the daughter of a 2M female is 1.73 times as likely to be 2M herself, while the daughter of a 2F females is 0.60 times as likely to be 2M as 2F. Through this mechanism, the hormonal influence provided by a particular generation of brothers goes on to increase the number of nephews they will have in the subsequent generation.

One might think that these trends would eventually lead to overwhelmingly male sex ratios in the gerbil, but this does not occur. The high production of females in the 2F line is thought to assist in keeping things better balanced. Furthermore, the later onset of puberty experienced by 2M females also leads to fewer litters over their lifetimes than the number of litters produced by the earlier maturing 2F females.

The mechanism by which this second-generation effect is caused is not known. Speculation hinges on the original hormone effect somehow changing the nature of the 2M gerbils' eggs, perhaps making them more susceptible to penetration by the lighter sperm carrying the smaller Y chromosome.

In humans, the effect of testosterone on the female fetus with a twin brother has been variously suggested to cause more left-handedness or greater mathematical ability, but studies have not really been conclusive in either regard.

The Case of the Calculating Canary

The effect of testosterone on aggression is found throughout much of the animal kingdom. It is the same in birds as it is in mammals, for example. Testosterone levels account for many of the changes one observes in birds as the seasons progress. In at least one instance (and probably in others not studied) the effects of testosterone in the egg, which can be considered to be the bird's equivalent to the womb, are evidently designed by evolution to be useful in preserving the species.

The species that has been studied by Hubert Schwabl of the Rockefeller University's Field Research Center in Dutchess County, New York, is the common canary(*Serinus canaria*), a partly domesticated songbird whose wild form is found in the Canary Islands and other islands off the northwest coast of Africa. Schwabl found that the female canary controls the amount of testosterone each egg supplies to the developing chick in a manner calculated to equalize chances of survival.

Birds minimize their weight in various ways. Among these ways is the practice of laying a clutch of eggs over a period of days, so that the mother bird does not have to carry about a half dozen or more fully developed eggs at any one time, which might prove too heavy to allow flight. The tradeoff of this practice is that eggs laid a day or even two days apart will hatch in the order they are laid. By the time the last chick has hatched, the first one might be a week old. This older chick could then demand more food and attention or could crowd its youngest sibling out of the nest. It would be helpful to level the playing field among the chicks that share a nest.

Mother canary arranges for such a leveling. As with hyenas, a high testosterone level before birth results in a newly hatched chick that is especially aggressive. So the first egg is formed with just a thin layer around the yolk that, as the chick develops, will produce only a small amount of testosterone. The first-hatched has no need for great aggressiveness, since there are no other chicks as yet. The next egg to be formed gets a thicker layer, which then produces more testosterone. By the time the fifth and usually final egg is laid, the level of testosterone received by the developing chick may be as much as 20 times the amount produced in the first egg to be laid. Thus, the aggressiveness of the chicks increases with the order of their hatching, which reflects the greater threat to survival posed by the older chicks.

The Case of the Mutating Male

The preceding cases show that sex is powerful, but the question baldly asked by James Thurber and E. B. White—*Is Sex Necessary?*—has not been answered by science, although for most species Mother Nature has answered "yes." Mammals are so committed to sex that they have evolved barriers prohibiting parthenogenesis (the production of offspring with no male involvement). Yet even among phyla with no such barriers, parthenogenesis, when it arises, is not especially successful in evolutionary terms. Most theories which attempt to explain why sex is desirable in a world of evolution by natural selection end up foundering upon close examination or mathematical analysis.

In 1994, Rosemary Redfield of the University of British Columbia in Vancouver used computer modeling to test some of the effects of sex on mutation rates. It had already been established that male animals and plants produce sperm or spores with much higher mutation rates than the eggs of females, possibly because males generally produce vastly more sperm cells or spores than females do eggs. Some studies have show that the mutation rate increases directly with the total number of reproductive cells, or gametes, produced. In humans, the rate of mutations in the DNA of sperm is thought to be 10 times that of the DNA of eggs.

Nearly all mutations in DNA either cause no change in the individual, or else cause harm. By itself, a high mutation rate for male gametes would seem to be harmful. Even when Redfield took into account such factors as the suppression of the action of a harmful male gene by the functioning female allele, the computer models showed that parthenogenesis is apparently the sounder evolutionary course. Redfield did show that sex could be desirable under some conditions. If a small number of mutations increases fitness and a large number causes death, then for a moderately high mutation rate there is a mathematical advantage to sex. It is not clear, however, that this scenario resembles reality. Redfield calls for more research into actual rates and conditions.

Perhaps the most popular theory to explain why sex is useful states that it speeds up evolutionary change. Most parasites, including bacteria, protists, and worms, are capable of rapid change. Parasites can shift their outer coats, for example, to avoid detection by the immune system or to evolve quickly in the face of such challenges as antibiotics. Slower breeding plants and animals, which are the kind most likely to practice sexual reproduction predominately, can use sex to increase their own mutation rate. Robert Vrigenhoike of Rutgers University in Brunswick, New Jersey, studied fish called topminnows, which can be either sexual or asexual. These topminnows were found to be vulnerable to their chief parasite (a flatworm) to the extent that all topminnows are genetically alike. The most common asexual form in a given pool is attacked more often than are other topminnows in the pool. Generally, sexual forms are less likely to be parasitized. Only when a pool population has been founded by a small group of inbred sexual fish is the situation reversed, but the sexual population quickly becomes less likely to be parasitized if a few females are added from a different genetic group, overcoming the inbreeding.

The theory which suggests that sexual reproduction exists because it provides an advantage in the war against parasites is related to another theory. This other theory addresses the advantage of sexual selection for traits, citing the peacock's tail. Although the peacock's tail reduces the survivability of the individual male's possessing them, the tail also increases the probability of heirs through sexual selection. The theory goes on to suggest that males who can survive with such impediments must have improved resistance to parasites.

In 1994, Stephanie J. Schrag of Emery University and coworkers demonstrated that a species of snail grows up to be a self-fertilizing hermaphrodite in conditions that favor low levels of its main parasite, while it turns to sexual reproduction if it is developing in conditions that usually lead to high levels of the parasite. Thus, the idea that sex is primarily a weapon in the fight against parasites has some experimental backing, although even Schrag thinks that more research is needed and that parasites are probably not the whole story.

The Case of the Voluptuous Vole

Testosterone has a key role in determining an individual's sex, while estrogen mediates many processes in women, particularly those relating to reproduction.

Other hormones also play a number of different roles in each sex, and greatly influence sexual behavior—certainly in rodents, and probably in humans.

While the hormonesvasopressin and oxytocin are known primarily for their other functions (vasopressin for kidney and blood-pressure control, oxytocin for contractions in labor and milk production during lactation), they are clearly involved with issues of sex in some fundamental way. Studies of male humans have shown that both vasopressin and oxytocin are released during orgasm, with oxytocin responsible for pleasurable sensations, according to some studies. Vasopressin is known to play a role in male territorial behaviors, such as scent marking in rats.

Thomas R. Insel of the National Institutes of Mental Health and C. Sue Carter of the University of Maryland in College Park are among the scientists who have chosen the mouselike vole as an experimental animal in which to study the functions of vasopressin and oxytocin in regard to sex and sexual behavior. Interest in the vole was incited by the fact that some vole species, such as the prairie vole (*Microtus orchogaster*), are notable for monogamy and for involvement of both parents equally in raising the offspring (pups). An adult male prairie vole bonds for life to the first female with whom he copulates. After several days of copulation, often reaching 50 or so episodes, the male will aggressively fight off any potential rivals for his mate's interest. When pups are born to this union, they are cared for by both parents.

Research into vole reactions to hormones has been facilitated by the use of antagonists to vasopressin and oxytocin developed by Maurice Manning and coworkers at the Medical College of Ohio in Toledo. Hormone antagonists are chemicals that bind to the receptors for a specific hormone but do not have the action of the hormones. When most of the receptors are occupied by the antagonist, the effect of a hormone is greatly reduced even when the normal amount of the hormone is produced.

Insel and Carter and coworkers showed that a previously unmated male vole injected with a vasopressin antagonist behaved with a receptive female in the same manner as a control male, coupling intensely and repeatedly. However, the male whose vasopressin had been blocked failed to bond with the female and did not defend her against other males. The vasopressin-blocked male also failed to show any other signs of preference for his recent partner. When an unmated male that had been injected with vasopressin was put in with a female who had been made surgically incapable of participating in sexual intercourse, the male defended her and bonded with her even though physical mating did not occur. Thus, in prairie voles it appears that release of vasopressin is a direct factor in monogamy. Experiments with the montane vole (*M. montanus*), a closely related vole species that does not normally pair-bond, showed that vasopressin injections in males had no effect in that species' behavior patterns. Presumably, the montane vole does not possess the receptors that cause pair-bonding in the prairie vole.

Similar research by Geert De Vries and Zuoxcin Wange of the University of Massachusetts at Amherst have show that a vasopressin antagonist injected into a prairie vole father reduces such typical father-vole behaviors as huddling over pups, grooming them, and even carrying them to safety in the face of a threat.

Similar experiments with males and oxytocin showed no effect. While oxytocin seems to have no behavioral effect on male voles, it appears to help female voles develop better social bonds, including a willingness to mate and to care for the pups. In rats and mice, oxytocin released during lactation has shown to change the wiring in the brain.

No one knows for sure if any of the bonding effects of vasopressin or oxytocin occur in humans, but the hormones themselves are produced, so scientists are creating experiments to learn more about this apparent second set of sex hormones.

(Periodical References and Additional Reading: *New York Times* 9-1-92, p C1; *Nature* 8-19-93, pp 671 & 712; *New York Times* 8-24-93, p C4; *Nature* 10-7-93, p 545; *New York Times* 11-2-93, p C1; *New York Times* 1-25-94, p C1; *Nature* 5-12-94, pp 99 & 145; *New York Times* 5-17-94, p C1; *Science* 1-5-95, p 30; *Natural History* 8-95, p 12)

PLANT BEHAVIOR

Botanists often complain that people, even fellow scientists, concentrate on animals too much, missing all the excitement of plants. If you think of a plant as something that does not move, fight, make love, or communicate with other plants or with animals, then you might be forgiven for finding plants dull. Yet plants actually do all those things. The difference is that plants behave in plantlike ways, which differ from animal ways.

A Gene Called *S*

In 1994, scientists learned about the way that some plants fend off unwanted pollen produced by their own flowers, while a plant with a different agenda takes steps to see that it *must* fertilize itself. Most orchardists know that some fruits, including all apples and many cherries, cannot fertilize themselves. Two apple trees of different varieties are needed if apple-blossom time is to lead to fruit production. If you just have one apple tree, better hope that your neighbor has a tree as well. It is less obvious that some garden plants are also unable to fertilize themselves, because one seldom grows a single tomato plant. The family *Solanaceae,* which includes tomatoes, tobacco, and petunias, has this trait. If a plant from this family discovers that the pollen grain growing toward its eggs comes from itself, the plant takes steps to halt this unnatural union.

In February 1994, Bruce McClure at the University of Missouri in Columbia and coworkers reported on an experiment revealing how a *Solanaceae* plant deals with its own pollen, while another team led by The-hui Kao at Pennsylvania State University at State College described their separate experiment, which also confirms this plant's method. A theory of how *Solanaceae* stay self-incompatible is based on the observation that when pollen begins growing in the pistil of a flower, a single gene, called the *S* gene, is activated. There are many different alleles of *S,* each one producing a version of an enzyme that, if it can bind to a structure in RNA corresponding to its own allele, acts to destroy the RNA. Thus, when the *S* gene is activated in the pistil, it will shred the RNA of a growing grain of pollen from one of its own flowers, killing the pollen cell, but will have no effect on RNA from another plant possessing a different allele of the gene. In some ways, the variable S gene resembles those variable parts of the immune system of animals that function in recognizing self and nonself.

McClure's team tested this theory by inserting an S gene into a plant that lacked the gene, a plant that thus was self-compatible. The resulting transgenic plant became self-incompatible. Kao's team added a third allele of the S gene to a plant already possessing two different alleles (plants, unlike mammals, often have triplets or other multiple sets of chromosomes), thereby producing a plant that rejects pollen containing any one of the three alleles.

Making Sure of a Little Help from Its Friends

Flowering plants that rely on animal intermediaries for pollen distribution generally have to pay them for the task. Sometimes the compensation is very direct, as when a bee collects a salary of nectar for each visit. Often, however, the situation is somewhat more complex. Some plants find that they must lure pollinators by various foul means. The "foulest" of these techniques may be the smell of carrion that some South American plants use to attract beetles or other carrion eaters.

Even when the reward is honestly provided by the plant, the plant may still have to take measures to prevent being taken advantage of by the pollinator. One example of how a plant keeps the upper hand was observed by Walter Whitford of the Environmental Protection Agency and coworkers. They already knew that the flower of the desert yucca plant is pollinated by a small white moth, the yucca moth, that goes from blossom to blossom, depositing in each flower a ball of pollen taken from the previous flower it has visited. Earlier researchers had determined that the pollen ball is accompanied by a single egg laid in each flower by the moth. The egg hatches into a larva, which lives on seeds that develop in the pollinated fruit. Since there are more seeds in a yucca fruit than the larva needs, uneaten ones are enough to complete the task of reproduction. As with many plants that are pollinated by moths, the yucca depends on the moth as its sole pollinator. Similarly, the moth depends completely on the yucca as a combination playground and cafeteria for its young. At first glance, the arrangement seems to be a clear-cut example of beneficial mutualism.

Yet it is not that simple. What prevents the moth from exploiting the yucca by laying more than a single egg? If the moth laid all her eggs on one blossom, the larvae that hatched would eat all of the yucca seeds; but that is no concern of the moth. One does not imagine a moth looking into the future consequences of its actions, thinking, "If there are too many larvae, they might cause the yuccas to become extinct; then what will my grandchildren do for food?" From an evolutionist point of view, the plant has to be doing something that directs the moth to spread out its eggs among as many flowers as possible, pollinating the flowers in the process.

Whitford and his team took to the desert for a month, watching the flowers develop, the moths pollinate and lay eggs, and the pollinated flowers grow into fruit. Nearly all of the action takes place at night, so they observed by flashlight. In July 1994, they announced that they knew how the plant keeps the moth on the move. Some 9 out of 10 flowers die before producing fruit, even if pollinated. Thus, the moth is confronted with a classic dilemma. If the moth lays all its eggs in one flower and quits, that flower will not develop seeds because it will not have been pollinated, and

THE BIRDS AND THE BEES AND THE FLOWERS

Traditionally, parents describe how flowers are pollinated by bees and how birds raise chicks from eggs to explain human reproduction to their children. The story of human reproduction is actually simpler than the story of how flowering plants reproduce, however. For our kind, there are only two humans of opposite sex involved, while flowering plants have available to them several different sexual reproductive procedures, each of which involves an intermediary that is not even a plant. (Some flowering plants also rely on asexual reproduction most of the time, but that is a different story.) Thus, it might be easier to use human reproduction to explain the birds and the bees.

Human reproduction begins when a large number of small motile cells (**sperm**) produced by the male are placed (via an organ for which this appears to be the primary, but not sole, purpose) into a receptacle in the female that periodically contains a single large cell (**egg**). The stage called **fertilization** is completed when one of the sperm succeeds in penetrating the outer membrane of the egg. After this point, a new human can begin developing.

Reproduction in flowering plants begins when a large number of two-celled grains called **pollen** are produced and loosely collected on an organ (the **anther**) whose only purpose is to hold the pollen in an appropriate place. Some other agent, usually the wind or an insect, but sometimes birds, bats, or other animals, shakes a few of the grains from the anther and transports these grains away at random. Fertilization will occur when a pollen grain reaches a receptive plant egg of the same species. One cell from the pollen fuses with the egg, and the other fuses with an unusual cell with two nuclei called the **endosperm mother cell**. The egg and the mother cell are part of an apparatus of seven cells that is most often found at the base of a flower, connected to the outside world by a tube or stem. The whole apparatus, including the tube or stem, is called the **pistil**.

One part of the pistil, the **stigma**, collects any pollen that is transported to it. Since pollen is not self-propelled as is human sperm, a pollen grain moves in the way that plants move, by growing. It can grow inside the tube or column, often keeping the actual pollen cells at its growing tip. When these cells have been transported into contact with the seven-cell egg complex, the whole nine-cell combination is the newly fertilized plant.

As if the basic process were not complicated enough, plants have evolved numerous variations on this theme. Many plants (for example, wheat and peas) contain all the needed parts within a single blossom, and pollen from that flower's anther needs only to be transported to the same flower's stigma. More commonly, a plant will only accept pollen from a different blossom, although both flowers may be on the same plant. In other cases, such as apple trees and tomato vines, the flowers need to be from different plants. The plant reproduction schema which most closely resembles human reproduction occurs in plants with two sexes, such as holly and asparagus. In these plants, pollen from an anther-bearing flower found only on the male plant must be transferred to a pollen-accepting flower found only on the female. Such plants can be recognized because only the female produces fruit, such as the bright-red berries found on both female holly and asparagus.

the larvae will starve. If the moth lays two or three batches of eggs, the chances are still high that the larvae will starve, since most of the pollinated flowers die before producing seed. To be assured of at least one larva surviving, the best strategy is to lay one egg in each flower, which is what the moth does.

A similar study was conducted in less dramatic surroundings at the Spring Grove Arboretum in Cincinnati, Ohio, where, for over a century, there has been a large population of the yucca known as Adam's needle. Like all yuccas, Adam's needle is pollinated by a yucca moth in the classic mutualistic relationship. In the case of Adam's needle, the moth often has to make several attempts before laying an egg, and proceeds to lay several eggs in the same flower, yet not so many that there will be no seeds left over for reproduction after the larvae have had their fill. The moth follows each egg-laying attempt, successful or not, with a trip up the pistil to apply to the stigma some of the pollen it has brought with it. If moths were rational creatures, one might think that the moth actually recognizes the necessity of the pollen it applies as allowing the flower to produce seeds as food for the moth's young.

As with all yuccas, flowers frequently fail to produce fruit, which scientists have attributed to a process of selective abortion. Still, several different theories exist to explain why Adam's needle "selects" certain flowers to abort and others to bear fruit. The theories were tested through experiments performed by Olie Pellmyr and Chad J. Huth of Vanderbilt University in Nashville, Tennessee, and their conclusions reported in the November 17, 1994 *Nature*. As many scientists had expected, those flowers with the greatest numbers of eggs laid in them were more likely to abort than those with smaller numbers, encouraging the moth that pollinates Adam's needle to work toward the "one flower, one egg" scheme used by the moth pollinating the desert yucca. Yet there was apparently another factor at work, since the lowest rate of flower abortion did not correlate well with the lowest number of eggs. Further study showed that the factor that best predicts flower retention followed by fruit develop-ment is the presence of 4-12 scars from egg-laying attempts. Thus, since each attempt to lay an egg is followed by a trip to the stigma to leave pollen, these flowers have been pollinated at least four times, assuring an ample supply of pollen. By aborting most flowers that have insufficient pollen or too many eggs, Adam's needle selects for moths that will provide high-quality pollen, but not lay too many eggs.

Plants have other ways of controlling reproduction that involve changing the behavior or even the physiology of animals, almost all of which are ultimately dependent on photosynthesizing plants as the source of energy. In the June 1994 *Ecology*, K. Greg Murray and Kathy Winnett-Murray reported that they had demon-strated for the first time that one long-suspected way in which plants manipulate those animals depending on them involves controlling digestion. The researchers, both from Hope College in Holland, Michigan, experimented with fake fruits intended to closely resemble those of the Puerto Rican shrub *Witheringia solanacea*. They exposed these fruit imposters to birds called black-faced solitaires, known to eat *Witheringia* fruit. Half of the sweet fakes included an extract from the real fruit, while the other half did not. All of the fake fruits contained real seeds.

The seeds from fruit imposters containing real *Witheringia* extract passed through the bird's gut and were deposited twice as fast as those from imposters containing no extract. Furthermore, germination was impaired in those seeds that had spent a longer time in the solitaire's digestive tract. The viability of seeds that had passed through a bird fell from a 70% germination rate in those imposters containing extract to 20% in those without the extract, a significant effect. Furthermore, since seeds from both the extract-containing imposters and from real fruit both passed in about 15 minutes, the birds were likely to have carried the seeds away from their parents, but not very far away.

Fighting Back

Plants may form the food base for most of the animal world, but it is not out of charity. Just as there must be balance between the needs of the yucca moth and yucca plant, between the needs of the *Witheringia* and the black-faced solitaire, there must also be some way for plants to protect themselves against excess predation. If animals could successfully consume all individuals of a plant species, the species would become extinct, which would ultimately harm the animal feeders as well.

While becoming extinct may be the ultimate revenge of a plant species on its animal predators, most often there is some other outcome. Plants fight predation in various ways, while predators continue to develop new ways to overcome plant protection. Here are a few of the discoveries of plant retaliation from the early 1990s, as well as ways that predators get around the defense.

- Piney-smelling chemicals called terpenes are used by pine trees to poison not only plant tissues, but even the air close to trees. When a beetle attacks a pine tree, the tree causes its own cells to burst. Each dead cell pours out sticky gums loaded with terpenes to jam up beetle paths and burrows and to poison the air inside the tunnels used by the beetles.
- Some species of pine beetles consume a small amount of the pine resin and its terpenes, so that the aromatic terpenes can be converted by beetle body chemistry to a signal calling other beetles of the same species. Soon, thousands of beetles are attacking a single tree in an apparent attempt to overcome the tree's defense.
- The Mexican desert shrub called the bursera also uses aromatic resins to deter beetles such as the flea beetle, but Judith X. Becerra of the University of Colorado in Boulder found that the bursera is not content just to allow the resins to ooze out of cut cells. Instead, it stores the resins under pressure in specialized canals which, if broken into, actively squirt the poisons for distances as great as 6 ft (2 m). The gum shoots into the mouth of a beetle grub that bites through the canal and can even cover the grub completely.
- After several exposures to resins from the bursera, the flea beetle grub learns to begin at the base of the leaf and systematically open all the canals. The grub can then proceed to the outer edge of the leaf, munching the whole leaf in safety. However, since it takes over an hour to prepare each leaf for a few moments of dining pleasure, the flea beetle grub cannot do serious damage to the bursera.
- The coyote tobacco plant can tell the difference between loss of leaves from wind or rain and loss from the bites of caterpillars. In the latter case, but not in the former, the plant quickly concentrates large amounts of poisonous nicotine in the

part of the plant being attacked. It is thought that the plant recognizes a chemical in caterpillar spit.

- Caterpillars often use poisons from plants to kill their own predators. The case of the monarch butterfly becoming unpleasant to eat by feeding on milkweed plants is famous, although perhaps not clearly established. Some caterpillars that feed on tobacco plants or on tomatoes take nicotine or tomato alkaloids, intended by the plants to deter the caterpillars, and recycle the poisons to deter wasps that otherwise parasitize the caterpillars. Similarly, gypsy moth caterpillars use tannin from tree leaves to defend themselves against viruses.

- Fire blight is a bacterial disease that often infects fruit trees (especially pear trees) causing the ends of branches to turn black and die. In 1992, Steven V. Beer and coworkers at Cornell University in Ithaca, New York, found that the bacterium contains a virulent molecule called *hairpin* that by itself will set off a reaction in which any cell touching a *hairpin* molecule will commit suicide. Trees also build walls of dead cells at the base of infected branches, keeping the infection from spreading within the tree.

- Corn, according to a study by Ted Turlins and James Tumlinson of the U.S. Department of Agriculture, responds to a caterpillar attack on one leaf by releasing terpenoid chemicals (similar to those of the pine-tree defense) from every leaf. Wasps have learned that terpenoids are a signal that the corn is under attack by their favorite prey, the caterpillar. Without the plant's signal, it is unlikely that wasps would be able to locate caterpillars in a cornfield.

Sharing

Some years ago, theoretical biologists analyzing how water flows through plants and ecosystems determined that water found in shallow plant roots as a result of the plant's active transport of water from deep in the ground would tend to seep into dryer ground near the surface. During the day, however, the flow of water vapor through leaves would keep the internal water moving upward and out. At night, when the process of vapor passing through leaves shuts down, the difference in water pressure between wet roots and dry soil would reverse the normal course of a plant circulation, allowing water to flow from the roots into the soil. Most biologists had not expected this to actually happen, believing that plants would have some mechanism to prevent water flow out of their roots.

Scientists, however, eventually check their predictions and hunches. John Baker of the U.S. Department of Agriculture was the first to do so in this case. He looked at Bermuda grass, which is adapted to a dry environment, and found that the theoreticians had been correct. Water does, in fact, escape from the roots at night. Later researchers found that sagebrush, in an even dryer environment, releases enough water each night to dampen the desert soil around the base of the plant. Grasses that grow near sagebrush could be presumed to take advantage of the situation.

In 1991, Todd Dawson of Cornell University in Ithaca, New York, observed that plants around sugar maples fared better under drought conditions than did those further away from the maples. Dawson calculated that a maple tree 40 ft (12 m) tall might deliver as much as 40 gal (150 L) or more of water from the damp depths where its deepest roots reach to the shallow soil where its lateral roots are located. If some of this water then escaped from the lateral roots into the soil, it would be available to the plants surrounding the base of the maple. Such lateral roots could easily reach out a distance of 16 ft (5 m) from the trunk of the maple. Dawson analyzed deuterium content (present in different concentrations in rainwater than in ground water) to determine the source of water in herbs, shrubs, and other trees growing near sugar maples. He found that the plants near the sugar maples were using ground water that could only have been released by the maples' lateral roots.

Such water sharing results in strong influences by water-transporting plants on the environment in which they grow. The plants that are able to use water around a maple's lateral roots and therefore thrive may not be exactly the same plants found elsewhere in the forest. Similarly, other studies have shown that some trees, such as the black walnut, release chemicals from their roots into the soil that prove toxic to many plants growing adjacent to them. Some seeds, such as sunflower seeds, also control their immediate environment with toxic chemicals. This is just another way in which plants are more active and less passive than people once thought.

(Periodical References and Additional Reading: *New York Times* 6-12-92, p C1; *New York Times* 9-12-92, p C2; *New York Times* 8-18-93, p C1; *New York Times* 10-26-93, p C1; *Science* 1-14-94, p 183; *New York Times* 7-12-94, p C4; *Nature* 11-17-94, p 257; *New York Times* 12-6-94, p C13; *Discover* 1-95, pp 88 & 89)

FINDING THEIR WAY

Animals often conduct seasonal migrations (many birds in North America and Europe fly south in autumn and return north in spring), or perform lifespan migrations (Pacific salmon travel from their birth upriver to the sea, and years later return upriver to spawn and die), or conduct longer-than-lifespan migrations (monarch butterflies and milkweed bugs may pass through several generations during a seasonal migration).

The mystery of how birds and other animals navigate as they migrate over long distances may never be completely resolved. It has become increasingly evident that animals use many different means to locate their paths and goals, sometimes employing several such techniques at once. Unaided humans also use several navigational tools at once, ranging from the sense of how long a journey may be and a very general sense of direction—which may or may not be influenced in a given individual by the position of the Sun—to specific recognition of important landmarks along the way. Other animals employ all these means and more.

Here are some of the ways animals are known to find their way about, as illuminated by research of the early 1990s.

Turtle Travels

We tend to think of birds as the master migrators. Yet, in terms of complexity of routes and potential hazards faced along the way, other creatures surpass birds in their migration behavior, particularly those who migrate to distant breeding grounds after a number of years away from them. The green turtle (*Chelonia mydas*) breeds on mid-Atlantic islands, such as Ascension island in the South Atlantic. Its young somehow travel from 1,200–2,800 mi (2,000–4,500 km) to the shores of Brazil to feed on turtle weed. After two or three decades, a green turtle develops the urge to mate, and returns to its birth island, a tiny speck in the ocean. Turtles live a long time, and therefore the long journey between the mid-Atlantic and the Brazil Coast is repeated many times, probably annually.

Other sea turtles, such as the loggerhead (*Caretta caretta)*, also migrate long distances between birth and feeding grounds. Atlantic loggerhead turtles hatch on beaches in the Caribbean or along Florida and immediately begin migrating. They take a circuitous route to their feeding ground in the middle of the Atlantic Ocean, where they travel about for several years. Then they return to their native beaches to breed, before heading back to the Sargasso Sea in the middle of the Atlantic.

In 1994, Kenneth Lohmann of the University of North Carolina at Chapel Hill reported on his study of how the loggerhead manages its travels. Since 1985, it has been known that turtles, like many other creatures, have magnetic minerals in their brains. Lohmann was especially interested in finding whether magnetic fields actually guide turtle migrations, so he devised a giant water-filled dish which allowed magnetic fields to be applied to swimming turtles.

Lohmann's experiments confirmed that turtle hatchlings just out of the shell head for the nearest source of dim light. Because turtles hatch at night, such a light source would be the direction of starlight or moonlight reflected from the sea. After that, the turtles' magnetic sense takes over. In the experimental tank, this keeps the turtles heading in the same direction as that initially begun. In real oceans, however, Lohmann found that hatchlings always swim into the waves. Swimming toward oncoming waves is an effective way to swim away from a beach. Lohmann thinks that further out, when the waves in the open ocean begin to become swells coming from every direction, the turtles begin to rely on their magnetic compass.

The turtles in the giant dish orient themselves using the inclination of the magnetic field. When wandering about Earth, inclination of the magnetic field is one indicator that tells how far north or south one is. In the tank, with the magnetic inclination set for the latitude of Florida, where the turtles were hatched, they swam east. When the slowly changing magnetic field reached 60°N, the turtles abruptly began to head south.

Checking the map of the turtle's regular migration pattern showed that the habit of swimming east at low latitudes would put the turtles into the Gulf Stream, which would then carry them north. When the Gulf Stream reaches 60°N, it branches. The northern branch would take the turtles to England where they would find it too cold to live, while the southern branch heads toward the mid-Atlantic. Swimming south brings the turtles to their feeding ground in the Sargasso Sea, where they spend the next several years. When they reach breeding age, Lohmann thinks, the adolescent turtles use their magnetic sense to head back to Florida.

Other biologists are impressed by Lohmann's studies, but some also think that turtles need more than a compass to find small islands or beaches. One popular alternative idea suggests that turtles, like other underwater migrators, may use taste or smell to recognize native waters.

The first creatures to use magnetite to orient themselves and the first known to science were certain bacteria. Since the early 1970s, when Richard P. Blakemore of the University of New Hampshire at Durham discovered magnetite-producing bacteria, other creatures have been added to the list. The discovery in the mid-1980s of magnetic crystals in some animal brains and the appearance of strange black spots on certain images produced by magnetic resonance imaging led a biologist to look for such crystals in humans. In 1992, Joseph Kirschvink of the California Institute of Technology at Pasadena and coworkers reported in *The Proceedings of the National Academy of Sciences* on small crystals of the mineral magnetite found in human brains. These are extremely tiny crystals, but there are a lot of them. According to Kirschvink, there are some seven billion crystals per brain. However, Kirschvink, who has a reputation for far-reaching theories based on little evidence, does not claim that the magnetite is used in navigation by humans. Indeed, the amount of magnetite is thousands of times less than that found in brains of migrating animals.

Birds Do It

Birds are among those animals who have been found to use Earth's magnetic field as a navigational aid, although this use is primarily a backup system when other means are not available. Experiments show that navigation in birds involves a sophisticated interaction of various senses. Just as a baby turtle must first set its magnetic compass with the light from the sea, some birds appear to keep their magnetic compasses on track or to reset them with light from the sky.

Unlike humans, who sometimes navigate by noting the position of the Sun, Moon, and stars, some birds have been shown to use a band of polarized light that is invisible to the human eye. This band always appears at a right angle to the position of the Sun, so it is equivalent to seeing the Sun. To a bird, the polarized light appears as a dark band across the sky any time the sky is clear and the Sun is up. Various tests have shown that homing pigeons and some other species of birds use this polarized band as a navigational tool rather than the disc of the Sun. In 1993, Kenneth P. and Mary A. Able of the State University of New York at Albany conducted experiments demonstrating that a migrating bird called the Savannah sparrow (*Passerculus sandwichensis*)uses polarized light to maintain the direction of its magnetic compass. The Ables used various combinations of shifted magnetic fields, depolarizing filters, and natural conditions on birds they raised themselves. A bird that was exposed to a shifted magnetic field but had been raised in depolarized light based its migration directions on the magnetic field alone, aiming directly for magnetic north. The most dramatic change occurred in birds that could see the normal sky, but were provided with a magnetic field shifted so that magnetic north was directly west. These birds reoriented their migration direction to match the E-W field. Thus, the polarized light clue overcame the magnetic clue.

The same August 5, 1993 *Nature* that contained the Ables' report also contained a letter by Wolfgang and Roswieth Wiltschko of the University of Frankfurt in Germany and coworkers providing evidence that some birds can see the magnetic field directly. The Wiltschkos work built on a set of 1992 experiments by J. B. Phillips and S. C. Borland that had demonstrated that the color of light affected migratory behavior in red-spotted newts (*Notophthalmus viridescens*). The Wiltschkos demonstrated that a migratory bird, the silvereye (*Zosterops lateralis*), did not know how to orient itself in red light, although it had no difficulty when presented with a nearly full spectrum from an incandescent bulb, or green and blue lights produced by light-emitting diodes. Although competing theoretical explanations for this difference have been put forth, the bottom line appears to be that the birds are somehow able to use higher frequency light to perceive the magnetic field in some way. Newts, however, were not tested under low-frequency light, but did orient differently when exposed to blue-green light.

It's In Their Genes

Many experiments with birds have shown that birds hatched in laboratory incubators have inbred migration routes, as well as senses attuned to the sky or magnetic field with which to follow those patterns. Similarly, those insects that go through several generations during migration must have genetically programmed maps. A change in the route among birds or insects who have not been subjected to experimental alteration of directional clues would seem to result only from a mutation affecting the hereditary paths.

In 1993, A. J. Helbig of the University of Heidleberg, Peter Berthold, and coworkers reported that a genetic mutation had, in less than 40 years, altered the migratory behavior of a kind of European wood warbler called the blackcap. Normally, birds from Central Europe migrate to Spain and the Mediterranean during the winter, but these birds had been genetically rewired to migrate to England in the region around Weston-super-Mare. Climate changes and the English habit of setting out bird feed in winter (still a relatively new practice) are thought to have brought about the change. Blackcaps that winter in Weston-super-Mare return healthier and better able to breed because of the richer feed put out for them in England.

Another Migration Mystery Solved

Since the 1920s, scientists have known that European and American eels are born, mate, and die in the Sargasso Sea, home of the adult loggerhead turtle. The eels then travel to freshwater rivers on the appropriate continent, where they spend some years before returning to the Sargasso to mate and die. In the case of European eels, it takes three years for a young eel to travel from its birthplace to the mouth of a freshwater river. Before this discovery, the appearance of sub-adult eels with no apparent newborn form in the vicinity was a major biological mystery.

With the example of the two Atlantic species of eel in mind, it was not difficult to suspect that the Japanese eel might also find a quiet place in the Pacific to reproduce. At last, in 1991, Matsumi Tsukamoto and coworkers from the University of Tokyo tracked the elusive eel to its lair, some 1,200 mi (2,000 km) east of the Philippines. Like the Sargasso, this part of the Pacific is encircled by currents.

(Periodical References and Additional Reading: *New York Times* 5-12-92, p C1; *Science News* 5-16-92, p 331; *Discover* 1-93, p 39; *Nature* 8-5-93, pp 491, 523, & 535; *New York Times* 9-28-93, p C1; *New York Times* 12-22-93, p C1; *Science* 4-29-95, p 661)

Background

Known methods for navigation by animals include almost all the navigational systems and tricks used by human navigators. While there are so far no known instances of navigation among other animals that is based on using radio signals from any source, human navigators increasingly rely on radio-based location using fixed transmitters and Earth-orbiting satellites (the latter, the Global Positioning System, appears to be becoming a new standard for all forms of human navigation). Yet most of the other basic methods of navigation are shared by both humans and other animals.

Celestial navigation Birds evidently orient their general direction using the Sun, and there is some evidence that salmon at sea also use the Sun. In planetarium experiments, migrating birds have been shown to maintain a fixed direction by observing stars as well.

Dead reckoning European garden warblers change headings on their migration around the end of September as a result of an internal clock. This is a form of dead reckoning, since they migrate to the southwest for about two months, then turn and migrate to the southeast for about three months, arriving in December at the correct destination on the west coast of Africa.

Echolocation Dolphins and porpoises can determine their distance from objects including the ocean floor using sound waves. Humans use sonar and charts of ocean or lake depths for navigation, but it is not completely established whether dolphins also navigate with this kind of piloting.

Follow the leader Many birds and some other animals migrate in flocks or in lines. This is particularly true of birds such as the North American bobolink and the Pacific long-tailed cuckoo, which migrate with pinpoint accuracy over long distances. It is natural to assume that birds who have previously made the trip act as guides for the whole flock. Yet these birds are also known to use celestial or other means, even when raised in situations providing no contact with previous migrators.

Geomagnetic navigation Since 1965, when magnetic direction finding was demonstrated in the European robin, it has been known that birds can use a built-in compass to plot paths on the basis of Earth's magnetic field. Humans have been using magnetic compasses for nearly 2,000 years. Since 1980, however, various experiments have shown that humans, like many other animals, can also sense Earth's magnetic field directly. While most animals appear to depend on magnetic crystals that sense Earth's magnetic field in the same way that a compass does, some sharks and rays appear to observe the field from the electrical currents set up as they swim through lines of force, the same principle as an electric generator.

Inertial guidance Although similar in some ways to dead reckoning, an inertial guidance system depends on sensors, such as gyroscopes, to recognize changes in direction or speed. Use of such systems is well established in insects, but apparently only for short journeys. There has been speculation that inertial guidance might be used in long-range navigation, but proof is yet to be gathered.

Piloting During their annual migrations, hawks, eagles, and gray whales follow natural landmarks, such as mountain ranges and shorelines. The whales are often seen with their heads raised from the water, in what appears to be a look around for landmarks on the shore.

Smell or taste Although seldom used by modern humans, animals are often thought to be able to find their way using smells carried by the wind, or the taste of particular waters.

Waves and swells Both fish and birds use the constant direction of waves on the ocean for long-range migration, a trick formerly employed by humans during long-distance canoe travel on the Pacific.

Winds and currents Fall migration over the Atlantic by songbirds from eastern North America is apparently accomplished primarily by maintaining a constant course with respect to winds, nearly always resulting in landfall at an appropriate destination after three or four days. Columbus is famed for crossing the Atlantic using a similar system.

Speculative theories proposed uniquely as techniques of animal navigation and not used by humans include gravitational fields, barometric pressure, temperature of water, electrical potentials in water, the Coriolis force, and low-frequency sounds that travel easily through air but are inaudible to humans. Polarized light is definitely used by insects as well as by birds (see above). Some humans also have a few unique tricks. For example, canoe navigators in the Pacific report using underwater flashes, probably bioluminescence, that point out islands well over the horizon.

Evolution UPDATE

The theory of evolution was preceded intellectually by the new geology of the late eighteenth and early nineteenth century. This began with James Hutton's 1785 work describing geological features as formed by uniform processes working slowly over long periods of time, a theory that was fully developed by the publication in 1830-33 of Charles Lyell's *Principles of Geology*. The essence of this concept is its emphasis on gradual change, in contrast to the theories of abrupt catastrophes preceding it. The theory of evolution likewise was a gradualist theory from its beginning.

It is fitting, then, that acceptance of a new catastrophism also began with geology — namely the 1980 discovery of the iridium-rich layer at the boundary between the Cretaceous and Tertiary Periods. The roots of the concept that living organisms may not undergo change gradually can be dated to eight years earlier, when mathematician René Thom brought catastrophe into a mathematics dominated by continuity, and biologists Niles Eldredge and Stephen Jay Gould proposed that species appear quickly in the geologic record and then enter a long period of stability. Throughout the 1980s, while the chaos theory of mathematics developed as an extension of Thom's work, biologists also became more receptive to ideas of sudden change. In the 1990s, many of those biologists who had clung to gradualism reviewed new evidence which convinced them to fling it aside.

One of the bastions of gradualism that for many remains firmly entrenched involves the origin of life itself. Nearly all theories, which are varied in detail, share at core the idea that slow changes occurring over eons have led to life on Earth. Yet there have always been a few biologists who found the gradual evolution of life unsatisfactory, since these theories fail to explain certain mysteries. If, for example, there is an equal probability of two different ways for life to have originated, is it possible to find some gradual process that caused life to all be one way and not the other? Or must one invoke some sudden event to explain why life is the way it is today?

Although there is considerable evidence supporting that other animals can think and feel in much the same way as humans, it has long been established that much of the behavior of these animals that seems purposeful is actually the result of evolved patterns of rigid behavior—instinct, or in a metaphor that is perhaps more meaningful to the computer generation, hard-wired behavior. That is, instinctive behavior is that built into the hardware from the start, unlike creative or learned behavior. However, unlike the computer program that disappears when the machine is turned off, some behavior in animals gradually wears its path into later behaviors, becoming preserved by natural selection. In the 1990s, the investigation of when and how this happens in nonhuman animals has taken great strides forward. It may be too soon, however, to apply the new insights to human behavior, although that has not stopped biologists, anthropologists, and just plain folks from trying to do so.

Origin of Life

A marching band or army needs to know which foot to put forward first in order to allow the entire body to stay in step. The organic molecules composing all living things keep together in somewhat the same way.

The molecules of DNA are what chemists call right-handed. To keep in step, all proteins made from the DNA blueprint are also right-handed (*see* Background). In the late 1980s, chemists showed that single strands of DNA must be in a homochiral solution (*homochiral* means either all right-handed or all left-handed) even to form a double helix. In other words, a mixture of approximately half right-handed and half left-handed DNA results in a tangle like that formed when right-handed honeysuckle entwines with left-handed morning glory.

While there are practical consequences to homochirality, the origin of homochirality presents more of an intellectual than a practical challenge, albeit one with implications for general theories about the origin of life. One needs to know, as Winnie-the-Pooh did, where to begin. Pooh knew that one of his hands was the right and the other the left, but his knowledge was ultimately of no help, because he had forgotten where to start. Similarly, when on February 15-17, 1995, researchers interested in the origin of life gathered in Santa Monica, California, to debate the issues and share newly found knowledge, they found that the main impediment to discussion was that they did not know where chirality could have begun.

No conclusions were reached. There seemed to be an unbridgeable gap between those who thought that some physical process caused molecules of a uniform chirality to form, which then led to the origin of life, and those who thought that life originated in some helter-skelter fashion, with some event allowing one chiral form of organic molecules to win out over the other. Among the latter, Stanley Miller of the University of California, San Diego, suggested that life was originally based on a smaller and tougher nucleic acid than DNA, one called peptide nucleic acid (PNA) that was discovered in 1991. Unlike DNA and RNA, PNA is not initially a chiral molecule by itself. In 1994, however, it was found that adding lysine to the end of a PNA strand fixes it as either right or left. Miller's idea is that PNA-plus-lysine might have existed in both forms early in life's history, but one of the two forms had some evolutionary advantage that is not obvious to us from this vantage point.

The scientists who disagreed with Miller, believing that some physical process was at work in the origin of chirality, tended to look to outer space as the origin of the molecules of life, if not of life itself. Polarized light from a neutron star might, for example, cause amino acids to form in one chiral form and not the other. Mayo Greenberg of the University of Leiden in the Netherlands told the conference that he had experimental evidence showing that polarized light from a laboratory imitation of a neutron star could direct the chirality of the amino acid tryptophan. Observations of meteorites, thought to represent matter in outer space, show no chirality, however. Some look to the first mission that will sample matter from a comet, considered to be made from a more primitive material than that of a meteor, in hopes that the sample will reveal some form of homochirality.

Background

Chiral forms are mirror images of each other, like hands or gloves. The two in a pair are exactly alike, except that a right-handed glove will not go on the left hand. Proteins are made from 20 subunits called amino acids. Amino acids, because they are based on carbon and are sufficiently complex, can generally exist in two chiral forms (the amino acid glycine is an exception). Only one of the two chiral forms of amino acids, the left-handed one, occurs in living organisms. As a result, proteins formed by connecting long strings of amino acids are made from subunits that are helixes (corkscrew shapes). Since the components are almost all left-handed, the helixes are right-handed.

Similarly, the five nucleotides that form RNA and DNA (**nucleotides** are combinations of a base such as cytosine, a sugar, and phosphoric acid that connect to form the RNA or DNA molecule) are all left-handed, so connecting them in strings also forms right-handed helixes. When a protein is made following a DNA blueprint, by combining RNA and attendant preexisting proteins (*see* "Genetics and Beyond," p 336), all of the molecules are right handed from beginning to end. Indeed, inserting a left-handed molecule into the middle, such as a left-handed RNA, would throw a left-handed monkey wrench into the whole works.

On the other hand, everything would work properly if all the molecules were reversed. Suppose DNA and RNA were made from right-handed bases. Then each molecule would twist itself into a left-handed double helix. If RNA's helper proteins were made from right-handed amino acids, they also would contain left-handed helixes. The protein-making machine would be set to pick up right-handed amino acids and make left-handed proteins out of them.

Scientists have been working to understand why life has been based on right-handed molecules made from left-handed parts since the discovery of molecular chirality by Louis Pasteur in 1847. Pasteur himself had suggested that it might result from Earth's magnetism, or from the apparent path of the Sun from east to west, but he proved by his own experiments that these ideas were not possible. In 1989, chemist Francis Robert Japp of the University of Aberdeen in Scotland proposed that the presence of chirality should be taken as proof of God's existence, since there is no other way to explain it.

Mysteries of right and left handedness go beyond molecules, of course. It is not known why various other examples of handedness exist throughout life, from the twist of the tusk of the narwhal to the right-handed pattern common to humans. Handedness was even found in Cambrian seas, as reported in 1993 by Loren E. Babcock of Ohio State University and Richard A. Robison of the University of Kansas. They established that more than 70% of trilobites that had been bitten, based on a sample of more than 300 fossils exhibiting the effects of healed bites, had been bitten on the right side. Whether the trilobites themselves most often escaped to the right, or the unknown predator generally attacked from the left, or both, continues to be a mystery.

Evolution Also Takes Quantum Jumps

A principal change in the world view of twentieth-century scientists, compared to those of the nineteenth century, has been the recognition that matter, motion, and probably even time are discrete and not continuous. The "quantum leap," or instantaneous passage of a particle from one place to another without traversing the in-between, has become a popular metaphor for any abrupt change. Despite the internalization of the quantum point of view among physicists, however, biologists have clung to continuity. Popular language perpetuates this distinction: an idea that *evolves* is one that results from a continuous process, not from a quantum leap of the imagination.

Biologists do recognize discrete change, however. Even Darwin saw evolution as caused by small differences between discrete individuals. Still, the steps were so tiny that the result was pictured as a continuous process. In 1901, a year after the quantum was born, Hugo De Vries introduced the concept of mutation, allowing for a large jump from more-or-less similar members of a species to a mutant that was quite different. In the long run, this concept proved to be of more interest to science-fiction writers than to scientists.

However, in 1972, Niles Eldredge of the American Museum of Natural History and Stephen Jay Gould of Harvard University proposed that the larger process of evolution also consisted of discrete steps, quantum jumps, so to speak. They called their theory "punctuated equilibrium," meaning that a species stays in equilibrium with its environment most of the time, but that this state is "punctuated" with discrete changes from one species to another. Although the theory of punctuated evolution arose from careful studies of how evolution occurs in trilobites and in land snails, biologists have been slow to accept it. They prefer to think, as Darwin did, that evolution proceeds as a slow, steady series of changes. What appear in the evidence to be quantum jumps are attributed instead to gaps in the fossil record caused by geological processes.

Careful study of two other genera of animals, reported in 1994, substantiates the theory of punctuated equilibrium for those genera, at least. Gradualist Alan Cheetham of the Smithsonian Institution started out in 1986 to demonstrate that the small colonial animals called bryozoa, or "moss animals," evolved the way Darwin had expected, not in the Eldredge and Gould manner. Instead, his data showed that over a 15 million year span, 17 species of the genus *Metrarabdotos* each evolved in periods of time that were less than 160,000 years, and then stopped changing for periods ranging from 2 million to 6 million years. Punctuated equilibrium, indeed! His results were challenged in a friendly way by Jeremy Jackson of the Smithsonian Tropical Research Institute in Panama. Jackson was primarily concerned that there might be some flaw in the methods used for species identification among the bryozoa, who all look pretty much alike, even to biologists. Species separation is easier in living animals, so Jackson arranged for Cheetham to demonstrate that the methods he used for fossils would also correctly recognize separate species among known living forms of bryozoa, working only with skeletons. Several different kinds of tests, including ecological and molecular, were used to make sure that Cheetham's classification system based on skeletal features, the only features available in fossils, is completely accurate, which it proved to be. Then, as a further reality check, Jackson and Cheetham worked together to study the evolution in a second genus of bryozoa, *Stylopoma*. In this case, the 19 species identified all evolved according to punctuated equilibrium.

About the same time that Jackson and Cheetham's work was published, a similar look at evolution over 20 million years of the snail genus *Nucella* also found punctuated equilibrium. This study was conducted by Timothy Collins of the University of Michigan and coworkers.

There remain a number of similar studies that did *not* show punctuated evolution in snails or trilobites or other phyla. Although some species do appear to evolve gradually, the ratio of those known to have evolved quickly to those known to have evolved gradually continues to increase as more studies are done.

One of the most dramatic cases of swift evolution discovered in the 1990s involves the whale, a familiar mammal instead of an obscure invertebrate. The discovery of intermediate species of whales by J. G. M. Thewissen of Northeastern Ohio Universities College of Medicine in Rootstown and coworkers (announced in the January 14, 1994 *Science*) and by Phillip D. Gingerich of the University of Michigan in Ann Arbor and coworkers (announced in the April 28, 1994 *Nature*) demonstrates that the whole process took place with extreme rapidity in geological terms. The intermediate species of whales evolved from deer-like ancestors to the fish-like forms known today. The species *Ambulocetus natans* discovered by Thewissen and coworkers is somewhat like a large seal (about 10 ft or 3 m long), and was probably an efficient swimmer that could walk awkwardly on four limbs on land. *Rodhocetus kasrani* found by Gingerich and coworkers, dating about 3 million years after *A. natans,* has lost its hind flippers and long tail, but gained the ability to use its tail in swimming. Both fossils were found in Pakistan. Although there is not yet a complete set of fossils showing the evolution of whales, the gaps are being rapidly filled in. It appears that a complete evolution from land to sea took place in less than 10 million years.

Even swifter evolution has occurred in cats, although the effects are subtler than the loss of limbs and development of flukes. In the January 1993 *The Journal of Neuroscience*, Robert Williams of the University of Tennessee College of Medicine in Memphis reported that his investigations showed that in just 20,000 years the brain of the domestic cat has evolved to be far different from that of its wild ancestor, as represented by two adult and one fetal specimen of *Felis silvestris tartessia*, the Spanish wildcat. Not only are the adult brains of the wildcats almost a third larger than that of a domestic tabby cat of the same body mass, but the visual system of the wildcats is also far more sophisticated than that of the tabby. Most dramatically, wildcats have the apparatus for color vision, but domestic cats do not. Williams thinks that wildcats hunt in daylight, when color vision is helpful, while domestic cats were driven by humans to hunt at night, when color is not apparent in any case. The skull of the domestic cat is twice as thick as that of its wild counterpart, which is one way to have a larger brain in the same sized head.

Williams thinks that the evolution of many of the brain changes were produced by reprogramming apoptosis so that those cells not needed in domestication but vital to the wild cousins committed suicide in the tabby (*see also* "Development Developments," p 342). The evidence for programmed cell death is that the brain of the wildcat fetus appears to be the same size and complexity as that of a fetus of the domestic cat. Evolution into domesticity results in the pruning away of cells not needed by an adult house cat.

Watching Species Evolve

Evolution can proceed even faster than that evidenced in the cat. Starting a quarter of a century ago, in a series of major experiments with and studies of fish and birds, biologists have observed evolution in the face of changes in environment. These results have been criticized by people who do not believe that evolution exists because the changes caused by environment have not resulted in new species. These critics imply that the experiments are essentially the same as animal breeding which, although resulting in lines within species displaying dramatically different traits from the wild type, has gone on for 10,000 years without producing a new species. Although each scientific experiment in evolution may take years to complete, the scientists involved think that they are seeing the origin of species, if not speciation itself.

- In the 1970s, John Endler of the University of California, Santa Barbara, showed that within months, guppies evolve over 9 or 10 generations to match their background. The difference between these experiments and breeding is that natural selection, not breeder selection, produces the change.
- In 1993 (reported in the November 4, 1994 *Science*), Dolph Schulter showed that a species of threespine sticklebacks who naturally live in only one part of their pond will outcompete a species of sticklebacks who naturally live throughout their home pond, pushing the generalist group to become specialists. The intent of the experiment was to demonstrate that competition for resources could produce speciation, and the outcome supported that hypothesis.
- In 1987, Peter R. Grant of Princeton and H. L. Gibbs reported that changing climate conditions on Daphne Major, one of the smaller Galápagos islands, have resulted in an increase in the number of plants with small seeds and a decrease in those with large seeds. One consequence is that a finch species (*Geospiza fortis*) quickly evolved to have a smaller bill, better adapted for the small seeds. This change in bill size probably happened because smaller-billed finches survive better and leave more descendants in a changed environment than do those with large bills. Another factor may have been the recent arrival of a breeding population of a different

species of finch (*G. magnirostris*) that was even better adapted to eating large seeds.

- In 1992, Peter R. Grant and Rosemary Grant, also of Princeton University, reported on another experiment of nature. *G. fortis* on Daphne Major has begun violating the common definition of "species" by interbreeding with two other species of finch found on the island. The resulting hybrids are fertile and more fit in terms of survival than either parent. The hybrids are gradually becoming a new species.
- In 1993, A. J. Helbig of the University of Heidleberg, Peter Berthold, and coworkers reported that in less than 40 years a genetic mutation had altered the migratory behavior of a kind of European wood warbler called the blackcap (*see* "Finding Their Way," p 407).

Frozen Evolution

Scientists studying behavior have sometimes turned to the frozen behavior of the spider web. Each species of web-spinning spider spins a characteristic pattern that has evolved along with other behaviors that enable a spider to capture its prey. While the web is not a physical part of the spider like its size, color, or shape, there is no doubt that the web is as much a product of evolution as are the cells and hormones of the spider. In 1994, Catherine L. Craig of Yale University in New Haven, Connecticut, showed that the white zigzags we see on webs woven by the common garden spider of the genus *Argiope* appear much brighter to insects, because they reflect ultraviolet light, while the main radii and circles of the typical orb web do not reflect much in the ultraviolet. Bees and other insects are attracted to ultraviolet light (an attraction also exploited by flowering plants), luring them toward the zigzags, but trapping them in the other strands. Craig also observed that the *Argiope* spiders not only had evolved the strategy of tricking insects this way, but had also developed the habit of building a different zigzag every day, making it more likely that the web will catch an insect that may have seen through this trick and turned away the day before. Furthermore, the spiders even benefit from bees that become trapped but get away. The trapped bee leaves some edible pollen on the web. Although all spiders are carnivores, they also consume their web each night for its raw materials. The next day, the spider has the protein needed to spin a new web.

Another orb-web-weaving spider of the genus *Nephila* has evolved a yellowish web. Craig showed that yellow webs are more confusing to bees than are other colors, perhaps by looking somewhat like yellow flowers to the bee eye.

Sophisticated web-building behavior has its evolutionary roots in spiders living in forests and trapping insects in the dark, similar to a human fishing with a trawler net in the deep sea, collecting fish and other sea creatures mechanically and indiscriminately. The advanced spiders, like *Argiope* and *Nephila*, have gone far beyond using their webs as simple nets.

The Genetic Basis of Evolution

Now that we know much more about genetics than De Vries did in 1901, the concept of a mutation can be pinned down more precisely. Today, a mutation refers to a change in a single gene and doesn't necessarily result in a new species as De Vries had suggested. A major change, such as that between one species and another, would seem to require many mutations in many different genes. This kind of speciation, involving a number of independent changes in specific proteins (*see* "Genetics and Beyond," p 336), seems, on initial viewing, to fit better with the concept of gradualism than with that of punctuated equilibrium.

Today, however, we have tools that enable scientists to go beyond speculation on such matters, and to conduct directed experimental studies. H. D. "Toby" Bradshaw Jr., Sibylle M. Wilbert, Kevin G. Otto, and Douglas W. Schemske of the University of Washington in Seattle asked the question, "How many gene mutations does it take to change one species into a different, but closely related species?" They conducted the necessary studies to find the answer in a particular situation, taking advantage of the rule that sufficiently close species can be made to breed in a laboratory situation, even when normally reproductively isolated in the wild (for example, a zoo can breed a lion with a tiger). The species they used in the experiment are two different kinds of monkey flowers that do not ordinarily interbreed, because their blossoms attract different pollinators. The blossom of Lewis' monkey flower (*Mimulus lewisii*) is pink with a yellowish "landing pattern" and rich nectar, exactly the combination that attracts bees, with a blossom shape that causes a bee visitor to emerge from deep in the flower covered with pollen. The blossom of the other species, the cardinal monkey flower (*M. cardinalis*), is bright red and filled with large amounts of a dilute nectar, features that attract hummingbirds. This blossom is shaped so that a feeding hummingbird is dusted with pollen on the back of its head, from which the plant also acquires pollen obtained during the bird's prior visit to another cardinal monkey flower (*see also* "Plant Behavior," p 401).

Altogether there are eight major differences between the two species that affect pollination: relative amounts of two kinds of pigment, distribution of the pigments, shape of the blossoms, size of the blossoms, amount of nectar, concentration of nectar, length of stamens, and length of pistils. In fact, the pollen transportation system is such that even if a bee should feed at a cardinal monkey flower, it will not pick up pollen, and if a hummingbird feeds on a Lewis's monkey flower, it will transfer very little pollen on its smooth beak.

The University of Washington researchers found that when they hybridized the two species in the laboratory, the hybrids ranged in traits all the way from one closely resembling a parent plant to various combinations in between. They then used a new technique to determine the location on the flower's genome of those genes for the eight traits, treating corolla width and petal width as representative of size and shape. The evidence pointed to a single gene at work for a large part of the difference in each case. A single mutation could account for about one-fifth to one-third of the change from a pink flower (bright in the ultraviolet region used by bees) to a bright red flower (dull-looking to bees but a stand-out for birds). A single gene makes the yellowish stripes appear or disappear; the allele responsible for no stripes is recessive. Half the variation in the volume of nectar is accounted for by one gene.

Although the genetic evidence does not explain all speciation (especially since it only applies to the flowering plants studied), it is suggestive. The University of Washington researchers hypothesize that species creation in this case began with a Lewis's monkey flower already in existence. A single mutation results in a change in blossom color from pink to red, which attracts hummingbirds and lessens attractiveness to bees. Although pollination is not efficient at first, enough reproduction occurs to sustain the mutant gene in the monkey-flower population. Nectar amounts are variable at first, but because red flowers with larger stores of nectar enjoy longer visits from birds, the high-volume plants bearing red flowers are more often pollinated than those that contain only a sip or two of nectar. Also, plants whose red flowers are lucky enough to have long sexual organs that brush the birds' necks reproduce more often than plants with short ones tucked into the blossom. Thus, the original color mutation tends to foster traits that contribute to changes in the entire pollination system. Not many genes are involved, and speciation occurs in a relatively short time, as predicted by punctuated equilibrium. Furthermore, the two species, once the punctuation occurs, tend to stay separate for as long as the environment continues to provide bees and hummingbirds, allowing equilibrium.

The University of Washington scientists are now conducting further experiments to see if their hypothesis works. They are breeding hybrid monkey flowers with just one of the eight traits mutated from the original, perceived to be the Lewis's monkey flower. These plants will be placed in the wild and observed to see which animals pollinate them.

Evolution's Big Bangs

Even if one accepts the theory of punctuated equilibrium for individual species, there is another sort of punctuated equilibrium that appears in the fossil record. Against a background rate of approximately steadily changing species, there are short periods of time in which higher classifications (families, orders, or phyla) evolve. Often, such periods of explosive evolution follow a sudden mass extinction, an association that seems to have a ready explanation. When many species become extinct all at once in a mass extinction, many ecological niches become unfilled. Existing species then spin off or "radiate" radically different new species to take advantage of gaps. These are sufficiently different to form a higher level of taxonomy. Occasionally, there is also a burst of family or order radiation that appears to have no known cause.

One of the most dramatic examples of radiation at the level of orders and families appears among birds. Until the 1990s, evolutionary ornithologists presumed that the class of birds originated as an offshoot of the dinosaurs in the Jurassic Period, gradually evolving into the present day orders, families, genera, and species. However, in 1995, Z. Zhou recognized that the birds of the Cretaceous Period were mostly what Alice might call "looking-glass birds," with their internal bone structure the mirror image of modern birds. Only one lineage of birds possessing the modern structure is known from that period, a group called "transitional shorebirds," similar to the present-day orders of waterfowl and stilt-legged birds. During the Cretaceous-Tertiary (or K-T) mass extinction, all the looking-glass birds, the major orders of birds prior to the K-T boundary, ceased to exist. There are no descendants of the looking-glass birds known today. The amazing part, however, concerns the fate of those descendants of the shorebirds. Within 5 million to 10 million years of the K-T event, the shorebirds are known to have evolved into all but one of the 23 modern orders of birds. The remaining order, the only one not known in early Tertiary fossil beds, is the passerine birds, the songbirds. Passerines also radiated rapidly almost as soon as the first species appeared, producing more than half the present species of all birds today. This second big bang of bird diversity occurred about the time of the boundary between the Oligocene and Miocene Epochs, which is not associated with a mass extinction.

What could cause periods of explosive speciation other than a preceding mass extinction? One possibility is the introduction of a major new way of living that proves vastly more successful than the previous lifestyle. For many years, scientists thought that evolution's Big Bang(with initial capital letters), also known as the Cambrian explosion, occurred when animals evolved hard outer shells, although this theory is no longer as popular as it once was. The Cambrian Big Bang deserves its capital letters, as that was the time about 600 million years ago when *all* the modern animal phyla developed in a short period of time (as well as a few phyla that did not last until the present). In addition to the old hard-shell theory, a prominent idea theorizes that rising oxygen levels may have resulted in the explosion.

A more contemporary theory, from fossil algae specialist Andrew Knoll of Harvard University in Cambridge, Massachusetts, postulates that the Cambrian Big Bang was exponential because it caused itself. In exponential growth, the rate of growth increases in step with the amount of growth that has taken place: the more growth, the faster the rate of increase. Knoll points out that diversity itself opens up new ecological niches, which then tend to be filled by newly evolved species. Thus, when diversity among animal phyla became great enough to ignite this process, it began to spawn both new species and even new phyla, which in turn resulted in more diversity and so on. The Big Bang only stopped when it ran out of possible new niches. Presumably, the major niches were all filled, so that although no new phyla have appeared since, new species continue to evolve.

Even earlier than the Cambrian Big Bang there was another major species explosion, which Knoll can document in certain one-celled algae whose outer coats are hard enough to form numerous fossils. The explosion of species among these algae, dated at about a billion years ago (some three billion years after the first forms of life are thought to have occurred) was most likely accompanied by a similar radiation among other forms of life, but the early animals were all soft bodied and did not leave much in the way of a fossil record. The algal record, however, shows not only an increase in the number of new species, including the first multicellular algae, but also a shortening of the length of time before any given species became extinct.

Knoll thinks this first known explosion took place when sex was invented. The first evidence for early algal sex comes from a reconstruction of the evolutionary history of algae based on analysis of DNA and RNA sequences in living species. The molecular record points to sexually reproducing algal groups evolving during or after the early explosion. By providing additional diversity from genetic recombination, sex could have caused the earliest big bang.

(Periodical References and Additional Reading: *Science* 4-10-92, p 193; *New York Times* 1-12-93, p C1; *New York Times* 6-15-93, p C1; *Nature* 11-18-93, p 223; *New York Times* 12-22-93, p C1; *Science* 1-14-94, pp 180 & 210; *Nature* 4-28-94, pp 807 & 844; *New York Times* 5-3-94, p C1; *Science* 11-4-94, pp 746 & 798; *Science* 1-6-95, pp 30, 33, & 87; *Science* 2-3-95, p 637; *Science* 3-3-95, p 1265; *New York Times* 4-19-95, p C1; *Science* 3-10-95, p 1421; *Nature* 8-31-95, pp 736 & 762; *Science News* 9-2-95, p 148; *New York Times* 9-5-95, p C1)

BIOLOGY LISTS AND TABLES

LIFE SCIENCE RECORD HOLDERS

Size

Largest structure made by living creatures	Great Barrier Reef off coast of Australia	Length: 1,260 mi (2,030 km) Area: 80,000 sq mi (200,000 sq km)
Largest living creature	Currently thought to be a clone of aspens in the Wasatch Mountains of Utah	Weight: About 13.2 million lb (6 million kg) Circumference: 114.6 ft (34.9 m)
Largest living animal	Blue whale	Length: 110 ft (34 m) Weight: 136 sht tons (150 m tons)
Largest invertebrate	Giant squid	Length: 55 ft (17 m)
Largest living land animal	African elephant	Height: 13 ft (4 m) Weight: 13 sht tons (13 m tons)
Longest animal	Bootlace worm, a sea worm	Length: 180 ft (55 m)
Largest land animal ever	*Argentinosaurus,* a dinosaur from 100 million years ago	Weight: 100 sht tons (90 m tons)
Largest bacterium	*Epulopiscium fishelsoni,* symbiot of the surgeonfish	Length: 0.02 in (0.57 mm) Diameter: 0.002 in (0.06 mm)
Smallest free-living creature	Pleuro-pneumonialike organisms of the genus *Mycoplasma*	Diameter: 0.000004 in (0.0001 mm)
Smallest creatures of any kind	Viroids—viruslike plant pathogens without coats	Diameter: 0.00000000007 in (0.000000002 mm)
Most teeth in a mammal	Toothed whales	Amount: 260 teeth
Longest trip by a shorebird	Semipalmated sandpiper	Length: 2,800 mi (4,500 km) in four days
Tallest flying bird	Eastern sarus crane (*Grus antigone*)	Height: 5 ft (2.4 m)
Largest egg ever	Laid by extinct elephant bird of Madagascar	Capacity: 2.35 gal (8.9 L) Weight: 27 lb (12 kg)
Largest seed	Seed of palm *Lodoicea maldivicia* of Seychelles in Indian Ocean	Length: 20 in (50 cm)
Largest nest	Built by bald eagle	Weight: 6,700 lb (3,000 kg)

Time

Oldest living creature	King Clone, a creosote plant in California desert	Age: 11,700 years
Longest recorded age of an animal	Ocean quahog (clam)	Age: 220 years

Oldest living bird	Wandering albatross	Age: Banded birds recovered after 40 years; estimated at 80 years in the wild
Longest gestation time	Alpine black salamander	Period: Up to 38 months
Speed		
Fastest flight	Peregrine falcon	While diving, 212 mph (340 kph)
Fastest swimmer	Cosmopolitan sailfish	Speed: 68 mph (110 kph)
Slowest mammal	Three-toed sloth	Speed 4 mi (6 km) a day on the ground

(Periodical References and Additional Reading: *Science* 12-16-94, p 1805)

THE NOBEL PRIZE FOR PHYSIOLOGY OR MEDICINE

Date	Name [Nationality]	Achievement
1995	Edward B. Lewis [American: 1918-] Christiane Nüsslein-Volhard [German: 1942-] Eric F. Wiechaus [American: 1947-]	Location and explanation of genes that guide development in fruit flies.
1994	Alfred G. Gilman [American: 1941-] Martin Rodbell [American: 1926-]	Independent discovery in the 1960s of G-proteins, chemicals that help cells convert extracellular signals into cascades of internal reactions by binding to guanosine triphosphate (GTP).
1993	Richard J. Roberts [British/American: 1943-] Phillip A. Sharp [American: 1944-]	Discovery that genes in higher organisms are not continuous; instead, they are fragmented by stretches of unrelated DNA known as introns.
1992	Edmond H. Fischer [American: 1920-] Edwin G. Krebs [American: 1918-]	Discovery in the 1950s of protein kinases and the key role they play in turning on cell switches through the process of phosphorylation (addition of phosphate groups) and discovery of phosphatases, which make cells inactive by removing phosphate groups.
1991	Erwin Neher [German: 1944-] Bert Sakmann [German: 1942-]	Analysis of ion channels in cells.
1990	Joseph E. Murray [American: 1919-] E. Donnall Thomas [American: 1920-]	First kidney transplant (Murray), and first successful bone-marrow transplant (Thomas)
1989	J. Michael Bishop [American: 1936-] Harold E. Varmus [American: 1939-]	Discovery of the cellular origin of cancer-causing genes found in retroviruses.
1988	Sir James W. Black [British: 1924-] Gertrude B. Elion [American: 1918-] George H. Hitchings [American: 1905-]	Development of artificial variations on DNA that block cell replication, the basis for many new drugs.

Date	Name [Nationality]	Achievement
1987	Susumu Tonegawa [American: 1939-]	Discovery of the genetic principle of antibody diversity.
1986	Stanley Cohen [American: 1922-] Rita Levi-Montalcini [Italian/American: [1909-]	Discovery of growth factors.
1985	Michael S. Brown [American: 1941-] Joseph L. Goldstein [American: 1940-]	Discovery and analysis of cholesterol receptors.
1984	Niels K. Jerne [Danish: 1911-1994] Georges J.F. Köhler [German: 1946-] César Milstein [British/Argentinian: 1927-]	Pioneering work in immunology (Jerne) and the invention of monoclonal antibodies (Köhler and Milstein).
1983	Barbara McClintock [American: 1902-]	Discovery of mobile genes in chromosomes of corn.
1982	John R. Vane [British: 1927-] Sune K. Bergstrom [Swedish: 1916-] Bengt I. Samuelsson [Swedish: 1934-]	Studies on formation and function of prostaglandins, hormonelike substances that combat disease.
1981	Roger W. Sperry [American: 1913-1994] David H. Hubel [American: 1926-] Torsten N. Wiesel [American: 1924-]	Studies on the organization and local functions of brain areas.
1980	George D. Snell [American: 1903-] Baruj Benacerraf [American: 1920-] Jean Dausset [French: 1916-]	Discovery of antigens useful for making the immune system accept transplanted organs.
1979	Allan McLeod Cormack [American: 1924-] Godfrey N. Hounsfield [English: 1919-]	Invention of computed axial tomography, or CAT scan, now often called the CT scan.
1978	Daniel Nathans [American: 1928-] Hamilton O. Smith [American: 1931-] Werner Arber [Swiss: 1929-]	Use of restrictive enzymes in gene splicing, to produce mutants in molecular genetics.
1977	Rosalyn S. Yalow [American: 1921-] Roger C. L. Guillemin [American: 1924-] Andrew V. Schally [American: 1926-]	Advances in the synthesis and measurement of hormones.
1976	Baruch S. Blumberg [American: 1925-] D. Carleton Gajdusek [American: 1923-]	Identification of and tests for different infectious viruses.
1975	David Baltimore [American: 1938-] Howard M. Temin [American: 1934-1994] Renato Dulbecco [American: 1914-]	Discovery of the interaction between tumor viruses and the genetic material of host cells.
1974	Albert Claude [American: 1898-1983] George E. Palade [American: 1912-] C. René de Duvé [Belgian: 1917-]	Advancement of cell biology, electron microscopy, and structural knowledge of cells.
1973	Karl von Frisch [German: 1886-1982] Konrad Lorenz [German: 1903-1989] Nikolaas Tinbergen [Dutch: 1907-1988]	Study of individual and social behavior patterns of animal species for survival and natural selection.

Date	Name [Nationality]	Achievement
1972	Gerald M. Edelman [American: 1929-] Rodney R. Porter [English: 1917-1985]	Determination of the chemical structure of antibodies.
1971	Earl W. Sutherland Jr. [American: 1915-1974]	Work on the action of hormones.
1970	Julius Axelrod [American: 1912-] Ulf von Euler [Swedish: 1905-1983] Sir Bernard Katz [English: 1911-]	Discoveries in the chemical transmission of nerve impulses.
1969	Max Delbrück [American: 1906-1981] Alfred D. Hershey [American: 1908-] Salvador E. Luria [American: 1912-1991]	Discoveries in the workings and reproduction of viruses within human cells.
1968	Robert Holley [American: 1922-1993] Har Gobind Khorana [American: 1922-] Marshall W. Nirenberg [American: 1927-]	Understanding and deciphering the genetic code that determines cell function.
1967	Haldan K. Hartline [American: 1903-1983] George Wald [American: 1906-] Ragnar A. Granit [Swedish: 1900-]	Advanced discoveries in the physiology and chemistry of the human eye.
1966	Charles B. Huggins [American: 1901-] Francis Peyton Rous [American: 1879-1970]	Research on causes and treatment of cancer.
1965	François Jacob [French: 1920-] Andre Lwoff [French: 1902-1994] Jacques Monod [French: 1910-1976]	Studies and discoveries on the regulatory activities of human body cells, identifying how one gene controls others.
1964	Konrad Bloch [American: 1912-] Feodor Lynen [German: 1911-1979]	Work on cholesterol and fatty acid metabolism.
1963	Sir John C. Eccles [Australian: 1903-] Alan Lloyd Hodgkin [English: 1914-] Andrew F. Huxley [English: 1917-]	Study of the mechanism of transmission of neural impulses along a single nerve fiber.
1962	Francis H. C. Crick [English: 1916-] James D. Watson [American: 1928-] Maurice Wilkins [English: 1916-]	Determination of the molecular structure of DNA.
1961	Georg von Békésy [American: 1899-1972]	Study of auditory mechanisms.
1960	Sir Macfarlane Burnet [Australian: 1899-1985] Peter Brian Medawar [English: 1915-]	Study of immunity reactions to tissue transplants.
1959	Severo Ochoa [American: 1905-1993] Arthur Kornberg [American: 1918-]	Artificial production of nucleic acids with enzymes.
1958	George Wells Beadle [American: 1903-1989] Edward Lawrie Tatum [American: 1909-1975] Joshua Lederberg [American: 1925-]	Beadle and Tatum for genetic regulation of body chemistry, Lederberg for genetic recombination.
1957	Daniel Bovet [Italian: 1907-1992]	Synthesis of curare.

Date	Name [Nationality]	Achievement
1956	Werner Forssmann [German: 1904–1979] Dickinson Richards [American: 1895–1973] André F. Cournand [American: 1895–1988]	Use of catheter for study of the interior of the heart and circulatory system.
1955	Hugo Theorell [Swedish: 1903–1982]	Study of oxidation enzymes.
1954	John F. Enders [American: 1897–1985] Thomas H. Weller [American: 1915–] Frederick C. Robbins [American: 1916–]	Discovery of a method for cultivating poliomyelitis virus in tissue culture.
1953	Fritz A. Lipmann [American: 1899–1986] Hans Adolph Krebs [English: 1900–1981]	Discovery by Lipmann of coenzyme A, and by Krebs of citric acid cycle.
1952	Selman A. Waksman [American: 1888–1973]	Discovery of streptomycin.
1951	Max Theiler [S. Africa: 1899–1972]	Development of 17–D yellow fever vaccine.
1950	Philip S. Hench [American: 1896–1965] Edward C. Kendall [American: 1886–1972] Tadeusz Reichstein [Swiss: 1897–]	Discovery of cortisone and other hormones of the adrenal cortex and their functions.
1949	Walter Rudolf Hess [Swiss: 1881–1973] Antonio Egas Moniz [Portuguese: 1874–1955]	Hess for studies of middle brain function, Moniz for prefrontal lobotomy.
1948	Paul Müeller [Swiss: 1899–1965]	Discovery of effect of DDT on insects.
1947	Carl F. Cori [American: 1896–1984] Gerty T. Cori [American: 1896–1957] Bernardo A. Houssay [Argentinian: 1887–1971]	Coris for discovery of catalytic metabolism of starch, Houssay for pituitary study.
1946	Hermann J. Muller [American: 1890–1967]	Discovery of x–ray mutation of genes.
1945	Sir Alexander Fleming [English: 1881–1955] Sir Howard W. Florey [English: 1898–1968] Ernst Boris Chain [English: 1906–1979]	Discovery of penicillin, and research into its value as a weapon against infectious disease.
1944	Joseph Erlanger [American: 1874–1965] Herbert Spencer Gasser [American: 1888–1963]	Work on different functions of a single nerve fiber.
1943	Henrik Dam [Danish: 1895–1976] Edward A. Doisy [American: 1893–1986]	Dam for discovery and Doisy for synthesis of vitamin K.
1942	No award	
1941	No award	
1940	No award	

Date	Name [Nationality]	Achievement
1939	Gerhard Domagk [German: 1895–1964]	Discovery of first sulfa drug, prontosil (declined award).
1938	Corneille Heymans [Belgian: 1892–1968]	Discoveries in respiratory regulation.
1937	Albert Szent-Györgyi [Hungarian: 1893–1986]	Study of biological combustion.
1936	Sir Henry Dale [English: 1875–1968] Otto Loewi [Austrian: 1873–1961]	Work on chemical transmission of nerve impulses.
1935	Hans Spemann [German: 1869–1941]	Discovery of the "organizer effect" in embryonic development.
1934	George R. Minot [American: 1885–1950] William P. Murphy [American: 1892–] George H. Whipple [American: 1878–1976]	Discovery and development of liver treatment for anemia.
1933	Thomas H. Morgan [American: 1866–1945]	Discovery of chromosomal heredity.
1932	Sir Charles Sherrington [English: 1857–1952] Edgar D. Adrian [American: 1889–1977]	Multiple discoveries in the function of neurons.
1931	Otto H. Warburg [German: 1883–1970]	Discovery of respiratory enzymes.
1930	Karl Landsteiner [American: 1868–1943]	Definition of four human blood groups.
1929	Christiaan Eijkman [Dutch: 1858–1930] Sir Frederick G. Hopkins [English: 1861–1947]	Eijkman for antineuritic vitamins, Hopkins for growth vitamins.
1928	Charles Nicolle [French: 1866–1936]	Research on typhus.
1927	Julius Wagner von Jauregg [Austrian: 1857–1940]	Fever treatment, with malaria inoculation, of some paralyses.
1926	Johannes Fibiger [Danish: 1867–1928]	Discovery of Spiroptera carcinoma.
1925	No award	
1924	Willem Einthoven [Dutch: 1860–1927]	Invention of electrocardiograph.
1923	Sir Frederick Banting [Canadian: 1891–1941] John J. R. MacLeod [English: 1876–1935]	Discovery of insulin.
1922	Archibald V. Hill [English: 1886–1977] Otto Meyerhof [German: 1884–1951]	Archibald for discovery of muscle heat production and Meyerhoff for oxygen–lactic acid metabolism.
1921	No award	
1920	Shack August Krogh [Danish: 1874–1949]	Discovery of motor mechanism of blood capillaries.
1919	Jules Bordet [Belgian: 1870–1961]	Studies in immunology.
1918	No award	
1917	No award	
1916	No award	

Date	Name [Nationality]	Achievement
1915	No award	
1914	Robert Bárány [Austrian: 1876-1936]	Studies of inner ear function and pathology.
1913	Charles Robert Richet [French: 1850-1935]	Work on anaphylaxis allergy.
1912	Alexis Carrel [French: 1873-1944]	Vascular grafting of blood vessels and organs.
1911	Allvar Gullstrand [Swedish: 1862-1930]	Work on dioptrics, refraction of light in the eye.
1910	Albrecht Kossel [German: 1853-1927]	Study of cell chemistry.
1909	Emil Theodor Kocher [Swiss: 1841-1917]	Work on the thyroid gland.
1908	Paul Ehrlich [German: 1854-1915] Elie Metchnikoff [Russian: 1845-1916]	Pioneering research into the mechanics of immunology.
1907	Charles L. A. Laveran [French: 1845-1922]	Discovery of the role of protozoa in disease generation.
1906	Camillo Golgi [Italian: 1843-1926] Santiago Ramón y Cajal [Spanish: 1852-1934]	Study of structure of nervous system and nerve tissue.
1905	Robert Koch [German: 1843-1910]	Tuberculosis research.
1904	Ivan P. Pavlov [Russian: 1849-1936]	Study of physiology of digestion.
1903	Niels Ryberg Finsen [Danish: 1860-1904]	Light ray treatment of skin disease.
1902	Sir Ronald Ross [English: 1857-1932]	Work on malaria infections.
1901	Emil von Behring [German: 1854-1917]	Discovery of diphtheria antitoxin.

Chemistry

THE STATE OF CHEMISTRY

Chemists have been excited throughout the early 1990s. After a period of time in which many in the field thought that chemistry had matured to a point approaching senility, chemists suddenly are enjoying new elements, idiosyncratic classes of compounds, unlikely types of reactions, and even a new phase of matter (*see* "The Newest Forms of Matter," p 645). All of these were unexpected or merely predictions until a few years ago. The most dramatic of the chemists' new toys are described here, but the excitement actually extends well beyond the big stories and into the very pores of the field.

BASIC DEFINITIONS

A fundamental understanding of the basic concepts of any science requires more than the brief descriptions of interlocking concepts that can be offered here. The definitions below may be used, however, either as reminders or as introductions to the mysteries. Further details on electrons, neutrons, and protons can be found in the article "Subatomic Particles" on p 666.

Atomic number The number of protons in an atom of an element is called the atomic number. A specific element is characterized by its atomic number, which ranges from 1 for hydrogen to 111 for an as-yet unnamed element.

Bond Any of several types of attachment that can form between atoms are called bonds. Typically, bonds are thought of as forming molecules, although ionic bonds can produce crystals in which there is no specific molecule.

Catalyst Any chemical that increases the rate of a reaction between two other chemicals without itself being changed is a catalyst. Catalysts that are proteins are generally known as enzymes.

Electron The electron is a small, negatively charged fundamental particle which, in its travels from atom to atom, constitutes the main actor in chemistry, forming the various kinds of bonds.

Element A chemical that cannot be broken down into simpler substances by chemical processes because it is consists of atoms all of which possess the same atomic number. Traditionally, the "building blocks of chemistry" are elements, not atoms *per se*.

Ion An atom that has become electrically charged because it has gained or lost an electron is called an ion. A molecule or other group of atoms which is electronically charged in this way is called a polyatomic ion.

Matter   Ordinary matter is any chemical made from atoms or ions. Some subatomic particles, including but not limited to electrons, neutrons, and protons, are also considered to be matter.

Molecule   A particle consisting of two or more atoms connected to each other by bonds.

Nucleus   The part of an atom consisting of the protons and neutrons (or in the case of one form of hydrogen, just the proton) is called the nucleus. For the most part, the nucleus is not a part of chemistry except for its role in determining the atomic number.

Neutron   A neutral particle that is found in most atoms. It closely resembles a proton and often decays into a proton, releasing its charge and other energy in the form of smaller particles to produce one kind of radioactivity.

Phase   A phase (also known as a state) of matter is largely a description of how the particles of that matter have arranged themselves. Traditionally, the three states of matter are the solid, liquid, and gas. However, more recently the plasma phase (which is to a gas what a gas is to a liquid) has been recognized. In the 1990s, a fifth phase, the Bose-Einstein condensation gas, has been added.

Plasma   A phase of matter in which the individual atoms have separated into their nuclei and their normally attendant electrons is called a plasma. Although rare on Earth, the plasma is perhaps the most common phase of matter in the universe as a whole.

Proton   A relatively heavy positively charged particle found in all atoms.

Chemists Take Over the Universe

A lingering confusion persists in the public mind regarding where chemistry stops and physics or biology begins. Chemists have no such difficulty: "Chemistry is the study of matter" is an old definition, and chemists take it more seriously than ever. If a scientist is studying matter in any way, he or she is doing chemistry. The American Chemical Society makes the point that "everything is a chemical." Some would add that chemistry, then, is the science of everything.

Some physicists fail to recognize this. They have taken to calling themselves "materials scientists," although a lot of what they do is actually chemistry. Physicists, however, think that physics is the true science of everything.

It might be said that a chemist is any materials scientist who cares to call himself or herself a chemist. Increasingly, chemistry merges seamlessly into physics and biology. This is far from saying that chemistry is disappearing. Just because there is a continuous spectrum from red to violet does not mean that one cannot identify green, red, and violet.

Chemists (or at least materials scientists) also lay claim to patterns of dust particles in the highly ionized state of matter called a plasma. In 1994, these particles were found by two different teams to form analogs of crystals, the dust particles arranging themselves in arrays that mimic the arrangement of atoms in crystals. When small plastic spheres about 0.0003 in (7 microns) in diameter are sprinkled into a plasma, the resulting "electric crystal" is even visible to the naked eye. Materials scientists hope to be able to use this model as a way for studying crystal growth and dissolution.

On the other side of the equation are the biochemists. While biochemistry is recognized today as a discipline in its own right, some of the syntheses of new molecules inspired by biology actually concern the basic chemistry of various types of bonds. Should synthetic molecules that are not direct imitations of naturally occurring products of living creatures really be classified within the field of biochemistry, even if they are designed to do some of the same things biological molecules do? The answer to this question is not obvious.

In addition to incursions into the neighboring sciences of biology and physics, chemists have also made a few forays into more distant disciplines. One such sortie has been in the area of computation. John Ross of Stanford University and coworkers at the Max Planck Institute in Göttingen, Germany, have proposed that oscillating chemical reactions be used as the basis of a computer that would not rely on electromagnetic interactions at all. Such a computer might be the basis of a neural network system, somewhat similar to the actual neural network believed to be employed in animal brains.

What Can One Look for from Chemistry?

First, one can expect progress in the specific fields discussed in this section of the book. Most of the topics covered here are exciting to chemists because these studies are near the beginning of major developments.

For some time, lasers have been a vital tool used for various purposes in the field of chemistry. Recently, they have become increasingly more important in new techniques for slowing down atoms and molecules. At room temperature, atoms and molecules are constantly moving, although in solids they do not move very far in any one direction. Lasers have been used to slow individual atoms or molecules almost to stopping, allowing them to participate in reactions that could not occur at the usual subatomic-particle speed. Lasers have also been vital in studying small clusters of atoms or molecules. The only other technique used to slow atoms and molecules down to near zero velocity involves bulk cooling to temperatures close to absolute zero. This latter method has not been especially useful in recent chemistry, although it remains extremely important to materials scientists.

There is a Monty Python television routine about a Society for Putting One Thing on Top of Another, the point of which is that its members are engaged in a perfectly useless activity. Chemists could organize a similar Society for Putting One Thing Inside Another. The work of such a society would consist not only of an effort to solve some theoretical problems, it would also have the potential to produce materials of considerable practical value. Putting an atom (or ion or molecule) inside a molecule or other small cage both alters the properties of the cage and protects the entity within the cage. The latter property has been exploited primarily in biological delivery systems, but it could have many far reaching uses (*see* "Ornamented and Filled Buckyballs," p 457).

Chemists have always been specialists in finding new ways to assemble molecules. In addition to putting one completely inside another, chemists have found ways to make molecular daisy chains. The recent work with self-replicating and self-assembling molecules suggests that some chemists are abdicating their traditional role to the molecule itself. Progress in self-replication could lead to many practical possibilities, and also could help people better understand how life evolved (*see* "Sporty Molecules," p 445).

Some new assembly techniques combine elements of self-assembly with methods which place individual atoms into the molecule a single atom at a time. Materials scientists commonly use a process called molecular beam epitaxy, in which beams of atoms are directed so that the beams interact first with each other, and then with a suitable solid called a substrate. This method has led to high-quality semiconductors by, for example, inserting foreign atoms into silicon or germanium crystals. In 1995, scientists from the University of Oregon demonstrated that the same method could be used to synthesize such materials as high-temperature superconductors.

In an even smaller-scale assembly, chemist Peter Schultz and physicist Paul McEuenof the University of California at Berkeley devised a way to cause two chemicals—azide and a high-hydrogen alcohol—to react with each other one molecule at a time. They used a scanning tunneling microscope that had a platinum tip. As it often does, the platinum acted as a catalyst (*see* "Catalyst UPDATE," p 463). Nestled next to the platinum atoms, one azide molecule cashes in two nitrogen atoms to obtain two hydrogen atoms supplied by one alcohol molecule. The result is that the azide molecule becomes an amine molecule. Since the platinum is unaffected, the tip can be moved to another azide molecule, where it can repeat the same procedure as often as desired. The most likely application of this technique is in the manufacture of very small parts (*see* "Nanotechnology UPDATE," p 732).

Another experiment in assembly of organic molecules also initially appeared to be a breakthrough on how life came to be made from molecules with handedness (*see* "Evolution UPDATE," p 412). This experiment was subsequently found to be a fraud. When German workers announced that a strong magnetic field influences the handedness of molecules formed in its presence, groups all over the world tried unsuccessfully to duplicate the startling results. They all failed, because unlike the original German research team they did not have a "post-doc" (recent doctoral graduate still working as an assistant) who doped the mixture to support the faulty conclusions of his doctoral thesis. When the original team repeated the experiment without this extra help, they too found that the claim of magnetically influencing handedness was faulty. This was a notable instance of science working the way it is supposed to, with the flawed result quickly expunged from the record.

Lasers are not the only way to exploit light. Chemists have begun to use various forms of electromagnetic radiation, especially light, to affect chemical reactions. In some cases reported in 1993, light of just the right frequency was used to break the stronger bond in a molecule, while leaving a weaker bond intact. In 1995, a group from the University of Illinois at Chicago controlled the distribution of products in a reaction by using quantum-mechanical effects produced by a pair of lasers. Of course, green plants and some other organisms have long used light to control and power the reactions of photosynthesis. There are continuing efforts by chemists to accomplish the same with greater efficiency than present solar cells.

The chemistry of short-lived elements, all of which are artificial and highly radioactive, is gradually becoming known. While this is largely of theoretical importance thus far, there is some reason to believe that elements just a few protons beyond the current high-proton champion of atomic number 112 may be more stable. If that is the case, then chemistry on the fly may also turn out to pave the way for practical applications as well.

Another field recognized as important in the early 1990s involves the analysis of biological substances. Although biologists or medical researchers usually discover the actions of specific substances in plants or other organisms, chemists are needed to find out what the substances really are and, if possible, how to synthesize them. This program is increasingly urgent, as many species near extinction.

In all forms of organic chemistry and biochemistry, students in the United States get off to a slower start than do their European counterparts, although after later years of schooling they frequently catch up or surpass the achievements of scientists from other countries. The slow start comes because American high school students receive very little introduction to organic chemistry, while organic chemistry is integrated into the study of chemistry from the beginning in most European school systems. Despite this disadvantage, American high school students rank fairly high in international chemistry competitions, which implies that the best American high school students are about as good as the best from other nations. There is a much greater international disparity, however, when average students are compared, with American students ranking much lower on the average.

Most of the developments discussed in the previous paragraphs are phenomena that are being studied in the laboratory. They are a long way from being the polyvinyls, the high-octane gasolines, the synthetic rubbers, the 12-hour cold capsules, the water-based house paints, or similar advances that have changed the way we live. Practical progress continues to be made, however.

While the only true artificial enzyme took two years to design and two months to manufacture in laboratory-sized amounts, inorganic catalysts of great promise are progressing quickly from laboratory synthesis to factory production (*see* "Catalyst UPDATE," p 463).

(Periodical References and Additional Reading: *New York Times* 4-12-92, p C7; *New York Times* 7-22-92, p A17; *Science* 5-13-94, p 908; *Science* 7-1-94, p 21; *Science News* 8-6-94, p 84; *New York Times* 9-11-94, p III-7; *Science* 10-6-95, p 77; *Science* 11-17-95, pp 1157 & 1181; *Discover* 1-96, p 72)

TIMETABLE OF CHEMISTRY TO 1992

450 B.C.	Greek philosopher Leucippus [490 B.C.-?] of Miletus is the first to introduce the concept of an atom, expanded upon about 430 B.C. by his pupil Democritus [Greek: 470-380? B.C.].
1250	Albertus Magnus [German: 1193-1280] is probably the first to describe the element arsenic, although it may have been known to earlier alchemists.
1662	Robert Boyle [British: 1627-1691] announces Boyle's Law, which states that for a gas kept at a constant temperature, pressure and volume vary inversely.
1669	Hennig Brand [German: c. 1630-c. 1700] discovers phosphorus, the first element definitely not known to ancient chemists and alchemists.

1670	Robert Boyle produces hydrogen gas, but fails to recognize it as an element.
1735	Georg Brandt [Swedish: 1695-1768] discovers cobalt, the first element in a cascade of new discoveries over the next half century.
1751	Axel Fredrik Cronstedt [Swedish: 1722-1765] discovers and isolates nickel.
1753	Claude J. Geoffrey (the Younger) recognizes that bismuth is distinct from lead.
1755	Joseph Black [British: 1728-1799] discovers carbon dioxide.
1766	Henry Cavendish [British: 1731-1810] is the first to make extensive investigations of hydrogen, and is generally credited with its discovery.
1772	Joseph Priestly [British/American: 1703-1804] discovers nitrous oxide, the fourth recognized gas (after air, carbon dioxide, and hydrogen). He soon discovers many other gases, including ammonia, sulfur dioxide, hydrogen chloride, and oxygen.
	Daniel Rutherford [British: 1794-1819] and several other chemists discover nitrogen.
	Karl Wilhelm Scheele [Swedish: 1742-1786] discovers oxygen, but does not announce his discovery until after its independent discovery by Joseph Priestly in 1774. He also discovers chlorine, but does not recognize that it is an element.
	Antoine-Laurent Lavoisier [French: 1743-1794] demonstrates that diamond is a form of carbon.
1774	Johann Gottlieb Gahn [Swedish: 1745-1818] isolates manganese, recognized as an element as early as 1755 by Joseph Black [British: 1728-1799].
1778	Lavoisier discovers that air is mostly a mixture of nitrogen and oxygen.
	Scheele recognizes molybdenum as an element, although it is not isolated until 1782, by Peter Jacob Hjelm [Swedish: 1746-1813].
1781	Lavoisier states the law of conservation of matter.
1783	Fausto D'Elhuyar [Spanish: 1755-1833] and his older brother Juan José are the first to isolate tungsten, although its existence had been recognized a few years earlier by various chemists, including Scheele.
1784	Henry Cavendish [British: 1731-1810] announces that water is a compound of hydrogen and oxygen.
1789	Martin Heinrich Klaproth [German: 1743-1817] discovers oxides of the element uranium in the ore pitchblende and of zirconium in zircon.
1791	Jeremias Richter shows that acids and bases always neutralize each other in the same proportion.
	William Gregor [English: 1761-1817] discovers titanium.
1794	Johan Gadolin [Finnish: 1760-1852] investigates the "earth" yttrium, which in 1843 yields three separate elements. (Gadolin is often credited with discovery of the element yttrium, although he did not specifically isolate or identify it.)
1797	Louis-Nicolas Vauquelin [French: 1763-1829] isolates chromium in a mineral from Siberia.
1801	Charles Hachett discovers that an ore sent from North America a hundred years earlier contains a previously unknown metal, although the metal is not isolated for another 63 years. Today, the official name of the metal is niobium,

but in commerce it is nearly always called columbium because of its origin in the lands discovered by Columbus.

Andrés del Rio [Spanish/Mexican: 1764-1849] discovers vanadium, but is persuaded that he has made a mistake. Vanadium is finally rediscovered in 1830.

1802   Anders Gustaf Emeberg [Swedish: 1767-1813] discovers tantalum.

1803   John Dalton [British: 1766-1844] develops the atomic theory of matter.

Smithson Tennant [British: 1761-1815] discovers iridium and osmium.

Klaproth, Jön Jakob Berzelius [Swedish: 1779-1848], and Wilhelm Hisinger [Swedish: 1766-1852] all discover the element cerium.

William Hyde Wollaston [British: 1766-1828] discovers palladium and rhodium.

1807   Amedeo Avogadro [Italian: 1776-1856] proposes that equal volumes of gas at the same temperature and pressure contain the same number of molecules (Avogadro's law).

Humphry Davy [British: 1778-1829] begins using electricity to separate elements, and quickly discovers potassium and sodium using this new method.

1808   Davy uses an improved electrolysis method to isolate barium, boron, calcium, magnesium, and strontium, although he is beaten to the discovery of boron by nine days by Joseph-Louis Gay-Lussac [French: 1778-1850] and Louis-Jacques Thénard [French: 1777-1857].

1810   Davy is the first to recognize that chlorine is an element and not a compound.

1811   Bernard Courtois [French: 1777-1838] discovers iodine in seaweed, although he fails to recognize that it is a new element. It is so identified by both Davy and Gay-Lussac in 1814.

1813   Berzelius proposes the system of chemical abbreviations that is still used today.

1817   Friedrich Strohmeyer [German: 1776-1835] discovers cadmium.

J. A. Arfvedson discovers lithium.

1818   Berzelius discovers selenium.

1824   Gay-Lussac discovers chemical isomers—chemicals with the same formula but different structures.

Berzelius discovers silicon.

1825   Hans Christian Oersted [Danish: 1777-1851] is the first to prepare aluminum metal.

1826   Antoine-Jérôme Balard is the first to recognize that a brown liquid extracted from seaweed is a new element, bromine.

1828   Friedrich Wöhler [German: 1800-1882] prepares an organic compound from inorganic chemicals, showing that life is basically the same as other matter. Wöhler is the first to isolate beryllium, which had been recognized as an element in 1798 by Vauquelin.

1829   Berzelius discovers thorium.

1830   Nils Gabriel Sefström [Swedish: 1787-1845] rediscovers vanadium, first isolated by del Rio in 1801, and becomes the one whose name for the metal is accepted.

1839   Carl Gustav Mosander [Swedish: 1797–1858] discovers lanthanum.

1843   About this time, Mosander isolates the rare earths yttrium, erbium, and terbiumfrom the yttrium first investigated in 1794. Mosander also prepares a mixture of praseodymium and neodymium that he calls didymium, which will not be separated into its two elements until 1885.

1859   Gustav Kirchhoff [German: 1824–1887] and Robert Bunsen [Germany: 1811–1999] introduce the use of the spectroscope to identify elements by the light they give off when heated or burned.

1860   Kirchhoff and Bunsen use their new technique to discover a new element, cesium.

1861   Kirchhoff and Bunsen discover their second new element, rubidium.

            William Crookes [British: 1832–1919] recognizes a new element in a spectrogram, which he names thallium.

1863   Ferdinand Reich [German: 1799–1882] and Hieronymous Theodor Richter [German: 1824–1898] discover indium, confirming the discovery by its spectrum.

1868   Pierre-Jules-César Janssen [French: 1824–1907] and Sir Joseph Lockyer [British: 1836–1920] discover helium by observing the spectrum of the Sun.

1869   Dimitri Mendeléev [Russian: 1834–1907] publishes his first version of the periodic table of the elements.

1875   Paul-Emile Lecoq de Boisbaudran [French: 1838–1912] discovers gallium, the first discovery of a predicted element (predicted by Mendeléev on the basis of his periodic table).

1876   Lars Fredrik Nilson [Swedish: 1840–1899] discovers the element scandium, predicted by Mendeléev as eka-boron.

1879   Lecoq de Boisbaudran discovers the rare earth samarium.

            Per Teodor Cleve [Swedish: 1940–1905] discovers the rare earths holmium and thulium.

1880   Jean-Charles Galissard de Marignac [Swiss: 1817–1894] discovers gadolinium.

1885   Karl Auer, Baron von Welsbach [Austrian: 1858–1929] separates the rare earth Mosander called didymium into two elements, which he names praseodymium and neodymium.

1886   Although fluorine had been recognized from the time of Humphry Davy, efforts to isolate this most active of all elements fail until this year, when on June 26 Ferdinand-Frédéric-Henri Moissan [French: 1852–1907] succeeds in liberating it by using all-platinum equipment.

            Clemens Alexander Winkler [German: 1838–1904] discovers germanium.

            Lecoq de Boisbaudran discovers dysprosium.

1894   William Ramsay [British: 1852–1916] discovers argon.

1896   Eugène-Anatole Demarçay [French: 1852–1904] discovers europium.

1898   William Ramsay and Morris William Travers [British: 1872–1961] discover the noble gases neon, krypton, and xenon.

            Marie Sklodowska Curie [Polish-French: 1867–1934] and Pierre Curie [French: 1859–1906] discover polonium and radium.

1899	André-Louis Debierne discovers actinium.
1906	Mikhail Tsvett [Russian: 1872–1919] develops paper chromatography, the beginning of modern methods of chemical analysis.
1907	Georges Urbain [French: 1872–1938] isolates lutetium and ytterbium, as do several other independent chemists about this time.
1908	Fritz Haber [German: 1868–1934] develops a cheap process for making ammonia from nitrogen in the air.
1918	Otto Hahn [German: 1879–1968] and Lise Meitner [Austrian: 1878–1968] discover protactinium.
1923	Observing that element 72 had not yet been found, György Hevesy [Hungarian/Danish/Swedish: 1885–1966] and Dirk Costner [Dutch: 1889–1950] set out to find it. Upon succeeding, they name the new element hafnium, after the Latin name for Copenhagen.
1937	Emilio Segrè [Italian/American: 1905–1989] synthesizes technetium, the first previously unknown element to be made by deliberate nuclear bombardment.
1939	French scientist Marguerite Perey isolates element 87, francium, which is among the rarest of naturally occurring elements because all of its isotopes are quite unstable. It is the least stable element naturally occuring in any amount.
1940	Glenn T. Seaborg [American: 1912–] and coworkers create element 94 (plutonium), and also produce the element astatine from bismuth. Astatine also exists in small amounts in nature, and plutonium may also be produced in very small amounts from natural nuclear processes.
	Edwin McMillan [American: 1907–] and Philip Abelson [American : 1913–] create element 93, neptunium.
1943	Albert Hofmann [Swiss: 1906-] discovers that LSD is hallucinogenic.
1944	Seaborg and coworkers discover elements 95 (americium) and 96 (curium).
1945	J. A. Marinsky, L. E. Glendenin, and C. D. Coryell make the first chemical identifications of the rare earth promethium, element 61, which had probably been first produced in nuclear reactions in 1941 at Ohio State University. This radioactive element is probably not found on Earth, but has been identified in star spectra.
1949	Glenn T. Seaborg and coworkers create elements 97 (californium) and 98 (berkelium).
1951	Russian chemist B. P. Belousov discovers chemical reactions that repeatedly oscillate between two states.
1952	Albert Ghiorso and coworkers create elements 99 (einsteinium) and 100 (fermium).
1955	Ghiorso and coworkers create element 101 (mendelevium).
1961	Ghiorso and coworkers create element 103 (lawrencium).
1962	Neil Bartlett [British: 1932–] shows that the noble gases can form compounds by creating a compound of xenon.
1969	Ghiorso and coworkers create element 104, whose name is still in dispute (see "Heavier Heavy Elements and More Naming Feuds," p 439).
1970	Ghiorso and coworkers create element 105, whose name remains in dispute.

1974   Frank Sherwood Rowland [American: 1927-] and Mario Molina [American: 1943-] warn that chlorofluorocarbons (freons) are destroying the ozone, that layer in the atmosphere which serves to protect life on Earth from harmful ultraviolet radiation.

1982   German scientists produce one atom of element 109.

1984   German scientists produce one atom of element 108.

1985   Richard E. Smalley [American: 1943-] and Harold W. Kroto discover that carbon forms hollow balls of 60 atoms. Named buckminsterfullerene (after the inventor of the geodesic sphere that possesses the same structure as one of the balls), the individual particles come to be nicknamed "buckyballs," while other carbon particles with similar structures are labeled fullerenes.

1986   Herbert Naarmann and N. Theophilou develop a plastic that conducts electricity more effectively than copper.

1987   Scientists confirm that a hole in the ozone formed in August and September is caused by chlorofluorocarbons.

        Frank Kilisko [American: 1942-] develops the first electo-rheological fluid, which changes viscosity in the presence of an electrical field.

1989   On March 23, B. Stanley Pons and Martin Fleischmann [British: 1927-] announce that they have discovered how to fuse hydrogen nuclei at room temperature, a claim that has not been confirmed.

1990   Donald Huffman, Lowell Lamb, Wolfgang Krätschmer, and Konstantin Fostiropoulos discover a way to manufacture large amounts of fullerenes, touching off a continuing wave of fullerene chemistry.

        David Lawrence and coworkers from the State University of New York at Buffalo develop complex molecules that assemble themselves from component parts.

        John Stewart, Karl W. Hahn, and Wieslaw A. Klis of the University of Colorado Medical School create the first artificial enzyme.

        Fumihiro Wakai and coworkers at Japan's Government Industrial Research Institute in Nagoy develop a stretchable ceramic.

1991   Henri Kessler of the National Superior School of Chemistry in Mulhouse, France, and coworkers develop an artificial zeolite that, because it has pores whose cross-sections resemble a four-leaf clover, is named cloverite.

NEW FORMS OF MATTER

HEAVIER HEAVY ELEMENTS AND MORE NAMING FEUDS

Filling out the periodic table by adding a new element has always been newsworthy, even from the days before there was a clear notion of the definition of an element. By the 1940s, however, all the elements known to occur naturally on Earth had been found, a couple of gaps between 1 and 92 had been filled, and the only way to add an element to the periodic table was at the end. Making such additions, while not central to either chemistry or physics, became the province of three small groups of scientists, one in California, one in Germany, and one in the Soviet Union. Each group has created new elements, but considerable controversy surrounds who gets credit for particular elements first created during the 1950s, 1960s, and 1970s, when cold-war tensions fed the rivalry between the Berkeley, California, group and that at the Joint Institute for Nuclear Research in Dubna, Russia. In the 1980s and 1990s, the German group (Gesellschaft für Schwerionenforschung, or Center for Heavy Ion Research, in Darmstadt) has become the acknowledged leader. The Germans have produced a few atoms each of several short-lived elements, all of which are extremely radioactive, and therefore decay with great rapidity. After reaching element 109 in 1984, however, there was little progress made for 10 years. Then, in 1994, the Darmstadt group succeeded with element 110 in November and 111 on December 8. About a year later, on February 9, 1996, the team produced a single atom of element 112. All the Darmstadt heavy elements were produced by bombarding targets made of one moderately heavy atom (such as bismuth or lead) with another (such as nickel or zinc) and looking for occasions when the bomb and the target briefly melded into a single atom. For element 111, three such superheavy atoms were observed before they decayed over a period lasting about two-thousandths of a second.

On the way out of existence, the new elements decompose into various isotopes of other superheavy elements. Element 111, for example, decays into element 109-268, which then decays into element 107-264 (the number after the hyphen corresponds to the atomic mass of the atom, identifying the isotope of the element). Similarly, element 112 begins its decay chain with element 110 followed by element 108. If elements 107, 108, and 110 had not been previously created by direct bombardment, they could have been made as decay products. However, those isotopes formed as decay products differ from those produced by direct union of two atomic nuclei.

There was some controversy following announcement of the Darmstadt production of element 110, because scientists from the Lawrence Berkeley Laboratory thought that they had produced one atom of element 110 in 1991, although it had taken them until 1994 to recognize their own accomplishment. Similarly, the Dubna group had been working on the synthesis of element 110 and could perhaps, after further analysis of data, also claim to have been first. For element 111, however, Darmstadt clearly led the field.

HOW HEAVY IS "HEAVY"?

The fundamental entities of chemistry are elements, substances formed entirely from atoms that have the same number of protons. The element hydrogen, for example, consists mostly of atoms that have one proton (and one electron), although one atom out of each 7,000 contains a proton and a neutron (and one electron), and certain conditions produce atoms of hydrogen that contain a proton and two neutrons (and one electron). Furthermore, any of these atoms can be ionized, which means that the atom can lose its electron or gain an extra electron, although the resulting ion is still hydrogen because of the single proton. Similarly, atoms that contain two protons are helium, whether they also contain one neutron or (more commonly) two, and whether or not the helium is ionized.

Since each element is characterized by an integral number of protons per atom, this **atomic number** identifies the element, also allowing one to arrange elements in order by number. Hydrogen, with an atomic number of 1, is the first, since anything without a proton in its atom would not be called an element. Helium is number 2. There is no way to squeeze in another element between the 1 of hydrogen and the 2 of helium, since the proton is chemically indivisible. Furthermore, the element after helium must be one possessing atomic number 3, with three protons (which happens to be lithium).

Since the protons and neutrons contribute nearly all the mass to an atom, most atoms of element atomic number n are heavier than most of number $n - 1$ and lighter than most from element $n + 1$. One has to say "most," because an atom of element number n that has few neutrons may be lighter than an atom of element number $n - 1$ that has a lot of neutrons. The actual weight of a particular amount of an element in liquid or solid form depends on how the atoms are arranged, not just on how much each atom weighs. Therefore, the designation "heavy element" is used to describe the weight of each atom, averaged over the various different atoms with the same number of protons, not on the weight of a particular volume.

Normally when chemists speak of **light elements**, they mean only those elements with very small atomic numbers, up to, say, element 11 (sodium) or 13 (aluminum). Similarly, the **heavy elements** (such as those often referred to in discussions of toxic wastes) are those starting about element 42 (molybdenum) or 48 (cadmium). In the context of making artificial elements, the heavy elements are much heavier, beginning traditionally with element 93 (neptunium). There are also elements less than atomic number 93 that are only known as artificially created forms, such as number 61 (promethium), as well as elements greater than 93, such as element 94 (plutonium), that exist naturally in small amounts.

The Promised Islands of Stability

Different isotopes of the same element have different lifetimes. Therefore, even though all the heaviest elements are unstable, some are more unstable than others. It is very difficult to study an element that exists for only two thousandths or five ten-thousandths of a second. The atoms are gone so fast that the only way to verify their presence is to look for the products of their decay. Chemistry of such elements is based largely on inference and guesswork.

The theory of nuclear stability for superheavy atoms is far from being complete, but chemists have made inferences based on the behavior of lighter atoms. Chemists have foreseen that some isotopes would be longer-lived than others. Elements 110 and 114 were predicted to be more stable than the other elements with atomic numbers greater than 106. In the 1990s, some of these predictions were tested by production of the nuclei in question for the first time.

In November 1993, scientists announced that a collaborative effort between two former rival groups, Lawrence Livermore Laboratory and Dubna, had located two isotopes of element 106, both of which were more stable than the first form produced. The first form of element 106, which had an atomic mass of 263 and is therefore called element 106-263, existed for less than a second before decaying into its daughter elements.

Element 106-263 has 106 protons and 157 neutrons. The theory of nuclear stability suggests that an isotope whose protons and neutrons exist in close to equal numbers will be more stable than an isotope with a proton-to-neutron ratio which is further from a one-to-one ratio. According to this, one would expect that a form of element 106 with fewer than 157 neutrons would be more stable, while one with a greater number would be less stable. Yet Polish theorist Adam Sobiczewski calculated that for element 106, some heavier isotopes would be more stable. Sobiczewski was correct, and the standard rule-of-thumb was wrong. The new isotopes, each with a half-life of over a minute, have 159 and 160 neutrons.

The isotopes, known as element 106-265 and element 106-266, were physically produced at Dubna, but their identities were detected with instruments developed and made at Lawrence Livermore Laboratory in California. Credit for these discoveries has therefore been shared between the two groups.

Sobiczewski has gone on to predict that other isotopes of superheavy elements will be even more stable.

Others have also predicted that certain isotopes of superheavy elements would be particularly stable. The best-known prediction was put forth in the 1960s by Georgy N. Flerov of Dubna, who suggested that element 114-298, with 184 neutrons, might itself be stable, or at least much more stable than its neighbors. This prediction has yet to be verified, since the highest atomic number created so far is 112. That island of stability may exist, however, and it appears to be within the reach of scientists during the next few years.

Another prediction suggested that element 110 would be more stable than those elements with atomic numbers close to 110. The first isotope to be produced, and the only thus far, lived for only about 0.0005 second. In fact, element 111 lived longer, around 0.004 second. Note that for these elements, one speaks of a lifetime instead of a half-life. The concept of a half-life concerns a large collection of atoms, in which some will unpredictably live longer than others, and the average time for half of them to have existed can be determined. When there are only three atoms to consider (the total amount of element 111 produced), the actual lifetime of the atoms can be considered.

Even though element 110 is not an island of stability, that is no reason to think that element 114 will similarly fizzle. If element 114 is stable because it has a magic number, then it is logical that element 111 lives longer than 110, since 111 is closer to 114 in terms of number of protons.

Background

Nuclear theory predicts that some atomic nuclei will be more stable than others. The most easily observed effect, obvious from a table delineating the abundance of particular elements in the universe arranged in order of atomic number, is that nuclei with even numbers of protons exist in a greater abundance than those with odd numbers of protons. For example, tin (atomic number 50) is about 4 times as abundant as antimony (atomic number 51), while platinum (atomic number 78) is 10 times as abundant as gold (atomic number 79). This is a result of how nuclei are produced in fusion and of what stable forms emerge. The correspondence between evenness and stability is not exact, however. An example of its non-exactness is that mercury (atomic number 80) is not much more common than odd-numbered gold.

A factor which tends to influence stability more than just the *number* of protons is the *ratio* of protons to neutrons in the nucleus. Except for the most common form of hydrogen, all atoms need both protons and neutrons in their nuclei in order for the nuclei to be stable. This is essentially because the strong force (that force which keeps the positively charged nucleus from flying apart— *see* "The State of Physics," p 625), works by a process that transforms protons into neutrons and vice versa, as long as the different nucleons are sufficiently close to each other. The strong force is usually explained as resulting from the exchange of subatomic particles called pions, which are mesons—particles that contain exactly two quarks (*see* "The Subatomic Particles," p 666). For very small separations, the strong force deserves its name, being vastly stronger than the electromagnetic force that pushes the positively-charged protons apart. Another result of this process of transmuting neutrons concerns the neutron which, within a nucleus, becomes a quasi-stable particle. This stability comes from the fact that such a neutron is usually turned into a proton before it has a chance to decay, while outside the nucleus the neutron decays in a matter of minutes. One form of radioactive atom results from the occasional decay of a neutron within a nucleus, a neutron that was not changed into a proton in time to avoid such decay.

Common hydrogen, which has only a single proton in its nucleus, is stable because there are no neutrons to cause difficulties. Yet elements with a given number of protons also exist in different isotopes possessing different numbers of neutrons. Deuterium, the isotope of hydrogen with one proton and one neutron, is also stable, since the two particles can easily change back and forth by exchanging a pion between them, producing the strong force. Tritium, the isotope of hydrogen possessing one proton and two neutrons, is rather unstable, since there is no second proton with which the second neutron can interchange.

Add another proton to tritium and you get helium, stable because both neutrons have partners. Also note that it is impossible to make a stable atom whose nucleus contains two protons and only a single neutron.

As atomic nuclei become bigger, this simple need for a one-to-one ratio is no longer the whole story. Even if there are enough protons and neutrons to go around, the shape of the large nucleus may keep the actors too far apart for frequent and completely successful interchanges. Thus, as nuclei become heavier, other factors come in to play. The nuclei will need more neutrons to keep the protons together at all, yet some nuclei will remain unstable because their characteristic shape keeps the interacting pairs too far apart. Thus, most very heavy elements are inherently radioactive, although their instability is not a direct result of size. Uranium-230, with 92 protons and 138 neutrons, is rather unstable. Half of uranium-230 decays into another element in only 21 days. Uranium-238 (with 92 protons and 146 neutrons) is also radioactive, but has a half-life of billions of years, while uranium-240 (with 92 protons and 148 neutrons) has a half-life of only 14 hours.

Clearly, some combinations are more stable than others. Nuclear chemists investigating this phenomenon call the most stable combinations of protons and neutrons **magic numbers**. This number refers to the number of either protons or neutrons, with the ratio of the two kept close to one to one. The first few magic numbers proposed by theorists, based upon such concepts as the Pauli exclusion principle and nuclear spin, are 2, 8, 20, 50, 82, and 126. For example, calcium-40, with 20 protons and 20 electrons, is especially stable. Lead, with the magic number of 82 protons and the magic number of 126 neutrons, is often thought of as the most stable heavy element, although bismuth-209, with 126 neutrons, is also quite stable.

Magic number theory suggests that an element with 126 protons and 126 neutrons might be relatively stable, but scientists are far from producing an element with atomic number 126.

Meanwhile, nuclear chemists test magic-number theory by producing isotopes of various elements which possess magic numbers. For a long time, chemists pursued tin-100, an isotope with 50 protons and 50 neutrons, a doubly magic isotope with a one-to-one ratio. Despite all this magic, when Darmstadt finally won the race by making tin-100 early in 1994, the seven atoms they produced existed for only a few seconds each before decaying. Still, a lifetime of a few seconds is thousands of times as long as other nearby isotopes of tin. The next doubly magic isotope with a one-to-one ratio would be lead-164, with 82 protons and 82 neutrons, but no one thinks that it will be possible to create an atom of that isotope.

Name Calling

No long-running dispute in current science has less scientific and more nuisance value than does the quarrel over what to name the superheavy elements. In the 1990s, the apparently official names of elements greater than 103 have been changed with alarming regularity. In the meantime, books and papers are published using the name of the moment. Only readers with a firm grasp on the whole controversy from the beginning have a chance of keeping the names straight when consulting older publications that use now-outmoded names. Fortunately, the current importance of the superheavy elements in the big picture of applications or even theory is quite small. Serious consequences of the name confusion seem unlikely.

TIMETABLE OF SUPERHEAVY NAMES

1964	A team from the Joint Nuclear Research Institute at Dubna, Soviet Union (Russia), led by Georgy N. Flerov claims to have produced element 104. They propose to name the element kurchatovium (Ku), in honor of Igor Vasilevich Kurchatov [Russian: 1903–1960].
1967	The Dubna team claims to produce a few atoms of element 105.
1969	A team led by Albert Ghiorso and coworkers at Lawrence Berkeley Laboratory in Berkeley, California, claim to produce element 104, and also note that they cannot duplicate the Dubna process of 1964. The Berkeley team takes full credit, and proffers the name rutherfordium (Rf), after Ernest R. Rutherford [British: 1871–1937].
1970	The Dubna team from the Soviet Union and Ghiorso's group at Lawrence Berkeley Laboratory in the U.S. both claim to have produced element 105 in quantities, but the Berkeley groups claims that they cannot duplicate the methods used by the Soviets. Berkeley proposes the name hahnium (Ha), after Otto Hahn [German: 1879–1968]. Dubna later counters with the name nielsbohrium, after Niels Bohr [Danish: 1885–1962].
June 1974	The Dubna team announces that they have synthesized element 106.
September 1974	Ghiorso's group announces that it has produced element 106.
1976	Flerov's group at Dubna produces element 107, announcing this in 1977.
1980	The International Union of Pure and Applied Chemistry rejects earlier names for elements 104 and 105 and chooses to call 104 *unnilquadium* (Unp), for "one-zero-four element," and 105 *unnilpentium* (Unp) for "one-zero-five element." They also christen element 106 *unnilhexium* (Unh) for "one-zero-six element."
1982	A team from Gesellschaft für Schwerionenforschung (Society for Heavy Ion Research) in Darmstadt, (West) Germany, produces a few atoms of element 109.
1984	The Darmstadt team synthesizes a few atoms of element 108.
1991	Scientists in Berkeley believe that they have produced element 110, but this is not confirmed, and is generally discounted.
1992	The International Union of Pure and Applied Physics/Chemistry, headed by Sir Ernest Wilkinson, announces that element 107 will be named nielsbohrium, element 108 hassium (from Hassia, the Latin name of the German state of Hesse), and element 109 meitnerium, for Lise Meitner. These are the names proposed by the Darmstadt team that first synthesized each element.
October 1993	Albert Ghiorso, head of a team that claims first production of element 106, proposes that it be named alvarezium, after Luis Alvarez.

March 1994	The American Chemical Society announces that a group of scientists, including Ghiorso, who had been invested with the task of naming element 106 have decided upon seaborgium (Sb), after the still-living American physicist Glenn T. Seaborg [1912–], who discovered several other superheavy elements.
October 1994	The International Union of Pure and Applied Chemistry (IUPAC) announces that its nomenclature committee decided in August to rename all but one of the elements heavier than 103. The new names were to be as follows: element 104, dubnium, for Dubna; 105, joliotium, for Frédéric Joliot-Curie [French: 1900–1958]; 106, rutherfordium, for Ernest Rutherford; and 107, bohrium, for Niels Bohr; 108, hahnium, for Otto Hahn. Element 109 is to remain meitnerium.
November 1994	Peter Armbruster of the Darmstadt group announces that although the group has successfully produced element 110 on November 9, it will not propose a name for it in protest of the action by IUPAC committee of the previous month that ignored the names Darmstadt had proposed for elements 107 and 108. The Darmstadt scientists also announces that the group wrote to IUPAC to inform them that it will do everything it can to get the group to agree to return to the names Darmstadt had proposed.
March 1995	Ghiorso reports a rumor that the Darmstadt group plans to hold element 111 hostage until the other names are settled.
February 1996	The Darmstadt group, which has produced an atom of element 112, fails to propose a name for it.

(Periodical References and Additional Reading: *New York Times* 10-12-93, p C1; *New York Times* 3-15-94, p C6; *Science* 5-6-94, p 777; *New York Times* 10-11-94, p C10; *Nature* 11-25-94, p 306; *Science* 11-25-94, p 1311; *Science News* 11-26-94, p 356; *Science* 12-2-94, p 1479; *New York Times* 12-22-94, p A5; *Science* 1-6-95, p 29; *Scientific American,* 3-95, p 21)

SPORTY MOLECULES

Throughout the 1990s, the molecule most famous for its shape has been the buckyball (*see* "Ornamented and Filled Buckyballs," p 457), but chemists have also produced other shapes of interest. Such shaped molecules are termed advanced inorganic materials. In 1991, for example, French and Swiss chemists produced cloverite, which in its solid form exhibits a structure filled with molecule-sized holes, each shaped like the leaves of a four-leaf clover. Molecules whose shapes and electrical properties have been carefully controlled are important in catalytic and molecular sieving, which is what cloverite is used for. Other uses for these architecturally engineered molecules include improvement of lasers and wave guides; high-temperature superconductors; advanced materials, especially ceramics; electronic and magnetic applications; and even energy storage.

However, the shaped-molecule emphasis in the 1990s seems to have been in finding molecules with a sports association.

Molecules in Chains

A particularly difficult challenge has been to develop molecules with rings that interlock. Three interlocking rings make a famous symbol, once best known in the United States as a logo designating a brand of beer. The interlocked rings have the almost magical property that although the three are linked, no two of the rings themselves are joined. Therefore, removal of any one of the rings causes the other two to fall apart. This is different from the way three rings would be if linked in a chain, which would fall apart only if the middle ring were removed. Another distinct way of linking rings involves forming a ring of rings, by linking one end of a chain to the other end. For the short chain with three links, removing any one ring leaves the other two still attached. In the branch of mathematics called topology, the study of the differences among the three ways that three rings can be linked has been formalized and extended.

When the modern Olympic games looked for a symbol, they used the same idea as the old beer symbol, but with five rings to symbolize five continents (Africa, Asia, Europe, Australia, and the Americas treated as one continent). In the way this is sometimes shown, the symbol consists of two of the old beer symbols that share one ring. If the shared ring is removed, the others all separate, while removing any other ring produces two single rings and one group of three that are still linked.

The exact chemical equivalent to the Olympic symbols has not been developed, although in 1994 a team from the University of Birmingham in England led by J. Fraser Stoddart and David Amabilino succeeded in linking five rings together in a chain. Being chemists instead of topologists, the British chemists named their chain compound olympiadane, because it has five linked rings. They subsequently produced molecules incorporating seven linked rings, although these rings form a branched chain rather than a single line. The general class of molecules consisting of linked rings in straight or branching chains has been given the name catenanes, from the Latin for *chain*.

Another group of molecules uses rings that are not interlinked, although the whole material is filled with adjacent rings. These compounds, which may have as many as 10 to 12 windows formed by rings, are important in the conversion of petroleum to various specific oil fractions (such as gasoline) or even the conversion of the smaller molecules of methane to the large molecule of gasoline.

Ball Molecules

The buckyball, or buckminsterfullerene molecule, has been compared to a soccer ball, although certain variations on it are closer in some respects to the patterned ball used in European football. The elongated shape used in rugby, or the even longer one used in American football, can also be found in carbon fullerenes possessing 70 atoms instead of the 60 found in the soccer ball fullerene.

However, while most balls are round, the characteristics which separate various ball-shaped molecules are dictated not by the shapes themselves, but by the way the sphere is put together. This is true, for example, of the aforementioned buckyball molecule and the regulation soccer ball. Both the molecule and the ball are essentially spheres, but are distinguished by being formed from pentagonal pieces. In 1992, A.

THE DREAM AND THE RING

Near the dawn of chemistry, when Dalton's atomic theory of 1803 was gradually being accepted, the science had not advanced sufficiently for anyone to think seriously about how molecules might look, although it was recognized that molecules were formed in some way by atoms that tended to stick together in predictable numbers. Slightly more than 40 years later, Louis Pasteur was the first to demonstrate that shape matters. He showed that two different substances sharing the same chemical formula were different from each other because their molecules were assembled differently, albeit from the same parts. In 1860, the chemist Friedrich August Kekulé explained how the ability of carbon to bind with itself and other elements in several different ways accounts for the different shapes of molecules.

Kekulé is most famous for a dream that he interpreted as revealing the shape of the benzene molecule, which was puzzling chemists studying carbon compounds. In 1865, Kekulé dreamed of a snake swallowing its own tail, which led him to realize that the atoms in benzene form in the shape of a ring. Since that time, the chemistry of compounds based upon or including rings has been an important part of chemistry.

Welford Castleman Jr. and coworkers at Pennsylvania State University accidentally discovered yet another soccer ball molecule, composed of 20 titanium and carbon atoms arranged somewhat like the characteristic pentagons and 12-region structure of the balls. Since each pentagon contains two carbons and three titaniums, there are only four pentagons in all, as opposed to the dozen pentagons of the buckyball. Thinking of carbon as black, as it is in graphite and soot, and of titanium as white (since it is a white metal and an important brightener for white paint) enhances the soccer ball comparison. The official name for this kind of chemical soccer ball is a metalo-carbohedrene. After the first accidental discovery, Castleman and coworkers found that similar metalo-carbohedrenes, or met-cars, could be made by replacing the titanium with vanadium, zirconium, or hafnium. Furthermore, the soccer balls could be linked to form much larger molecules, just as one can stack icosahedrons, the basic shape of these molecules.

A tennis ball is formed from two mirror-image half spheres, although neither is a simple hemisphere. Instead, each half sphere has a curved structure similar to one side of the famous ying-yang symbol of China. René Wyler of Massachusetts Institute of Technology, working with Julius Rebeck Jr., succeeded in developing the molecular equivalent to a tennis ball, a molecule in which the two mirror images are stitched together with weak hydrogen-oxygen attachments called polar interactions. One advantage of this structure is that the bonds can easily be removed or replaced, making it possible to open and close the tennis-ball shaped molecule. For many purposes, a molecule which has another molecule trapped within it is useful, so this zipping and unzipping feature is valuable. Open it up, tuck in a molecule, and glue the edges together again. The tennis-ball molecule, however, is too small to contain large molecules, which are often the molecules of particular interest and usefulness. Thus

Rebeck and Wyler are now trying to develop a larger version of the tennis-ball molecule, which they refer to as a softball molecule.

A group of Australian chemists led by Richard Robson of the University of Melbourne reported in the May 17, 1995 *Journal of the American Chemical Society* that they had produced a giant molecular structure consisting of cagelike compartments interspersed with rings. Instead of being like a ball, this structure is more like a squash or handball court. The chamber is large enough to trap as many as 20 large organic molecules, each one too large to fit inside a tennis-ball molecule. The discovery came by accident as the researchers were trying to develop a new kind of zeolite (*see* "Catalyst UPDATE, p 463). The four different building blocks they used, when put into a solvent together, self-assembled into the squash-court cages (*see* further discussion of molecules that self-assemble below).

Self-Replication

The goals of linking molecules together to form specific shapes involve more than just finding solutions to chemical puzzles. Cloverite and the other zeolites are molecular sieves that permit only certain shapes and sizes of molecules to pass, and similar designs can be used to sieve other designated molecules. Linked rings can be used to store information, just as DNA is a molecular memory "chip." Molecules such as the tennis ball molecule are capable of forming copies of themselves when given suitable chemicals as "food."

If all this sounds like the processes used by living organisms to separate molecules and reproduce, it is intended by chemists to do so. The skill that cells have for passing substances from one side of a membrane or cell wall to the other is well worth imitating, for example.

In the 1950s, the mathematician John von Neumann suggested that technology could produce devices which, given a suitable supply of parts, could reproduce themselves. Computer scientists continue today to work on ways that robots could build more robots without human intervention, and many think that this will someday be possible. In fact, chemists have already begun to produce compounds that reproduce themselves when given a suitable mix of molecules to use as parts.

Rebeck has been working on this problem for several years. The principal challenge has been to fine-tune the amount of attraction in the weak bonds. Early versions tended to be too strong. One version caused the template to grab its own ends and effectively stop the reaction before it could make a new molecule. Another held onto the first new molecule it produced, stopping the reaction at that point. In 1990, Rebeck's team had its first major success with a molecule that could, in a suitable medium, reproduce using means similar to those employed by living systems. The molecule developed by Rebeck's team is officially known as amino adenosine triad ester, or AATE, although it is usually called the J-molecule, after its shape. Placed in a chloroform solution of its two components, an amine and an ester, AATE grabs one of each, bonding an amine to its own ester end and an ester to its own amine end. The ester and amine are positioned to form what is called an amide bond between them. An amide bond is the same kind of bond that forms between amino acids in proteins. The result of the assembly process is two J-molecules that are loosely bonded to each other by hydrogen bonds. Thermal agitation is enough to break the hydrogen bonds, so the two J-molecules can float apart and seek partners to make new copies. Chemists think of it this way:

$$AATE + amine + ester \longrightarrow 2\ AATE$$

Because there is a geometric progression in the number of copies, AATE can duplicate itself a million times a second if it has the proper "food." If no AATE is added to the solution as a template, the same reaction will occur, but a hundred times slower. A major breakthrough would be the production of a self-replicating molecule that can assemble two or more components in a particular sequence *only* with the template present.

The practical importance of a self-replicating molecule would involve a situation wherein one of the components is a useful chemical that is difficult to manufacture by ordinary means. After the molecule was allowed to duplicate itself enough times, it should be easy to break off the useful part and separate it from the rest, subsequently returning the separated parts to solution as part of the "food" for the larger molecule. This somewhat resembles the way that living cells make difficult-to-produce substances. It is also somewhat like catalysis, in that some of the chemicals involved could be recycled without using them up.

An interesting scientific question related to this issue involves how closely creating such self-replicating molecules resembles the origin of life. A prominent theory of the origin of life involves self-replicating molecules of ribonucleic acid (RNA) floating in a solution containing suitable bases. The RNA of this hypothesis uses the bases to make copies of itself. Self-assembly is recognized as the foundation for production of a number of different helical and multiple-helical molecular parts in living organisms.

Two groups of chemists have already developed chemicals that assemble into large molecules, given the right ingredients combined in the right order. There is no template, as required by the J-molecule. Instead, the large molecules put themselves together using electrostatic interactions and hydrophobic (water avoiding) regions, coupled with the proper shapes for self-assembly.

Think of the hydrophobic regions as directed by forces comparable to those between similar poles of a bar magnet (likes repel) and the electrostatic interactions as directed by forces comparable to those between the opposite poles of magnets (opposites attract). If you put into the same jar a cylinder and a long piece that could thread through the opening in the cylinder and then shake the jar, you can use such forces to cause the long piece to stick with its middle in the cylinder and the two ends outside. Then, if you put in a couple of wing nuts whose magnetic orientation opposed that of the ends of the long piece and shook the jar some more, you would end up with a large piece in which the long piece is held in place both by the internal situation with the cylinder and the wing-nut caps at each end. In effect, this is what David Lawrence and Tata Venkata S. Rao of the State University of New York at Buffalo did at the molecular level. Their cylinder was a starchlike molecule called a cyclodextrin. The long piece was a salt whose middle section avoids contact with water. The salt, therefore, benefits by hiding where it is protected by the cyclodextrin. The ends of the salt are electrostatically attracted to molecules resembling four-bladed wing nuts. Thus, the complex molecule assembles itself.

In one application of self-assembly, methods have been used to produce a new class of solids that are not quite porous, but not quite impervious either. In this new class, the distinguishing characteristic is that all of the medium-sized pores are approximately the same size.

Further experiments along these lines have imitated the assembly of shells and other structures produced by marine algae such as diatoms and marine protozoans such as radiolaria. The natural structures produced by these small organisms are famous for their diversity and complexity. Chemists from the University of Toronto in Canada and from the University of Bath in the United Kingdom have developed methods that *almost* persuade inorganic molecules to assemble themselves into similar shapes. With further work, it is thought that these new inorganic materials could be used in industrial processes in ways similar to zeolites. Furthermore, a long-range goal would be to be control the self-assembly such that even more intricate patterns could be produced, perhaps even patterns which could be useful as parts of computer chips.

(Periodical References and Additional Reading: *New York Times* 3-17-92, p C5; *New York Times* 5-26-92, p C2; *Nature* 10-22-92, p 710; *Nature* 2-3-94, p 441; *Science* 3-25-94, p 1698; *Science* 5-6-94, p 795; *Scientific American* 7-94, p 48; *New York Times* 8-20-94, p C8; *Science* 6-23-95, p 1698; *Nature* 9-28-95, p 320; *Nature* 11-2-95, pp 17 & 47; *Science* 11-24-95, p 1299)

HARDER HARD STUFF

Science fiction stories from before World War II and even a few years afterwards often involved the invention of a new material that, in the story, is harder, stronger, and lighter than previously known materials. Today we recognize that the assembly of various desirable properties into one substance is more likely to be accomplished by combining two or more materials. Still, one longs for the magic cloth of *The Man in the White Suit* that would never wear out or get dirty. Chemists continually seek such a super substance.

In 1989, Marvin L. Cohen of the Lawrence Berkeley Laboratory at Berkeley, California, and graduate student Amy Y. Liu calculated that it might be possible to develop a material harder than any previously known. Diamond has been the hardest material, natural or synthetic, for the entire history of the universe (*see* "Diamond Discoveries," p 453). The hardness of diamond results from the extreme compactness and tight bonding of the arrangement of carbon atoms in its crystal. The theoretical new material would be more like another hard substance, silicon nitride, than like diamond. While silicon is a part of several more familiar hard substances as well, including quartz (silicon dioxide) and carborundum (silicon carbide), silicon nitride is the hardest, almost as hard as diamond.

In many ways, silicon more closely resembles carbon than does any other element. For example, the two of them are the only nonmetals in Group IVA of the Periodic Table. However, since silicon is a larger atom than carbon, Cohen and Liu thought that carbon nitride would be even harder than silicon nitride. The hardness of both crystals is due to the very short bonds between atoms of the molecules, which are even shorter between carbon and nitrogen than between silicon and nitrogen. The calculations showed that not only would carbon nitride be harder than silicon nitride, it would be harder than diamond. One arrangement of carbon nitride, called the beta phase, was predicted to be stable.

No one knew how to produce beta-carbon nitride in 1989. It was not until July 1993 that Charles M. Lieber and Chunming Niu of Harvard University in Cambridge, Massachusetts, announced the production of carbon nitride in the form of a thin film deposited on glass or silicon. Their method used a laser, which blasted carbon atoms from a graphite target into an atmosphere of nitrogen molecules. This atmosphere was then disrupted into atomic nitrogen by artificial lightning. The carbon and nitrogen would then combine, forming a solid film on the surface of the chamber. The resulting film was identified by Lieber and Niu as beta-C_3N_4 at the time, although the amounts of carbon nitride in the film were too small to be certain. Using similar methods, other chemists also produced something that might be beta-carbon nitride, but there was no convincing demonstration of a material harder than diamond.

The problem is that carbon nitride molecules, like the carbon atoms of diamond, find more comfortable ways to arrange themselves than in the tight matrix of the beta form. It is as if you threw a child's blocks into a pile, expecting the blocks to stack themselves into a cube. Luck might produce tiny regions where several blocks stack, but the majority of the blocks would settle into a stable-but-loose configuration—a heap rather than a stack.

In 1995, Yip-Wah Chung and coworkers at Northwestern University in Evanston, Illinois, tried a totally different approach for getting the carbon nitride molecules to arrange into the beta form. They used steel as a substrate on which they sputtered (electronically pushed molecules onto) alternating layers of titanium nitride and individual molecules of carbon nitride.

The composite material that results from such a technique is a coated steel that is extremely hard, almost as hard as diamond, and much harder than either a heap of carbon nitride molecules or uncoated steel. Chung and his group think that they can detect layers of beta-carbon nitride in the coating which account for the hardness, but, as usual, the amounts are too small for certainty. Nevertheless, the hard coating is a significant discovery in itself, since the sputtering method for producing such a coating is already in industrial use, and there is always a demand for harder coatings. Furthermore, the same approach might possibly be used with other molecules, such as zirconium nitride, to produce thicker layers of beta-carbon nitride (if that is indeed what is being assembled). Work on the method continues.

Meanwhile, Nature's diamond is still the victor.

Harder and Bouncier

Hardness is not by itself the most desirable trait for a substance, since many hard materials (such as ceramics and glass) are brittle. They are hard, but easy to shatter. A little "give" would improve this kind of hardness.

On the other hand, the materials with the most "give" are the flexible polymers, notably rubber and various synthetic rubbers. Their flexibility precludes shattering, but they are not hard enough for many applications.

The really hard materials are different from the bouncy ones in several ways. Most of the hard materials are inorganic, while the stretchy ones are organic. Hard materials are either crystalline or amorphous, but always feature three-dimensional structures that link atoms in all directions. Materials with "give" on the other hand are typically linear polymers: long chains of molecules, loosely bound, with interconnections between the chains to form a sort of two-dimensional structure.

Late in 1994, at a meeting of the Materials Research Society, Ken Sharp of DuPont reported that he and DuPont had patented a hybrid they call a star-gel that is hard like glass but flexible like rubber. The method involved opening up the connections in a three-dimensional array to make them looser, like those in a polymer. Some of this opening up is accomplished by inserting shock-absorber structures between bonds. These shock absorbers are short hydrocarbon chains connected to each other with silicon atoms. Another way to open up the star-gel is to have each molecule link up with only three of those surrounding it, rather than four as is common in glass.

The star-gel can take five times the strain that would cause glass to break. Instead of breaking, the star-gel bends. When the strain is removed, the star-gel bounces back to its original shape.

(Periodical References and Additional Reading: *New York Times,* 7-20-93, p C10; *Science* 7-20-93, p 334; *Science* 12-16-94, p 1808; *Science* 2-24-95, p 1089)

CARBON CONFIGURATIONS

DIAMOND DISCOVERIES

It is not clear whether or not chemistry is everything, but the chemistry of hydrogen, nitrogen, oxygen, and carbon is virtually everything there is to life. The remarkable chemical consequences which allow the simple compound of one oxygen atom with two hydrogen atoms to form water, dissolving almost everything, attracting and repelling specific ends of molecules, floating on itself when frozen, and so forth are well known. Yet it is carbon that allows the other three elements to combine into the vast array of organic compounds that form materials, particularly the materials of life. It has been calculated that 94% of all known compounds contain carbon.

Even carbon alone is a remarkable substance. It can be relatively structureless (as amorphous carbon), structured in sheets as graphite, wrapped into spheres as fullerenes (*see* "Ornamented and Filled Buckyballs," p 457), or bonded tightly with itself into diamond.

The carbon bonds in a diamond are so strong that a crystal of diamond behaves almost as if it were a macroscopic molecule. As a result, diamond has the useful and ornamental properties that make it so valuable: extreme hardness and incompressibility, a high index of refraction, and extreme transparency.

Easy Diamonds?

Nature's way of making a diamond is thought to be via great heat and pressure in volcanic vents. Humans succeeded in duplicating that process, also using heat and pressure, more than 40 years ago. The diamonds were not as good as those from nature, but they were diamonds, and could be used where hardness was needed in industrial processes.

Soon after these initial techniques were established, an easier method was found to produce some bits of diamond film from methane gas (made up of carbon and hydrogen). This method involved the gentle removal of hydrogen, so that the carbon had time to arrange itself comfortably into a diamond configuration, although initially the carbon atoms were as likely to settle into graphite as into diamond. By changing the surface on which the carbon settled in order to more effectively nudge the atoms into a diamond's tetrahedral crystal, and by improving the process in other ways, better and better diamond films have been produced. Eventually, such diamond films have also moved into commercial use, mainly at first to reduce wear.

Both artificial diamonds and diamond films have been expensive to produce, but not as expensive as natural diamond. Thus, the 1993 news that scientists at Pennsylvania State University had produced a diamond by baking a commercially available plastic at temperatures any kitchen oven can reach resulted in astonishment and disbelief among diamond scientists. Patricia A. Bianconi, leader of the Penn State group, began by calculating that it would be possible to produce a diamond by means she had already used to make tetrahedrally-linked crystals of silicon and germanium compounds. The idea involved the use of a similar compound, with carbon replacing the silicon or germanium. Upon removal of those atoms that were not carbon, the carbon atoms could settle peacefully into diamond.

It was not quite that easy. Although a suitable tetrahedral polymer of carbon existed in the form of trichlorotoluene, heating it produced amorphous carbon, not diamond. Bianconi and her coworkers made it easier for the noncarbon atoms to boil off by making an alloy of sodium, potassium, and trichlorotoluene. When this alloy is heated to moderate temperatures, the sodium and potassium boil off, taking the other noncarbon atoms with them. The residue is a transparent substance that the Penn State scientists believe to be diamond.

One of the early hints that the material actually is or contains diamond came when a graduate student tried to grind up a sample for use in x-ray diffraction studies. The material was so hard that it wore out the mortar and pestle, instead of grinding up nicely. Eventually, however, enough powder was prepared, and the diffraction patterns showed diamond as well as less desirable forms of carbon.

The first production of the material occurred 18 months before announcement of the process, because the Penn State group wanted to verify the presence of diamond, given the unorthodox nature of their method. The time was also used to apply for a patent on their product.

TIMETABLE OF ARTIFICIAL DIAMOND

The ability to grow diamond in various forms has existed for a long time. In 1772, Antoine Lavoisier [French: 1743–1794] had demonstrated that diamond was nothing but an expensive form of carbon. The virtues of diamond were such that the first attempts by modern chemists to make artificial diamonds started in the nineteenth century, although they were unable to produce the heat and pressure needed. Recognition of actual diamond has proved difficult, and many of the announced successes in the field evaporated under later scrutiny.

1893	Henri Moissan [French: 1852–1907] announces that he has used heat and pressure to produce artificial diamond. Later studies show, however, that he did not have sufficient heat or pressure to accomplish this, and that the diamonds he exhibited had probably been placed in his equipment as a hoax by unknown perpetrators.
1905	Percy Bridgman [American: 1882–1961] develops the first devices which can produce pressures higher than 100 atmospheres.
1952	William G. Eversole [American: 1898–] of Union Carbide grows diamond seed crystals from carbon monoxide and hydrocarbon gases at moderate pressures.
1953	A Swedish group synthesizes diamond using high pressure, but keeps their process and results secret.
1955	Bridgman and coworkers at General Electric use his high-pressure devices to produce artificial diamonds, which are immediately offered for sale. Many years later, members of the original team discover that the first diamond they announced was actually a natural one that had been misidentified. This "success" leads to true artificial diamonds on the next try.
1958	A method for producing diamond from methane is patented in the United

States, but the diamond is mixed with graphite, making it of limited use for commercial purposes.

1965	John C. Angus demonstrates that Eversole's 1952 method of growing diamond works, by growing a blue semiconducting diamond.
1976	Researchers in the Soviet Union find a way to make diamond from methane without also producing undesirable graphite impurities.
1981	The Soviet Union diamond researchers can produce single-crystal diamond films on previously existing diamonds and can deposit multiple-crystal diamonds on metal. The single-crystal diamonds prove much more desirable for many purposes.
1983	Japanese scientists at the National Institute for Research in Inorganic Materials in Tsukuba-shi demonstrate methods they have developed for growing diamond films rapidly at very low pressures.
1990	In July, workers in the United States report that they can grow pure carbon-12 diamond films that conduct heat 50% better than natural diamond, which contains 1% carbon-13. Pure carbon-12 diamond can also withstand laser radiation much better than natural diamond.
1991	On April 19, 1991, Jagdish Narayan and Vijay Godbole of North Carolina State University and Carl White of the U.S. Oak Ridge National Laboratory announce in *Science* that they have succeeded in growing thin single-crystal diamond films on metal surfaces.
1993	Workers at General Electric use modern spectroscopic techniques to analyze their first "artificial diamond" from the early 1950s, and find that it is a natural diamond (used to seed crystallization) that contaminated the experiment.

New Work for Diamond

Because of its hardness, diamond has long had a place in industry. Its incompressibility has made it the key part of the diamond anvil, the high-pressure generator of choice. The prospect of diamond films and cheaper artificial diamond has suggested to many engineers and designers that diamond may become an integral part of various familiar devices. These all use other properties of diamond that are unique. Surprisingly, although diamond itself has long been known as a material, some of these properties were little known or misunderstood until recently.

The properties of interest arise directly from the structure of a diamond. Because the bonds between carbon atoms in a diamond tie up almost all available electrons at room temperature, diamond is exceptionally insulating for electric current. Yet heat and other forms of electromagnetic radiation pass through diamond easily for the same reason. Because the electrons are all closely bound, they hardly interact with passing photons. Diamond at room temperature is the best known conductor of heat, and is one of the most transparent to light. The same tight bonding makes diamond hard, stiff, and slick. Diamond is even the material in which the speed of sound is the fastest.

Although diamond is an electrical insulator, breaking into its structure with a foreign atom can provide an extra electron or gap where there is no electron, depending on what specific atom is inserted into the network. Adding an extra electron with its negative charge produces an *n*-type (for "negative") region. Inserting a "hole" where an electron would normally be creates a p-type (for "positive") region. Thus, diamond can be made into a transistor, one possessing also all the useful properties described above. Carbon, the stuff of diamond, is the element most like silicon, which is used in most semiconductor devices. However, the shorter bonds in diamond would mean that (if technical problems could be resolved) diamond semi-conductor devices could pack more electronics into a smaller space. Alas, the process of inserting the appropriate doping atom into diamond on an industrial scale is a long way from development at this time. Boron, which can provide electron-less "holes" when substituted for carbon, makes a suitable p-type doping agent, with the added advantage that a diamond doped with boron is a better semiconductor at high temperatures (between 212–932°F, or 100–500°C) than it is at room temperature, the opposite of commonly used silicon semiconductors. Unfortunately, there is no similar suitable element that can be used in producing the n-type region, or electron donor, needed to complete manufacture of a transistor or of any other conventional elec-tronic device. Nitrogen, which would seem to be the best electron donor, holds on to the free electrons too tightly. Phosphorous might work, but fails to dissolve in diamond. There has been limited success based on using a "gun" to shoot ions of phosphorus or other electron donors into diamond, but this approach is far from commercial realization.

It is possible, but not very practical, to make devices that just use p-type semiconducting regions, thus bypassing the problem of electron donors. Diamonds combined with silicon can be used to solve some chip problems, but practical applications are still limited to using diamond substrates to provide heat conduction, not as semiconductors.

Other applications, however, are now being realized, or appear to be practical and even common for the immediate future. So far, most of these use the process of carbon-vapor deposition, a method for making diamond developed in the 1970s and 1980s that uses high temperatures and a carbon-containing gas such as methane. Carbon-vapor deposition produces diamond films, but the temperature of production can interfere with possible applications. Because the films do not adhere well to many surfaces, the result can be far from smooth. Despite these problems, tests of machine tools that use a diamond coating on cutting edges have been impressive. The least resolved difficulty involves the tendency of the film to chip away, especially in tools containing cobalt, which does not bond well with diamond.

A process developed at Lawrence Livermore Laboratory in California called "ca-thodic arc deposition" produces a somewhat different kind of film called amorphous diamond. This film does not require high temperatures for manufacture, adheres well to tungsten carbide and silicon substrates, and is extremely smooth. However, amorphous diamond is not truly diamond, since it has no crystal structure whatso-ever. As a result, some desirable properties generally derived from the crystal struc-ture (including transmission of heat and light) are not present. Amorphous diamond films can be used for coating cutting tools, both in industrial and medical applications such as scalpels; as a lubricant, even in outer space where many lubricants fail; and to reduce wear in all sorts of situations.

Sony produced one of the first commercial applications of diamond films for home use. Sony manufactures audio speakers in which the diamond film serves to stiffen the speaker cone, allowing better reproduction at high frequencies.

Nalin Kumar and coworkers at Microelectons and Computer Technology Corporation of Austin, Texas, discovered that a thin film of synthetic diamond emits electrons when exposed to a weak electric field, an effect quite different from the effective insulation of true diamond. This occurs because diamond film is mostly carbon, mixed with both hydrogen picked up in manufacture and diamond where appropriate bonds have formed. One possible application for the electrical properties of diamond film would be in a new form of display device, different from the cathode-ray tubes and liquid-crystal devices that are now commonly used. A diamond-film display would be a "flat-panel" display, the most desirable type. Other devices that use treated diamond as electron emitters might include discharge lamps, cathodes, or electron multipliers, all important parts of modern electronic devices.

John D. Hunn and coworkers at Oak Ridge National Laboratories have used diamond to make parts for nanomotors (see "Nanotechnology UPDATE," p 732), microscopic motors made by etching silicon with techniques similar to those used in fabricating semiconductor devices. A diamond gear is vastly harder and stronger than a silicon gear. The same technology can be employed to develop other devices, such as pressure sensors that can be used in contact with liquids that would be corrosive to silicon and other materials.

(Periodical References and Additional Reading: *New York Times* 9-23-92, p D19; *Scientific American* 10-92, p 84; *New York Times* 6-4-93, p A19; *Nature* 9-2-93, p 19; *New York Times* 9-28-93, p C1; *Nature* 8-25-94, p 601; *Scientific American* 4-95, p 35)

ORNAMENTED AND FILLED BUCKYBALLS

Buckyballs are molecules that look like soccer balls, are ubiquitous and superconducting, and may have various other tricks. They were unknown prior to artificial production in 1985. In 1992, they were recognized by Semeon Tsipursky of Arizona State University as occurring naturally in the mineral shungite from northern Russia, and they were also found later in minerals formed when lightning strikes soil. Certainly, no other molecule but DNA has been as newsworthy as buckminsterfullerene and its close relatives, the fullerenes. According to a survey in the July/August 1994 *Science Watch*, papers concerning this group of carbon molecules and structures were cited much more often than others. For example, in 1993, 16 papers attracted 10 or more citations, of which 10 were concerned with fullerenes and carbon tubules. It is unlikely, however, that the fullerenes will come close to DNA in true significance. Indeed, there has been far more fuss than fulfillment as yet.

Early in 1991, Sumio Iijima of the NEC Corporation discovered that carbon can form long open cylinders, called nanotubes, as well as spheres. These were first produced in large quantities a year later, in July 1992, when Thomas Ebbesen and P. M. Ajayan at NEC used graphite rods in helium gas as the anode and cathode through which a current was passed. Positively charged carbon ions left the positively charged anode and deposited themselves in nanotubes on the cathode. The tubes, of mixed lengths, formed a sort of powder heap that increased by a few tenths of a millimeter (hundredths of an inch) each minute.

In 1995, Charles M. Lieber of Harvard University in Cambridge, Massachusetts, and coworkers added to the mix what they call nanorods. Although the nanorods are not pure carbon (as are buckyballs and nanotubes), the nanorods are structurally quite strong and could be used in constructing tiny devices, or as reinforcements in other materials.

The practical benefits of buckyballs, if any, may still be a long way off. There is some reason to believe, as is discussed below, that nanotubes may have practical applications even before buckyballs do. Although it is easy to produce enough of the materials for study, it is not yet possible to make fullerenes and related compounds on the scale needed for mass markets.

Filling the Ball

Success in putting foreign atoms into the interiors of buckyballs was first achieved with helium and neon, two gases that do not enter into chemical compounds in normal ways. It is debatable whether a filled fullerene should be called one molecule or two. Robert J. Poreda of the University of Rochester, who first observed filled fullerenes, and coworkers proposed using the @ symbol, well known today from Internet addresses, to describe the combination. A molecule of buckminsterfullerene filled with a helium atom could be designated as $He@C_{60}$. Creation of the "compound compound" occurred when a bond in a heated fullerene molecule slipped, opening the ball wide enough to let the helium or neon atom in. Cooling then caused the bond to reassert itself, and the noble-gas molecule was securely trapped.

Soon chemists trapped a lanthanum atom inside a buckyball, then a yttrium atom, and then other metals. This process has become an especially rich form of chemistry in a short time. One of the outstanding unresolved issues, however, involves how to separate the filled molecules from the empty ones. The methods used for enclosing the atoms have been somewhat random in nature, and more empty than filled balls exist within any sample. Surprisingly, soot heated to high temperatures consists largely of filled carbon balls, but these are onion-layered fullerenes: a 60-atom buckyball is encased in a 240-atom fullerene, surrounded by a 540-atom version, and so on up to more than 60 some layers. Soot that has not been given enough energy to reformulate itself into these structures is amorphous.

In 1995, researchers at the University of California at Santa Barbara developed a systematic way for filling a buckyball. They first ornament each target ball with a nitrogen-methoxyethoxymethyl (N-MEM) group, which weakens the bond nearest the point of attachment. Putting the sample of ornamented balls in a container of monatomic oxygen for about three hours causes the weakened bond to open completely, leaving a hole in the ball. While the hole is open, the ball can be filled with an atom or even a small molecule. The space is not large enough for a large molecule. Putting the filled balls in an acid bath removes almost all of the N-MEM group and the adjacent carbon atoms, leaving a nitrogen atom behind. The nitrogen fits into the space left by the carbon atom and closes up the ball, which is now $C_{59}N$ instead of C_{60}.

The interior of a buckyball tends to carry a negative charge, so it may be easier to insert positive ions rather than neutral atoms or small molecules. If all the balls in a sample have holes, and are placed in contact with a solution of positive ions, the ions would naturally sidle into the hole and linger there. Closing the hole would keep them in place thereafter.

Although buckyballs are a dramatic form of cage, other "container molecules" have been developed, many by Donald J. Cram of the University of California at Los Angles. The process of chelation, known to many as an approach of alternative medicine, uses one molecule to trap an atom, although the trapped atom is simply grabbed, instead of caged. The non-buckyball container molecules have already proved their worth in biochemistry and other applications. It is widely believed that container buckyballs will have various applications. In addition to possible high-temperature superconductivity, more traditional applications might include using the buckyball to hide a radioactive atom. This would allow the radioactive atom to be traced while preventing it from reacting chemically with other atoms. Furthermore, a filled buckyball might be used as part of, or a step toward, other more complex structures that could not be made any other way.

THE NAMES OF THE GAME

The chemical literature often makes important distinctions using names that to the uninitiated seem so similar that they must mean about the same (nitrate and nitrite come to mind) despite important differences in chemical properties. The names of the various entities connected to "buckychem" also make significant distinctions.

buckminsterfullerene   The most stable fullerene, with sixty carbon atoms, or C_{60}.

buckyballs   The soccer ball of buckminsterfullerene and the rugby balls of C_{70}, as well as larger fullerenes. Fullerenes smaller than C_{60} are sometimes called baby buckyballs.

fullerenes   Any carbon structure with exactly 12 pentagons and enough hexagons to form a closed geodesic dome with carbon atoms at the vertices.

fullerites   Solid forms of carbon in which the atoms are arranged to form fullerenes.

heterofullerenes   Buckyball-like structures that have one or more carbon atoms replaced by another atom, such as nitrogen.

metallofullerites   Solid materials built from fullerenes, each enclosing an atom of some metal, such as potassium. These have not yet been made.

nanorods   Solid rods composed of carbon and another element, such as silicon or tin, that are 2-30 nanometers (0.00000008-0.0000012 in) in diameter and a micrometer (0.00004 in) long.

nanotubes   Structures of pure carbon formed in long hollow tubes, about a nanometer (a billionth of a meter, or 0.00000004 in) in diameter, but perhaps 10 to 100 nanometers (0.0000004 to 0.000004 in) long.

Background

Carbon is unusual among elements in the way it can bond with itself. Not only can carbon form as many as four bonds with atoms of other elements, it can form single, double, or triple bonds with itself. It has long been known that carbon can form long chains in which each carbon atom is bonded to another carbon atom, sometimes for as many as 70 carbon atoms in a row. Another characteristic of carbon's versatility is that two carbon-based chemicals (known as **organic** chemicals), each containing the same number of atoms of the same types, can have very different properties. The geometric arrangement of atoms in a molecule often determines what it does, and in three dimensions a carbon-based molecule can be left-handed or right-handed, or differ in other ways from another molecule constructed of exactly the same atoms. For example, both glucose and fructose, two different sugars, share the formula $C_6H_{12}O_6$, but a key difference exists in the atoms' arrangements.

Even by themselves, carbon atoms can be arranged in varying geometric configurations possessing vastly different properties. The example of the black, greasy graphite used for pencil leads and the transparent hard diamond is well known. In diamond, carbon atoms bond very closely to each other in the pyramidal shape that mathematicians call a tetrahedron, so that each carbon atom is bonded to exactly four other carbon atoms in a rigid structure. In graphite, each carbon atom is bonded to three other atoms in a plane, making for a structure that consists of hexagons existing in layers that are loosely bound to each other. Before 1985, it was widely believed that this was the whole story, although earlier studies showing that soot, charcoal, and other forms of carbon are basically the same as graphite were not very convincing. Nevertheless, textbooks comfortably stated, "All the apparently amorphous forms of carbon are found to possess the graphite structure."

However, in 1985, Richard E. Smalley of Rice University in Houston, Texas, and Harold W. Kroto of the University of Sussex in the United Kingdom discovered that some of the "apparently amorphous" forms of carbon possess, in fact, a radically different structure. At first, there was considerable controversy over whether or not Smalley and Kroto had found anything new. Their claim was that, under various conditions (including ordinary combustion), carbon atoms tended to form in clusters of 60 atoms, represented in chemist's shorthand by the formula C_{60}. In other words, 60 is a magic number for carbon clusters. When Smalley and Kroto tried to determine what sort of structure might cause this, they recognized that a geodesic sphere, the most stable structure made from pentagons and hexagons, has sixty vertices. Apparently, some pentagons are capable of inserting themselves into the basic hexagonal form of graphite. This causes the flat layers found in graphite to warp. If exactly a dozen pentagons are present, it is possible for the warping to be sufficient to produced a closed surface, essentially a sphere (although technically a truncated icosahedron). The carbon structure and its cousins were named after the famous architect, inventor, and philosopher Buckminster Fuller [American: 1895–1983], who developed the half-spheres called geodesic domes.

Leonhard Euler established in the eighteenth century that all closed three-dimensional structures made entirely from pentagons and hexagons must have exactly 12 pentagons. Since buckminsterfullerene is the smallest such molecule, its structure is the most symmetric structure of this type possible in ordinary three-dimensional space.

In May 1990, Donald Huffman and Lowell Lamb of the University of Arizona and Wolfgang Krätschmer and Konstantin Fostiropoulos of the Max Planck Institute for Nuclear Physics in Heidelberg, Germany, found a way to make large amounts of nearly pure samples of the material (the C_{60} buckyballs are mixed with elongated C_{70} balls). Actually, they had found the basic step in this manufacturing process (vaporizing a graphite rod in helium at one seventh of an atmosphere pressure) as early as 1983. However, because that was before Smalley and Kroto had discovered buckminsterfullerene, neither team of chemists knew what it was they were producing. The two teams realized they had something interesting and announced their discovery in September 1990.

Chemists quickly learned the basic properties and possible applications of the newly discovered form of carbon. Joel M. Hawkins of the University of California, Berkeley, and coworkers were able to make x-ray diffraction studies of buckminsterfullerene that confirmed for the first time that the proposed geodesic shape was correctly identified. Other researchers demonstrated that the stable geodesic-dome structure makes buckyballs tough enough to survive collisions with targets at 20,000 mph (32,000 kph), impacts that would rip other molecules to pieces. C_{60}, however, reacts vigorously with various organic molecules.

Another property may even be more useful. Arthur F. Hebard and coworkers at AT&T Bell Laboratories announced at the April 1991 Atlanta, Georgia, meeting of the American Chemical Society that buckyballs doped with potassium ions are superconducting at temperatures below 18K ($-428°F$ or $-255°C$). Forms of buckminsterfullerene were soon developed that showed superconducting properties at higher temperatures, but not as high as the so-called high-temperature superconductors (see "Superconductivity UPDATE," p 662). As experimental results were announced or rumored, however, many came to believe that large fullerenes (perhaps with as many as 540 carbon atoms) offer the most hope for the long-sought room-temperature superconductor. Surprisingly, the temperature of the transition to superconductivity lowers when metallofullerites are put under high pressures. In most superconductors, it works the other way: high pressure increases the temperature at which superconductivity begins. Because buckyball superconductors conduct equally well in all three directions, they may turn out to be more useful than the high-temperature superconductors, which are strongly two-dimensional.

Related Progress

Several chemists have found ways to produce polymers based on buckyballs. One of the early successes was the work of Fred Wudl of the University of California at Santa Barbara. He began by ornamenting each buckyball with an extra carbon or extra nitrogen atom. The ornament could then be used to attach the buckyball to a conventional polymer. A polymer made by linking buckyballs to each other tends to cross-link, making the polymer difficult to process.

In 1992, scientists showed that buckyballs could be compressed into diamonds just as graphite can, but that the pressure needed to accomplish the task is only about two thirds the amount needed for graphite.

The electric arc-discharge process used to produce both fullerenes and carbon nanotubes can be adapted to production of a number of other substances. The electric-arc process is the main method for filling and ornamenting buckyballs or filled nanotubes. In 1994, French chemists from the Université Paris-Sud in Orsay and the Université Montpellier II Sciences at Techniques du Languedoc in Montpellier reported that the method could be used to dope graphite or carbon nanotubes, much as a silicon semiconductor chip is doped. Because of their atomic structures, the elements boron and nitrogen are most suitable for this purpose (see "Diamond Discoveries," p 453). In theory, the electric arc-discharge would not work, because the temperature is so high that the boron and carbon would be expected to form compounds, expelling the nitrogen. However, perhaps because the process takes place very fast, the boron and nitrogen atoms are able to replace carbon atoms in the nanotube or graphite structures that are being produced.

Unfilled nanotubes may have an important application in display devices, although it is early for this idea. An international team lead by Walt de Heer and André Châtelain of Switzerland and Daniel Ugarte of Brazil reported in May 1995 that they had found a way to produce an electrically conducting plastic film that includes thousands of tiny nanotubes aligned perpendicular to the plane of the film. By November, they had learned that applying layers of mica and copper to the film allowed them to push electrons into the film and out through the nanotubes. The result is like a shower head for electrons, with all the electrons streaming forth from the flat surface at once and aimed in the same direction. If the researchers can now find a way to control which patch of nanotubes emits electrons at a particular time, then the electrons can be aimed at the desired spots on a phosphor screen, producing an image in exactly the same way that a cathode ray does. If successful, the nanotube display would have two advantages over the cathode ray version. First, it could be quite flat, since it does not depend on an evacuated tube through which the electrons must be shot. Second, the picture would not have to be refreshed one line at a time, as does the cathode ray. There is also reason to believe that once an assembly line was started, nanotube displays would be inexpensive to produce.

(Periodical References and Additional Reading: *New York Times* 7-10-92, p A15; *Nature* 10-22-92, pp 670 & 707; *Discover* 1-93, p 72; *New York Times* 3-5-93, p A18; *Science* 9-2-94, p 1357; *Science* 12-9-94, p 1683; *Scientific American* 10-95, p 34; *Science* 11-17-95, pp 1119 & 1179; *Science News* 12-9-95, p 397)

APPLICATIONS AND MATERIALS

Catalyst UPDATE

Catalysts make chemical reactions happen faster. Therefore, they are the backbone of both industrial chemistry and the chemistry of life. Indeed, one place that chemists are searching for new catalysts is among the biological catalysts, traditionally called enzymes. Since many industrial processes take place at some amount of heat or in other extreme conditions, an interdisciplinary group of some 160 chemists, biochemists, and biologists met in 1994 to discuss ways of obtaining enzymes from creatures that live under such conditions. Their aim was to find enzymes that can be used in high temperatures, high salinity, or in organic solvents. Several promising high-temperature enzymes were already known to cleave proteins or otherwise react with biological molecules. The hope, however, was to learn how the trick is done so that it can be converted to industrial processes.

New Tools

A different approach to catalysis is, in fact, based upon a failed idea from biology. In the 1950s, Linus Pauling had proposed that antibodies might mold themselves around antigens. Although this is not the way that antibodies form, it is a reasonable way to form catalysts. Some of the researchers who have been using the idea, with polymers that form in specific shapes to match molecules, are Günter Wulff of the University of Düsseldorf in Germany, Warren Ford of the University of Oklahoma, and Klaus Mosbach of the University of Lund in Sweden. In 1994, Mosbach and coworkers demonstrated that the method could be used to make a molecule that could pick out a specific protein from a group of similar molecules. One of the applications already in use for this method involves sorting left-handed molecules from right-handed ones for the pharmaceutical industry. Kenneth Shea of the University of California, Irvine, is using this method to develop molecules that will behave more like traditional catalysts, holding two molecules in positions where they can interact in a specific way. In 1994, Shea showed that the method could be used to develop a simple catalyst which removes a hydrogen fluoride group from another molecule.

Masahiro Murakami, Hideki Amil, and Yoshihiko Ito of Kyoto University in Japan invented a new method of catalysis that opens the tight bonds between carbon atoms so that some other atom or group of atoms can be tucked in between them. Previous methods for accomplishing this had been ineffective. The new method relies on inserting a rhodium complex in between carbon atoms that are next to a carbonyl group, a common arrangement in organic molecules. The process can be used either to remove the carbonyl group or to catalyze desirable reactions. The Kyoto chemists think that the insertion method may open up new avenues in organic synthesis.

Gerard von Koten of Utrecht University and coworkers in the Netherlands discovered that dendrimers—fractal molecules resembling complex snowflakes that were first recognized in 1985—can be used as catalysts. The active sites can be placed at the ends of the spidery rays that emanate from the center of the molecule. Thus, more of the active sites can be involved in the reaction at any one time. The process is similar to using a polymer to hold the active sites, but the location of the active sites is much more easily controlled in a dendrimer than by simply studding a polymer with sites in hopes that they will be pointing in the right direction to ensure catalysis. The main advantage of the dendrimer molecule is that the dendrimers can easily be filtered from the reaction products. The Dutch chemists announced their first experiments with dendrimers in the December 15, 1994 *Nature*.

Wood Alcohol Advance

Methanol has many industrial uses. At one time it was a common antifreeze, but few persons outside of chemical workers directly encounter methanol today. This situation could change in the future if a cheaper way of producing methanol becomes available, since methanol is suitable either as a fuel additive to gasoline (a role now occupied largely by ethanol produced from corn) or even as an actual replacement for gasoline. Burning methanol produces less of the pollution that participates in photochemical smog, one of the least desirable results of burning gasoline.

Methanol is still known as "wood alcohol" because it was once made from wood, but most methanol today is produced from simple gases such as methane. Production of methanol by any method other than distilling wood requires the use of catalysts. There are vast reserves of natural gas which is mostly methane, but the catalysts used in converting the methane in natural gas to methanol are not powerful enough to efficiently remove one of the hydrogen atoms from a methane molecule and replace it with molecules of hydrogen and oxygen.

ALCOHOLS WE KNOW AND LOVE

Although chemists class hundreds of different compounds as alcohols, most people are familiar with at best two or three of them. By far the best known is familiarly called grain alcohol, due to its production by fermentation of wheat, barley, and corn (although fruit and potatoes are also common sources). Today, the name **ethanol** is widely used for grain alcohol, especially when grain alcohol is to be added to or burned instead of gasoline. One presumes that ethanol manufacturers want to separate themselves from the distillers, brewers, and vintners who also make a product containing ethanol, referred to in that context simply as alcohol. Another name for grain alcohol is ethyl alcohol.

Distilling wood in the absence of air produces an alcohol that is similar to ethanol in many ways, but poisonous and a famous cause of blindness if ingested. This "wood alcohol" is seldom made from wood any more, and generally goes by its chemical name of methyl alcohol, or **methanol**. Methanol is mostly manufactured from carbon monoxide and hydrogen, or produced from natural gas. Methanol is the simplest of the alcohols, all of which consist entirely of carbon, hydrogen, and oxygen atoms arranged in particular configurations.

The third familiar alcohol is, of course, **isopropyl**, better known as rubbing alcohol from its application as a skin cooler and conditioner.

In 1993, however, scientists from Catalytica Inc. in Mountain View, California, developed a new catalyst based on ionized mercury, which operates at a lower temperature than earlier catalysts. If it can be scaled up to commercial size, it would greatly reduce the cost of production. Although sulfur dioxide and sulfuric acid appear during the process, they are consumed by the time the methanol and the method's only by-product, carbon dioxide, is produced. Increasing the amount of carbon dioxide in the air would contribute to the greenhouse effect, but otherwise carbon dioxide is not considered a form of air pollution.

(Periodical References and Additional Reading: *New York Times,* 1-17-93, p I-26; *Science* 3-4-94, p 1221; *Science* 7-22-94, p 471; *Nature* 8-18-94, p 540; *Science* 10-14-94, pp 215 & 259; *Nature* 12-15-94, p 617)

CHEMISTRY LISTS AND TABLES

THE NOBEL PRIZE FOR CHEMISTRY

Date	Name [Nationality]	Achievement
1995	Paul Crutzen [German] Mario Molina [American: 1943-] Frank Sherwood Rowland [American: 1927-]	Crutzen for discovery that nitrous oxide can reach the stratosphere and reduce the ozone layer; Rowland and Molina for demonstrating that chlorofluorocarbons rise into stratosphere and also destroy ozone.
1994	George A. Olah [Hungarian/American: 1927-]	Use of superacids and cryogenics to create and stabilize carbocations, important fragments of hydrocarbons that normally occur in creation of fuels and other chemicals from crude oil. These could not be studied without the techniques developed by Olah in the early 1960s.
1993	Kary B. Mullis [American: 1945-] Michael Smith [English/American: 1930-]	Mullis for development of the polymerase chain reaction, the key to making millions of copies of a specified fragment of DNA, which he developed in 1983; Smith for development of a technique called oligonucleotide-based site-directed mutagenesis, which is used to alter specific coding sites on DNA.
1992	Rudolph A. Marcus [Canadian/American: 1923-]	Mathematical analysis of the oxidation-reduction process, in which an electron is transferred from one molecule to another, conducted between 1956 and 1965.
1991	Richard R. Ernst [German: 1933-]	Improvements in nuclear magnetic resonance techniques.
1990	Elias James Corey [American: 1928-]	New ways to synthesize organic molecules.
1989	Sidney Altman [American: 1939-] Thomas Cech [American: 1947-]	Discovery of catalytic properties of RNA.
1988	Johann Deisenhofer [German: 1943-] Robert Huber [German: 1937-] Hartmut Michel [German: 1948-]	Determination of the structure of molecules involved in photosynthesis.
1987	Charles J. Pedersen [American: 1904-1989] Donald J. Cram [American: 1919-] Jean-Marie Lehn [French: 1939-]	Making artificial molecules perform the same functions as natural proteins.
1986	Dudley R. Herschbach [American: 1932-] Yuan T. Lee [American: 1936-] John C. Polanyi [Canadian: 1929-]	Herschbach and Lee for the crossed-beam molecular technique; Polanyi for chemiluminescence for studying chemical reactions.
1985	Herbert A. Hauptman [American: 1917-] Jerome Karle [American: 1918-]	Work on equations to determine the structure of molecules.

Date	Name [Nationality]	Achievement
1984	R. Bruce Merrifield [American: 1921-]	Method for creating peptides and proteins.
1983	Henry Taube [American: 1915-]	New discoveries in basic mechanism of chemical reactions.
1982	Aaron Klug [South African: 1926-]	Developments in electron microscopy and study of acid-protein complexes.
1981	Kenichi Fukui [Japanese: 1918-] Roald Hoffmann [American: 1937-]	Application of laws of quantum mechanics to chemical reactions.
1980	Paul Berg [American: 1926-] Walter Gilbert [American: 1932-] Frederick Sanger [British: 1918-]	Berg for development of recombinant DNA; Gilbert and Sanger for methods to map the structure of DNA.
1979	Herbert C. Brown [British/American: 1912-] Georg Wittig [German: 1897-1987]	Brown for study of boron-containing organic compounds; Wittig for phosphorus-containing compounds.
1978	Peter Mitchell [British: 1920-]	Study of biological energy transfer by mitochondria.
1977	Ilya Prigogine [Russian/Belgian: 1917-]	Nonequilibrium theories in thermodynamics.
1976	William N. Lipscomb Jr. [American: 1919-]	Study of bonding in boranes.
1975	John W. Cornforth [Australian: 1917-] Vladimir Prelog [Yugoslav/Swiss: 1906-]	Cornforth for work on structure of enzyme-substrate combinations; Prelog for study of symmetric compounds.
1974	Paul J. Flory [American: 1910-1985]	Study of long-chain molecules.
1973	Ernst Otto Fischer [German: 1918-] Geoffrey Wilkinson [British: 1921-]	Work on the structure of ferrocene.
1972	Christian B. Anfinsen [American: 1916-] Stanford Moore [American: 1913-1982] William H. Stein [American: 1911-1980]	Pioneering research in enzyme chemistry.
1971	Gerhard Herzberg [German/Canadian: 1904-]	Study of geometry of molecules in gases.
1970	Luis F. Leloir [Argentinian: 1906-1987]	Discovery of sugar nucleotides and their biosynthesis of carbohydrates.
1969	Derek H. R. Barton [British: 1918-] Odd Hassel [Norwegian: 1897-1981]	Determination of three-dimensional shape of organic compounds.
1968	Lars Onsager [American: 1903-1976]	Theoretical basis of diffusion of isotopes in a gas.
1967	Manfred Eigen [German: 1927-] Ronald G. W. Norrish [British: 1897-1978] George Porter [British: 1920-]	Study of high-speed chemical reactions.
1966	Robert Mulliken [American: 1896-1986]	Study of atomic bonds in molecules.
1965	Robert B. Woodward [American: 1917-1979]	Synthesis of organic compounds, including quinine, cholesterol, cortisone, reserpine, and chlorophyll.

Date	Name [Nationality]	Achievement
1964	Dorothy Crowfoot Hodgkin [British: 1910-1994]	Analysis of the structure of vitamin B_{12}.
1963	Giulio Natta [Italian: 1903-1979] Karl Ziegler [German: 1898-1973]	Synthesis of polymers for plastics.
1962	John C. Kendrew [British: 1917-] Max F. Perutz [British: 1914-]	Kendrew for location of the position of the atoms in myoglobin; Perutz for hemoglobin using x-ray diffraction.
1961	Melvin Calvin [American: 1911-]	Work on chemistry of photosynthesis.
1960	Willard F. Libby [American: 1908-1980]	Invention of radiocarbon dating.
1959	Jaroslav Heyrovsky [Czech: 1890-1967]	Polarography for electrochemical analysis.
1958	Frederick Sanger [British: 1918-]	Discovery of structure of insulin.
1957	Sir Alexander Todd [British: 1907-]	Study of chemistry of nucleic acids.
1956	Sir Cyril Hinshelwood [British: 1897-1967] Nikolai Semenov [Russian: 1896-1986]	Parallel work on kinetics of chemical chain reactions.
1955	Vincent Du Vigneaud [American: 1901-1978]	Synthesis of polypeptide hormone oxytocin.
1954	Linus C. Pauling [American: 1901-1994]	Explanation of chemical bonds.
1953	Hermann Staudinger [German: 1881-1965]	Study of polymers.
1952	Archer J. P. Martin [British: 1910-] Richard L. M. Synge [British: 1914-]	Separation of elements by paper chromatography.
1951	Edwin M. McMillan [American: 1907-] Glenn T. Seaborg [American: 1912-]	Discovery of plutonium and research on transuranium elements.
1950	Otto Diels [German: 1876-1954] Kurt Alder [German: 1902-1958]	Synthesis of organic compounds of the diene group.
1949	William Francis Giauque [American: 1895-1982]	Methods of obtaining temperatures very close to absolute zero.
1948	Arne Tiselius [Swedish: 1902-1971]	Research on blood serum proteins using electrophoresis.
1947	Sir Robert Robinson [British: 1886-1975]	Studies of plant alkaloids such as morphine and strychnine.
1946	James B. Sumner [American: 1887-1955] John H. Northrop [American: 1891-1987] Wendell Stanley [American: 1904-1971]	Sumner's first crystallization of an enzyme; Northrop's crystallization of enzymes; Stanley's crystallization of tobacco-mosaic virus.
1945	Artturi Virtanen [Finnish: 1895-1973]	Improvement of fodder preservation.
1944	Otto Hahn [German: 1879-1968]	Discovery of atomic fission.
1943	György Hevesy [Hungarian/Swedish: 1885-1966]	Use of isotopes as tracers.
1942	No award	
1941	No award	

Date	Name [Nationality]	Achievement
1940	No award	
1939	Adolf Butenandt [German: 1903-1995] Leopold Ruzicka [Croatian/Swiss: 1887-1976]	Butenandt for study of sexual hormones (declined award at Hitler's direction); Ruzicka for work with atomic structures and terpenes.
1938	Richard Kuhn [Austrian/German: 1900-1967]	Carotenoid and vitamin research (declined award at Hitler's direction).
1937	Walter N. Haworth [British: 1883-1950] Paul Karrer [Swiss: 1889-1971]	Haworth for work on carbohydrates and vitamin C; Karrer for work on carotenoids, flavins, and vitamins.
1936	Peter J. W. Debye [Dutch/American: 1884-1966]	Study of dipolar moments of ions in solution.
1935	Frédéric Joliot-Curie [French: 1900-1958] Irène Joliot-Curie [French: 1897-1956]	Synthesis of new radioactive elements.
1934	Harold C. Urey [American: 1893-1981]	Discovery of heavy hydrogen.
1933	No award	
1932	Irving Langmuir [American: 1881-1957]	Study of surface chemistry of monomolecular films.
1931	Karl Bosch [German: 1874-1940] Friedrich Bergius [German: 1884-1949]	Invention of high-pressure methods for producing ammonia from nitrogen in the atmosphere.
1930	Hans Fischer [German: 1881-1945]	Analysis of the structure of heme.
1929	Arthur Harden [British: 1865-1940] Hans von Euler-Chelpin [German/Swedish: 1873-1964]	Harden for discovery of the first known coenzyme (involved in fermentation by yeast); Euler-Chelpin for analysis of that coenzyme's structure.
1928	Adolf Windaus [German: 1876-1959]	Study of cholesterol
1927	Heinrich O. Wieland [German: 1877-1957]	Analysis of the structure of steroids.
1926	Theodor Svedberg [Swedish: 1884-1971]	Development of and work with the ultracentrifuge.
1925	Richard Zsigmondy [German: 1865-1929]	Study of colloid solutions.
1924	No award	
1923	Fritz Pregl [Austrian: 1859-1930]	Microanalysis of organic substances.
1922	Francis W. Aston [British: 1877-1945]	Development of the mass spectrograph and discovery of the whole-number rule of atomic weights.
1921	Frederick Soddy [British: 1877-1956]	Discovery of isotopes.
1920	Walther Nernst [German: 1864-1941]	Discovery of the third law of thermodynamics, the impossibility of obtaining absolute zero.
1919	No award	
1918	Fritz Haber [German: 1868-1934]	Synthesis of ammonia from atmospheric nitrogen.
1917	No award	

Date	Name [Nationality]	Achievement
1916	No award	
1915	Richard Willstätter [German: 1872–1942]	Research on chlorophyll in plants.
1914	Theodore Richards [American: 1868–1928]	Determination of atomic weights.
1913	Alfred Werner [German/Swiss: 1866–1919]	Discovery of coordination bonds (secondary valence).
1912	Victor Grignard [French: 1871–1935] Paul Sabatier [French: 1854–1941]	Grignard for discovery of reagents; Sabatier for hydrogenated compounds.
1911	Marie Curie [Polish/French: 1867–1934]	Discovery of radium and polonium.
1910	Otto Wallach [German: 1847–1931]	Work with terpenes.
1909	Wilhelm Ostwald [Russian/German: 1853–1932]	Work on catalysis, chemical equilibrium, and reaction rates.
1908	Ernest Rutherford [British: 1871–1937]	Study of alpha wave disintegration of elements.
1907	Eduard Buchner [German: 1860–1917]	Discovery of noncellular fermentation.
1906	Henri Moissan [French: 1852–1907]	Isolation of fluorine and introduction of electric furnace.
1905	Adolf von Baeyer [German: 1835–1917]	Work on organic dyes.
1904	Sir William Ramsay [British: 1852–1916]	Discovery of inert gas elements, and placement in periodic table.
1903	Svante Arrhenius [Swedish: 1859–1927]	Theory of electrolytic dissociation.
1902	Emil Fischer [German: 1852–1919]	Sugar and purine synthesis.
1901	Jacobus van't Hoff [Dutch: 1852–1911]	Laws of chemical dynamics and osmotic pressure.

Earth Science

THE STATE OF EARTH SCIENCE

Earth science is really a group of sciences with a single focus: our planet and how it works. As a result of that focus, which is different in many ways from that of other broad areas of science, earth science encompasses a bit of virtually all the major branches of science. For example:

- Since Earth is a planet, astronomy and space are both involved with earth science. Indeed, some earth scientists prefer to be called planetary scientists. Like astronomy, earth science requires an understanding of "deep time."
- Earth is made from minerals, some of which (from our surface-dwelling point of view) are under unusual stresses. Since mineralogy is allied to chemistry, many earth scientists can be considered geochemists.
- For now at least, Earth is our environment, and it is often difficult to determine the boundary between earth science and environmental science. For example, where exactly does the study of soil erosion fit?
- Paleontologists are classed as earth scientists, even though they specialize in studying the remains of living creatures, the subject of evolutionary biology.
- Increasingly, an earth scientist or two is required for any anthropological or archaeological expedition.
- Earth scientists who study earthquakes or meteorology use some of the most sophisticated systems of mathematical analysis in science.
- Geophysics is becoming the heart of earth science, although the term is used very broadly by earth scientists.
- The largest source of employment for earth scientists around the world is technology, in particular the petroleum industry.

Understanding Earth's Interior

Earth science is relatively young when compared with biology and physics. It was not until the eighteenth century that important results were achieved, and these were mostly connected with easily visible processes, such as weathering and erosion, deposition, crystallization, fossilization in sedimentary rock, and so on. Great progress was made in understanding these surface phenomena throughout the nineteenth century.

Understanding what goes on beneath the surface did not start until the twentieth century. Large structures, such as the mantle and core, were detected with earthquake waves. Studies of meteorites and materials ejected by volcanoes also helped scientists understand the chemistry of Earth's interior.

At the same time, a combination of factors (residual magnetism on the ocean floor was the main key) led earth scientists to the theory of plate tectonics (*see* "Advances in Plate Tectonics Studies," p 542). This theory took shape in the 1950s and 1960s. The mechanism that moves tectonic plates is an important part of the theory. Convection currents caused by differences in temperature are thought to drive the plates. If a material is all of the same composition, heating the bottom makes the warm lower region less dense than the cooler upper part. If the material is fluid, the cool part from the top flows downward under gravity, pushing the warmer stuff up. The result is a continuous movement of the material in a circular pattern called a convection current.

Convection currents require a heat source at the bottom, a fairly uniform material, and some fluidity. Plate tectonics theory requires that the interior of Earth contain a region with the appropriate characteristics. Originally, it was thought that the large region between Earth's crust and core, known as the mantle, meets the criteria. As more data were collected, however, problems began to develop. During the 1980s, earth scientists began to take sides among the rival theories that tried to reconcile the data with the concept of convection currents as the driving force for tectonic activity on the planet. As the 1990s began, opinions were sharply divided, but new tools and new studies seem likely to resolve the problem during this decade.

Among the most direct new tools being employed are superdeep holes drilled into continental crust. These greatly extend the range of samples available from drilling. Many oil wells are no more than 3,250 ft (1,000 m) deep. The deepest water well is 7,320 ft (2,231 m) deep, and oil wells of more than 10,000 ft (3,000 m) are rare. The superdeep German borehole project started drilling with a pilot hole of somewhat more than 12,800 ft (4,000 m) that was completed in April 1989. On September 8, 1990, the project began its main hole about 650 ft (200 m) east of the pilot near the Bavarian village of Windischeschenbach. Geologists call the drill holes the KTB holes after the jawbreaker name of the project: *Kontinentales Tiefbohrprogramm der Bundesrepublik Deutschland,* roughly translated as German Continental Drilling Project. The main KTB borehole had been planned to reach 33,000 ft (10,000 m) by 1994, but on October 12, 1994, the project stopped at approximately 30,000 ft (9,100 m)—short of its planned depth, although still strongly productive of new scientific insights.

Geologists have developed a concept of the rock layers underlying the center of Europe based on various indirect means of measurement. One of the goals of the KTB holes has been to test whether these concepts are accurate. The superdeep borehole provided the greatest percentage of its new data from below 25,000 ft (8,000 m), a region that is still largely unexplored. An expected boundary, thought to have formed when two tectonic plates met some 320 million years ago, failed to show up, for example. Other results from the boreholes have also been surprising already, notably the high temperatures encountered and the presence of cracks filled with very salty and gassy water, even near the greatest depth reached. There is also some indication that the gases in the water were caused by living organisms.

One of the goals of the KTB project was to reach a zone where rock is so hot and under such great pressure that it flows. Although such a zone had not been expected for another 0.6 mi (1 km), the KTB staff thinks that they were forced to stop drilling because they had already reached such a layer. In any case, they gave up drilling, because every time they pulled the drill up to change a worn bit, the hole already drilled closed up.

The record holder for superdeep holes remains the one drilled by Russia in the Kola Peninsula, which is nearly 50,000 ft (20,000 m) deep.

Better International Climate for Earth

Direct observation, however, is not needed to deduce much that is happening far below the deepest holes. In 1755, John Winthrop, observing an aftershock of the Boston earthquake of that year, noted that the shock produced an actual wave in the bricks as it passed through his fireplace. Earth scientists have been using such waves to deduce facts about the interior of Earth since 1906, when Richard D. Oldham speculated that patterns in earthquake waves reveal the Earth to have a partially liquid core. The amount of information that can be obtained from earthquake waves has increased dramatically in recent years. Mathematics has been one key, and the computer revolution the other.

The first sophisticated mathematical analysis of earthquake waves was conducted by Beno Gutenberg [German/American: 1889–1960], who refined the speculations of Oldham and others, correctly locating the boundary between the core and mantle, called the Gutenberg discontinuity. Yet such calculations could reveal only the general outlines of regions hundreds or thousands of kilometers beneath's crust.

In 1956, however, mathematician Ronald N. Bracewell developed the projection-slice theorem, the essential tool for recovering the shape of an object from various projections (shadows) of the object. Initially, the mathematics required was too daunting for the theorem to have many practical applications. However, cheaper and faster computers changed the picture. By 1972, the first application of the theorem to medicine, computerized axial tomography (familiar as a CT scan or CAT scan), was in use in the United Kingdom. Earth scientists put the idea into practice about 10 years later. Since earthquake waves are used, this method is called seismic tomography.

At first seismic tomographers did not have records from enough seismographs or enough earthquakes to obtain very good resolution of the layers in Earth's interior. During the same time period that new seismograph networks were set up and new techniques for producing waves other than earthquake waves were introduced, the ratio of computer power to cost was skyrocketing, enabling more and more geophysicists to use seismic tomography and refine it. Today, with ultrapowerful supercomputers, seismograph networks around the globe, modern communications, and Bracewell's indispensable theorem, geophysicists can not only analyze evidence for actual shapes within the Earth, they can also simulate how these shapes change over time according to various models. This means that for the first time geophysicists can test their ideas using a simulated Earth.

All this became easier with the end of the cold war. A network of seismological stations was partly set up in 1995 as part of the International Seismic Monitoring System required by a treaty, still not ratified, to ban all nuclear weapons tests. The stations are intended primarily to observe covert weapons tests and to report over continuously dedicated telephone lines to a central location. Now that cold-war secrecy is no longer needed, the information can also be siphoned off by scientific agencies, such as the U.S. National Earthquake Information Center in Golden, Colorado, or by anyone with access to the Internet. By the end of 1995, there were 37 stations spanning the world with 13 more stations planned, along with a hundred auxiliary sites.

Other parts of the treaty will aid other branches of Earth science as additional monitoring is set up and the results made available to scientists. Oceanographers can benefit from a network of hydroacoustic sensors listening for sounds in the oceans. Meteorologists will be able to benefit from two sets of recorders, one intended to observe low-frequency air waves that might be caused by nuclear blasts (volcanologists might also use this one), and another that measures radioactive particles that float in the air, which could be used for establishing regular air currents as well.

Fossil Fights

The trend of aboriginal peoples to keep their remains and holy relics out of museums (see "Return of the Heritage," p 113) has recently been matched by efforts to shut down some recovery operations for dinosaurs. Although there are no known dinosaur ancestors making the protests, the U.S. Federal Bureau of Investigation (FBI) has taken up the cause. The first major raid was on the Black Hills Institute of Geological Research in Hills City, South Dakota, on May 14, 1992. The FBI was acting for U.S. Attorney Kevin Schieffer and the U.S. Bureau of Indian Affairs, which claimed that a fossil *Tyrannosaurus rex* had been illegally removed from lands belonging to the Cheyenne River Sioux.

The case began in 1990, when a collector from the commercial Black Hills Institute spotted dinosaur bones on a cliff face on the ranch of Maurice Williams, a Cheyenne River Sioux. The Institute paid Williams $5,000 for what proved upon excavation to be an unusually complete fossil *T. rex*, which was soon christened Sue. After Sue was excavated, however, Williams denied that he had given permission (although not that he had been paid $5,000). Meanwhile, the Cheyenne River Sioux claimed that Williams did not own the land in any case—it belonged to the tribe.

Paleontologists from all over became involved in the issue. One group, the Society of Vertebrate Paleontologists, chose to make an issue of the commercialization of fossil collecting, pushing for a congressional bill to limit collection on federal lands and to make any fossils so collected permanently under government ownership. This would be the same rules by which archaeological treasures are maintained. Other paleontologists feel that such a rule might reduce the flow of fossils, since there is now a commercial value put on them. At present, only recognized paleontologists or their students can, with a permit, collect fossils on federal land. A bill prepared for Congress in 1995, the Fossil Preservation Act, has completely different aims. It would permit anyone (including amateurs and commercial collectors) to take fossils from the surface—provided the fossils are not "scientifically unique"—and to obtain permits to excavate buried fossils. Only scientifically unique fossils would have to go to museums instead of private collections. This is not what the Society of Vertebrate

Paleontologists has in mind at all; instead, it is being pushed by a group whose aim is greater commercial exploitation of federal lands.

At the end of 1995, the new bill had not been acted upon, Sue's bones remained in a government warehouse, and the controversy continued.

(Periodical References and Additional Reading: *New York Times* 5-24-92, p 117; *Science* 10-16-92, p 391; *Discover* 1-93, p 80; *New York Times* 2-5-93, p A12; *New York Times* 7-27-93, p C6; *New York Times* 11-25-93, p A17; *Science* 10-28-94, p 155; *Science* 12-8-95, p 1559; *Science News* 12-9-95, p 391)

TIMETABLE OF EARTH SCIENCE TO 1992

c. 15,000 B.C. A map of region around what is now Mezhrich, Russia, is made on bone.

c. 1000 B.C.   Duke of Chou [Chinese] builds a "south-pointing carriage," which may be an early magnetic device.

A map on a clay table shows Earth with Babylon at its center.

c. 575 B.C.   Anaximander of Miletus [Greek: c. 610–c. 546] in Ionia (Turkey) states that fossil fish are the remains of early life.

c. 530 B.C.   Pythagoras of Samos [Greek: c. 560–c. 480] argues on philosophical grounds that Earth is a sphere.

c. 500 B.C.   Hecataeus of Miletus [Greek: c. 550–c. 476] makes a map showing Europe and Asia as semicircular regions surrounded by ocean.

c. 350 B.C.   Aristotle [Greek: c. 384–22] lists observations that support the Pythagorean idea that Earth is a sphere.

The explorer Pytheas of Massalia (Marseilles, France) [Greek: c. 330–?] suggests that the Moon causes tides.

c. 300 B.C.   Dicaearchus of Messina (Sicily) [Hellenic: 355–285] develops a map of Earth on a sphere using lines of latitude.

The Chinese *Book of the Devil* contains the first clear reference to the magnetic properties of lodestone.

271 B.C.   In China, the first form of magnetic compass is used.

c. 240 B.C.   Eratosthenes of Cyrene (Shahat, Libya) [Hellenic: c. 276–c. 196] correctly calculates the diameter of Earth.

c. 79 B.C.   The explorer Hippalus [Hellenic] discovers the regularity of the monsoon.

79 A.D.   Pliny the Elder [Roman: 23–79] is killed by the eruption of Mt. Vesuvius that buries Pompeii and destroys Herculaneum. His nephew Pliny the Younger writes a detailed account of disaster.

132   Inventor Zhang Heng [Chinese: 78–139] develops the first seismograph.

c. 1110   Chinese sailors begin to use magnetic compasses for navigation.

c. 1180   *De naturis rerum* by Alexander Neckham [English: 1157–1217] contains the first known Western reference to a magnetic compass.

1600   William Gilbert [English: 1544–1603] suggests that Earth is a giant magnet.

1643   Evangelista Torricelli [Italian: 1608–1647] invents the barometer.

1667   Robert Hooke [English: 1635–1703] invents the anemometer for measuring wind speed.

1669	Nicolaus Steno [Danish: 1638-1686] correctly explains the origin of fossils as once-living organisms preserved in stone.
c. 1672	Isaac Newton [English: 1642-1727] shows that the rotation of Earth causes a bulge at the equator and flattening at the poles.
1701	Edmond Halley [English: 1656-1742] publishes a chart of variations in Earth's magnetic field.
1735	George Hadley [English] describes how temperature controls the major air movements around the globe.
1736	Pierre de Maupertuis [French: 1698-1759] proves by measurement Newton's theory that Earth is flattened at the poles.
1761	Chronometer Number Four, designed by John Harrison [English: 1693-1776], successfully demonstrates a method of finding longitude at sea.
1774	Abraham Gottlob Werner [German: 1750-1817] introduces standard ways to classify minerals.
1777	Nicolas Desmarest [French: 1725-1815] proposes that basalt is formed from lava.
1785	James Hutton [Scottish: 1726-1797] proposes that geologic features of Earth result from tiny changes taking place over very long periods of time.
1795	Georges Cuvier [French: 1769-1832] shows that giant bones found in the Meuse River are the remains of an extinct giant reptile.
1797	James Hall [Scottish: 1761-1832] shows that melted rocks form crystals on cooling.
1803	Luke Howard [English: 1744-1864] classifies cloud types.
1816	William Smith [English: 1769-1839] publishes the first geological map in which rock strata are identified by fossils.
1821	Ignatz Venetz [Swiss: 1788-1859] proposes that glaciers once covered Europe.
1822	According to Gideon Mantell [English: 1790-1852], he and his wife Mary Ann are the first to discover and recognize dinosaur bones this year. External evidence suggests that Mantell may have found the first bones as early as 1818.
	Friedrich Mohs [German: 1773-1839] introduces Mohs' scale of hardness for minerals.
1830	Charles Lyell [Scottish: 1797-1875] begins to publish *The Principles of Geology*, the work that convinces geologists that Earth is at least several hundred million years old.
1835	Gustave Coriolis [French: 1792-1843] shows that the rotation of a sphere causes the displacement of paths of objects moving on its surface.
1837	Louis Agassiz [Swiss/American: 1807-1873] introduces the term "ice age."
1839	Karl Friederich Gauss [German: 1777-1855] uses data gathered by Paul Erman [German: 1764-1851] and his son Georg Adolf [German: 1806-1877] to develop a mathematical theory of Earth's magnetic field.
1842	Charles Darwin [English: 1809-1882] classifies coral reefs and explains how atolls are formed.
1850	Matthew Maury [American: 1806-1873] charts the Atlantic Ocean.
1855	Luigi Palmieri [Italian: 1807-1896] builds the first Western seismometer.

1859	Edwin Drake [American: 1819-1880] drills the first oil well.
1866	Gabriel-August Daubrée [French: 1814-1896] suggests that Earth has a core of iron and nickel.
1880	John Milne [English: 1850-1913] invents the modern seismograph.
1896	Svante Arrhenius [Swedish: 1859-1927] discovers the greenhouse effect of carbon dioxide in the atmosphere: global temperatures rise for higher levels of carbon dioxide.
1899	Clarence Dutton [American: 1841-1912] describes vertical movements of Earth's crust in terms of isostasy, the equilibrium between the crust's weight and forces that cause parts of the crust to rise.
1902	Léon-Phillipe Teisserenc de Bort [French: 1855-1913] discovers the stratosphere.
1906	Richard D. Oldham [Irish: 1858-1936] uses earthquake waves to show that Earth's core exists.
	Bernard Brunhes shows that Earth's magnetic field can be recorded by cooling rock.
1907	Bertram B. Boltwood [American: 1870-1927] shows that the age of rocks containing uranium can be determined by measuring the ratio of uranium to lead.
1909	Andrija Mohorovičić [Croatian: 1857-1936] discovers the boundary between Earth's crust and the mantle, now known as the Mohorovičić discontinuity, or "Moho."
	Charles Doolittle Walcott [American] and his wife discover the Burgess Shale in British Columbia, a fossil bed from Cambrian times that contains rare fossils of soft parts of animals.
1912	Alfred Wegener [German: 1880-1930] proposes his theory of continental drift.
1913	Charles Fabry [French: 1867-1945] discovers the ozone layer in the atmosphere.
1920	Milutin Milankovich [Serbian: 1879-1958] begins work on a theory linking the ice ages to complex cycles of changing geometric relations between the Sun and Earth.
1925	The German *Meteor* ocean expedition discovers the mid-Atlantic ridge.
1929	Motonori Matuyama [Japanese: 1884-1958] shows that Earth's magnetic field reverses polarity every few hundred million years.
1935	Maurice Ewing [American: 1906-1974] begins study of the ocean floor using refraction of waves caused by explosions.
	Charles Richter [American: 1900-1985] develops a scale for measuring the energy of earthquakes.
1938	A British mining engineer named Callendar predicts that burning fossil fuels will cause global warming as a result of the greenhouse effect of carbon dioxide in the atmosphere. Two years later global *cooling* begins, and lasts for about a quarter of a century.
1943	A Mexican farmer discovers a volcano (later named Mt. Parícutan) growing in his cornfield.
1944	U.S. B-29 bombers attacking Tokyo on Nov. 24 find they are flying faster than

the theoretical top speed of their aircraft, resulting in the discovery of the jet stream.

1946   Vincent Schaefer [American: 1906–1993] discovers that dry ice can be used to cause clouds to release rain.

1950   Hannes Alfvén [Swedish: 1908–] shows that the solar wind interacting with Earth's magnetic field causes auroras.

1953   Maurice Ewing discovers a rift that runs down the middle of the mid-Atlantic ridge.

1957   The Soviet Union (Russia) establishes the Vostok ice station in the heart of the Antarctic ice cap.

1958   James Van Allen [American: 1914–] discovers belts of radiation that surround Earth in space, now known as the Van Allen belts.

1960   Harry Hess [American: 1906–1969] develops his theory of seafloor spreading.

1963   J. Tuzo Wilson proposes that the Hawaiian islands and other island chains are caused by stationary, long-lasting features beneath the tectonic plates that move above them.

1967   Independent of Callendar's 1938 concept, S. Manabe and R. T. Wetherald predict that rising carbon dioxide levels will cause global warming. When global temperatures begin to increase in the 1980s, their prediction is taken seriously.

1968   The U.S. National Science Foundation's Deep Sea Drilling Project, using the ship *Glomar Challenger*, begins. Between this year and November 1983, it takes 20,000 core samples from 624 sites.

1969   John Ostrom [American] proposes that some dinosaurs may have been warm-blooded.

1977   On February 19, John Corliss [American: 1936–] and two crew mates aboard the research submersible *Alvin* discover living organisms near undersea volcanic vents. The food pyramid for bacteria, worms, clams, and crabs is based on sulfur, not sunlight.

1978   John R. Horner [American] and Robert Makela [American] discover that some dinosaurs nested in colonies.

1980   Walter Alvarez [American: 1940–] and coworkers discover a layer of iridium at the K-T boundary identified with the demise of dinosaurs. Alvarez attributes both the iridium and extinction to the impact of a large comet or meteorite.

1981   Oil geologists working on the Yucatan peninsula in Mexico discover the giant crater that will later be identified as the site of the impact that deposited a worldwide iridium layer and probably caused the K-T mass extinction.

1984   Hou Xianguang [Chinese] discovers in Yunnan Province, China, the Chengjiang bed of fossils from early in the Cambrian period.

1985   The British Antarctic Expedition detects a hole that forms annually in the ozone layer over Antarctica.

1986   Using theories developed in wake of the very strong 1982-83 El Niño, the U.S. National Weather Service successfully predicts the 1986–1987 El Niño.

1987   The U.S. National Aeronautics and Space Administration (NASA) measures the movement of continents; the results are consistent with the theory of plate tectonics.

1988   Harmut Heinrich [Germany] discovers six (later expanded to seven) layers of stones dropped by massive flotillas of icebergs as they passed from the Labrador Sea to the coast of France. Later study suggests that the fresh water introduced into the North Atlantic by these periodic Heinrich Events has caused abrupt changes in worldwide climate.

John Delaney and Milton Smith detect light given off by the type of undersea geothermal vent known as a black smoker.

1989   Eirik J. Krogstad and Gilbert N. Hanson [American: 1936–] of the State University of New York, Stony Brook; S. Balakrishnan and V. Rajamani of Jawaharlal Nehru University, India; and D. K. Mukhopadhyay of Roorkee University, India, find evidence that tectonic plates of 2.5 billion years ago clashed in what is now the Kolar schist belt in India, the earliest indication that current geologic processes were under way that long ago.

A mapping project by scientists from the University of Washington in Seattle and Woods Hole (Massachusetts) Oceanographic Institution reveals a string of 16 spreading centers that make up the mid-Atlantic ridge. Each center is over a bull's-eye-shaped region of low gravity that may be caused by hot mantle rising toward the center.

William F. Ruddiman of the Lamont-Doherty Geological Laboratory and John E. Ketzbach of the University of Wisconsin-Madison use a computer model to demonstrate that tectonic uplift of the Tibetan plateau and the Rocky Mountains caused global cooling and weather patterns that could have set off ice ages.

Quarry workers uncover the most extensive set of dinosaur tracks in North America in Culpepper, Virginia. About a thousand footprints of a carnasaur, a coelurosaur, and an unknown quadruped dating from 210 million years ago are well preserved.

In December, Philip D. Gingerich [American] and B. Holly Smith [American] discover the first whale with feet, *Basilosaurus isis,* which lived about 40 million years ago.

1990   Kathleen Crane [American] and a Soviet-American team discover hot vents in the floor of Lake Baikal in Russia, suggesting that this is a "spreading center," or widening crack, in the crust of the Eurasian landmass.

Michael T. Clegg, Edward M. Golenberg, and coworkers clone DNA from a magnolia that lived 18 million years ago.

Soviet scientists make available a long core of Antarctic ice taken at the Vostok ice station, providing a climate key that extends back 160,000 years.

Alan R. Hilderbrand and William B. Boykton of the University of Arizona establish that an impact by a large meteorite at Chicxulub in Yucatán, Mexico, is the probable cause of the extinction of the dinosaurs.

1991   Volcanologists successfully predict major eruptions of Mount Unzen in Japan and Mount Pinatubo in the Phillippines, saving thousands of lives.

William M. Gray finds a strong positive correlation between the amount of summer rain falling in the western part of the Sahel region of Africa and the number and strength of hurricanes reaching the coast of North America.

THE EVOLUTION OF LIFE

INSTANT EVOLUTION

Most people agree with Charles Darwin's belief that vast reaches of time were needed for the gradual change of one species into another, but many scientists today have come to accept the evidence that in most cases evolution proceeds rapidly (*see* "Evolution UPDATE," p 412). For no time is such evidence richer than for the brief 10-million-year period between 530 million years ago (mya) and 520 mya.

The Cambrian revolution—a sudden burst of evolution near the beginning of the Cambrian Period, which is now agreed to have begun about 543.9 mya—was perhaps the most astonishing event in biology. Before the Cambrian "explosion," a few bacteria or related blue-green algae, and some small, odd, soft creatures existed. Within a few million years (a span measured at a site in Siberia in the early 1993 as only eight million years long, and beginning about 530 mya) all the basic forms of modern life suddenly came into being. Since that time only one new phylum, known as the Ectoprocta, has appeared, and it has not been especially successful. Insofar as one has heard of them at all, we know the ectoprocts as "moss animals," or by their former official name, the phylum Bryozoa. Furthermore, there are good reasons to think that ectoprocts actually were already in existence during the Cambrian revolution, although paleontologists have yet to find evidence of this.

In 1994, two more of the approximately three dozen main animal phyla were found to have taken shape during the Cambrian revolution. One of these phyla, the Pentastomida, or tongue worms, had long been thought a candidate for more recent evolution, especially since all tongue worms are parasites on vertebrates, and vertebrates had not been thought to have evolved until post-Cambrian times (*but see* "Chordates and Mammals," p 503). Also, the Pentastomida, along with the Tardigrada (water bears) and the Onychophora (velvet worms), were thought to have evolved from the Arthropoda, a phylum first sighted during the Cambrian revolution. A new interpretation of fossils from the famous Burgess Shale, laid down during the Cambrian revolution, had shown in 1991 that several of the fossils were definitely in the phylum Onychophora, however. This is evidence that the Onychophora existed alongside Arthropoda from early on.

Theories as to Why

The original explanation for the Cambrian revolution theorized that it was an artifact brought about by the invention of hard parts. Ultimately, however, fossils of earlier life forms were discovered, and the original explanation was found wanting. Here are a few of the recent theories that purport to explain why the explosion occurred.

Feces   Graham Logan from Australia has proposed that life in the sea was limited before Cambrian times because excreta from organisms fed large numbers of surface-dwelling bacteria, thereby depleting oxygen from what otherwise would have been even more productive surface waters. Cambrian animals developed guts, and with them produced fecal pellets that sank to the bottom. With less nutrient material at the surface, the bacteria died out, leaving more oxygen. The Cambrian revolution was then fueled by that oxygen. Evidence in favor of this theory comes from isotopic analysis of rock that formed at the bottom of pre-Cambrian and post-Cambrian oceans. **Volcanoes**   Geerat Vermeij from the University of California at Davis thinks that the Cambrian revolution and other bursts of evolutionary activity have been set off by the eruption of undersea volcanoes. Such eruptions could add carbon dioxide to the atmosphere without at the same time spewing sulfates or tiny particles into the stratosphere. Volcanic eruptions on land are often followed by slight global cooling as a result of the haze of sulfates and dust. Vermeij thinks that the undersea volcanoes would have produced global warming as a result of the greenhouse effect (*see* "The Future of Climate," p 525). The warmer temperatures then sparked the revolution. **Food**   Ronald Martin of the University of Delaware thinks that erosion of land or some other factor provided nutrients that moved up the food chain. This prompted some organisms to take advantage by reproducing in great numbers. Such increases in reproductive activity are usually followed by population collapses. When such a collapse occurs, it somehow triggers a reciprocal burst of evolutionary activity that serves to fill up the now available niches.

Background

As long ago as 600 B.C., the Ionian scholar Xenophanes reported that he had seen fossils of sea creatures high on mountaintops. Other scientists continued to report ancient shells and bones (fossils) with varying interpretations of their origins until well into the seventeenth century, when finally Athanasius Kircher suggested that some fossils might be remains of animals that had missed Noah's Ark and therefore become extinct. By the nineteenth century, it was generally recognized that most fossils were the bones, shells, and hard parts of extinct animals or plants. Although a few fossils with preserved soft parts, such as the feathers of *Archaeopteryx,* were found as early as 1861, it was generally believed that early fossils that revealed soft body parts were virtually nonexistent.

This belief was used to explain the "Cambrian revolution," a name given to the sudden appearance of life in the geologic record about 530 mya. The nineteenth-century geologists could find no fossil traces of life before this revolution, although more recent work has revealed traces of bacteria and single-celled algae. After this revolution, practically every currently surviving animal phylum is represented in the fossil record. It was thought that there must have been a long period of evolution of multicellular life prior to 530 million years ago, but scientists could not find traces of it because there were no hard parts to record this development.

In 1909, the Burgess Shale was found in British Columbia by Charles Doolittle Walcott. This fossil bed preserved the soft parts of animals as well as the hard ones, probably because the shale is itself the remains of mud slides that had buried complete undersea communities. Buried in thick mud, the organisms were protected from oxygen, and therefore from decay, throughout the long period of time it took the mud to turn to rock. Although the fossils of soft animals were immediately recognized as important clues, the pivotal place of these fossils in understanding early evolution was not fully grasped until the 1980s. About 1984, at a site in China called Chengjiang, Hou Xianguang discovered a fossil field that proved as rich in soft-bodied fossils as the Burgess Shale. Furthermore, Chengjiang was determined to be even older than the Burgess Shale.

BEDS OF RARE PRESERVATION

Sources of ancient or soft-bodied fossils	Organisms preserved	Geological time period
Ediacara Hills, Australia; Nama Groups outcrops, southern Namibia; Mistaken Point, Newfoundland	"Quilted" creatures that some scientists think have no separate cells, others think are early lichens, and still others consider the immediate ancestors of the animals.	Vendian period (610–543 mya)
Mt. Stephen, British Columbia, Canada; Burgess Shale, British Columbia, Canada; Chengjiang, Yunnan Province, China; Kangaroo Island, Australia; Lancaster, Pennsylvania; Peary Land, Greenland; House Range, Utah	Sponges, brachiopods, polychaete worms, priapulid worms, trilobites and other arthropods, mollusks, echinoderms, chordates, and onychphorans, along with many creatures of no known phylum, including *Anomalocarus*, *Opabinia*, and *Microdictyon*.	Cambrian period (543–510 mya)
Beecher's Bed, New York; Soom Shale, South Africa	Trilobite (limbs) in Beecher's Bed and giant conodont in the Soom Shale.	Ordovician period (510–425 mya)
Lesmahagow, Scotland; Waukesha, Wisconsin	Early marine uniramian that resembles a centipede; possible leech; *Ainiktozoon*, of no known phylum; faint conodont animal.	Silurian period (425–409 mya)
Gilboa, New York; Hunsrck, Germany	Trigonotarbids, mites, centipedes, spiders, and trilobite (limbs).	Devonian period (409–363 mya)
Granton Shrimp Bed, Edinburgh, Scotland	Conodont animals, primitive chordates.	Mississippian (363–323 mya) subperiod of Carboniferous
Mazon Creek, Illinois	Priapulid worms and *Tullimonstrum* (the Tully Monster) of no known phylum, along with hundreds of animals and plants.	Pennsylvanian subperiod of Carboniferous (323–290 mya)
Solnhofen, Bavaria, Germany; Christian Malford, England	*Archaeopteryx lithographica*, an early dinosaurlike bird; squid.	Jurassic period (208–146 mya)
Santana, Brazil	Fish with muscles, guts, and gut contents.	Cretaceous period (146–65 mya)
Green River, Wyoming	Freshwater fish, insects, and plants.	Paleocene epoch of Tertiary period (65–56.5 mya)

Sources of ancient or soft-bodied fossils	Organisms preserved	Geological time period
Messel River, Germany		Eocene epoch of Tertiary period (56.5–35.5 mya)

Before the Revolution

Some revolutions wipe away the immediate predecessors, but most incorporate them into the new structure in some way. Sometimes, that which is hailed as a revolution is simply the natural outcome of processes that were actually ongoing but not necessarily recognized. Scientists are not at all sure in which of these categories the Cambrian revolution belongs.

Quick evolution may be natural in Earth's early history. When J. William Schopf of the University of California at Los Angeles found fossils of single-celled organisms from 3.485 billion years ago, very near the beginning of life, he was able to classify them into 11 different major groups. This suggests a very fast rate for early evolution. Single-celled organisms continued to be the only life forms on Earth until shortly before the Cambrian revolution.

The immediate predecessors of the Cambrian animals were strange creatures themselves. Scientists still argue over whether they are animals or something else. Notable theories include those of Adolf Seilacher of Tübingen University in Germany, who thinks that a seventh kingdom of life needs to be set up to accommodate the creatures. Gregory Retallack of the University of Oregon thinks that the pre-Cambrian creatures were more like lichens, fungi and algae that live together on one basis or another (see "Fungi Surprises of the 1990s," p 388). Still others agree with Simon Conway Morris of the University of Cambridge in the United Kingdom, who thinks the creatures were animals, and therefore the ancestors of the Cambrian revolutionaries. One reason that many scientists believe that these creatures, variously known as Ediacaran fauna, Vendian-Period fauna, or vendobionts, might *not* be ancestors of later animals is that there seemed to have been a 20-million-year gap between the last Ediacaran life forms and the first Cambrian animals.

New evidence related to the controversy appeared in 1995. Scientists from M.I.T. in Cambridge, Massachusetts, led by John Grotzinger and Samuel Bowring, made an expedition to a rock formation in southern Namibia known to contain pre-Cambrian and early Cambrian fossils. They found many of both types of fossils, including one previously unknown species ("goblet-shaped fossils"). They also discovered crystals of zircon that could be used to pin down the dates of fossil-bearing strata.

Perhaps the major outcome of the expedition involved disproving the existence of the presumed 20-million-year gap, at least in Namibia. The last Ediacaran fossils virtually overlap with the first Cambrian examples. This redating greatly strengthened the case for those who preferred to think that the vendobionts were ancestors of animals, since there was no longer an unexplained time period to be bridged. On the other hand, those who believed that the Vendian life forms became extinct after the Cambrian animals ate them all were also comforted. Thus, while removing the gap did not resolve the controversy, it did make it somewhat easier to claim the vendobionts as ancestral.

Developmental biologists (*see* "Development Developments," p 342) Kevin Peterson of the University of California, Los Angeles, and Eric Davidson and R. Andrew Cameron of Caltech argue that the multicelled animals did evolve in pre-Ediacaran times but were microscopic, as are many marine larvae today. These creatures would have been too small to have left observable fossils. Today's marine larvae use a different developmental mechanism to reach the larval stage than they do for the transition from larva to adult. The California biologists hypothesize that very slow evolution of a new approach to development gradually prepared these microscopic life forms for the great leap forward of Cambrian times. That they find the clues for this in development of present day organisms seems rather like the idea "ontology recapitulates phylogeny." Such an approach, while sometimes providing helpful clues, has also led to conclusions later proved incorrect.

The Algae Revolution

Along with the sudden evolution of animal phyla in the first third of the Cambrian Period, there was a similar explosion in another life form. The diversity of the single-celled algae doubled during the same period. Furthermore, the species turnover increased by a factor of 10. Any explanation of the changes in Cambrian fauna needs also to explain this change in Cambrian flora.

The early history of the true algae is the history of humans as well, for these one-celled planktonic creatures are, as far as we know, the first organisms with nucleated cells. Except for the monerans and archaebacteria (*see* "The Kingdom of the Archaea," p 378, and "Bacteria: Giant, Deep, and Social," p 382), all present-day life forms have cells with nuclei and are classed as eukaryotes. Since the algae were the first eukaryotes, some 1.8 billion years ago, they are presumed to be the ancestors to all others.

Andrew Knoll of Harvard University in Cambridge, Massachusetts, has been making a census of algae species over time. He found that very little change occurred in the first 5 or so species of eukaryote algae during the first 800 million years or so. The same species continued on, with little evolution or extinction taking place. During that long period of time, the total number of species managed to double from about 5 to about 10. About a billion years ago the pace began to pick up, so that in only about 100 million years the number nearly tripled and then continued slowly to rise for a couple of hundred years. Not only did the total number of species existing at any one time rise, but the turnover in species also rose rapidly, with many becoming extinct and being replaced by others.

About 600 million years ago, following a major ice age, the number of species of planktonic algae crashed. The total number of species was cut in half. Surprisingly, this was followed by a major spike in which the number of algal species briefly doubled to what had been its previous high. This, however, did not last, and the number of species again dropped as the Cambrian revolution was about to begin.

When the explosion in animal phyla began, it was accompanied by another spike in the number of algal species, which reached nearly 80 during the early part of the

Cambrian revolution. This last explosion was probably precipitated by the complexity of the new environment resulting from all these new phyla and species. People tend to think that the number of new animal phyla caused the algae to respond with new species, although it is possible that the spike in algal species had something to do with the increased pace of animal evolution. Perhaps both were the result of some outside force, although many think that the complex interactions of new species would by themselves trigger other new species in a kind of chain reaction, similar to that which triggers a nuclear explosion.

The earlier changes in algal speciation may have another cause. Knoll thinks that the algae evolved sex for the first time, which would have increased the number of new species. This idea is partly confirmed by gene analysis, which suggests that the very first sexual revolution may have taken place more than 500 million years ago.

(Periodical References and Additional Reading: *New York Times* 4-30-93, p A10; *New York Times* 9-7-93, p C1; *Science* 9-3-93, pp 1274; *Natural History* 1-95, p 6; *Science* 1-6-95, p 33; *Science* 10-27-95, pp 580 & 598; *Science* 11-24-95, pp 1300 & 1319; *Time* 12-4-95, p 66; *Discover* 1-96, p 42)

Mass Extinction UPDATE

Events in which a great many species (or even families or orders) ceased to exist over a very short period of time have only recently been identified as an important area of study. Today, these mass extinction events are seen as one of the ways in which evolution progresses. Perhaps our present view of the significance of great die-offs comes in part from the still newish idea that the dinosaurs were destroyed by a large meteorite or comet that struck Earth some 65 million years ago. Until the past 10 or 15 years, nearly everyone believed that the causes of dinosaur extinction were rooted within the dinosaurs themselves, not the result of a *Deus ex machina* from outer space. When we believed that great size, or a small brain, or the rise of the mammals caused the dinosaur line to disappear, the actual cause of their death was more of an intellectual puzzle than a serious line of inquiry. When, as we now believe, it becomes apparent that our own ascendance as Lords of the Universe resulted from what an insurance company would term an Act of God, and not from our own marvelous worth, then the subject of mass extinctions become considerably more personal. What happened to the dinosaurs, after all, could happen to us.

With these thoughts in the background of science, the subject of the cause and extent of mass extinctions has developed into a topic on which a considerable number of scientists dwell.

TIMETABLE OF MASS EXTINCTIONS

700 mya	**Late Paleozoic:** One-celled algae called acitarchs become extinct. Seventy percent of the algae disappear.
505 mya	**Late Cambrian:** Various trilobite species are lost, with three separate episodes of extinction preceding this one.
435 mya	**Final Ordovician:** Trilobite species, cephalopod species, crinoid

species, and other marine species, totalling about 24% of marine families and 45-50% of marine genera, become extinct.

367 mya
Late Devonian: Many corals, brachiopods, placoderm fish, and trilobites, totalling about 22% of marine families and 47-57% of marine genera, are lost. Echinoderms nearly disappear from fossil record. Evidence in the form of tektites in Europe, perhaps caused by an impact in Siljan Ring in Sweden or Charlevoix Crater in Quebec, suggest that the extinctions resulted from such an impact.

250 mya
Final Permian: End of trilobites and many land species, as well as more than 50% of marine families; 76-80% of marine genera and about 70% of all vertebrate families become extinct. Some estimates are that 96% of all marine species were lost.

198 mya
Late Triassic: Many cephalopod species and shelled invertebrates disappear, as well as final extinction of conodont animals, along with losses of mammal-like reptiles and other land animals; 24% of marine families and 45-58% of marine genera become extinct. Some evidence (e.g., shattered quartz crystals in Italy) of an impact as the cause.

144 mya
Final Jurassic: Many marine invertebrates and dinosaur species disappear from the fossil record.

65 mya
Final Cretaceous: Extinction of all dinosaurs, ammonites, and many other animals. 16% of marine families and 38-46% of marine genera are lost. Almost certainly the result of an impact.

11,000 years ago
End of last glacial period: Large mammals, including giant beavers, mammoths, mastodons, giant bisons, ground sloths, and giant kangaroos, are eliminated. Some think that humans are the cause.

The Big Impact: End of the Cretaceous

The mass extinction responsible for the death of the dinosaurs, as well as many other creatures on land and sea, is sometimes called the K-T event. The name is derived from the fact that this extinction marked the boundary between the Cretaceous (K) and the Tertiary (T) periods. At 65 mya, it was also the boundary between the Mesozoic (middle life) Era and the Cenozoic (recent life) Era. In the Cenozoic, most of the organisms alive would easily be recognized as similar to those of today. The Mesozoic biological landscape, however, was entirely different from that of today.

Except for a very few paleontologists, most researchers believe that the Mesozoic came to an abrupt end when a large object smashed into the Yucatán Peninsula. The dissenters think that the Mesozoic was winding down anyway, and that whatever happened when the object hit was not decisive. Circumstantial evidence supporting the impact explanation continues to build, however. In 1992, a team led by Walter Alvarez of the University of California at Berkeley dated the Yucatán crater as having been formed 64.98 mya, for example. One study in 1993 found the diameter of the impact crater to have been 185 mi (300 km), considerably larger than original estimates, making it the largest known impact crater on Earth. A notable first occurred in February 1994, when a major conference on catastrophes in Earth's history revealed no one with a paper attacking the impact hypothesis. At that same conference, Robert Ginsberg of the University of Miami arranged a blind test of samples containing fossils of marine plankton called forams to determine whether extinctions were gradual or abrupt. Most geologists thought that the results of the test favored abrupt extinction, with as usual a hold-out or two. At the same conference, another blind test led four out of five sediment experts to conclude that sand layers in Mexico had been the result of giant tidal waves.

Still, given that everyone recognizes that a large object landing on the edge of North America would cause considerable damage, there is no single accepted theory as to how that damage could result in a mass extinction that not only surgically removed the dinosaurs, leaving behind the birds and the other beasts, but also ended the evolutionary life of nearly half the species found in the sea. Here are a few notions as to how the impact could cause such an extinction.

Darkening and Freezing   Dust from the impact would have reached the stratosphere, where it would circulate for many years. During this time, the surface of the Earth would be dark, breaking the food chain by killing all the green plants on land and in the sea. Furthermore, even plants that could go dormant for a time might be lost because of freezing. The lack of food and the cold temperatures would have caused the mass extinction. Small and furry or feathered creatures would be better able to live on slim pickings in a deep freeze than would giant, scaly ones.

Burning and Scalding   Debris from the impact would have risen high into the stratosphere or even into nearby space. Upon falling back to Earth, it would have been like the reentry of a billion space shuttles all at once. The red-hot debris would have heated the atmosphere enough to kill off most life on Earth.

Corrosion and Chemistry   The Yucatán Peninsula is mainly limestone. When struck by a fiery meteorite, the limestone and ocean water would have reacted to form sulfuric acid. The chemical would rise into the atmosphere where it would be transmuted into super-acid precipitation. This would rain down on Earth, killing not only many of the animals directly, but also destroying plant life. In another version, the impact in Yucatán would send shock waves through the Earth to trigger volcanoes on the other side, which would have rained down deadly gases and aerosols as well as volcanic ash.

Burning and Freezing   The debris would be hot enough to set forests on fire all over the globe. Not only would these fires be destructive themselves, but the fires would also inject vast amounts of soot high into the stratosphere. There, the soot would cut off light, resulting in the lower temperatures and subsequent death of green plants already discussed.

Drowning   Since the impact was partly in the ocean, it might have caused worldwide tsunamis (tidal waves), which would have contributed to the mass extinction, although not sufficient to cause it by themselves.

The less well-known K-T disaster in the oceans has long been considered to have been much more extreme than anything that happened on land. In 1995, one report showed that the ocean was apparently devoid of all life for about two million years, a condition known among paleo-oceanographers as "the Strangelove ocean," after the movie *Dr. Strangelove* that ended with a doomsday machine destroying all life on Earth. Steven D'Hondth of the University of Rhode Island determined that a core from the South Atlantic showed no signs of life—not even a carbon-13 to carbon-12 ratio that is considered the signature of life in the ocean—over a time period representing two million years after the K-T event.

But D'Hondth and coworkers at Rhode Island think that it is impossible for all ocean life to have been lost for that long a period. Instead, they argue, larger life forms may have become extinct, leaving only tiny ones. The remains of tiny organisms would dissolve before reaching the ocean floor, so there would be no evidence of their existence. Larger organisms would fall all the way to the ocean floor after dying, and even their fecal pellets (*see* "Instant Evolution," p 480) would be large enough to reach bottom. This scenario makes more sense in relation to certain other evidence than does the Strangelove ocean. Also, it might relate in some way to the apparent loss of large animals and survival of small ones on land.

The Big One: The Permian

The largest known mass extinction as measured by the number of species, genera, families, and orders that were affected came at the end of the Permian period, about 250 mya. About 90% of all genera in the oceans were wiped out at this time. This was the only mass extinction to have had any significant effect on insects, which have been able to crawl, creep, or fly through other mass extinctions nearly unscathed. The Permian wiped out 65% of all insect families alive at the time.

No one knows for certain what caused the Permian mass extinction, although new theories appear every few years.

In 1991 and again in 1992, separate groups of geologists proposed that giant volcanic eruptions in what is now Siberia served to lace the atmosphere with sulfuric acid, raise and lower sea levels rapidly, and block sunlight from reaching Earth's surface for several years. The lava from such eruptions now forms the Siberian Traps. In 1995, Paul Renne of the Berkeley Geochronology Center in California announced that he and his coworkers had established the precise date of the great Siberian eruptions, and this date coincided with that of the Permian mass extinction.

In 1992, Michael R. Rampino of the Goddard Space Flight Center in New York City and Verne Oberbeck of the Ames Research Center proposed that a giant asteroid caused both the breakup of Pangaea (*see* "Advances in Plate Tectonics Studies," p 542) and the Siberian volcanoes, leading to the Permian extinctions.

In 1995, Andrew Knoll of Harvard University in Cambridge, Massachusetts, Richard Bamback of Virginia Polytechnic Institute and State University, and John Grotzinger of M.I.T. in Cambridge proposed that the Permian extinction might have been caused by a mechanism similar to that by which Lakes Monoun and Nyos killed more than 1,700 people in Africa during the 1980s: carbon dioxide that had been building up in lower waters of these lakes was suddenly released when the lakes "turned over" their layers of water of different density. This theory, based upon evidence for carbonates precipitated from seawater during Permian times and upon climate studies, states that ocean circulation in Permian days was weak because there was only one continent. Thus, organic material could build up at the bottom of the single sea. Decay of this organic material would produce a layer of carbon-dioxide-laden water at the bottom of the ocean. An ice age with attendant glaciers could then cause the oceans to "turn over," bringing the carbon dioxide to the surface layers of the sea where it could wreak havoc.

In July 1994, Kun Wang of the University of Ottawa reported that he and his workers had determined that ocean plankton died at the end of the Permian in one very short pulse, perhaps as short as even a few days, although perhaps as long as a few thousand years (still very short for geologic time). His dates depend on a sharp change in isotope ratios of carbon that occurs exactly at the Permian-Triassic boundary. He believes these isotope ratios to be consistent with some violent end, such as an impact or giant volcanic eruptions.

(Periodical References and Additional Reading: *New York Times* 8-14-92, p A12; *New York Times* 8-16-92, p IV2; *New York Times* 8-21-92, p D16; *New York Times* 12-15-92, p C1; *New York Times* 8-3-93, p C1; *New York Times* 9-17-93, p A1; *Science* 3-11-94, p 1371; *Science* 10-7-94, p 28; *Science News* 7-16-94, p 38; *New York Times* 12-27-94, p C1; *Science* 10-6-95, p 27; *Science* 12-1-95, p 1441)

ONTO THE LAND, THEN BACK TO THE SEA

The transition of mammals from land to full-time existence in water has been compared to the transition of the first fish/amphibians from the sea to land, or early reptiles from land to sea and from land to air. Other vital passages took insects, birds, and mammals into the air. Finding organisms part way along the trajectory of such a transition is one of the keys to understanding evolution. In the early 1990s, a series of major discoveries cast new light on how whales moved from the land back into the water.

Sea to Land

When we picture the movement from the sea to the land, we tend to think of that time somewhat before 370 million years ago (mya), when the first amphibians learned to live part of their lives on land (although one may presume that, like today's amphibians, they began life in water). The land had been colonized long before that, or the only reason to go on land would have been to escape from enemies at sea. Mosses or similar low-growing plants were among the first creatures to find a way of life on land, although even they had been preceded by the one-celled blue-green algae, a moneran more closely related to bacteria than to plants. In 1994, Robert J. Horodyski of Tulane University in New Orleans, Louisiana, and L. Paul Knauth of Arizona State University in Tempe reported on fossil algae dating from 1.2 billion years ago in what is now Arizona. These lived on land as well. The algae are thought to have had the land to themselves for about 800 million years, at which point some of them may have evolved into mosses, although mosses might well have evolved from other one-celled algae. Soon after the moss began to grow on land, centipedes and other invertebrates found a way to live on land. About 50 million years later, when insects, spiders, and forerunners of horsetails and ferns, as well as the earlier inhabitants, had formed a pleasant ecosystem, the ancestral amphibian found the land both attractive and safer than the water.

Although we tend to picture the first amphibian as crawling onto an ocean beach and gasping for air, it seems more likely that the transition to land from the ocean was made via fresh water rivers and ponds.

The earliest amphibians were identified in 1991 from scattered remains found in

Scotland and dated at about 370 mya. In 1993, Neil H. Shubin and graduate student Edward B. Daeschler of the University of Pennsylvania found a fossil amphibian they named *Hynerpeton bassetti*. *Hynerpeton* had lived about 365 mya in marshy rivers about 100 mi (160 km) from an equatorial sea. Today, this region is the village of Hyner, Pennsylvania, in the Allegheny Mountains. The report of this find appeared in the July 29, 1994 *Science*. Because *Hynerpeton* is advanced as a land creature, the discovery suggests that the first amphibians existed earlier than previously thought, arising perhaps 380 mya. The ancestors of the first amphibians are thought to have been fish that could breathe air part of the time, similar to present-day lungfish.

Land to Sea

In December 1989, husband and wife team Philip D. Gingerich and B. Holly Smith of the University of Michigan and Elwyn L. Simons of Duke University unearthed a fossil of a whale that still possessed hind limbs complete with feet, the first ever found on a species of whale. The whale, *Basilosaurus isis*, is a long, slender sea creature that has been known since before 1835, although no complete skeleton has ever been assembled. The main purpose of the 1989 expedition to a place paleontologists call Zeuglodon Valley ("Zeuglodon" is a former name for early whales) was to locate more *B. isis* bones in hopes of reconstructing still unknown parts, notably the front flipper. Five days before the end of the expedition, Gingerich located a bone showing the place where the pelvis attached to the vertebrae, indicating the position along the 50 ft (15 m) length where leg bones might be found. Some other early whales are known to have small remnants of former hind limbs, so it seemed worthwhile to look for those of *B. isis*. Returning to other partially exposed fossils, Gingerich, Smith, and Simons started searching, and in the remaining days of the trip located the legs, from the femur down to the toes. After studying the bones back in Ann Arbor, Gingerich reported that the legs were about 20 in (50 cm) long—not very big for a 50 ft (15 m) whale, but much longer than expected. Furthermore, these were clearly functional legs, not just rudiments.

The whale, which lived 40 million years ago in a shallow sea where Egypt is today, apparently did not use the legs in swimming. Speculation about their use has ranged from helping the whale work its way out of mud (by Lawrence G. Barnes of the Natural History Museum of Los Angeles County to clasping the opposite sex during mating (by Gingerich). Gingerich has pointed out that since the pelvis is 35 ft (11 m) from the brain along a highly flexible body, legs may have been needed to keep the proper alignment for copulation. Barnes notes that sexual organs are often different sizes in males and females. If more limbs are found, and if they differ greatly in size, he will concede the case to Gingerich.

The four toes found on *B. isis* suggest that it is more closely related to the even-toed ungulates, or Artiodactyla, than to the odd-toed ungulates. In other words, whales descended from the ancestors of cattle, deer, pigs, and so forth. In 1992, J. G. M. Thewissen of Northeastern Ohio University's College of Medicine in Rootstown and coworkers discovered a seal-like ancestor of the whales that existed part way along the transition. *Ambulocetus natans* used its feet instead of its tail to propel itself. Unlike the Artiodactyla species thought to be its ancestors, but similar to seals and

WHALE FAMILY TREE

The ancestors of whales were land mammals until about 60 million years ago (mya). The early whale ancestors, the mesonychids, were carnivorous relatives of the ungulates, the herbivores that became today's horses, cattle, pigs, and so forth. Although the mesonychids became extinct on land about 40 mya, those that moved into the water gradually evolved into today's whales and dolphins. The oldest known whale, *Pakicetus,* discovered by Gingerich in 1978, lived about 50 mya and may still have ventured onto land from time to time, perhaps like a modern sea otter.

In 1989, Gingerich discovered *Prozeuglodon,* a small but recognizable whale about 15 ft (4.5 m) long with 6 in (15 cm) vestigial legs.

B. isis is not the only fossil whale found in Zeuglodon Valley. Gingerich and Smith also found what they now believe are leg bones from a small whale called *Dorudon,* the probable ancestor of modern whales. It seems increasingly likely that all sufficiently early whales had legs, probably for the first 10 million years after they had begun to live full time in the water.

Until recently it was thought that sperm whales, which have teeth for catching giant squid and other creatures, are more closely related to toothed dolphins than they are to the large group of baleen whales, which strain small organisms from ocean water using specialized horny plates called *baleen.* The baleen takes the place of place of teeth. In 1993, DNA studies showed the sperm whales to be recent relatives of the baleen whales, with a common ancestor some 10–15 mya that is not shared with the dolphins, porpoises, and narwhals. It was not clear from the analysis where the beaked whales fit into the picture, although they may be a third, separate branch of the cetacean family tree.

modern whales, *Ambulocetus* was a carnivore.

Whales are, of course, not the only land animals to recolonize the sea. They are not even the only mammals. Manatees, dugongs, and the now extinct Steller's sea cows collectively form the order Sirenia, which experienced a return to total marine and aquatic life entirely separate from that of whales. The Sirens have not been very successful, possessing the fewest individuals according to most estimates of any order of mammals. Their closest relatives are elephants and hyraxes.

Some other animals have taken to the sea, at least in part. Sea otters, for example, are born and live almost entirely at sea, although they seldom move into deep water. In 1995, Greg McDonald at the Hagerman Fossil Beds National Monument in Idaho and coworkers discovered fossils of a large sloth in Peru. The sloth had returned to the sea some three to seven million years ago. Although not so well adapted to marine life as a sea otter, the sloth was much better adapted to swimming and to eating seaweeds than the land sloths. Evidently, however, it was not adapted well enough, as it became extinct.

Among the invertebrates, a 1994 surprise candidate for land-to-sea transfer is the lobster. Almost any body of fresh water of any permanence in the United States contains what appear to be small lobster relatives, which are the various species of crayfish (also known as crawfish or "crawdads"). Some crayfish also live away from surface water in places where the water table is sufficiently high. Until 1994, the earliest known crayfish had been fossilized in shallow brackish or salt water at the ocean margins dating from about 130 mya. Steven Hasiotis of the University of Colorado, however, located crayfish from 220 mya, all of which lived in fresh water streams or ponds or in deep burrows on land, burrows that reached into the water table to become flooded at their base. Since the 220 mya crayfish had already evolved in various complex ways by that time, Hasiotis thinks that the ur-crayfish may have been in existence 300 mya, placing it before the origin of its big lookalike, the lobster. A return to the sea by the crayfish would make the crayfish the lobster's ancestor, instead of, as people simply assumed, the other way around.

(Periodical References and Additional Reading: *New York Times* 1-28-93, p A18; *New York Times* 1-14-94, p A25; *Science* 1-14-94, p 210; *New York Times* 2-22-94, p C7; *Science* 1-14-94, pp 180 & 210; *Science* 7-29-94, p 639; *New York Times* 8-2-94, p C7; *Discover* 1-95, pp 82 & 84; *Discover* 1-96, p 45)

DINOSAURS AND BIRDS

It may seem odd to discuss dinosaurs and birds in one section, but many paleontologists believe that the two kinds of dinosaur and the modern birds are more closely related to each other than either dinosaurs or birds are to such reptiles as snakes, lizards, and turtles. Crocodilians may also be in the dinosaur-bird group, although some paleontologists think that birds and crocodilians share a common ancestor different from the dinosaurs. In this section, we will not only review recent information related to this taxonomic question, but also look at some of the other recent discoveries about dinosaurs and the earliest birds.

What Were Dinosaurs *Really* Like?

Reevaluation of the place of dinosaurs in the history of life continues. Increasingly, young paleontologists are espousing new, controversial ideas, and new finds are also changing what we know about dinosaurs in ways that may provoke little controversy, but much revision.

The following paragraph moves from more controversial ideas to new ideas that are now accepted as fact:

Dinosaurs were not really reptiles at all. They were warm-blooded creatures that did not become extinct but evolved into birds. Like birds they swallowed stones to grind their food in a gizzard. Dinosaurs ran fairly quickly and carried their tails high. Carnivorous dinosaurs were highly intelligent and hunted in packs the way wolves do. Some did not abandon their offspring, the way most modern reptiles do, but cared for them in colonies that resemble seabird enclaves. They did not live in tropical climes only, but ranged far to the north and south.

Warm Blooded Like Us?   One of the main arguments used to claim that birds are just dinosaurs with feathers is that both are warm blooded. That is, they are *endotherms* who use internal means of heat regulation, rather than *ectotherms* who use environmental methods. Regulating internal temperature by generating excess internal heat is not, however, a characteristic that evolved only with the birds and mammals. Some modern fish, notably tunas, are also endotherms. Most fish and other ocean creatures, however, are cold blooded, because sea temperatures are relatively constant and it is energetically easier to "go with the flow" of heat. Land animals, however, face extremes of heat and cold. Reptiles use external means (mostly sunlight, but also sometimes rotting vegetation) to provide heat needed for swift movement. Without the external heat, reptiles move slowly, or conserve heat by not moving at all.

Nearly all animals, however, use a combination of external and internal means of regulating heat. You are probably wearing clothes right now, for example. Ectothermic insects that need heat to fly will generate some by exercising wing muscles. Endothermic cats like to sun themselves. Large creatures of all kinds generally retain heat because of the mathematical relationship between volume and surface area. A large ectothermic lizard might need an organ such as a crest to increase its surface area, thus allowing itself to get rid of excess heat.

Throughout the early 1990s, evidence accumulated both for and against the hypothesis that dinosaurs regulated their internal temperature in much the same way as birds and mammals.

In 1992, Reese E. Barrick of the University of Southern California and William J. Showers of North Carolina State University measured the ratio of oxygen-16 to oxygen-18 in various fossil bones of late Cretaceous dinosaurs, those dinosaurs that were the last to become extinct. The scientists' idea was based upon the fact that oxygen-16 more readily incorporates into warmer bones. Therefore, the ratio of oxygen-16 to oxygen-18 in bones deep in the body and those at the extremities should be alike in warm-blooded animals and dissimilar in cold-blooded ones. This hypothesis was confirmed by measurements on one modern ectotherm and two modern endotherms. Measurements by Barrick and Showers indicate that the late Cretaceous dinosaurs were endotherms—that is, warm blooded. In 1994, further measurements of tyrannosaurus bones confirmed Barrick's and Showers' original diagnosis of endothermy in dinosaurs.

In 1995, on the other hand, John Ruben of Oregon State University used a small cartilage or bone construct found in the noses of modern warm-blooded animals, and not thought to have been observed in dinosaur fossils, to proclaim that the creatures were as cold blooded as lizards. Not only that, but Ruben and coworkers found that some very early birds, including *Archaeopteryx,* did not have the construct. Therefore, according to this criterion, they could not have been endotherms. Among the disputants is John R. Horner of the University of Montana, who thinks he has evidence for the cartilage structure in a duck-billed dinosaur.

Complicating the issue was a 1994 study by Anusuya Chinsamy and Peter Dodson of the University of Pennsylvania and Luis M. Chiappe of the American Museum of Natural History in New York City. They found annual growth rings in the bones of some of the enatniornithe birds as well as in an Argentine fossil bird from 20 mya named *Patagopteryx defarrariisi.* Growth rings do not form in warm-blooded animals because internal conditions are kept more or less constant. An animal exposed to seasonal internal changes, however, might develop such rings. Crocodiles, for example, show such rings, as do some dinosaurs. Thus, some now-extinct birds might have been cold blooded. Chiappe, however, thinks that these birds had a metabolism somewhere between that of an ectotherm and an endotherm.

More and more of the researchers involved have come to believe that perhaps some dinosaurs were ectotherms and some were endotherms. The same may have been true of very early birds. This is not an unheard-of situation; among the fish, for example, nearly all are ectotherms, but tuna are both fish and endotherms.

The Dinosaur Look   Part of the new view of dinosaurs is that they were more erect than pictured in early dioramas. It was clear from early on in dinosaur studies that the legs of dinosaurs were beneath the body, not splayed out as in amphibians and most reptiles, but recent reconstructions emphasize the mammal-like stance of dinosaurs. Earlier views had also shown dinosaurs with typical reptilian scales, while more recently they have been drawn or sculpted with smooth, leathery skin, rather like elephants and hippos.

In 1992, a new set of fossils suggested that the sauropods, including *Apatosaurus,* often known as *Brontosaurus,* the *Barosaurus* featured in the new Dinosaur Hall at the American Museum of Natural History, and *Diplodocus,* might have looked more reptilian than their most recent reconstructions. The newly found fossils included many spikes, which paleontologist Stephen A. Czerkas claimed would have run down the back and tail of the giant sauropods, rather similar to those spikes found on modern iguanas.

The upright posture assumed in reconstructions of dinosaurs such as *Utahraptor* may have sometimes been too much of a good thing for very large dinosaurs. Cardiologist Daniel Choy has calculated that for a large *Barosaurus* to get blood from its body to its brain, located some 25 ft (7.6 m) apart, it might have needed eight hearts, especially if it were to rear up on its hind legs as is depicted in a reconstruction at the American Museum of Natural History's Dinosaurs Halls, redone and reopened on June 2, 1995. Conservative paleontologists could interpret this calculation as indicating that dinosaurs may not have been as athletic as some of the recent reconstructions suggest.

Watch Out for the 'Raptors   In the book and movie *Jurassic Park*, the most fearsome dinosaurs were the velociraptors, fairly small representatives of the dromaeosaurs that were intelligent and hunted in cooperative groups. Author Michael Crichton could claim considerable evidence in favor of this concept. In one find, for example, three dromaeosaurs of a related species, *Deinonychus*, were found in a group with a herbivorous dinosaur that the three *Deinonychus* seemed to have been hunting when disaster overtook the whole group. In 1992, James Kirkland, a dinosaur expert, identified a dromaeosaur called *Utahraptor*. *Utahraptor* is the earliest known member of the 'raptor clan, and the largest and meanest looking, with sharp claws that reached 1 ft (30 cm) in length. *Utahraptor* was about 20 ft (6 m) long and weighed some 1,500 lb (800 kg). Kirkland thinks that it was a fast runner and brighter than your average dinosaur. The biggest mystery involves why it allowed itself to become extinct some 125 mya, long before the famous meteor that many now believe to have permanently ended the reign of the dinosaurs.

Herd of Dinosaurs?   Studies of dinosaur tracks indicate social behavior in dinosaurs. Tracks from 100-million-year-old limestone in what is now Texas show what is probably a herd of sauropods being stalked by a pack of three theropods. Another site has been thought to be of a herd gathered around its young, but more recent measurements dispel this notion. It is simply a herd of sauropods, led by the largest members of the herd, as they move across the region. Like the sauropods, ornithopods show by their tracks that they were herd animals. Some of the track pathways of dinosaurs appears to be migration routes.

Care-Givers Like Birds   That dinosaurs laid eggs was not really surprising, since most reptiles lay eggs. The discovery in the 1920s of what appeared to be preserved eggs was surprising. As later finds were nearly always close to and in the same strata as dinosaur fossils, there seemed no doubt about the identity of the eggs. Some of the herbivorous dinosaurs were thought to have been the parents, while others, notably a group named oviraptorids, were thought to have been present near eggs with the intention of making breakfast of them. The first one, *Oviraptor philoceratops*, was named "egg eater that loves ceratopsians," in the belief that it hung around the nests of the ceratopsian *Protoceratops* waiting to eat its eggs and/or young.

In 1983, the first embryos still in dinosaur eggs were located in the Gobi desert, and they proved to be herbivores. Other embryos were found in Utah in 1992, and proved to be of 150 million-year-old *Camptosaurus*, another herbivore. Then in 1994 Amy Davidson of the American Museum of Natural History painstakingly scraped out a fossil embryo from a 75-million-year-old egg and found it to be one of the oviraptorids. Apparently, the presence of the oviraptorid fossils near the egg fossils was because they defended their nests, a behavior common in birds but rare in reptiles.

The egg in which Davidson located the embryo was one of a half dozen that had been placed in a circle, apparently so that the parent or parents could roost on them. Such roosting would make more sense, of course, if the parents were endotherms. The nest also contained two skulls of baby dinosaurs from another carnivorous species. While these could be the remains of a parent's meal, skulls of modern-day baby birds are often found in the nests of modern raptors such as hawks. The parents bring back the young birds of other species to feed their own young, who devour everything except the heads. Thus, the oviraptorid parents may also have captured baby dinosaurs of other species to feed to their young, care-giving behavior common in birds.

An even more startling find was made in 1993, although it took two years for the block of sandstone to be chipped away from the fossils. Then, paleontologists could recognize the importance of what was attached to the pair of claws that Mark A. Norell of the American Museum of Natural History had noticed protruding from the stone in the Gobi. The claws belong to an oviraptor that died while brooding its clutch of 15 or more eggs, each egg carefully placed in a circle with the wider part of the egg toward the center. Speculation is that the oviraptorid parent was caught in a sudden sandstorm some 80 mya. The posture and the position of the legs was exactly as is found in modern brooding birds.

Reptiles are not without modern-day care-giving parents. Crocodilians commonly take care of young, although they do not brood the eggs with their own bodies. Crocodilians use rotting vegetation as a source of heat. A few lizards, but no snakes, are also care-giving parents. A few snakes, however, brood their eggs, using muscular contractions to raise their own internal temperature when necessary to keep eggs warm. In any case, dinosaur parenting is not by itself enough evidence to connect dinosaurs more closely to birds than to reptiles, although it is another clue. Because birds evolved well before the time of the brooding oviraptors, it would appear most likely that an early common ancestor developed egg brooding, which provided an advantage for natural selection. The other possibility, as always, is that this is a striking example of convergent evolution.

Roaming the Earth   When one hears the phrase "when dinosaurs roamed the Earth," one pictures them as not roaming very far from the tropics, or at least not outside tropical climates (which at that time may have been more widespread than today's tropical climes). Discoveries in the late 1980s and early 1990s, however, put dinosaurs inside the Arctic circle, where it is dark for six months of the year.

In 1991, William Hammer of Augustana College in Rock Island, Illinois, also located fossils of several dinosaurs about 400 mi (650 km) from the south pole. By 1994, he had identified one of the fossils as having belonged to a previously unknown carnivorous dinosaur that he named *Cryolophosaurus ellioti*, which means "Elliot's frozen, crested reptile," after D. Elliot, a discoverer of the site (Hammer was not referring to the *Cryolophosaurus* as being frozen when alive, but upon collection).

During the heyday of *Cryolophosaurus* and three other already-identified dinosaurs from the site, the climate at the site where the fossils were found was somewhat milder, since movements of tectonic plates have carried that site to a new location in the 200 million years since the *Cryolophosaurus* died and became a fossil. Fossil tree trunks were also found. That is not to say that the middle of the Antarctic continent had drifted to the equator. Indeed, Antarctica has not strayed far from the South Pole for hundreds of millions of years. The place deep in the interior of the continent where *Cryolophosaurus* was found would probably have been just north of the Antarctic circle 200 mya. However, generally warmer conditions worldwide meant that the climate then at the Antarctic circle might have been like that of the northwest coast of the United States rather than the ice cap and glaciers currently present at that latitude. Still, the coast of Washington state is not exactly tropical.

Is It a Bird? Is It a . . .

Although Roy Chapman Andrews' famous expedition of the 1920s to the Gobi Desert in Central Asia was intended to find traces of early humans, the expedition's principal success was the discovery of the first known clutch of dinosaur eggs. Since that time, paleontologists have returned to that remote and inhospitable land whenever politics and war have allowed, finding large numbers of dinosaur and other fossils, but no early humans. In 1993, one of the main finds concerned a creature termed *Mononykus olecranus*, a common fossil in the Gobi.

In reconstruction, *Mononykus* looks rather like an emu or ostrich with a long tail, instead of plumes; two short, heavily clawed arms projecting from its breast, instead of wings; and a head that looks more like that of a lizard than that of a bird. Cover it with feathers, and it is a bird that looks like a dinosaur. Give it scales, and it is a dinosaur possessing a birdlike breastbone and legs. Mark Norell thinks that the keel on the breastbone could only have evolved for flight. Therefore, perhaps the ancestors of *Mononykus* were flyers, just as the ancestors of ostriches and penguins are thought to

Background

In the 1820s, Gideon Mantell, aided by his wife Mary Ann, was collecting fossil bone fragments and teeth from English gravel beds. He concluded that he had found the remains of a giant plant-eating reptile. These fossils are now recognized as the first evidence for dinosaurs to be uncovered. Georges Cuvier and Richard Owen were among the paleontologists who studied these and other fossils of large reptile-like creatures.

In 1841, Owen grouped many of the fossils into a new taxon called Dinosauria, a name that in its anglicized form, **dinosaur**, has stuck until today. By the late nineteenth century, however, paleontologists had determined that there were two groups of "dinosaurs" that were no more closely related to each other than they were to crocodilians or to pterosaurs (such as pterodactyl). These groups are the Saurischia, which included the sauropods and theropods, and the Ornithischia, which included stegosaurians and ceratopsians. Despite this, members of both groups are still generally known as dinosaurs. However, both groups differ from other reptiles. For example, reptiles sprawl and dinosaurs stood upright (*see* below for a discussion of whether dinosaurs actually are reptiles). To further complicate matters, in 1990 Rolf E. Johnston of the Milwaukee Public Museum and John H. Ostrom of Yale University made the controversial claim that *Triceratops* and related Orithischian dinosaurs sprawled. Dinosaurs used their upright stance to walk the Earth from Late Triassic times until the end of the Cretaceous period—about 140 million years.

Traditional views of the dinosaurs as slow-moving, stupid creatures with lifestyles similar to those of crocodilians began to be attacked in 1968, when Robert T. Bakker proclaimed their upright stance made dinosaurs swift-moving, not sluggish. The following year, John Ostrom suggested that dinosaurs, or at least some of them, might have been warm blooded (technically, endothermic). Bakker took up the cause, first at Yale (where he was Ostrom's student) and then at Harvard, eventually going far beyond Ostrom's suggestion to argue throughout the 1970s in favor of dinosaur endothermy. Indeed, another set of revisionist ideas concerns structures such as the plates along the back of stegosaurus and the frills found on Triceratops and related dinosaurs. These have been described by some as devices used to cool large dinosaurs, whose immense size would make their blood warm without the need for any mechanism such as those found in endotherms.

Another recent issue concerns the dinosaurs' relationship to birds. As early as 1867, Thomas Henry Huxley proposed that birds are descended from some branch of the dinosaurs. Until the 1970s, however, most scientists believed the dinosaurs and birds shared a common ancestor that was neither one nor the other. Starting in 1973, Ostrom argued with considerable success that Huxley had been right. He based his claim largely on a careful reexamination of *Archaeopteryx*, which, aside from the feathers, he found almost identical to small dinosaurs of the time. In 1975, Bakker took the idea one step further, proposing that dinosaurs and birds together formed one class of vertebrates while reptiles were a separate class from which the dinosaur/bird class had evolved.

have flown. Norell thinks that the keel and other bones on *Mononykus* are more like those of modern birds than they are like the famous early bird *Archaeopteryx*. Other paleontologists are not so sure. The keel may have been a purely dinosaur device that evolved to anchor muscles used in digging or other ripping activities. In any case, *Mononykus* is either a birdlike dinosaur or a dinosaurlike bird.

Speaking of *Archaeopteryx lithographica,* April 1993 also brought forth the seventh known fossil of that dinosaurlike bird, and the first to reveal its breastbone. The breastbone, a key adaptation for flight, had been missing in the earlier *Archaeopteryx* fossils, which were otherwise exceptionally well preserved. Previously studied fossils suggested that despite its wings and feathers, it could not have been much of a flyer. For example, a 1994 analysis of the three *Archaeopteryx* fossils with well-preserved feathers shows the typical feather pattern of flightless birds. Similarly, an analysis of *Archaeopteryx* wrist mobility in 1992 and muscle strength in 1993 showed poor adaptation for flight.

The 1993 *Archaeopteryx* fossil may represent a different species from earlier ones. It is noticeably smaller, which would account for breastbone differences and the flight capability it suggests. Because of the size difference and the breastbone, Peter Wellhofer of the Munich Museum in Germany has proposed that the new specimen be given a new species name: *A. bavarica* instead of *A. lithographica.* It has long been recognized that if it were not for the feathers the earlier specimens of *Archaeopteryx* would have been classed as theropod dinosaurs. The new find provides in the breastbone another difference between the theropods and at least one sort of *Archaeopteryx.* However, another dinosaur characteristic was found on *A. bavarica*: a series of bony plates in the mouth that are found in theropod dinosaurs but not known in birds.

University of North Carolina ornithologist Alan Feduccia analyzed the *Archaeopteryx* feet in February of 1993 and concluded that it was a tree percher and not a ground walker. Feduccia concludes from this that *Archaeopteryx* could not be descended from dinosaurs, since dinosaurs did not climb trees. Although some paleontologists then claimed that it was possible that small dinosaurs, unable to fly, still climbed trees and perched on the limbs, others who saw less likelihood of tree-climbing dinosaurs accepted the analysis of *Archaeopteryx's* perching as proof of birdness. The close resemblance of *Archaeopteryx* to theropods suggests, however, that either some dinosaurs did climb trees or that *Archaeopteryx* developed perching feet after it learned how to fly up into the trees.

There are only seven *Archaeopteryx* fossils ever found, one of which is missing, having apparently been stolen in 1991.

The First Dinosaur

Paul Sereno of the University of Chicago has made a number of important discoveries related to early dinosaurs and birds, based on fossils found in Argentina and in Africa. He has been one of the principal investigators of *Herrerasaurus* from the Argentine Andes, generally believed to be the oldest known dinosaur. The first fossils of *Herrerasaurus* were found in 1959, but aside from their age (established eventually as 228 mya from an analysis of volcanic ash) and dinosaurness, little could be learned from just a leg bone and a pelvis. In 1988, researchers from the University of Chicago, including Sereno, located five nearly complete skeletons. From these, Sereno was able to show that *Herrerasaurus* was a carnivore and a likely ancestor of the line that led to *Tyrannosaurus rex.* The description of the *Herrerasaurus* was published in *Science* in 1992, and described a creature with powerful hind legs, short forearms, recurved claws on its fingers, and a jaw filled with inwardly curved teeth.

About a mile away from the Andes site where the five *Herrerasaurus* fossils were found, the University of Chicago scientists worked with scientists from the National University of San Juan, Argentina. In October 1991, they located remains of another early dinosaur. *Eoraptor* lived about the same time as *Herrerasaurus* but has more primitive features. For example, although a theropod like *Herrerasaurus*, *Eoraptor* lacked the specialized jaw that *Herrerasaurus* and later members of the theropod line utilized in seizing and devouring their prey. It is thought that a common dinosaur ancestor would most resemble *Eoraptor.*

Early Birds

The early bird *Sinornis,* found in China in 1992, has been called a missing link between the dinosaurs and the birds, although this remains a controversial idea. Paul Sereno, who reconstructed the fossil from two separate pieces, thinks that *Sinornis* is more birdlike than *Archaeopteryx,* which he views as a theropod dinosaur with feathers. *Sinornis* is only 135 million years old (compared to the 147-million-year-old *Archaeopteryx*), so Sereno sees a progression from theropod dinosaur to *Archaeopteryx* to *Sinornis* to modern birds.

Some serious discrepancies argue against this progression.

In 1995, Z. Zhou recognized that the birds of the Cretaceous Period were mostly enantiornithines—what Alice would have called "looking-glass birds"—with their internal bone structure the mirror image of modern birds. Only one lineage of birds with the modern structure is known from that period: a group called "transitional shorebirds," which were like the present-day orders of waterfowl and stilt-legged birds. During the Cretaceous-Tertiary (K-T) mass extinction , all the looking-glass birds (the major orders of birds before the K-T boundary) ceased to exist (*see* "Mass Extinction UPDATE," p 485). *Sinornis* was a looking-glass bird, and so could not have been the ancestor of any modern bird.

One of the looking-glass birds, *Confuciusornis sanctus,* lived from about 132-142 mya and was the earliest bird to have had a beak. Because it was an enantiornithine, it could not have been the ancestor of modern beaked birds. It is thought that beaks evolved at least twice in modern birds as well as in the enantiornithines.

Because (geologically speaking) *Confuciusornis* lived only a short time after *Archaeopteryx* but differed significantly (although like *Archaeopteryx* it had claws on its wings and a bony tail), there are no doubt many other early birds from this time whose fossils have not yet been found. Because of their hollow bones and other adaptations of weight reduction for flight, birds do not form fossils under most conditions. The beak, which we now know evolved at least three times in birds, is thought to be another weight-reducing adaptation.

(Periodical References and Additional Reading: *New York Times* 7-21-92, p C1; *New York Times* 9-1-92, p C2; *New York Times* 11-3-92, p C1; *New York Times* 11-17-92, p C10; *Discover* 1-93, pp 48, 49, 50, & 66; *New York Times* 1-6-93, p A17; *New York Times* 2-5-93, p A12; *New York Times* 4-15-93, p B7; *New York Times* 4-20-93, p C1; *New York Times* 5-7-93, p A17; *Discover* 1-94, p 51; *New York Times* 3-17-94, p B10; *New York Times* 5-6-94, p A17; *Science* 75-6-94, p 828; *New York Times* 7-8-94, p A15; *Science* 7-8-94, pp 188 & 222; *Nature* 8-18-94, p 514; *New York Times* 10-18-94, p C6; *New York Times* 11-4-94, p A1; *Discover* 1-95, p 85; *Natural History,* 6-95, pp 30, 41, 47, 52, 56, 59, & 62; *Science* 11-3-95, p 735; *New York Times* 12-21-95, p A22; *Discover* 1-96, p 50)

THE *JURASSIC PARK* SYNDROME

The 1993 movie *Jurassic Park,* like the 1990 novel by Michael Crichton on which the movie was based, utilized the supposition that it would be possible to reconstruct living dinosaurs from their DNA. The question of whether or not dinosaur DNA would ever become available was partly answered in the early 1990s, although whether the answer was "yes" or "no" depended on which paleontologist was asked.

The Reports of Success   Although there was originally some considerable skepticism over whether DNA could survive for over 65 million years (that is, since the last dinosaurs lived), various scientists have reported success in finding, isolating, and copying DNA from that time or even earlier. In 1992, scientists obtained DNA from an extinct termite which had been preserved in amber and dated back to 25 to 30 million years ago (mya). The DNA indicated that termites are not descended from cockroaches. In September 1992, George O. Poinar Jr., Raúl J. Cano, and coworkers reported on 40-million-year-old DNA from a stingless bee in amber. In June 1993, the same group reported that they had cloned DNA from a 130-million-year-old insect that had been trapped in amber. Mary Schweitzer claimed to have recovered DNA in the summer of 1993 from the bone marrow of a *Tyrannosaurous rex* fossil bone.

Finally, on November 18, 1994, Scott R. Woodward of Brigham Young University in Provo, Utah, made the ultimate claim. He had not only located a dinosaur gene, but was able to identify it as the mitochondrial cytochrome b gene. He could not identify the dinosaur species involved, but it or they (there were two bones used) came from a layer of coal in Utah dating back to 80 mya. Woodward had looked for bones in coal, as he thought that coal might act as a preservative for DNA by keeping oxygen out. The fragments of bone that he found did not, upon microscopic examination, appear to have been fossilized by replacement of tissue with minerals. Instead, they appeared to consist of organic material with the cell structure preserved. As part of his effort to make sure that the dinosaur genes were not contaminated, Woodward carefully checked the coding sequence of the dinosaur gene against that of human mitochondrial DNA and found significant differences. In fact, the sequences he found did not resemble those of any living creature.

One encouraging set of results for ancient DNA comes from abundant mammoth bones and flesh, some dating from as long ago as 50,000 years. Moderately numerous samples from more than half a dozen soft-tissue fossils show clearly the expected relationship between the mammoths and modern elephants and are consistent with each other.

The Case Against   Immediately after Woodward reported capturing dinosaur DNA, critics began to pick the idea apart. In the beginning, though, all that they had as ammunition was to say that it was too old to have survived and that it really ought to look like bird DNA (*see* "Dinosaurs and Birds," p 492). By May 1995, Hans Zischler of the University of Munich in Germany had shown that humans contain copies of the gene in question outside the mitochondria, making human genes a more likely contaminant. At the same time, S. Blair Hedges of Pennsylvania State University in University Park reported that his analysis showed the sequences of purported dinosaur DNA to be more like those of humans than those of either reptiles or birds. A few months later, in November 1995, Carol-Beth Steward and Randall V. Collura of the State University of New York at Albany independently confirmed that in various primates, including humans, parts of the cytochrome b gene had jumped from the mitochondria to the nucleus more than 30 mya, resulting in nonfunctional inserts in human nuclear DNA. Their study was devoted to problems in unraveling primate lineages using genetic analysis, but along the way they realized that the gene fragments they found in humans, and only in humans, matched some of the fragments of Woodward's "dinosaur genes."

Why Stop at DNA?

At the end of 1992, George O. Poinar Jr. and a graduate student, Benjamin M. Waggoner, found pollen grains and single-celled microorganisms in amber dating from 230 mya. One of the pollen grains had grown after it had been trapped, or just as it was trapped. This demonstrated that amber can preserve the microorganisms.

In May 1995, Raúl J. Cano of California Polytechnic State University in San Luis Obispo carefully removed the stomach contents of a bee that had been trapped in amber since at least 25 mya and perhaps as much as 40 mya. Instead of fishing for any DNA, he simply transferred the contents to a Petri dish of nutrients. Being very careful to avoid contamination from modern sources, and gradually increasing the nutrients from a simple batch of electrolytes and amino acids to heartier fare, Cano and graduate student Monica K. Borucki managed to grow bacterial colonies in the dishes. Their first success was a variety of *Bacillus sphaericus,* common in bee stomachs today. Eventually, using bits and pieces of various insects from amber, Cano and workers have produced a couple of thousand different growing cultures from million of years ago, including yeasts as well as various bacteria.

The idea behind this program is that many bacteria or other microorganisms have phases in which they cover their exteriors and suspend animation for long periods of time. These phases are generally produced in response to adversity, such as drying out or running out of food. It is well known that some of these phases, called *spores* (not to be confused with plant spores, which are entirely different), can survive for many years under suitable conditions. Despite this known ability of bacterial spores to survive under difficult conditions, many biologists think that Cano is mistaken, and must have contaminated his samples with modern microorganisms. Contamination is hard to disprove, but Cano claims that the DNA sequences he finds in his ancient microbes differ significantly from those found in contemporary relatives.

Background

George O. Poinar of the University of California at Berkeley proposed the same way of obtaining genes from dinosaurs that became the basis of *Jurassic Park*. So far, his proposal remains a movie fantasy, partly for the reasons outlined below.

DNA can be preserved for longer periods of time when it is still in cells that are protected by some coating. Ancient magnolia leaves from 17 million years ago, for example, have been found preserved in airtight shale, which kept the cells intact and preserved the DNA. Some very old fossils that are also remarkably well preserved are small creatures that have been trapped in tree resins that, through geologic time, became the gemlike substance we know as amber. Like the original resins, amber is sufficiently transparent that one can see the creature trapped inside. Most often, it is an insect or spider, but occasionally a frog or lizard is found. Such insects and other creatures are so well preserved that they appear to have been engulfed recently, even when they are millions of years old. Fossils in amber would seem to be good candidates for DNA analysis through the technique of polymerase chain reaction, or PCR.

PCR is a method for amplifying genes or complete strands of DNA. The method involves separating the paired strands of DNA and then using each strand as a template to create other strands. Appropriate chemicals are added to the mixture as a feedstock. With PCR, part of a DNA molecule can be reproduced in as great a quantity as one needs for analysis. Often, instead of using the entire DNA molecule, a single gene is isolated. Then the copies of the gene can be used in genetic engineering. With additional copies to insert into a bacterium or an egg cell, the chance for the success of such engineering projects is increased.

Poinar had an even more ingenious idea. Many insects obtain some of their energy by sucking the blood of other creatures. For example, we are most familiar with various mosquito species in which the female obtains the blood from a mammal, often a human, to get the energy needed to produce and deposit eggs. Biologists think that mosquito species using blood in this way first preyed on cold-blooded animals, reptiles, and amphibians, as so some species of mosquitoes today. Whether dinosaurs were cold-blooded or warm-blooded, they probably were preyed upon by mosquitoes.

Suppose a female mosquito is found trapped in amber from a time before 65 mya. Such a mosquito is likely to have had a blood meal. A little of the DNA from this blood could be extracted from the mosquito's digestive system. With PCR, an unlimited supply of the DNA could be obtained, enabling one to study the genes of whatever creature the mosquito had bitten. Given the number of dinosaur fossils, it would seem entirely possible that the DNA might be from a dinosaur. Other biting insects or arachnids might also be used for such an experiment.

There are several hitches in this plan. For one thing, perfectly preserved specimens in amber are extremely valuable to collectors and are even used as exotic jewels. Poinar once found a female lizard preserved in amber whose eggs he would have liked to test for well-preserved DNA, but a Japanese collector bought it for $20,000, far beyond what Poinar could pay. Secondly, it has not been proved conclusively that DNA will be preserved in amber for tens or hundreds of millions of years. Finally, there is the difficulty of locating a biting insect or arachnid that had recently fed on a dinosaur.

In *Jurassic Park,* complete dinosaur DNA is located and produced in quantity. Complete DNA contains the entire genetic makeup of the creature. To activate the dinosaur DNA, crocodile DNA is first removed from crocodile eggs, then dinosaur DNA is used to replace it. Various species of dinosaurs hatch from the eggs. This is fiction.

What does one do with ancient microbes? Cano has located what he believes to be a new antibiotic, brewed some "Jurassic Ale" beer, and formed a corporation to exploit the concept commercially.

(Periodical References and Additional Reading: *New York Times* 9-25-92, p A14; *New York Times* 1-8-93, p A14; *New York Times* 6-10-93, p A1; *New York Times* 6-13-93, p IV2; *Discover* 1-94, p 55; *Nature* 8-4-94, p 333; *New York Times* 11-18-94, p A26; *Science* 11-18-94, pp 1159 & 1229; *Science News* 11-19-94, p 324; *Science* 5-19-95, pp 977 & 1060; *Nature* 11-30-95, p 485; *Science News* 12-3-96, p 373; *Discover* 1-96, p 49)

CHORDATES AND MAMMALS

Being the creatures we are, we tend to be particularly interested in the evolution of vertebrates, mammals, primates, and hominids. Discussion of the latter two categories is reserved for anthropology (*see* "News About Early Humans and Our Other Ancestors," starting on p 29), while the early evolution of vertebrates and mammals tends to be part of paleontology, an earth science.

First Vertebrates

Until recently, scientists believed that the first vertebrates were the armored fish. The armored fish (now extinct) preceded the two modern groups we think of collectively as fish: the true fish and the sharks and rays. However, unknown to the paleontologists, hundreds of thousands of fossils of earlier vertebrates were common in limestones from around the world. Indeed, the fossil vertebrates are so common that they have become one of the standard ways used to date limestones formed between about 520 and 205 million years ago (mya). The fossils were called *conodonts* because they appeared to be tiny conical teeth, although different theories as to what these anatomical features actually were have abounded. The problem that promoted multiple hypotheses about these "teeth" involved the fact that no one could clarify what kind of creature might have possessed such teeth. Without the rest of the body known, the "teeth" could actually be spines, internal hard parts, horns, or even parts of plants, and not associated with animals at all.

In 1982, however, Euan Clarkson of the University of Edinburgh observed a fossil of a small (2 in or 5 cm) eel-like creature that possessed conodonts in the mouth region. He also observed a rodlike part of the creature's body that appears to be a notochord, the ancestor of the backbone. Vertebrates are a subclass of the class of chordates, a class that includes lampreys, hagfish, and tunicates (none of which posses true backbones) as well as the backboned vertebrates. If the creature possessed a noto-chord, then it might be a chordate.

Paleontologists were intrigued but not convinced. Even if the apparent conodont creature had conodonts in its mouth, they were not necessarily teeth. For example, they could be mouth supports. Or perhaps the eel-like creature died trying to eat the true conodont animal!

In May 1992, new evidence on the issue was published by Ivan J. Sansom of the University of Durham in the United Kingdom and coworkers. They examined some 23 different types of conodonts by dissolving parts of them in chromium sulfate, which brought out the internal structure. It was clear to Sansom's team that the conodonts were not all that different from chordate teeth found in the primitive hagfish today. Thus, the apparent teeth in Clarkson's conodont animal are real teeth, strongly suggesting that the conodont animal was a chordate. Even so, doubters remain.

Further evidence appeared in 1995. Mark A. Purnell and coworkers at the University of Leicester in England used photographs from several angles to reconstruct a three-dimensional model of the location of the conodonts (hard parts) in the mouth of the conodont animal. This told them where the conodonts would touch each other when in use. They then used a scanning electron microscope to examine conodonts, and found that they have wear patterns consistent with teeth which are used to grasp and chew other animals in the same size range as the conodont animals themselves.

Other University of Leicester paleontologists Sarah E. Gabbott and R. J. Aldridge, working with J. N. Theron of the South African Geological Survey Geological Survey, located a single "giant conodont animal" in an Ordovician shale found in mountains north of Capetown, South Africa. About 10 times as long as Clarkson's conodont animal, the giant was only partly preserved, but would have been about 15 in (40 cm) long when alive. The Gabbott group also used the scanning electron microscope on their specimen and determined that the eyes and muscles appear to be those of vertebrates. Combined with all the other evidence, this last finding greatly increases the probability that conodonts were chordates.

An even earlier chordate ancestor was identified in 1995, although it had been discovered in 1991. The ancestor, *Yunnanozoon lividium*, was found in the Chenghiang fossil field in China and dates from about 525 mya. *Yunnanozoon* is not a conodont but is related to one of the three main branches of Chordata, the subphylum Celphalochordata, best known for a small marine burrower called the lancelet or *Amphioxus*. Although at one time *Amphioxus* was touted as likely being the common ancestor to all the chordates, it is now seen as far too specialized to have been the root. Thus, the presence of *Yunnanozoon* in Cambrian times should not be taken to mean that *Yunnanozoon* is the earliest ancestor. Instead, its affiliation with *Amphioxus* suggests that the chordates had already split into its main branches at that time. If so, there should be a true vertebrate ancestor somewhere in the Cambrian seas, although it is yet to be identified.

The Rise of the Mammals

One small branch of contemporary mammals is often considered to be extremely primitive because of its retention of the characteristic of egg laying, whereas the vast majority of mammals are viviparous, or give birth to "live" babies. Today, the only monotremes (as the egg-layers are known) are one species of platypus, found only in Australia, and two species of echidnas, one from Australia and the other from New Guinea. The platypus was more successful before 15 mya, when various species flourished, all of which had teeth. One theory is that loss of teeth led to the decline into a single species, although neither the fossil record nor evolutionary logic exists to support this concept. In 1992, one of the teeth from an early platypus was found in Patagonia, in South America, dating from about 63 mya. At that time, South America, Antarctic, and Australia were joined into the single continent called Gondwana (or Gondwanaland). It is not clear why monotremes survived in Australasia and not in the Americas.

(Periodical References and Additional Reading: *New York Times* 5-2-92, p A16; *Discover* 1-93, pp 66 & 67; *Nature* 4-27-95, pp 761 798, & 800; *Science News* 4-29-95, p 261; *Nature* 10-26-95, pp 682 & 720; *Discover* 1-96, p 45)

WEATHER AND CLIMATE, NOW AND THEN

LIGHTNING AND RELATED PHENOMENA

Lightning is one of the most common phenomena on Earth. From space, it can be observed that at any moment about 2,000 storms are occurring on Earth, and many of these are producing lightning bolts. Ancient cultures tried to explain these long bright streams of light and their accompanying loud noises as the weapons of a god. They felt quite certain that lightning was a weapon since it killed people and animals and set trees on fire. In the United States alone, lightning continues to set about 10,000 forest fires and to kill 100–200 people each year.

Despite humanity's uncomfortable familiarity with lightning, its nature was not known with certainty until Benjamin Franklin's famous kite experiment of 1752. Since that time, although it has been clear that lightning is a phenomenon involving an electric current, its cause has still been debated among scientists, who have failed to convince each other that any of the proposed mechanisms for generating great currents is correct in detail. Furthermore, lightning other than bolts between clouds or between a cloud and the Earth has only recently been accepted by the majority of specialists, although ordinary people observing weather have often seen "ball lightning" and have watched it behave in ways quite different from bolt lightning.

Strange Reports and Sightings

Since ordinary lightning is still not well understood, it should be no surprise that science continues to discover other forms of lightning. In the 1990s, researchers have come to accept that giant flashes of energy head upwards into the stratosphere, producing visible light, radio wave, and gamma rays. Even though some of the visible-light versions have been reported for many years by various observers, the reports were for a long time treated somewhat the way reports of meteorites were handled early in the history of science, or more recently the reports of ball lightning. In the early 1990s, however, a number of investigations produced what scientists would deem hard evidence of the stratospheric equivalent to lightning, which is to ordinary lightning what Jupiter is to Earth.

The first evidence came from military satellites, which detected mysterious gamma-ray bursts (*see* "Gamma-Ray Burster Mysteries and Solutions," p 202) from space while trying to detect nuclear explosions on Earth, and also confirmed that these bursts from Earth were not part of nuclear explosions. Since the Compton Gamma Ray Telescope was launched in 1991, in part to learn more about the bursts from space, it has also found dozens of Earth-based gamma-ray bursts, each lasting a few milliseconds. These are not caused by lightning, but appear to be a phenomenon caused by energy releases into the stratosphere. Even if lightning in the lower atmosphere, or troposphere, produced gamma rays, the energy would be absorbed long before it reached a satellite in space. The altitudes at which the gamma rays form are the same as those of some of the odd visual phenomena described below, however. As Gerald Fishman of the Marshall Space Flight Center in Huntsville, Alabama, who wrote the report on the gamma ray bursts in the May 27, 1994, *Science* commented, if there is no connection between the optical phenomena and the gamma rays then "we have two mysteries instead of one."

Also in 1991, a group of scientists led by Umran S. Inan from Stanford University in Palo Alto, California, detected something odd above thunderstorms while studying the ionosphere with radio signals from Earth. The group proposed that their odd phenomenon might be due to heating of the ionosphere caused by lightning in the storms, but the following year a video camera aboard the space shuttle observed a giant flash above a thunderstorm, a phenomenon that appeared to be more complicated than mere heating caused by underlying lightning. In 1993, however, Stanford scientist Inan and coworker Yuri Taranenko detailed a theoretical explanation of how a lightning bolt below or in a thunderstorm could heat the ionosphere enough to make a visible flash. Observations seem to support their theory.

Another satellite, *ALEXIS*, was launched by the U.S. Department of Energy in 1993 to detect nuclear explosions by collecting high-energy radio waves. Like the previous efforts with gamma rays, the high-energy radio waves were soon found to come from something happening in the stratosphere, clearly not nuclear tests. The flashes of radio waves carried 10,000 times the energy of those produced by lightning, and dozens of the radio flashes happen every year. Like the gamma ray bursts, the radio flashes are associated with thunderstorms. Since the discovery of optical phenomena occurring above thunderstorms, it has become clear that at least some of the radio waves are linked to these.

Starting in 1993, Walter A. Lyons of the Mission Research Corporation in Fort Collins, Colorado, set up a video camera at the edge of the Rocky Mountains where it could photograph the sky above several hundred miles of the nearby Great Plains. The cameras were soon recording giant and strange-looking light shapes high in the stratosphere. Like the radio waves and gamma rays, the colored sheets of light are associated with thunderstorms. These structures exist as high as 50-60 mi (80-100 km) above the storms, however, and measure as much as 12 mi (20 km) wide.

Sprites, Elves, and Blue Jets

The newly found forms of light displays are now sufficiently well known that they have received individual names. While a name such as *blue jets* for one type of upward bolt is self-explanatory, one may need help in understanding the terms *red sprites* and *Elves*. Apparently, the word "sprite" was applied to the low-intensity colored phenomena somewhat whimsically to indicate that the colored lights were evanescent and poorly understood, as are the small elusive creatures called sprites in some folk cultures. The name *Elves* is a partial acronym, derived from "Emissions of Light and Very Low Frequency Perturbations Due to Electromagnetic Pulse Sources" which could have been, one supposes, *Eolavlfpdteps,* although that is a bit harder to remember than *Elves.* Elves and sprites are pretty much the same thing in folk mythology.

Experiments suggest that the red color of most sprites, as well as of the Elves, comes from ionized nitrogen. It is not clear, however, how the storm ionizes the nitrogen, although one theory suggests that lightning produces an electric field that serves to accelerate upward any loose electrons which incoming cosmic rays have knocked off molecules. These upward heading electrons produce a chain reaction of ionization that eventually affects the nitrogen molecules that constitute 80% the atmosphere. A similar electric field has been reproduced in the laboratory by Earle R. Williams of M.I.T. in Cambridge, Massachusetts. Further tests are needed to see if such a mechanism is really at work in the stratosphere.

Sprites are not only noted for their red color but also for their ability to reach up to about 55 mi (90 km) above thunderclouds, while the narrower blue jets only reach half as far. Some accounts report sprites as being salmon, green, yellow, or pink instead of red. Sprites also are connected with radio signals while blue jets are not. Sprites often appear in small groups (five being a common number) and seem to appear all at once for their whole length. Blue jets are alone and can be seen to propagate just as lightning does. The jets are narrow at the base and widen as they flash upward. They travel at about 60 mi (100 km) per second, 300 times the speed of sound, but still slower than the speed of light. Elves are vast donuts, appearing so briefly that they cannot be seen by humans, even though they may be 250 mi (400 km) in diameter. They are also higher than sprites or blue jets, appearing at altitudes of 40 to 60 mi (70 to 100 km). Ordinary lightning which occurs between clouds or between clouds and the Earth is limited to altitudes below 12 mi (20 km). Because of their short duration, Elves are the most likely candidates as the cause of the gamma-ray bursts from Earth.

Reports of sprites go back a long time, because it is possible to see a sprite with the naked eye, provided that the eye is completely dark-adapted and is looking at an appropriate angle above a thunderstorm. One reason sprites are not seen as often as lightning is that a sprite lasts only about a thousandth of a second while lightning often produces flashes lasting a second or more. Pilots have often reported seeing something above thunderstorms, but scientists were not much interested. In 1989, a sprite was accidentally photographed from the ground, occasioning a lot of puzzlement and starting the current research effort. In 1993, a directed effort by Davis D. Sentman and Eugene Westcott of the University of Alaska in Fairbanks captured definite images in black and white. They took many more images in June and July of 1994. These were captured in color, recording both the red sprites and blue jets with specially designed low-light video cameras aboard jet airplanes. By then, many scientists recognized that they had interesting and possibly important newly accepted phenomena to study. By the end of 1994, one sprite investigator, Bill Boeck of Niagara University, estimated that one sprite per minute occurs somewhere on Earth.

(Periodical References and Additional Reading: *Science* 5-27-94, pp 1250 & 1313; *EOS* 6-7-94, p 259; *Science* 8-5-94, p 740; *Science News* 8-6-94, p 87; *Science News* 12-17-94, p 405; *EOS* 12-20-94, p 601; *Discover* 1-95, p 67; *New York Times* 1-17-95, p C1; *New York Times* 1-22-95, p IV-1; *New York Times* 12-12-95, p C1; *Science News* 12-23&30-95, p 421)

PATTERNS OF WEATHER

We have understood the basic causes of climate and weather for a long time. Even ancient peoples knew that climate changes with altitude and latitude. Early humans had ways, some still valid, to predict the next day's weather and those annual changes called seasons. Recently, we have made progress concerning the problem of predicting and understanding longer-term changes in climate and medium-term changes in weather. The keys to such progress involve knowledge of the interactions between the oceans and the atmosphere and recognition of the ways that water and heat are transported around the globe. In 1995, the U.S. National Weather Service took this progress into account and began issuing its first regular forecasts for three-months of weather at a time.

Rivers in the Air

One of the newer insights into the movement of air and water began when Reginald Newell of M.I.T. in Cambridge, Massachusetts, began looking at trace gases in the atmosphere. As early as 1985, he observed carbon monoxide unexpectedly appearing in the air above the Indian Ocean. Newell thought that it must have been transported to this region, far from any significant automobile traffic or other source of carbon monoxide, by water vapor. To check his hypothesis, he started a program of combining into a single chart information on water vapor from balloon soundings, satellite studies, and any other sources he could find. To his surprise, Newell observed water vapor forming a vast transportation network, some places in sheets, and other places in definite rivers through the atmosphere. The carbon monoxide that had originally started Newell on this quest had been transported by sheets and rivers from Africa (where the gas had been produced by burning forests to make grassland) all the way to the Indian Ocean.

By 1993, Newell and his coworkers had found that the water vapor forms vast rivers in the atmosphere, carrying as much as 350 million lb (160 million kg) of water past a given point each second. This capacity is possible because the rivers tend to be about 150 mi (240 km) wide although just 1 mi (1.5 km) deep. The longest such river is about 4,000 km (6,500 km). Most sky rivers flow basically from the tropics toward the poles, although their course is deflected to the east in the Northern Hemisphere and to the west in the Southern Hemisphere. These deflections are caused by the Coriolis force, secondary to the Earth's rotation. Thus, the rivers in the air mimic the rivers (such as the Gulf Stream) within the sea. The chief difference between the two kinds is that atmospheric rivers come and go, while the ocean currents seem to be permanent (*but see* "Climate Clues to the Past," p 515). There is some evidence suggesting that sky rivers are the main source of water for large storms such as hurricanes.

A Conveyer Belt in the Atlantic Causes Weather Cycles

Climatologists studying both present-day climate variation and those of the past are fond of the metaphor of a conveyer belt for the main ocean currents. These are the currents that stir up most of the ocean, moving cold water toward the equator and warm water toward the poles. The main characteristic that the currents have in common with a conveyor belt is that they carry heat from one part of the world to another. Furthermore, the currents come in pairs (with one often atop the other), each moving in opposite directions, like the top and bottom part of a continuous conveyor belt. The conveyer belt of the North Atlantic has been of special interest, both because it is thought to have been one of the main actors in recent ice ages (*see* "Climate Clues to the Past," p 515) and because it may cause an important cycle in weather today.

In February 1994, Michael Schlesinger and Navin Ramankutty of the University of Illinois developed a model that combined the effects of greenhouse gases with those of aerosols that promote cooling by reducing sunlight. Running the model over the past 130 years showed that there seems to be an underlying 70-year cycle of warming and cooling. The greatest effect seems to be in the region dominated by the North Atlantic. A different simulation, developed by Thomas Delworth and coworkers at the Geophysical Fluid Dynamics Laboratory in Princeton, New Jersey, reveals how the conveyor belt of currents in the North Atlantic gradually slows itself down for a time and then speeds up again. This cycle results from natural feedback as the currents interact with fresh water from the rivers of North America. This cycle seems to take from 40-60 years, which means that it may be the same as the one found at the University of Illinois. In this kind of study, 60 is close enough to 70 to think they may be the same, and in any case the University of Illinois has fewer than two complete cycles to measure if the period is 70 years.

Not to be outdone by a smaller ocean, the North Pacific may also have an underlying cycle. Mojib Latif of the Max Planck Institute for Meteorology in Hamburg, Germany, and Tim P. Barnett of the Scripps Institute for Oceanography in La Jolla, California, ran a computer model that dealt primarily with the interactions of ocean and atmosphere over the past 100 years. They were surprised to observe a 20-year period in changes in sea surface temperature in the North Pacific that occurred independent of any other changes. Checking with other computer modelers at the Max Planck Institute and at the Geophysical Fluid Dynamics Laboratory in Princeton, New Jersey, showed the same period of about two decades. The proposed cycle may be based upon a natural frequency that stems from the amount of time it takes wind to travel in a vast circle about the North Pacific.

The main problem with the models and the theory, however, is that it is difficult to find the changes in actual temperature records to support the models, although another study suggests that perhaps changes in Siberian land temperature might correlate with the proposed sea-surface temperatures. A hundred years from now these models may be able to be tested against sea-temperature variations calculated from satellite data, but at present the amount of North Pacific data from the past is quite small. The *Topex/Poseidon* satellite from the United States and France, for example, can detect water temperature from space, but it was not launched until August 1992.

El Niño and Climate

Although the El Niño was well known to people interested in climate and weather or even in the anchovy fisheries of Peru, it did not become famous worldwide until an unusually great El Niño began in 1982. By 1984, most people assumed that the 1982 event was over. Gregg Jacobs and coworkers from the U.S. Naval Research Laboratory reported a dozen years later that in some ways the great El Niño was still around. They tracked the wave of warm water that started the process. After it warmed up the coast of South America, the wave gradually spread north and south, raising water levels along the entire west coasts of North and South America. Unable to travel further in that direction, the wave bounced back and headed west, back toward Asia and Australia. The Pacific is enormous, and waves travel at speeds that are determined by the ratio of the wavelength to the period, so that the long-period El Niño wave moved slowly across the sea. It was not until 1991 that El Niño encountered the Kuroshio Current, which flows between New Guinea and Japan and past Japan. Even in 1991, the wave still had enough strength to push the current some 200 mi (300 km) out of its usual path. The last remnants of the giant wave were still detectable in the Bering Sea as late as 1994.

Meanwhile, the early 1990s got repeated doses of the El Niño climate medicine, as typical El Niño warming in the South Pacific appeared in three out of the first four years of the decade (keep in mind that the *decade* starts with 1991, although the *1990s* include 1990). Indeed, the El Niño pattern persisted until early 1995. What is more, the South Pacific has essentially been in its warm mode since 1976. During the past 20 years, there have been four strong, warm El Niños and only one cool La Niña, although in earlier periods the warm and cool tended to alternate. Furthermore, the effects of moderate warm El Niños in 1986-87 and 1991-95 were superimposed upon the lingering waves of the strong El Niño of 1982-83.

El Niño alters the worldwide climate in mysterious ways. In 1994, it was discovered that harvests in Zimbabwe are more closely correlated with changes in El Niño than they are with changes in rainfall in Zimbabwe itself. A more typical link connects El Niño to unusually warm winters throughout the northern United States and unusually cool and wet conditions in the southwestern United States.

One of the reasons that El Niño has such global effects concerns its relationship to similar weather patterns in the central Indian Ocean and to related effects in the Atlantic. The El Niño in the Indian Ocean takes place at the same time as the one in the Pacific, but the event in the Atlantic trails the other two by about a year or year and a half. Each event in the Indian and Pacific Oceans begins as a warm pool of water in the west, gradually moving eastward across the Indian and Pacific oceans in phase. A year or so later, a similar warm pool develops off the coast of Africa, occurring at about the same latitude as the initial pools in the other two oceans. However, the warm water in the Atlantic has no place to go, since it has the African continent directly to the east.

Although the cause of an El Niño is not known for sure, one interesting speculation was proposed in 1993 by Ken Macdonald of the University of California at Santa Barbara, who suggested that undersea volcanic eruptions might trigger the warm water that touches off the whole pattern.

Background

El Niño was first recognized as a change in conditions off the coast of Peru and Ecuador, during which normally dry inland areas received torrential rains while, simultaneously, fish and the birds that fed upon them died. The phenomenon was named El Niño (the boy child), after the Christ child, because the phenomenon usually occurred around Christmas time. Long ago, scientists determined that both the peril to fish and the local weather changes could be explained by a shift in the position of the cold Humboldt Current, which usually washes the coast.

In 1982-83, the strongest El Niño of the century occurred, spurring intensive study of the phenomenon. During the 1982–1983 event, weather seemed to be abnormal worldwide. India and Australia experienced droughts that were thought to be caused by the event. Animal and kelp die- offs around the world were also related to the unusual weather. During this period, meteorologists realized that a central Pacific phenomenon they had named the Southern Oscillation was behind the changes in the Humboldt Current. They also suspected that this Southern Oscillation caused global changes. Among meteorologists, the event came to be called the ENSO for "El Niño-Southern Oscillation," although most people continue to say "El Niño." A similar cooling trend that often alternates with ENSO is logically called "El Viejo," the old one, but sex has triumphed over logic once again, and the cooling trend is called "La Niña" (the girl child).

Studies of historical records coupled with examinations of weather preserved as ice in South American glaciers have shown that an ENSO typically takes 12–15 months from start to finish and occurs every 4–7 years.

The Sun and Weather

Since the later 1970s, most climate researchers admit to a cyclical solar influence on global climate changes (see "Climate Clues to the Past," p 515), but there is considerably more controversy about cycles that are longer than the familiar day-night and yearly patterns and shorter than the 26,000-year precession of the equinoxes that probably plays a role in creation of ice ages. At the end of the eighteenth century, William Herschel was the first to propose that weather responded to the observed cycle of sunspots, then thought of as having a period of 11 years, although now considered to be twice as long because magnetic orientation is taken into effect. In the late 1980s, Karin Labitzke of the Free University of Berlin and Harry Van Loon of the U.S. National Center for Atmospheric Research in Boulder, Colorado, presented evidence that convinced most climatologists that Herschel had been right in stating that there was a distinctive 11-year pattern of low-pressure systems over the North Atlantic. In 1994, Labitzke and Van Loon proposed that the actual cycle was more like 10–12 years and that it stemmed from changes in air circulation between the tropics and the poles, air movement known as Hadley circulation.

Various astronomers before and since have calculated the actual change in the energy available to Earth from sunlight and have found it short of what would be needed to make any difference in weather or climate. In 1994, J. Lean and David Rind of the Goddard Institute for Space Studies calculated that the change in solar output between a solar maximum and a solar minimum is only about one part in a thousand, for example. A random string of changes of that magnitude over a period of ten 22-year cycles, however, could account for as much as a change as one part in 400. That amount is enough variation to cause a period in which the temperature is lower or higher than at the other extreme. Although Lean and Rind think that the amount of solar energy cannot account for changes in weather, they also recognize that the weather variations do follow the same 11-year cycle as the sunspot cycle and that long-range weather forecasting can safely be based on the solar cycle.

If there is not enough variation in energy output, then what does cause the change in weather? Several ideas have been proposed. In August of 1994, Brian A. Tinsley of the University of Texas reported that his calculations show that the solar wind, which also varies with sunspots, has enough influence on the electrical field surrounding Earth that it can cause changes in precipitation. Such changes can by themselves alter temperature. This idea has actually been championed by Tinsley for a number of years, but he did not previously have the data needed to back up the hypothesis.

A similar modulation has been proposed for changes in the ozone layer. J. D. Haigh of Imperial College of Science, Technology, and Medicine in London, England, developed a simple computer model to determine if stratospheric ozone could influence the heat balance of the lower atmosphere by screening out or failing to screen out ultraviolet radiation. Haigh was especially interested in whether the greater variability of the 22-year solar cycle at ultraviolet wavelengths might have an effect on climate that was modulated in part by the ozone. The Haigh model explains how the greater amount of change at ultraviolet levels combined with seasonal changes in the ozone layer could tie the global weather pattern more closely to the sunspot cycle. David Rind has calculated that the ultraviolet flux will heat up the stratosphere, which he thinks might modulate the weather through changes in cloud cover or winds.

Recently, researchers working on lasers confirmed an effect postulated by Italian scientists Roberto Benai, Alfonso Sutera, and Angelo Vulpiani, which explains the mysterious 100,000-year cycle in past ice ages (*see also* "Climate Clues to the Past," p 515). They observed that the generally "noisy" background of climate changes might amplify a weak cycle caused by changes in the eccentricity of Earth's orbit. The same effect could be at work for the sunspot cycle. Even though the periodic change in energy is too weak to cause changes in weather, it can be amplified by the "noise" of irregular weather changes that happen all the time.

Hurricane Helpers

The observed correlation between the weather off Peru's coast, in Australia, and in South Africa suggests that other worldwide patterns must also exist. Understanding long-term weather patterns is important, but most of us are more concerned about predictions of short-range weather. For example, it has long been recognized that the typical pattern for hurricanes in the North Atlantic is for them to increase in number in late summer, to disappear in winter, and to strike the Caribbean and Gulf of Mexico more often than New England. This knowledge is only slightly helpful to people who live along the western coast of the North Atlantic. What people really want to know is exactly when a hurricane is on its way, how big it will be, and where it will go.

A possible advance in the tricky art of hurricane prediction occurred in 1990, when weather forecasters reviewed the performance of the trial run of a new Yugoslavian-designed computer model called ETA (named after the Greek letter that was its model designation). ETA takes data on existing weather conditions and forecasts what the conditions will become after 24 and 48 hours. A hurricane is powered by heat produced as clouds release rain. ETA apparently beats other models in hurricane prediction partly because it assumes more abundant rainfall under given conditions, more accurately reflecting what happens during a hurricane. Additionally, because it is devoted solely to predicting for North America, it has more computational power available for that task than programs that deal with the whole world. With more power, ETA can use a smaller 50 mi (80 km) grid of data points, improving the sharpness of its predictions. This precision will be increased in future versions.

Overall trends also help in making specific predictions. In 1990, William M. Gray of Colorado State University in Fort Collins reported on a newly discovered weather pattern that has become another tool for hurricane predictors. He recognized that a well-known weather pattern in the western Sahel region of Africa, just south of the Sahara, is directly related to a previously unobserved pattern in hurricanes reaching the southeastern United States. The Sahel, for reasons that are not currently known, alternates between wet and dry periods on a cycle of about 20 years. The evidence for such a cycle stretches back to the seventeenth century. Although tropical storms that develop into hurricanes can form over any stretch of warm water, those that form off the coast of West Africa and remain coherent enough to reach the east coast of North America are generally stronger and more dangerous than those that form closer to North America. During rainy weather in the western Sahel, also known as a serious monsoon, conditions are excellent for forming strong, coherent tropical storms. The easterly jet stream is slower, giving storms a longer time to form. It propels the storms slowly but surely toward the Caribbean and the Gulf. Consequently, some weeks after a storm forms off Africa, a strong hurricane arrives in North America. The difference in frequency of hurricanes can be astonishing. During the wet Sahel cycle from 1947-69, 13 strong hurricanes struck the United States, compared with just one during the dry cycle from 1970 through 1987.

Gray has been issuing hurricane-season predictions since 1984. He revised his method considerably in 1990 to factor in sea-surface temperatures, atmospheric pressure over the ocean, shifts in major wind currents, and, of course, rainfall in Africa. Every factor that influences hurricanes according to Gray's system came together in 1995. The rains came in the Sahel. El Niño was missing for the first time in several years. There was a warm spring in the Caribbean region, producing a low air pressure cell that would strengthen storms as they reached the west side of the Atlantic. Gray predicted a heavy hurricane season, but he did not quite expect the final outcome, which included 11 hurricanes, 5 of them intense, as well as 8 tropical storms. During the twentieth century, only 1933 proved to be a more severe hurricane season in North America. An average season is 6 hurricanes and 4 tropical storms.

With El Niño gone, Gray predicted a normal hurricane season for 1996.

The pattern of strong and frequent hurricanes in 1995 was only one more example of the 1990's more-violent-than-usual weather. Many computer models of global warming (*see* "The Future of Climate," p 525) forecast sharp changes in weather from month to month after worldwide temperatures reach a certain level. Yet the temperature during the early 1990s, when the extreme weather conditions occurred, had not reached a level that should cause such conditions. Thomas Karl and coworkers at the National Climactic Data Center in Asheville, North Carolina, have developed an index based on weather extremes, however, including high temperatures, heavy rainstorms, droughts, and long periods of similar weather of any kind. They find that recently the weather has tended more toward those extremes.

In case you have forgotten the extreme weather, you might not want to be reminded of

- the extra-cold summer in the northeastern United States in 1992
- Hurricane Andrew in August 1992
- the violent Nor'easter of March 1993 on the northeastern Atlantic Coast of the United States
- the great Midwestern floods of the summer of 1993
- the endless winter and continuous snows in the northeastern United States in the winter of 1993-94
- the strangely warm winter of 1994-95
- the heavy rains on the West Coast of the United States in the winter of 1994-95 and the autumn of 1995
- the succession of record snows across the eastern United States in the winter of 1995-96 and the continued pounding of heavy rain on the West Coast
- the El Niño and hurricanes described above.

One explanation is a shift in the jet stream, but there seems to be nothing to account for that shift. Some blame everything on El Niño. A simple explanation calls it a run of really bad luck.

(Periodical References and Additional Reading: *New York Times* 2-14-93, p A1; *New York Times* 2-28-93, p I-23; *Discover* 1-94, p 34; *New York Times* 5-24-94, p C1; *New York Times* 7-26-94, p C4; *EOS* 8-9-94, P 369; *New York Times* 8-9-94, p C4; *EOS* 8-16-94, P 377; *Science* 10-28-94, pp 544, 632, & 634; *Scientific American* 11-94, p 14; *EOS* 11-22-94, P 585; *Discover* 1-95, pp 68 & 77; *New York Times* 1-14-95, p 1; *New York Times* 1-24-95, p C6; *Science* 4-21-95, p 363; *Scientific American* 8-95, p 66; *Scientific American* 11-95, p 62; *Science News* 12-16-95, p 415; *Discover* 1-96, p 92)

CLIMATE CLUES TO THE PAST

Scientists have found ways to measure entities that once seemed beyond their reach, including the interior of the active human brain; the once invisible world of the atom; the temperatures of stars; the composition of Earth's interior; and the date a long-dead human passed away. One of the trickiest of these indirect measures concerns finding the global temperature of the Earth in the past or attempting to predict it in the future (*see also* "The Future of Climate," p 525). Although some easy inferences can be made about past climate (leaf impressions in coal suggest that the coal probably formed in a temperate or warmer zone, for example), most knowledge about past climates derives from less direct clues.

Before the nineteenth-century revelation of the existence of the recent ice age, scientists and people in general thought that Earth's climate stays fairly much the same, although older people have probably always claimed that it was colder or hotter or wetter when *they* were young. Once Louis Agassiz had popularized the notion that an ice age had produced a dramatic climatic change, other scientists began to devise ways to examine past climates. As more of the fossil record became known, it was easy to recognize that in some periods, vegetation or animal life corresponded to that now associated with life prevalent in the tropics or in cool climate areas. Geologists were also able to recognize characteristic scars and deposits left by glaciers from a variety of times much earlier than the recent ice age. These clues produced a general impression of Earth's climate history as consisting of long warm periods intermingled with a number of ice ages. Scientists, however, crave details and not general impressions, so they have worked to develop the means of obtaining such details.

Sometimes the initial observations used in obtaining these facts seem to have little to do with climate. Agassiz and the eighteenth-century geologists who preceded him initially diagnosed the cold of the ice ages from scratches on rocks—scratches, they deduced, must have been caused by glaciers. One of the key clues to recent swings in climate has come from the work of Harmut Heinrich of the University of Göttingen, who observed in the early 1980s that several odd layers of sediment were found in the North Atlantic. Fragments of rock, half a meter thick toward North America, but thinning toward Europe, contain few shells of foraminifera, the generally ubiquitous protozoa of ocean surface waters. Heinrich identified the cause of these odd layers as six (later increased to seven) separate flotillas of giant icebergs that had each been launched into the Labrador Sea from an icecap covering Hudson Bay. The ice cap had contained limestone and dolomite from the Hudson Bay regions brought to the edge of the sea by glaciers. As the icebergs slowly melted on their passage across the Atlantic they dropped the rock debris, which sank to the bottom and remains there to this day.

Very Recent Climate Changes

Written accounts provide some information about the past few hundred years, although important devices used for measuring even such a basic quantity as temperature were not invented in the most primitive form until the seventeenth century. Furthermore, many places of climate interest are not represented in written accounts, which are generally available only for northern temperate zones on land. Past changes in climate for particular regions can be inferred from various environmental changes that left permanent records, however. For approximately the past 10,000 years, carbon-14 can be used to date plant and animal remains, indicating which species were growing at a particular time. Furthermore, tree trunks contain in their rings a yearly weather record that can show drought or other severe climate conditions. Pollen grains can also be used to indicate which species were plentiful at a given time. Even oxygen ratios (*see* below) can provide global data or data on local rainfall. Advances or retreats of lakes can also be used to ascertain general rainfall patterns.

Written records indicate that there was a warm period in Medieval times in northern Europe, at least. Although the name "Greenland" was a bit of puffery on the part of Eric the Red designed to attract settlers in the tenth century, it is clear that Greenland actually was much warmer during the period of Viking settlers. Iceland, too, was warm, growing wheat and barley. This warmest period is generally identified as from about 1100-1300. During most of that time, English farmers were able to grow and harvest wine grapes, which no longer thrive in Britain. There is also some evidence for a cold snap early in the twelfth century. Later, a longer cold wave (referred to as the Little Ice Age) struck Europe and North America during the period from 1550-1800. Warming after the Little Ice Age persisted until about 1940, when a cooling period set in that lasted for about a quarter century. Recently, on the average, the climate has been growing warmer.

In 1993, climate investigations by Charles R. Ortloff from the FMC Corporate Technology Center in Santa Clara, California, and Alan L. Kolata of the University of Chicago in Illinois suggested that a massive drought in South America coincided roughly with the warm Medieval times in Europe. Prior to 1000 A.D. there had been a civilization centered on Lake Titicaca and now called the Tiwanaku (better known, perhaps, by its older spelling of Tihuanaco). Ortloff and Kolata attribute the collapse of that civilization in the year 1000 to the effects of the drought, which may have started as early as 950 and not concluded completely until 1410. The main evidence used by scientists to identify the drought includes pollen and oxygen ratios. Another 1993 study by J. Q. Fang demonstrated drought in Eastern China during the first part of this period, from 950-1250.

In the June 16, 1994 *Nature*, Scott Stine of California State University at Hayward reported on his use of a combination of carbon-14, tree rings, and lake levels to identify two similar superdroughts in California. The first drought was from 892-1112, a period of 220 years, while the second was from 1209-1350, a period of 141 years. Water levels fell in the lakes, which had no outlets, as much as 50 ft (15 m) during the drought periods. In between, conditions were wet and lake levels high. A similar finding by the same author, refining in 1994 the results of a 1990 study, demonstrates that at approximately the same time as the first California drought there were similar drought conditions interspersed with periods of wetness in the Patagonian Andes of Argentina.

Because the droughts in both North and South America occurred at about the same time as the known warm spell in Medieval Europe, Stine has proposed that the droughts were one result of global warming. He thinks that present-day global warming will bring back drought conditions in California.

Ice Cores

When more snow falls each year than melts, glaciers or ice caps gradually build up. Because the regions near the poles and on some high mountains have received more snow than has melted for thousands of years, these glaciers and ice caps have often been in place for that entire time. A "core" refers to a column of sediment removed from its surroundings so that it can be examined by Earth scientists. In this context, glacial ice is much like any other sediment. Recently, scientists have been using ice cores from sites in Greenland and Antarctica to study the climate and even weather patterns of the past. Russian scientists in the period 1980-85 prepared a long core of Antarctic ice taken at their Vostok ice station. This core holds the record of the past 160,000 years. A later Russian ice core from the Vostok sampled much deeper and older ice. A U.S. ice-core operation called the Greenland Ice Sheet Project 2 (GISP2) and a European one called the Greenland Ice Core Project (GRIP) each drilled about 19 mi (30 km) apart at the summit of the Greenland ice cap. The Greenland ice sheet where they are drilling is about 2 mi (3.2 km) thick and represents a record that goes back about 250,000 years. Greenland was chosen because its mean annual temperature is too low for summer melting, while its annual accumulation of 10 in (0.25 m) of the water equivalent of snow and ice is large enough to allow analysis of the record year by year.

The GRIP core hit bedrock at a depth of 9,938 ft (3,029 m) in 1992. The GISP2 project also finally ran out of ice, hitting bedrock at a depth of 9,984 ft (3,044 m) in July 1993. Bedrock at the Vostok site is about 11,350 ft (3,456 m) below the surface but represents a much longer record than the Greenland cores because annual ice layers in Antarctica are thin. The air in Antarctica is too cold to hold much water, and so not much snow falls each year.

The main surprise from the Greenland ice cores came from an analysis of individual years, which is not possible in the Vostok Antarctic core due to the thinness of the annual layers. Astonishingly, the Greenland ice cores demonstrate that about 10,000 years ago the annual amount of snowfall in central Greenland doubled over a space of just three years. A temperature increase of about 12°F (7°C) would be needed to sufficiently increase moisture content in air enough for that much snow to fall. The sudden increase in temperature is also marked by a change in the ratios of oxygen isotopes.

The oxygen isotopes reveal that such sudden temperature changes are common. At least 10 times in the past 40,000 years there was a similar abrupt warming, which then lasted for 500–2000 years. These warm periods are called Dansgaard-Oeschger events, after Willi Dansgaard of Denmark and Hans Oeschger of Switzerland, scientists who had major roles in identifying the warm spells in ice cores. Preliminary reports from the GRIP core showed sudden cold spells in the previous interglacial, as well as the brief warm periods during the past glacial. Later analysis, however, demonstrated that the apparent cold snaps were an artifact caused by breaks in the core record resulting from ice flow or changes in ice composition.

In addition to measuring oxygen isotopes, which are thought to represent worldwide temperature because of atmospheric mixing, scientists also measured the amount of methane in bubbles in the ice. Methane levels change from time to time, although the causes for such changes are not very well understood. Since the amount of methane is thought to remain the same throughout the atmosphere at any one time, a particular methane signature marks a given moment in Earth's history. The methane amounts can be correlated with oxygen isotopes to check on possible errors and, by correlating the GISP2 and GRIP cores, eliminate problems caused by ice flow. Furthermore, the combination of methane and oxygen isotopes can be used to correlate air bubbles in Greenland ice cores with bubbles in Antarctic ice, effectively linking northern and Southern Hemisphere climate changes. Methane is a greenhouse gas

Background

When snow accumulates year after year, the accumulated weight first compresses the snow crystals into a mixture of snowy ice and icy snow called **firn**, from the German for "last year's (snow)." Air and dust trapped in the snow become part of the firn. During this stage, some gases in air can chemically react with the snow or otherwise become altered, but dust and other gases stay as they were. The gases form tiny bubbles that become less mobile as the weight of the firn and snow above turns the bottom layer of firn into glacial ice. Gases in glacial ice may become pressurized even more as ice layers build higher, although the gas cannot escape until the ice melts. Indeed, when people use glacial ice in drinks, it snaps, crackles, and pops as it melts, releasing the pressurized gases trapped within.

The temperature in the snow, firn, and ice is almost always below freezing from the time the snow falls, and certainly at or below freezing thereafter. So, one might ask, how do scientists use air or dust trapped in ice to determine past global temperatures?

The classic oceanographer's method for dating ice ages comes from studies of two isotopes of oxygen. Most elements found naturally are a mixture of different isotopes (*see also* "Heavier Heavy Elements and More Naming Feuds," p 439). A given element, such as oxygen, always has the same number of protons in its nucleus—eight in the case of oxygen. The number of neutrons in the nucleus can vary, however, producing the different isotopes of the element. An isotope is identified by a number that is the sum of the protons and neutrons in the nucleus. Thus, the most common form of oxygen, which has eight neutrons as well as the obligatory eight protons, is called oxygen-16, symbolized $_8O^{16}$. Adding neutrons to a nucleus can make the atom unstable, or radioactive. Oxygen-16, oxygen-17, and oxygen-18 are stable, but oxygen-19 and oxygen-20 are radioactive. In a natural sample of oxygen, about 99.759% is oxygen-16, about 0.204% is oxygen-18, and only about 0.037% is oxygen-17.

Because of the two extra neutrons in oxygen-18, that isotope is slightly heavier than the common oxygen-16. The heavier oxygen-18 is also slower and, one might say, lazier than ordinary oxygen. In evaporation, for example, oxygen-16 is more likely to make it from the liquid to the gaseous state. This effect was used in the classic studies dating the ice ages during the 1970s. That work used the oxygen-18/oxygen-16 ratio in the sediments formed by the shells of tiny sea organisms. The idea behind the method was that during an ice age, as oxygen-16 evaporated, fell as freshwater rain, and was locked up in glaciers, oxygen-18 was left behind in seawater. The more oxygen-18 in the sediments, therefore, the more ice on the continents.

Analysis of the air in ice also produces other results of interest. Glacial ice contains a higher amount of oxygen-16 as a regular course. The ice, however, has air bubbles trapped in it. Oxygen in the air comes from photosynthesis, most of which takes place in the ocean. Plants and protists in the ocean get the oxygen from seawater, which, as we have already seen, is enriched in oxygen-18 during an ice age. Thus, an increase in glaciation is marked by more oxygen-18 in the air bubbles trapped in ice.

(*see* "The Future of Climate, p 525), which makes its connection to climate changes doubly interesting.

Methane levels measured in Greenland ice cores by Jerome A. Chappellaz of the Laboratory of Glaciology and Geophysics of the Environment near Grenoble, France, show that cold periods also result in lower methane levels in ice bubbles. Since methane is produced mostly in swamps and bogs, and since swamps and bogs near the poles are generally frozen and produce little methane, a drop in atmospheric methane is thought to be caused by drier conditions in the tropics. Thus, this drop is taken to mean that cold spells in the region around the North Atlantic have been accompanied by worldwide changes in climate.

Another form of ice-core analysis seems almost too easy to work: it consists of comparing measurements of ice-core temperatures at different depths. Although there is considerable loss of detail because of heat conduction, ice-core temperatures do reveal the broad changes in climate at the site of the core. In 1995, an analysis of the GISP2 core by Kurt M. Cuffey of the University of Washington in Seattle and coworkers demonstrated that the ice-core temperatures independently confirm the temperatures derived from analysis of oxygen isotopes. However, the profile also showed that temperature changes at the end of the last ice age were double what had been expected in Greenland. Cuffy and coworkers attribute this change — an increase of 25-29°F (14-16°C) — to greater warming near the poles than is predicted by many climate models. A different research group using similar methods reached essentially the same conclusions.

Lonnie G. Thompson of Ohio State University and coworkers have been studying two ice cores taken from long-lived glaciers in the high Andes Mountains of Peru. The deeper parts of these cores, dating from 14,500-20,000 years ago, show large amounts of dust, suggesting that much of South America was dry during the glacial periods. Also, the glaciers at that time descended about 0.6 mi (1 km) down the slopes of the mountains, suggesting a colder climate. This is one of many indications that what happened in the polar regions was mirrored in events all over the world.

Ocean Sediment Cores

Geologists take cores in the ocean floor for many reasons. One goal is to analyze climate change. Not only can they use ratios of different isotopes in such an analysis, but examination of the species present in cores provide significant climate clues. For this purpose, the best animals and algae to use are the tiny creatures that build hard shells, for the empty shells that rain on the ocean floor form well-defined layers of ooze that eventually hardens into clay and later into stone.

In September 1993, Gerald Bond of the Lamont-Doherty Earth Observatory used one especially cold-loving species of foraminiferan to identify sudden cold snaps and warm spells in the North Atlantic. The results he obtained from his mud core matched those just released for the two Greenland ice cores. Furthermore, six of the warm spells appeared to be timed just following a Heinrich event, when a flotilla of icebergs crossed the North Atlantic.

There is a theory to explain how a Heinrich event results in warming of the North Atlantic region. During a fairly warm period, like the present interglacial, the ocean currents in the North Atlantic form a sort of conveyor belt that brings heat to Canada, Greenland, and northern Europe (*see also* "Patterns of Weather," p 509). During a glaciation, however, icebergs flow into the North Atlantic, where they melt into liquid freshwater. The freshwater interferes with the conveyor belt, which is powered primarily by the sinking of cold, salty water near the poles. Thus, during the glacial periods of an ice age, the conveyor belt no longer brings warm water up to replace the cold, and polar regions become colder and colder. As a result, the ice sheet extending out from North America becomes thicker and thicker. Eventually, however, a huge chunk of it breaks off from its own weight and floats across the Atlantic. This is the Heinrich event. As that chunk melts, it continues to hold back the conveyor belt. When it is gone, there are no further icebergs released for a time. Thus, the surface waters of the far north become salty and dense, sinking to the floor of the sea and allowing warm water from the south to move up. Very suddenly, the climate in the northern North Atlantic region becomes warm. After a time, when considerable warming has taken place, the air up north becomes more moist, allowing for more freshwater rain and snow, which shuts down the conveyor belt again. As it gets cold, icebergs begin to float across the sea, keeping the conveyor belt shut down until another Heinrich event.

This theory, invoked by a number of specialists in past climates, fails to account for all the data, although it does seem to explain why there is warming after a Heinrich event. One glitch in the theory is that while there are only seven Heinrich events, there are many more episodes of sudden warming or cooling. One of these has an obvious cause, similar to that of a Heinrich event. When the last ice age was coming to an end, the melting ice formed a great lake in the middle of the North American continent. At first, water from the lake (now called Lake Agassiz) flowed into the Gulf of Mexico. Later, when the ice receded a little farther, a new opening to the sea was found, where the St. Lawrence River is today. The fresh water from Lake Agassiz flowed into the North Atlantic with the same effect on the conveyor belt as a Heinrich event: it stopped the conveyor belt and started the sudden period of chill temperatures known as the Younger Dryas when the currents resumed flowing.

A second problem with the theory involves shutdowns of the conveyor belt which would be expected to last for hundreds or thousands of years. Yet the cold spells have not lasted for such long periods without interruption. Stefan Rahmstorf of the University of Kiel has proposed that there is a shallow, secondary conveyor belt that springs into place a few years after the primary one is shut down. His theory does not explain, however, how the system returns from this state to the primary conveyor belt that is in operation today. One of the attractive points of the Rahmstorf mechanism is that it would better agree with the record of radioactive carbon in ocean sediments and as found in dating past events. Shutdowns of the primary conveyor belt for hundreds or thousands of years would interfere with the normal mechanisms removing carbon-14 from the atmosphere, where it is continually formed by cosmic rays.

Also, we know for certain from all sorts of evidence that the climate has been relatively warm for about 11,000 years. Why hasn't this warm period shut down the conveyor belt? No one has a clue.

Another discovery from ocean cores concerns the longer cycle, or repeating period, of ice ages. A well-known theory explaining the ice ages is that of Milutin

Milankovich, who proposed that cyclical changes in the shape of Earth's orbit (90,000–100,000 years), changes in the angle the Earth's axis makes with the plane of its orbit (41,000 years), and wobbling of Earth's axis (26,000 years) together alter the amount of energy from sunlight reaching Earth. Among the various Sun-Earth cycles that have controlled weather in the past, geologists have long recognized that the signal at 100,000 years is especially strong. This period corresponds to changes in the eccentricity, or lack of circularity, of Earth's path around the Sun. The problem is that such changes should not affect temperature as much as some other changes in the relation between the position of Earth with respect to the Sun. Still, this particular change is the one that over the past 800,000 years has been associated with frigid ice ages. In July 1995, K. A. Farley of CalTech in Pasadena, California, observed that helium-3 in ice cores can be used to measure changes in the amount of dust from space falling into Earth's atmosphere and observed that the amount changes over time. The following month, physicist Richard A. Muller of the University of California at Berkeley and Gordon J. MacDonald of the University of California, San Diego, reported that they could connect the 100,000 year cycle to changes in the inclination of Earth's orbit. Furthermore, such changes could influence climate if, in some inclinations, Earth passes through clouds of cosmic dust. This possibility caused Farley and his associate D. B. Patterson to look for a 100,000-year periodicity in the amount of helium-3, which they found and reported before the end of 1995. Surprisingly, the high levels of cosmic dust, which might cause ice ages by dimming sunlight, are associated with the wrong part of the orbital inclination for the Muller-MacDonald theory. Farley and Patterson also point out that it is possible that the ice age somehow causes change in the helium-3 record, rather than both the ice age and helium-3 being the result of extraterrestrial dust.

Challenge from Devils Hole

Devils Hole (no apostrophe) is a 425 ft (130 m) deep crack in Nevada about 75 mi (115 km) northwest of Las Vegas. Over the past 560,000 years or so, rainwater has seeped into the crack, filling it much of the time, and depositing the mineral calcite, just as water does in limestone caves and caverns. The calcite builds up slowly, so a 14 in (35 m) core represents more than a half-million years of accumulation. Isaac J. Winograd of the United States Geological Survey obtained such a core and analyzed its isotopes in two different ways. One way involved using the ratio of oxygen-18 to oxygen-16 to determine global temperatures, as has been done for ice cores and other geological records. The second analysis used isotopes of uranium to date the parts of the core.

The Devils Hole record quickly became notorious because it seemed to provide impeccable evidence that the data from ocean cores was wrong. In particular, the ocean-core data support the Milankovich cycles, but the Devils Hole core record, while matching earlier cycles, no longer matches from about 110,000 years ago until about 60,000 years ago. During that period, the temperature from Devils Hole oxygen ratios is warmer for a period of several thousand years, implying a longer interglacial than shown by the other data.

Various other factors might have influenced the Devils Hole record, but it is difficult

ISOTOPES AND RADIOACTIVITY

Radioactive isotopes change into other forms in various ways until they find a form that is stable. Radioactive carbon-14 (with six protons) becomes stable nitrogen-14 (with seven protons) when one of the neutrons in carbon-14 emits an electron and turns into a proton. Other radioactive isotopes may go through several different forms before becoming stable elements.

Most methods of dating based on decay of radioactive isotopes rely on ratios comparing the original amount of an isotope to either the stable isotope representing the end product of the decay or to an isotope with a relatively long half-life. For example, the first such method used the ratio of radioactive uranium to its ultimate product: lead. It is assumed that when the rock was formed the radioactive isotope was present and the stable decay product was not. If there is no other gain or loss in either isotope (or any other radioactive isotope that has the same end-product), then the ratio provides an unambiguous date for the rock formation. Unfortunately, the date can easily be compromised by intrusions of one kind or another. In addition, many rock samples formed without containing a suitable radioactive isotope that could be used in dating.

Carbon-14 dating is based on a slightly different process. Cosmic rays are subatomic particles that bombard Earth from space. The energy from cosmic rays converts some nitrogen in air to carbon-14, producing throughout the atmosphere a definite ratio of radioactive carbon-14 to the much more common stable carbon-12. This ratio is kept relatively constant by absorption of excess carbon into ocean water and sediments. Atmospheric carbon interacts with oxygen to become carbon dioxide; plants incorporate carbon dioxide during photosynthesis; some animals acquire the carbon by eating plants, others by eating animals that eat plants. As long as all the organisms are living, they continue to incorporate carbon-14 into their tissues in a specific ratio to carbon-12. When an organism dies, however, this process stops. Since carbon-14 is radioactive, it decays at a steady rate, going back to nitrogen. Half of the amount of carbon-14 present will be returned to nitrogen after about 5,730 years. Thus, the amount of carbon-14 that remains, as compared with the amount originally present, can be used to date anything that once was alive. The method can be used only for fairly recent dates (less than about 40,000 years), as the amount of carbon-14 will eventually become too small to measure.

to prove that they did. Seepage could have altered the amount of uranium in some layers, for example. The main argument against Devils Hole, however, is that it represents only one source, contradicted by many other sources.

Other Clues

Many other climate indicators have been used in determining past climates, some obvious and many rather unexpected.

Levels of Lakes Clearly, lakes with no outlet rise during wet climate periods and have lower water levels in dry periods. For example, about 14,000 years ago, immediately after the last Heinrich event, Lake Lahontan in Nevada grew to be 10 times the size it is today. This implies that the southwestern United States was undergoing the kind of weather that it normally gets during a strong El Niño—yet this El Niño seems to have lasted a thousand years.

Pollen in Lakes Pollen from trees and other vegetation floats through the air for much of the year, especially in warmer climates. If it lands on an appropriate flower, the pollen fertilizes the flower. Most airborne pollen, however, fails to connect with the appropriate flower and falls onto one surface or another. For climate studies, the pollen that falls on the surfaces of ponds and lakes is most useful. It mixes with other sediment and drifts to the bottom, where annual layers called *varves* form, each embedded with that year's pollen. Pollen from a particular plant is easily recognized under the microscope, and most pollen does not decay easily. The record held by the oldest pollen found in the United States is at the bottom of Lake Tulane in Florida, which goes back 50,000 years.

Recognizing climate from pollen is straightforward. In Florida, wet climates during the past 50,000 years show up as varves with mostly pine pollen. In drier times, the varves are filled with oak, ragweed, and other pollen. According to the pollen records of Lake Tulane, the damp, piney periods occurred during the same time frame as the last five Heinrich events, which shows that these events' effect on climate (or the effect of whatever caused the events themselves) extended beyond the North Atlantic region.

Tree Rings Special x-ray studies have demonstrated that wood forming tree rings is denser at the end of warm summers than at the end of cool ones. Studies by Gordon Jacoby and Rosanne D'Arrigo of the Lamont-Doherty Earth Observatory used this concept, reporting that tree rings sampled in Alaska indicate that this region is warmer now than at any time in the past 300 years. In theory, the same method can be used with tree-rings from earlier times as well.

Coral Reefs An interglacial is marked by higher ocean levels, since less of Earth's water is tied up in glaciers and ice caps. As a result, an interglacial can be identified by ancient coral reefs that now are above the water line, since coral does not grow unless it is underwater. Improved forms of uranium/thorium dating were used by various oceanographers in 1990 to date the top layer of ancient coral reefs in Barbados. These dates confirmed that the last long interglacial (as opposed to a Dansgaard-Oeschger event) was between 122,000 and 130,000 years ago.

Noble Gases in Ground Water According to the laws of physics, cold water dissolves more of such noble gases as neon, argon, krypton, and xenon from the atmosphere than does warm water. If ground water can be found that has not been in contact with the atmosphere, the amount dissolved should reflect the temperature at which it was last in such contact. According to Martin Stute of the Lamont-Doherty Earth Observatory in Palisades, New York, the ground water he found about 7° S. latitude in Brazil, not far from the equator, has not interacted with the atmosphere for 18,000 years. The amount of noble gases measured in this water would correspond to a climate about 10°F (5.4°C) cooler than it is today. This is further evidence that the last ice age affected the climate all over the Earth. Pollen records also indicate cooling in Central and South America during this period, for example.

Bones in Fish Ears   The calcium carbonate "stones" that form in fish ears, generally a half-dozen within the internal ear, record weather as well as climate. Because these form in thin layers, the layers can be used to date periods in terms of weeks or sometimes even days. Such "stones," called otoliths, can also be analyzed for oxygen isotopes to obtain the general temperature of the water at the time that the otolith was formed. By correlating these data, scientists can relate long summers or winters to global weather changes.

(Periodical References and Additional Reading: *New York Times* 12-1-92, pp C1 & C6; *Nature* 7-15-93, p 218; *New York Times* 7-15-93, p A1; *New York Times* 9-7-93, p C5; *Nature* 12-2-93, p 443; *New York Times* 12-9-93, p B12; *Discover* 1-94, p 67; *New York Times* 2-1-94, p C1; *Science* 2-17-95, p 961; *Science* 3-17-95, p 1595; *Nature* 7-13-95, p 153; *Nature* 6-16-94, pp 518 & 546; *Science* 7-7-94, pp 32 & 46; *New York Times* 7-19-94, p C1; *Science* 7-21-94, p 379; *Nature* 9-14-95, p 107; *Science* 10-20-95, pp 444 & 455; *Scientific American* 11-95, p 62; *EOS* 11-21-95, p 477; *Nature* 12-7-95, pp 558 & 600)

THE FUTURE OF CLIMATE

Knowing about past climates helps solve many scientific problems, but the quality of human life and our environment in general will be defined by what happens to Earth's climate in the future. Robert Frost's well-known poem *Fire and Ice* describes the two extremes, each of which has been foreseen by various scientists as the ultimate fate of the Earth (*see* "Stories on the Solar Wind," p 146, for the latest theory on the final days of Earth). More to the point, what can be expected in the lifetimes of those who are alive today? The most common prediction is global warming—an overall increase in temperatures around the world, although differing in amount from place to place.

Since 1938, with a fresh start in 1967, scientists have predicted that a rise in atmospheric carbon dioxide would trap solar heat, causing a global increase in temperatures. The climate *has* warmed on average during the twentieth century, but global climate is like the weather: sometimes it gets hotter, sometimes it gets colder, and we often don't know why. The climate cooled after the 1938 prediction of global warming and changed very little in the first few years after the 1967 warning. Although the climate seems to have warmed more in the past few years and overall has risen about 1°F (0.5°C) since 1900, such recent warming is no guarantee of a continuing rise.

The most authoritative group of scientists to claim that global warming in the twentieth century has been caused by carbon dioxide and other greenhouse gases is the United Nations Working Group I of the Intergovernmental Panel on Climate Change, which included 2,500 members. Its report, made available late in 1995, states that "the balance of evidence suggests that there is a discernible human influence on global climate." That is, we are responsible at least in part for global warming. The key element in this conclusion was the recognition of the roles sulfate aerosols have in climate change, as reflected in new computer models (*see* subsection "Computer

Models" below in this article). The report also predicts future warming over at least the next century that would be somewhere between 2-9°F (1-5°C). Earlier in 1995, an extremely careful study by researchers at the Max Planck Institute for Meteorology in Hamburg, Germany, also concluded that the odds were 40 to 1 against global warming having been caused by natural variability.

It's Not Just the Heat, But . . .

Carbon dioxide is not a bystander in the ecology of Earth; it is an active player in many ways. Since it is used by green plants in photosynthesis, the size of the world's population of green plants affects the amount of carbon dioxide in the air. Trees, especially, tie up large amounts of carbon in their woody parts for years—as long as thousands of years in the case of a few species. They take the carbon dioxide from the air, use it to collect energy from the sun, release much of the oxygen and keep the carbon. The carbon returns to circulation when the wood rots or is burned.

Sometimes carbon locked up by green plants does not return to the atmosphere for even longer. If it is buried or formed into peat in bogs, it can become fossilized as coal, oil, or natural gas. Certainly in the past, this process removed vast amounts of carbon from the atmosphere. It seems less likely that creation of fossil fuels is a major factor today.

The ocean also collects carbon dioxide. Some carbon dioxide is merely dissolved in ocean water (where it ceases to contribute to the greenhouse effect), and some is incorporated into calcium carbonate, the principal component of most sea shells. After the organism that made the shell dies, the shell is either buried, where it may later be converted to limestone or chalk, or it dissolves in ocean water. In the latter case, the calcium carbonate becomes available for other organisms to use in their cells. Mechanisms such as these remove large amounts of carbon dioxide from the air and keep it out of the air for long periods of time. Eventually, limestone or chalk may be thrust up from the ocean floor by geological processes. Only then does some of the carbon trapped in sediments return to the atmosphere.

It is believed that changes in oceans and in ocean currents caused by movements of tectonic plates have caused variations in carbon dioxide resulting in vast climate changes in the past. Changes have also resulted from carbon being locked up in vast forests or as fossil fuels. The biological storage of carbon may also have been caused in part by movements of tectonic plates. When all the continents are joined in the tropics, and the polar regions are all sea, there would be vast forests; but the forests gradually remove carbon and cool the climate. When a continent is directly over a pole, as Antarctica is today, it can develop an ice cap and store essentially no carbon. This would gradually tend to warm the climate.

These factors make it difficult to predict the amount of carbon dioxide over long periods of time. Furthermore, cloud cover and the presence of snow change the climate in ways that are hard to predict. Thick clouds screen out solar radiation, but if the radiation is emanating from the earth, they trap it. Snow is cold, but is also white, which means that it reflects most light back into space.

With the difficulties of making accurate projections in mind, it is still useful to list some of the possible consequences of the greenhouse effect.

Background

Recently, whenever there is a hot summer or a warmer-than-usual winter, people have tended to blame the greenhouse effect. The greenhouse effect is a real phenomenon, but its actual impact on Earth's climate under present circumstances is not clear.

The cause of the greenhouse effect is a change in the mixture of gases that make up air. Gases tend to be transparent to electromagnetic radiation at visible wavelengths, because light does not react easily with electrons in isolated molecules. At different wavelengths, however, gases are generally not transparent. One example is ozone, which is less transparent to ultraviolet radiation than are other gases in the atmosphere. Another example is carbon dioxide, which is less transparent to infrared radiation, or heat, than it is to light. Both of these reactions between a gas and electromagnetic radiation have proved to be involved in worldwide environmental problems.

Carbon dioxide (and several other gases) in the atmosphere permits solar radiation in the form of light to reach the surface of Earth. There it is absorbed by solids or liquids, although some is reflected in all directions. The light that is absorbed heats the material that absorbs it. This heat is emitted from Earth's surface as infrared radiation. Gases that are not transparent to infrared radiation, such as carbon dioxide, collect this heat and keep it in the atmosphere. If all gases in the atmosphere were transparent to infrared radiation, or if there were no atmosphere, the heat would escape into space. Because a similar process traps infrared radiation in a greenhouse, this is known as the greenhouse effect, and gases that are less transparent to infrared radiation are called greenhouse gases. Besides carbon dioxide, the principal greenhouse gases in the atmosphere are methane, chlorofluorocarbons, nitrogen oxides, and low-level ozone. Methane is produced primarily by natural sources, such as digestive processes of cattle and termites; chlorofluorocarbons are synthetics; and the other two gases are forms of air pollution caused mainly by automobile exhausts and burning of wood or fossil fuels (coal, oil, and natural gas).

Mars, although its atmosphere is 95% carbon dioxide, has a total amount of carbon dioxide that is not sufficient to trap very much heat (its air pressure is about 0.007 times that of Earth's air pressure). As a result, the surface temperature of Mars is $-67°F$ ($-55°C$). Venus also has a 96% carbon dioxide atmosphere, but its air pressure at the surface is 90 times that of Earth. Consequently it retains much more heat, resulting in a surface temperature of about 854°F (307°C).

Earth has been comfortably in between these extremes. Our atmosphere is only 0.035% carbon dioxide, but that, combined with other greenhouse gases, traps 88% of the Sun's energy, some of which is reradiated toward Earth's surface while the rest is radiated into space. This process then repeats, with some reradiated heat collected by the gases and some escaping into space. Eventually, 70% of the infrared radiation is emitted toward space while the remaining 30% stays in Earth's surface and atmosphere, giving Earth an average surface temperature of 59°F (15°C). If it were not for the greenhouse effect, this temperature would be much lower.

The greenhouse effect has become a possible cause of global warming because the greenhouse gases are increasing in the atmosphere. Carbon dioxide has increased by roughly 25% since about 1850. It is generally assumed that much of this increase is due to the burning of fossil fuels, but it is also clear from the geologic record that carbon dioxide levels have varied from time to time in the past, resulting in periods when most of Earth was tropical as well as periods where there were ice ages. A major source of carbon dioxide is deforestation due both to burning and to rotting of cut wood. Dr. Paul Crutzen, an atmospheric scientist at the Max Planck Institute for Chemistry in Mainz, Germany, has estimated that two-thirds of the carbon dioxide being added to the atmosphere comes from burning fossil fuels while the remaining third comes from deforestation. Other greenhouse gases are also increasing, in most cases clearly as a result of human activity. It remains less clear what is causing the rise in methane levels.

Scientists have fairly good data for global temperatures in the past 100 years, when carbon dioxide has increased from about 0.028% of the air to 0.035% and other greenhouse gases have also increased. Although the trend has not been constantly up, overall the increase in temperature during this period has been about 0.9° F (0.5°C).

Climate   If global warming occurs, local climates will change, but it is far from clear how specific regions will alter. In the United States, one of the concerns is how much water will fall in various regions. One study suggests that a local temperature increase in the neighborhood of 5°F (2.7°C) might reduce runoff in the Colorado River Basin by as much as 10%. This would affect water use over much of the west.

The Seasons   It is clear that the onset of true winter conditions would be later and true summer conditions earlier in a warmer world, although the equinoxes and solstices would not be affected. As it is now, people in northern temperate zones tend to think of summer as starting at the end of May, with winter beginning sometime early in December or at the end of November. An engineer from AT&T, David J. Thomson, looked at this phenomenon historically and reported his findings to the American Geophysical Union in December 1994. He showed that in the Northern Hemisphere, the seasons in central England, where long-term records are available, stayed about the same until 1940, when dates for the seasons began to shift. The cause appears to be greenhouse warming, which produces greater heat in June, July, and August in the Northern Hemisphere even though Earth is farther from the Sun at that time. No seasonal shift is observed in the Southern Hemisphere, however, because the greenhouse warming comes at a time when Earth is closer to the Sun.

Violent Weather   Computer models of global warming generally predict that weather will have more extremes. Thomas Karl and coworkers at the National Climactic Data Center in Asheville, North Carolina, have developed an index based on weather extremes which shows that situations predicted by the computer models are already taking place. Although the year with the highest rating recently was in the 1930s before present-day global warming, most recent years have had indices that are high, and the trend seems to be toward more extremes in the future.

Sea Level Rise   Ice caps will melt faster than they are melting now. This, combined with the increase in the volume of water that occurs with increasing temperature, will cause sea levels to rise around the world. Predictions are that this rise will be from 1.5-5 ft (0.5–1.5 m) over the next 50–100 years.

Vegetation   Plants that spread easily can adopt to new climates, but trees do not migrate (as scientists call it) very fast. If climates change faster than southern forests can migrate north, much of the temperate forest might be lost. In this instance, as in many others, the speed of the climate change is important. Most scientists think the climate will change faster than it has in the past 5,000 years, so civilization does not possess the necessary experience to predict how well it will be able to react.

One study of trees in northern forests, where warming is greater than at lower latitudes (4°F or 2°C since the 1880s), showed that although forests initially responded with faster growth, the growth slowed down with further warming. This may be because the heat dried out the air. Also, insect parasites increased in populations because more larvae or eggs survived the winters. In bark beetles, warm temperatures cut the reproductive cycle from two years back to one year per generation.

Reduction of rainfall, or increased heat, could change crop patterns all over the United States. Some fruits, such as apples, need a certain amount of winter cooling to flower and fruit, for example. Corn needs a lot of rain at just the right time. Winter wheat needs the water that comes with snowmelt. Furthermore, rising temperatures would permit insect and fungal pests from the south to migrate into northern farming regions.

Fish and Birds   Since the mid-1970s, water flowing into the ocean from rivers around the world has already become warmer. Various studies have shown that the ocean may also be warming. A parallel set of analyses, primarily of salmon in the Atlantic and Pacific oceans, shows that as waters become warmer, fish populations rise while the size of individual fish declines. Studies of fish other than salmon, such as Pacific rockfish, have shown that the reduction in size stems from a decrease in the plankton that thrive in cooler water. Not only are the prolific fish declining in size, but birds that live on marine plankton, who produce small numbers of eggs per individual compared to fish, are declining in number. Between 1987 and 1994, one ornithologist found that his counts of marine birds had dropped by 40%. It is not clear whether these changes result from global warming or natural cycles, however.

Disease   As the climate warms, tropical diseases are likely to spread, including malaria, dengue fever, yellow fever, and other diseases carried by insect vectors that thrive in warm climates. One 1995 estimate predicted an increase in malaria of 50 million to 80 million cases by 2100. The changes caused by warming are not alone among those that might affect disease. Many attribute the sudden 1993 appearance of a new hantavirus infection in the United States (variously called Four Corners hantavirus and *hantavirus sin nombre*) to the El Niño of 1991–1993. The same El Niño may have actually reduced the incidence of Lyme disease in California, simply because cool, rainy weather in that state kept people indoors more of the time. Mosquitoes that carry malaria, dengue fever, or yellow fever all find more places to breed in rainy conditions. If global warming occurs, some regions will become wetter and some drier than they were previously.

Sea-Level Indications

The idea of measuring sea level is somewhat daunting, given constant waves and tides that change the average level twice a day. Yet we continue to measure land heights in terms of a conventional sea level, for which waves are stilled and tides averaged out. Global warming would include the warming of regions around the poles where ice caps are, and also the mountain heights in other parts of the world that are now capped with snow and glaciers. If any significant amount of this ice melts, sea level will rise.

A rise in sea level would affect coastal regions around the world, but especially coastal plains such as in Bangladesh. The U.S. Environmental Protection Agency (EPA) estimates that if sea level rises 3 ft (1 m), the United States would lose an area the size of Massachusetts despite any efforts to build dikes and protect critical shorelines. A 6 in (15 cm) rise in sea level would lead to the North American Atlantic shoreline moving landward by an average of 100 ft (30 m); a 30 in (80 cm) rise would lead to an average takeover by the sea of 500 ft (150 m) of land all along the Atlantic coast.

One factor determining sea level is water temperature. Although ice is providentially less dense than water, at temperatures between freezing and boiling water behaves like nearly all other substances, with warmer temperatures producing expanded volume, which is the same as a decrease in density. A study conducted in 1992 to commemorate the first voyage of Christopher Columbus measured deep sea temperatures in the Atlantic at 101 stations along the 24th parallel N. of latitude (approximately the route of Colombus). Similar measurements had been taken in 1957 and 1981 (although not in 1492, which was more than a hundred years before the first crude thermometer). The comparisons revealed that the deep ocean was warming at a rate of about 2°F (1°C) a century, at least for that third of a century between 1957 and 1992. The temperature change could reflect long-term global warming. Furthermore, warming could by itself account for some sea level rise (in addition to that caused by melting ice) from the expansion of water.

In April 1994, as a preliminary to the problem-plagued Acoustic Thermometry of Ocean Climate experiment (see "Oceanography UPDATE," p 536), Peter Mikhalevsky of Science Applications International in MacLean, Virginia, and coworkers used the speed of sound to measure the temperature of the Arctic Ocean. In comparison with temperatures found in the past by ocean-going ships, the data from the sound experiment suggested that the northern ocean had warmed by 1°F (0.5°C) over the past decade alone. The experiment was corroborated by temperatures measured recently from shipboard in the same polar seas.

Is the Ice Melting Yet?

Other factors affecting sea levels are also already occurring. For no known reason, the Antarctic Peninsula is warming much faster than the rest of the world, and the ice shelf around it is melting away. In 1995, three large parts of the ice shelf broke off and floated away into the ocean. The largest pieces will take several years to melt. Ice that is already floating in water does not raise levels when it melts, since the mass of the water displaced by the ice is the same as the mass of the floating ice. Ice that is on land, however, such as that on the peninsula itself, would raise sea levels either by melting or by flowing from the land into the water as giant icebergs. Geologists are currently in disagreement regarding how stable the ice sheet has been in the past. Many believe the evidence shows that it has not melted for millions of years, while others claim that the evidence shows the continent to have been an archipelago of tree-covered islands just three million years ago (mya).

Kurt M. Cuffey of the University of Washington in Seattle says that ice-core records (*see* "Climate Clues to the Past," p 515) reveal that "the Greenland ice cap is in a precarious position climatologically. A modest warming would make the whole ice sheet melt if the warming is not accompanied by changes in snowfall rate." Cuffey's research shows that at the end of the last ice age the temperature in Greenland rose significantly more than it did at lower latitudes. It needs to be noted, however, that for small changes in warming, in a climate like that of Greenland more snow falls than when the air is colder.

Despite such concerns, a group of specialists convening in Japan in November 1995 concluded that, for the most part, there was no really reliable indication that polar ice was melting faster than usual. Natural variability could account for all the observed changes in polar ice.

Teams of glacier specialists have been studying trends in mountain glaciers all over the world, with data collected at the World Glacier Monitoring Service in Zurich, Switzerland. Reports are mixed indeed.

- **Alps:** In 1995, estimates are that as much as a third to a half of all the ice in the Alps has melted since 1900. Alpine ice in general reveals the ups and downs of climate history. Historical records of ice in this region go back to the Middle Ages, when a warm spell kept open many of the mountain passes that are clogged with ice and snow year-round today. In the seventeenth and eighteenth centuries the ice advanced, only to begin its current retreat in mid-nineteenth century. The rate of melting has increased since 1980.
- **Andes:** Recently the glaciers in the Andes have melted more in the summer than they have grown in the winter.
- **Caucasus:** In essentially the same climate zone as the Alps, the Caucasus mountains of southern Russia, Georgia, Armenia, and Azerbaijan are also shedding ice.
- **Greenland:** Snow is contributing to growing glaciers, even though temperatures are warmer (in fact, *because* temperatures are warmer).
- **Iceland:** Like Greenland's glaciers, those of Iceland grow each year.
- **New Zealand:** Climatologists are not at all sure how the changes in the Northern Hemisphere match up with changes in the Southern Hemisphere. In New Zealand, glacier watchers report increases in glacial ice.
- **Scandinavia:** Like other northern regions, Scandinavia reports growing glaciers. In Norway, glaciers have grown an average of 30 ft (9.1 m) in length during the early 1990s. Before 1950, however, glaciers in Scandinavia had been shrinking for half a century.
- **Urals:** Like other mid-latitude mountain ranges in the Northern Hemisphere, the Ural Mountains of Russia contain melting glaciers.

Computer Models

In the continuing debate over whether global warming is occurring, most of the information available has not been anything so tangible as temperature measurements at any one place on the globe, nor measurements of changes in sea level, nor reports on glaciation. Instead, the controversy nearly always comes down to dueling computer models.

As early as 1961, Edward N. Lorenz used a computer model to try to predict long-term weather changes. He discovered that small changes in initial conditions soon produced large changes in weather, making long-term prediction essentially impossible. Climate, however, is thought to be easier to predict than weather. When sufficiently powerful computers became available and the idea of greenhouse warming was growing in popularity, climatologists took the computer models that they were using to derive Earth's climate from first principals and plugged in an increase in carbon dioxide in the atmosphere to see what would happen.

The first global climate models had very few variables as inputs, and most scientists knew that predictions of global warming (or lack thereof) derived from these computer models would not be reliable. A lot of fine-tuning and the addition of other important variables would be needed to obtain models with truly trustworthy results. The best way to test such models is to use information known for the variables of the past 50–100 years and then see if the temperatures that result are the ones that actually occurred on Earth. Often the word "predict" is used for these tests, wherein the models use past data to predict the present, which then indicates something about the reliability of the models.

The latest generation of computer models includes simulations of world climate developed and improved over many years at Lawrence Livermore National Laboratory in California and at the British Meteorological Office in the United Kingdom. In its 1995 update, each included as a new variable the effect of the release of sulfates into the atmosphere, best known as the principal cause of acid rain. Sulfates in the air tend to cool the Earth by reflecting energy back into space. In May 1995, Thomas Karl of the U.S. National Climatic Data Center released word of findings by his team of researchers. Their research made the first definite connection between regional cooling and regional production of aerosols, primarily sulfates. Karl pointed out that summer cooling caused by aerosols was "significant enough to have an impact on global temperatures." He estimated a cooling over the entire Northern Hemisphere of about 0.9°F (0.5°C). It was this analysis that spurred the new computer models, especially since older computer models tended to predict present-day climate about 0.9°F (0.5°C) warmer than it actually measures.

The Lawrence Livermore model was preceded by five different versions that failed to produce results resembling reality. The new model concentrated on the way that sulfates change chemically over time, and on the location of sulfate-producers, essentially the industrial nations. The British model simply used a factor to compensate for sulfates, while being more realistic regarding the transport of heat by ocean currents.

Both groups found that their models were accurate for the past few years, but less accurate 50–100 years ago. Although the Livermore model is essentially a description of present-day Earth, the British one can be projected into the future, where it predicts a global temperature rise of 2.3°F (1.3°C) by the year 2050.

Neither of these models includes all the factors known to effect global temperature. For example, neither includes dust or haze other than sulfates. Gradually, however, known factors are being installed into the models. In part, this is becoming increasingly possible because of the continually improving ability of supercomputers, especially those called massively parallel computers. If the models fail to predict past climates with great accuracy after all the known factors are taken into account, we will know (as Sherlock Holmes might have noted) that there is some unknown factor to be found. A model run by James Hansen of the Goddard Institute for Space Studies in New York City included not only the rise in greenhouse gases and the cooling by aerosols but also an increase in cloud cover caused by warming. Hansen's was the first model that could predict warmer nights, an observed phenomenon.

All the computer models that included clouds were incorrect before January 1995, when two separate sets of researchers discovered that clouds absorb more energy from sunlight and reflect back less than had previously been believed.

A model run by the EPA and used as the basis of a report in October 1995 called "The Probability of Sea Level Rise," authored by Jim Tituds and Vijay Narayanan, used 35 different atmospheric and oceanic variables as well as corrective input from two dozen climatologists. The report concluded that it was most likely that global temperatures would rise by 1.8°F (1°C) by 2050 and 3.6°F (2°C) by 2100. Furthermore, the

---

THE ODDS ARE . . .

In August 1994, J. D. Mahlman, the director of the Geophysical Fluid Dynamics Laboratory at the U.S. National Oceanic and Atmosphere Administration, provided a personal assessment of the probabilities of various climate events. Here is a summary of his predictions.

Safe Bets   Greenhouse gases will rise, producing a rise in the heat content of Earth, although a combination of ozone depletion and increased carbon dioxide will lower temperatures in the stratosphere.

Almost Safe   The global rise of about 1°F (0.5°) over the past century was caused by the greenhouse effect, but no one will be able to definitively prove this.

Good Bets   The average temperature at ground level, considering Earth as a whole, will rise 2–6°F (1–3°C) by the middle of the twenty-first century, producing increased precipitation when averaged over the Earth as well as melting of some sea ice in the Northern Hemisphere, resulting in a sea level rise of 4–12 in (10–30 cm).

Even Better Bet   Although soil moisture and temperature is likely to rise in the summer at temperate latitudes, no one can be sure how any one region will be affected or the timing of such an effect.

Hang on to Your Money   It is possible that there will be more tropical storms and that the salinity and density of sea water in the North Atlantic, Arctic, and North Pacific oceans will both decrease. However, the complexity of interactions, especially with regard to cloud cover, means that such conclusions are at best tentative.

reported cited a 1 in 10 chance that the amount of increase might actually be double these predictions. Since temperatures at the poles appear to go up faster than those at lower latitudes, the report includes a 1 in 40 chance that temperatures in Greenland and Antarctica will rise so much that sea level will rise by 10 ft (3 m).

Computer modelers will have an easier time of it as more data is collected. One important new source is the Global Emissions Inventory Activity, for which several hundred atmospheric scientists from more than 30 countries are contributing data on half a dozen gases that are regularly added to the atmosphere, including chlorofluoro-carbons, methane, nitrous oxides, and sulfur emissions. Starting in 1994, the Inventory has worked toward building a close grid showing emissions from both natural and artificial sources. In 1994, the best computer models used data points that were each about the size of Poland. The size of the grid has gradually decreased as more data becomes available.

Responses to Global Warming

Although it was not clear in 1992 that global warming would take place in the near future, there was still enough evidence to provoke official alarm. At the Rio de Janeiro United Nations conference on the environment that year, 140 nations signed a nonbinding treaty in which they promised to reduce emissions of greenhouse gases slightly. Specifically, the treaty stated that by the year 2000 emissions would be reduced to the 1990 level existing in each country. Although this was only a slight reduction from the 1992 levels, it was a major commitment to make for the year 2000 because of the high rate of growth in emissions at that time. The United States committed itself further to that goal on Earth Day (April 21) in 1993.

By 1995, it was apparent that the goal of the treaty would be very difficult to implement. Although some industrial nations, including Germany and the United Kingdom, were on target to meet that goal in 1995, others, including the United States, were far behind. In the United States, steps had been implemented that would originally have met the goal. The plan, however, was sabotaged by faster economic growth than expected, coupled with lower-than-anticipated energy costs. Further-more, the Republican congress elected in 1994 appeared to lack sympathy with the goal in any case. Republican proposals were mostly aimed at reducing restrictions rather than reducing greenhouse gases.

In the United States, the EPA has proposed the following ways to mitigate the greenhouse effect: raise prices on fossil fuels; increase use of alternative energy sources, especially those such as solar and nuclear power that do not produce greenhouse gases; grow new forests around the planet; stop use of chlorofluorocar-bons (an action already agreed to because of the effect of these chemicals on the ozone layer); capture gases now released by landfills (primarily methane); and change ways of raising rice and cattle to reduce production of methane. In 1995, the United Nations panel reported that they believe that "significant decreases in greenhouse gas emissions are technically and economically feasible." Unfortunately, the EPA says that if all this were done worldwide, the measures taken would result in a rate of gas build-up that would not begin leveling off until sometime in the twenty-second century. The worse news in 1995 involved a bill passed by the house which banned the EPA or the National Oceanic and Atmospheric Administration from any further studies of the problem. The bill did not become law in that form, but concern exists that the EPA might in the future be reluctant to make reports that could cut off its funding.

(Periodical References and Additional Reading: *Science* 2-7-92, p 683; *New York Times* 5-10-94, p C10; *EOS* 8-12-94, p 353; *Nature* 8-18-94, p 544; *Scientific American* 9-94, p 18; *Science* 9-9-94, p 1528; *Science* 10-28-94, pp 544, 632, & 634; *EOS* 11-22-94, P 585; *Science* 12-23-94, p 1947; *Science* 1-27-95, pp 454, 496, & 499; *Scientific American* 2-95, p 13; *Science* 2-3-95, p 612; *Science* 2-17-95, p 957; *Science* 3-17-95, p 1595; *Science* 3-31-95, p 1911; *Science* 4-21-95, p 363; *Scientific American* 5-95, p 21; *Science* 5-12-95, p 802 *Science* 6-9-95, p 1436; *Science* 6-26-95, p 1567; *Science* 10-20-95, pp 444 & 455; *Scientific American* 11-95, p 18; *Science* 11-3-95, p 731; *EOS* 11-7-95, p 452; *EOS* 11-21-95, p 478; *New York Times* 11-28-95, p A1; *Science* 12-8-95, p 1565; *New York Times* 12-19-95, p C4; *Discover* 1-96, pp 78 & 93)

OCEANS AND OCEANOGRAPHY

Oceanography UPDATE

Perhaps the most startling news from oceanography was a single photograph made in 1992 and published in *Nature* in 1994. It showed a dark green line that extended straight along the waters of the Pacific Ocean for hundreds of miles. Taken from the space shuttle *Atlantis,* it depicted a phenomenon that was easily observed from space and had always been partly visible to sailors who crossed the line in the open ocean. Similar photographs had been obtained from research planes, but they had enhanced the effect with laser-induced fluorescence. The immediate cause of the line is a vast profusion of diatoms in a region about 300 ft (100 m) wide. In turn, the abundant growth results from increased nutrients, which are the result of a sharp border between the cold South Equatorial Current and the warm North Equatorial Counter-current. As the cold, nutrient laden water meets the warm waves, the cold water sinks below the warm, with nutrients diffusing across the boundary.

Such a line has existed near the equator for millions of years. One result is that the shells of the diatoms and other plankton produce a "chalk line" on the ocean floor, which eventually becomes buried by other sediments. Cores drilled from the ocean floor show the line at depths indicating how long ago it was formed. Evidence of one such line has been pushed almost to Japan by sea-floor spreading, but at the Japan trench the remnants of the line will plunge down into the mantle along with that edge of the Pacific plate.

Rubber Duckie Science

On May 27, 1990, a severe storm knocked 21 containers off a cargo ship in the mid-Pacific. Five of the containers were filled with Nike products, including 61,280 shoes made from materials lighter than sea water, although one container sank. At least 1,600 shoes were later recovered, mostly by beachcombers on the northern West Coast of North America. It is thought that floating objects from other containers were also recovered but not reported. The drifting shoes provide oceanographers with details on the flow of currents in the North Pacific, which they had previously also sought by deliberately releasing thousands of drift bottles at various locations in the Pacific.

In January 1992, another storm washed a cargo container containing 29,000 plastic bathtub toys into the North Pacific near the International Date Linefloating. These included yellow ducks, red beavers, green frogs, and blue turtles, each about 2.5–5 in (6–12 cm) long. By November 16, 1992, the first batch of six toys was found by a beachcomber near Sitka, Alaska. Scientists considered these toys a real boon because they can remain buoyant for extremely long times and are easily identified. Further-more, the exact time and location of the spill is known. More than 400 toys had been recovered in 1994, when data from the spill was used to refine computer programs simulating circulation in the North Pacific. Currents were expected to carry some of the toys through the Bering Strait into the North Atlantic. Furthermore, because some currents are essentially large gyres, or wheels, the toys that do not land elsewhere will be brought back to the region west of the initial spill. Oceanographers are expecting to receive reports of toys found all over the Northern Hemisphere early in the twenty-first century, where they will continue to aid in correcting our knowledge of ocean surface currents.

Satellite Maps

When results about land or oceans obtained by a research airplane or Earth-orbiting satellite are checked with the data obtained by close-up examination, the result is a comparison with "ground truth." With increasing sophistication, however, such checks are less needed. Ocean currents can be mapped by changes in water color or temperature that are observed from satellites. In some ways, these results are better than anything that can be obtained by floating drift bottles, shoes, or plastic bathtub toys. Satellite data can also penetrate beneath the surface of both soil and sea.

The discovery that a satellite could map the ocean floor from space was made by civilians in 1978 with the *Seasat* earth-orbiting satellite. That satellite, however, gathered only three months of data. In 1985, the U.S. Navy launched *Geostat*, which provided even more detailed information. However, since *Geostat* was a secret project, scientists got only a glimpse of what it could do, for most of the data was classified until 1990. *Geostat* was able to map the ocean floor primarily by measuring the changes in gravitational attraction caused by undersea mountains, canyon, and anomalies.

Civilian scientists knew enough about *Geostat* to campaign for more access to the maps, and in 1990 they got the first information since shortly after the 1985 launch. That year the Navy released information from the Antarctic. Two years later, the Navy agreed to free up anything south of 30° S. latitude. Finally, on July 19, 1995, the Navy released all the data from the period of *Geostat* operation: March 1985 through October 1986.

One factor in this release of information involved the launching of a satellite by the European Space Agency, which provided ocean maps using many of the same techniques as *Geostat*, reducing any reason to keep *Geostat* data secret. The *Earth Resources Satellite* (*ERS-1*) was launched in 1991, and gathered data through April 1995.

A second factor in release of *Geostat* data involved pressure exerted by a group of scientists calling themselves Medea (Measurements of Earth Data for Environmental Analysis) and by U.S. Vice-President Al Gore. Both Medea and Gore are especially concerned with all data that can be used to track long-term environmental changes, not just the ocean maps. The Navy is expected to release additional information that Medea thinks will be useful, including surveys of seabed magnetism, sediment thickness and composition, ice cover, ocean temperature, salt content, and light transmission in sea water.

Both *ERS-1* and *Geostat* measured changes in the ocean floor by using the height of the ocean surface. A large submerged mountain attracts water from the region around it, producing a slight bulge in the surface of the sea, not noticeable to a ship passing over it, but observable in precise measurements. Combining data from the two satellites improves reliability and resolution. The combined resolution shows any patches of raised or lowered sea level that are at least as large as the area of the District of Columbia.

Among the main new scientific results from the satellite data is a re-evaluation of chains of subsea volcanoes and volcanic islands once thought to have been caused by the passage of a tectonic plate over a plume of rising magma (*see* "Plumes," p 554). Some of these changes may, in fact, have other causes. Ground truth for one such change, the Pupapuka Ridges, reveals that the feature appeared all at once. Volcanoes caused by a plate moving over a "hot spot" where a plume rises are always of different ages, with the older ones at the trailing edge of the track.

Another advantage of the newly available data is that it is accessible by geologists all over the world on CD-ROM or on the Internet. Thus, oceanographers who cannot afford the time or expense of a trip to the South Seas can still weave and test theories about how the ocean crust in those regions has come to be the way it is.

Underwater Truth

Although satellite studies can provide a better large-scale picture of the ocean bottom than soundings or other spot checks, there is much information that can only be obtained by closer examination. Humans can travel to ocean depths if they are encased in strong submersibles, but such dives are dangerous and limited in many ways by the human cargo aboard. A better method of close examination for ocean depths involves the use of robot vehicles. Today, there are dozens of such vehicles probing the seas, with more to come.

A typical advanced robot is *Jason*, which can travel to depths of more than 1 mi (2 km) while tethered to a ship by cable. *Jason* is about 7 ft (2 m) long and contains two video cameras, a still camera, a manipulator robot arm, and devices for measuring such quantities as temperature and pressure. All of the information is not only cabled to the ship but is then sent by electromagnetic waves to a satellite. From there, the data is sent to receivers that use telephone lines and the Internet to transmit the data all over the world within moments of the information having been collected from the ocean floor.

Another type of ocean robot is untethered. An example of this type is the *Autonomous Benthic Explorer* (*ABE*), which like *Jason* is controlled by the Woods Hole Oceanographic Institution of Cape Cod, Massachusetts. Slightly smaller than *Jason*, *ABE* can travel deeper in part because it has no cable to drag behind it. Furthermore, once it is some 4 mi (6.5 km) deep on the ocean floor, it can lie there for a year, transmitting data from a suitable location. After its mission, a signal from a research ship brings it back to the surface. *ABE* is not nearly as expensive to make as *Jason* ($1 million as opposed to $5 million) and is of course less expensive to operate as well. The most famous human-carrying research submersible, *Alvin*, cost $50 million to build in 1964, and costs about $25,000 a day when operating. Even cheaper than *ABE* is *Odyssey*, launched in 1993, which can dive to more than 3 mi (5 km) but was built from only $50,000 in parts.

Japan, along with the United States, has been a leader in the submersible movement in the 1990s. On March 1, 1994, the tethered Japanese robot *Kaiko* sent back television pictures from the depths of the Challenger Deep in the Mariana Trench in the Pacific, the deepest known part of the ocean. *Kaiko* did not succeed in its goal of landing, however, because of equipment failure. In 1960, a manned submersible set the record by landing on the bottom of the Challenger Deep at a depth of 35,800 ft (10,912 m). *Kaiko* was only 3 ft (1 m) short of this record when it aborted its mission.

In 1990, Japan built what is the current leader in manned submersibles for depth. *Shinkai 6500*, which carries a crew of three, set the depth record for human travelers in the Atlantic Ocean in 1994, dropping down 3.74 mi (6 km).

Many other untethered robots are under construction or testing in both the United States and Japan and will contribute greatly to "ground truth" in the ocean.

News from the Spreading Centers

There are nearly 600 volcanoes on land, including very many that are actually in oceans but have (like the Hawaiian Islands) formed their own land. Still, much of the volcanic activity on Earth is on ocean bottoms, and many volcanoes have not risen high enough to produce islands. In many cases, the activity does not come from what one thinks of as a volcano at all, but instead consists of long cracks, called rifts, that pour forth lava and other materials at frequent intervals. We now know that such rifts between tectonic plates on ocean bottoms produce new ocean floor, and are one of the engines of plate-tectonic motion through the mechanism of sea-floor spreading. There are also more conventional volcanoes on the ocean floor, which is generally less than half the thickness of continental crust. All in all, according to Haraldur Sigurdsson of the University of Rhode Island, 90% of Earth's volcanic activity takes place on the ocean floor instead of on continents.

The greatest concentration of volcanoes ever found in one region was located on the floor of the Pacific Ocean in 1993. The research vessel *Melville* mapped 1,133 seamounts and volcanic cones, of which two or three might be erupting at any given time. The volcanoes are clustered in a region about the size of New York State, some 600 mi (1,000 km) northwest of Easter Island, west of the boundary where the Pacific and Nazca tectonic plates are moving apart. This region is known as the East Pacific Rise.

The Atlantic Ocean is not to be outdone entirely, although the most spectacular activity is located in the Pacific or Indian oceans. In 1993, *Alvin* located a new field of seven undersea vents near the Azores. This is the third such vent group located in the Atlantic and proves to be the largest of the three. Evidence of the vent's existence was obtained by dredging in 1992, so *Alvin* was dispatched to investigate. The first such vents were found by *Alvin* in the Pacific, in 1977.

Although it has long been clear that submarine volcanoes must exist, none was caught in active eruption until 1991, when a volcano was found in the Pacific on the Juan de Fuca ridge off the coast of Oregon. Another undersea volcano of 1991 erupted in the Pacific off the coast of Costa Rica, although the eruption just preceded an April visit to the site by *Alvin*. Nearly a year later when *Alvin* returned, the site was still depopulated. By December 1993, on the third visit from *Alvin*, giant tube worms had grown there, reaching lengths of 5 ft (1.5 m). This was the fastest growth ever recorded for a marine invertebrate.

Ocean Acoustic Experiment Controversy

Sound travels better in water than in air, and humans and various other animals have taken advantage of how well sounds propagate in water. For humans, the most familiar application is sonar, using echoes to locate objects. Marine mammals also use sonar. In fact, there is considerable evidence suggesting that many marine mammals also use underwater sound for communication. Among the mammals thought to use sound in one or the other of these ways are humpbacked whales and other similar baleen whales, sperm whales, and elephant seals.

Underwater sound can be used in other ways as well. Temperature and pressure affect the speed of sound waves through the ocean (water composition contributes also, but the slight variations in composition of seawater do not make much difference). Small changes in the speed of sound waves bends the path of the waves, just as small changes in the speed of light between air and water cause a pencil sticking out of a glass of water to appear bent.

The first experimental work using these effects was conducted in 1981 by Robert C. Spindel of the University of Washington and Peter F. Worcester of the Scripps Institute of Oceanography and colleagues. By measuring how sound traveled through water and then using a computer to reconstruct the patterns of temperature and currents, Spindel was able to produce a three-dimensional picture of part of the ocean. The basic idea in such a reconstruction is to use tomography, the same technique that is more familiar in the computerized x-ray tomography used in medical diagnosis (the familiar CT or CAT scan). Spindel's method is called ocean acoustic tomography.

Since 1981, much experimentation has been conducted, with steady improvement in the technique. Primarily, ocean acoustic tomography has demonstrated that most of the variability in the ocean is not caused by large slow currents, but rather by much smaller patterns that can persist for months. These are the ocean equivalents of weather in the atmosphere. That is to say, previous techniques allowed scientists to study only the ocean's "climate," but tomography allows the study of "weather."

Low-frequency sound waves are used in ocean acoustic tomography because these frequencies travel great distances through the water. It occurred to climatologists that they could use the same idea to capture the average temperature of the entire Pacific Ocean, perhaps the most important factor in worldwide weather in the atmosphere. Sound travels slower in warm water than in cold. With only two deep sound emitters and a network of 13 listeners, the whole Pacific could be sampled all at once, providing data that would not be available from either ocean ship nor satellite. The feasibility of this method had been successfully tested in 1991.

Originally the measurement, called Acoustic Thermometry of Ocean Climate (ATOC), was scheduled to begin in the spring of 1994. Protests concerning the effect of sounds (loud enough to be heard from California to New Zealand) on marine mammals caused the experiment to be put on hold. The first such concerns had been raised on the marine-mammal forum on the Internet. The main problem would involve a whale or seal diving within about 600 ft (200 m) of the speakers, one located off Point Sur, California, and the other near Kauai, Hawaii. Only sperm whales or elephant seals could dive that far. The sound frequency proposed is not the one used by sperm whales to communicate, although it is within the range used by elephant seals. One species of elephant seal is found along the California coast. The scientists in favor of the experiment also argue that the sounds are not big blasts like explosions but come on slowly, and would not damage ears even within the 600 ft (200 m) range.

Eventually, a compromise was reached. Instead of sending sound blasts every four hours, the signals will be sent every four days. Furthermore, the first phase of the program, lasting 10 months, will be devoted entirely to observing the biological effects of the signals. With that understanding, the first signal from the Point Sur transmitter went out on December 2, 1995. Preliminary results released in 1996 showed no damage to marine mammals.

Assuming no biological problems arise, the experimenters will need to apply for new funds to begin measuring temperature. Then, if ATOC temperature measurements are a success, scientists plan to follow that experiment with a similar acoustic mapping of global ocean temperatures.

(Periodical References and Additional Reading: *New York Times* 9-22-92, p C4; *New York Times* 2-14-93, p A1; *New York Times* 3-9-93, p C1; *New York Times* 6-29-93, p C5; *New York Times* 3-2-94, p A8; *Science* p 339; *EOS* 9-13-94, p 425; *New York Times* 10-18-94, p C1; *Nature* 10-20-94, pp 663 & 689; *New York Times* 10-20-94, p B13; *New York Times* 10-25-94, p C7; *New York Times* 12-11-94, p I-40; *Discover* 1-95, p 75; *Science* p 727; *Science* p 1122; *New York Times* 11-28-95, p C1; *Science News* 12-16-95, pp 410 & 415)

THE DYNAMIC EARTH

ADVANCES IN PLATE TECTONICS STUDIES

Although our pre-human ancestors taught themselves the differences among various rocks, geology as a modern science is younger than physics (seventeenth century), biology (early eighteenth century), and chemistry (late eighteenth century). Geology did not progress very far until well into the nineteenth century, and truly modern ideas about the interior of the Earth and geological processes had to await the twentieth century. Acceptance of the ideas surrounding plate tectonics in the 1950s and 1960s revolutionized all that had gone before. Today, to understand any of the important processes that take place in and around our planet, it is necessary to know something of the main ideas of plate tectonics. As these ideas are refined and improved, new insights can be found regarding everything from the ephemeral weather on the planet to the longest lasting rock formations.

Where the Continents Drift

In 1912 and in subsequent publications and talks, German meteorologist Alfred Wegener described and promoted his theories of continental drift and the super-continent he named Pangaea. His ideas had little impact until aided in the 1960s by the then-new theory of plate tectonics. Although plate tectonics made continental drift a familiar idea accepted by most people, the tectonic theory of the 1960s also replaced Wegener's original idea of drifting continents with the more sophisticated idea of moving plates. The new construct included most plates containing ocean crust as well as continental patches, although the largest plate, the Pacific, has nary a continent riding on it.

Despite the modern concept's transcending continental drift, past peregrinations of plates are nearly always thought of in terms of the continents. There are several reasons for this. Perhaps the most important is that while ocean floor is continually being created and destroyed, the lighter rock of the continents floats above this fray. All the oldest rock is that locked up in continents, which persist even as the plates move beneath them and carry them on their backs. Thus, as geologists reconstruct the past, it appears as if the continents really were drifting about, combining and separating from time to time, but still recognizably the rock formations that exist today.

The combination of continental rock that Wegener called Pangaea consisted of all the present-day continents arranged about 250 million years ago (mya) into a single supercontinent which had been gradually formed from bits and pieces. When Pangaea broke apart, the present-day seven continents did not all break off at once, either. Although the continental rock has been maintained, the continents themselves are always slowly rearranging themselves.

Just how the rearrangement took place is deduced largely from the orientation of Earth's magnetic field that becomes "frozen" in rock. Another source of data is the recognition of similar rock formations that at some time in the past were continuous.

In 1994, for example, T. H. Torsvik of the University of Michigan and coworkers D. Roberts and B. A. Sturt from Trondheim, Norway, assembled a set of such clues, using the clues to revise the then-accepted picture of continental rearrangements. Two continental regions—Siberia (today's Siberia, but not attached to the rest of Eurasia) and Baltica (today's Scandinavia and northern Europe, but not attached to the rest of Eurasia)—circled about each other in the period from about 500 mya to about 425 mya. This new version of opening and closing of oceans between Baltica and Siberia accounts for certain sediments and individual pebbles found today in Norway that appear to have originated in Siberia. The main tool was a map of "frozen" magnetic fields, called paleomagnetism, which provided information of latitudes at various times in the past. Some part of the reconstruction of movements is simply informed guessing. Putting the movement of early continents together from this sort of data is rather like assembling a jigsaw puzzle in which the pieces are only approximately the correct shapes.

TIMETABLE OF DRIFTING CONTINENTS

750 mya	All the continents have come together to form a supercontinent that Eldridge Moores of the University of California, Davis, in 1991 named Rodinia, with Australia-Antarctica on the west coast of the ancestor of North America, which he calls Laurentia (a name also used for Greenland).
490 mya	Laurentia, the precursor of North America, has broken from Rodinia and is swinging around it. Along the way, it leaves some of its rock that had been part of what is now the Gulf Coast attached to what is now South America, where it winds up as part of the Andes. At this point, in the traditional view, Laurentia includes bits and pieces of Baltica and Siberia, as well as the core of North America. As reconstructed by Torsvik and his colleagues from Trondheim, however, Siberia and Baltica are in the Southern Hemisphere, while Laurentia is on the equator.
437 mya	In the reconstruction by Torsvik, Roberts, and Sturt, Siberia has moved north of the equator with Baltica just below it. Laurentia is also on the equator, and all are about to collide.
425 mya	In the Torsvik reconstruction, this is the time at which Baltica and Laurentia meet. Siberia's location is somewhat uncertain, but this would have also been the occasion for materials from Siberia to have been left behind in Baltica.
380 mya	There are three large continents, Laurentia, Eurasia, and Gondwana (sometimes known as Gondwanaland; named in the nineteenth century by Eduard Suess after a district in India), which includes what are now South America, Africa, Antarctica, Australia, and India.
340 mya	Laurentia has drifted into contact with Gondwana, but Eurasia remains separate.
250–200 mya	All the continents are grouped into the single supercontinent Pangaea ("all Earth"). From north to south these were Eurasia, North America to the west, the equator, Africa to the east, South America west of and

	nestled into Africa, the subcontinent of India to the southeast, and Antarctica. Australia was a great peninsula on the eastern flank of Antarctica.
135 mya	Laurasia (a name coined in the 1920s by Alexander Du Toit from Laurentia and Asia, but consisting mostly of North America and Eurasia) is in the Northern Hemisphere separated from the other continents, which are grouped together as Gondwana.
110 mya	Greenland and North America separate from each other.
100 mya	Gondwana splits and South America is separated from Africa as the South Atlantic Ocean forms. India and Australia also part from Gondwana at this time, with India beginning to head north.
65 mya	Eurasia and North America, although still joined at the top, are largely split apart with Greenland between them.
45 mya	India on its trek north passes the equator with Australia trailing behind, but leaving Antarctica near the south pole.
40 mya	India begins to plow into Asia, raising the Himalayas.
24 mya	Africa swings about to meet Eurasia at the Middle East.
3.4 mya	Volcanic activity raises the Isthmus of Panama and connects North and South America.

Set 13 Plates, Martha

From early in the development of plate tectonics theory, a dozen large plates were found that make up almost all Earth's crust, although it was soon determined that there are also a number of smaller plates tucked into the interstices. In July 1995, however, persuasive evidence was put forward that the number of large plates is actually 13 instead of 12. James Cochran of the Lamont-Doherty Earth Observatory and coworkers confirmed a suggestion first proposed by scientists from Northwestern University: India and Australia, instead of occupying two ends of the same tectonic plate, are actually going their own separate ways.

The break between the Indian and Australian plates is not one of the types known earlier. All the other plate boundaries are sharp, involving two plates which move away from each other or past each other, or one plate moving under the other. The Indian-Australian boundary is wide and diffuse, with the Australian subplate slowly swirling into the Indian subplate and the Indian subplate slowly swirling away. The resulting boundary region measures hundreds of kilometers across, cutting across most of the Indian Ocean between Sumatra and Africa. East of that boundary, the larger part of the Australian subplate has more conventional borders with the Eurasian plate at Indonesia and with the Pacific plate along New Guinea and the islands of Melanesia. In the eastern region of the Australian subplate, one plate pushes under the other to form a sharp boundary marked by deep trenches and rising volcanoes. At the newly found wide border with the Indian subplate, the crust is heavily fractured with faults, but the trenches and tall volcanoes are missing.

Along another boundary, the west side of the Pacific plate crashes into the east side of the Australian subplate in the process forming the Tonga Islands. In 1995, this collision produced a new island starting on June 6 and continuing into July, when the island was about 900 ft (300 m) across and 140 ft (40 m) high. Like other recent Tonga islands formed at the same spot, this one is expected to erode away sometime in the near future.

One of the first great successes of plate-tectonics theory was its explanation that the Himalaya Mountains and associated Tibetan Plateau were caused by the collision of the Australian-Indian plate and the Eurasian plate (now evidently just the breakaway Indian subplate and Eurasia). In this explanation, the Himalaya are the crumpled edge of the Indian subplate, while the Tibetan Plateau is the edge of the Eurasian plate lifted up either by the force of the collision or by part of the Indian subplate slipping below the Eurasian plate.

However, a glance at a relief map of Asia shows that the crumpling and folding does not stop with Tibet but goes on to the Altun Mountains, the Tian Shan, and even the Altai Mountains that extend along northern Mongolia and southern Siberia in Russia. This great region of uplift is too far from the point of collision (the base of the Himalaya) for conventional plate tectonics to explain. In 1994, however, Sean D. Willett and Christopher Beaumont of Dalhousie University in Halifax, Nova Scotia, developed a computer model that would account for these long-distance effects. In their model, the collision causes the light crustal part of the Eurasian plate to split off from the cool and heavy slab of mantle that is the base of the plate (*see also* "Able to Move Great Continents," below). The mantle portion bends downward, moving under the Indian subplate while the crust simply continues pushing into the crust of the Indian subplate, causing the long-range crumpling. A basin between the Tibetan plateau and the Altun Mountains is the surface evidence of this plunging mantle.

Able to Move Great Continents

When Wegener first proposed his theory of drifting continents, the two main objections were that continents could not "drift" in solid rock, and even if they could, no power on Earth would be strong enough to shove them from place to place. Replacing the concept of drifting continents with theories of sea-floor spreading and of rotating and sinking patches of crust stilled the first objection. It has been more difficult to solve the problem of what impels the motion of the plates.

Sea-floor spreading itself appeared to be a good candidate, at least where two plates were moving away from each other with new ocean crust being built between them. The new crust could push the plates away from it. Yet many plate motions are in other directions, not simply moving away from sea-floor spreading centers. Also, it appeared to many geologists that the new ocean floor was rising up to fill a widening crack, rather than pushing up to make the crack widen. That kind of ocean basin creation would not provide a force that would move the plates.

More and more, geologists came to treat the mantle as a fluid instead of as the dense rock envisioned earlier, although it was long recognized that there must be some fluidity to very hot rock. With the mantle treated as fluid, most geologists and geophysicists in the 1970s came to believe that the true source of tectonic power was the same as in any fluid heated from below. Such heating makes the lower levels of the material less dense than upper levels. The denser upper levels are pulled downward by gravity, displacing the lower levels, which have no place else to go but up. A

Background

Geologists could not develop the theory of plate tectonics until they had some notion about what might lie too deep below the surface for direct observation. In 1909, Andrija Mohorovicic observed that earthquake waves (also called seismic waves) showed a definite change some tens of kilometers below Earth's surface. Eventually, this change was identified as the boundary between Earth's thin rocky outer sphere, known as the **crust**, and the bulk of the planet, labeled the **mantle** (it is not clear why geologists use the word *mantle*, which means "outer covering," for the body of the planet). The boundary between crust and mantle is now known as the Mohorovicic discontinuity, or "Moho." About the same time, geologists employed seismic waves to locate another boundary deep in the heart of the Earth, which is recognized as the beginning of Earth's **core** .

Plate tectonics is largely a theory of Earth's crust and its interaction with the mantle. The original theory of plate tectonics, as developed by a number of earth scientists in the 1950s and 1960s, described Earth's crust and upper mantle as composed of a dozen giant plates with a few smaller subplates filling in the larger cracks. Some plates are diverging as new warm material is added at their ocean-ridge edges (called **rifts**), while cooler heavy material at the opposite edges plunges into the mantle in ocean **trenches**. Such plunging occurs when two plates meet under the oceans, or where one plate carries a continent at its leading edge while the other one does not. A plate that is moving under another plate travels at least part way into the mantle, a process called **subduction**. If both plates that come together are carrying continents where they meet, something different and less well understood occurs.

Although overwhelming circumstantial evidence for this theory piled up in the next quarter century, it was not until 1987 that the theory was definitely established by actual measurements from space showing that the plates are slowly moving with respect to one another. For example, the Pacific plate moves to the northwest about 1/5 in (48 mm) a year.

circular current is set up in this fashion; specifically, a pattern called a convection current. This is a well-known effect. A convection current that causes a thin film to move about can be seen in your kitchen whenever a simmering soup moves islands of scum across its surface. The simmering soup or convection current model was considered a logical answer.

However, unresolved problems began to be raised about that model as well. For example, convection currents are caused by uneven heating. Earth's interior is thought to be kept hot by decay of radioactive elements, which would tend to cause nearly uniform heating throughout the Earth. Perhaps the core is hotter than the mantle, which also seems likely for other reasons (*see* "At the Core," p 558). The core could then provide heating from the bottom of the mantle, like heating a soup on the stove. Yet even with heating from below, the amount of energy used in plate motion seemed to some geologists to be too great for the convection currents to carry, although many continued to believe in the currents as the main engines of plate tectonics. Furthermore, many models of the convection currents limited the currents to the upper layer of the mantle, which also reduced the amount of energy the currents could carry.

A subsequent hypothesis is based upon measurements of stresses and strains, which suggest that some plates are being pulled instead of pushed. The new theory begins with the observation that relatively hot rock forms the ocean floor near mid-ocean rifts. This rock rides high on the mantle because of its low density, which results from its heat content. As the sea-floor spreads away from the rift, it cools down. Cooler rock is denser rock. Eventually the edge of the plate becomes cool enough so that it is dense enough to sink into the mantle. Gravity pulls on the edge of the plate, which geologist now call a *slab*, perhaps thinking of the well-known "cold slab" in a morgue. The slab is still attached to the warmer part of the tectonic plate, so as it descends into the mantle, the slab pulls the rest of the plate with it. In the cold slab hypothesis, it is this pull that opens up spreading centers rather than convection currents.

In this hypothesis, gravity appears to be the engine rather than heat. One should recognize that convection currents are powered by gravity also, and that warm and cold regions are needed to make the cold-slab hypothesis function as well. Only a push from sea-floor spreading provides a tectonic force that does not depend directly on gravity, and even then gravity is needed to make the sea-floor spread.

The main problem with the cold-slab hypothesis is similar to one of the difficulties with the sea-floor-spreading-push hypothesis: some of the plates move in the wrong direction. Descending slabs are plausible in the Pacific region, where slabs subduct and form deep trenches. It is difficult to see how they are involved in moving the North American plate, which does not subduct at any edge.

Ignoring this objection, Mark Richards of the University of California, Berkeley, with Caroline Lithgow-Bertelloni of the University of Göttingen, Germany, and, independently, Vincent Deparis of the University of Strasbourg, France, and coworkers, developed computer models of tectonic motion over the past 200 million years based on power from cold-slab descent into the mantle. The models are surprisingly accurate in reproducing the motions as deduced from geomagnetism and other clues. The modelers think that their results show that about 95% of the power for tectonic movements comes from descent of cold slabs, while the rest may come from sea-floor spreading. These modelers do not mention the once ubiquitous explanation of convection currents.

Yet there remain unexplained motions. Aside from observed plate movements in the wrong directions some of the time, the main unresolved difficulty is the apparent shift of plate motions in the past. For example, the history of the motion of the Pacific plate is thought to be shown by the line of seamounts and islands that end at Hawaii (*see* "Plumes," p 554). This line has a bend that is believed to correspond to a change in the direction of the Pacific plate amounting to about 60° some 43 mya. It is not clear how the descent of cold slabs could have caused this shift.

Another place where the cold-slab hypothesis runs into a difficulty is at the western margin of South America. Plate tectonics has been invoked to explain the formation of the Andes Mountains. The explanation states that the Nazca plate is pushing under the South American plate. As the cold slab of the Nazca plate warms in the mantle below South America, the leading edge thickens and also sends up blobs of hot magma, which rise to form the very high Andes plains and even higher volcanic Andes Mountains. However, Paul Silver of the Carnegie Institution of Washington and coworkers say that their seismic studies of South America show that something different is happening. They find that the region just below the Andes and near the descending Nazca plate is cold, viscous mantle, not the partly fluid stuff on which the tectonic plates are thought to travel. Silver and geophysicist Raymond Russo of the University of Montpellier in France have worked out a different theory to explain the Andes. They note that the South American plate is moving westward at about 1.4 in (3.5 cm) a year. As it hits the big chunk of viscous mantle near the edge of the Nazca plate, it buckles and bends. The bend is clearly visible on a map of the west coast of South America, represented by a great continental-border concavity with its deepest point just where Peru and Chile meet. Just behind that concavity lies the Altiplano, the highest part of the continent.

Still, the cold-slab theory cannot find a way to push the South American plate into the block of viscous mantle. Another force is needed, and sea-floor spreading in the Atlantic is just not powerful enough. Silver opts for the convection-current idea. He also suggests that perhaps the plates that meet along the mid-Atlantic ridge are all powered by convection currents that rise at the ridge and then flow east and west, while the plates that cover the Pacific are all powered by cold-slabs descending into the mantle. In this theory, one size does not fit all.

Some geologists object to the concept proposed by Silver and Russo on the grounds that the observed cool region in the mantle is too far below the continental rock of South America to have much influence. Silver plans further studies to locate unambiguously the mantle flows he thinks are there.

Wandering from Continent to Continent

In addition to continents breaking up and coming together, smaller bits and pieces are carried by spreading sea floors from one continent or from the middle of an oceanic plate to become stuck to the sides of the larger continental chunks. Such smaller pieces that are stuck on like a piece of spearmint gum on the bedpost are called *terranes*. This concept was first suggested in the early 1970s by Myrl Beck of Western Washington University in Bellingham. By the 1980s, geologists recognized that much of the West Coast of North America consists of terranes carried by the Pacific plate and left behind.

In truth, geologists are a conservative breed. *Geophysicists* were the first to accept plate tectonics and have been the most enthusiastic believers in wandering terranes. *Geologists* view any long-range motion of parts of the crust with some skepticism. Until about a quarter of a century ago, nearly all geologists would only accept the crustal motions *up* and *down*, with the exception of small amounts of slippage along obvious cracks called faults. Geophysicists, who spend much of their time mapping traces of relic magnetism in rock formations, are willing to believe anything their measurements tell them.

Thus, when measurements of fossil magnetism in parts of the northwestern United States and the west coast of British Columbia, including Vancouver Island, showed that these areas had solidified from molten rock at about the latitude of Baja California, geophysicists reported that the rock formations had been carried by close relatives of today's San Andreas Fault northward for about 2,000 mi (3,000 km) from where they were 95 mya. Geologists, however, said that the evidence for long-range movement was not good enough. The geologists counter that the obviously foreign terranes had been islands off the coast of Laurasia, offshore bodies that had then been jammed onto the coast by a little eastward jog of the Pacific Plate. If fossil magnetism indicates a more southern latitude, the geologists said, the rocks containing the magnetic evidence must have been folded and otherwise moved about. So geophysicist Jane Wynne of the Geological Survey of Canada and coworkers went back to the original terrane and showed how the rock formations could be unfolded into a flat plane from which more accurate magnetic remnants could be read. The unfolded terrane still appeared to come from down south, according to Wynne's report on April 10, 1995. Geologists still want to see some physical evidence of the fault along which the terrane could have slid, but more are willing to grant some credence to the geophysicists' ideas.

An earlier study by Scott Bogue of Occidental College in Los Angeles, California, presented in the fall of 1994, showed that a different terrane, now Duke Island off the coast of Alaska, also started far to the south. In the case of Duke Island, an unusual crystal pattern makes it more likely that the rocks are in the same orientation today as they were at the time of formation. If Duke Island could do it, why not Vancouver Island?

Some terranes have undergone an even stranger migration than Vancouver Island. In November 1995, several geologists found that their different discoveries were related, and together formed the story of how a terrane from North America became a part of the Andes in Argentina. William Thomas of the University of Kentucky explained his discovery that about 540 mya, a terrane 500 mi (800 km) long had disappeared from what is now the Gulf Coast of the United States. A related discovery reported at the meeting had been made by Ricardo Astini of the National University of Córdoba, Argentina, who determined that rock formations in a part of the Argentine Andes closely resembled rock from the southeastern United States. Ian Dalziel of the University of Texas in Austin connected the two reports, postulating that the Gulf Coast terrane somehow parted from North America and stuck to South America, while Laurasia, the predecessor of North America, swung past Gondwana.

Tales of Subduction

The Philippine plate does not contain the Philippine Islands, but is located to the east of them and south of Japan. All of the Philippine plate is covered with part of the Pacific Ocean, except for a few tiny islands along its eastern border. Since it is all ocean, the plate is sometimes called "the Philippine sea plate," which can be confusing, since there is no "Philippine Sea." This small tectonic plate is entirely bordered on the west by the giant Eurasian plate, which is plunging below it, or subducting. The Philippine plate is bordered on the east almost entirely by the giant Pacific plate, which is also subducting it. On the southeast, some geologists identify a small plate containing the Caroline Islands and known as the Caroline plate, although it is often considered a part of the Pacific plate. In either case, this region of crust is also plunging below the Philippine plate.

Herein lies a mystery. The two largest tectonic plates on Earth are both sliding under the relative tiny Philippine plate, and no one knows why. The movements of these plates are of considerable significance to people in the region around the Philippine plate, the island nations of Japan and the Philippines, both of which are plagued by earthquakes and volcanoes as a result of the plate movements. Geologists have set up a special set of survey markers in hopes of determining how the Philippine plate is itself moving with respect to the mantle.

In the November 1995 issue of the *Journal of Geophysical Research*, Chris Scholz of the Lamont-Doherty Earth Observatory and Jaime Campos of the University of Santiago in Chile proposed that there are two different ways that an oceanic plate can subduct. If the upper plate is moving toward the descending plate, the combination produces frequent great earthquakes. If the upper plate is moving away from the lower plate, many fewer and milder earthquakes result, even when the subducting lower plate is moving faster than the receding upper. Scholz and Campos think that the receding upper plate turns the descending mantle downwards, so the slab plunges deep into the mantle before it can cause major quakes. Because the angle of plunge is so steep, the crust beyond the deepest trench is also pulled down, forming the beginning of a new ocean.

One Mantle, or Two?

Although everyone agrees that there is clear-cut evidence for some edges of plates descending into the mantle, forming deep trenches in the ocean floor, the fate of these descending slabs is involved with the complex problem of what difference exists between the upper and lower mantle. The evidence suggests that at about 415 mi (670 km) below Earth's surface the nature of the mantle changes in some way. The mantle begins just below the crust at depths between 3 mi (5 km) and 55 mi (90 km), depending on the part of the crust it is below, and extends to about 1,790 mi (2,890 km) from the surface to where the outer core begins. Thus, if there are two parts to the mantle, the inner one by itself is a hollow sphere with internal radius 2,180 mi (3,510 km) and external radius 3,550 mi (5,730 km), while the outer part is a hollow sphere with an internal radius of 3,550 mi (5,730 km) and an external radius of about 3,900 mi (6,350 km).

In 1992, two different conclusions were reached about the fate of descending slabs into subduction zones as they reached the lower mantle. Peter M. Shearer and T. Guy Masters of the University of California at San Diego found that in the region off Japan the descending plate makes a 15 mi (25 km) dent in the boundary between the upper and lower mantle before becoming compressed at the boundary. Their calculations suggest that if the upper and lower mantle were made from different minerals, the dent would be more profound, perhaps 100 mi (150 km) instead of 15 mi (25 km). About the same time, Shun-Ichiro Karato and Ping Li of the University of Minnesota calculated that descending slabs at the Marianas trench actually pass through the boundary because the rock grains in the slab are crushed by the overlying mantle and crust, making it easier to enter the lower region. This debate over whether the plates descend through the entire mantle to the core or just dissolve and disappear at the boundary with the inner mantle is still unresolved.

The same situation exists with regard to convection currents. Some scientists believe that the two layers of the mantle never interact, while others believe that the apparent differences between the layers are artifacts and that there is really only a single mantle. According to some geologists, convection currents and plumes of magma rising to form volcanoes and other lava flows exist only in the upper mantle. In this vision, since the lower mantle does not interact with the upper, no one knows or cares much what is going on in the lower mantle, so long as it is a source of heat to power the convection currents. The extreme of the alternative view is that the convection currents and plumes go all the way to the core.

Paul J. Tackley of the University of California at Los Angeles and coworkers propose that the truth embodies both ideas. Their computer models show that the magma behaves like a deep lake that has periods during which it is separated into two layers and others in which the upper layer becomes cool enough to fall through the bottom layer, causing the lake to "overturn." In temperate regions, such an overturn is an annual event in the spring after the ice breaks up. For the Earth, an overturn by the mantle might initiate a new period of plate movements, continent growth, and volcanic activity. For geologists, who prefer to find explanations for life's mysteries in the Earth rather than the stars, such an overturn with associated great lava outpourings can explain events such as mass extinctions, and even the apparent periodicity of such extinctions. (*See also* "Mass Extinction UPDATE," p 485, and "Plumes," p 554.)

(Periodical References and Additional Reading: *New York Times* 3-17-92, p C10; *New York Times* 3-24-92, p C6; *New York Times* 6-15-93, p C1; *Nature* 6-23-94, p 642; *Science News* 7-2-94, p 15; *EOS* 10-4-94, p 461; *Nature* 11-3-94, p 43; *Scientific American* 3-95, p 30; *Science* 5-5-95, p 635; *Science* 12-8-95, p 1567; *Discover* 1-96, pp 93 & 94)

MAKING MINERALS

For historical reasons, geologists use the word *minerals* to refer to chemical compounds or elements. (For different historical reasons, dietitians use the same word to refer to chemical elements only.) Minerals are often mixed together, as in granite, but are sometimes found in great concentrations, or even in pure forms. Often, a rich source of minerals is a winding path through some other rock formation, which is called a vein. Minerals that are commercially valuable are often found concentrated in such veins as ores, concentrations high enough to exploit for profit, or even in pure form, such as a vein of "native" copper. It is not surprising that geologists are interested in determining which processes cause the formation of pure minerals, high concentrations in ores, or veins of ores.

Not all such cases of ore formation or concentration have the same cause. In the 1990s, however, progress was made toward finding the reasons behind some instances of mineral formation.

From Volcanic Vents

A vent is an opening in the crust through which lava flows from time to time. Lava is the liquid rock formed either when hot magma from Earth's interior melts the crust or when the magma itself emerges and is changed chemically through the release of tons of gases kept previously in the rock by pressure. Lava may also be a mixture of melted crust and degassed magma. Volcanoes are essentially mountains produced by vents that spew forth lava and by debris from explosions caused by the expansion of gases dissolving out of the magma. Long vents are called rifts. Rift valleys, where tectonic plates are spreading apart, have many individual vents, nearly always under the sea.

In the 1980s, gold was found at some undersea volcanic vents, precipitating a great deal of interest in underwater mining, although very little in the way of actual operations. One of the largest mounds of minerals found at an undersea vent is called the TAG mound, because it was located during the Trans-Atlantic Geotransverse, a survey of the mid-Atlantic. Preliminary study revealed the TAG mound to consist largely of sulfides of copper, iron, and zinc. Since 1990, plans have been underway for drilling into the TAG mound to determine more about how vents concentrate minerals, the degree of concentration, and the types of minerals to be located there. An extensive operation began in June 1994 and continued into the following year, including dives by two different submersibles and drilling directly into the mound to obtain a core.

The same process that can be observed live at modern-day volcanoes and vents is thought to have created veins of mineral in the past. Heat enables fluids to bring into solution more minerals than the fluids could absorb at cooler temperatures. Under the great pressure of the interior of the Earth, water can remain liquid at very high temperatures. It is thought that this superheated water dissolves large amounts of

---

Background

Geothermal vents are found where two tectonic plates are splitting apart (although similar vents are caused by another mechanism in the Gulf of Mexico). Such places are called **rifts**. Most rifts are found under the oceans, although there are rifts on land in Africa and Iceland. The crust is thin and cracked under the rifts, and magma is not far from the surface. For underwater rifts, this means that there are places where water fills the cracks and is heated to high temperatures by the magma. The first geothermal vents were found in the Red Sea in the 1960s. In 1977, John B. Corliss and Robert D. Ballard, using the research submersible *Alvin*, discovered the geothermal vents at a mid-ocean rift in the Pacific near the Galápagos Islands. This rift location made scientists begin to realize how odd and important the vents are.

In the oceanic rifts, temperatures may reach 650°F (340°C), but the pressure is so great that the water does not boil. The superheated water does dissolve minerals from the crust, especially sulfur compounds. Eventually, the super-heated water is transported back into the ocean. When the hot mineral-laden water emerges into the cold ocean, the compounds are precipitated. It is thought such a scenario was responsible for the formation of the copper compounds that made the island of Cyprus an early center of mining activity and trade.

minerals from the rocks through which it flows, with some minerals more easily absorbed than others, resulting in high concentrations of a particular mineral in the superheated water. When the water cools or perhaps evaporates, depending on local conditions, it leaves behind the minerals that had been dissolved. The winding mineral paths through the rock are called the veins.

From Volcanic Pipes

Pipes are mysterious openings in Earth's crust that are sometimes called volcanoes, although they are not much like the undersea volcanoes implicated in production of minerals, nor are they like other familiar volcanoes.

Volcanoes, Rifts, and Other Vents	*"Volcanic" Pipes*
Found largely at plate margins where plates are separating or colliding.	Found only in plate interiors, especially in the oldest continental crustal blocks, called cratons.
Of varying sizes, but usually tens to hundreds of yards (meters) across, and sometimes much larger.	Extremely narrow, sometimes only a few yards (meters) across.
Have shallow roots, perhaps 50 mi (80 km) at most, even shallower on thin ocean crust.	Extend for great lengths through the crust, perhaps as far as 300 mi (500 km).
Produce molten rock known as lava. At ocean vents, the lava tends to flow smoothly from the vent. Sometimes can be explosive when gases dissolve.	No lava associated with pipes, although there is evidence that material bursts explosively from the top of the pipe during eruptions.
Volcanoes are often capped with tall mountains.	The openings of pipes are in craters below the level of the surrounding crust.
Eruptions on land familiar to everyone, although undersea eruptions were not recognized until recently.	No human is known to have ever witnessed the eruption of a pipe.
Volcanoes produce mostly sulfur compounds, but undersea vents appear to be involved in creation of various ores.	Pipes produce an unusual mineral called kimberlite that contains diamonds.
Volcanoes continually form. Some are known to have appeared unexpectedly in the past 50 years.	Pipes are dated as having formed, ranging from about 45 mya, to as much as 2.6 billion years ago.

The production of diamonds from other forms of carbon generally occurs when carbon is heated to high temperatures under great pressure, which allows the carbon atoms to move into a tight crystal structure (*see* "Diamond Discoveries," p 453). For natural diamonds, the only place possessing the correct conditions is far down in the crust, or even at the very base of volcanic pipes. Whatever process drives material up the pipes carries the diamonds along with it, tossing them about the openings of the pipes, along with garnets and the minerals pyrope and chromium diopside. Most natural diamonds are very old, some as much as 3.3 billion years.

There are very few old crustal cratons through which diamond-bearing pipes pass. The best known is in South Africa, but diamonds have been found in other old parts of continents, including South America, Asia (India and Siberia), and North America (Arkansas). The oldest and largest craton in North America is the greater part of Canada. The first diamonds were found in the Northwest Territories of Canada in the 1980s by geologist Charles Fipke. Word of his discoveries did not leak out until 1991, when the news started a "diamond rush" that is still ongoing.

Along the 2.5 billion-year-old Canadian craton, magma in cracks in the bedrock cools, forming swarms of a geologic feature called dikes. The dikes, which are seen on the surface as rows of long low hills, were formed in the already old craton well before the pipes formed. Bruce Kjarsgaard of the Canadian Geological Survey in Ottawa reports that evidence not very well reported by prospectors suggests that the pipes are always located near dikes. He believes that the magma intrusions weakened the craton enough to allow the pipes to form, although the last magma intrusions are dated at 1.2 billion years ago and the pipes are from about 50 mya. There are still more mysteries than solutions regarding natural diamond formation.

(Periodical References and Additional Reading: *New York Times* 2-15-94, p C1; *New York Times* 0-9-94, p C1)

PLUMES

The essential idea of plumes began in 1963 when J. Tuzo Wilson pointed out that the then-new theory of plate tectonics could explain certain features of the Hawaiian Island chain and its associated chain of undersea volcanoes called seamounts. Wilson proposed that there was a stationary "hot spot" in the mantle. As the Pacific plate moved, the hot spot produced periodic volcanoes above it. In the 30-some years since Wilson's idea was proposed, the concept has been strengthened, expanded, refined, and occasionally derided. Although geologists still sometimes speak of "hot spots," the most common language today reflects an underlying concept: the spot is hot because material rises through the mantle at that location. Although the rising material could have been called a "fountain," a "jet," or some other name, the name "plume" (apparently in reference to a rising column of smoke) has stuck.

Some 60 plumes are generally recognized, although there continue to be disputes about some of them. Plumes under ocean crust are more likely to break through and cause volcanoes because the crust is thinner than continental crust.

Origin of Plumes

The main question currently being debated about plumes is "Where do they start?" If the mantle is relatively homogeneous, plumes may start as low as the core and traverse the entire mantle. If the mantle is layered, then it is most likely that plumes originate in its upper layer, thought by most proponents to start around 415 mi (670 km) below Earth's surface. At the beginning of the 1990s, a sizable majority of earth scientists appeared to believe plumes start at the core-mantle boundary. At this time, it was assumed that the plumes represented mantle material.

In 1993, however, a series of reports suggested an entirely different origin for the plumes. The German research ship *Sonne* dredged volcanic glasses from a volcanic hot spot near Pitcairn Island in the South Pacific. Analysis of isotopes of oxygen and carbon in the glasses matched ratios expected of crustal rocks, not ratios thought to exist in the mantle. The international crew of scientists aboard came to believe that the plume beneath the hot spot was continental crust returning to the surface after it had plunged into the mantle at a plate boundary. At later conferences back on shore, the geologists put forth various ways that such an event could happen.

- When two plates meet and one dives under the other, the one diving down is halted by the lower mantle, which is somehow different from the upper mantle (*see* "Advances in Plate Tectonics Studies," p 542). The diving plate heats from the collision, and bounces back to become a plume.
- Plates diving down reach the lower mantle and are halted there, but do not bounce back. As the forces of the diving plate compress it, the lower edge of the plate becomes heated and breaks off, rising through the upper mantle as a bubble.
- There is no real difference between the upper and lower mantle. Plates descending through the mantle become hotter and hotter as they travel deeper into the mantle. Eventually, parts break off and rise as bubbles through the mantle.
- The plates keep descending until they reach the core, which is now known to be a lumpy ball. The bumps on the core knock of parts of the plate, which then rise all the way back through the mantle as bubbles.
- As the plate descends, heat from the core melts the leading edge, which then bubbles up to form the plume.

As the above scenarios suggest, the original concept of a "plume like a plume of smoke" has necessarily been modified. Now the rising material forms puffs, like those of a cigar smoker. Geologists generally call the puffs *bubbles,* despite the fact that bubbles usually denote containers of gas, and these puffs are either solid or slightly plastic when they start to ascend. Some geologists, however, call the puffs *blobs,* which is closer to what they all really mean.

In March 1995, Kaj Hoernle of GEOMAR in Kiel, Germany, reported that the characteristic chemical signature of the plumes that formed both the Canary and the Madeira island groups in the Atlantic Ocean also is found at various sites in Europe, including Mount Etna in Sicily and sites in Germany. Hoernle suggests that in this case the "plume" is actually a large "sheet" of molten material from which plumelets arise at various places, resulting in volcanic activity in the Atlantic basin and in Europe. The sheet might be halfway down in the mantle between the crust and the core. Partial confirmation of this idea comes from a tomographic reconstruction of the interior of that part of the mantle prepared by Yu-Shen Zhang of the University of California at Santa Cruz. His picture seems to show a faint sheetlike structure that could give rise to the plumelets.

In November 1995, Phillip D. Ihinger of Yale University in New Haven, Connecticut, suggested another kind of mantle structure that could affect plumes. He believes that the fluid part of the mantle near the crust must contain currents, similar to ocean currents, and like them caused by differences in density dictated by temperature or chemical composition. One such current could account for some anomalies in the Hawaiian volcanoes and associated islands. Sometimes adjacent volcanoes, such as Mauna Loa and Mauna Kea, both on the big island of Hawaii, produce lava that differs

Background

In 1963, J. T. Wilson proposed that chains of oceanic islands, such as the Hawaiian chain, were caused by stationary long-lasting features below the moving tectonic plates. In 1971, W. Jason Morgan of Princeton University extended this idea by suggesting that hot material rising from the core through the mantle at about 20 different locations accounts for several noticeable features on Earth's crust, including the Hawaiian Islands, Iceland, and the Yellowstone Park region. This concept has come to be accepted by most earth scientists, and the number of plumes has been expanded to about 60. The columns of rising heat and rock are called **plumes**, and the places where they appear on the surface are known as **hot spots**.

A hot spot is caused by a plume that is relatively stationary with respect to the moving crust above it. When magma from the hot spot breaks through the crust, it forms a volcano that grows into a seamount or island. As the plate moves away from the hot spot, the volcano becomes extinct, and the island or seamount begins to erode and subside. The "big island" of Hawaii is currently over the largest hot spot known. This hot spot is now forming what will become a new island in the Hawaiian chain (already named Loihi) while still causing eruptions on the big island. The Hawaiian hot spot has left behind a chain of extinct volcanoes that stretches from Hawaii to the northwest, reaching almost to the Kamchatka Peninsula, and including the Emperor seamounts as well as the Hawaiian Islands themselves.

Earth is not a perfectly smooth sphere. In addition to the obvious mountains and river valleys, there are large regions located further from the center of Earth than average, even when the flattening at the poles and bulging at the equator predicted by gravitational theory is taken into account. Some regions, such as continents, are higher than others because they are made of lighter rock that floats in a plastic or partly fluid region called the asthenosphere. The asthenosphere includes the bottom of the crust and the top of the mantle. For raised regions smaller than continents, other explanations must be found. For example, the Tibetan Plateau and the Himalaya Mountains are both caused by one tectonic plate pushing under another. Other high regions are harder to explain. Since the 1960s, many of these have been thought to be caused by plumes of material from the mantle pushing the crust up, sometimes erupting through it to form new layers of crust above the old.

Seismic waves are observed to speed up in columns below regions that are higher than the average level of Earth's surface. For the past 30 years, it has been generally assumed that such fast waves are caused by hot plumes of mantle material. As early as 1980, however, J. G. Schilling, M. B. Bergeron, and R. Evans suggested that the data might have another interpretation. They theorized that lighter material, not hotter, might be rising to produce the higher elevations and the accompanying volcanoes.

There are two possible ways to produce a plume of material rising through the mantle to reach the surface, both of which require that some parts of the mantle be less dense than other parts. One way is for the less dense material to be hotter than the material around it. The other is for it to have a different composition, one with more light elements than are found in the surrounding material. Both ways occur in currents that rise and fall in the atmosphere and oceans. In all these cases, the heavier material actually pushes the lighter material up, with gravity the force supplying the energy.

In the atmosphere or in the ocean, it is fairly easy to tell which of the two mechanisms (temperature or composition) is the cause of a rising current. One can measure the temperature of the air or the salinity of the water. The mantle, however, is inaccessible directly except for a few rock samples believed to have been brought up from the mantle by deep-rooted volcanoes. Most of what we know of the mantle comes from studies of seismic waves. Unfortunately, seismic waves travel faster in both hotter and lighter rock. Thus, the presence of faster waves in a portion of the mantle tells that the portion is hotter or less dense, but not which of the two.

significantly from volcano to volcano. Ihinger suggests that the mantle current disrupts a plume that is rising from further down in the mantle, causing different discrete blobs of plume material, which makes for differences in the lava of particular volcanoes.

Episodes of Superplumes

In the early 1990s, various geologists discovered evidence that there were periods in Earth's past when giant floods of lava poured forth through the crust, far in excess even of the great lava floods of the well-known Columbia plateau in North America and the Deccan Traps in India. About 120 million years ago (mya), the Ontong-Java plateau in the western Pacific Ocean and the Kerguelen plateau in the Indian Ocean erupted, producing the characteristic volcanic rock basalt which is also formed at ocean trenches and makes up the floor of the ocean. The plateaus, however, are formed from younger basalts that have risen through the ocean floor. The large regions of younger basalt beneath the oceans are vastly larger even than the better known flood basalts on the continents. The largest, the Ontong-Java plateau, is 25 times the size of the Deccan Traps, the largest continental flood basalt. The Ontong-Java plateau covers an area of Earth's crust about two-thirds the size of the continent of Australia, and possesses a volume such that if it were spread like icing over the lower 48 states of the United States it would cover them in 16 ft (5 m) of basalt.

Geologists began to explain these massive outpourings in terms of superplumes, giant blobs of magma that had risen through the crust. One reason that the flood basalts are thought to have been superplumes is that each of them emerged quickly by geologic standards. From beginning to end, an entire lava outpouring was accomplished over the span of 2-5 million years at most. This period is very short when compared to the tens of millions of years involved in building mountains or other geologic processes. Thus, the lava floods must have originated from isolated, or discrete, events in Earth's history, such as the arrival of a superplume at the surface.

In the November 3, 1994, *Nature*, Mordechai Stein of the Hebrew University in Jerusalem, Israel, and Albrecht W. Hofmann of the Max Planck Institute for Chemistry in Mainz, Germany, proposed that such superplumes arise during a brief period in Earth's history as a result of a phenomenon in the mantle similar to the mixing of layers in deep lakes in the spring, called overturn. During the mantle's overturn, plumes can establish themselves throughout the mantle from core to crust, but when the mantle is layered, as it is today, the smaller plumes associated with hot spots are confined entirely to the upper layer. There are then no superplumes so long as the layered mantle persists.

The arrival of a superplume greatly affects the environment of the whole Earth, not just the part immediately covered in molten rock. For example, the Ontong-Java plateau would have raised sea level by about 30 ft (10 m) over the 3 million years that it took to erupt. The carbon dioxide released 120 mya, when both the Ontong-Java plateau and the Kerguelen plateau were rising, would have raised global temperatures by 13.7-22.5°F (7.6-12.5°C). (*See* "The Future of Climate," p 525, for explanation of the influence of carbon dioxide on global temperatures.) Many of the flood basalts, especially those on the continents, coincide with mass extinctions (*see* "Mass Extinction UPDATE," p 485). Stein and Hofmann envision continents growing rapidly during the mantle overturns, with little or no continental growth during the layered phases of the mantle.

Roger L. Larson of the University of Rhode Island in Kingston notes that superplumes have emerged often in Earth's history. It has been 120 million years since the great mid-Cretaceous superplume that produced the giant undersea igneous provinces, 65 million years since the smaller event of the Deccan Traps, and only 16 million years since the Columbia flood basalts of the U.S. northwest. The next superplume could arrive at any time, with little or no warning.

(Periodical References and Additional Reading: *New York Times* 6-15-93, p C1; *Science* 4-29-94, p 662; *Scientific American,* 10-94, p 42; *Nature* 11-3-94, pp 43 & 63; *Scientific American,* 2-95, p 82; *Scientific American,* 3-95, p 30; *Science News* 11-25-95, p 357; *Discover* 1-96, p 91)

AT THE CORE

The conventional picture of the Earth's interior has not changed in its basic outline since 1936, when the diffraction of earthquake waves established that in addition to a crust and mantle differing in composition, Earth has a liquid outer core and a solid inner core. In recent years, however, this simple picture has been refined in various ways (*see* "Plumes," p 554, for example). Although much of this fine tuning has come from a more detailed analysis of earthquake waves, a large part also derives from experiments on materials subjected to high pressure—experiments made possible by the invention of the diamond anvil press.

Before the recent reanalysis of the core and its interaction with the lower mantle, the principal effect of the core on life on the surface was thought to derive from Earth's magnetic field, probably generated by the core. Some also thought that heat from the core reached the surface, but that was less certain. Now it appears that the core may also influence the rotation of the planet and drive the convection currents thought to be the engines of plate tectonics (*see* "Advances in Plate Tectonics Studies," p 542).

The Strange Core-Mantle Boundary

When geologists want to investigate the structure of Earth's crust and no handy earthquake supplies waves of the right sort in the right place, the geologists set off explosions to make their own waves. The same idea would work for the core, except that the explosions would have to be huge. When China detonated a very large nuclear device on May 21, 1993, about 1 mi (2 km) below Earth's surface, geologists were able to use the blast as a probe deep into the mantle and even into the core. Seismometers in the southwestern U.S. produced the greatest amount of information,

WAVES OF INFORMATION

In the nineteenth century, scientists calculated that an earthquake must produce two kinds of waves that travel through the Earth. One kind, identical in nature to sound waves, consists of a movement of a high-pressure region and a low-pressure region through the material that makes up the Earth. These waves are called pressure (or compressional) waves, abbreviated as **P** waves. The type of wave, a three-dimensional version of the two-dimensional ripple waves at the surface of a liquid, are called shear waves, abbreviated as **S** waves. The modern type of seismograph was invented in 1900, and both of the predicted types of waves were detected. Because **P** waves travel faster than **S** waves, some geologists translate the abbreviations as meaning Primary and Secondary waves.

The different characteristics of **P** and **S** waves are vital in understanding the composition of the inner Earth. The arrival times of the two types of waves, for example, are used to determine the distance to an earthquake origin, or focus. Because **P** waves are weakened and slowed by Earth's core, the core must be somehow different from the mantle, the main body of Earth. The mantle itself was recognized by a similar transition in speed below Earth's thin outer layer, or crust. **S** waves do not travel through liquids. Therefore, the outer core, which fails to transmit **S** waves, must be liquid, despite the huge pressures deep in Earth's interior.

Experiments with materials at high pressure have suggested how fast **P** waves should travel in various materials. The likely identity of minerals that make up the mantle has been derived primarily from this information, with some clues from volcanoes and other sources of deep minerals. The composition of the core is thought to be mostly iron and nickel, as previously suspected, but firmly established by F. Birch in the 1950s and 1960s. On the basis of experiments measuring how sound waves travel through various minerals, Birch also found that the iron core is partly combined with some nonmetal such as silicon or oxygen. In 1994, experiments with a diamond anvil suggested to Yingwin Fei and Ho-kwang Mao of the Geophysical Laboratory and Center for High-Pressure Research in Washington, D.C., that the nonmetals in the core, often known to geophysicists as light elements, were likely to be oxygen and sulfur. That is, part of the outer core might consist in large part of iron oxides and iron sulfides.

because they received two separate sets of **P** waves seconds apart. Since there was only a single explosion, the pair of signals meant that part of the **P** waves had traversed a stiffer material that conducted these waves faster than the rest of the mantle. The location of this large body of stiffer material was just above the outer core. It could be calculated from the times of arrival at various seismometers that the body must be about 200 mi (320 km) wide and 80 mi (130 km) thick.

This discovery set geologists looking more closely at the boundary between the core and the mantle. Since the outer core is liquid and exists under the gravitational attraction of the inner core, scientists before 1993 had assumed without much thought that the outer surface of the core was a sphere. In 1994, a number of researchers performed careful studies of the core-mantle boundary using computer-assisted seismic tomography (*see* "The State of Earth Science," p 471), revealing that the core-mantle boundary is covered with hills and valleys as much as 3.5 mi (6 km) high or deep. Part of this hilly topography may come from chemical interactions between the lower mantle and the core, facilitated by the high pressure and temperature. Data, however, is outstripping theory; that is, observations are piling up that no one knows how to make convincing sense from as yet.

The Great Crystal

Seismic waves travel through Earth's inner core about 4% faster when they are moving parallel to Earth's axis than when they are perpendicular to the axis. Usually, a difference in wave speed in Earth's interior is attributed to a difference in composition, although in some cases differences in pressure, temperature, or state may be suspected. Yet none of those conditions could be involved in producing the different speeds perpendicular to each other in the core, since the core consists of the same material, in the same state, at the same temperature and pressure. Still, the measured speeds are different. There has to be some difference that is solely due to direction of transmission.

In 1995, Ron E. Cohen of the Carnegie Institution of Washington, District of Columbia, and Lars Stixrude of the Georgia Institute of Technology found a plausible explanation. Using a computer, the two calculated various possible crystal arrangements for iron. Materials under high pressure can assume configurations that do not normally occur (*see* "Diamond Discoveries," p 453, for example). Cohen and Stixrude found a way that iron atoms could form interlocking hexagonal prisms, a crystal pattern classed as "hexagonal close packed," or hcp, which produces in computer simulation the kind of difference in speed of seismic waves that has been observed.

The properties of large single crystals are generally different from those of a solid made from many small crystals. Distinctions tend to be lost in the jumbles of smaller crystals. The hcp iron crystals are no exception. If small crystals had formed with random orientations in the inner core, the difference in speed would vanish. Thus, the crystals need to be either very large and aligned or, if smaller, very closely aligned for the effect to occur. Cohen goes further than that, suggesting that the iron may have formed essentially into a single giant crystal; that is, the individual atoms are all aligned. Perhaps Earth's rotation or its magnetic field is enough to have caused iron

Background

It has long been known that rock under great pressure and extremely high temperatures does not have the same properties as rock under what people at the surface view as normal conditions. It has been difficult to speculate about what these differences caused by pressure and temperature mean to our interpretation of events in Earth's interior, where the observable information is the behavior of seismic waves.

The development in the 1970s of the diamond anvil by Charles Weir, Alvin van Valkenburg, and coworkers at the National Bureau of Standards revolutionized the way earth scientists study the interior of Earth. A diamond anvil applies pressure to a small sample held between the surfaces of two specially cut gem-quality diamonds. A fundamental law of physics states that pressure equals force divided by area of application. Since the diamonds are tiny and the anvil's hydraulics, levers, and screws multiply force to a high level, pressure between the points is great. Because diamond is very hard, the points do not break or bend. Using this device, scientists have been able to produce pressures in the laboratory comparable to those in Earth's interior. Diamond anvils routinely develop pressures 2 million times air pressure at sea level (2 megabars). The record for a diamond anvil is 4.16 megabars, set by Arthur L. Ruoff and coworkers at Cornell University. The maximum theoretical strength for a diamond anvil is about 8-10 megabars, the pressure at which diamond becomes a metal and too malleable to be useful. Theoretical pressure might be increased by replacing diamond with buckminsterfullerene, since buckminsterfullerene is expected to become twice as stiff as diamond under high pressures (see "Ornamented and Filled Buckyballs," p 457).

A few other devices produce even higher pressures than diamond anvils. The two-stage light-gas gun is a 105 ft (32 m) cannon whose first stage is a gunpowder explosion that pushes a piston that compresses a light gas, almost always hydrogen. At high pressure, the hydrogen bursts a barrier and propels a plastic projectile. The plastic projectile travels about 23 in (60 cm) through a vacuum to its target. When the plastic projectile hits a target, it transfers its momentum to the target, producing pressures as great as 50 megabars.

Hydrogen is often used in this device because molecules of light gases accelerate faster than those of heavier gases. Although momentum is a product of both mass and velocity, it is easier to increase momentum by accelerating particles of low mass to a high velocity than by accelerating heavy particles to a low velocity. Since acceleration equals force divided by mass, for a given force, lower mass means higher acceleration. After a short period of acceleration, the greater velocity of the light gas molecules contributes more to the momentum than the low mass of the molecules takes away. In practice, only about 10 of the lightest molecules are useful in gas guns (or for rocket fuels). Plastic projectiles are used to reduce the loss of momentum (and therefore pressure) that would be produced by bouncing off the target. The projectiles do not bounce and plastic does not compress, also contributing to the transfer of momentum.

Even higher pressures have been produced by nuclear explosions (158 megabars) and implosion of a capsule using lasers (about 1,000 megabars).

The advantage of the diamond anvil over devices that have produced higher pressures is that impact and implosion methods create pressures that last for very short times, while diamond anvils increase pressure slowly. All methods for producing high pressures suffer from the small size of the samples that can be used and the difficulty of measuring physical properties while maintaining the pressure. In some ways, the ability to measure what they have achieved, often by bouncing laser light off the samples, is as remarkable as the pressures themselves.

atoms in the core to arrange themselves in this regular way, as the solid settled slowly out of the liquid. The liquid iron that is now left, the outer core, may still be cooling and depositing layers of aligned iron atoms on the exterior of the inner core, growing the crystal even further.

The giant-crystal concept might also explain the alignment of Earth's magnetic field, which is also different in one direction, although not exactly along the axis. Cohen and Stixrude think that there may be two components to the field. The main one would be caused by movements of the liquid outer core, while the secondary one would be trapped in the crystalline inner core. If these two are not aligned, the result may be the askew field that is observed at Earth's surface (*see also* "Geomagnetism UPDATE," p 563).

How Hot is the Core?

Some theories of what propels plate tectonics or causes plumes in the mantle (*see* "Advances in Plate Tectonics Studies," p 542, and "Plumes," p 554) incorporate heat emanating from Earth's core. To some degree, these theories are fueled by experiments with diamond anvils and shocks produced by gas-powered guns, which since 1987 have indicated that Earth's core is much hotter than anyone expected. The other possibility is that the experiments have been giving the wrong results—that the core is not quite that hot.

Part of the problem has always been that the pressure at the core is simply too high to reproduce with a diamond anvil or even with high-pressure shock-wave experiments. It is relatively easy to calculate the pressure at the core, which is a direct result of Earth's mass (known since 1798) and density at various levels (which must be estimated). The pressure is thought to be 330 gigapascals, where a pascal is a derived SI unit of pressure equal to one newton per meter squared (roughly 0.00001 pound per square inch) and a gigapascal is 1,000,000,000 pascals (roughly 140,000 pounds per square inch). Therefore, 330 gigapascals is somewhat more than 45 million pounds per square inch. Another way of thinking of the equivalence is in terms of bars, also called "atmospheres." A gigapascal is about 10,000 bars or 10,000 times air pressure at sea level. In 1987, the best that could be gotten from a diamond anvil was about 100 gigapascals, from which the temperature was extrapolated.

The way that temperature is derived from pressure is as follows: Earth's interior maintains a delicate balance between temperature and pressure. Solid or nearly solid rock from the interior becomes flowing lava when at the surface because its temperature is hot enough to melt but the high pressure keeps it from doing so. Earthquake waves have been used to establish that the liquid core solidifies in the center where the pressure is greater. Thus, the temperature of the core at the boundary where it begins to solidify is the temperature at which iron (or perhaps iron combined with some other elements) melts when it is at a pressure of 330 gigapascals. This number cannot be predicted from first principles but must be derived from experimentation.

In the original 1987 experiments, Raymond Jeanloz and Quentin Williams of the University of California, Berkeley, found that iron heated to 4,000K (almost 7,000°F or 3,700°C) melted at 100 gigapascals, from which they calculated that at 330 gigapascals the iron would need to be heated to about 7,600K (about 13,000°F or 7,300°C) to melt. Their construct, however, depended on a core temperature of only 4,000K, a little more than half the experimentally derived temperature.

Reinhard Boehler of the Max Planck Institute for Chemistry in Mainz, Germany, has also been working with diamond anvils and has consistently found a melting point of only 3,000K at 100 gigapascals. This difference translates to a temperature at 330 gigapascals that more closely approximates the expectations of theorists. In 1995, Thomas Ahrens and Kathleen Gallagher of CalTech in Pasadena, California, reanalyzed the 1987 experiments, finding that they had been contaminated by the effect of an aluminum oxide casing used to hold the iron samples. With that correction, Boehler's work makes more sense. Further experiments by Boehler have now demonstrated iron melting at 4,000K at 200 gigapascals, which translates to 4,850K (8,250°F or 4,550°C) at 330 gigapascals, although the original Jeanloz experiment, even when corrected, is still about 6,500K (11,250°F or 6,200°C).

Jeanloz continues to base his core-mantle interaction theories on a core that supplies about one-fifth of the heat energy of the planet and that heats the mantle from below.

(Periodical References and Additional Reading: *New York Times* 2-11-93, p B16; *EOS* 11-29-94, p 563; *Science* 12-9-94, pp 1662 & 1678; *Discover* 1-95, p 90; *Science* 3-17-95, p 1597)

Geomagnetism UPDATE

Sailors around the world had been using a magnetic compass to locate north or south for several hundred years before 1600, when William Gilbert made his dent in the history of science by pointing out that Earth behaves like a large magnet. More than 200 years later, Michael Faraday began to explain magnetism and electricity in terms of fields, regions in space for which there is a definite force in a definite direction at each point. Lines of force in a given direction define the field; similar lines appear when iron filings are shaken on a sheet of paper placed over a small iron magnet. Today, the magnetism of Earth is called Earth's magnetic field, and it is identified by the lines of force that pass from the north magnetic pole to the south magnetic pole. These are the lines of force that are pictured as lining up the compass needle in much the same way as iron filings are lined up by a small magnet. These lines of force are also pictured as lining up magnetic domains in liquid rock or in some kinds of sediments. When the rock or sediment turns to stone, the lines are preserved as they had been, since the domains can no longer adjust to changes in the magnetic field.

Identifying Earth's Magnet

It is generally accepted that Earth's magnetic field begins with the iron and nickel core, since these metals are known for their ability to act as magnets. Specifically, most scientists today believe that the fields originate in the interaction of electric currents and moving physical currents in the outer core. Just as a dynamo generates electricity by moving iron through a magnetic field, the process can be reversed to generate magnetism, as in an electromagnet. One problem with this concept concerns finding a cause for the motion in the fluid outer core.

In 1995, J. F. Poirier, reviewing the kind of light elements that might be mixed with iron in the outer core, suggested that compounds of these light elements might be expelled from the precipitate as the iron solidifies at the inner core. If this process occurs, the migration of the lighter elements, or compounds from the solidifying margin of the inner core through the outer core, might provide the convection currents in the outer core needed to create geomagnetism. The 1994 iron oxide experiments of Yingwin Fei and Ho-kwang Mao of the Geophysical Laboratory and Center for High-Pressure Research in Washington, D.C., show that the compound can enter a hexagonal crystal state at high pressures and temperatures, becoming a conductor, like a metal. Thus, moving iron oxide could generate a magnetic field.

Peter L. Olson of Johns Hopkins University in Baltimore, Maryland, and Roger L. Larson of the University of Rhode Island in Kingston have connected geological periods with few magnetic reversals to the emergence of superplumes (see "Plumes," p 554). They picture the plumes as stemming from "boiling" in Earth's core. When the core is boiling furiously, it changes Earth's magnetic orientation relatively fast. This boiling is the process that sends up plumes through the mantle, and the plumes take away heat from the lower mantle, allowing the core to boil faster by creating a greater difference in temperatures at the core-mantle boundary. By the time the plumes arrive at the surface, the boiling is down to a simmer for a while and heat builds up in the lower mantle. Thus, when the plumes arrive, geomagnetic reversals are at a minimum.

There are observations, however, that seem to contradict Olson and Larson's hypothesis. In 1985, Michel Prévot of the University of Montpellier in France and Robert S. Coe of the University of California, Santa Cruz, discovered surprising evidence that a sudden geomagnetic reversal had taken place while the lava was flowing during the Columbia flood-basalt episode (see "Plumes," p 554). This discovery was met with skepticism. In the April 20, 1995, *Nature,* however, Coe, Prévot, and Pierre Camps of the University of Montpellier report on further study of the Columbia lava flows in Washington state, which they say captured the magnetic field rotating at a rate of 6° a day. This rate is about a thousand times faster than any changes in the magnetic field observed today. The authors of the report carefully examine alternate hypothesis, but conclude that the rapid shift of the field must really have happened.

One problem with rapid shifts in geomagnetism, however, is that it seems probable that Earth's mantle, or at least the inner mantle, is a semiconductor. A semiconductor would dampen down the change, slowing down a rapid reversal as it propagated through the nearly 1,800 mi (3,000 km) from the core to the surface. Coe, Prévot, and Camps assume that only the lower mantle has this effect, and are just barely able by

Background

The state of Earth's magnetic field is captured by any molten rock that solidifies, or by grains of magnetite in sedimentary rock. The magnetic record in rocks reveals that from time to time (usually every few hundred thousand years or so) the magnetic field of Earth reverses. The north and south magnetic poles interchange, usually taking about 5,000 years to accomplish the flip. These reversals have happened about 300 times in the past 170 million years. Although this process was first noticed by Bernard Brunhes in 1906, it was so surprising that most geologists did not accept the phenomenon until the 1960s. Even today, no one knows for sure why the reversals occur. One theory suggests that the flips are caused when a larger than usual meteorite (which might be called an asteroid under other circumstances) bumps into Earth, changing the flow of the outer core, which is where the magnetic field is thought to originate.

Some periods have almost no reversals. The Cretaceous and Permian periods are among those in which the poles stayed the same for most of the time. Scientists who favor impact theories for mass extinctions and who also favor impacts as affecting Earth's magnetic field may be cheered by that information, since both the Cretaceous and the Permian ended with mass extinctions followed by long periods of geomagnetic reversals.

Normal polarity is considered to be what we experience today, with reversed polarity positioning the north magnetic pole near the geographic South Pole, and the south magnetic pole near the geographic North Pole. For the past 2,000 years or so, Earth's magnetic field has been getting weaker, but it is not clear whether this portends another reversal. The intensity peaked some 2,000 years ago, and is now about 40% of what it was at that time.

Magnetic reversals have become useful tools for dating past events. Knowledge of the fossil magnetic orientation of a rock along with that of rocks above and below it (for which the magnetic orientation is often different) can be combined with other information to determine a date within a range that averages about half a million years (long by some standards, but short by geologic standards). Since geomagnetic reversals are worldwide events, there is no problem correlating events on one continent with those on another.

that assumption to reduce the amount of damping enough to agree with their experimental results. They think that the kind of swift change they measured in Washington state occurs as brief excursions during a longer period of slower change, just as a baby grows in phases, with periods of no growth in between. Or, one could call their theory "punctuated geomagnetism."

Geomagnetic Oddities Explained?

There are many unresolved questions about Earth's magnetic field. Lars Stixrude of the Georgia Institute of Technology, working with Bradford Clement of Florida International University in Miami and Ron E. Cohen of the Carnegie Institute of Washington in the District of Columbia, has developed a hypothesis that perhaps can answer some of them. The mysteries include why Earth's magnetic equator is 4° away from the horizontal level expected by theory; why the magnetic poles do not replace each other symmetrically during a geomagnetic reversal; and why the poles as they move from one hemisphere to the other during a reversal seem to always follow the same path, lingering at favored locations. The hypothesis that Stixrude proposes suggests that the inner core of the Earth has its own highly aligned magnetic field that causes all of these effects through its interactions with the more variable and more powerful magnetic field of the outer core.

Cohen and Stixrude have developed a giant-crystal concept for the inner core (*see* "At the Core," p 558). According to Clement and Stixrude, this crystal would have an alignment that could tilt the stronger field of the outer core by 4°, as is observed at the magnetic equator. The same tilt could cause the North and South Magnetic Poles to move to different locations when they have been interchanged in a geomagnetic reversal. Finally, as the magnetic field caused by the outer core weakens during a geomagnetic reversal, the permanent field from the inner core would become the preeminent one. Thus, the permanent field could always direct the poles to the same places, since the inner core's field is thought not to wander the way that of the outer core does.

The basic concepts of this hypothesis seem to make sense, but calculations proving that the inner core could account for the observations have yet to be made. Furthermore, even if the calculations show that the "two magnet" hypothesis explains the events, they will not prove that there are two separate magnets. The circumstantial evidence will be very strong indeed, however.

(Periodical References and Additional Reading: *Nature* 11-18-93, pp 205, 234, & 238; *Science* 12-9-94, pp 1662 & 1678; *Scientific American,* 2-95, p 82; *Science* 3-31-95, pp 1910 & 1972; *Nature* 4-20-95, pp 674 & 687; *Science News* 4-22-95, p 244)

VOLCANOES AND EARTHQUAKES

VOLCANO STUDIES AND SURPRISES

Although many vocanologists (as scientists who study volcanoes are known) are still at work around the world, they have already learned much about the basic mechanisms that cause volcanoes. The answers to many unresolved questions are more in the province of geophysics and plate tectonics (*see* "Advances in Plate Tectonics Studies," p 542, and "Plumes," p 554) than in pure volcanology. Volcanologists have become remarkably good at predicting both when volcanoes will erupt and the possible consequences of eruptions, although sometimes warnings are overlooked and many volcanoes are in regions where they receive little study. Despite all this success, volcanoes continue to produce surprises, not only to those living near the "smoking mountains," but also to the volcanologists themselves. On January 14, 1993, the Galeras volcano in Colombia erupted suddenly while it was being studied by a group of visiting scientists, killing six and severely injuring another.

Interactions of Volcanoes and the Environment

Much that is being learned new about volcanoes today comes from the study of how volcanoes change the environment about them and even around the world. Volcano-induced change can happen suddenly during an eruption or more slowly during the relatively quiet periods of a still-active vent. Furthermore, in some instances the change works the other way, such that events in the environment can change what is happening to volcanoes.

Fire and Ice   Today, Iceland is known as "The Land of Fire and Ice" because of its landscape of active volcanoes and glaciers. In the past, however, the relationship between glacial ice and volcanic heat was even more intimate than mere proximity. Gregory Zielinski of the University of New Hampshire in Durham and coworkers analyzed the interactions of volcanoes and glaciers around the time of the recent ice age using the GISP2 ice core from Greenland (*see* "Climate Clues to the Past," p 515). They were primarily looking for dust from volcanoes in the recovered ice. The scientists found confirmation of an old theory which had stated that during the glacial phase of the ice age, pressure from 1 mi (2 km) of ice overhead would have been enough to cap any volcanoes that the ice covered. Thus, when the ice melted, all the volcanoes that had been prevented from erupting would go off at once. The ice-core record suggests that this scenario actually occurred at the end of the most recent ice age. Furthermore, the volume of sulfate found in the ice suggests that each erupting volcano produced about three times the amount that would be produced by the same volcano under today's conditions.

Another old theory about the relation between volcanoes and ice postulates that one or more giant volcanic eruptions, known to cool the Earth by screening out sunlight with sulfates and dust, might have tipped the climate scales and initiated an ice age. Zielinski and his coworkers also found evidence tending to support this theory. About 68,000–75,000 years ago, just at the start of a major glacial advance, there is a thick layer of sulfates. Zielinski connects this layer to a previously identified eruption of the volcano Toba on Sumatra.

Modern Cooling   A 1993 study by scientists from the U.S. National Aeronautics and Space Administration (NASA) and the University of Maryland in College Park used satellite information to calculate sulfur dioxide emissions, which are thought to be the main contributor to cooling after large volcanic eruptions. The researchers found that volcanoes on average contribute a tenth or less of the amount of sulfur dioxide that humans contribute to the atmosphere. During explosive volcanic eruptions for the period 1979-1992, however, the amount of sulfur dioxide thought to have reached the stratosphere was about 33 million tons. Virtually no sulfur dioxide produced by human factories every reaches the stratosphere. It is sulfur dioxide and dust in the stratosphere that is often blamed for global cooling after powerful volcanic eruptions.

TIMETABLE OF VOLCANOES AND CLIMATE

5675 B.C.	Eruption of Mt. Mazama in what is now Oregon (date from GISP2 ice core) produces volume of sulfates similar to that of historical eruption of Tambora in A.D. 1815. Therefore, it may have created lower-than-normal global temperatures.
1773-74 A.D.	Rain, volcanic ash, poison gases, and darkness from the Laki fissure eruption on Mt. Skaptar not only kills crops and livestock, but the clouds prevent fishing. About 24% of the population starves.
1815	The eruption of Tambora on Sumbawa in Indonesia is the largest since written accounts of eruptions started. The following year, there is widespread cold and crop failure in Europe and the United States, with 1816 named in New England as "The Year Without a Summer."
1883	Eruption of Krakatau in East Indies causes tsunamis that kill 34,000. Sunsets are reddened around the world for several years.
1982	Global temperature is lowered slightly by eruption of El Chichón in Mexico, the first modern documentation of such an effect.
1991	Eruption of Mt. Pinatubo in Philippines stops global warming trend for two years.

Volcano Hazard Prediction

In recent years, one of the methods used by volcanologists to predict eruptions involves measuring changes in the shape of volcanoes. This is accomplished by combining lasers with markers placed on geological features. In June 1995, Didier Massonnet of the French National Center of Space Studies and coworkers demonstrated that the same kinds of measurement of shape can take place from space. The European satellite *ERS-1* is engaged in using radar to make detailed maps of Earth's features. The maps are accurate to a fraction of a centimeter. When the same feature is mapped in a second pass, one computerized image can be superimposed over the other. Any difference, positive or negative, in height can be shown by a "false color" image that highlights any part of the topography that has moved. Color is also used to indicate in which direction movement has taken place. Checking this idea, Massonnet and coworkers found that the satellite images could easily detect the deflation in

Mount Etna caused by the eruption of 1991-93. The team is now using the technique to monitor various volcanoes around the world, in search of movement in the opposite direction. Earlier experience with ground-based measurements has shown that inflation of a volcano often signals eruption.

Another important indicator concerns changes in the gases emitted. These gases were used in the successful but ignored prediction of the 1991 eruption of Ruiz in Colombia, the successful prediction of the 1990 eruption of Redoubt in Alaska, and the life-saving prediction of Pinatubo in the Philippines. A change in the gases emitted by Mammoth Mountain (*see* "Other Geophysical Events," p 580) in Long Valley in California may also signal a forthcoming eruption. The most recent eruption of Mammoth Mountain was several hundred years ago.

The most successful recent prediction concerned the eruption of Rabaul in Papua, New Guinea. After an eruption of Rabaul in 1937 that killed 500 residents of the city (also named Rabaul) in its caldera, the government set up a volcano observatory. In 1994, seismologists from the observatory recognized that after a period of quiet, a series of small earthquakes had merged into a kind of continuous earthquake called a "volcanic tremor." Such an earthquake is an almost certain indication of an eruption, so the observatory called for 30,000 people to be evacuated. On September 19, 1994, the day after the evacuation, the eruption began. For the next five days, volcanic ash poured down on the town, covering everything to a depth of about 3 ft (1 m) and destroying a quarter of the buildings with its weight. The seismologists have since returned to the volcano observatory from which they closely monitor Rabaul for signs of an even larger event. Along with Long Valley in California and the Phelgreaen Fields in Italy near Naples, Rabaul is a giant caldera with a history of violent explosions— larger than any that have been observed in historic times.

One result of the Galeras disaster of 1993 was a new realization. Apparently, when an active volcano becomes suddenly quiet, it can mean that a blockage has developed that may suddenly blow out, leading to violent eruptions. In particular, if a gas such as sulfur dioxide stops flowing through the cracks, the gas may be building up inside the volcano. Similarly, very small and increasingly far-apart tremors may suggest that an explosion is about to occur.

Trips into the Volcano

Dante 1 was a robot designed by scientists from Carnegie Mellon University in Pittsburgh, Pennsylvania, and primarily intended to aid in research which would lead to robots designed for visits to other planets. Its test on Earth, however, was to explore the crater of active volcano Mt. Erebus in Antarctica. The robot was directed through and powered by a long umbilical cord, called its tether. After traveling 28 ft (9 m) into the crater, it failed on January 1, 1993, when its fiber-optic communications cable broke.

Dante 2 is an eight-legged robot weighing 770 kg (1,700 lb) that was Carnegie Mellon's next try. Again, the test was entry into an active volcano, this time one that had just erupted the year before.

Dante entered the crater of Mt. Spurr in Alaska in late July and spent eight days traveling some 660 ft (200 m) into the crater, climbing over rocks and snow and down slopes as steep as 30°. Large rocks dislodged from the crater walls fell all about the 10 ft (3 m) long robot. In the crater, it analyzed gases escaping from several small vents called fumaroles by volcanologists. Mission accomplished, Dante began the climb back up, but after 250 ft (75 m) a misstep on August 5 caused it to role over. A helicopter tried to lift the robot out of Mt. Spurr by its tether, but the tether snapped in the attempt. Finally, two humans hiked into the crater, where they were able to truss up Dante so that the helicopter could lift it out. Between the fall and the rescue, Dante broke seven of its eight legs.

Despite the slip on a 35° slope, the mission was considered a success. Not much was really learned about volcanoes, but in a field where several human volcanologists have been lost in the past few years because of unexpected eruptions, there could be a place for a robot volcanologist.

(Periodical References and Additional Reading: *Nature* 9-29-93, p 327; *New York Times* 3-10-94, p B11; Science 5-13-94, pp 913 & 948; *Science* 8-5-94, p 731; *Science News* 8-13-94, 101; *Science News* 8-20-94, p 117; *Science* 8-26-94, p 1171; *Science* 9-30-94, p 2005; *Science News* 10-15-94, p 254; *Science* 1-13-95, p 256; *Nature* 8-24-95, pp. 644 & 675; *Science* 9-1-95, p 1223; *Discover* 1-96, p 91)

EARTHQUAKE MYSTERIES

Both the cause and some of the effects of earthquakes are less well understood than most people assume. Although it is clearly established that cracks in Earth's crust called faults are associated with earthquakes, many earthquakes occur in locations where such faults either have not been located or else do not exist. In some cases, it is known that an earthquake has a cause other than a fault. For example, earthquakes associated with volcanoes are caused by movements of the molten rock called magma, not by movements of rock strata along fault lines.

The timing of earthquakes may be largely at random or may be influenced by particular forces. A popular but unproven idea assigns some responsibility to the phases of the Moon, which are already known to be directly related to tides (*see also* "Earthquake Prediction UPDATE," p 573).

Many people have reported earthquake effects that are unexplained or at best poorly understood. Although scientists sometimes doubt that visible waves can be seen in solid objects during an earthquake, such waves have been reported by many reliable observers. Earthquakes produce strange sounds, lights, and unusual behavior in animals. Some people associate earthquakes with certain types of weather, deemed "earthquake weather."

New Ways to Cause Earthquakes

If one earthquake closely follows another and there seems to be some connection between the two, the second earthquake is termed an *aftershock*. However, if the second earthquake is larger than the first and largest in the series, the first is called a *foreshock* and the second is *the* earthquake. Often, aftershocks are caused by something other than the slippage along the fault that results in the main earthquake.

After the 1992 Landers earthquake in California's Mojave Desert, which had a moment magnitude of 7.3, geologists were puzzled by the great number of apparent aftershocks that occurred all over the West, some as many as 759 mi (1,200 km) from Landers. One clue to the origin of these unusual small quakes concerned their location in regions where, at some time in the past, volcanoes had erupted, such as the Long Valley caldera in California and Yellowstone National Park in Wyoming. Alan T. Linde of the Carnegie Institution of Washington in the District of Columbia and coworkers reported in the September 29, 1994, *Nature* that their measurements at Long Valley showed that the aftershocks were caused by bubbles in the magma chambers still existing in ancient volcanoes. The shaking caused by the Landers quake would be like shaking a bottle of soda pop, causing the bubbles to free themselves from the magma and exert pressure on the walls of rock chambers. A mild explosion as parts of the chamber collapsed would result in the observed small earthquake.

Surprising Earthquake Effects

Scientists have known for a long time that the ionosphere is affected by earthquakes. The ionosphere is the ionized layer of the upper atmosphere that partly reflects AM and shortwave radio signals back to Earth, making it possible to receive radio messages beyond the curve of the horizon. The motion of Earth's surface sends a wave similar to a sound wave through the atmosphere with great power, although with very low frequency, so the waves are not heard as sound by humans. As the waves rise through regions of decreasing atmospheric density, the energy must be transmitted by molecules that are farther and farther apart. This has the effect of increasing the amplitude of the waves. The lower surface of the ionosphere moves far enough when the waves arrive that the movement can easily be detected by ground-based radar.

The Northridge earthquake in January 1994 moved the earth a distance of about 10 in (40 cm) which propagated as atmospheric waves, reaching and moving the ionosphere about 120 mi (200 km) away. Eric Calais and J. Bernard Minster of the Scripps Institution of Oceanography in La Jolla, California, showed that one can detect such waves using the Global Positioning System (GPS; *see* "The State of Technology," p 679) based on the disturbances the waves cause to the signal as it passes from the satellites to the receiver on Earth. Thus, the GPS becomes a sort of auxiliary seismometer.

What Causes Deep Earthquakes?

The earthquake that struck beneath Bolivia on June 9, 1994, had a magnitude of 8.3 and was felt all the way to Toronto, Ontario. It was the largest earthquake recorded since March 27, 1964, when an earthquake in Southern Alaska registered 8.5. Yet unlike the Alaskan earthquake, which killed 131 and caused widespread damage, the quake below northern Bolivia killed no one and caused virtually no damage. The main reason for the difference was that the 1994 earthquake occurred 390 mi (630 km) below Bolivia, the deepest large earthquake ever recorded.

Earthquakes that deep are like the famous flying bumblebee: proved by science not to occur. At a depth of over 40 mi (70 km), rock should be too soft and ductile to crack. By the time the depth reaches 390 mi (630 km), the rock is further softened by a temperature of about 3,000°F (1,650°C). Near the surface, earthquakes occur at slippage along cracks, but at a deep-earth pressure of 3,000,000 lb per sq in (650,000 newtons per sq cm), all the cracks should have been squeezed shut. Given the conditions, it is surprising to learn that earthquakes have been known to occur to a depth of 415 mi (670 km).

Deep earthquakes may be caused by processes that differ from the various causes of shallower quakes, which are principally interactions of tectonic plates at their boundaries. From a few kilometers deep to about 200 mi (350 km), the number of earthquakes gradually decreases, but below a depth of 200 mi the number begins to increase again.

In all cases of very deep earthquakes, the quake takes place where a block of crust is plunging down into the mantle at the juncture of two tectonic plates. The Nazca plate is sliding under the South American plate, raising the Andes and lowering the depth at which earthquakes such as the deep Bolivian blast occur. One theory suggests that the deep earthquakes originate in parts of the crust that have not yet been melted and melded. By the time they reach 415 mi (670 km), these slabs are being incorporated into the mantle. Other evidence, however, suggests that slabs of crust maintain their integrity far deeper into the mantle.

Two deep earthquakes in 1994—the Bolivian quake described above and a magnitude 7.6 earthquake beneath Tonga in March—produced seismic waves that indicated huge faults as the cause. The Bolivian quake had a fault region about 10 by 30 mi (30 by 50 km), while the Tonga fault was about 30 by 40 mi (50 by 65 km). Such deep earthquakes occur only in cooler slabs, ones that formed a longer time ago than did other slabs.

Some geologists have proposed that the cause of deep earthquakes is heat from the mantle causing rock to give up water loosely bound to minerals in the rock. This water greases the fault slippage, which is otherwise much like that in shallower earthquakes, also thought to be enhanced by wetness. Critics say that if this effect were to cause deep earthquakes, the depths of all such earthquakes should be about the same, which they are not. Also, there is no evidence that water is released from rocks in slabs as they descend.

Some geologists think that a phase transition caused by pressure is the cause of deep earthquakes. Just as pressure can turn the mineral graphite into diamond by rearranging the relationship of the atoms to each other, pressure can also change the mineral olivine (a main component of a crustal plate) into the structure called spinel. The spinel is slipperier than the olivine, so latent earthquakes start when the transition occurs. Critics of this approach say that the temperature profiles of actual deep earthquakes vary too widely for this effect to explain what is happening.

A third idea is that deep earthquakes occur because the descending slab hits some sort of a barrier when it reaches the lower mantle and pieces break off. The weak spots that developed when the slab was younger have been preserved and become the places where the block breaks. A version of this theory has been used by Raymond Russo and Paul Silver of the Carnegie Institution of Washington in the District of Columbia to explain the Bolivian deep quake. They believe that there is a viscous region of the mantle that is bending South America (*see* "Advances in Plate Tectonics

Studies," p 542). Their seismic profile shows the Bolivian quake took place where the descending tectonic plate encountered mantle turbulence, which could have caused the break.

(Periodical References and Additional Reading: *Scientific American* 9-94, p 64; *Nature* 9-29-94, p 408; *Science News* 10-15-94, p 254; *Science News* 12-17-94, p 415; *Discover* 1-95, p 65; *Science* 4-7-95, pp 49 & 69)

Earthquake Prediction UPDATE

Although earth scientists have developed fairly reliable means of predicting the imminent eruption of a volcano, the problem of short-time-scale earthquake prediction, despite some reported successes in the past, remains unsettled. Medium-time-scale prediction (stating, for example, "the part of the San Andreas Fault south of Palm Springs will have a magnitude 7.5 earthquake in the next 30 years with a probability greater than 40%") has not really proved itself either, although geologists are somewhat confident that such predictions are valid. Most reliable of all are long-time-scale predictions (such as "the eastern United States will experience a serious earthquake on the average of once in every 500 years"). Of course, the longer the time scale, the less the practical advantage of such a prediction.

A lot of ideas appear initially to work, but while they get some press, they fail on closer examination. For example, a geyser in California known as the Old Faithful Geyser of Calistoga changed its period before earthquakes in 1975, 1984, and 1989, according to historical records, but has not ever been used to predict an earthquake. Nor does a change in period tell anything about where a quake might occur, as the quakes involved are all more than 120 mi (200 km) from the geyser and have occurred in two separate directions.

In Mexico City, where a destructive earthquake in 1985 killed 20,000, officials installed an earthquake warning system in August of 1993. The system uses seismometers along the west coast of Mexico, where earthquakes are most likely to occur, and triggers alarms in radio stations that automatically broadcast a siren sound. The cutoff is intended to be a quake that registers at least 6 on one of the coast seismometers. However, in October, the system failed to alert people to a magnitude 6.8 earthquake. The next month, the system produced a false alarm, with no earthquake in sight. In December of 1993, the whole system was turned off because it upset too many people. It was later turned back on. On September 14, 1995, the alarm system proved its worth by providing two-and-a-half minutes of warning when a large earthquake struck.

Some scientists, despairing of successful earthquake predictions, suggest doing geological surveys of soils to find which ones will lead to destruction of buildings or highways during quakes. Mapping hazardous soils and then attempting to ensure that any structures are earthquake proof is entirely feasible with today's knowledge and could in the absence of reliable short-term earthquake predictions save many lives and much property damage. Indeed, it would prevent damage to property even if short-term prediction ever should become feasible.

Another major help, although not a solution, is better prediction of tsunamis, which is currently possible (*see* "Other Geophysical Events," p 580).

It's on the Radio Right Now

One possible short-term earthquake predictor utilizes the appearance of electromagnetic changes caused by stresses in rock. In 1989, the Stanford University physicist Antony C. Fraser-Smith accidentally detected ultra-low frequency radio signals shortly before the Loma Prieta earthquake that struck just south of San Francisco. Fraser-Smith then set up detectors for such waves along the San Andreas fault in regions such as the area around Parkfield, California, where geologists have made medium-term predictions of California earthquakes. Although he got no signal from the devastating Northridge earthquake of 1994, that quake did not occur where his instruments were positioned. He did receive a signal a week prior to a small magnitude 5.0 earthquake struck Parkfield, in November 1994.

On January 17, 1995, a different kind of radio signal also preceded an earthquake. Astronomer K. Maeda was monitoring high-frequency radio signals from Jupiter when his instruments picked up a steady and unexpected signal from Earth. Just 40 minutes later and 50 mi (75 km) away, a 6.9 earthquake devastated the city of Kobe, Japan.

The most systematic search for electromagnetic signals preceding earthquakes is also the most controversial. Since 1983, Panayiotis Varotsos, a physicist from the University of Athens, has been monitoring various locations in Greece for changes in the electromagnetic field of Earth. His inspiration was laboratory observation of electric currents generated in squeezed crystals. If this happens on a small scale, he reasoned, it should also happen on a large scale when rocks begin to move at the beginning of an earthquake. In 1995, he felt confident enough of his predictions to fax them to about 30 institutions around the world. Each of his predictions, he claims, on May 4, May 13, and June 15, was followed by an earthquake. The June 15 earthquake measured 6.5 in magnitude; it killed 26 persons and caused great damage. Critics, however, say that Varotsos' predictions are so broad in terms of time and magnitude that they are no better than random. On the other hand, a study of the predictions by Stephen Park of the University of California at Riverside found that the issued predictions revealed a success rate of 60-70%, far better than chance in this field. Still, none of Varotsos' predictions has been exact in either location or timing.

Medium-Term Tremors

Predicating earthquakes in the earthquake-prone town of Parkfield, California, has not been very successful, but in December 1994 Robert Nadeau of the University of California, Berkeley, went out on a limb. Nadeau predicted that a baker's dozen of quakes would hit between then and the end of 1996. More specifically, he predicted that each of 13 different places along faults in the Parkfield region would produce a small earthquake, less than magnitude 1, within specific time windows. Nadeau's prediction approach is based on the idea that earthquakes repeat in the same location at definite intervals. Despite the history of small earthquakes at those spots, as 1995 progressed the first 2 of the predicted 13 had failed to materialize within their given

Background

Medium-scale predictions of earthquake periodicity are based on the pattern of past earthquakes. In California and other regions where the historical record of earthquakes is short, the record of past earthquakes is found using geology. Geologic evidence of past earthquakes of magnitudes greater than about 6.2 occurs because such earthquakes shake water-saturated sediment, changing it briefly from a damp solid to a fluid. The suddenly liquefied sediment, usually called sand, can now flow through any cracks that are available. The earthquake itself may open some new cracks, either permanently or temporarily. If a crack is completely in the crust, the result of its filling by the liquefied sand is what geologists call a **dike** (a dike is any crack in the crust that is filled with some material different from the crust around it). If a crack leads to the surface, the liquid sand flows out and forms a cone around the crack's opening, similar to the cone of a volcano. When the shaking stops, the water-drenched sediment becomes solid again and remains in its new location. The dike or cone of sand may later be exposed when excavations are made for highway cuts or irrigation ditches. The presence of such formations is taken as an indication of a past earthquake. Dating the rock or soil layer where the formation is found (usually by carbon-14 dating of trapped organic materials) is used as evidence of the date of the past earthquake.

The first widely noted medium-time-scale earthquake prediction came in 1979, when B. A. Bold and R. H. Jahns proposed that the odds of California getting a major earthquake before 1990 were even. People were not impressed, since California is a big state, and 50-50 odds do not make disaster seem inevitable.

Formal medium-time-scale earthquake prediction in the United States did not get started until 1985, when the U.S. Geological Survey (USGS), sanctioned by the National Earthquake Prediction Evaluation Council, predicted a 20-1 chance that a magnitude 6 earthquake would strike the Parkfield segment of California's San Andreas Fault before 1993. Parkfield is a tiny town of 34 people located in the hills that border California's central valley, roughly halfway between San Francisco and Los Angeles. Since the 1985 prediction, Parkfield has undergone two C-level alerts (both in the fall of 1990). C-level is the lowest level at which the USGS notifies the public of what they believe to be increased danger. Neither alert produced an earthquake, although, despite efforts to prevent panic, each managed to frighten a few folks. As the deadline approached, some USGS scientists found evidence in old documents that serious earthquakes on the Parkfield segment have diminished in number since 1930 and that different quakes broke different segments of the fault. Both pieces of evidence suggest that a Parkfield quake may be harder to predict than expected. A small Parkfield quake did occur in 1994, but it was only a tenth the size of that predicted.

The most famous successful short-range earthquake prediction came in 1975 when Chinese seismologists correctly predicted a 7.3 earthquake on the basis of small earthquakes that preceded it, but this represents the only time that this method has been successful.

time windows. It is beginning to appear that the apparent repetition of Parkfield quakes is either a mirage, or else whatever caused the periodicity might have passed. Perhaps, as in quantum physics and much folk magic, simply looking at Parkfield by scores of geologists has prevented earthquakes from happening there.

Historical views of earthquakes at Parkfield and other places along the San Andreas fault do not seem to suffer from magical disappearance when observed. William L. Ellsworth, Lynn Dietz, and John Vidale of the USGS in Menlo Park, California, along with Robert Wesson of the USGS in Reston, Virginia, have found strikingly periodic historical earthquakes at Bear Valley, California, at Parkfield, and at other places on the San Andreas fault, all apparently associated with places where a stretch of very hard rock keeps the fault from slipping for a time, ultimately giving way to the strain. The earthquakes involved in this repetitive pattern are all small, since there are always more small earthquakes than large ones. Seismic tomography has confirmed that for some specific large earthquakes (notably the 1994 Northridge, California, quakes) the fault at the heart of the quake matches this pattern of harder rock at the epicenter compared to rock at other places nearby.

The usefulness of this periodicity in prediction remains in question. For example, although Bear Valley quakes have a periodicity of about 15 years, there have also been a couple of quakes that occurred during periods that were supposed to be inactive.

A method thought to offer better chance for predictive ability involves obtaining more direct information about the nature of rock along the sides of faults. Plans are being made to drill into some of the faults in California to see whether or not the predicted hard rock under stress can be found.

Predicting Size

Ellsworth and Gregory C. Beroza of Stanford University in Palo Alto, California, have found a way to predict which earthquakes will become very large as much as five seconds before such earthquakes release large amounts of energy. Their 1994 study of 51 earthquakes found that little quakes (such as a magnitude 1) start quickly, reaching their peak in tiny fractions of a second. However, the great earthquake of magnitude 8.0 in their study took five seconds to reach peak size, and other large earthquakes were also found to be slow starters. Thus, since radio waves travel at the speed of light and earthquake waves at the speed of sound, an early warning might be issued in time to provide several seconds of lead time. This would not allow time for people to leave buildings, but it could enable utility workers to shut off gas or electric lines or to mobilize utility crews.

Another study identifying a possible source for earthquake warnings appeared in *Science* on December 2, 1994. Pierre F. Ihmlé of the Institut de Physique du Globe in Paris, France, and Thomas H. Jordan of M.I.T. in Cambridge, Massachusetts, analyzed the records of 107 earthquakes. They found indications that in 20 instances (19 of these on oceanic tectonic plate boundaries), a long period "slow" or "quiet" earthquake preceded the ordinary observable earthquake by dozens of seconds. The one continental earthquake to show this pattern was also an earthquake on a plate boundary, but the boundary was the incompletely separated rift system of Africa. The low frequency waves of the "quiet" precursors cannot be detected at a distance, however, limiting their use as a warning system. Even up close, no such precursors have been observed before any of the well-monitored California earthquakes.

(Periodical References and Additional Reading: *New York Times* 9-4-92, p A8; *New York Times* 10-11-92, p I-37; *New York Times* 12-19-93, p I-12; *Science* 12-2-94, p 1656; *Science News* 9-3-94, p 159; *Science* 12-2-94, p 1547; *Science News* 12-3-94, p 374; *Science News* 12-17-94, p 415; *Science News* 10-21-95, p 260; *Science* 11-10-95, p 911; *Science News* 12-13&30-95, p 531)

EARTHQUAKES IN THE UNITED STATES

Although Americans traditionally feel safer from disaster than other people from around the world, this perception is not necessarily rooted in reality. In fact, wealth does provide some protection to the citizens of the United States against such natural disasters as famine, and two oceans have helped keep the nation from feeling the worst effects of war. However, the climate of the United States, taken as a whole, is the worst for disasters of all the nations of the world (think of hurricanes and floods, as some recent examples). Furthermore, although recent volcano eruptions have caused less loss of life than in other places on the globe, some of the big blow-ups of the past can match those of Indonesia.

When it comes to earthquakes, the United States has also held its own, although the types of construction popular in America has limited damage to less than that occurring in many other nations. The mud and stone houses of Turkey and Algeria tend to collapse much more easily than the frame house that is standard throughout the United States. Luck has also limited damage, as some of the worst earthquakes in the United States struck in regions before they became heavily populated. Luck, however, eventually runs out.

Faults in Los Angeles

The San Andreas fault in California is easily the most famous crack in Earth's crust. The west side of the fault is moving jerkily to the north at an *average* rate of 2 in (5 cm) per year. At each jerk, California suffers an earthquake. Yet motions of the San Andreas fault are not the only cause of earthquakes in California. Indeed, all of the most recent events stem from other faults.

On January 17, 1994, the most destructive earthquake to hit California since 1989, and the most destructive in southern California since 1971, hit Los Angeles at 4:31 a.m. Pacific Time. The epicenter, under the city of Northridge, was not on the San Andreas fault. Instead, it was on a hidden fault that had not been suspected previously. Northridge is only a few kilometers from the epicenter of the 1971 San Fernando quake, which was not caused by the San Andreas fault either. Furthermore, the magnitude 7.4 earthquake of 1992, which did less damage, although larger than the Northridge and San Fernando earthquakes, was not on the San Andreas fault either, occurring east of Los Angeles, in Landers, California.

Here are a few unfamiliar names with which to toy: Elysian Park fault, Hollywood fault, Malibu Coast fault, Newport-Inglewood fault, Palos Verdes fault, San Cayetano fault, Santa Monica fault, Santa Susana fault, and Whittier fault. All are in the urban regions immediately surrounding Los Angeles, much closer to the city than the San Andreas fault. All in all, there are more than a hundred active faults in the Los Angeles region. On October 15, 1994, a major project was begun. Expected to take several years, researchers are attempting to map all the underground faults in Southern California, using compressed-air blasts and underground explosions to produce seismic waves that can then be analyzed as if they came from small, precisely located earthquakes.

Although the largest Los Angeles earthquakes in the past 50 years have been magnitude 6.7, the geological evidence of past earthquakes suggests that it has been more common for several faults to rupture together, producing a 7.7 to 7.6 moment magnitude quake. Such an earthquake, which has not occurred in the Los Angeles region for at least 200 years, would be vastly more destructive than the Northridge earthquake of 1994. An analysis of earthquake probabilities by Susan E. Hough, based on a mathematical approach not unlike betting the stock market from charts instead of company performance, suggests that there may be an earthquake in the 7.2 to 7.5 magnitude range every 140 years in the Los Angeles region. She also calculates that over a 300-year-period there might be six quakes similar to Northridge (magnitude 6.7) and one much larger—magnitude 7.4 or 7.5. Since fewer than that number have occurred recently, Los Angeles is suffering from what some call an earthquake "debt," a debt that might have to be paid soon. An analysis by David D. Jackson of the University of California, Los Angeles, and coworkers suggests that the debt has an 80-90% chance of being called in before 2025.

In the December 8, 1994 *Science*, Ralph H. Archuleta and Kim B. Olsen of the University of California, Santa Barbara, along with Joseph R. Matarese of M.I.T. in Cambridge, Massachusetts, reported that a computer simulation of a San Andreas fault earthquake indicates that damage to Los Angeles could be much greater than anyone had previously expected. The simulation assumed that movement of the 105 mi (170 km) stretch of the fault from Gorman to San Bernardino, California, causing a 7.75-magnitude earthquake. Even though the simulation was simplified by eliminating short-period waves and was performed by a massively parallel supercomputer, it took 23 hours to run. The result showed ground waves amplified by the shape of the Los Angeles basin moving the surface at speeds several times as fast as had been expected, particularly threatening to tall buildings and long bridges. It was not clear what effect there would be on small buildings, which are commonly found in most of the Los Angeles region.

Haresh Shah of Stanford University in Palo Alto, California, utilized information from the Northridge quake to upgrade predictions of damage from a magnitude 8 quake, one like the 1906 San Francisco disaster. He found that fewer people would be killed in such known earthquake targets as San Francisco, Los Angeles, or Tokyo than had been predicted using older data, but that the financial impact would be far greater than expected. His estimates ranged from a low of 2,000 dead in either of the California cities to a high of as many as 60,000 deaths in Tokyo. Damage might range from as low as $115 billion in San Francisco or $125 billion in Los Angeles to as high as $1.2 trillion in Tokyo. The 6.7 magnitude Northridge quake killed 61 persons, but caused an estimated $15 billion in damages.

Danger in the Northwest

If you drive north along the Pacific Coast of the United States from San Diego to Seattle, the famous earthquake region is encountered first, that region that follows the San Andreas fault north until it passes into the Pacific Ocean at Point Area about 100 mi (160 km) north of San Francisco. The famous fault, which is the boundary between the Pacific and the North American tectonic plates, does not then head out to sea toward Japan. Instead, the edge of the Pacific plate continues for a long time in a more-or-less north-south direction, paralleling the coast of North America.

Another small, narrow plate interposes itself between the Pacific plate and the North American plate along the northwest coast of the American continent. It runs along the coastline from about Cape Mendocino in California to Vancouver Island off British Columbia. This strip, known as the Juan de Fuca plate after the strait separating Vancouver Island from the rest of North America, is not rotating past the North American plate as does the Pacific plate along the San Andreas. Instead, it is wedging itself into and under the North American plate at a rate of nearly 4 in (10 cm) per year. As a result, a line of tall volcanoes, including such famous mountains as Hood, Rainier, and St. Helens, has risen in Washington and Oregon. As in Chile, where a similar plate collision is occurring, the movement in this region can be expected to produce great earthquakes from time to time.

Geologists studying the tectonic past of the Cascadia Subduction Zone (where the plate collision in the northwestern United States is taking place) discovered a sheet of sand that appeared to have been deposited by a tsunami about A.D. 1700. The sand deposits showed that the wave must have been about 30 ft (10 m) high. A quake producing such a great wave would have to measure at least a magnitude 8 and perhaps a magnitude 9.

Kenji Satake of the Geological Survey of Japan looked at his nation's historical records for reports of a tsunami around 1700. He found that on January 27, 1700, a tsunami more than 7 ft (2 m) high struck the Pacific Coast of Japan, which implies that the Cascadia earthquake must have been huge, with a magnitude of 9 to match those great earthquakes known from Chile.

Data previously kept secret by petroleum companies was released in 1994, revealing that the fault structure underlying Seattle, Washington, is similar to that under Los Angeles, California. One large horizontal fault which is not likely to cause earthquakes itself connects many shallower faults, the kind that do cause quakes. The horizontal fault is about 9 mi (15 km) below Puget Sound. The layer of crust above the fault slides northward, causing earthquakes from time to time as it buckles, like a blanket lying on a smooth floor which is pushed from one edge.

Thus, it is not surprising that a study of earthquake hazards paid for by the state of Oregon concludes that earthquakes of a magnitude 8 are a virtual certainty every 450 years (plus or minus 200 years) in the region from Vancouver Island down to northern California. That is, they would expect at least one truly major earthquake every 250-650 years. They also concluded that a magnitude 9 earthquake, involving the whole fault where the Juan de Fuca plate is subducting under the North American plate, has only 1 chance in 20 of occurring during that time frame. In view of the evidence for a magnitude 9 earthquake only 300 years ago, that conclusion seems a bit doubtful to many geologists. Also, in other similar geologic regions, such as northern Alaska and Chile, very large earthquakes are more common than are merely large quakes.

(Periodical References and Additional Reading: *Discover* 1-93, p 61; *Science News* 9-3-94, p 159; *Science* 9-23-94, p 1802; *New York Times* 10-16-94, p I-34; *Science* 10-21-94, p 389; *Science News* 12-24&31-94, p 442; *Discover* 1-95, pp 64 & 65; *Science* 1-13-95, pp 176, 199, 206, & 211; *Science News* 1-21-95, p 37; *Scientific American* 4-95, p 14; *Science News* 12-16-95, p 404; *Discover* 1-96, p 94)

OTHER GEOPHYSICAL EVENTS

Movements of the atmosphere that cause destruction are called storms, but people have many different ways to refer to the movements of the hydrosphere and crust. Depending on where it came from and how solid it is, a hydrosphere movement can be a flood, a wave, a tsunami, or an avalanche. In many cases, the movement of water begins as a movement of crust, since earthquakes have caused floods by damming rivers or breaking dams, as well as by producing giant harbor waves and tsunamis. An avalanche (the snow equivalent of a landslide) can be initiated by less than an earthquake, but some of the most deadly are those caused by a mountain's shaking. Movements of molten rock, called magma within the crust and lava when it spills out, also trigger earthquakes, avalanches, and landslides. In more complex ways, movements of magma can also initiate harbor waves or tsunamis.

On October 18, 1992, in northwestern Columbia, 36 people were injured when an earthquake triggered the explosion of a geyser, showering the area around it with boiling mud.

Hazards of the Volcano

The classic picture in everyone's mind of an erupting volcano involves a great column of ash heading toward the stratosphere, large hot rocks that volcanologists call bombs flying from the crater, and molten lava streaming over the edge and forming a river of fire down the side. When Mount St. Helens erupted, however, one side of the mountain gave way with a blast of ordinary soil and rock, followed by and then mixed with extremely hot gases. On Hawaii, the eruption of Kilauea, which has continued unceasing for over a dozen years and poured forth about 0.3 cu mi (1.3 cu km) of lava, has not caused any deaths directly, but has the potential to destroy the populous city of Hilo at any time. Kilauea is unlikely to do so with anything like the eruption of the type described in the first sentence of this paragraph, in part because it is not that kind of volcano (*see* "Active Volcanoes and Major Volcanic Events," p 598). It might, however, provoke a volcanic earthquake. In 1868, when part of the island slipped along the base, the earthquake reached an estimated magnitude of 8, while a 1975 slip caused a measured magnitude 7.2 quake. Each of these earthquakes, in turn, generated destructive tsunamis. However, the most devastating event associated with Kilauea and other volcanoes on Hawaii concerns the precipitation of a giant landslide into the sea. One prehistoric landslide into the sea caused a tsunami that reached a height of 1,200 ft (365 m) on the neighboring island of Lanai.

Indeed, a radar study of Kilauea from the space shuttle *Endeavor* showed that part of the mountain is sliding toward the sea now, but only at the speed of 4 in (10 cm) a year.

Even volcanoes that no longer erupt can pose hazards of various kinds, in part because they are quite steep and high. When compared to other kinds of mountains, tall, cone-shaped volcanoes are much weaker, since such volcanoes are either cinder cones built from either heaps of volcanic ash and rubble or composite volcanoes in which such heaps are somewhat strengthened with rock left from lava flows. Nonvolcanic tall mountains have cores of bedrock. Volcanoes are usually geologically younger than ordinary mountains, and so have had less time to erode into rounded shapes.

A recent eruption can add vast amounts of unconsolidated material to the mountain sides. This can then form rivers of ash called lahars that flow rapidly down the slopes. More than four years after the April 1991 eruption of Mount Pinatubo in the Philippines, lahars continued to plague the local region, displacing thousands. Even those whose homes have not been destroyed by lahars must live elsewhere until the threat finally subsides.

Another hazard of tall, inactive volcanoes can be a lake of water in the crater which can break through the weak walls. Snow and ice at the top of a volcano can form an avalanche more easily, because the sides of tall volcanoes are steeper than those of most mountains. Cinder cones often have slopes of 35°, for example. Heating by magma movements can also melt the underside of the snow or ice, thus precipitating an avalanche.

As an example of the kind of problem that nonerupting volcanoes can produce, consider the case in the Philippines of the volcano named Parker. At the top of Parker, some 6,044 ft (1,842 m) above sea level, the crater filled with water is known as Maughan Lake. In 1995, a passing typhoon added so much water to Maughan Lake that it overflowed and sent additional water down the Alah River flowing down the mountain. Furthermore, the rain-soaked sides of Parker gave way in two or three places, with the resultant landslides damming the Alah. A sheet of water flowed down the flank of the mountain, killing more than 60 people and displacing about 50,000 from their homes, 300 of which were destroyed in the flood. Parker is thought to have erupted last about 250 years ago, although that eruption in 1640 might have been a different volcano on a nearby island.

Even more people were killed when an earthquake triggered mudflows on Nevado del Huila in southern Columbia on June 6, 1994. The Paez earthquake, with a magnitude of 6.4, occurred on the southwest flank of the volcano itself. Part of the south slope of the mountain at an altitude of about 10,000 ft (3,000 m) was dislodged and great earth slides developed. Because of heavy rainfall in the days preceding the earthquake, the soil was very damp and liquefied into a kind of mud. Even though the main body of the slide was only about 6 ft (2 m) thick when it reached level ground, it continued with great force for several kilometers, killing more than 500 persons, displacing 20,000 from their homes, and destroying 5 bridges and more than 60 mi (100 km) of roads.

Events similar to the mudslides at Nevado del Huila are far from being rare. One instance in Ecuador occurred in 1987, for example, and killed about a thousand people. In the United States, Mount Rainier has the kind of unconsolidated steep slopes that would make such a disaster possible in a large earthquake (*see also* "Earthquakes in the United States," p 577).

Underground Gas Attack   Often, the magma reservoir of a volcano or a volcanic region releases carbon dioxide gas as the magma cools and is relieved of some of the pressure that contains the gas. Carbon dioxide, the waste product of respiration, is a gas that cannot be used any further by living organisms. In the long run, it is recycled both by plants and geologic processes, but in the short run, too much carbon dioxide can kill. This is what happened at two lakes in Africa in 1984 and 1986, when carbon dioxide that had built up in them was suddenly released, poisoning the animals and plants in the immediate vicinity.

The Long Valley and Mammoth Lakes region in California is among the volcanic regions of the United States, of which the most famous is Yellowstone National Park. Small gas-releasing vents called fumaroles have long been known to exist on the volcano Mammoth Mountain in that region. In 1990, studies of the gases at these fumaroles began to reveal the presence of a new magma intrusion into the base of the mountain. Christopher Farrar of the United States Geological Survey in Carnelian Bay, California, and coworkers report in the August 24, 1995 *Nature* that trees observed to die from previously unknown causes on Mammoth Mountain are actually being poisoned by air containing as much as 90% carbon dioxide in some locations. The main problem seems due to suffocation of tree roots in carbon-dioxide saturated soil. Since high atmospheric carbon dioxide levels could also harm humans, the U.S. Forest Service has closed one campground near the carbon-dioxide zone. Potentially lethal levels of the gas were found to have built up in some buildings at the campsite.

George King of the University of Michigan and coworkers have been looking at the two killer lakes, Lake Monoun and Lake Nyos, which between them asphyxiated about 1,750 people with carbon dioxide. King says that the gas is once again filling lower layers of the lakes, and by 2005 or sooner Lake Monoun could be dangerous again, while it might take Lake Nyos to about 2025. Unless action is taken to drain away the carbon dioxide, a lake overturn (in which the bottom water layer mixes with the top) will happen again, releasing another cloud of deadly gas.

Earthquakes Trigger Tsunamis

The Nevado del Huila mudslides are only one of the many kinds of problems that earthquakes can launch, in addition to the obvious hazards of falling buildings, cracking of the earth, collapsed bridges and highways, and so forth.

The best-known hazard of earthquakes, apart from damage caused by shaking, is the tsunami, often referred to as a "tidal wave." In the early 1990s, there were four instances in which modest-sized earthquakes set off tsunamis that were unexpectedly high upon reaching land, calling for a new analysis of the relationship between earthquakes and the ocean waves they trigger.

Costas Synolakis of the University of California, like most oceanographers, had assumed that accounts of dry land appearing along beaches before a tsunami were folk myths despite the wealth of firsthand accounts reporting the phenomenon. The physical description of how a tsunami moves through the ocean would not permit the wave trough to arrive before the crest. While interviewing survivors of the 1992 Nicaraguan tsunami, however, he came to believe that there had to be truth in these accounts. With his colleague Srinivas Tadepalli, he reanalyzed the formation of tsunamis, developing a model in which waves could be generated by earthquakes close to shore in such a way that the trough does precede the crest. Not only did this model accord with frequent observations of dry beaches followed by high waves, but it also predicted a higher crest for tsunamis with forerunning troughs. Thus, the new model predicts the higher-than-expected tsunamis, provided that the waves are generated close to shore. This model can explain some of the higher tsunamis of the 1990s but not all of them, since not all originated with near-shore earthquakes.

One remarkable case of tsunami destruction occurred in the earthquake in Indonesia on December 12, 1992. The wave was most damaging on the part of Babi Island that faced away from the source of the wave. Babi Island is a roughly circular island about 1.2 mi (2 km) in diameter. When the tsunami hit the island, part of it washed up on the shore while the two parts beyond the edges of the island were each refracted (bent). By the time the part of the wave near the island passed beyond it, the two wave fronts were traveling toward each other. The fronts collided outside the harbor that was on the far side of the island, creating a swift-moving wave that shot into the harbor and onto the beach, killing 238 persons in two villages on the shore. Although the wave that struck the island first was perhaps 1.7 times as high as the one that destroyed the villages, it was not as swift and powerful. Analysis of the Babi Island disaster was conducted by Harry Yeh of the University of Washington in Seattle and coworkers with the aid of the U.S. Army Corps of Engineers.

There has long been a successful tsunami warning system for waves traveling a long distance, such as from Alaska to Hawaii, but most tsunamis cause their greatest damage close to the earthquake or volcanic collapse that starts them. This was the case with all the serious tsunamis of the early 1990s, including Nicaragua and Indonesia in 1992; Okushiri Island of Japan in 1993; and Irian Jaya in Indonesia and Chimbote, Peru, in 1996. Hiroo Kanamore of CalTech in Pasadena, California, and

---

BIG WAVES

A **tsunami** is caused when an earthquake raises or lowers a section of the seabed. This produces a wave that, while not generally noticeable at sea, can reach great heights as it approaches land. High waves are also caused by volcanic explosions or collapses, such as the explosion and collapse of Krakatau. These giant waves from volcanoes are generally called tsunamis also, since they originate because of a sudden change in the sea bed and demonstrate wave lengths similar to those of tsunamis caused by earthquakes. Tsunamis of either type are capable of traveling great distances across the sea at speeds of about 500 mph (800 kph). In some instances, if suitably aimed, one could circumnavigate Earth.

Related to the tsunami is the **harbor wave**. Usually, harbor waves are caused when a landslide falls into a bay or strait. Because of the confines of the land around the sea, a harbor wave can be more destructive and even higher than a tsunami.

Kasayuki Kikuchi of Yokohama City University in Japan observed that the 1992 tsunami that killed 116 people in Nicaragua was preceded by an unusual seismic wave that could perhaps be used to predict which undersea earthquakes will be destructive. The U.S. National Oceanic and Atmospheric Administration has proposed putting seismometers at various points under the Pacific Ocean to provide better tsunami warnings.

(Periodical References and Additional Reading: *New York Times* 10-19-92, p A5; *New York Times* 2-25-93, p D23; *New York Times* 1-3-94, p A11; *Science* 4-1-94, pp 26 & 46; *Science* 4-29-94, p 660; *Bulletin of the Global Volcanism Network,* 7-94, p 7; *New York Times* 7-14-94, p A1; *EOS* 7-19-94, p 322; *Nature* 11-24-94, p 353; *Science* 11-25-94, p 1327; *New York Times* 11-29-94, p C5; *Science News* 4-8-95, p 216; *Scientific American* 7-95, p 24; *Bulletin of the Global Volcanism Network,* 9-95, p 5)

EARTH SCIENCE LISTS AND TABLES

MEASURING EARTHQUAKES

The size of an earthquake is generally reported in the United States using the Richter scale, a system developed by Caltech geologist Charles Richter [American: 1900–1985] in 1935, intending to deflect questions from inquisitive reporters asking "How big was it?" The scale is based on measurement of the waves on a seismograph, which directly reflect the disturbance of ground motion. Each whole number on the scale, called a magnitude, represents a seismograph wave that is 10 times as large. Thus, a magnitude 6 earthquake produces a ground wave 10 times as large as a magnitude 5, given that both earthquakes are the same distance away and at the same depth.

This does not mean that a magnitude 6 earthquake has 10 times the energy of a magnitude 5. Measuring the actual energy requires instruments placed at the site of the earthquake. Various methods have been developed for extrapolating the energy from the magnitude. These show that a difference of one magnitude corresponds to an energy change of 30-60 times. The energy of a magnitude 8 earthquake, a very serious event, can be as much as 1 million to 10 million times as much as that of a magnitude 4 earthquake, one that can be felt but that causes almost no damage.

A very different scale, developed by Giuseppe Mercalli (Italian: 1850–1914) in 1902 and modified by Harry Wood and Frank Neumann in the 1930s, is used for some purposes in the United States and on a regular basis in certain other countries. This scale, the Modified Mercalli scale, is based on the effects of the earthquake at a point above the earthquake. Comparing the Modified Mercalli scale with the Richter scale helps one to understand the relative energy of earthquakes.

RICHTER SCALE	MODIFIED MERCALLI EARTHQUAKE INTENSITY SCALE
1.0 About the energy of 6 ounces of TNT.	**I** Not felt except by a very few under specially favorable circumstances.
2.5 Generally not felt, but recorded on seismometers. ***Very minor*** *below 3.0.*	**II** Felt by only a few persons at rest, especially on upper floors of buildings.
3.5 Felt by many people. ***Minor*** *from 3.0 to 3.9.*	**III** Felt noticeably indoors, especially on upper floors of buildings, but not necessarily recognized as an earthquake.
	IV Felt during the day indoors by many, outdoors by few. Sensation is like a heavy truck striking a building.
	V Felt by nearly everyone; many awakened; disturbances of trees, poles, and other tall objects sometimes noticed.
4.5 Some local damage may occur. ***Light*** *from 4.0 to 4.9.*	**VI** Felt by all; many frightened people run outdoors; some heavy furniture moved; few instances of fallen plaster or damaged chimneys; damage slight.

5.0 About the energy of a thousand tons of TNT.

5.5 *Moderate from 5.0 to 5.9.*

VII Most everybody runs outdoors; damage negligible in buildings of good design and construction; slight to moderate in well-built ordinary structures; considerable in poorly built or badly designed structures.

6.0 A destructive earthquake. *Strong from 6.0 to 6.9.* About as much energy as the Hiroshima atomic bomb.

VIII Damage slight in specially designed structures; considerable in ordinary substantial buildings with partial collapse; great in poorly built structures; chimneys, factory stacks, columns, monuments, and walls fall.

IX Damage considerable in specially designed structures; buildings shifted off foundations; ground cracks conspicuously.

7.0 A major earthquake; about 10 occur each year. *Major from 7.0 to 7.9.*

X Some well-built wooden structures destroyed; most masonry and frame structures with foundations destroyed; ground badly cracked.

7.2 About the energy of a million tons of TNT.

8.0 Great earthquake; these occur once every 5–10 years. *Great at 8.0 and above.* About the amount of energy of the eruption of Mount St. Helens.

XI Few, if any, masonry structures remain standing; bridges destroyed; and broad fissures in ground.

XII Damage total; waves seen on ground surfaces; objects thrown upward into air.

9.5 Greatest earthquake ever recorded had a moment magnitude (see below) this high, about the energy of 3 billion tons of TNT (Chile in 1960).

Since 1935, various improvements have been made in the Richter scale used by geologists, although these refinements are not generally reported by the press. The original Richter scale measures waves at the surface from nearby earthquakes. It was specifically designed with California in mind. Two variations were developed by Richter and Beno Gutenberg, also of Caltech. In the first, designated by either Ms or M_s, a different set of surface waves is measured. This method can be used for earthquakes that are much farther away than the original version. A later scale, called either Mb or M_b, works better with deep earthquakes because it measures waves that travel through the Earth (body waves, hence the "b") instead of surface waves. The original scale is sometimes designated M_L ("L" meaning "local"). Another popular measure, the moment magnitude (sometimes called M_w), is calculated later and quantifies the total energy of the earthquake. Although these different scales do not translate exactly from one to another for any given earthquake, they result in similar, if not exactly the same, numbers. For example, the January 7, 1994, Northridge,

California, earthquake had a surface magnitude (Ms) of 6.6, which later was supplanted by a moment magnitude (M_w) of 6.7.

(Periodical References and Additional Reading: *Science News* 10-15-94, p 250)

MAJOR EARTHQUAKES

Earthquakes have been among the most destructive forces of nature throughout time. Most earthquakes are caused when rock on one side of a crack (a fault) in Earth's crust moves with respect to rock on the other side of the fault. The motion sets up vibrations in the crust that travel as waves through the rock. When the waves reach Earth's surface, they cause it to move in various ways. Small earthquakes that accompany volcanic eruptions are caused by the motion of liquid rock (magma).

Unlike volcanoes, which at present seem to be beyond control, the amount of damage an earthquake does to human life and property can be largely controlled by proper construction of buildings. Note that the energy of an earthquake is poorly related to the destruction it causes.

In the following table, all of the magnitudes before 1880 (when John Milne developed the first modern way of measuring the power of an earthquake) are estimated, based on various forms of evidence such as contemporary accounts or geologic changes. Where available, moment magnitude (based on the actual slippage of rock along a fault) instead of surface magnitude (based on the size of earthquake waves at a seismograph station) is reported. For large earthquakes, the moment magnitude may be slightly greater than the more traditional surface-wave measurement. For example, the greatest earthquake ever recorded by modern seismographs, the Chilean earthquake of 1960, is 9.5 on a moment magnitude scale, but 8.3 on the surface-wave scale. Moment magnitudes are marked with an asterisk (*) in the table.

Date	Location and Remarks	Estimated Deaths	Magnitude
May 20, 526	Antioch, Syria (now Turkey)	250,000	N.A.
856	Corinth, Greece	45,000	"
1036	Shanxi, China	23,000	"
1057	Chihli (now Hopeh), China	25,000	"
1170	Sicily	15,000	"
1268	Cilicia (now Turkey)	60,000	"
September 27, 1290	Chihli (now Hopeh), China	100,000	"
May 20, 1293	Kamakura, Japan	30,000	"
1356	Basel (Switzerland)	N.A.	7.4*
January 26, 1531	Lisbon, Portugal	30,000	N.A.
January 24, 1556	Shanxi, China	830,000	"
1604	Taiwan Straits	N.A.	7.7*
1605	Hainan Island, China	N.A.	7.3*
February 5, 1663	St. Lawrence River	N.A.	N.A.

Date	Location and Remarks	Estimated Deaths	Magnitude
November 1667	Shemakha, Azerbaijan	80,000	N.A.
June 7, 1692	Port Royal, Jamaica	30,000	"
1693	Naples, Italy	93,000	"
January 11, 1693	Catania province, Sicily, Italy	60,000	"
January 27, 1700	Northwest coast of North America; tsunami reaches Japan	N.A.	9
1707	Tsunami hits Japan	30,000	N.A.
1727	Boston, Massachusetts	N.A.	
December 30, 1730	Hokkaido Island, Japan	137,000	"
1731	Peking (Beijing), China	100,000	"
October 11, 1737	Calcutta, India	300,000	"
June 7, 1755	Northern Persia	40,000	"
November 1, 1755	Earthquake with tsunami hits Lisbon, Portugal	60,000	8.7
November 18, 1755	Boston, Massachusetts (centered at Cape Ann)	0	6
February 4–5 and March 28, 1783	Calabria, Italy	35,000	N.A.
February 4, 1797	Quito, Ecuador, and Cuzco, Peru	41,000	"
December 16, 1811	New Madrid, Missouri	fewer than 10	8.1*
January 23, 1812	New Madrid, Missouri	0	8.2*
February 7, 1812	New Madrid, Missouri	0	8.3*
December 21, 1812	Traverse Range in California	N.A.	N.A.
1819	Kutch, India	N.A.	7.8*
September 5, 1822	Aleppo (Syria)	22,000	N.A.
December 28, 1828	Echigo, Japan	30,000	"
June 1838	San Francisco, California	N.A.	"
December 24, 1854	Tokai, Japan	3,000	8.4
October 1855	Edo (now Tokyo) Japan	more than 2,000	N.A.
1857	East of Naples, Italy	more than 10,000	N.A.
January 9, 1857	Fort Tejon, California	2	8.3
1858	Portugal	N.A.	7.1*
November 18, 1867	Virgin Islands	N.A.	"
1868	Hayward Fault, California	N.A.	6.8
April 2, 1868	Hawaiian Islands	N.A.	N.A.
August 13–15, 1868	Peru and Ecuador	40,000	"
March 26, 1872	Owens Valley, California	60	"
May 16, 1875	Venezuela and Colombia	16,000	"
August 31, 1886	Charleston, South Carolina	83	7.6

Date	Location and Remarks	Estimated Deaths	Magnitude
October 28, 1891	Central Japan	7,300	N.A.
June 15, 1896	Tsunami hits Sanriku and Kamaish, Japan	26,000	"
June 12, 1897	Assam, India	N.A.	"
September 3, 1899	Yakatanga, Alaska	0	8.3
September 10, 1899	Yakatanga, Alaska	0	8.6
April 4, 1905	Kangra, India	20,000	N.A.
1906	Exmouth Plateau, Indian Ocean northwest of Australia	N.A.	7.2*
April 18, 1906	San Francisco, California	667	8.3
August 16, 1906	Valparaiso, Chile	20,000	8.6
1907	Tadzikistan, Russia	40,000	N.A.
December 28, 1908	Messina, Italy	75,000	7.5
1909	Anegawa, Japan	41	6.8
1914	Senpoku, Japan	94	7.1
January 13, 1915	Avezzano, Italy	30,000	N.A.
October 2, 1915	Pleasant Valley, Nevada	0	7.8
January 13, 1916	Avezzano, Italy	29,980	7.5
October 11, 1918	Mona Passage in Caribbean Sea (Nanai)	116	7.4*
December 16, 1920	Gansu, China	more than 180,000	8.6
September 1, 1923	Tokyo and Yokohama, Japan	143,000	7.9
1925	Kita Tajima, Japan	428	6.8
June 27, 1925	Helena, Montana	0	6.8
1927	Offshore of San Luis Obispo, California	N.A.	7.7
March 7, 1927	Kita Tango, Japan	2,935	7.3
May 22, 1927	Nan-Shan, China	200,000	8.3
1928	Northwest coast of California	N.A.	7.2
1929	Newfoundland (Grand Banks), Canada	27	7.4*
1930	Kita Izu, Japan	272	7.3
August 16, 1931	Mt. Livermore, Texas	0	6.4
December 20, 1932	Cedar Mountain, Nevada	0	7.3
December 26, 1932	Kansu, China	70,000	7.6
1933	Baffin Bay, between Greenland and Baffin Island, Canada	N.A.	7.7*
March 2, 1933	Tsunami hits Sanriku, Japan	3,064	8.9
March 10, 1933	Long Beach, California	120	6.3
January 15, 1934	India, Bihar, and Nepal	10,700	8.4
March 12, 1934	Great Salt Lake, Utah	0	6.6
May 31, 1935	Quetta, India (now Pakistan)	50,000	7.5

Date	Location and Remarks	Estimated Deaths	Magnitude
October–November 1936	Helena, Montana	2	N.A.
January 24, 1939	Concepción, Chile	30,000	8.3
December 27, 1939	Erzincan, Turkey	30,000	7.9
May 18, 1940	Imperial Valley, California	9	7.1
1943	Tottori, Japan	1,083	7.2
1944	Higashi Nankai, Japan	998	7.9
1945	Mikawa, Japan	1,961	6.8
April 1, 1946	Earthquake at Unimak Island, Alaska, causes tsunami in Hilo, Hawaii	173	7.2
December 21, 1946	Nankai on Honshu Island, Japan	1,330	8.0
June 28, 1948	Fukui, Japan	3,769	7.1
October 1948	Ashkhabad, Turkmenistan	fewer than 20,000	N.A.
1949	Olympia, Washington	8	7.1
July 10, 1949	Tadzhikistan, Tajikistan	120,000	7.5
August 5, 1949	Pelileo, Ecuador	6,000	6.8
August 15, 1950	Assam State, India	1,500	8.7
1951	South Tasman Rise, Indian Ocean southwest of Tasmania	N.A.	7.0*
July 21, 1952	Bakersfield, California	12	7.7
March 18, 1953	Northwest Turkey	1,200	7.2
September 9–12, 1954	Orléansville, Algeria	1,660	N.A.
December 16, 1954	Frenchman's Station, Nevada	0	7.1
June 10–17, 1956	Northern Afghanistan	2,000	7.7
July 2, 1957	Northern Iran	2,500	7.4
December 13, 1957	Western Iran	2,000	7.1
July 9, 1958	Lituya Bay, Alaska	3	7.9
August 17, 1959	Hebgen Lake, Montana	28	7.1
February 29, 1960	Agadir, Morocco	12,000	5.8
May 21–30, 1960	Southern Chile; on May 22, a tsunami strikes various Pacific islands, including Hawaii, where 61 are killed in Hilo; greatest earthquake ever recorded	5,700	9.5*
September 1, 1962	Northwest Iran	12,403	7.1
February 21–22, 1963	El Marj, Libya	260	5.0
July 26, 1963	Skopje, Macedonia	1,011	5.5
March 27, 1964	Southern Alaska	131	8.5
March 28, 1965	Central Chile	420	N.A.

Date	Location and Remarks	Estimated Deaths	Magnitude
August 19, 1966	Eastern Turkey	2,520	6.9
July 27, 1967	Caracas, Venezuela	more than 250	6.5
December 11, 1967	Konya, India (caused by water filling a reservoir; the extra pressure resulted in the earth shifting)	117	6.5
August 31, 1968	Khurasan, Iran	12,000	7.8
July 25, 1969	Eastern China	3,000	N.A.
January 5, 1970	Yunnan Province, China	10,000	7.7
March 28, 1970	Gediz, Turkey	1,086	7.4
May 31, 1970	Yungay, Ranrahirca, Huarás, and other cities in Peru	66,794	7.7
February 9, 1971	San Fernando Valley, California	64	6.6
April 10, 1972	Ghir, Iran	5,057	6.9
December 23, 1972	Managua, Nicaragua	more than 10,000	5.6
April 26, 1973	Hawaii	0	6.2
August 28, 1973	Puebla, Mexico	527	6.8
December 28, 1974	North Pakistan	more than 5,200	6.3
February 2, 1975	Near Islands, Alaska	0	7.6
February 5, 1975	Liaoning Province, China (successfully predicted)	300	N.A.
September 6, 1975	Lice, Turkey	2,312	6.8
November 29, 1975	Hawaii	0	7.2
February 4, 1976	Guatemala City, Guatemala	22,778	7.5
May 6, 1976	Northeast Italy	946	6.5
June 26, 1976	New Guinea and Irian Java	8,000	7.1
July 28, 1976	Tangshan, China	750,000	8.0
August 17, 1976	Earthquake and tsunami hit Mindanao, Philippines	8,000	7.3
November 24, 1976	Eastern Turkey	4,000	7.9
March 4, 1977	Bucharest, Romania	1,541	7.5
August 19, 1977	Indonesia	200	8.0
November 23, 1977	Northwest Argentina	100	8.2
December 20, 1977	Central Iran	more than 500	N.A.
January 14, 1978	Tokai region, Japan	25	7.0
June 12, 1978	Sendai, Japan	more than 27	7.5
September 16, 1978	Northeast Iran	25,000	7.7
November 29, 1978	Eastern Iran	200	6.7-7.5
January 17, 1979	Oaxaca, Mexico	0	7.8
April 15, 1979	Meshed, Iran	500	6.7

Date	Location and Remarks	Estimated Deaths	Magnitude
November 14, 1979	Macedonia and Albania	129	7.2
December 12, 1979	Colombia and Ecuador	800	7.9
October 10, 1980	Northwest Algeria	4,500	7.3
November 23, 1980	Southern Italy	4,800	7.2
July 28, 1981	Kerman Province, Iran	8,000	N.A.
December 13, 1982	Yemen	2,800	6.0
1983	Tsunami hits Akita, Japan	104	7.7
March 31, 1983	Popayán, Colombia	200,000	5.7
October 28, 1983	Challis, Idaho	2	6.9
October 30, 1983	Erzurum, Turkey	1,330	7.2
March 3, 1985	Algarrobo, Chile	177	7.8
September 19 & 21, 1985	Mexico City	4,200	8.1
January 31, 1986	Northeast Ohio	0	4.9
1986	Hokkaido, Japan	52	7.9
October 10, 1986	San Salvador, El Salvador	more than 1,000	7.5
March 5-6, 1987	Ecuador	2,000	7.0
October 1, 1987	Whittier Narrows, California	0	5.9
August 20, 1988	Nepal and India	more than 700	6.5
November 6, 1988	Yunnan Province, China, and Burma	144	7.0
November 25, 1988	Chicoutimi, Quebec	0	6.0
December 3, 1988	Pasadena, California	0	4.9
December 6, 1988	Mamasani, Iran	7	5.6
December 7, 1988	Armenia, U.S.S.R.	28,854	6.9
January 22, 1989	Tadzhikistan, U.S.S.R. (Tajikistan)	274	5.5
May 23, 1989	Macquarie Ridge, 300 mi southeast of New Zealand	0	8.2-8.3
October 17, 1989	Loma Prieta in Santa Cruz Mountains of California, but destructive in San Francisco and Oakland as well	63	7.1
February 28, 1990	Southern California	0	5.2
June 20, 1990	Northern Iran	50,000	7.6
July 16, 1990	Luzon Island, Philippines	1,700	7.7
January 31, 1991	Hindu Kush mountains of Afghanistan and Pakistan	600	6.8
June 28, 1991	Los Angeles, California	2	5.9
October 20, 1991	India and Nepal	more than 2,000	7.1

EARTHQUAKES OF 1992

February 13	No casualties are reported for a 6.8 magnitude shallow earthquake about 200 mi (320 km) northwest of Port Vila in Vanuatu.
February 27	No damage results from a 6.7 magnitude shallow earthquake near the coast of eastern New Guinea.
March 2	No damage or casualties result from a 6.8 magnitude earthquake with a focal depth of 27 mi (44 km) centered about 56 mi (90 km) east of Petropalvlosvsk-Kamchatskiy on the Kamchatka peninsula of Russia.
March 4	No damage is caused by a 6.6 magnitude earthquake in the Admiralty Islands off New Guinea, focal depth 30 mi (48 km).
March 4	A shallow 4.5 magnitude earthquake in the Lordegan-Ardal area of southeastern Iran kills six, destroys many houses, and blocks roads with landslides.
March 13	A 6.8 magnitude earthquake with a focal depth of 17 mi (28 km) about 37 mi (60 km) north of Erzincan, Turkey, kills more than 600, making it the worst earthquake disaster in Turkey since 1983.
April 25	An earthquake measuring 6.9 in magnitude strikes 35 miles south of Eureka, California, near Ferndale, causing considerable damage to buildings and injuring 80 persons.
June 28	Two large earthquakes, the first a 7.5 magnitude jolt that strikes the Yucca Valley near Landers, California, at 4:58 a.m., and the second a 6.5 magnitude quake at Big Bear City, California, at 8:05 a.m., are just 19 mi (30 km) apart. The first quake kills a three-year-old child when a chimney falls on him, and injures at least 70 other persons, while both quakes caused serious damage to highways, houses, and water systems throughout the area. Some effects of the quakes are noticed as far away as Seattle.
August 19	A 7.5 magnitude earthquake strikes Kyrgyzstan in Central Asia, killing 60.
August 21	An earthquake measuring about 4.1 in magnitude strikes near Charleston, South Carolina, doing little damage.
September 2	An earthquake off the coast of Nicaragua, measuring 7.2 in magnitude, kills 170.
September 11	Nine are killed by a 6.8 earthquake centered in Kabala, Zaire.
October 12	A 5.9 magnitude earthquake in Egypt kills 651, injures 6,512, and destroys 3,239 buildings in Cairo.
October 18	A 7.2 magnitude earthquake at Murindo, Colombia kills eight and triggers a geyser that injures 36.
December 13	A 7.5-magnitude earthquake with an epicenter 20 mi (32 km) southeast of the town of Maumere on the island of Flores in Indonesia kills at least 2,500, many as a result of tsunamis as high as 80 ft (24 m) that sweep as much as 1,000 ft (300 m) inland, wiping out whole coastal villages, although many buildings collapsed in Maumere and in Larantuka, another town on Flores.

EARTHQUAKES OF 1993

January 15	Two are killed by a 7.0 magnitude earthquake at Kushiro, Japan.
March 20	One person dies and about a thousand houses are destroyed as a result of a shallow 6.0 magnitude earthquake in China.
April 18	Five die as a result of a 6.2 magnitude earthquake with a depth of 56 mi (91 km) in central Peru.

May 13	A surface magnitude 6.9 earthquake centered 36 mi (60 km) deep strikes Alaska near Sand Point, 1,450 mi (900 km) from Anchorage, where it is strong enough to knock items off store shelves.
June 8	A 40 mi (70 km) deep 7.1 surface magnitude earthquake strikes the Kamchatka Peninsula on the east coast of Russia, but does no damage.
July 12	More than 200 people are killed and more than 300 injured when 7.8 earthquake and two major aftershocks, causing a large tsunami over 30 ft (10 m) high, hit Japan's large northern island of Hokkaido and Okushiri Island. The main cause of death and destruction is the tsunami on Okushiri.
August 1	Two are killed by a 6.2 magnitude earthquake in Sudan.
August 8	An earthquake with a magnitude of 8.0 occurred in the Mariana Islands, causing no fatalities.
August 9	A 6.4 magnitude earthquake strikes the Hindu-Kush region of Afghanistan, killing six.
September 10	One person dies as a result of a 7.2 earthquake in Chiapas State, Mexico.
September 20	A 5.4 magnitude earthquake in Klamath Falls, Oregon, claims one life.
September 29	Although the earthquake measures only 6.4 in magnitude, it kills an estimated 30,000 people when it strikes India's Latur District, which includes Bombay. Some 11,000 bodies are actually removed from the wreckage, but nearly twice that many are thought to be beyond recovery. This is a rare earthquake, occurring in stable continental crust far from plate boundaries and any known fault.
October 11	One person is killed by an earthquake that measures 6.5 in magnitude near Tokyo, Japan.
October 13	More than 63 people are killed by a 7.2 magnitude in Eastern Papua, New Guinea.
December 4	The second 5.4 magnitude earthquake of 1992 strikes Klamath Falls, Oregon, also killing one person.

EARTHQUAKES OF 1994

January 17	An earthquake measured at a 6.7 moment magnitude (6.6 surface magnitude, often reported as 6.8) strikes the Northridge section of Los Angeles, California, killing 61, injuring 8,000, and damaging more than 92,000 buildings—3,000 so badly that they are subsequently declared unsafe for occupancy. More than $7 billion in insurance claims are filed.
January 21	A shallow earthquake of surface magnitude 7.3 on the sparsely populated Halmahera Island near Sumatra in Indonesia kills nine and injures 300, destroys 16 villages, and produces tsunamis that drove waves as far as 0.6 mi (1 km) inland.
February 5	An earthquake of surface magnitude 6.2 in the East African Rift Valley of Uganda causes landslides that kill two and injure several others.
February 12	A 150 mi (250 km) deep earthquake of surface magnitude 7.2 near Erromango Island in Vanuatu causes little damage.
February 15	In Indonesia, a shallow 7.2 surface magnitude quake in the Liwa regions of southern Sumatra kills at least 217 people, injures about 3,000, and leaves 75,000 homeless.

February 23	A shallow earthquake with a surface magnitude of 6.1 strikes eastern Iran near the border with Pakistan and Afghanistan, killing six and injuring many others. It is followed by several aftershocks of nearly the same size, but they produce no additional casualties.
March 1	An earthquake with a surface magnitude of 6.0 kills two in southern Iran.
March 2	A 36 mi (58 km) deep earthquake in Haiti with a body magnitude of only 5.3 kills four people, and because of its depth is felt as far away as Cuba and the Dominican Republic.
April 18	A shallow earthquake measuring 6.8 in surface magnitude strikes the Solomon Islands, causing no damage.
May 25	A shallow earthquake of surface magnitude 6.6 strikes the Irian Jaya regions of Indonesia on New Guinea, causing considerable damage but no reported loss of life.
June 3	An earthquake with a surface-wave magnitude of 7.2 sets off 10-43.3 ft (3-13.2 m) tsunamis in Java and Bali in Indonesia, killing at least 218 persons, injuring 400, and destroying about a thousand homes.
June 6	An earthquake with a magnitude of 6.8 causes landslides of mud from the sides of Nevado del Huila volcano in Colombia, killing more than 500 people in the Paez river valley as mudslides sweep away all or parts of about a dozen villages.
June 9	The largest deep earthquake ever recorded strikes some 390 mi (630 km) below the Amazon rain forest of northern Bolivia, registering a moment magnitude of 8.3. Although damage is relatively minor and there is no loss of life, the quake is felt in such distant places as Seattle, Washington; Chicago, Illinois; Toronto, Canada; and throughout the Caribbean islands.
June 18	A shallow earthquake with a body magnitude of 7.1 is felt throughout the South Island of New Zealand and into the North Island, collapsing some structures and damaging roads but causing no fatalities.
June 20	An earthquake with a surface magnitude of 6.0 in Iran kills two and produces landslides that block a road.
July 4	A shallow earthquake west of Mexico with a body magnitude of 6.0 collapses a stone wall, killing two.
July 13	A large earthquake with a surface magnitude of 7.3 strikes the Vanuatu Islands, but causes no reported damage in Port Villa, where it is strongly felt.
August 10	Building damage but no casualties result from a shallow earthquake with a body magnitude of 4.7 in southern Iran.
August 18	An earthquake 90 mi (146 km) below northern Algeria kills 164 persons, injures almost twice that many, and leaves perhaps 10,000 persons homeless. The quake has a moment magnitude of 6.0.
August 19	A large earthquake with a body magnitude of 6.6 occurs 350 mi (565 km) below Argentina, but causes no surface damage.
August 31	A 4.9 body magnitude shallow earthquake centered near Jhansi, India, causes the collapse of eight houses in Bhopal, killing two and injuring dozens.
September 1	Although there is no damage from a 7.0-surface-magnitude earthquake off the coast of California, a small tsunami with 5.5 in (14 cm) waves is recorded in Crescent City on the north coast of the state.
September 12	An earthquake of magnitude 6.3 strikes the Lake Tahoe region of northern California and Nevada, but causes little damage because it affected a region of sparse population.

September 13	An earthquake of magnitude 4.6 strikes western Colorado, rolling some boulders into a highway and breaking a water pipe but causing little overall damage.
September 16	A shallow earthquake with a surface magnitude of 6.6 centered in the Taiwan Strait causes at least 400 injuries and one death by trampling. Most of the injuries occur in stampedes at schools and other public buildings in Guangdon Province in China, although some injuries are caused by collapsing walls of buildings as well.
October 1	A shallow earthquake with a surface magnitude of 6.5 takes place under the ocean amid the islands of Vanuatu, causing no damage.
October 4	A large earthquake with a moment magnitude of 8.2 under the ocean along the Japan-Kuril trench kills 10 in the Kuril Islands and injures 180 in Japan. At Russian military installations on South Kuril Island, 89 out of 380 buildings collapse. Tsunamis are measured as high as 11.5 ft (3.5 m) from peak to trough but cause little damage, although a five-ton fishing boat is sunk.
October 8 & 13	Two earthquakes near Halmahera Island in Indonesia with surface magnitudes of 6.9 and 6.4 kills one person, injures 71, and damages hundreds of buildings.
October 25	An earthquake 153 mi (247 km) below the Hindu Kush region of Afghanistan produces a body magnitude of 6.1, shaking a wide area but causing only minor damage.
November 14	A large, shallow earthquake with a surface magnitude of 7.2 kills at least 78 and injures 130 more on Mindoro, a large island in the Philippines. An associated tsunami carries off about 50 houses in Calapan on the north shore of Mindoro.
November 20	A shallow 6.3 surface magnitude quake beneath Cenderaisih Bay injures 28 people on the Irian Jaya Region of Indonesia, the western part of the island of New Guinea.
December 28	A large earthquake measuring 7.4 or 7.5 in magnitude centered above a point in the Pacific Ocean northeast of the island of Honshu, Japan, kills three people and injures more than 200 others. Small tsunamis—as high as 22 in (56 cm) at Hachinote, Japan, about 90 mi (60 km) west of the epicenter—strike Honshu and elsewhere in Japan.

EARTHQUAKES OF 1995

January 17	A massively destructive 7.2 body magnitude shallow earthquake strikes the island of Honshu, Japan. Because the worst-hit region is the city of Kobe, just 20 mi (32 km) north of the epicenter on Awaji Island, the shock and its more than 1,300 aftershocks become popularly known as the Kobe quake, although geologists call it the Hyogo-ken Nanbu earthquake. Not only are about 5,300 people killed and about 35,000 injured, but also nearly 170,000 houses are destroyed, adding greatly to the misery.
January 19	About seven persons are killed and 35 injured in and around Bogota, Colombia, by a shallow earthquake with a surface magnitude of 6.5 centered near the town of Paez.
January 27	A 6.9 surface magnitude earthquake strikes the Irian Jaya region of Indonesia on the island of New Guinea.
January 28	A small 5.0 magnitude earthquake centered 10 mi (16 km) south of Seattle, Washington, is felt from Vancouver Island in Canada to Salem, Oregon. Damage is minor.

February 5	A deep earthquake with a surface magnitude of 7.5 occurs in the Pacific off a relatively uninhabited region on the North Island of New Zealand, causing minimal damage because of a depth of about 60 mi (100 km).
February 8	At least 36 persons are killed and 250 injured when a 6.5 body magnitude earthquake with a depth of 45 mi (70 km) and an epicenter below Pereira topples government buildings there.
February 18	An earthquake at a depth of about 6 mi (10 km) and a surface magnitude of 6.6 strikes off the California coast near Cape Mendocino but causes little damage.
February 23	A rockslide caused by a 6.2 surface magnitude quake in northern Taiwan kills two passengers on a bus and injures 10 others when a rockslide is dislodged upon the vehicle. A separate 5.8 surface magnitude in the eastern Mediterranean also kills two when a house collapses on them in their village of Milou on Cyprus.
March 19	A shallow earthquake of surface magnitude 7.2, preceded and followed by other large quakes, strikes the southern coast of Irian Jaya, the Indonesian part of the island of New Guinea, but there is no reported damage.
April 7	A large shallow earthquake with a surface magnitude of 7.9 but a body magnitude of 6.8 occurs in the South Pacific southwest of Samoa.
April 14	A 5.7 surface magnitude earthquake strikes near Alpine, Texas, in the Glass Mountains, the largest Texas quake since 1931.
April 21	Samar Island, one of the Philippines, is struck by a 7.3 surface magnitude earthquake which results in some structural damage to buildings but no injuries.
May 16	A shallow surface magnitude 7.7 undersea earthquake near the island of New Caledonia in the Loyalty Island provokes tsunami warnings for most of the South Pacific, but no great waves are caused.
May 27	A shallow surface magnitude 7.6 earthquake on Sakhalin Island in Russia kills about 2,000 persons.
June 15	An earthquake near the gulf of Corinth in southern Greece measuring 6.5 in magnitude kills 26, injures more than 60, and destroys about 1,500 buildings, while damaging many more.
June 17	An earthquake on Flores Island in Indonesia measuring 5.4 in magnitude kills 1, damages about 75 buildings, and causes landslides.
June 25	A landslide caused by a 5.9 magnitude earthquake on Taiwan kills one and injures four; a second person is killed and three more injured elsewhere on the island.
July 3	A 7.1 magnitude earthquake near the Raoul Island volcano in the Mermadex Islands northeast of New Zealand, followed by four large aftershocks, causes landslides but no fatalities.
July 11	A Burmese earthquake of magnitude 7.2, just on the Myanmar side of the border with China, kills two in China and results in damage as far away as Thailand.
July 21	A small 5.7 magnitude earthquake in Gansu province in China causes 14 deaths, 60 injuries, and damage to many buildings.
July 30	Along the coast of northern Chile, a 7.8 earthquake felt as far away as La Paz, Bolivia, and Buenos Aires, Argentina, kills three people in Antofagasta, Chile, and causes damage and injuries for more than 60 mi (100 km) from the epicenter.

August 16	A shallow 7.8 surface magnitude earthquake near Bougainville Island in the Solomon Islands produces the collapse of a few buildings in Rabaul, New Britain, but no serious damage. It is followed that day and the next by four major aftershocks ranging from surface magnitude 6.5 to 7.2.
September 14	A shallow 7.2 surface magnitude earthquake centered along Mexico's Pacific Coast near the border of Guerrero and Oaxaca states kills several people and injures dozens. An alarm system in Mexico City provides about 150 seconds of warning before buildings there shake for a minute or more with the arrival of the earthquake waves.
October 1	An earthquake in Dinar, Turkey, kills at least 55, and perhaps as many as 100.
November 22	A shallow earthquake with a surface magnitude of 7.7 strikes the Gulf of Aquaba, killing 40 in Elat, Israel, and four more in Nuwebia, Egypt.
December 3	A shallow earthquake with a magnitude of 8.0 strikes the Kurile Islands, but causes no serious damage.

EARTHQUAKES OF 1996

January 1	A shallow magnitude 7.7 earthquake on the Minahassa Peninsula, Sulawesi, Indonesia, raises a 3 ft (1 m) high tidal wave that kills nine and causes considerable damage to buildings and roads.
February 3	More than 300 people were killed and about 17,000 injured when a shallow magnitude 6.4 earthquake struck near K'un-ming, China, the capital of Yunnan province.
February 17	A severe magnitude 8 earthquake and associated tsunamis struck the New Guinea trench in the Pacific, killing over 100 Indonesians. Some tsunamis reached heights of 23 ft (7 m) along the coast of Irian Jaya.
February 21	A surface 6.7 magnitude earthquake in the Pacific off the coast of Peru causes a 20 ft (6 m) tsunami in the town of Chimbote some 80 mi (130 km) southwest of the epicenter. At least 10 people were killed.

(Periodical References and Additional Reading: *Science* 12-10-93, p 1666; *EOS* 6-28-94, p 289; *New York Times* 9-14-95, B8; *New York Times* 12-2-94, p A7; *Discover* 1-95, p 64; *New York Times* 1-16-95, p A13; *New York Times* 1-17-95, p A1; *New York Times* 10-3-95, p A6)

ACTIVE VOLCANOES AND MAJOR VOLCANIC EVENTS

A volcano is an opening in Earth's crust that emits melted rock (lava), hot gases, and solid rocks of various sizes. It is also the mountain that forms as solidified lava and ejected rocks pile up around an opening—called a vent if it resembles a crack, or a crater if it is fairly circular.

Volcanoes do not occur at random across the world. Almost all are found at plate boundaries, such as the famous Ring of Fire around the Pacific Ocean. A few, such as the volcanoes of Hawaii and the volcanic region (without a volcano) of Yellowstone Park, appear to be over a "hot spot"—a place where molten rock flows upward with sufficient force to burn through Earth's crust.

About 600 volcanoes are now active. The following list contains more than a third of all the known active volcanoes, with emphasis on volcanoes that have been active in recent years, and volcanoes that have had famous eruptions.

Other Vocabulary Terms

Active: Describes volcano that has erupted in historic times. Thus Tambora, which has not erupted since 1815, is considered active. Sometimes used for volcanoes that produce gases or earthquakes even when there has been no historical eruption.

Ash: Ejected rocks, larger than dust, but smaller than about a quarter inch (half a centimeter).

Bombs: Ejected rocks larger than several inches (centimeters) across. Some may be as heavy as 100 tons.

Caldera: A large depression formed by the collapse of a volcano when its magma reservoir flows back into the crust. A caldera may have several vents or craters on its floor.

Cinder cone: A mountain, usually regular and steep sided, produced by debris tossed from a volcano's central crater. This is the classic image of a volcano.

Cinders: Ejected rocks larger than ash, but smaller than bombs; also called scoria.

Composite volcano: A volcano in which both lava and debris ejected from vents combine in layers to produce a mountain. Nearly all familiar volcanoes are composite in structure, with the shield volcanoes of Hawaii and Iceland as the main exceptions.

Dormant: Not active, but showing signs, such as hot springs or mild earthquakes, that it may become active. Sometimes a dormant volcano is labeled active.

Extinct: Neither active nor dormant, but showing signs that it is actually a volcano, not some other feature. It is usually not clear whether a volcano is dormant or extinct. Some volcanoes thought to be extinct become active.

Hot spot: A place where magma continually flows upward, sometimes causing a volcano by burning a hole in the crust. Yellowstone Park, Iceland, and Hawaii are currently over hot spots. Since tectonic plates move over the hot spot, it may appear that the hot spot moved in the past, although in fact the spot stayed still while the plates moved.

Lahar: A landslide or mudflow of volcanic debris, often caused when the dam to a lake is opened or overflows, or when a glacier or snowcap melts.

Lava: Molten rock that has flowed from a volcano's vent or crater. Because it loses gases to the air and quickly changes composition as it nears the surface, lava is not the same substance as molten rock that remains deep below the surface.

Magma: Molten rock that remains beneath Earth's surface, where it powers all of the phenomena associated with a volcano.

Nuée ardente: A mixture of hot dust or ash and very hot gases that is ejected from a volcano and flows downhill at great speed.

Phreatic explosions: Violent steam blasts.

Pyroclastic flow: Same as a *nuée ardente*.

Shield volcano: A volcanic mountain formed almost entirely from lava, especially less viscous lava that flows easily out of vents and down the sides of the giant, gently sloping mountain. Such volcanoes are formed at hot spots, or where rifts are found on land.

AFRICA AND THE INDIAN OCEAN

Volcano	Location	Height in ft (m) above sea level	Last reported eruption	Remarks
Cameroon Mt.	Cameroon	13,435 (4,095)	1982	October-November eruption causes evacuation of two towns as lava flows 7.5 mi (12 km) down mountain.
Erta-Ale	Ethiopia	2,011 (613)	1973	
Fournaise, Piton de la	Réunion Island	8,631 (2,631)	1992	Brief eruption starts August 27 and lasts about a month. There have been more an a hundred such eruptions recorded in the past 300 years.
Karthala	Grande Comore, Cormoro Islands	8,000 (2,438)	1991	July 11
Kimanura	Zaire		1989	New cone in the Nyamuragira-Nyirangongo complex.
Ol Doinyo Lengai	Tanzania	9,482 (2,890)	1994	September 18, 1994. Present eruption cycle began in 1983.
Nyamuragira	Zaire	10,033 (3,058)	1994	Starts July 4, 1994, and is a continuation of active period that since 1976 has produced more than seven eruptions.
Nyirangongo	Zaire	12,381 (3,469)	1994	June 23. Part of Virunga chain with Nyamuragira, its permanent lake of fluid lava flowed through a fissure and killed 70 people on January 10, 1977.

ANTARCTICA

Volcano	Location	Height in ft (m) above sea level	Last reported eruption	Remarks
Big Ben	Heard Island	9,007 (2,745)	1986	
Deception Island	South Shetland Islands	1,890 (576)	1970	
Mt. Erebus	Ross Island	12,447 (3,794)	1990	The southernmost active volcano. It was erupting when first sighted in 1841.

Volcano	Location	Height in ft (m) above sea level	Last reported eruption	Remarks
Unnamed	Ross Ice Shelf	2,100 (640)	?	Donald Blankenship of the University of Texas and Robin Bell of Lamont-Doherty Earth Observatory discovered in 1993 an active volcano in large caldera under more than a mile of ice.

ASIA

Volcano	Location	Height in ft (m) above sea level	Last reported eruption	Remarks
Agung	Bali, Indonesia	10,308 (3,142)	1964	May 17-21, 1963, eruption kills 1,584.
Akan	Hokkaido, Japan	4,918 (1,499)	1988	
Akita Komagatake	Japan	5,449 (1,661)	1970	
Alaid	Kuril Islands, Russia	7,674 (2,339)	1972	
Anak Krakatau	Indonesia	330 (100)	1993	New volcano at the site of Krakatau (see below).
Anak Ranakah	Flores Island, Indonesia		1989	December 28, 1987, eruption begins in area of no previously known volcanic activity.
Amburomrribu	Indonesia	7,051 (2,149)	1969	
Asama	Honshu, Japan	8,399 (2,560)	1983	April 8 eruption produces ashfall as far away as 120 mi (200 km).
Aso	Kyushu, Japan	5,223 (1,592)	1995	Vents at Nakedake cone eject rocks (September 12, 1994) and mud (May 2, 1994); Nakedake has erupted more than 165 times since 553 A.D., which is first documented historic Japanese volcano.
Avachinsky	Kamchatka, Russia	9,026 (2,751)	1991	January 13 eruption not so large as February 1945 eruption nor as the huge explosion about 3400 B.C.
Awu	Sangihe Islands, Indonesia	4,350 (1,326)	1968	Eruption begins August 12, 1966, and kills 39 people.
Azuma	Honshu, Japan	6,640 (2,024)	1978	

Volcano	Location	Height in ft (m) above sea level	Last reported eruption	Remarks
Baitoushan	Jilin, Chine		1702	One of the world's largest eruptions took place here about 1000 A.D., followed by small, probably phreatic, eruptions in 1413, 1597, and 1668.
Banda Api	Banda Sea, Indonesia	2,247 (685)	1989	May 9-15, 1988, eruption forces evacuation of 7,000 people.
Bandai	Japan	6,204 (1,891)	1888	July 15 eruption causes mudslides that kill nearly 500 people.
Barren Island	Andaman Islands, India	1,000 (305)	1995	Eruption that starts in April, 1991, is first since 1803; followed by December 20, 1994, eruption that continues into 1995.
Batur	Bali, Indonesia	5,633 (1,717)	1994	First eruption in 18 years begins on August 7; frequent eruptions have taken place since first historic eruption in 1804.
Bezymianny	Kamchatka, Russia	9,456 (2,882)	1995	October 6. The October 21, 1993, eruption is largest since May 20, 1956, which was among the most energetic of the twentieth century.
Bulusan	Philippines	5,135 (1,565)	1995	Mild phreatic and ash eruptions begin on November 27. Similar eruptions have been recorded periodically since 1852, with more aggressive episodes only in the period 1918-22; nevertheless, people were warned against approaching the volcano.
Canlaon	Negros Islands, Philippines		1992	June.
Catarman (Hibokihibok)	Mindanao, Philippines	4,364 (1,330)	1951	December 4 eruption kills 500.
Chikurechki	Kuril Islands, Russia		1986	November 18.
Chokai	Honshu, Japan	7,300 (2,225)	1974	

Volcano	Location	Height in ft (m) above sea level	Last reported eruption	Remarks
Dukono	Halmahera, Indonesia	3,566 (1,087)	1995	First of five historical eruptions in 1550. Dukono has been erupting almost continuously since 1933.
Ebeko	Kuril Islands, Russia	3,734 (1,138)	1989	Starts February 3.
Fukutoku-okanoba	Volcano Islands, Japan	−46 (−14)	1986	January eruption builds a temporary island.
Galunggug	Java, Indonesia	7,113 (2,168)	1982	Eruptions on October 8 and 12, 1822, cause mudslides that kill 4,000 people.
Gamalama	Ternate Island, Indonesia	5,625 (1,715)	1994	No serious damage from several eruptive episodes in 1994.
Gamkonora	Indonesia	5,365 (1,635)	1981	
Gerde	Indonesia	9,705 (2,958)	1949	
Ijen	Java, Indonesia	7,828 (2,386)	1952	Three eruptions so far in twentieth century.
Ilewerung	Nusa Tengarra Islands, Indonesia		1993	Submarine eruption 165 ft (50 m) below sea level.
Iliboleng	Nusa Tengarra Islands, Indonesia	5,443 (1,659)	1991	Three small eruptions in 1991; at least 20 moderate eruptions since 1885.
Ivan Grozny	Kuril Islands, Russia	3,799 (1,158)	1990	
Karangetang (Apu Siau)	Sangihe Islands, Indonesia	5,853 (1,784)	1995	Earlier eruption in 1992 kills 6 on May 11. Eruption in 1994 produces a number of hot lahars; 40 reported eruptions in past 300 years.
Karymsky	Kamchatka, Russia	4,875 (1,486)	1996	Starts January 1, 1996. First historical eruption was in 1771; period of 11 eruptions between 1945 and 1970.
Kawah Ijen	Java, Indonesia			Phreatic (gas) emissions in 1994 were the fourth period of such emissions in the twentieth century.
Kelimutu	Indonesia	5,381 (1,640)	1968	June 3 eruption features 300 ft (100 m) geyser; in 1995, Dutch tourist disappears after falling into crater.

Volcano	Location	Height in ft (m) above sea level	Last reported eruption	Remarks
Kelut	Java, Indonesia	5,679 (1,731)	1990	February 10 eruption kills 32, with four more killed by lahars in November. In May 1919, the crater lake spills over, killing more than 5,000 people.
Kirishima	Japan	5,577 (1,700)	1959	Eruption in the 1700s kills six people. Five historic eruptions.
Kliuchevskoi	Kamchatka, Russia	15,584 (4,750)	1994	Large eruption that begins September 8 produces lava bursts as high as 14,750 ft (4,500 m) above crater rim and ash column estimated at 12 mi (20 km) high on September 30, disrupting Pacific air traffic for three days. There have been more than 80 historic eruptions since 1697.
Komaga-take	Hokkaido, Japan	3,740 (1,140)	1996	Starts March 6, 1996. A June 17, 1929, eruption kills two and damages houses.
Korayksky	Russia	11,339 (3,456)	1957	
Kozu-shima	Izu Islands, Japan	1,883 (574)	840	Although last eruption was over a thousand years ago, earthquake activity in the early 1990s suggests renewed activity.
Krakatau	Indonesia	2,667 (813)	1994	August 26-27, 1863, eruption blows up island, producing giant waves that kill 36,000 people.
Kusatsu-Shirane	Honshu, Japan	7,139 (2,176)	1989	January 6 eruption is small.
Lewotobi Lakilaki	Flores, Indonesia	5,217 (1,590)	1991	Eruption starts May 11 and continues into June. Previous eruption was in January 1971.
Lokon-Empung	Sulawesi, Indonesia	5,184 (1,580)	1992	Main eruption begins May 17-18, 1991, and continues into 1992; 11,000 villagers evacuated, but volcanologist Vivianne Clavel killed by eruption. Previous eruption was in 1988.
Mahawu	Sulawesi, Indonesia		1958	July eruption causes lahars that kill one and injure 10.

Volcano	Location	Height in ft (m) above sea level	Last reported eruption	Remarks
Marapi	Sumatra, Indonesia	9,485 (3,045)	1994	Has been active since 1987, mostly with moderate ash explosions; not to be confused with Merapi on Java.
Mayon	Philippines	8,077 (2,462)	1993	Unexpected explosion on February 2 causes *nuée ardente* that kills 68.
Me-akan	Japan	9,846 (3,001)	1966	
Merapi	Java, Indonesia	9,550 (2,911)	1995	Collapse of summit dome sends *nuée ardente* down slope on November 22,1994, killing more than 40; a smaller collapse occurs on January 5, 1995.
Mutnovsky	Kamchatka, Russia		1961	
Nasu	Japan	6,210 (1,893)	1977	
Niigata-yake-yama	Honshu, Japan	8,064 (2,458)	1987	
On-Take	Kyushu, Japan	10,049 (3,063)	1979	Volcanic earthquake on September 14, 1984, starts a landslide that kills 29.
Oshima	Izu Island, Japan	2,487 (758)	1990	October 4 eruption is the first since January 1988.
Papandayan	Java, Indonesia	8,743 (2,665)	1772	Explodes on August 11-12, killing more than 3,000.
Pinatubo	Luzon, Philippines	5,250 (1,600)	1991	Explosion on April 2, 1991, signals a renewal of activity. A major eruption starts on June 9, which injects ash 18.6 mi (30 km) into the air; ranks as second largest volcanic eruption of twentieth century (after 1902 eruption of Santa Maria in Guatemala). 754 person are killed or missing, including 358 from disease in evacuation camps. The previous major eruption was about 1380 A.D.

Volcano	Location	Height in ft (m) above sea level	Last reported eruption	Remarks
Raung	Java, Indonesia	10,932 (3,332)	1991	September-October eruption produces vigorous ash eruptions and plumes. Six eruptions in period between A.D. 1586 through 1817 result in a small number of fatalities. More then 50 eruptions in nineteenth century.
Rinjani	Lombok Island, Indonesia	12,224 (3,726)	1995	Eruption begins in June. On November 3, a cold lahar kills 30 or more people. First reported historical eruption was in September 1847, and Rinjani has been fairly active since.
Sakura-jima	Kyushu, Japan	3,665 (1,117)	1996	Eruption begins in 1955; 55 separate eruptions in August 1994, including more than 25 explosive ones. Falling cinders damage automobile windshields in area.
Sangeang Api	Indonesia	6,351 (1,936)	1988	
Sarychev	Kuril Islands, Russia	5,115 (1,559)	1986	
Semeru	Java, Indonesia	12,060 (3,676)	1995	Highest and perhaps most active volcano on Java. First historical eruption in 1818 but nearly continuous eruptive state since 1967. Lahars kill 252 on May 14, 1981; 1994 eruption kills six.
Sheveluch	Kamchatka, Russia	11,138 (3,395)	1993	Strong eruption on April 22, 1993. Crater forms in 1964 in explosion that along with Sheveluch eruption in 1864 counts among the largest historical eruptions of any volcano.
Siau	Indonesia	5,853 (1,784)	1976	
Sinila	Indonesia	7,000 (2,134)	1979	February 21 eruption kills 175.
Slamet	Java, Indonesia	11,260 (3,432)	1989	Weak eruptions in July 1989 preceded by tephra eruption to 1,640 ft (500 m).
Soputan	Sulawesi, Indonesia	5,853 (1,784)	1989	April 22-24 eruption damages 500 houses.

Volcano	Location	Height in ft (m) above sea level	Last reported eruption	Remarks
Sorikmarapi	Sumatra, Indonesia	7,037 (2,145)	1986	July eruption emits ash cloud to 2,300 ft (700 m) above summit.
Suoh	Sumatra, Indonesia			In 1993, a gas emission produces a small burst of hot mud.
Suwanose-jima	Japan	2,621 (799)	1995	Frequent explosive eruptions since 1956.
Taal	Luzon, Philippines	1,312 (400)	1970	Has erupted 24 times since 1572. 1911 blast kills 1,300; September 28, 1956, eruption kills 200; and 1965 event kills at least 150. Volcano has been restless since 1991.
Tambora	Sumbawa, Indonesia	9,354 (2,851)	1815	Eruption starting April 5 causes immediate death of 10,000 locally; additional 80,000 thought to die worldwide as a result of climate changes caused by eruption, including "Year without a Summer" in New England.
Tangkuban Prahu	Java, Indonesia	6,637 (2,023)	1967	
Tengger Caldera	Java, Indonesia	7,641 (2,329)	1984	May 12-31 eruption; dark smoke to 0.6 mi (1 km).
Tiatia	Kuril Islands, Russia	6,013 (1,833)	1973	
Tjarme	Indonesia	10,098 (3,078)	1938	
Toba	Sumatra, Indonesia			Eruption in 71,500 B.C. is second largest in history and may have caused global temperatures to fall as much as 9°F (5°C), causing the human population to decline to very low levels.
Tokachi	Hokkaido, Japan	6,814 (2,077)	1989	Eruption begins December 16, 1988, after 26 years of dormancy; 1926 eruptions killed 144.
Tolbachik	Kamchatka, Russia		1976	The Great Tolbachik Fissure Eruption takes place from 1971 through 1976. Has erupted 30 times since 1740.

Volcano	Location	Height in ft (m) above sea level	Last reported eruption	Remarks
Unzen	Kyushu, Japan	4,921 (1,500)	1996	Eruption starts on November 18, 1990, and is largest Japanese eruption in 50 years; current eruption kills 43, including three volcanologists, destroys more than 2,000 buildings, and causes the evacuation of 12,000 from vicinity. Last previous eruption in 1782 produced harbor wave that killed 15,000; first historic eruption was in 1663 A.D.
Usu	Japan	2,390 (728)	1978	
Yake Dake	Kyushu, Japan	8,052 (2,455)	1963	Steam explosions killed a tourist on February 11, 1995, although such phreatic eruptions are not counted as volcano eruptions.
Zhupanovsky	Kamchatka, Russia	9,490 (2,958)	1959	Preceded by six explosive eruptions since 1776.

CENTRAL AMERICA AND THE CARIBBEAN

Volcano	Location	Height in ft (m) above sea level	Last reported eruption	Remarks
Acatenango	Guatemala	12,992 (3,560)	1972	
Arenal	Costa Rica	5,436 (1,657)	1995	The first recorded eruption begins in 1968, killing 78 at that time, and continues.
Cerro Negro	Nicaragua	2,214 (675)	1995	May 28 eruption is minor compared to large eruption in April 1992. Volcanic cone formed in an eruption in April 1850.
Concepción	Omtepe Island, Nicaragua	5,106 (1,556)	1986	First historical eruption was in 1883.
Conchagua	El Salvador	4,100 (1,249)	1947	
El Viejo (San Cristóbal)	Nicaragua	5,840 (1,780)	1987	
Fuego	Guatemala	12,582 (3,835)	1987	January eruption produces incandescent ash.

Volcano	Location	Height in ft (m) above sea level	Last reported eruption	Remarks
Ilopango	El Salvador	1,476 (450)	260	Destroys early Maya civilization.
Iraz	Costa Rica	11,260 (3,432)	1965	First historical eruption in 1722. Several dozen people killed by lahars in 1963-65 eruptions.
Izako	El Salvador	7,749 (2,362)	1966	
Kick-'em-Jenny	Subocean, off Grenada	− 160 (− 49)	1990	March 26 eruption not so definite as the preceding eruption on December 29-30, 1988.
Las Oukas (El Hoyo)	Nicaragua	3,444 (1,050)	1954	
Masaya	Nicaragua	2,083 (635)	1989	Probably on June 2.
Miravalles	Costa Rica		1946	Steam explosion only.
Mombacho	Nicaragua		N.A.	No historical eruptions have been recorded, only smoking fumaroles. A sixteenth-century avalanche that killed 400 may have originated with this volcano.
Momotombo	Nicaragua	4,127 (1,258)	1905	First historical eruption was 1524, although volcano began forming about 4,500 years ago.
Pacaya	Guatemala	8,373 (2,552)	1994	June 1 eruptions sends plume 4 mi (6 km) high. First historical eruption in 1565; current activity cycle began in January 1990.
Pelée	Martinique	4,500 (1,372)	1930	May 8, 1902, eruption wipes out city of St. Pierre, killing 29,000.
Poás	Costa Rica	8,885 (2,708)	1994	August 5–22; acid rain from 1986–90 eruption damages crops and causes reported human health problems.
Rincon de la Vieja	Costa Rica	6,286 (1,916)	1995	Eruption starts November 11 and ends two days later. At least 16 eruptions noted since first historic activity in 1851.
San Cristóbal	Nicaragua		1977	October small ash emission may also have occurred in November 1987.
San Miguel	El Salvador	6,988 (2,130)	1987	Weak eruption ends in February.

Volcano	Location	Height in ft (m) above sea level	Last reported eruption	Remarks
San Salvador	El Salvador	6,187 (1,886)	1923	
Santiaguito (Santa Maria)	Guatemala	12,362 (3,768)	1993	Continued strong pyroclastic explosions since 1922 eruptions of Santa Maria. 1902 eruption is largest of century. Four hikers killed July 19, 1990.
Soufriére	St. Vincent and the Gredadines	4,048 (1,234)	1979	Eruption May 8, 1902, kills 1,500-2,000.
Soufriére Hills	Montserrat, West Indies	3,002 (915)	1996	On August 21-23, 1995, more than 6,000 people are evacuated, but the eruptions of gas and steam caused no serious problems and all evacuees were allowed to return by September 6. Lava starts flowing in November 1995, and 4,000 are evacuated from its path the following month.
Tacaná	Guatemala	12,400 (3,780)	1988	
Telica	Nicaragua	3,477 (1,060)	1994	Starts July 31 and explosions continue for about two weeks. First historical eruption in 1527-29, but geologic evidence suggests heavy ash falls earlier as well as lava flows about a thousand years ago.
Turrialba	Costa Rica	10,958 (3,340)	1866	

EUROPE AND THE ATLANTIC OCEAN

Volcano	Location	Height in ft (m) above sea level	Last reported eruption	Remarks
Askja	Iceland	4,594 (1,400)	1961	
Beerenberg	Jan Mayen Island, Norway	7,470 (2,277)	1985	
Eldfell	Iceland	327 (99)	1973	

Volcano	Location	Height in ft (m) above sea level	Last reported eruption	Remarks
Etna	Italy	10,991 (3,350)	1996	Eruption starts August 2, 1995, and continues into 1996. Previous eruption in 1991–92 produces lava flows that damage houses and orchards despite efforts of Italian government and U.S. Air Force to redirect or block the flow. An eruption on January 11, 1683, kills more than 60,000.
Fogo	Cape Verde Islands	9,281 (2,829)	1995	Eruption begins on April 2, causing the evacuation of 5,000 and some destruction of buildings. Fogo, which built the island, erupted almost continuously until about 1760 and has erupted intermittently since, with the previous large eruption in 1951.
Heimay	Iceland		1972	Fissure eruption causes city to be evacuated.
Helka	Iceland	4,892 (1,491)	1991	January 17 eruptions lasts until March 11.
Ischia	Italy	2,589 (789)	1883	Earthquakes from July 28 eruption kill more than 2,000.
Krafla	Iceland	2,145 (654)	1984	September.
Leirhnukur	Iceland	2,145 (654)	1975	
Skaptar (Lakagigar)	Iceland	1,640 (500)	1783	Fissure emits gases that kill crops and livestock and interfere with fishing, causing famine that kills 9,800.
Stromboli	Italy	3,038 (926)	1996	In almost continuous state of eruption for about 2,000 years.
Surtsey	Iceland	568 (173)	1967	Creates a new island in the Atlantic during eruptions that start in 1963.
Teide	Spain			Selected for study by the European Laboratory Volcano program and the Decade Volcano effort.

Volcano	Location	Height in ft (m) above sea level	Last reported eruption	Remarks
Thera	Mediterranean near Crete; also known as Santorini	430 (131)	1550 B.C.	Explosive eruption destroys island and affects whole Mediterranean, possibly helping end Minoan civilization on Crete. Date often given as 1643, but recent recalibration of record suggests that 1550 B.C. date is probably valid.
Tristan de Cunha	St. Helena	6,760 (2,060)	1961	
Vesuvius	near Naples, Italy	4,203 (1,281)	1944	Eruption in 79 A.D. destroys Pompeii and Herculaneum, killing more than 2,000 people; earliest identified eruption was in 15,000 B.C. Eruption in 1631 killed 3,000.
Vulcano	Aeolian Islands, Italy	1,640 (500)	1890	Frequently eruptive during nineteenth century; breaks a submarine cable with eruption in 1892.

NORTH AMERICA

Volcano	Location	Height in ft (m) above sea level	Last reported eruption	Remarks
Akutan	Alaska	4,275 (1,303)	1992	Small eruptions March 8–April 9.
Amukta	Alaska	3,490 (1,064)	1963	
Aniakchak	Alaska	4,450 (1,356)	1931	
Atka Island	Alaska	5,030 (1,533)	1987	Two volcanoes on island: Korovin, which erupted in 1987, and Kliuchev.
Augustine	Alaska	3,995 (1,218)	1986	Eruption March 27-31 interrupts international air traffic. Eruption in 1883 and subsequent collapse produces landslide into Cook Inlet that causes massive harbor waves.
Bogoslof	Alaska	150 (45)	1931	
Carlisle	Alaska	5,315 (1,620)	1838	
Cerberus	Alaska	2,560 (780)	1873	
Chinginagak	Alaska	7,985 (2,434)	1929	
Cinder Cone	California	7,985 (2,105)	1851	

Volcano	Location	Height in ft (m) above sea level	Last reported eruption	Remarks
Cleveland	Alaska	5,675 (1,730)	1994	May 25 eruption produces ash cloud that rises 35,000 ft (10,700 m). Last previous eruption was in 1987.
Colima	Mexico	12,650 (3,850)	1994	Lava extrusion begins March 1, 1991, and increases on April 16–17 when dome collapses, producing avalanches.
El Chichón	Mexico	3,478 (1,060)	1983	Eruptions March 28, 1982, and April 3–4, 1982, kill 2,000 locally and send a cloud of ash around the globe.
Fisher	Alaska	3,545 (1,081)	1826	
Gareloi	Alaska	5,370 (1,637)	1982	
Great Sitkin	Alaska	5,775 (1,760)	1974	
Iliamna	Alaska	10,140 (3,091)	1978	
Isanotski	Alaska	8,185 (2,495)	1845	
Kagamil	Alaska	2,945 (898)	1929	
Kanaga	Alaska	4,288 (1,307)	1994	About a dozen low-level eruptions in the past 20 years. 1994 eruption starts in December 1993.
Katmai	Alaska	7,540 (2,298)	1974	1912 eruption is one of third largest in twentieth century.
Keniuji	Alaska	885 (270)	1828	
Kiska	Alaska	4,025 (1,227)	1990	The 1990 eruption is the first since the first known eruption in 1962.
Lassen Peak	California	10,453 (3,186)	1914–21	Signs of restlessness, noted in 1982, soon die down.
Little Sitkin	Alaska	3,945 (1,202)	1828	
Mageol	Alaska	7,295 (2,224)	1946	
Makushin	Alaska	6,680 (2,036)	1987	First historical eruption was in 1763; there have been 17 minor eruptions since. Steam cloud observed January 30, 1995, but eruption not confirmed.
Martin	Alaska	6,102 (1,860)	1960	Steam plumes reported on March 15, 1995, but no real eruption.

Volcano	Location	Height in ft (m) above sea level	Last reported eruption	Remarks
Mt. Baker	Washington	10,778 (3,427)	1870	Begins steaming in 1875, but activity soon ceases.
Mt. Hood	Oregon	11,245 (4,392)	1801	Shaken by earthquakes in 1980.
Mt. Rainier	Washington	14,160 (4,316)	1882	Eruption today could loose floods on Tacoma or Seattle.
Mt. Saint Helens	Washington	8,368 (2,549)	1991	May 18, 1980, eruption kills 61. Activity since has been off and on.
Mt. Shasta	California	14,160 (4,316)	1855	Earthquake swarms in 1981 and 1982 do not lead to eruption.
Novarupta	Alaska	2,759 (841)	1912	The eruption of Katmai, one of the three largest of the twentieth century, also gives birth to Novarupta.
Okmok	Alaska	3,540 (1,079)	1988	
Parícutin	Mexico	1,500 (457)	1953	Volcano grows in a cornfield, starting in 1943. Lava overruns two villages before eruption terminates.
Pavlof	Alaska	8,960 (2,731)	1988	
Pavlof Sister	Alaska	7,050 (2,149)	1786	
Peulik	Alaska	5,030 (1,533)	1852	
Pogromni	Alaska	7,545 (2,300)	1964	
Popocatépetl	Mexico	17,930 (5,465)	1996	In 1994, more than 50,000 people are evacuated when the volcano begins smoking and producing small earthquakes. Similar bursts of light activity had occurred in 1943 and 1947, with a more significant eruption as recently as 1922. Twenty million people live on the sides of the volcano or in the valleys at its base.
Redoubt	Alaska	10,265 (3,129)	1990	Eruption starts December 14, 1989, but largely subsides in late spring of 1990.
Sarichef	Alaska	2,015 (614)	1812	
Sequam	Alaska	3,465 (1,056)	1993	Observed from a distance on August 19 by U.S. Coast Guard.

Volcano	Location	Height in ft (m) above sea level	Last reported eruption	Remarks
Shishaldin	Alaska	9,430 (2,857)	1995	Eruptions sighted by satellite and aircraft on December 23, 1995.
Spurr	Alaska	11,070 (3,374)	1992	July 9, 1953, eruption drops ash on Anchorage, 75 mi (120 km) away.
Tanaga	Alaska	7,015 (2,138)	1914	
Tobert	Alaska	11,413 (3,478)	1953	
Trident	Alaska	6,830 (2,082)	1974	
Veniaminof	Alaska	8,225 (2,507)	1995	Erupted at least five times since 1990; this eruption begins July 30, 1993, after a six-year quiet period.
Vsevidof	Alaska	6,965 (2,123)	1880	
Westdahl	Alaska	5,055 (1,541)	1992	Previous eruptions occur in 1978 and 1979.
Yunaska	Alaska	1,980 (603)	1937	

OCEANIA—AUSTRALIA, NEW ZEALAND, AND THE PACIFIC ISLANDS

Volcano	Location	Height in ft (m) above sea level	Last reported eruption	Remarks
Agrigan	Mariana Islands	3,166 (965)	1990	Residents evacuate island in August, but volcanic activity poses no danger.
Ambrym	Ambrym Island, Vanuatu	4,376 (1,334)	1994	Almost continuously active since discovery by James Cook in 1774.
Anatahan	Mariana Islands	2,592 (790)		No known historical eruptions, but island was evacuated in 1990 following earthquake activity.
Aoba (Ambae)	Aoba Island, Vanuatu	4,908 (1,496)	1870	Some evidence of a return to activity starting in 1991 and continuing. Previous eruptions destroyed villages on the island. On March 8, 1995, the government of Vanuatu started an evacuation of nearby villages.
Bagana	Bougrainville Island	5,742 (1,750)	1991	Among the most active volcanoes in Melanesia.

Volcano	Location	Height in ft (m) above sea level	Last reported eruption	Remarks
Gaua	Santa Maria Island, Vanuatu	2,615 (797)	1777	
Haleakala	Hawaii	10,025 (3,056)	1790	
Hualalai	Hawaii	8,251 (2,515)	1801	
Karkar	Karkar Island, Papua New Guinea	4,920 (1,500)	1979	Summit almost completely denuded of vegetation by 1979 eruption. Nine eruptions known since 1643; some seismic activity in 1994.
Kavachi	Solomon Islands	82 (25)	1991	Temporary island built by eruption first observed May 8; eight previous eruptions have formed such islands since 1950. Height given is for temporary island.
Kilauea	Hawaii	4,009 (1,222)	1995	Continuous eruption begins January 3, 1983; since then has destroyed houses and cut off roads with lava flows that continue streaming into the sea.
Lamington	New Guinea	5,840 (1,780)	1951	January 15 eruption kills 3,000-5,000.
Langila	New Britain, Papua New Guinea	4,364 (1,330)	1996	Small explosions and light ashfalls observed almost daily in 1994 and into 1996; more-or-less continuous activity since 1973.
Lohoi	Hawaii	−3,215 (−980)		
Long Island	Papua New Guinea		1993	A major eruption in 1660 was as large as the famous eruption of Krakatau in 1883.
Lopevi	Lopevi Islands, Vanuatu	4,636 (1,413)	1982	
Macdonald Seamount	Central Pacific		1989	
Manam	Off coast of Papua New Guinea	5,928 (1,807)	1996	Continuous eruption since 1974.

Volcano	Location	Height in ft (m) above sea level	Last reported eruption	Remarks
Mauna Loa	Hawaii	13,678 (4,170)	1984	During eruptions, lava flows from Mauna Loa present a danger to the city of Hilo.
Metis Shoal	Tonga	141 (43)	1995	Shoal was 13 ft (4 m) below sea level before eruption, first observed on June 9; three to five previous eruptions also produced islands, which then eroded away. Most recently the island Late Iki existed for several months in 1979.
Motmot	Long Island, New Guinea	4,278 (1,304)	1974	
Ngauruhoe	North Island, New Zealand	7,516 (2,291)	1975	February.
Pagan	Mariana Islands	1,870 (570)	1993	Islanders who were evacuated during May 1981 eruption are still unable to return to their homes.
Rabaul	New Britain, Papua New Guinea	2,257 (688)	1996	Earthquake-based prediction leads to evacuation of 30,000 people before September 19, 1994, eruption, which is about the size of the 1980 eruption of Mount St. Helens in U.S.; a similar-sized eruption in 1937-43 caused more than 500 deaths. 1994 eruption ends December 23, and 1995 eruption begins on February 13.
Ruapehu	North Island, New Zealand	9,177 (2,797)	1995	Large eruptions produce lahars, starting September 23. A 1945 eruption creates a dam, which finally gives way on December 25, 1953, killing 150 on a passing train. Eruptions nearly every year since 1966.
Rumble III	Off Kermadeo Island, New Zealand	− 660 (− 200)	1986	Erupted every year from 1966 through 1973.
Soretimeat	Vanua Lava Island, Vanuatu	3,054 (931)	1966	

Volcano	Location	Height in ft (m) above sea level	Last reported eruption	Remarks
Tarawera	North Island, New Zealand	3,645 (1,111)	1886	
Ulawun	New Britain, Papua New Guinea	7,657 (2,334)	1994	April–June, with thick ash emissions. Number of eruptions has gradually increased each century since the first historic eruption in 1700.
White Island	New Zealand	1,053 (321)	1994	Mud and large rocks on July 28. The January 17, 1992, explosion is largest seismic event recorded at the volcano. The current eruption started in December 1976.
Witori	New Britain, Papua New Guinea	2,375 (724)	1933	
Yasur	Tanna Island, Vanuatu	1,148 (350)	1991	Continuous activity since first observed by James Cook in 1774. Three tourists killed in 1994–95 by volcanic activity.

SOUTH AMERICA

Volcano	Location	Height in ft (m) above sea level	Last reported eruption	Remarks
Alcedo	Galápagos Islands, Ecuador	3,707 (1,130)	1993	Unobserved eruption is recognized from craters that developed between 1991 and 1994 visits. Guides indicate activity was in 1993.
Cotachachi	Ecuador	16,204 (4,939)	1955	
Cotopaxi	Ecuador	19,347 (5,897)	1975	
Cumbal	Colombia	15,630 (4,764)	1926	In addition to 1926 eruptions, an eruption occurred in 1877
Fernandina	Galápagos Islands, Ecuador	4,905 (1,495)	1995	Eruption that starts on January 25 is the first flank eruption (vent in the side of the mountain, with lava pouring out) since 1968. Eruption concludes on April 8.

Volcano	Location	Height in ft (m) above sea level	Last reported eruption	Remarks
Galeras	Colombia	14,025 (4,276)	1993	Last of six eruptions in one year occurs on June 8, 1993. Sudden beginning of eruption on January 14, 1993, kills nine and injures six. Many historical eruptions since sixteenth century.
Guagua Pichincha	Ecuador	15,696 (4,784)	1993	Two scientists on rim are killed in the most recent in a decade-long series of occasional phreatic explosions.
Guallatiri	Chile	19,882 (6,060)	1960	Produces strong plumes in 1985 and 1987.
Hudson	Chile	8,580 (2,615)	1991	August 8 eruption melts glacier. Larger eruption on August 12–15 produces a cloud that travels completely around world in a week; locally, ash blankets about 25,000 sq mi (65,000 sq km) of Argentina; smaller eruption in October. Previous eruption in January 1973 causes 11 deaths from flooding by melting ice.
Láscar	Chile	18,346 (5,592)	1995	Starts February 28, 1994. Lascar has had more than 15 historic eruptions; largest ever was April 19-20, 1993.
Llaima	Chile	10,250 (3,125)	1994	Starts May 17, 1994. There have been more than 30 eruptions since 1640, the last previous being in 1957.
Lonquimay	Chile	9,400 (2,865)	1990	Thirteen months of eruptive activity ends in January.
Marchena	Galápagos Islands, Ecuador	low island	1991	September 25.
Nevado del Huila	Colombia	17,274 (5,265)	1555	On June 6, 1994, mudflows from volcano triggered by an earthquake kill hundreds, although volcano itself does not erupt.

Volcano	Location	Height in ft (m) above sea level	Last reported eruption	Remarks
Nevados Ojos del Salado	Chile	22,590 (6,887)	dormant	World's tallest active volcano, although technically not "active," since there are no known historical eruptions.
Planchón-Peteroa	Chile	Planchón: 13,048 (3,977) Peteroa: 13,455 (4,101)	1991	February 9 eruption kills large numbers of fish in Teno and Claro rivers.
Puracé	Colombia	15,604 (4,757)	1977	Has popular hot springs that bring many visitors.
Reventador	Ecuador	11,434 (3,485)	1976	January eruption produces lava flows over 4 mi (2.5 km) long.
Ruiz	Colombia	17,720 (5,401)	1991	November 13 eruption produces mud slides that kill 25,000.
Sabancaya	Peru	19,577 (5,967)	1995	Eruption starts May 28, 1990, and kills at least 20 when earthquakes cause landslides on July 23-24, 1991. Last previous known eruption in July 1784.
Sangay	Ecuador	17,159 (5,230)	1976	
San José	Chile	19,127 (5,830)	1895	Erupts four times in nineteenth century.
Shoshuenco	Chile	7,743 (2,360)	1960	
Tinguiririca	Chile		1994	About January 15; phreatic eruption only. Previous small event was 1917.
Tupungatito	Chile	18,504 (5,640)	1986	
Villarrica	Chile	9,338 (2,847)	1985	October 1984 to November 1985. Has erupted more than 50 times since 1558, with four eruptions resulting in fatalities from mud flows.
Yanteles	Chile		1835	

(Periodical References and Additional Reading: *Bulletin of the Global Volcanism Network* 1992–1995; *Science* 5-13-94, p 948; *Science News* 10-1-94, p 213; *New York Times* 12-26-94, p 3; *Science* 1-13-95, p 256; *Scientific American* 3-95, p 30)

RECORD BREAKERS FROM THE EARTH SCIENCES

The Earth is thought to be about 4.6 billion years old.

BY AGE OF ENTITY

Record	Holder	Age in Years
Oldest minerals	Zircons from Australia	4.3 billion
Oldest rocks	Granite from northwest Canada found in 1989 by Samuel Bowring of Washington University in St. Louis, Missouri	3.96 billion
Oldest fossils	Single-celled blue-green algae or bacteria from Australia	3.5 billion
Oldest slime molds (bacteria)	Slime molds, also called slime bacteria, are one-celled creatures that gather together to form multicellular bodies for reproduction	2.5 billion
Oldest true alga and oldest eukaryote	Alga, probably *Grypania,* found in banded iron formation near Marquette, Michigan, by Tsu-Ming Han and Bruce Runnegar in 1992	2.1 billion
Oldest petroleum	Oil from northern Australia	1.4 billion
Oldest life on land	Blue-green algae (cyanobacteria), traces of which were found in Arizona by Robert J. Horodyski and L. Paul Knauth in 1985	1.2 billion
Oldest multicelled organism	Small creatures resembling jellyfish, found as fossils in Mexico by a team led by Mark McMenamin of Mount Holyhoke College in Massachusetts	590 million
Oldest fish	*Sacabambasis,* found in Bolivia	470 million
Oldest land plant	Moss or algae	425 million
Oldest land animals	Trigonotarbids and centipedes found in the Ludlow Bone Bed at Ludford Lane, Ludlow, Shropshire, England	414 million
Oldest upright land plant	*Cooksonia pertoni,* a forerunner of horsetails and ferns	400 million
Oldest insect	Bristletail (a relative of the modern silverfish)	390 million
Oldest spider	Known only from a spinneret used to make silk, found in Gilboa, New York	380 million
Oldest tetrapod (ancestor of amphibians that walked on land)	Known only from scattered fossils collected in Scotland before 1950, and thought until 1991 to be fish	370 million

Record	Holder	Age in Years
Oldest reptile	"Lizzie the Lizard" (*Westlothiana curryi*), found in Scotland in 1988 by Stan Wood	340 million
Oldest dinosaur	*Herrerasaurus,* found in Argentina by Osvaldo Reig in 1959	228 million
Oldest dinosaur tracks	Those of a turkey-sized ornithiscian found in New Mexico by Russell Dubiel of the U.S. Geological Survey in 1983	225 million
Oldest frog	*Prosalirus bitis* found in 1995 by Neil Shubin of the University of Pennsylvania and Farish Jenkins of Harvard University in Arizona	190 million
Oldest bird	Despite other candidates, the winner by popular acclaim remains *Archaeopteryx lithographica*	147 million
Oldest flower	An unnamed tiny dicot similar to a pepper plant found in Koonwarra, Australia	110 million
Oldest mushroom	*Coprinites dominicana,* found in Haitian amber from the La Toca mine	35 to 40 million

BY SIZE OF ENTITY

Record	Holder	Size
Longest dinosaur	*Seismosaurus,* who lived about 150 million years ago in what is now New Mexico	About 150 ft (45 m) long
Heaviest dinosaur	*Argentinosaurus,* a sauropod who lived in South America about 130 million years ago	About 90 sht tons (more than 80 m tons) and perhaps as many as 110 sht tons (100 m tons)
Largest carnivorous dinosaur	*Giganotosaurus carolinii,* whose fossil remains were discovered in Argentina by Ruben Carolini in 1995	About 8 sht tons (about 7 m tons)
Largest land animal skull	The skull of *Chasmosaurus,* found by University of Chicago senior Tom Evans in Big Bend National Park in Texas in March 1991	5–6 ft (1.5–1.8 m) long
Largest land mammal	*Indricotherium,* also known as *Baluchitherium,* an ancestor of the rhinoceros	About 18 ft (5.5 m) high at shoulder; weighed five times as much as largest elephant

Record	Holder	Size
Largest exposed rock	Mt. Augustus in western Australia	At 1,237 ft (377 m) high, 5 mi (8 km) long, and 2 mi (3 km) wide, it is twice the size of the more famous Ayers Rock
Largest island	Greenland (also known as Kalaallit Nunnaat)	About 840,000 sq mi (2,175,000 sq km)
Largest ocean	Pacific	64,186,300 sq mi (166,236,100 sq m)
Deepest part of ocean	The *Challenger* Deep in the Marianas trench in the Pacific Ocean	35,640 ft (11,000 m), or 6.85 mi (11 km)
Greatest tide	The Bay of Fundy between Maine and New Brunswick	47.5 ft (15 m) between high and low tides
Greatest flood	In Altai Mountains of southern Siberia in Chuja River valley about 12,000 B.C.	Water about 1,500 ft (500 m) deep traveling at 90 mph (150 kph)
Highest wind in a tornado	Near Red Rock, Oklahoma, on April 26, 1991	Between 280–290 mph (450–465 kph) as measured by Doppler radar
Coldest temperature	Verkhoyansk, Russia in 1887	−90.04°F (−67.8°C)
Largest geyser	Steamboat Geyser in Yellowstone Park	Shoots mud and rocks about 1,000 ft (300 m) feet in the air
Longest glacier	Lambert Glacier in Antarctica (upper section known as the Mellor Glacier)	At least 250 mi (400 km) long
Longest known cave	Mammoth Cave, which is connected to the Flint Ridge Cave system	Total mapped passageway of over 340 mi (550 km)
Largest single cave chamber	Sarawak Chaper, Lobang Nasip Bagus, on Sarawak, Indonesia	2,300 ft (700 m) long, 900 ft (300 m) wide on average, 230 ft (70 m) or more high
Deepest cave	Réseau Jean Bernard in France	5,256 ft (1,602 m)
Highest volcano	Cerro Aconagua in the Argentine Andes (extinct)	22,834 ft (6,960 m) high
Most abundant mineral	Magnesium silicate perovskite	About two-thirds of the planet
Highest artificial pressure by a diamond anvil	Arthur L. Ruoff and coworkers at Cornell University	4.16 megabars
Deepest hole in oceanic crust	Hole 504B in Pacific, 500 mi (800 km) west of Panama	1.2 mi (2 km)

Record	Holder	Size
Deepest gorge (a gorge is a cut in the surface narrower than its depth)	Great Gorge on Mount Denali (Mount McKinley) in Alaska, found in 1992 by Keith Echelmeyer of the University of Alaska	9,000 ft (2,750 m) and twice as deep as it is wide
Largest canyon on land	Grand Canyon of the Colorado, in northern Arizona	Over 217 mi (350 km) long, from 4–13 mi (6.5–31 km) wide, as much as 5,300 ft (1,615 m) deep
Deepest continental canyon	El Cañon de Colca in Peru	10,574 ft (3,223 m)

(Periodical References and Additional Reading: *New York Times* 7-14-92, p C5; *New York Times* 8-2-92, p I-30; *Discover* 1-94, p 36; *New York Times* 9-3-94, p2; *New York Times* 6-7-94, p C6; *Science* 12-16-94, p 1805; *The Sciences* 3/4-95, p 26; *New York Times* 4-4-95, p C12; *Natural History* 6-95, p 41; *Discover* 1-96, pp 44, 48, & 50)

Physics

THE STATE OF PHYSICS

While modern chemists think that their study is the "science of everything," the Theory of Everything is a part of physics, albeit an unrealized part. The goal of some physicists is to explain matter and energy using a few basic principles, somewhat the way Euclid arranged the geometry of his time on the basis of five postulates. If that goal is reached, it would be the Theory of Everything. Some think such a theory is close to being realized, but others point out the difficulties involved. In any case, it would be a mistake to think that more than a few physicists are actively seeking such a theory. Furthermore, if the theory were found tomorrow, the chances are that it would consist of a mathematical construct called a symmetry group, from which one could calculate (at the most) the properties of some two or three dozen subatomic particles. It would be a great intellectual achievement, but it is difficult to see how it would help physicists solve real problems very well.

Physics is a discipline in which getting down to basics is important. These basics differ, however, among the different kinds of physicists, each of whom may be devoted entirely to a small part of physics: optics, acoustics, rheology (the deformation and flow of matter), crystallography, astrophysics, the vacuum, geophysics, particle physics, nuclear physics, cosmology, or materials science. Several topics of considerable concern to physicists have been dealt with earlier in this book, including astrophysics and cosmology in the **Astronomy** section, and geophysics in **Earth Science.** The remainder will be handled here.

The four recognized big divisions of science—biology, chemistry, earth science, and physics—are each separated into multiple smaller subdisciplines. Furthermore, each also has at least one great divide of its own: biology into whole-organism biologists and parts-biologists; chemistry into pure and applied disciplines; earth science into geology and everything else. Physics has the deepest split, into experimental and theoretical physics. The big stories in physics, the ones that leap out of science journals and onto the front pages of journals, are almost always experimental. New theories tend to take time to sink in, and therefore usually are not news until a theorist wins a Nobel Prize. The early 1990s had several front-page experiments. Physicists established experimentally a previously undetected particle, the top quark (*see* "Particle Physics UPDATE," p 633), and they created two new forms of matter (*see* "The Newest Forms of Matter," p 645). The newest theories, however, failed not only to make front pages, but even failed to win Nobel Prizes. Indeed, there has not been a Nobel Prize for a significant new theory in physics since the end of the 1970s, which might be taken as an indication of how far away physics really is from illuminating the Theory of Everything.

Almost the Theory of Almost Everything

Physicists generally recognize four forces in nature, although some prefer to call these forces "interactions" for a variety of reasons.

The most familiar force (or interaction) is gravity. In general relativity theory, which is the main theory of gravity today, gravity is not actually a force, since it is a side-effect of curved space-time. Instead, it acts like a force between any two entities that have mass, as well as between some, such as photons, that have no mass in the ordinary sense of the word.

General relativity theory, although sometimes criticized, is an excellent theory of everything about gravity. In 1919, Arthur Stanley Eddington made relativity theory famous by performing an experiment showing that relativity is better than Newton's theory of gravity in predicting how much the Sun's gravity will bend a beam of light from a star. In 1995, astronomers from Harvard University, M.I.T., and the Haystack Observatory made an extremely accurate measurement using essentially the same experimental setup. They measured how much the Sun's gravity bends a radio wave from a quasar. If the predicted value from relativity theory is set to 1, the astronomers obtained a value of 0.9996 ± 0.0017. Relativity remains an excellent theory.

Another important measure of gravity was also redone in 1995, although with less satisfactory results. Because gravity is the weakest of the four fundamental interactions, it is not easy to measure exactly. In 1798, Henry Cavendish was the first to measure G, which is the constant of proportionality in Newton's law that gravitational force between two bodies is directly proportional to the product of the masses of the bodies and inversely proportional to the square of the distance between them. In 1995, two groups used versions of the same experimental setup as Cavendish had, which was two masses suspended from a torsion balance, and a third group used a pair of pendulums to remeasure G. Not only did they all obtain constants that differ from each other, but all three differ from the currently accepted value. The results varied from $6.6656 (6) \times 10^{-11}$ to 6.71540×10^{-11}, on both sides of the currently accepted $6.67259(85) \times 10^{-11}$. Further work will be done to resolve the differences.

The second recognized force is charge, which accounts for electricity. Careful physicists, however, deny that electricity exists as an actual entity, stating instead that there is an electric *current* produced when particles with charge move from one place to another. Furthermore, charge produces observable forces of either attraction or repulsion for particles that possess it, including the electron and the quarks (*see* "Particle Physics UPDATE," p 633, and "Subatomic Particles," p 666). The electron is also subject to the third force, generally known as the weak interaction since it does not very closely resemble a force. The weak interaction and charge are both considered manifestations of a deeper force called the electroweak interaction. The electroweak interaction can only be observed at very high energies, since at lower ones it splits into charge and the weak interaction.

The fourth generally recognized interaction is referred to as the strong force, because for particles existing very close to each other it is much stronger than the other three forces. In the 1930s, Hideki Yukawa developed a theory explaining the strong force on the basis of the interchange of then unknown particles between particles that feel this force, which at that time were only known to include the proton and neutron. The postulated particles, which we now call pions, were found in the 1940s and would seem to explain the strong force. Still, when quark theory was developed, it became apparent that the strong force had to operate between quarks and not between protons and neutrons, which are made from quarks. Pions are also made from quarks. So the strong force was replaced by another, called the color force, which is caused by exchange of undetected (and theoretically undetectable) particles called gluons. Although quark-gluon theory and the color force are advances, the older idea of protons, neutrons, pions, and the strong force is much easier to understand and is still used in calculations by nuclear physicists.

The closest that physicists have gotten to a true Theory of Everything is the development of some mathematical ideas that would unify the electroweak and the strong force, although none of these have been completely accepted yet. Physicists are even further away from a theory unifying these forces with gravity (*see* "Strings and Things UPDATE," p 652).

A Really Good Theory of Nothing

The standard account of forces presented above also fails to take into account other forces that are just as real (or perhaps more real) than the fundamental four. One reason is that some of these forces can, it is agreed, be explained in terms of the fundamental four, just as the strong force is explained in terms of the color force. However, even taking that idea into account, there remain forces that are unrelated to the fundamental four in any real way. Among these are the forces of the vacuum. If two conductive plates are placed parallel to each other at a distance that is relatively close to the wavelength of moderately energetic electromagnetic waves, the energy fluctuations of the vacuum will produce forces. For example, there will not be enough room between the plates for low-energy electromagnetic waves, which have a wavelength too long to fit in the space, so longer waves cannot create themselves between the plates; only short wavelength waves are able to create themselves there, but short waves have high energies and cannot last very long. Outside the plates, there are no restrictions, so both long and short wavelength waves are briefly in existence. With more waves outside the plates than between them, the outer waves create a force that tends to push the plates together. Such a force was measured experimentally long ago.

In 1948, Hendrick Casimir and Dik Polder predicted that the vacuum would create another force. None of the longer created waves can be very close to either one of the plates because there is not room. Very short waves have such high energy that very few create themselves. Thus, the vacuum exerts another force from the middle of the plate toward the edges. This Casimir-Polder force would cause an atom traveling between the plates to veer toward one of the plates and eventually crash, since the lowest energy region is immediately adjacent to the plates. In February 1993, Ed Hinds and Charles Sukenik and coworkers at Yale University in New Haven, Connecticut, became the first to measure the Casimir-Polder force directly, establishing that the theoretical force actually exists. Notice that neither of the forces that results from the vacuum is among the four fundamental forces referred to in physics textbooks.

THE BUSY VACUUM

Aristotle wrote that the vacuum (a total absence of matter) could not exist, and this conclusion was generally accepted throughout the Middle Ages. Torricelli, however, was able to produce a vacuum in 1643, and a great deal of excitement followed. Production of vacuums and their application became one of the main demonstrations and tools of early science. Better vacuums led to new discoveries right up to the time of William Crookes, whose method of producing excellent vacuum tubes made possible the discovery of the electron in 1897 by J. J. Thompson.

It is easy to trace the development of the modern physics of quantum mechanics and relativity theory directly from the discovery of the electron. There have been many surprises. Over the course of slightly more than 30 years, quantum theory found that Aristotle had been partially correct after all. It is not so much that the vacuum cannot exist, however, as it is that the word "vacuum" no longer means a region totally devoid of matter in the way it did between 1643 and 1930.

Although the quantum vacuum is empty space, it is never truly empty (*see also* "Quantum Stuff," p 649). There is always a finite probability that any point in empty space will be suddenly occupied with matter, because a particle can rise out of the vacuum like a catfish jumping in a pond in summer. We don't get much of a chance to observe either the creation of a particle or the jumping catfish, because both return to their vacuum or pond very quickly. Nevertheless, both events happen, and they (or their effects) can be observed. For example, you can hear the catfish land in the water and see the ripples, even though you usually do not see the actual fish.

Similarly, an electron-positron pair can appear briefly and then annihilate each other. This can be viewed as a random fluctuation of energy that is permitted to break the laws of conservation of energy because it is so brief. In fact, it is like successfully writing a check. You know that on Tuesday you can write a check at the grocery store, even when you have no money in your bank account, provided you deposit money in the account on Wednesday. Similarly the electron-positron pair can violate the law of conservation of mass-energy, as long as it makes the energy up soon enough that nobody finds out.

Note that you have to look fast to observe the particle. Over a moderately long amount of time, such as a tenth of a second, the particle is not there. If, however, you reduce the time of observation to a very short period of time, you are almost always sure to find a particle. In both quantum theory and life in general, one should expect paradoxes; that is, don't be surprised that events are always surprising.

Not only is there a small probability that such particles will appear, there is strong evidence that they always appear. Forces produced by the particles can be detected in experiments, just as the ripples can be seen and the sound heard when the unseen catfish jumps. The quantum vacuum is seething with energy, and energy, as Einstein discovered, is the same as matter. Thus, the modern vacuum is not a vacuum at all in the old sense of the word. There is no empty vacuum, and Aristotle is vindicated.

Unresolved Questions of Physics

Despite the quest for a Theory of Everything, there continue to exist many unresolved issues in physics:

• Where does the proton get its spin? Since protons are made from three quarks, one might think that it derives its spin ½ from the two ups and one down that form it, each also with a spin ½. Experiments, however, show that this is not the case, and theorists are hard pressed to understand the experimental results. This remains a major area of investigation in basic particle physics.

• Exactly how big is the proton? Different experiments give different results. One experimental measurement of the proton's diameter is a size that could be predicted by quantum electrodynamics, but another measured size is outside the predicted range. New measurements appear to confirm the second result, which might mean that a new theory is needed.

• Why do particles of sand stick together to form "fingers" that move together when a single layer of sand in a pan is shaken so that the particles move at random?

• There are 21 different numbers measured by particle physicists that the standard model fails to predict; among these are the rest mass of the electron and the size of the unit of charge. What theory accounts for these parameters?

(Periodical References and Additional Reading: *Discover* 1-94, p 102; *Scientific American* 9-94, p 40; *Science* 6-2-95, p 1277; *Science News* 4-29-95, p 263; *Physics Today* 9-95, p 24; *Physics Today* 10-95, p 9; *Physics Today* 11-95, p 9)

TIMETABLE OF PHYSICS TO 1992

1586	Simon Stevinus [Belgian/Dutch: 1548-1620] shows that two different weights released at the same moment will reach the ground at the same time.
1604	Galileo [Italian: 1564-1642] discovers that a body falling freely will increase its distance as the square of the elapsed time of the fall.
	Johannes Kepler [German: 1571-1630] shows that light diminishes as the square of the distance from the source.
1663	Pascal's law by Blaise Pascal [French: 1623-1662] is published posthumously. (It was probably discovered around 1648.) The law states that pressure in a fluid is transmitted equally in all directions.
1675	Ole Römer [Danish: 1644-1710] measures the speed of light.
1676	Robert Hooke [English: 1635-1703] formulates Hooke's law: the amount a spring stretches varies directly with its tension.
1678	Christiaan Huygens [Dutch: 1629-1695] develops the wave theory of light.
1687	The *Principia* of Isaac Newton [English: 1642-1727] is published; it contains Newton's laws of motion and the theory of gravity.
1746	At least two experimenters in Holland invent a method for storing static electricity. The device becomes known as the Leyden jar, after the location of its invention.
1752	Benjamin Franklin [American: 1706-1790] performs his kite experiment, demonstrating that lightning is a form of electricity.

1787	Jacques Charles [French: 1746-1823] discovers Charles' law, which states that all gases expand the same amount with a given rise in temperature.
1791	Luigi Galvani [Italian: 1737-1798] announces his discovery that two different metals that touch each other produce an effect similar to that of an electric charge. Later this will be recognized as electric current as opposed to static electricity.
1798	Count Rumford [American/British/German/French: 1753-1814] shows that heat is a form of motion.
	Henry Cavendish [English: 1731-1810] determines the gravitational constant and the mass of Earth.
1799	Alessandro Volta [Italian: 1745-1827] invents the electric battery.
1800	William Herschel [German/English: 1738-1822] describes his discovery of infrared light.
1801	Johann Ritter [German: 1776-1810] discovers ultraviolet light.
1802	Thomas Young [English: 1773-1829] develops his wave theory of light.
1819	Hans Christian Oersted [Danish: 1777-1851] discovers that magnetism and electricity are two different manifestations of the same force (not published until 1820).
1820	André-Marie Ampère [French: 1775-1836] formulates the first laws of electromagnetism.
1830	Michael Faraday [English: 1791-1867] in England and Joseph Henry [American: 1797-1878] in the United States independently discover the principle of the electrical dynamo.
1842	Julius Robert Mayer [German: 1814-1878] is the first scientist to state the law of conservation of energy.
1850	Rudolf Clausius [German: 1822-1888] makes the first clear statement of the second law of thermodynamics: energy in a closed system tends to degrade into heat.
1851	William Thompson, later Lord Kelvin [Scottish: 1824-1907], proposes the concept of absolute zero, the lowest theoretically possible temperature.
1873	James Clerk Maxwell [Scottish: 1831-1879] publishes the complete theory of electromagnetism, which includes his prediction of radio waves.
1887	Albert Michelson [German/American: 1852-1931] and Edward Morley [American: 1838-1923] attempt to measure changes in the velocity of light produced by the motion of Earth through space. The negative result of the Michelson-Morley experiment is later interpreted as helping to establish Einstein's special theory of relativity.
1888	Heinrich Hertz [German: 1857-1894] produces and detects radio waves.
1895	Wilhelm Konrad Roentgen [German: 1845-1923] discovers x rays.
1896	Antoine-Henri Becquerel [French: 1852-1908] discovers natural radioactivity.
1897	Joseph John Thomson [English: 1856-1940] discovers the electron.
1900	Max Planck [German: 1858-1947] explains the behavior of light by proposing the quantum, the smallest step a physical process can take.
1905	Albert Einstein [German/Swiss/American: 1879-1955] shows that the

photoelectric effect can be explained if light has a particle nature as well as a wave nature.

Einstein shows that the motion of small particles in a liquid (Brownian motion) can be explained by assuming that the liquid is made of molecules moving at random.

Einstein develops his theory of relativity and the law $E = mc^2$ (energy equals mass times the square of the speed of light).

1911	Heike Kamerlingh Onnes [Dutch: 1853–1926] discovers superconductivity.
1914	Ernest Rutherford [New Zealander/Canadian/English: 1871–1937] discovers the proton.
1915	Einstein completes his general theory of relativity, a theory of gravity more accurate than that of Sir Isaac Newton.
1919	An expedition led by Arthur Eddington [English: 1882–1944] confirms that Einstein's theory of gravity is more accurate than Newton's in predicting the effect of gravity on light.
1923	Louis-Victor de Broglie [French: 1892–1987] theorizes that particles, such as the electron, also have a wave nature.
1925	Wolfgang Pauli [Austrian/American: 1900–1958] discovers the exclusion principle.
	Werner Heisenberg [German: 1901–1976] develops the matrix version of quantum mechanics.
1926	Erwin Schrödinger [Austrian: 1887–1961] develops the wave version of quantum mechanics.
1927	Heisenberg develops his uncertainty principle.
1932	James Chadwick [English: 1891–1974] discovers the neutron.
	Carl Anderson [American: 1905–1991] discovers the positron.
	John Cockcroft [English: 1897–1967] and Ernest Walton [Irish: 1903–] develop the first particle accelerator.
1937	Anderson discovers the muon.
1938	Otto Hahn [German: 1879–1968] splits the uranium atom, opening the way for nuclear bombs and nuclear power.
1947	Quantum electrodynamics (QED) is born from the labor of many parents, notably Richard Feynman [American: 1918–1988], Julian Schwinger [American: 1918–1995], Shin'ichiro Tomonaga [Japanese: 1906–1979], Willis Lamb Jr. [American: 1913–] (all of whom received Nobel Prizes), and Hans Bethe [German/American: 1906–] (whose Nobel Prize was for other work).
	Cecil Powell [English: 1903–1969] and coworkers discover the pion.
1952	A group led by Edward Teller [Hungarian/American: 1908–] develops the hydrogen bomb.
1955	Owen Chamberlain [American: 1920–] and Emilio Segrè [Italian/American: 1905–1989] produce the first known antiprotons.
	Clyde Cowan Jr. [American: 1919–] and Frederick Reines [American: 1918–] are the first to observe neutrinos.
1957	Experiments performed by a group led by Chien-Shiung Wu [Chinese/

American: 1912-], and quickly confirmed by others, show that the law of conservation of parity does not hold for the weak interaction.

John Bardeen [American: 1908-1991], Leon Cooper [American: 1930-], and John Schrieffer [American: 1931-] develop a theory that explains superconductivity.

1961   Murray Gell-Mann [American: 1929-] and, independently, Yu'val Ne'eman [Israeli: 1925-] and others develop a method of classifying heavy subatomic particles that comes to be known as the eightfold way.

1964   Gell-Mann introduces the concept of quarks as components of heavy subatomic particles such as protons and neutrons.

1967   Steven Weinberg [American: 1933-], Abdus Salam [Pakistani/English: 1926-], and Sheldon Glashow [American: 1932-] independently develop a theory that combines the electromagnetic force with the weak force.

1986   Alex Müller [Swiss: 1927-] and Georg Bednorz [German: 1950-] discover the first warm-temperature superconductor.

Nova, an experimental device at the Lawrence Livermore National Laboratory, achieves nuclear fusion by using a laser to heat hydrogen pellets.

1990   Following a trend that has seen almost all crafted basic measurement standards replaced by standards derived from physical phenomena, the volt and ohm are redefined in terms of the behavior of atoms and electrons. Previously, such measures as length and time had been defined on the basis of natural phenomena. The most resistant to such an approach continues to be mass, still defined in terms of a standard cast block of metal.

1991   Stuart Shapiro and Saul Teukolsky demonstrate that it is possible to receive information from a black hole even though electromagnetic radiation cannot escape it.

PARTICLE AND NUCLEAR PHYSICS

Particle Physics UPDATE

In one sense, particle physics reached an important point of closure in 1994. According to the standard model of particle physics, the last predicted particle of ordinary matter was found, and its measure taken. At the same time, there were many different notions of how to extend the standard model of particle physics (*see* "Strings and Things UPDATE," p 652). Each alternative model arrived accompanied by a whole host of described particles known only to theory. In that sense, particle physics could not be more wide open. Even if none of the weird particles of the alternative models actually exists, particles of matter and antimatter are not the whole story in any case. In addition to the two forms of matter, the universe is filled with forces of one kind or another, and in all models these forces are also particles. Just as quantum theory has particles as waves and waves as particles, so all of the fundamental forces are represented by particles in particle physics. Some of the particles connected with these forces are yet to be detected.

In the 1990s, particle physicists tended to a somewhat bleak view of what was going on in their branch of the profession. This pessimism stemmed from the fact that accomplishing what they deemed the next task seemed to require the construction of larger and therefore more expensive particle accelerators. Only national governments have enough money to build such devices, and national governments were beginning to think that they would be better off spending their money elsewhere (*see* "The End of Big Science in the United States?" p 13, for the demise of the Superconducting Supercollider).

The Top Quark at Last

Nevertheless, the top media story in physics in the early 1990s was the top quark, once also known as "truth" (*see* "The Newest Forms of Matter," p 645, for what was probably the true lead story). The search for this elusive particle ran from the end of the 1970s well into the 1990s. In 1985, scientists from CERN announced that they had succeeded in producing the top quark, but it was not to be. A preliminary rumor of success in May 1992 from Fermilab, denied by Fermilab physicists at that time, was indeed premature. Solid evidence for the top quark began to pile up at Fermilab in 1994. By 1995, it was clear that the top quark had been definitely produced, and an announcement was made on March 2 that Fermilab had done the trick.

The main obstacle to this success had not been the hidden nature of the top quark. All six quarks (*see* "Subatomic Particles," p 666) are hidden from direct view by the laws of physics. The real difficulty concerned the mass of the top quark—the amount of energy bound up in the particle. The equivalence of mass and energy that everyone knows as $E = mc^2$ becomes extremely important for subatomic particles. Even in a vacuum, particles create themselves by stealing a little energy for a few trillionths of a second, after which they give it back. When real energy is provided without a need for it to be returned, particles will create themselves for much longer. The way that particle physicists look for new particles is by providing the appropriate amount of energy, most often as a result of a crash between two existing particles. If one particle is the antiparticle of the other, the collision will result in pure energy that is the total of the both energies combined. This energy will then reform into particles. There is considerable chance involved as to what the new particles will be. Sometimes the original particles can reform, but more often a group of different particles, together possessing the same energy as the originals, appear. The probability of a single very massive particle appearing is low, although not zero.

In the beginning, the top quark was thought to be less massive than it has subsequently emerged as being. Therefore, the quark was initially sought among the collision debris of smaller and less energetic particles than were needed. When particle physicists did not find it, they raised their estimate of the least possible mass that the top quark could possess. When the quark was finally found in collisions of protons and antiprotons at Fermilab, just outside Chicago, Illinois, it weighed in at 200 times the mass of a single proton, or 100 times the amount of energy provided by the masses of proton-antiproton pair that had been collided to produce the top quark. A common comparison states that the top quark weighs nearly as much as an atom of gold. To make it even more difficult, production of a top quark by this method only works when one makes a top-antitop quark pair, which results in a need to double the amount of energy again. The rest of the energy—beyond the masses of the two colliding protons—has to come from the energy of motion; that is, each particle has to accelerate to a speed close to the speed of light to produce that much energy.

That did not solve the entire problem, however. The top quark is far from stable, decaying into a shower of less massive particles in a trillionth of a trillionth of a second. The lifetime is far too short for any kind of detector to register the top quark directly, although some of the smaller particles can be uniquely recognized. The problem, however, is that these particles could have arisen in some other way without the presumed top quark having been there at all.

The laws of particle probability have to be invoked at this point. At Fermilab, the physicists examined the results of trillions of proton-antiproton collisions between 1991 and 1994. They found a dozen that could have involved a top quark. While this is not a large number, it is twice as many as probability would allow without the top quark. So the probability is that the top quark was produced six times during those years, although the precise six occasions cannot be identified.

When this somewhat ambiguous result was first announced, there were physicists who doubted that the deed had finally been accomplished. Further study, however, convinced the doubters. By mid-1995, the two separate detection teams working at Fermilab had about 75 top quarks identified between them. Furthermore, the teams both obtained the same characteristics, even though they were using two different approaches for identification of the particle.

Searching for Higgs

One of the main aims of the cancelled Superconducting Supercollider was to have been to seek a particular large particle known as the Higgs particle. As the Europeans proceed with their plans for a giant synchrotron at CERN, the announced goal of that project is also to find the Higgs particle. Such large accelerators are needed, because the Higgs particle is expected to weigh in at more than 100 giga-electron-volts (GeV) and less than 1,000 GeV, with 600 GeV a common estimate. (The elusive top quark is only about 190 GeV.)

Proving the existence of the Higgs particle seems to be an essential step toward a continued validation of the standard model of particle physics. While such validation would be pleasant, it would not contribute much to the advancement of knowledge. Not finding the Higgs particle, or finding something quite different, would be far more interesting.

There are physicists, however, who seriously doubt the existence of the Higgs particle. For one thing, it has been calculated as having 0 spin, unlike any known fundamental particle except for some versions of the graviton, which is also an unobserved particle.

The Higgs is needed, however, to account for the existence of mass, which is otherwise unexplained. The analogy is with charge, also a mysterious quantity. The problem with mass is that it cannot be calculated from first principles but only measured. The problem with charge is similar, although there are only multiples of one-third to consider at this time. In the standard theory, charge arises from an

---

Background

When the structure of atoms was first envisioned, atoms were thought to be made of two components: the negatively charged electrons and a positive component that balanced out the charge, thus allowing atoms to be electrically neutral. Electrons are easy to spot because they can rub off (producing static electricity), steam off when substances are heated (the Edison effect), or be produced continuously by various elements (the beta decay of many radioactive elements). The positive charge was harder to identify and more complicated to understand. Historians generally credit Ernest Rutherford [New Zealand/UK: 1871–1937] with the 1914 discovery of the proton, the particle of positive charge. Yet he did not produce individual protons until 1919, 22 years after recognition of the electron. It was another 13 years before a scientist demonstrated the existence of another particle in the atom, the neutron, although some scientists predicted neutrons soon after isolation of the proton.

All atoms except the simplest form of hydrogen consist of a core made up of shells of protons and neutrons surrounded by shells of electrons. Atoms cannot have cores consisting only of protons, because the force that holds the core together (the strong force) depends upon the exchange of particles between neutrons and protons. A nucleus of only protons would not work because the particle exchange of the strong force changes protons to neutrons and vice-versa (most of the time). Neutrons also need to be changed into protons to stay stable.

electric field, a field that has particles associated with it. According to a 1964 theory put forth by Peter Higgs, if mass arises from a field, the field should also have particles. Although one usually speaks of *the* Higgs particle, in some versions of the theory there may be several Higgs particles involved with the field, all of which serve to produce mass.

One may think that gravity produces mass, which would make the Higgs particle unnecessary. However, although mass is affected by gravity, mass exists separately from gravity. In relation to other forces, mass can be recognized as what we often call inertia, for example.

Furthermore, even if the Higgs accounts for the origin of mass, its existence (if it does exist) does not explain why different particles possess their specific masses. Some other theory, with no doubt some other undiscovered particle, is needed for that. Nevertheless, Europe's Large Hadron Collider, which some physicists calculate has one chance in three of making a Higgs, is now going to be put on the case (*see* "Advanced Accelerators," p 657).

Neutrino Mass and Identity

Even before the discovery of neutrons, scientists recognizing that in certain atoms that spontaneously decay by emitting electrons, the energy lost and the change in spin did not make sense in terms of what was then known. We now know that in some atoms a neutron decays spontaneously into a proton, emitting an electron to achieve the change in charge. This is termed beta decay. The electron accounts only for the charge, however. An electron does not carry away enough energy to produce a particle possessing the mass of a proton. For this, more than two electrons would be needed. At the same time, the electron carries away too much spin—all of it, in fact.

In 1930, Wolfgang Pauli [Austrian/American: 1900-1958] proposed that there must be another particle involved in beta decay, a particle possessing the opposite spin and carrying small amounts of energy. The new particle would have no mass, since there is no inexplicable loss of mass in the transition from the neutron to the proton. This idea came into acceptance around 1933, when Enrico Fermi [Italian/American: 1901-1954] worked out the details of what has come to be known as the weak interaction, coining the name *neutrino* for the Pauli particle in the process. The following year, Pauli and Victor F. Weisskopf quietly proved that particles with spin ½ have near twins called antiparticles. Later, the Pauli particle was classed as an *antineutrino* to fit into this scheme.

A neutrino or its antineutrino twin (which has an opposite spin) is essentially spin and energy. It is rather like a dust devil, traveling at about the speed of light. The spin of an electron, a particle that can travel at any speed, can be reversed by stopping the electron and turning it around, just as you could change the spin of a top from clockwise to counterclockwise by turning it upside down. A neutrino, however, is always found with the same spin, and an antineutrino with the opposite spin. This implies that they cannot be stopped, which many physicists in the past assumed meant that they always traveled at exactly the speed of light. In turn, applying Einstein's theory of special relativity, this would mean that neutrinos have no mass.

More recently, particle physicists have discovered that particles and their antiparticles can come in mixed states. This idea is hard to picture when one thinks of particles as particles but is easier to understand when one thinks of particles as waves. Mixed states can impart mass to an otherwise massless particle. Using this idea, some theoretical physicists have suggested that the neutrino might have a small mass. Increasingly sophisticated experiments have ruled out any mass greater than 9 eV, however. A few holdouts still think that the mass of the ordinary neutrino and its antiparticle is exactly 0, as Pauli first indicated (Pauli cautiously said "vanishingly small").

Complicating this whole picture is the mystery of the three families of matter. For reasons that are not understood at all, in addition to ordinary matter there is a heavy family of matter, and an even heavier one. Furthermore, all three families are available in each of the two different kinds of stuff: the quark stuff that interacts with a strong interaction, and the lepton stuff that interacts with Fermi's weak interaction. The two heavier kinds of lepton stuff consist of the mu family (a muon, an antimuon, a mu neutrino, and a mu antineutrino) and the similarly constituted tau family. The three families are sometimes spoken of as three flavors, such that the mu neutrino would be termed "muon flavored." Particle accelerator experiments are said to prove that there is no fourth family.

A muon, discovered in 1937, is about 200 times as massive as an electron. There is no evidence that the mu neutrino has any measurable mass. Experiments in particle accelerators actually rule out much of a mass for the mu neutrino. The tauon is about 3,500 times as massive as an electron. It has only been known since 1975, and its properties are not fully investigated. Experiments also suggest that the tau neutrino has, at best, very little mass.

Two unrelated situations in astrophysics and cosmology have changed this fundamental picture of the three neutrinos. The experimental detection of neutrinos expected from fusion processes in the Sun is much lower than theory promises (*see* "Neutrino Astronomy Comes of Age," p 136), and the mass of the universe is lower than cosmologists think it should be (*see* "Dark Matters," p 226). There are various fixes possible for both of these shortfalls. In the case of the lack of solar neutrinos, the most popular idea has been a theory that, on the long trip from the Sun to Earth, some of the electron neutrinos become mu or tau neutrinos. The mu or tau neutrinos fail to be detected by neutrino telescopes, so there appear to be fewer solar neutrinos than would otherwise be expected. The cosmology problem can be solved if the neutrinos calculated to fill the universe possess some mass. Because so many neutrinos are thought to exist, even a very small mass per neutrino might solve the missing-mass problem.

These two problems and solutions are related by the observation that if one flavor of neutrino neutrino evolves into a different flavor as the particle travels about, then the two different flavors must have different masses. Suppose, then, that the mass of the electron neutrino is 0, as originally proposed. In that case, the masses of the other flavors of neutrino cannot be 0, and therefore both must be some different positive amount. Furthermore, if two of the three flavors of neutrino have a little bit of mass, then it seems likely that they all do. So, this line of reasoning goes, the shortage of neutrinos coming from the Sun might imply that there is enough mass to keep the universe from expanding endlessly. It is rather a lot to be obtained from a little.

In 1995, physicists of the Liquid Scintillator Neutrino Detector group at Los Alamos, New Mexico, released to the *New York Times* results of an experiment they had been conducting since 1993. The experiment consisted of a detector located at the end of an accelerator that produces a beam of protons and collides them to make mesons. Targets placed in front of the detector were intended to produce mu antineutrinos. Shielding was placed around the detector to exclude outside neutrinos, presumed to be electron antineutrinos. The reaction that the detector was looking for is called inverse beta decay. In this form of decay, instead of a neutron releasing an electron and an antineutrino to become a proton, an electron antineutrino and proton combine to produce a neutron and release an antielectron, or positron. This reaction requires an electron antineutrino to accomplish. If only mu antineutrinos are shot at the target, no reactions will take place, unless some of the mu antineutrinos turn into electron antineutrinos as they travel through the detector.

The detector consists of 167 tons of mineral oil combined with a gallon of a material that will flash a light when a high-energy antielectron or electron passes through it after such a reaction. This happens quite infrequently—only nine times in the many months that the Los Alamos group ran the experiment. However, if mu electrons fail to turn into electron neutrinos, the total number during that time would only have been two, so it appeared that seven of the flashes occurred as scientists caught an electron changing flavors.

Or did they? Although the experiment is shielded, some antineutrinos from the outside world enter anyway, mostly from the bottom where there is no shielding. Many physicists (including one who worked on the experiment) think that the signal is not strong enough to separate from the noise. A repeat of the experiments was run at Los Alamos in the period from August to November 1995. When analyzed, this run of experiments may help make the result more certain. The verdict on neutrino oscillation at this point is not proven, although circumstantial evidence continues to mount.

The same, then, is true for the neutrino mass. Although some physicists claim to have measured neutrino mass at greater than 0, they have still failed to convince everyone that they have pinned down the true number. The problem, as with any experimental result concerning neutrinos, is that the neutrino fails in most circumstances to interact with other matter at all. If the Los Alamos data is actually caused by neutrino oscillation, it would indicate that the difference between a mu neutrino and an electron neutrino would be about 2.4 electron volts, and that both neutrinos would possess masses between 0.5 and 5 electron volts. This mass represents less than a hundred-thousandth the mass of the electron, so it is not large. It could, however, be very significant if it is anything larger than 0.

(Periodical References and Additional Reading: *Discover* 1-93, pp 97 & 98; *New York Times* 4-26-94, p A1; *Discover* 1-95, p 45; *Science* 2-10-95, p 789; *Science* 3-10-95, p 1423; *Physics Today* 4-95, p 9; *Physics Today* 5-95, p 17; *Physics Today* 8-95, p 20; *Science* 9-1-95, p 1218)

FUSION PROGRESS AT LAST

Nuclear fusion remains a major although seemingly ever-receding hope for cheap and abundant energy. The principal attraction concerns the abundance of the fuel, ideally a mixture of atoms of heavy hydrogen (deuterium, in which a combination of one proton and one neutron called a deuteron replaces the single proton nucleus of ordinary hydrogen) and even heavier hydrogen (tritium, with two neutrons and a proton). Deuterium is found in ordinary water at a rate of about 1 oz (40 g) to the barrel. Tritium can be made relatively easily from the small amount of lithium found in the salt of sea water. Consequently, the top 1 yd (1 m) of the ocean would by itself contain enough fusion fuel to provide all the power needs of Earth for thousands of years, if fusion power were ever produced at all.

In the early 1990s, funding for the conventional form of fusion research in the United States waxed and waned, but progress in the field tended to occur just prior to congressional votes on appropriations bills, which helped. The Defense Department, which had been working on laser fusion as part of its weapons program, lifted its veil of secrecy in 1994. Plans were made for new non-defense facilities using both conventional and laser fusion. In 1995, the U.S. National Research Council issued a report saying that plasma physics should devote more time and money to basic research, and less to attempts aimed at achieving practical nuclear fusion devices.

Through all the trials and tribulations, predictions are issued from time to time as to when one or the other of the fusion methods will reach a point where it can be a source of energy. Although the predictions vary, none of the dates promised are any earlier than well into the next millennium.

Conventional Progress, But No Breakeven Yet

The conventional way to achieve fusion is to put very fast light atomic nuclei into very close confinement, using some form of magnetic field as the means of containment. The heat involved in such fast moving atomic nuclei is such that ordinary matter containers would melt or break. Magnetic fields, however, are not affected by heat. Until the early 1990s, the light atomic nuclei were usually deuterium. Tritium, which calculations had shown to be essential for significant fusion, is radioactive. Therefore, once a reactor starts using it, the reactor itself also becomes radioactive, greatly complicating the task of conducting experiments. The first tritium experiments at the Joint European Torus in the early 1990s were very cautious, using just 10% tritium and not obtaining much more than some new data.

In November 1993, the Tokamak Fusion Test Reactor (TFTR) at Princeton, which has been the main American fusion experiment since 1983, finally began using tritium as fuel, quickly working up to a 50-50 mixture with deuterium. In May 1994, the reactor set a new record for the amount of power it generated: some 9.3 million watts. It did even better in November 1994, reaching 10.7 million watts—enough power to light about 2,000-3,000 homes, although only for a brief moment.

While this seems like a lot of power, it is not nearly enough. For a fusion reactor to be even slightly successful, it needs to produce more power than the amount it takes to heat and contain the fuel. This has yet to be achieved. For a fusion reactor to become practical, it not only has to produce more power than it needs to start the reaction going (the breakeven point), but it also needs to produce enough heat to keep the reaction going (the self-sustaining point). In the November TFTR experiment, it appeared for the first time that the fuel was partly heating itself with the energy it was producing.

Based on this record-setting success, funding for TFTR (which had been scheduled to be shut down at the end of 1994) was renewed through 1998, but construction on a larger Tokamak Physics Experiment was dropped from the budget. By early 1996, physicists were discussing the complete shutdown of the TFTR and diversion of any money in the budget to a new International Thermonuclear Experimental Reactor, which would have Russian, Japanese, and European partners.

SIX MACHINES IN SEARCH OF FUSION

All fusion mechanisms are based on the concept of getting hydrogen isotopes into such close proximity that they will fuse into other hydrogen isotopes or, in most cases, helium isotopes, giving off energy in the process. Stars do it with gravity. Most of the common earthly methods involve putting the hydrogen isotopes into a plasma, where the nuclei are separated from the electrons, and then moving them around in a circular pattern. This pattern has an empty region in the middle, so it is the form that mathematician call a *torus* or *toroidal shape* (termed a *donut* by the lay world). In the problematical cases of cluster impact fusion and cold fusion, there is no donut because the hydrogen is contained without rapidly moving particles.

Cluster Impact Fusion   The hydrogen to be fused is in tiny heavy-water droplets that carry electric charges. Because of the electric charges, an electric field can be used to speed the droplets through a vacuum chamber, at the end of which is a target consisting of titanium that has absorbed heavy hydrogen. When the droplets hit the target at several hundred thousand miles per hour, the pressure and temperature at the target both become very great. This classic combination of high pressure and temperature is enough to cause fusion. This method has been deemed *cluster impact fusion*, because it only works when the droplets are clusters of 25–1,300 molecules of heavy water. The fusion route is that two heavy hydrogen atoms join to form the even heavier tritium, releasing protons in the process. Detailed physical explanations of what is happening, however, are lacking. Therefore, some theorists still believe that something other than cluster impact fusion is actually causing the apparent release of energy.

Cold Fusion   In 1989, two electrochemists claimed that they were achieving fusion by slowly packing heavy hydrogen into palladium using a simple electrical device. This process, known as cold fusion, is known to sometimes produce heat for chemists. Physicists who have tried it, however, find that their experiments fail to generate heat. Furthermore, the theory behind the process does not make sense to physicists. In general, most people believe cold fusion has been discredited, although some electrochemists still embrace the idea.

Compact Torus Fusion   In the compact torus device, theory predicts that a magnetic field will form a small donut shaped region that is intrinsically more stable than the larger donuts used in the most common apparatus, the tokamak. Although the theory appears plausible, the devices have yet to be built.

Reversed-Field Pinch (RFP) Fusion   The magnetic coils are inside the donut, set in such a way that their fields cause the plasma to twist as it rotates. Over time, the angle becomes great enough that the plasma twists away from the outer walls of the donut, containing itself, in effect.

Stellerator Fusion   Magnets creating fields outside the plasma are used to confine the plasma inside the donut where it can be heated.

Tokamak Fusion   Tokamak is a Russian acronym for toroidal magnetic chamber. Magnetic fields are created inside a moving plasma, causing the lines of motion of individual plasma particles to move in toroidal paths. Starting in 1995, a different setting on the tokamak (called reversed magnetic shear and proposed the previous year by Princeton theorist Charles Kessel) has been used to reduce instabilities that have plagued tokamak fusion in the past. The reversed magnetic shear configuration has generated high temperatures, dramatically greater density, and better containment at both the Princeton University tokamak in New Jersey and the General Atomics tokamak in San Diego, California. A better theoretical understanding of how this method works would likely suggest ways to improve it even further. Another possibility is that this method might be combined with a half-tritium fuel mix to achieve even higher energy output.

In addition to these various methods, several ideas for fusion based upon confinement of the plasma in other ways have been tried and since abandoned. An entirely different approach to fusion, inertial confinement, will be discussed in the next section.

Pelleted-Fuel Laser Fusion

"Inertial confinement fusion" is the proper name for fusion in which lasers or, less frequently, beams of ions are employed to ignite pellets containing a deuterium-tritium fuel. In other words, the concept involves bringing fuel nuclei close enough to fuse by inertial confinement instead of by magnetic confinement. If all the fuel particles are given a push in a direction toward the center of a pellet, then the particles (for the time that they are rushing toward the center) are "confined" in an ever-decreasing spherical boundary. Because of the push, the fuel particles continue to travel in a straight line toward the center. The cause of such motion is inertia. Many find it easier to think of this sort of confinement as an implosion. The same kind of implosion is used to ignite a plutonium bomb. Given that laser fusion was originally developed by the same Defense Department, the name "inertial confinement" may have a long but secret history.

There are two main types of pelleted-fuel laser fusion. In one, the target is really the fuel, which generally begins as liquid heavy hydrogen and is kept in a glass or plastic pellet primarily to give it a shape. In the other type, the target is the container, which for this indirect method is made from an element possessing a high atomic number.

Gold is a popular choice. In the indirect method, the laser or perhaps a beam of heavy ions vaporizes the container, producing a burst of high energy x rays. It is the x rays that inertially confine the fuel.

There have been various projects in laser fusion. The Defense Department Nova facility, completed at Lawrence Livermore National Laboratory in 1985, uses the indirect method with the most powerful and energetic laser ever built. In 1986, the Omega facility at the University of Rochester began a series of direct method experiments. Since then a number of other facilities have been built in Japan and Europe as well as in the United States. Some plans are underway for a U.S. National Ignition Facility for laser fusion, which would cost $1 billion if fully approved by Congress. President Clinton's 1997 budget called for an additional $59.2 million for design and development along with $131.9 million for preliminary operations. The French have a very similar fusion plant definitely in the works for a site near Bordeaux. The U.S. Naval Research Laboratory has an unusual fusion experiment called Nike, in which a gas laser is employed instead of a solid-state laser.

None of the laser fusion experiments are even as close to the break-even point as the magnetic-fusion tokamaks are. Many of the smaller laser facilities have not even progressed to the point of introducing fuel, and are instead just practicing on empty pellets. The Nova facility at Livermore has used fuel, but finds that the vaporized containers for the fuel get mixed with the confined fuel during implosion, thus preventing fusion from occurring.

From Experiment to Reactor

Many problems remain to be solved for both magnetic confinement or laser fusion. The initial plant necessary for magnetic confinement is larger and more complex, but once functioning, it should ultimately be easier to operate than a laser fusion facility. Laser fusion needs some system by which the fuel pellets can be made in a very precise manner and a method of putting them where they will be hit correctly by the lasers. The operation of laser fusion proceeds as a series of pulses, not continuously.

Both fusion and fission produce radioactivity. There is less radioactivity generated by fusion than fission, however, and fusion's radioactivity is a type that decays much faster as well. Neutrons bombard the walls of a fusion reactor chamber, making them brittle, just as neutrons embrittle the containment domes of fission reactors. In one plan, this neutron bombardment would be turned to a useful purpose. If the walls of the container are made from lithium (or rather, containing lithium, since pure lithium has a low melting point and is highly corrosive), the neutron attack will transmute the lithium to tritium, producing a good supply of the least available part of the fuel, just as a breeder nuclear reactor makes more of its fuel as it operates.

Furthermore, the heat produced by the neutron bombardment of the container walls will be the actual energy source used to power the reactor. Fusion power, like fission power from a nuclear reactor, is essentially a method of making heat. To get the energy into a form that can be used easily, the heat from either form of nuclear power must be converted into electricity, which is usually accomplished in much the same way as heat from a coal or oil furnace is employed to power a dynamo. One method proposed for a fusion reactor would use a gas confined to an outer wall around the hot reactor. The gas would be heated by contact with the wall, and would be carried off to drive a turbine that would turn the generator.

Spin-Offs

Usually one thinks of "spin-offs" as applications that develop adventitiously from basic research, but there is no reason why the development of applications cannot spin-off new scientific ideas. Indeed, it has happened regularly in the history of science and technology.

Although plasmas are rare on Earth (existing momentarily in lightning discharges, and perhaps nowhere else outside the laboratories of fusion researchers), they are common in the rest of the universe. Thus, fusion research is one way to perform experiments that duplicate conditions otherwise found mostly in or near stars, or perhaps in interstellar clouds. Furthermore, that interaction of magnetic fields and matter existing at the heart of most fusion devices produces results that may help to explain the general and somewhat mysterious creation and behavior of magnetic fields in space and, indeed, throughout the universe.

The RFP fusion process uses magnetic fields generated by moving plasmas as part of the containment of the plasma in a ring. Investigators had observed that there was more magnetic field than expected, even when all the observable factors seemed to have been taken into account. In 1995, Hantao Ji of the Princeton Plasma Physics Laboratory contemplated the possibility that the cause might be related to the movement of regions of high and low pressure in the plasma (essentially the same thing as sound waves in a gas). Measurements in denser plasmas showed that such changes in pressure do cause the waves. Ji thinks that perhaps the same mechanism might account for otherwise unexplained magnetic fields in space, although there is no definite proof that the sound waves are the cause of these otherwise unexplained fields.

Spin-offs of devices from fusion research also were suggested in 1995. Two entirely different kinds of thrusters for rockets using plasmas and failed fusion schemes were proposed at a meeting of plasma physicists in November 1995. In one, an abandoned system of using electricity instead of magnetism to confine a plasma could be adopted as a long-lived device for providing low-powered pushes needed to keep a spacecraft in proper orientation or in a geostationary orbit. The second concept was to use a failed fusion system of magnetic mirrors to propel an interplanetary spacecraft. Magnetic-mirror fusion failed because the mirrors leak. The idea of magnetic-mirror propulsion is based on their leakage, and is similar to one used in many lasers, in which two mirrors bounce back a light beam. One of the laser's mirrors deliberately reflects only part of the beam, so that the rest of the now-amplified beam passes out of the device as the laser light. For the interplanetary thruster, a magnetic-mirror fusion device weighing about 500 tons would confine a plasma that would leak from both ends. One end would be deliberately designed to be much leakier than the other, and the plasma streaming from that end would propel the spaceship. Calculations suggest that the system would be more powerful in the long run than chemical rocket fuel because the actual fuel would weigh little. Thus, the thruster could be kept going, producing acceleration (or deceleration) for the entire trip if needed.

(Periodical References and Additional Reading: *Physics Today* 9-92, p 32; *New York Times* 12-7-93, p C1; *New York Times* 12-12-93, p IV-4; *Physics Today* 9-94, p 17; *New York Times* 11-8-94, p C12; *Science* 12-2-94, p 1471; *Science* 6-23-95, p 1691; *Science* 7-14-95, pp 153 & 154; *Science* 7-28-95, pp 478; *Physics Today* 9-95, p 73; *Science* 12-8-95, p 1569; *Discover* 1-95, p 46; *Physics Today* 1-96, p 9)

BASIC THEORIES

THE NEWEST FORMS OF MATTER

Most people learn in elementary school that there are three states (or phases) of matter: solid, liquid, and gas. More recently, textbooks describe a fourth state of matter: a plasma (*see* "Fusion Progress at Last," p 639). Thus, production of yet a fifth form of matter in 1995 was one of the biggest news stories of the 1990s. A closer look at the phenomenon shows that the newest phase, called a Bose-Einstein condensate (BEC), is not quite that new. Satyendra Nath Bose and Albert Einstein worked out the theory behind its existence, and Einstein predicted BEC's existence in 1925. Since that time, what amount to BECs have been observed in certain supercooled states. For one reason or another, the observed examples were, for the most part, states of electrons or other particles within solids, where the BEC could not exist by itself and even then does not exist for more than a few milliseconds. The common form of helium becomes a partial BEC liquid when cooled sufficiently. A liquid, however, is too complicated to be a pure BEC because its particles are interacting in various ways.

Bose and Einstein predicted that a suitable gas could be cooled enough to form a substance in which the BEC characteristics are the most important part of the behavior of the substance. This was one feat accomplished in 1995.

The other 1995 production is not a form of matter, exactly; it is what is usually considered the opposite of matter. Yet antimatter, despite the name and reputation, is really a form of matter in the broad sense of the word. Only when one is contrasting antimatter to ordinary matter does it appear to differ from ordinary matter.

Supercool in the Orbit of Death

To a large degree, successful creation of the first Bose-Einstein condensate gas belongs to the whole BEC community. Many people worked on the problem, with each team building upon the ideas of others before them. Nevertheless, the first team that actually produced a BEC gas consisted of a number of physicists from different institutions in and around Boulder, Colorado, led by Eric Cornell of the National Institute of Standards and Technology, Carl Wieman of the Joint Institute for Laboratory Physics, and Michael Anderson of the University of Colorado. Graduate students Jason Ensher and Michael Matthews were part of the team as well, along with theorists John Cooper and Murray Holland.

The Colorado group used atoms of rubidium-87 as their bosons, particles that can form a BEC. Alkali atoms, such as sodium and rubidium, become metals when given the slightest chance, but a few atoms floating around can exist as a gas even at temperatures close to absolute zero. In such a gas, as in any gas, the speeds of individual atoms are distributed along a curve, with some atoms much slower than others. The average momentum of all the atoms is the temperature of the gas, although individual atoms can be slow (cool) or fast (hot).

First, the Colorado physicists slowed the atoms of the wispy rubidium gas with intersecting laser beams. When an atom absorbs a photon of the correct momentum, the atom then re-emits the photon in a random direction and, of course, recoils slightly in the opposite direction. As this process is repeated, the atom is gradually slowed down by the random recoils. This process accounts for the ability of a laser to cool an atom.

The Colorado team then used magnets to pin the slower atoms down in a trap—a method known as magnetic cooling. Magnetic cooling works because the hotter and therefore faster atoms move quickly enough to escape the trap. Therefore, only slow and cooler atoms are left. Essentially this is the same process as cooling your skin by letting alcohol evaporate from it. Not only can such a magnetic trap result in supercool atoms, it also concentrates them at the center of the trap. Furthermore, it aligns the spins of the atoms trapped, which is also necessary for a BEC to form.

A problem develops at the center of the trap, however. There is no magnetic field there, so the supercool atoms flip their spins over and leak away, leaving the density too low to produce a BEC gas. Cornell and his group solved this by adding a second magnetic field, one that rotates just enough to keep the spins from flipping. The zero point travels around the trap in the so-called "orbit of death." If a supercool atom is too energetic, it can make it from the center of the trap to the orbit of death, where its spin will flip and it will be thrown out of the trap. The result is a few atoms trapped inside the orbit which will be spin stabilized and close enough to each other that their waves will merge, just as light waves merge when they are in phase.

On June 5, 1995, the group got definite evidence of trapping a BEC gas of about 3,000 rubidium atoms. This is about the size of an average human cell. Because the atoms were essentially motionless, the BEC gas was as cold as any material ever produced: about 0.000000002K (about 0.000000036°F above absolute zero).

One way to think of the BEC gas is in terms of the Heisenberg uncertainty principle. This principle states that one cannot know the exact position and momentum of a particle at the same time. When an atom is that cold, its momentum is nearly zero. Therefore, its position becomes stretched out and wavelike. A BEC gas, however, is not just a collection of wavelike particles, it is the merging of the waves of all the atoms in the BEC gas.

By the end of the year, many other groups of physicists had also produced BEC gases using Cornell's method. While he was not the first to create a BEC gas, Wolfgang Kettele, who developed a way to keep atoms in the trap with a laser plug, got the best results. In as little as nine seconds, he was able to trap as many as 500,000 sodium atoms at once. In August 1995, a group at Rice University apparently succeeded with a third element: lithium. In principal, about 75% of all the elements have stable isotopes containing an even number of particles and so could be turned into BEC gases.

Might a BEC gas exist in nature? Nothing is impossible in nature, but a BEC gas is unlikely. Although it is cold in outer space, it is not as cold as a BEC gas. Furthermore, the appropriate atoms would have to be held sufficiently close with their spins aligned. Perhaps the best bet would be a cloud of monatomic hydrogen in deep space, but interactions between hydrogen atoms would be likely to interfere before the BEC gas could form.

Background

All particles have a mysterious characteristic called **spin**, so called because it behaves in some important ways like angular momentum (the momentum developed by spinning something around a central axis). For example, the same unit of measurement is used for spin and angular momentum. A big difference, however, is that angular momentum is a continuous quantity while spin is a quantum number, and therefore discrete. Every particle spin is a multiple of the number Planck's constant divided by 2π (a constant measured at about $1.05457266 \times 10^{-34}$ joule-seconds with an accuracy of about 0.60 parts per million). For largely historical reasons, this number is equated to ½. Thus, physicists call the spins of each particle either ½ or an integral multiple of ½, including 0, 1, and 2. For most of the familiar subatomic particles (including the electron, proton, neutron, neutrino, and quark) the spin is exactly ½. These are the particles that make up ordinary matter. Such particles are called **fermions** (*see also* "Subatomic Particles," p 666).

There is another large group of subatomic particles that have a spin of either 0, 1, or 2. These particles are called **bosons** after Bose, and include the particles that hold matter together by producing fundamental forces (*see* "The State of Physics," p 625), notably the pions and graviton (with a spin of 0) and the photon and gluons (with a spin of 1).

The behavior of fermions and bosons is very different. For example, fermions are distinguishable from their mirror images while bosons are not. From this simple premise, it is possible to show by largely mathematical reasoning that two fermions cannot occupy the same location in space. Of course, in quantum theory (*see* "Quantum Stuff," p 649) a particle does not occupy an exact location, but has a *probability* of being in a particular location. In quantum theory, the probability that two fermions will be in the same place is 0. Furthermore, in an atom this means that no two fermions can have the same set of quantum numbers. This is the Pauli exclusion principle, one of the great results of quantum theory.

However, two or more bosons can not only occupy the same place, but they will actually actively seek each other out. Given the location of one boson in space, any other boson that is near it will have a positive probability of being in the same exact location. This creates a sort of force that causes bosons to huddle together. Light waves, for example, can all be in phase, which viewed from a particle perspective means that they are acting as photons that have all come to occupy the same place.

Spins are additive. Thus, atomic hydrogen is a boson, since the ½ spin of the proton and the ½ spin of the electron must either add to 1 or cancel each other to 0. The calculations become more complex for larger atoms or molecules. The monatomic molecule of helium-4 is a boson with its two protons, two neutrons, and two electrons, none of which can be in the same state (since they are all fermions). Yet the helium-3 atom is a neutron shy and therefore a spin ½ short: it is a fermion. Physicists have long observed helium-4 becoming a 10% Bose-Einstein condensate (BEC) liquid when cooled to near absolute zero. The reason that the temperature must be so low is that the condensate phenomenon only occurs if the particles are within a wavelength of each other. Although both helium-4 and helium-3 become superfluids when sufficiently close to absolute zero, helium-3 behaves like the electrons in a metal (which are fermions) instead of like a BEC.

Photons, with 0 rest mass, form into a BEC at virtually any temperature. Atoms have definite masses and therefore rather short wavelengths. The heavier an atom is, the shorter the wavelength. Also, atoms often interact in other ways, which would prevent the BEC from forming. Thus, to obtain a BEC one needs a cool (nearly stationary) gas of bosonic atoms with their spins aligned so they are all the same. After all, two atoms can have the same fundamental spin, but one atom might be upside down relative to the other, which means that the spins are not exactly the same after all.

The First Antihydrogen Atoms

Antimatter in the form of antiparticles of various kinds appears spontaneously all the time, and physicists have for many years routinely created beams consisting of billions of antiparticles. Physicians use Positron Emission Tomography to observe soft tissues in a patient's body, a practical application of an antimatter particle. Thus, there is nothing really strange or unfamiliar about antimatter except that it is never encountered in bulk, only as subatomic particles. There are two reasons for this. In the first place, any time an antiparticle encounters its ordinary particle partner (which normally happens within instants of its creation) the two annihilate each other, shooting off an energetic photon of electromagnetic energy. Furthermore, antimatter's brief existence is only as an antiparticle, not as an antiatom. Even particles of ordinary matter are commonly only viewed when welded together into atoms, although there are some exceptions to this.

To a very limited extent, this situation changed in September 1995 when physicists at the European Laboratory for Particle Physics (generally known as CERN) created 11 atoms of antihydrogen, each lasting for about 40 billionths of a second. The atoms were created on the fly by shooting antiprotons at a xenon target. Some antiprotons passed right through the target, but others that hit it smashed out some antielectrons, also called positrons. On 11 occasions, one of the passing antiprotons and one of the positrons found themselves heading in the same direction at the same time and took the opportunity to travel together in the same conveyance. That is, the negatively charged antiproton and positively charged antielectron were attracted to each other by their equal charges and coupled to form an atom. In the familiar old Bohr model of an atom, the positron took up an orbit around the antiproton.

Because the antihydrogen atoms were all zooming quickly through space, they very soon struck the side of the container and annihilated themselves along with an ordinary proton and electron. The sudden flash of a high-energy photon that this annihilation produced enabled the CERN team, led by Walter Oelert of the Jülich Institute for Physics Research in Germany, to identify the antihydrogen atoms.

Very little could be learned from this experiment, except that an entity that most physicists believed possible was shown to exist. Other groups are trying to make some antihydrogen that is slower moving and longer-lived, which might then be studied to produce some profit. There is much that is unknown and unpredictable about an antimatter gas or other true antimatter substance.

Trapping antimatter is difficult. If slow moving positrons and antiprotons can be kept near each other, antihydrogen will form itself, as the CERN experiment demonstrates. Both positrons and antiprotons are charged particles that can be slowed and even trapped with magnetic fields. However, once they merge into antiatoms, their charges cancel each other and, like ordinary hydrogen atoms, they are no longer subject to magnetic forces. They quickly escape the trap and are annihilated. The other method of slowing small particles, with the use of intersecting laser beams, does not depend on charge and can be used. Thus, a procedure very similar to the method used for producing a BEC gas can in theory be used to make small amounts of antihydrogen gas. Physicists at Harvard, Penn State, Los Alamos, and elsewhere are using various versions of this method in the quest for bulk antimatter. For example, Harvard's Gerald Gabrielse and coworkers have trapped as many as 35,000 positrons in a small space, representing a step along the way. There is every reason to expect that these efforts will be successful. After this, the next step might be to attempt heavy antihydrogen, also called antideuterium, in which the particles that need to be placed in proximity include an antineutron as well as the antiproton and positron. Then, on to antihelium.

The main virtue of obtaining bulk antimatter will be to resolve the unanswered questions of how it behaves with respect to the forces that affect ordinary matter. Some have even suggested that antimatter will be repelled by gravity rather than attracted, although that concept is highly speculative. If there ever was a method developed for producing large quantities of antimatter and storing it until needed, or even simply producing antistuff on demand, the antisubstance would be a powerful source of energy. A common suggestion concerns an antimatter drive for space propulsion, one of the familiar science-fiction notions. No one speaks of the antimatter bomb.

(Periodical References and Additional Reading: *New York Times* 8-23-94, p C1; *New York Times* 11-15-94, p C1; *Science* 7-14-95, p 152; *Physics Today*, 8-95, pp 17 & BG9; *Physics Today*, 9-95, p 9; *Physics Today*, 12-95, p 9; *Science* 12-22-96, p 1902; *Discover* 1-96, p 58; *Physics Today* 1-96, p 22; *New York Times* 1-5-96, p A9)

QUANTUM STUFF

Bose condensation (*see* previous article) is "quantum stuff" in the most literal sense. However, although quantum theories of one kind or another are now nearly a century old, there continues to be a lot of new quantum stuff in a more figurative sense. For

Background

People in general have heard of quanta mostly in the expression **quantum leap**. The phrase "quantum leap" or an equivalent was well entrenched in common language even before its use as the title of a television series. In the television program, the hero moved instantaneously from one region of time and space to another. For a time, especially in the 1970s and 1980s, it became quite popular to describe any abrupt transition as a "quantum leap" or "quantum jump," especially if progress and science were in the background. "Smith's discovery resulted in a quantum jump in the treatment of the disease" or "the new Quince Computer is a quantum leap over any experience you may previously have encountered" are typical of this usage. Frequently, the implication is also of a great distance, figuratively at least, as well as of a sharp discontinuity. For example, an agnostic might require a leap of faith to accept an impersonal deity, but a *quantum* leap of faith to believe in earthly miracles.

While a physicist is more apt to speak of quantum *jumps* than leaps, he or she is not likely to mention either too often these days. In a strict sense, the quantum jump is part of a now abandoned view of electrons. Around the turn of the nineteenth century, the sum of various discoveries might be metaphorically termed a quantum leap in fundamental physics. The electron was discovered in 1897; Planck's idea that energy came in little bundles called **quanta** (singular, **quantum**) was announced in 1900; in 1905 Einstein showed light behaved as if it came in particles some of the time. When, in 1913, after several years of effort, Niels Bohr produced an explanation of the relationship between the hydrogen atom and light in terms of electrons, he thought of the electron as a physical particle in orbit about a proton. An interaction between the electron and light involved moving from one orbit to another with extreme suddenness, not passing through any intermediate state. This jump from one orbit to another occurred because energy was available only in the packets called quanta. Hence, the transition was a quantum jump. Note that the jump took place completely within the confines of the original atom. Therefore, the modern usage connoting a large transition is itself a quantum leap away from the original idea.

Ten years after the Bohr atom, Louis de Broglie suggested that electrons and light and all particles could be viewed as waves. The relation between the wave and the particle could be described by an equation in which the original size of the quantum appeared as a constant. Today, chemists and physicists think of electrons as behaving more like waves than like particles when they are in an atom. The sudden transition that Bohr described is the change from one wave shape to another one, so there is no jump in the ordinary sense at all.

Shortly after de Broglie's proposal, Werner Heisenberg (1925) and Erwin Schrödinger (1926) independently worked out two different versions of the mathematics needed to explain the behaviors described by Bohr and de Broglie. These mathematical treatments of the behavior of subatomic particles are known as **quantum mechanics**. The results of calculations using quantum mechanics appeared to be correct but were highly nonintuitive. Max Born, starting in 1926, provided an interpretation of quantum mechanics in terms of probability that helped make the odd conclusions comprehensible. Even with this interpretation (eventually named the Copenhagen school of quantum mechanics), some of the conclusions are difficult to accept by people who believe in ordinary cause and effect.

In 1947, a new quantum theory was born out of additional experimental data. This is called **quantum electrodynamics**, or **QED**, and is among the most successful theories of all time. It has been verified in its predictions often to as many as a dozen decimal places. QED is the theory of moving charges and the particles associated with them.

In 1964, when Murray Gell-Mann established that protons, neutrons, and mesons are not fundamental particles (*see* "Subatomic Particles," p 666), a new theory was needed to explain interactions of quarks. This theory, which is based on a concept known as the **color force** (*see* "The State of Physics," p 625), has come to be called **quantum chromodynamics** in obvious imitation of QED. It is sometimes abbreviated QCD. On May 11, 1993, a team of IBM scientists led by Donald Wiengarten of the Thomas J. Watson Research Center in Yorktown Heights, New York, reported that their year-long run of a specially-built parallel supercomputer accurately calculated the observed values of the ratios of the particles called hadrons to each other using QCD as a basis, suggesting strongly that QCD theory is correct.

Thus, there are several quantum theories, more even than those mentioned above, which are just the highlights. What they all have in common is their ultimate derivation from the now apparently indisputable data showing that energy is discrete and not continuous.

one thing, some physicists continue to argue the validity of the basic concepts behind the various components of quantum theory.

Atoms Make Waves

When one first encounters quantum theory, the idea that what one thinks of as particles (such as electrons and atoms) are also waves, and that what one thinks of as waves (such as light or radio) can often be better understood as particles, may seem quite strange. Although other quantum ideas are very hard to get used to, this one soon becomes second nature for dealing with the smallest particle-waves, such as electrons or light waves. It is, however, a bit more difficult to accept when it is applied to larger wave-particles. Even particle-waves as small as atoms seem in nearly all instances to behave as particles and not waves. In the 1980s and early 1990s, however, the wave nature of atoms was exploited in various applications. Quantum theory goes further, so that even particle-waves as large as bacteria or human beings or planets or stars have a wave nature as well as the particle nature that is apparent to us. For large particles, the associated waves are too small to be detected, since the wavelength of a particle is found by dividing the very small number known as Planck's constant (6.67259×10^{-34} joule-seconds) by the particle's momentum, which is the product of its mass and speed.

In February 1995, David Pritchard of MIT in Cambridge, Massachusetts, who had been working with the wave effects of atoms since 1988, demonstrated for the first time that atoms passed through a fine grating can be observed as waves. He and his coworkers were able to obtain diffraction patterns caused by interference between individual atoms passed though a vacuum and atoms that had been slowed by passing through a gas. Using these wave patterns, many specific measurements can be made with a precision not previously possible. Light-wave interferometers have similarly been used to make highly precise measurements for about a hundred years. The method uses sodium nitride crystals as the slits through which the atoms pass to create the diffraction patterns. This should not be confused with the familiar process in which x rays are passed through a crystal to reveal the structure of the crystal; indeed, in one sense, this is actually the reverse of that process.

In addition to sodium atoms, Pritchard has been able to find the waves of entire sodium molecules, which consist of two atoms when sodium is a gas. In theory, the same sodium-nitride-crystal technique could be used to obtain the diffraction pattern of a small bacterium treated as a wave. Quantum effects, however, very significantly slow the passage of such a wave through the crystal. It would take thousands of years for a bacterium to pass through the grating as a wave.

Other researchers are using the quantum properties of atoms to develop different tools. In June 1995, Edward Hinds and coworkers at Yale University announced that they had developed a concave mirror that reflects atoms to a single point, focusing atoms by a process similar to that used to focus the light from a distant star with a mirror. Atoms are increasingly used as waves in interferometry by various other methods as well. The first successful atom interferometer was created in 1991. By 1995, scientists were using atom interferometers to make extremely precise measurements of fundamental constants in physics.

Scientists at the University of Colorado have developed the atomic equivalent of fiber optics, combining hollow optical fibers and lasers to move the atoms around. Among many possibilities is that an atom with a wavelength comparable to the inside diameter of the fiber can be propagated through the fiber as a wave, and even split in two for interferometry experiments.

(Periodical References and Additional Reading: *New York Times* 5-11-93, p C6; *Physics Today* 7-95, p 17; *Physics Today* 9-95, p 9; *Discover* 1-96, p 58; *Physics Today* 12-95, p 9)

Strings and Things UPDATE

Although physicists base most experiments and devices on such theories as Newtonian dynamics, relativity theory, quantum theory, and the standard model of particle physics, many theoreticians feel strongly that there should be a deeper explanation for reality than any of these. Even Einstein looked for a unified field theory that would integrate gravity and electromagnetism. He met with little success and great mathematical difficulties.

In recent years, theoreticians have based the standard model of particle physics on that part of mathematics known as group theory, a discipline that is concerned with (among other things) how one object can be transformed into another, how structures can describe the interactions of objects, and what kinds of mathematical structures are possible. Group theory has proven to be a good home for particle physics, especially since group theory is also the mathematics of symmetry. Group theory has been at the center of such developments as electroweak unification (which combines charge and the weak force into one interaction) and the identification of quarks. It is only natural that the same mathematical tools should be used in proposed extensions of theory, such as supersymmetry. Interestingly, as has often happened in the past, the physicists trying to solve the difficult physical problems have emerged with new mathematical ideas that have greatly enriched mathematics.

Supersymmetric Successes

The basic idea of supersymmetry is that a fermion can become a boson (*see* "Subatomic Particles," p 666). The supersymmetric transformation of a boson into a fermion or vice-versa is generally added to other theories, sometimes using the acronymic prefix SUSY, so that there can be SUSY quantum theory, SUSY gravity or SUSY strings. In a SUSY theory, it is not that one familiar fermion, such as a muon, changes into a familiar boson of about the same mass, such as a pion. Instead, for each recognized and previously detected fundamental fermion or boson there is a twin, just as each fermion has an antiparticle twin. The SUSY twin of a fundamental fermion of a given mass is an undetected boson of the same mass, and so forth. It would seem, however, that if all these new particles with small masses exist, they would have been noticed previously. SUSY theory takes care of this problem with a symmetry-breaking operation that gives all the undetected partners very large masses, which explains why they have not been found. Indeed, *before* the symmetry breaking, the particle and its superpartner were one particle that, *with* symmetry breaking, became two particles of greatly unequal masses.

Common sense would suggest that it might be easier to locate something with a large mass than one with a small mass, as for example one could more easily find an elephant in a 10-acre field than a field mouse. Common sense seldom works for physics, however, even for old-fashioned Newtonian dynamics. Subatomic particles with large masses decay into smaller particles unless there is some powerful force keeping them from doing so. The reason that a particle elephant is difficult to locate is because it rapidly turns into its own weight in field mice. The more massive it is, the less time it takes to decay. To find a particle with a large mass, such as the top quark or presumed Higgs particle, one has to first create it and then recognize it from the "field mice" into which it decays. Thus, if the SUSY partner particles are very massive, this fact would explain why such particles have never been observed.

The supersymmetric transformation is an operation that subtracts one unit of spin from a particle while otherwise leaving it the same mass. The electron with spin ½ has as its superpartner a boson with spin 0. The photon with a spin of 1 has a superpartner with a spin of ½. If a particle has a spin of 0, as is the case for the pion, ½ is added to the spin to obtain a superpartner with a spin of ½.

There is a system for referring to the superpartners in terms of the names of the original particles. A superpartner of a fermion has the prefix *s-* added, so that the superpartner of the electron is the selectron, the superpartner of the quark is the squark, and so forth. The suffix *-ino* is used for superpartners of bosons, so one has the photino, gravitino, and gluinos. Composite particles, such as protons or pions, fail to have superpartners on their own, so no names such as sproton or pionino are needed.

In 1992, at the conclusion of a search using the Tevatron particle accelerator at Fermilab, the mass of the SUSY superpartners was established as being greater than 126 GeV for squarks and 141 GeV for gluinos. That is, interactions up to these levels produced no evidence of superparticles. The paper announcing the negative result on December 14, 1992, was signed by 315 scientists.

In 1994, Nathan Seiberg of Rutgers University in New Jersey and Edward Witten of the Institute for Advanced Study in Princeton, New Jersey, used a mathematical idea called a holomorphism, a powerful concept from complex number theory, to resolve various problems in supersymmetry theory. Among the many results obtained with the new technique was an explanation of symmetry breaking in terms of massless magnetic monopoles. The work brought a new recognition to the importance of duality in theoretical physics. The original idea of duality in mathematics occurs in projective geometry, where one can obtain a new theorem that is true from a previously proved theorem by interchanging the roles of point and lines. The new theorem with line for points and points for lines is the *dual* of the original. In physics, electromagnetism with magnetic monopoles added is a dual theory. That is, one can get a true theorem by interchanging electrons and monopoles while at the same time interchanging electricity and magnetism. For a dual physical theory, one can regard either part as fundamental, in which case the other part is derived.

Although the results of Seiberg and Witten are exciting to physicists, they only apply to a supersymmetric universe, which may or may not the one we inhabit. Mathematicians, however, are even more excited about the technique, because mathematicians (as Bertrand Russell once remarked) don't care whether what they are studying is real or not. The new technique is the key to resolving many unsolved problems concerning mathematical problems of the fourth dimension.

Difficulties in how mathematical space is constructed are complicated in an unexpected way. A one-dimensional space presents almost no problems. A two-dimensional space presents problems that can, with considerable work, be resolved. The apparently similar solutions to those problems in three-dimensional space have hidden or even obvious paradoxes, so there must be something wrong. The corresponding problems in four-dimensional space have been so difficult that there are not even self-contradictory solutions to most of them. After that, everything gets easy. Dealing with five- or six- or higher-dimensional space is as easy as one- and two-dimensional space. Thus, the discovery that the Seiberg-Witten approach works in four dimensions has had even more of an impact on mathematics than it has on physics.

Strings

Physical insights are important as well as mathematics. It is difficult to picture something that is a wave and a particle at the same time. In some ways, it is easier to envision that entity as a short length or loop of string, which can embody a wave in a physical entity. String theory has grown robustly from such elementary insights. Mathematics is once again part of the story, since it has been found easier to treat strings mathematically than it is to solve mathematical problems about particles and waves. In the case of string theory, the most relevant mathematical ideas have clustered around how objects can be embedded in spaces of various dimensions, with a 10- or 11-dimension space giving the best results for our real world, even though the universe seems more like a 3- or 4-dimensional entity to our senses.

One problem, however, is that a 10-dimensional string universe, which has come to be the accepted string universe, must become a 4-dimension perceived universe if it is to make any physical sense at all. The mathematics, however, offers no clues as to which of the tens of thousands of ways that 10 dimensions can be perceived as 4 is correct. As a result, no one has been able to devise a test for string theory that would establish that the 10-dimensional string universe exists at all.

Inspired by the Seiberg-Witten success with supersymmetry, Andrew Strominger of the University of California, Berkeley, one of the leaders in string theory, tried part of the approach on a new development that had appeared in his field of physics. In 1994, Chris Hull of Queen Mary's College in London and Paul Townsend of Cambridge University in the United Kingdom suggested that black holes, which are thought to "evaporate" mass, might (when most of their mass is gone) become something like elementary-particle strings. Strominger observed that the Seiberg-Witten theory works in part because it assumes that a massless particle appears, preventing equations from developing the infinite solutions known as singularities. When Strominger allowed the black holes to become massless, the same thing happened in string theory: as the mass of the black holes vanished, so did the singularities of the theory. With Cornell physicists David Morrison and Brian Greene working on the theory along with Strominger, the black holes were shown to be strings, or perhaps it would be more accurate to say that both black holes and strings were found to be manifestations of the essential reality. If all this is true, then, happily, there is only one way, not tens of thousands, to go from 10 dimensions to 4. However, unhappily, the mathematical description of the universe based on this idea is still unable to suggest a test that would tell whether the string picture or a more conventional idea is representative of the true reality.

And Things

Before supersymmetry and in addition to strings, there is the old Einstein idea of unifying the forces (although not the same forces that Einstein originally tried to unify). Grand Unified Theories, sometimes called GUTs or the grand unified model, extend the group theory of the standard model to unify the electroweak force and the color (strong) force (*see* "The State of Physics," p 625). When the transformations of the group for the standard model are extended, however, they predict that a quark can be transformed into a lepton. By quantum theory, if it *can* happen, then it sometimes *will* happen. Thus, if one waits long enough, and if the grand unified model is correct, then a quark will suddenly turn into a lepton. If the quark is in a proton, which is a likely place for a quark to be, the proton will suddenly turn into a meson and the newly born lepton (former quark). This transformation, although much sought after, has probably never been observed. A few physicists think that proton decay may have been observed in one of the giant experiments set up to find such a decay, but the event was not properly recognized. If supersymmetry is true, however, the quark-lepton transformation does not have to exist, since there is a fermion-boson transformation instead.

The standard model requires the as-yet undetected Higgs particle to imbue other particles with mass. Some unnamed variations on the standard model allow for more than one Higgs particle, while a version of "chromodynamics" called technicolor replaces the Higgs as a separate particle with two different particles of even greater mass. The undetected Higgs in technicolor is a blend of the two more massive and even less detected particles.

Although everyone else is hoping for new accelerators to find massive particles, John Jaros of the Stanford Linear Accelerator Center is using that particle accelerator to look for particles that have a charge in between a millionth and thousandth of that of the electron or proton. Such a millicharge is not ruled out by theory, although experimental results previously show that a particle with a charge between a thousandth of that of the electron and the 1/3 charge of some of the quarks cannot be found. The theoretical underpinning for a millicharged particle is a rather vague idea involving two parallel universes superimposed on top of each other, with a small amount of mass in the form of millicharged particles leaking from the otherwise unobservable other universe into this one. A search in 1994 and early 1995 did not reveal the millicharged particles, however.

(Periodical References and Additional Reading: *New York Times* 1-5-93, p C1; *Science* 3-10-95, p 1424; *Scientific American* 9-94, p 40; *Physics Today* 3-95, p 17; *Science* 6-23-95, p 1699)

NEW DEVICES IN PHYSICS

ADVANCED ACCELERATORS

The demise of the Supercooled Supercollider (SSC) in October 1993 (*see* "The End of Big Science in the United States?" p 13) was a great blow to the community of physicists whose work depends on experiments conducted by smashing speeded-up subatomic particles together, although it was far from the end of this kind of activity. Not only have grand plans continued to make progress in Europe, but projects only slightly less ambitious than the SSC have continued in the United States and around the world. Even if particle accelerators fail to grow as large as the SSC was intended to be, physicists have found ways to accomplish much with smaller systems.

Indeed, one of the smallest systems every built is an experimental laser accelerator developed by a consortium of laboratories in Japan. The powerful lasers were employed in short pulses to create a moving wave in a plasma. An electron inserted into the plasma is suddenly accelerated at a great rate of 30 GeV per meter, measured by physicists as the energy increase in the electron per distance traveled. If the process could be scaled up from its table-top operation to even a few meters, and if it could maintain that acceleration for those few meters, it would have the power of a 1 mi (1.6 m) long linear accelerator. A similar experiment was reported in April 1994 by scientists from the University of California at Los Angeles.

Background

From the beginning, the most useful information to be obtained about subatomic particles has come from smashing them into things and observing the results. A simple cathode-ray tube is a particle accelerator for electrons, boiling them off at one end and sending a stream down a vacuum tube to hit a target at the other. This first particle accelerator did all its accelerating at the start of the particle's journey, but was enough to establish the existence of the electron, lead to the discovery of x rays, and influence thinking in ways that ultimately led to the discovery of natural radioactivity.

A radioactive element is also a primitive particle accelerator. The processes of decay that change one element or isotope into another also accelerate electrons, positrons, photons, and helium nuclei (combinations of a pair each of protons and neutrons; such combinations are also known as alpha particles) in various mixtures. The alpha particles were of special interest to the first particle physicists because they are vastly heavier, and therefore make a much bigger hit than the other particles accelerated by radioactivity. An alpha particle is nearly 10,000 times as massive as an electron, sometimes called a beta particle in this context. With natural radioactivity as an accelerator, Rutherford and his assistants discovered the atomic nucleus and the proton. Even as late as 1932, James Chadwick used natural radioactivity as the accelerator in the discovery of the neutron.

The same year that Chadwick discovered the neutron, physicists began to employ small accelerators that used high voltage electricity, static electricity, electric fields, and magnets to accelerate particles to higher velocities. Soon, the best method proved to include a combination of high voltage to get the particles moving and magnets pulsing on and off to push them to move faster. Even the SSC, if constructed, would have used this basic idea with suitable refinements. One early problem was that effects predicted by the theory of relativity set limits on how fast charged particles could travel. These effects were transcended by the invention of the synchrotron in the 1940s. The most successful accelerators since that time have been refinements and improvements on synchrotons, in which charged particles are accelerated around a circular path by magnets. Synchrotrons, however, lose some power because the constant acceleration produced by travel in a curved path causes particles to radiate protons. An alternative idea that has been successful involves the use of very long straight accelerators, which do not lose energy from synchrotron radiation, but which are very expensive to build and finite in the lengths over which a particle can be pushed with a succession of shoves.

Although much of the progress in understanding subatomic particles and atoms has come from accelerating charged particles, most of the rest of the experimental knowledge derives from the opposite approach—slowing down a beam of neutral particles, generally neutrons. In 1995, the Technical University of Munich in German got approval to build a new neutron source, although the largest already exists at the Institut Laue-Langevin in Grenoble, France.

Although nuclear transmutation of lithium atoms was achieved by the first particle accelerator (built by Cockcroft and Walton in 1932 to accelerate protons), it was the nuclear fission of uranium by slowed neutrons that led to the development of atomic energy. This has proved to be the most practical direct application of particle research. Indirectly, however, the basic understanding of nature that has led to modern electronic devices of all types originated in experiments with particle accelerators.

Making Light

A distinct trend of the early 1990s was the rush to use synchrotron radiation for theoretical and practical purposes. A synchrotron is a particular design for a particle accelerator. Synchrotron radiation is, except for its synchrotron source, plain old electromagnetic radiation. Although physicists often call synchrotron radiation light, the most useful frequencies are much higher that those of visible light, existing in the realm of x rays. The synchrotron produces a continuous spectrum of electromagnetic radiation and can be tuned so that desired frequencies are dominant. Tuning allows narrow bands of a particular frequency to be selected out, just as a filter can pick out red light in the visible spectrum. One reason synchrotron radiation is valued is that it is extremely bright light.

The source is the energy that particles give off when they are accelerated. (In practice these particles are almost always electrons or positrons, although one Danish device called ASTRID, which started operation in 1990, uses heavy ions.) In a synchrotron, the particles travel in a circle, which means that they are in constant acceleration. The particles also travel at speeds near that of light, so their mass increases as predicted in relativity theory. Some of the energy involved is given off as electromagnetic energy.

When synchrotron radiation was first discovered in 1947 by physicists at General Electric Corporation, it was considered at best a nuisance, since it robbed energy from the particles. Physicists eventually began to find uses for that energy, however. By the 1980s, synchrotron energy was seen as an important source of energy for the manufacture of semiconductor chips, as well as for other purposes requiring very bright electromagnetic radiation (including a version of the x-ray microscope). By the early 1990s, plans were in place to build various new synchrotron sources. These include:

Name	Location	Year
Laboratorio National de Luz Sincrotron	Campenas, Brazil	1992
The Advanced Light Source	Berkeley, California	1993
European Synchrotron Radiation Facility	Grenoble, France	1994
The Advanced Photon Source	Argonne National Laboratory	1995
Pohang Light Source	Pohang, South Korea	1995
SPring-8	Harima Science Garden City, Japan	1998

The Advanced Photon Source at Argonne National Laboratory accelerates positrons until they reach a mass that is about 14,000 times as great as they would be if at rest. As they swing around the circle, the positrons emit hard x rays up to 100 keV with an unparalleled brilliance. The European Synchrotron Radiation Facility, which uses electrons at 6 GeV instead of positrons at 7 GeV, is about as brilliant. Japan's SPring-8 is so named in part because it will operate at 8 GeV when it is completed.

All three of the synchrotrons that went into operation in 1994–1995 are intended primarily for industrial use of one kind or another. Investigation of the properties of bulk matter are high on the list, including studies of polymers, superconductors, catalysts, genes, drugs, and viruses. One beam line at Pohang is dedicated to x-ray lithography for chip maker Gold Star Electron Company.

Factory Fever

In particle accelerator parlance, a factory is an accelerator producing a great amount of some kind of particle that can then be studied. Generally, the particle is the result of an impact of two other particles which together produce a short-lived complex that decays into a shower of particles. The factory is an accelerator that can make a large quantity of a specific particle among the various droplets in the shower.

Kaons, which are composed of a strange quark with an antiup or antidown quark, or an antistrange with an up or a down, are of particular interest because they are the simplest source of strange quarks, which, like all quarks, cannot exist by themselves. In Canada, the excitement revolves around a plan to produce a kaon factory at the Tri-University Meson Facility (TRIUMF) near Vancouver, British Columbia. On September

19, 1991, the Canadian government agreed to fund about a third of the new facility at a cost of $236 million. The TRIUMF backers are looking for additional funding from British Columbia as well as from sources outside of Canada.

In the United States, hopes have been high for so-called B factories that would produce many B mesons. A B meson is so called because it is a combination of a b, or bottom, quark and the b antiquark, a combination not common in nature, if in existence at all. Once created, however, B mesons do not decay very fast by exotic particle standards, so they can be produced in beams that contain large numbers of B mesons at any one time. Both Stanford and Cornell Universities wanted to build B factories, but on October 4, 1993, the grant was given to Stanford. The B factory, which will be operated in conduction with both Lawrence Berkeley National Laboratory and Lawrence Livermore Laboratory, was expected to begin operation in 1997. It will smash electrons and positrons together at energies great enough to generator B mesons. Other B factories are being built at the National Laboratory for High Energy Physics in Japan, a facility also known as KEK. At CERN, the Large Hadron Collider (*see* below) will be used to produce B mesons among its other activities.

Both kaons and B mesons are thought to be involved in processes that violate certain rules obeyed by other particles. These rules are called CP symmetry (for charge-parity symmetry), and kaons are the only particles thus far to be actually caught in the act of such a violation (in 1964 by Val Fitch and James Cronin). The Fitch-Cronin violation, or garden-variety violation, is not what the factory fans have in mind, however. They are looking for exotic violations of some parts of the standard model. In some scenarios, a B meson could decay into a long-sought-after Higgs particle or even one of the supersymmetry partner particles (*see* "Strings and Things UPDATE," p 652). Even the decay of the B meson into a tau-antitau pair would imply the existence of supersymmetric particles.

Other New Accelerators and Particle Beams

In addition to synchrotrons and particle "factories," various other facilities for dealing with subatomic or near-atomic particles are coming on line or being planned all the time. Some of the more significant new facilities and events around them are briefly described below.

In 1992, scientists at Brookhaven National Laboratory in New York began working on a reverse Cerenkov effect accelerator. In theory, any physical effect that can happen in one direction in time can also happen in the opposite direction. Normally, a particle produces a photon when it travels through a medium at a speed faster than the speed of light in that medium. This way of producing a glow is called the Cerenkov effect, and is a common way to detect and identify all sorts of particles. To reverse the Cerenkov effect, one needs to send the correct beam of light at a particle that is residing in some medium where it is not currently traveling faster than the speed of light. By careful synchronization of the light source and the speed of the particle, it is possible to accelerate the particle. The Brookhaven facility demonstrated acceleration of electrons at significant levels in January 1994, and work in this area is ongoing.

At the 2 mi (3 km) Stanford Linear Accelerator Center (SLAC), a new facility called the Final Focus Test Beam became operative in 1994 following three years of construction. It squeezes the final beam of electrons produced by the accelerator down to a width of 75 nanometers (roughly 4 hundred-millionths of an inch)—almost too pointlike to measure.

Building a large accelerator takes considerable time. The Continuous Electron Beam Accelerator Facility (CEBAF) at Newport News, Virginia, was first considered in the 1970s and approved in 1983; construction began in 1987, and it began operating in 1994. CEBAF is a high-powered device that works primarily in the basic physics of quantum chromodynamics, utilizing quarks, gluons, and their interactions. Instead of producing short bursts of electrons as the LEP does, CEBAF (as its name suggests) produces a steady stream of electrons at 4 GeV—not especially powerful by modern standards. It is hoped that its more gentle approach to particle smashing will reveal the secrets of the universe in ways hard to detect among the debris of the massive collisions induced at more powerful accelerators. CEBAF also produces many more collisions than do the pulsed accelerators.

One does not expect movie-like hostage dramas at particle accelerators, but in February 1995, there was such a scene at CERN's Proton Synchrotron and Proton Synchrotron Booster. It might not have made much of a movie, however. The hostage-taker took about 1,300 electronic components hostage and demanded that his ex-wife be fired from the project. He also cut about 200 cables and disconnected some 5,000. After a few days, the parts were found, the cables began to be reconnected, and all was returned to normal.

CERN was pleased to note that its Large Electron Positron (LEP) collider was still on schedule despite the sabotage. LEP has been its operation since 1989. In the beginning, the LEP beams of electrons and positrons began to wander in a mysterious way. Finally, in November 1992, the problem was traced to the Moon. Tides occurring in the land in which the large ring (with a 16.6 mi or 26.7 km circumference) is embedded were deforming the ring on a regular basis. The LEP scientists have learned to correct for the phase of the Moon.

For many purposes, beams of ions are more satisfactory than pure subatomic particles such as electrons or antiprotons. In 1995, the U.S. Oak Ridge National Laboratory introduced a new source, the Holifield Radioactive Ion Beam Facility, similar in some respects to beam facilities at CERN and the Cyclotron Researcher Center in Lougain-la-Neuve in Belgium. In such facilities, a target material is bombarded with accelerated protons to produce ions. The desired radioactive ions are separated and accelerated toward another target. The radioactive ions are useful in studying heavy nuclei that are close to breaking up of their own weight. Other facilities using this technique are located at Michigan State University, as well as in Japan, France, and Germany.

The accelerator everyone is waiting for is CERN's Large Hadron Collider (LHC), an extremely powerful (14-TeV) proton-proton collider backed by 19 countries and expected to be in action by about 2004. There are various design problems that the LHC must surmount, notably the large amount of hard radiation that will be produced when it is in operation. This radiation is more likely to damage the equipment than the researchers. Another problem involves finding ways to separate useful data from the vast number of particle collisions that will take place in each run.

While waiting for the LHC, physicists must be satisfied with the Tevatron at Fermilab, currently the most powerful accelerator. The Tevatron produces only 1.8 TeV, less than one-seventh of the planned power of the LHC. The Tevatron will itself be upgraded in 1999, after which it will come closer to the LHC's expected range.

(Periodical References and Additional Reading: *New York Times* 11-27-92, p A23; *New York Times* 10-5-93, p C7; *New York Times* 4-19-94, p C1; *Physics Today* 7-94, pp 22 & 33; *Scientific American* 9-94, p 40; *Physics Today* 11-94, p 80; *Science* 3-3-95, p 1266; *Science* 3-10-95, p 1423; *Science* 3-31-95, p 1904; *Physics Today* 5-95, pp 59, 60, & 61; *Physics Today* 8-95, p 9; *Science* 9-1-95, p 1218; *Physics Today* 10-95, p 21)

Superconductivity UPDATE

In the late 1980s, the promise of high-temperature superconductivity was a major story in newspapers and other mass media. When nothing superconducting and dramatic moved from the laboratory into the factory, home, or office, the story seemed to die down. That popular media impression was deceptive, however, as significant progress in the field of superconductivity has continued. In the early 1990s, however, that progress was reported mostly in science journals instead of newspapers.

High-temperature superconductivity is not the first form of superconductivity, as most people know, but the variety of different kinds of superconductors known today is surprising to people outside the group of physicists investigating the phenomenon. Traditional superconductivity at temperatures close to absolute zero occurs mostly in metals–originally lead and mercury, but more recently aluminum and zinc. High-temperature superconductivity is based on ceramics, especially those called cuprates, based on copper oxide. The most popular of the cuprates has become $YBa_2Cu_3O_7$ (known as YBCO) because it performs better in magnetic fields. Other cuprates in this field are usually named in full, such as thallium barium calcium copper oxide and thallium lead strontium calcium copper oxide. The thallium cuprates can be made into wires, but they fail when confronted with magnetism. In between the high and the low temperature superconductors are the fullerenes doped with appropriate ions. There are also organic superconductors and other variations, including what are called heavy-fermion superconductors that tend to be based on uranium. In 1994, Y. Maeno of Hiroshima University in Japan and coworkers discovered a form of perovskite (one of the main minerals in Earth's mantle) that can serve as a superconductor, although not a high-temperature one. Its great value to researchers is that it has the same structure as a cuprate without copper, allowing for a better determination of what makes cuprates work at high temperatures.

Better Superconductors and Theories about Them

The first ceramic, or oxide, superconductors were discovered in 1964, at the tail end of an important round of activity in the field. They languished until 1986, when one of them was found to break the old temperature record for superconductivity by a substantial amount. After that discovery, there appeared a flurry of well-publicized

Background

In 1908, when physicist Heike Kamerlingh Onnes [Dutch: 1853–1926] liquefied helium, a by-product of the process resulted in temperatures only a few
degrees above absolute zero. Experimenting with the cold liquid, Kamerlingh
Onnes discovered that an electrical current in frozen mercury or supercooled
lead could proceed without facing any resistance at such very cold temperatures,
although this property disappears in the presence of a sufficiently large magnetic
field. The newly discovered property was named **superconductivity**. In 1933, it
was discovered that a magnetic field not large enough to destroy superconductivity is excluded from the interior of superconductors. The cause of superconductivity was finally explained after 46 years by John Bardeen, Leon N. Cooper, and
John R. Schrieffer and termed the BCS theory. The BCS theory is explained in
terms of pairs of electrons bound together by small vibrations in the material
called phonons. These electron pairs are able to move through a material
effortlessly because they do not interact with the atoms.

BCS theory explains why superconductivity can take place only in materials
close to absolute zero, but it did not say how close. By 1967, the highest
temperature of superconductivity had passed 20K ($-423°F$ or $-253°C$). This
record was not to be substantially surpassed for almost 20 years, by which time
most researchers had stopped looking for higher temperatures. In 1986, when
Karl Alex Miller and Georg Bednorz discovered a ceramic oxide that is superconducting at 35K ($-396°F$ or $-238°C$), everyone was astonished. Soon other
"high" or "warm" temperature superconducting oxides were found. The record
of about 125K ($-234°F$ or $-148°C$) from the later 1980s still stands as the
highest temperature, despite occasional claims to have surpassed it. These
temperatures are all too high for BCS theory to explain, so it is assumed that high-
temperature superconductivity is based upon a different mechanism. Theorists
have not established a mechanism to explain high-temperature superconductivity satisfactorily, leaving experimenters to continue mostly on the basis of hints
and experience.

advances in the production of new materials that could reach higher and higher
superconducting temperatures. Today, the problems are more in the development of
materials that can carry higher electric currents, or that behave better in magnetic
fields, or that can be shaped into useful devices. Ceramics are not metals, and there
are problems in making wires, for example.

Another more fundamental problem is that high-temperature superconductors are
unable to conduct high electric currents. This has prevented the technology from
being used in most applications. Vortices, small magnetic whirlpools that appear
throughout the material in the presence of any substantial magnetic field, are the
cause. The tiny screwlike structures were first reported to the journal *Science* by a
research group from Los Alamos on December 27, 1990, and separately reported to
Nature by Georg Bednorz and coworkers from the IBM Research Division in Zurich,
Switzerland on February 15, 1991. Internal publishing schedules resulted in the
Nature report being published on March 28, 1991, and the *Science* report on March
29, 1991. These vortices interfere with the superconductivity, inducing resistance to

current. Since an electric current always produces a magnetic field, understanding the vortices and finding ways to deal with them has become one of the main goals of superconductivity research. The vortices show up not only in their effects, but can actually be imaged with scanning tunneling microscopes, magnetic decoration (essentially using iron filings to trace a magnetic field), and an adaptation of the atomic force microscope called the magnetic force microscope (*see also* "Nanotechnology UPDATE," p 732).

In 1993, Japanese researchers led by Yasuhiro Iijima developed a method of improving upon YBCO, the best high-temperature superconductor for performance in the presence of magnetic fields. They aligned crystals of YBCO by depositing it on a naturally aligned substance with a similar structure. In 1995, scientists for Los Alamos National Laboratory announced that they had found a way to make a superconducting tape that was considerably better. Their method, however, was simply to improve upon the Japanese idea by making the crystals in the substrate better aligned prior to depositing the YBCO on it. Critics said that the process was far too slow to become of use in any large-scale application. Another team working on the problem at Oak Ridge National Laboratory also reported progress using what they said was a faster method of improving the substrate.

A principal barrier to advances in superconductivity has been that the theorists have had difficulty agreeing on an explanation of the causes of the effect in the high-temperature materials. By 1993, however, most of the theorists had reached agreement, postulating that the mechanism was something like BCS superconductivity with the exception that the current is carried by pairs of electrons of opposite spin that do not need phonons to form pairs. The details of how they could do this were difficult to resolve at that time.

Good theories lead to good experiments, just as surely as the right experiments produce the best theories. High temperature superconductors were studied with almost every possible probe in 1994–1995 in the effort to resolve such basic issues as the angular momentum of the electron pair and what attracts one electron to the other in the first place. Most of the experiments found that the pair behaves like a boson with spin (the quantum angular momentum number) 2. Still, not all the experiments produce this result. The exact choice of measurement to make seems to make a significant difference. Either further experiments are still needed, or a different theory.

Meanwhile, German scientists from the Institute for Solid State Research in Jülich and from the Max Planck Institute in Stuttgart used a laser to detach individual electrons from molecules of buckminsterfullerene, or buckyballs (*see* "Ornamented and Filled Buckyballs," p 457), in an investigation of why fullerenes are superconducting at low temperatures. They interpreted their results as indicative that interactions between electrons and phonons cause superconductivity in fullerenes as they do in some metals when kept close to absolute zero. Doped fullerenes, however, are superconducting at about 30K ($-380°F$ or $-240°C$), much warmer than the requirements for superconducting metals, although some tens of degrees short of temperatures for the high-temperature cuprate superconductors.

The Future

Most progress in the 1990s has been in small increments, rather than in any basic new approaches, techniques, or discoveries.

In 1995, Arunava Gupta and coworkers at IBM's Thomas J. Watson Laboratory in Yorktown Heights, New York, produced much higher current densities in a high-temperature superconductor, reaching 100,000 amperes per cm^2 at temperatures of 110K ($-236°$F or $-163°$C). The team used a mercury cuprate superconductor, which produced results 10 times better than thallium- or bismuth-based materials, although it was necessary to keep any magnetic fields parallel to the current. Perpendicular magnetic fields disrupted the ability of the material to superconduct.

Applications of high-temperature superconductivity are beginning to reach certain specialized markets. Sensors for magnetometers were the first commercial products to be produced and sold in quantity. These are also known as SQUIDs (superconducting quantum interference devices) and have a multitude of scientific and medical applications. Other scientific and medical applications, especially improved magnetic resonance imaging devices, are in the pipeline, and one estimate has this market as high as $75 billion–100 billion a year by 2020. Military applications have been heavily funded, with mine detectors as one of the most promising applications. One of the main successes for experimental high-temperature superconducting devices has been in filters for separating microwave signals and cellular telephone signals. The same technology can also improve doppler radar systems used in weather forecasting. In 1996, a satellite filled with high-temperature superconductor devices for testing in space is expected to be launched. An earlier try with a similar satellite failed to make orbit.

Although slow, progress is being made toward the development of high-temperature superconducting wires that can be used in generating or transmitting electric power. If those applications become available, then there may be levitated trains or fast computers based on this technology sometime in the future. After all, the first high-temperature superconductor was only created in 1986, so the amount of progress in the first 10 years is quite significant.

(Periodical References and Additional Reading: *Physics Today* 5-93, p 17; *Nature* 12-8-95, pp 501, 502, & 532; *Physics Today* 2-95, pp 9 & 32; *Physics Today* 3-95, p 20; *Science News* 4-8-95, p 223; *Science* 5-5-95, p 644; *Physics Today* 11-95, p 19; *Physics Today* 1-96, p 19)

PHYSICS LISTS AND TABLES

SUBATOMIC PARTICLES

During the nineteenth century, most scientists came to believe that everything was made from atoms, even though there was no way then to detect atoms. We now know that they were right, and "photographs" of individual atoms are even available, thanks to images made with the scanning tunneling microscope. Early in the twentieth century, physicists discovered that atoms themselves are made from smaller pieces, so-called **subatomic particles**. At first, these smaller pieces seemed the ultimate limit of matter, although more and more of them kept appearing. Then, in the 1960s, physicists proposed that many subatomic particles are themselves made from smaller particles that are not detectable. Like the undetectable atoms of the nineteenth century, **quarks** (these undetectable smaller particles) have come to be quite commonly accepted. Perhaps in the next century there will be "photographs" of quarks.

Beside quarks, various other groups of particles are thought not to be made up of smaller pieces. Among these are the **leptons**. Together, the quarks and leptons form what we think of as matter. They are characterized by a spin of ½, as are the particles made from three quarks, the **baryons**. **Spin** is a number for each particle that behaves something like the number describing the rotation of an ordinary object. Other subatomic particles are similar to the **photon**, the particle that makes up light. Particles like the photon and other particles with a spin of either 0 or 1 are called **bosons**. Bosons act as the "glue" that holds matter together. They produce the four fundamental forces: gravity, electromagnetism, the strong force, and the weak force. Under certain conditions, electromagnetism and the weak force become a single force, the **electroweak** force.

Many subatomic particles carry a charge, which is a unit of electromagnetism. All charges are counted as either -1 or $+1$ or 0 (no charge), except for the quarks, which have charges in multiples of ⅓. No one knows why these charges come only in these particular amounts.

The masses of subatomic particles are measured in terms of the energy of the particle, because energy E is related to mass m and the speed of light c by Albert Einstein's famous equation $E = mc^2$. Since c is a very large number, the energy of a particle is much larger than its mass. Even so, the energy is expressed in a very small unit, the MeV, which is a million **electron volts**. An electron volt is the energy measured as 0.00000000000000000160217733 ($1.60217733 \times 10^{-19}$) joule (*see* "Units of Measure," p 786).

Every charged particle with spin ½ has an antiparticle whose charge is the opposite (negative particles have positive antiparticles) and whose spin is in the opposite direction. Even uncharged particles with spin ½ all have antiparticles, but the differences between the particle and antiparticle are more subtle. Only a few antiparticles are listed below, mainly as examples.

The information presented above and below and the theory from which it is derived have come to be known as the "standard model" of particle physics.

Recently, many physicists have proposed various other undetectable, or at least undetected, particles that are not part of the standard model. These, however, are omitted from the table.

Basic Particles of Matter (Fermions—all spin ½)

Leptons   The leptons are the "light particles," in contrast to the "heavy" and "middle-sized" particles described later in this compendium. There are six separate leptons.

The **electron** is generally considered to have been discovered in 1897 by British physicist Joseph John (J. J.) Thomson. Movement of electrons is the source of current electricity, while an excess or deficit of electrons causes static electricity. The properties of the electron form the basis of electronic devices such as computer chips. Electrons are found in all atoms, where they occupy several shells around the outside of the atom. Interactions between electrons account for all chemical reactions. Like all subatomic particles, the electron has a related particle called an antiparticle. The antiparticle of the electron is known as the **positron** (also called the **antielectron**). The positron is a mirror image of the electron with a charge of $+1$. The positron was the first antiparticle to be discovered. It was proposed in 1931 by Paul Adrien Maurice Dirac and discovered, serendipitously, by Carl David Anderson in 1932.

Muons and **tauons** are very poorly understood particles. The muon is often described as a "fat electron," since it has all the properties of an electron but its mass is 200 times as great. Similarly, the tau is a "fat muon," 17 times as heavy as a muon. No one predicted these particles, and no one knows what their role in the universe is. The muon was discovered in cosmic ray radiation by Carl David Anderson in 1937. The tauon was discovered by Martin Perl in 1975.

Particle	Charge	Mass (MeV)	Average Lifetime (sec)	Spin
Electron (e)	-1	0.51099906	Stable	1/2
Positron (ē)	$+1$	0.51099906	Stable	1/2
Muon (μ)	-1	105.658389	2.2×10^{-6}	1/2
Tauon (τ)	-1	1784	NA	1/2

The **neutrino family** is a group of particles associated with electrons, muons, and tauons. For each of the three charged particles there is a **neutrino** and an **antineutrino**. Aside from the fact of their associations, there would seem to be no difference among the three neutrinos. Originally, when the muon-associated neutrino, or mu neutrino, was found in 1961, it was said that the mu neutrino was "muon flavored," while the ordinary neutrino was "electron flavored."

They do not interact strongly with anything. Neutrinos are passing through your body all the time, and nearly all of them go on to pass through Earth and out the other side. Neutrinos were predicted by Wolfgang Pauli in 1930, and named neutrino ("little neutral one" in Italian) by Enrico Fermi in 1932. They were first found experimentally by Clyde Lorrain Cowan and Frederick Reines in 1955. Some evidence suggests that neutrinos have a very small mass (*see* "Particle Physics UPDATE," p 633).

Particle	Charge	Mass (MeV)	Average Lifetime (sec)	Spin
Electron neutrino (v_e)	0	Less than 2×10^{-5}	Stable	1/2
Mu neutrino (v_μ)	0	Less than 0.3	Stable	1/2
Tau neutrino (v_τ)	0	Less than 35	Stable	1/2

Quarks   Quarks were first proposed in 1964 by Murray Gell-Mann, and independently by George Zweig, to account for the relationships between various kinds of baryons. Each baryon is composed of three quarks, and each meson of two quarks. Baryons and mesons of ordinary matter are composed of quarks that are known as **up** and **down**. Even though their individual masses are small, the binding energy between quarks in a baryon produces most of the mass in the universe. Other baryons or mesons have a quality known as "strangeness," which is conferred by the **strange** quark, or a quality known as "charm," conferred by the **charm** quark. Two other quarks, known as **top** and **bottom** (once also called **truth** and **beauty**) complete the list of six quarks.

Each quark also comes in one of three "colors," although physicists disagree on what to call the colors (red, blue, and green comprise one set of popular choices). In a meson or a baryon, the colors must be combined so as to produce absence of color, so none can be detected directly. Furthermore, quarks are confined within the particles they make up and cannot be directly detected. Various experiments, however, have established that all actually exist, with the top quark the last to be confirmed in the early 1990s. Two different experiments so far have found slightly different values for the mass of the top quark.

Quark	Charge	Mass (MeV)	Average Lifetime (sec)	Spin
Up (u)	+2/3	5	Stable	1/2
Down (d)	−1/3	8	Stable	1/2
Strange (s)	−1/3	175	Stable	1/2
Charm (c)	+2/3	1270	N.A.	1/2
Top (t)	+2/3	$176,000 \pm 13,000$ or $199,000 \pm 30,000$	10^{-18}	1/2
Bottom (b)	−1/3	4250	Less than 5×10^{-12}	1/2

The six different quarks are sometimes, rather confusingly, called six different flavors of quark, a idea derived apparently from the "muon flavored" neutrino. It is more helpful to relate the three lepton families to three families of quarks. Each family has a heavy member and a light member among the leptons and among the quarks.

	Heavy member	*Light member*
First Family:	electron and up quark	electron neutrino and down quark
Second Family:	muon and charm quark	mu neutrino and strange quark
Third Family:	tauon and top quark	tau neutrino and bottom quark

Because experiments at the Large Electron Positron collider at CERN seem to have ruled out a fourth family of leptons, most physicists also believe that there is no fourth family of quarks.

The particles that consist of combinations of quarks are collectively called **hadrons**, because they are affected by the strong force. There are two kinds of hadrons, the "heavy particles" or **baryons**, and the "middle-sized particles" or **mesons**.

Baryons (All Spin ½)

Ordinary baryons   The **proton** was the first baryon to be discovered (by Ernest Rutherford in 1914). It is found in the nucleus of the atom. For about 20 years, it was assumed that atoms consist of a core of protons surrounded by electrons, although the actual situation is somewhat more complex. The proton appears to be stable, although some recent theories hold that protons may decay into pure energy after about 10^{31} years. So far, despite constant observation, no one has seen a proton decay. Every ordinary atom contains an equal number of protons and electrons, giving each atom a total charge of 0; but circumstances often add or remove an electron or two, producing a charged version of an atom called an **ion**.

The **neutron** is almost exactly like a proton, but because it has no charge it was much harder to detect. James Chadwick discovered the neutron, which had not been predicted, early in 1932. Neutrons in the nuclei of atoms are usually stable, but neutrons left to themselves soon decay into a proton, an electron, and an electron antineutrino. All atoms except hydrogen must have neutrons in their nuclei for the atoms to be stable. Each neutron has a mass of 939.6 MeV.

Strange Baryons   Other baryons include two **hyperons**, three **sigmas**, two **xis**, and an **omega**. Except for the omega, none was predicted. Instead, the baryons were found in cosmic-ray and particle-accelerator experiments in the late 1940s and 1950s. In 1961, Murray Gell-Mann predicted the omega on the basis of a theory that was preliminary to the quark theory. The omega was discovered in 1964. None of these particles is stable, decaying after much less than a second into other particles. They are not constituents of ordinary matter, and are "strange" because each contains a strange quark. The table also includes some charmed baryons, which are not strange.

In physics, a bar over the symbol for a particle indicates the symbol for the corresponding antiparticle, so p̄ means antiproton and ū refers to the anti-up quark.

Particle	Quark content	Charge	Mass (MeV)	Average Lifetime (sec)	Spin
Proton (p)	uud	+1	938.27231	Stable or 10^{31}	1/2
Antiproton (p̄)	ūūd̄	−1	938.27231	Stable or 10^{31}	1/2

Particle	Quark content	Charge	Mass (MeV)	Average Lifetime (sec)	Spin
Neutron (n)	udd	0	939.56563	9.18×10^2	1/2
Antineutron (n̄)	ūd̄d̄	0	939.56563	8.18×10^2	1/2
Lambda (Λ)	uds	0	1116	3×10^{-10}	1/2
Charmed lambda (Λ_c)	udc	+1			1/2
Positive sigma (Σ^+)		+1	1189.4	8×10^{-9}	1/2
Neutral sigma (Σ^0)		0	1192	6×10^{-20}	1/2
Negative sigma (Σ^-)		−1	1321	1.5×10^{-10}	1/2
Neutral xi (Ξ^0)		0	1315	2.9×10^{-10}	1/2
Negative xi (Ξ^-)		−1	1321	1.6×10^{-10}	1/2
Omega minus (Ω^-)	sss	−1	1672	1.3×10^{-10}	1/2

Mesons (Spin 0 or 1)

Pions were predicted in 1935 by Hideki Yukawa. They are the bosons that hold the nucleus of an atom together, and come in positive, negative, and neutral forms. When a pion is exchanged between a proton and a neutron, it can change each into the other particle. In the process it produces the strong force, which is needed to keep the positively charged protons from rushing apart due to the electromagnetic force (positive charges repelling each other). When the muon was first discovered in 1937, scientists thought it was the particle Yukawa had predicted, but by 1945 it was known that the properties of the muon were wrong for that role. In 1947, Cecil Frank Powell and coworkers located the pion in cosmic rays. Pions, like all mesons, are composite particles made from two quarks. Other bosons are thought to be elementary particles, not made up of other particles.

Other mesons. Various short-lived mesons heavier than the pion incorporate such quarks as strange, top, and bottom. None of them are constituents of ordinary matter, but the neutral K mesons, or **kaons**, have been very important in experiments that extended basic physical theories. Only the kaons are given in the table.

Particle	Quark content	Charge	Mass (MeV)	Average Lifetime (sec)	Spin
Positive pion (π^+)	ud̄	+1	140	2.6×10^{-8}	0
Negative pion (π^-)	ūd	−1	189.6	2.6×10^{-8}	0
Neutral pion (π^0)	mix	0	135	8×10^{-17}	0
Positive kaon (K^+)	us̄	+1	493.7	1.2×10^{-8}	0
Negative kaon (K^-)	ūs	−1	493.7	1.2×10^{-8}	0
K-zero-short (K^0_s)	mix	0	497.7	9×10^{-11}	0

Particle	Quark content	Charge	Mass (MeV)	Average Lifetime (sec)	Spin
K-zero-long (K^0_L)	mix	0	497.7	5.2×10^{-8}	0

The neutral mesons are blends or mixes of particles. For example, the neutral pion is represented as $(u\bar{u} + d\bar{d})/\int 2$.

Other Bosons

The **photon** is the agent of electromagnetic radiation, including ordinary light, radio waves, microwaves, x rays, and gamma rays. In the nineteenth century, Thomas Young demonstrated the wave nature of electromagnetic radiation. However, in 1905 Albert Einstein showed that light also has a particle nature. The particle came to be called the photon. Exchanging photons causes charged particles to be attracted (if the charges are unalike) or repelled (if the charges are alike). When an electron absorbs or emits a photon of sufficient energy, it can change into a positron. If a positron and an electron meet, they disappear, leaving an energetic photon. Electrons can emit and absorb photons in other ways as well.

Gluons. Although the Yukawa theory of the pion seemed to explain the strong force, the quark theory soon led to the understanding that pions and the strong force are both side effects of a more essential strong force, one carried by eight neutral particles called gluons. Exchanging gluons between quarks usually causes quarks to change from one color to another, keeping the quarks attracted to each other. Most gluons are named for the transformation they induce.

Particle	Charge	Mass (MeV)	Average Lifetime (sec)	Spin
Red to blue	0	0	Stable	1
Red to green	0	0	Stable	1
Green to red	0	0	Stable	1
Green to blue	0	0	Stable	1
Blue to red	0	0	Stable	1
Blue to green	0	0	Stable	1
Neutral (1)	0	0	Stable	1
Neutral (2)	0	0	Stable	1

Other Vector Bosons Like the photon, the **positive** and **negative W particles** and the **neutral Z particle** are vector bosons. These three were predicted by the electroweak theory and produced and detected by Carlo Rubbia and coworkers in 1983. They are very massive. Two other particles of this class have been predicted, but not detected. They are the **Higgs particle**, a massive particle that helps give mass to the W and Z particles, and the massless **graviton**, which should have the same relation to gravity as the photon does to electromagnetism.

Particle	Charge	Mass (MeV)	Average Lifetime (sec)	Spin
Photon	0	0	Stable	1
W^+	+1	80,410	10^{-20}	1
W^-	-1	80,410	10^{-20}	1
Z^0	0	91,160	10^{-20}	1
Higgs particle*	0	100,000 to 1,000,000	N.A.	0
Graviton*	0	0	Stable	2

* Predicted, but not yet observed.

(Periodical References and Additional Reading: *New York Times* 5-31-94, p C10; *Scientific American* 9-94, p 40; *Physics Today* 5-95, p 17; *Physics Today* 8-95, pp 9 & BG9)

THE NOBEL PRIZE FOR PHYSICS

Date	Name [Nationality]	Achievement
1995	Martin Perl [American:1926-] Frederick Reines [American:]	Perl for discovery of the unpredicted tauon in 1977, a particle in the same class as the electron and muon, but much heavier; Reines for discovering the neutrino with Clyde Cowan in 1956. The neutrino had been predicted in 1930 by Wolfgang Pauli and was generally accepted long before Reines and Cowan observed it. Cowan died before the prize was awarded.
1994	Bertram Brockhouse [Canadian:] Clifford D. Schull [American: 1915-]	The use of neutron scattering to determine atomic structure. Working with Ernest O. Wolman (who died before the prize was awarded), Schull studied elastic scattering, starting right after World War II. In the 1950s, Brockhouse studied inelastic scattering.
1993	Russell A. Hulse [American 1951?-] Joseph H. Taylor [American 1941?-]	Discovery in 1974 of the first binary pulsar, which furnished indirect proof of the existence of gravity waves. Hulse was Taylor's graduate student when they made the discovery using the giant radio telescope at Arecibo, Puerto Rico. Four year later, Taylor and other coworkers also found slowing of the pulses caused by the emission of the gravity waves.
1992	George Charpak [Polish-French: 1924?-]	Invention in 1968 of the multiwire electronic detector for high-energy subatomic particles, the basic tool used in determining what is happening in particle accelerators.

Date	Name [Nationality]	Achievement
1991	Pierre-Gilles de Gennes [French: 1933-]	Studies of the ordering of molecules, especially in liquid crystals.
1990	Richard E. Taylor [American: 1929-] Jerome I. Friedman [American: 1930-] Henry W. Kendall [American: 1926-]	Confirmation of the existence of quarks, in experiments between 1967 and 1973.
1989	Norman F. Ramsey [American: 1915-] Hans G. Dehmelt [American: 1922-] Wolfgang Pauli [German: 1913-1993]	Development of the separated oscillatory field method and its use in atomic clocks (Ramsey) and the ion trap technique (Dehmelt and Paul).
1988	Leon M. Lederman [American: 1922-] Melvin Schwartz [American: 1932-] Jack Steinberger [American: 1921-]	Development of neutrino-beam methods and discovery of the muon neutrino.
1987	Georg J. Bednorz [Swiss: 1950-] K. Alex Müller [Swiss: 1927-]	Discovery of high-temperature superconductivity.
1986	Ernst Ruska [German: 1906-1988] Gerd Binnig [German: 1947-] Heinrich Rohrer [Swiss: 1933-]	Development of the electron microscope (Ruska) and the scanning tunneling microscope (Binnig and Rohrer).
1985	Klaus von Klitzing [German: 1943-]	Discovery of the quantized Hall effect.
1984	Carlo Rubbia [Italian: 1934-] Simon van der Meer [Dutch: 1925-]	Detection of the W and Z intermediate vector bosons, which communicate the weak interaction.
1983	William A. Fowler [American: 1911-] Subrahmanyan Chandrasekhar [American: 1910-]	Investigations into the aging and ultimate collapse of stars, leading to the creation of heavier elements.
1982	Kenneth G. Wilson [American: 1936-]	Theory of phase transitions.
1981	Nicolaas Bloembergen [American: 1920-] Arthur L. Schawlow [American: 1921-] Kai M. Siegbahn [Swedish: 1918-]	Advances in technological applications of lasers for the study of matter (Bloembergen and Schawlow) and high-resolution electron spectroscopy (Siegbahn).
1980	James W. Cronin [American: 1931-] Val L. Fitch [American: 1923-]	Studies in the violation of symmetry by subatomic particles.
1979	Steven Weinberg [American: 1933-] Sheldon L. Glashow [American: 1932-] Abdus Salam [Pakistani: 1926-]	Electroweak theory, unifying electromagnetism and the weak force of radioactive decay into a single force.
1978	Arno A. Penzias [American: 1933-] Robert W. Wilson [American: 1936-] Pyotr L. Kapitsa [Russian: 1894-1984]	Penzias and Wilson for discovery of cosmic background microwave radiation; Kapitsa for work in low-temperature physics.
1977	Philip W. Anderson [American: 1923-] Nevill F. Mott [English: 1905-] John H. Van Vleck [American: 1899-1980]	Fundamental investigations into the electronic structure of magnetic and disordered systems.
1976	Burton Richter [American: 1931-] Samuel C. C. Ting [American: 1936-]	Discovery of J/psi subatomic particle.

Date	Name [Nationality]	Achievement
1975	James Rainwater [American: 1917–1986] Ben Mottelson [Danish: 1926–] Aage Bohr [Danish: 1922–]	Studies proving asymmetrical structure of the atomic nucleus.
1974	Antony Hewish [English: 1924–] Martin Ryle [English: 1918–1984]	Hewish for discovery of pulsars and Ryle for invention of the aperture synthesis technique of radiotelescopy.
1973	Leo Esaki [Japanese: 1925–] Ivar Giaever [American: 1929–] Brian Josephson [English: 1940–]	Theories and discoveries on tunneling effects in superconductors and semiconductors.
1972	John Bardeen [American: 1908–1991] Leon N. Cooper [American: 1930–] John R. Schrieffer [American: 1931–]	Explanation of low-temperature superconductivity.
1971	Dennis Gabor [English: 1900–1979]	Invention of holography.
1970	Hannes Alfvén [Swedish: 1908–] Louis Néel [French: 1904–]	Alfvén for plasma physics and Néel for work with ferromagnetics and antiferromagnetism.
1969	Murray Gell-Mann [American: 1929–]	The classification of elementary particles known as the eight-fold way.
1968	Luis W. Alvarez [American: 1911–1988]	Discovery of many resonance states of elementary particles.
1967	Hans A. Bethe [American: 1906–]	Study of energy production of stars.
1966	Alfred Kastler [French: 1902–1984]	Optical study of subatomic energy.
1965	Julian S. Schwinger [American: 1918–1994] Richard P. Feynman [American: 1918–1988] Shin'ichiro Tomonaga [Japanese: 1906–1979]	Development of the basic principles of quantum electrodynamics.
1964	Charles H. Townes [American: 1915–] Nikolai G. Basov [U.S.S.R.: 1922–] Alexander Prokhorov [U.S.S.R.: 1916–]	Development of maser and laser principles in quantum mechanics.
1963	Eugene Paul Wigner [American: 1902–1995] Maria Goeppert-Mayer [American: 1906–1972] J. Hans D. Jensen [German: 1907–1973]	Wigner and Goeppert-Mayer for mechanics of proton-neutron interaction; Jensen for theory of structure of atomic nuclei.
1962	Lev D. Landau [U.S.S.R.: 1908–1968]	Explanation of superfluidity in liquid helium.
1961	Robert Hofstadter [American: 1915–1990] Rudolf Mössbauer [German: 1929–]	Hofstadter for measurement of nucleons; Mössbauer for work on gamma rays.
1960	Donald Glaser [American: 1926–]	Invention of the bubble chamber for subatomic study.
1959	Emilio Segrè [American: 1905–1989] Owen Chamberlain [American: 1920–]	Demonstration of the existence of the antiproton.

Date	Name [Nationality]	Achievement
1958	Pavel A. Cherenkov [U.S.S.R.: 1904–1990] Ilya M. Frank [U.S.S.R.: 1908–1990] Igor Tamm [U.S.S.R.: 1895–1971]	Discovery and interpretation of principle of light emission produced by the motion of charged particles.
1957	Tsung-Dao Lee [Chinese: 1926–] Chen Ning Yang [Chinese: 1922–]	Prediction of violations of law of conservation of parity.
1956	William Shockley [American: 1910–1989] Walter H. Brattain [American: 1902–1987] John Bardeen [American: 1908–1991]	Studies on semiconductors and invention of the electronic transistor.
1955	Willis Lamb Jr. [American: 1913–] Polykarp Kusch [German/American: 1911–1993]	Lamb for measurement of the fine structure of the hydrogen spectrum; Kusch for measuring the magnetic momentum of electron.
1954	Max Born [German/English: 1882–1970] Walther Bothe [German: 1891–1957]	Born for work in statistical interpretation of quantum mechanics; Bothe for studies of particles in cosmic radiation.
1953	Fritz Zernicke [Dutch: 1888–1966]	Invention of the phase-contrast microscope.
1952	Felix Bloch [American: 1905–1983] Edward Mills Purcell [American: 1912–]	Measurement of magnetic fields of atomic nuclei.
1951	Sir John Cockcroft [English: 1897–1967] Ernest T. S. Walton [Irish: 1903–]	Transmutation of atomic nuclei by use of a particle accelerator.
1950	Cecil Powell [English: 1903–1969]	Invention of method for studying subatomic particles with photographic plates.
1949	Hideki Yukawa [Japanese: 1907–1981]	Prediction of mesons and explanation of strong nuclear force.
1948	Patrick Blackett [English: 1897–1974]	Discoveries of particles in cosmic radiation.
1947	Sir Edward Appleton [English: 1892–1965]	Discovery of ionic layer in atmosphere.
1946	Percy W. Bridgman [American: 1882–1961]	Development of tools needed for high-pressure physics and studies based on them.
1945	Wolfgang Pauli [Austrian: 1900–1958]	Discovery of exclusion principle of subatomic particles.
1944	Isidor Rabi [American: 1898–1988]	Invention of the resonance method for studying the magnetic properties of nuclei.
1943	Otto Stern [American: 1888–1969]	Discovery of the magnetic momentum of the proton.
1942	No award	
1941	No award	
1940	No award	
1939	Ernest Lawrence [American: 1901–1958]	Invention of the cyclotron.

Date	Name [Nationality]	Achievement
1938	Enrico Fermi [Italian: 1901-1954]	Discovery of new radioactive elements and nuclear reactions caused by slow neutrons.
1937	Clinton Davisson [American: 1881-1958] George P. Thomson [English: 1892-1975]	Discovery of electron diffraction by crystals.
1936	Victor F. Hess [Austrian: 1883-1964] Carl D. Anderson [American: 1905-1991]	Hess for discovery of cosmic radiation; Anderson for the detections of the positron.
1935	Sir James Chadwick [English: 1891-1974]	Discovery of the neutron.
1934	No award	
1933	Paul Dirac [English: 1902-1984] Erwin Schrödinger [Austrian: 1887-1961]	Discovery of new equations for atomic theory.
1932	Werner Heisenberg [German: 1901-1976]	Discovery of quantum mechanics.
1931	No award	
1930	Sir Chandrasekhara Raman [Indian: 1888-1970]	Laws of light scattering.
1929	Prince Louis de Broglie [French: 1892-1987]	Prediction of wave character of electrons and other subatomic particles.
1928	Sir Owen Richardson [English: 1879-1959]	Studies on effect of heat on electron emission.
1927	Arthur Holly Compton [American: 1892-1962] Charles Wilson [Scottish: 1869-1959]	Compton for effect of wavelength change in x rays; Wilson for demonstrating visible ion tracings in the cloud chamber.
1926	Jean Baptiste Perrin [French: 1870-1942]	Discovery of discontinuous structure of matter and equilibrium of sedimentation.
1925	James Franck [German: 1882-1964] Gustav Hertz [German: 1887-1975]	Discovery of the laws of electron impact on atoms.
1924	Karl Siegbahn [Swedish: 1886-1978]	X-ray spectroscopy.
1923	Robert Millikan [American: 1868-1953]	Measurement of elementary electrical charge.
1922	Niels Bohr [Danish: 1885-1962]	Analysis of atomic structure from spectrograms of light emitted by hydrogen.
1921	Albert Einstein [German: 1879-1955]	Explanation of photoelectric effect as a light behaving like a particle.
1920	Charles Guillaume [Swiss: 1861-1938]	Invention of special nickel-steel alloys.
1919	Johannes Stark [German: 1874-1957]	Doppler effect in canal rays and spectral lines in electrical fields.
1918	Max Planck [German: 1858-1947]	Quantum theory of light.
1917	Charles Barkla [English: 1877-1944]	Discovery of characteristic x rays of elements.
1916	No award	

Date	Name [Nationality]	Achievement
1915	Sir William Bragg [English: 1862–1942] Sir Lawrence Bragg [English: 1890–1971]	Analysis of crystal structure using x rays.
1914	Max von Laue [German: 1879–1960]	Discovery of diffraction of x rays by crystals.
1913	Heike Kamerlingh Onnes [Dutch: 1853–1926]	Low temperature physics and liquefaction of helium.
1912	Nils Gustaf Dalén [Swedish: 1869–1937]	Automatic gas regulators for lighthouses and sea buoys.
1911	Wilhelm Wien [German: 1864–1928]	Discovery of laws of heat radiation.
1910	Johannes Van der Waals [Dutch: 1837–1923]	Development of the equation of state for liquids and gases.
1909	Guglielmo Marconi [Italian: 1874–1937] Karl Ferdinand Braun [German: 1850–1918]	Contributions to wireless telegraphy.
1908	Gabriel Lippmann [French: 1845–1921]	Use of interference effects to produce color photography.
1907	Albert A. Michelson [American: 1852–1931]	Development of interferometer and use to measure velocity of light.
1906	Sir Joseph Thomson [English: 1856–1940]	Investigations of electrical conductivity of gases.
1905	Philipp Lenard [German: 1862–1947]	Work on cathode rays.
1904	John Strutt, Lord Rayleigh [English: 1842–1919]	Discovery of argon.
1903	Antoine Henri Becquerel [French: 1852–1908] Pierre Curie [French: 1859–1906] Marie Curie [French: 1867–1934]	Becquerel for discovery of spontaneous radioactivity and Curies for later study of radiation.
1902	Hendrik A. Lorentz [Dutch: 1853–1928] Pieter Zeeman [Dutch: 1865–1943]	Discovery of effects of magnetism on radiation.
1901	Wilhelm K. Roentgen [German: 1845–1923]	Discovery of x rays.

Technology

THE STATE OF TECHNOLOGY

One dictionary defines technology as "the body of knowledge available to a civilization that is of use in fashioning implements, practicing manual arts and skills, and extracting or collecting materials." While this is the anthropological meaning of the word, it seems a broad enough definition. Today, technology is usually linked to science—but technology is not only *not* science, it does not even depend upon science. Until Galileo's time (the early seventeenth century), science as we know it today did not really exist, while technology of a type not totally unlike that of today had been a part of human experience for over two million years, preceding even the evolution of modern humans.

Today, we expect scientific discoveries eventually to contribute to technology. Galileo himself was among the first to connect technology and science, but in his day, technology informed the newly emerging fields of science. In Galileo's hands, the technological advance of the telescope became a scientific tool. If a necessary tool had not been developed by technology, however, Galileo was not above inventing it himself, as was the case with the first crude thermometer.

Since the seventeenth century, the relationship between technology and science has remained close. By the middle of the nineteenth century, the flow of information was often reversed, with science informing technology. In our century, it is clear that the two disciplines reinforce each other in many ways. The phrase "science and technology" now seems like a single idea, like "horse and buggy" or "bagels and lox." As with other famous pairs, the two disciplines remain separate, but work much better together.

A new technology always has the potential to change a small part of the way people live, or life in general. The chain saw revolutionized work with trees, but did not revolutionize society. At a higher level of pervasiveness is an invention like Velcro, which in less than 50 years has become ubiquitous. The transistor led to microprocessors that are found embedded everywhere today, from credit cards to supercomputers. Often, the inventor hasn't even fully anticipated the ultimate uses of the invention. Bell Labs made the first practical lasers, but their lawyers hesitated to apply for a patent because it was unclear whether lasers would have any use in communications. The patent eventually went to a graduate student who developed the same basic technological idea. Meanwhile, the laser has proved important in hundreds of applications; in conjunction with fiber optics, laser has become the basic instrument of communication by telephone.

Today's Information Revolution

Those readers who are over 50 may remember the year of the last information revolution. Television had been invented in the early years of the twentieth century at nearly the same time as modern radio. The technology, however, was still primitive until after World War II. In the late 1940s, better cameras and receivers for black-and-white television made TV first a luxury item, then a common part of life for the middle class by the early 1950s. Commercial color television began in 1953. About that same time, television became a necessity in every home in the United States, and very soon throughout the world. Although screens have become larger, cable and satellites provide more channels and sharper pictures, and videotape has added a new dimension to television, the basic technological impact occurred in the 1950s, changing the way people live.

Forty years later—not very long by revolution standards—it appears that yet another revolution is at hand. Like the television revolution, the roots of the new information revolution are deep, going back to the first computers of the 1940s; the first modem in 1958, time sharing in 1961, the first computer networks in 1972, the first wave of personal computers in 1977, and the first fiber-optic telephone lines the next year. All of these individual innovations had a major impact on business and on some professions, especially the various sciences. Yet, like television before the 1950s, there was little impact on the world at large.

Computer hardware and software developers alike knew that the key would be what had become known as a "killer ap," the nickname for a computer tool that users everywhere would instantly recognize as necessary. The personal computer itself became indispensable to accountants with the spreadsheet as the killer ap; to writers with word processing; to graphics designers with desktop publishing. The true killer ap, however, would be the one that everyone had to have, no matter what they did for a living.

By its nature, the killer ap would be a surprise (or else someone would already have invented it). The real surprise was that when one was finally identified, a host of people had invented it, and almost everyone had thought of it. The killer ap was essentially the "information superhighway" that politicians had called for early in the 1990s, but when it arrived, it came with no government support. Astonishingly, it was free, or nearly so. The killer ap was the World Wide Web.

In 1993, the U.S. government was talking about committing $2 billion to building an information superhighway, with the money to be spent on fiber-optic networks and supercomputer-managed databases. That same year, people began to become aware of the Internet, mostly due to news stories about computer viruses spread by the Internet. There were then about 15 million users around the world, with about 5 million of them scientists and the rest business people and computer nerds. The Internet grew out of local area networks and links set up by the U.S. Department of Defense or the National Science Foundation to connect universities.

The very first part of the Internet was ARPANET, set up in 1969 to connect Defense Department contractors. Because the Defense Department wanted the system to be invulnerable to attack, command nodes were spread out over the entire network. A packet switching system is used to transmit data over the 'net, which means that each packet of information chooses its own path and millions of packets can be handled at once on the same system. Packet switching and distributed control is what makes the Internet accessible to millions of people all at the same time.

Even so, the Internet was hard to use until the graphics system known as the World Wide Web was attached to the Internet. In December 1990, software writer Tim Berners-Lee wrote the program for the World Wide Web in an attempt to make use of the Internet easier for physicists at CERN in Geneva where he was employed at the time. In addition to adding graphics and sound capability, he linked various parts of the Web with hypertext, a system in which key words or phrases that are highlighted in one text allow a user quick access to a different file which also pertains to the highlighted word or phrase. In the summer of 1991, Berners-Lee installed the World Wide Web on the Internet. In September 1994, as more and more came to depend on the Wide World Web for basic communications, a group of more than 100 software companies (including I.B.M., Microsoft, and Sun Microsystems) formed the World Wide Web Consortium to maintain and set standards for the Web, with Berners-Lee as the director. At the end of 1995, the World Wide Web consisted of more than 30 million electronic pages of information stored on tens of thousands of computers. These pages are accessible via various different "search engines," many of which will look into all the pages in their domain, calling forth any in which specified words appear. The general use of hypertext allows a Web user to leap from site to site effortlessly with a point and a click.

Recognition that this resource could be of general use to young and old alike made the World Wide Web the killer ap for ordinary people. That recognition appeared to occur during the latter part of 1995, spurring a sudden rise in stocks associated with the Web and inducing many to "get connected."

One of the main virtues of the Web is the ability to show realistic pictures (as realistic as color television, at least) in an environment where the user can interact directly with sight and sound. Apple Computer's Quicktime, which appeared in 1992, added movie-like pictures and sound to existing Macintosh programming. Microsoft added these capabilities to the Windows operating system in 1995. As 1995 progressed, new and better ways of handling sound and motion, such as the Java programming language, became a part of the Web.

Global Positioning for Consumers

While the new information superhighway represents the most prominent technological change of the early 1990s, another revolutionary system was also seeping into the fabric of our lives, although by the end of 1995 it had not yet effected the penetration of the Internet. While the key to the Internet is that it is not located anywhere, the key to the global positioning system (GPS) is that one always knows exactly where one is located.

On June 26, 1993, the U.S. Air Force launched the last of 24 satellites that transmit digital signals encoded in radio waves. These transmitted signals contain two pieces of information: the precise time and the location of the satellite in space. The 24 satellites are arranged in six orbits, with four satellites following each other equidistantly in each orbit about 11,000 mi (17,700 km) high. As a result, any point on Earth is in radio range of four of the satellites at any time.

To use this system, one needs a GPS receiver, available in various forms from sets for backpacks to complex installations on oceangoing freighters. Each GPS radio receiver includes a clock that it synchronizes to the atomic clocks on the satellites, as well as a small single-purpose computer. The radio receives the four signals revealing the locations of the four satellites. The clock is used to determine the length of time each signal took to arrive, from which the computer determines the distance to each satellite. Each such distance defines a giant sphere in space. Knowing the distance to two satellites means that one is located on the circle in space formed where the two spheres intersect. A third sphere then intersects that circle at two points. Knowing the distance to a fourth satellite identifies at which of the two points the GPS receiver is located.

More sophisticated receivers also contain devices that can determine how fast and in what direction the receiver is moving. Some of these simply take two positions and compare them, which of course takes some time. The best GPS receivers can measure the frequency of the radio waves exactly. If the frequency is higher than the emitted radio signal, the GPS receiver is traveling toward the satellite or the satellite is traveling toward the receiver. Eliminating effects caused by the motion of the satellites gives an instantaneous velocity. Altogether, one can obtain information regarding latitude, longitude, altitude, velocity, and, of course, the correct time.

The GPS was originally a military system, and the military continues to use it. When the U.S. military allowed civilian access to the system, it also coded the information so that it could retain the *exact* locations for itself, allowing civilian users only an accuracy of a few tens of meters. More recently, the military has allowed civilian systems to become more accurate. There is no fee charged for the GPS signal.

Although the GPS system is cheap enough that moderately well-off hikers or mountain climbers can use a small unit to keep from getting lost, the main impact so far has been on navigation at sea and in the air. Gradually, GPS is replacing LORAN and even the blind-flying-and-landing systems used in smaller airports. In the United States, civilian pilots have been able to use GPS since 1993 and commercial airlines since 1994. The United States has also granted that right to any other civilian air traffic around the world. According to the U.S. Federal Administration Association, the use of GPS will not only make airport landings safer all over the world but will also allow additional commercial flights over places such as the Atlantic Ocean (where there are few landmarks), since planes using GPS can fly closer together in safety.

A few automobiles have been equipped with GPS, with more being added to the list each year. Rockwell's GPS-based navigation system allows a user to type in an address, following which a display will show written directions on how to get there, even providing voice directions to turn left or right as needed. In one automobile system (also available separately), not only does the screen identify a user's location and provide a map, but the system's computer also taps into a popular travel guide to list the local sights and suggest where one might wish to dine. The U.S. Transportation Department wants to integrate automotive GPS with other new technology to produce IVHW—the intelligent vehicle highway system, a part of Intelligent Transportation Systems. The Intelligent Vehicle Highway Society of America is financed jointly by the Transportation Department and private industry, with more than a dozen well-known American corporations (including General Motors and IBM) working together on standards and new devices. Some trucking companies use GPS to monitor the locations of their trucks, and the same system can be used for taxi fleets, police, or emergency vehicles. Other parts of IVHW will sense road traffic, reporting both to traffic controllers and automobiles or trucks equipped for this service.

Information Everywhere

Both the Internet and GPS are big parts of an information revolution that has the potential to change life in so many little ways so very gradually that we sometimes fail to grasp the immensity of the overall change. Today, anyone who works with people who are in another office, whether in the same building or across the world, expects all written communications to be sent instantly, or at worst by overnight delivery service if the document contains something that cannot be faxed or sent by E-mail. Mailing lists and other information make everyone in the United States vulnerable to companies that think they know what you want before you do. A grandmother need only order one catalog toy for a grandchild to be deluged with catalogs from dozens of manufacturers of toys and children's clothing, for example. Grocery stores offer their customers cards that allow access to certain sales but are really intended to let the store know about what the shopper is buying. Cash registers are programmed so that if your purchase includes an item of a particular brand, your receipt may come with a discount coupon for a different brand of the same item. Bar codes are used not only for pricing and tracking goods in stores, but also for tracking mail or checks.

Standard bar codes (the Universal Product Code, or UPC) were not good enough for United Parcel Service, which wanted to have more information in machine-readable form regarding package location. United Parcel created Maxicode, which contains more information and which can be scanned from any direction or angle, even when a package is warped. Other companies may also license the new technology, although the UPC will likely continue to be used in retail stores.

Scanned product codes represent a new language that is spreading around the world—a language that our computers can read but most humans cannot. The scanner, on the other hand, is a cheap accessory to a personal computer which can to some degree read ordinary English and copy it into a computer. Optical character reading (OCR) is available for less than a hundred dollars, including an even greater ability to import drawings and photographs onto the computer's hard disk. The artwork can be scanned and later printed out in color. If one is patient, the image can be sent over the Internet.

Joining the latest wave of technology is usually incremental and relatively painless. If an old computer printer breaks down, you can buy a new one for less than the old one cost. The new one will print color and have a higher resolution (dots per inch) than the old. Most computers come with the latest devices and software as part of the package. Automobiles, too, sport improved technology with advancing model years.

When the day comes to replace an old television set, VCR, or oven, one takes another step into the new age of information, whether such an advance was intentional or not.

(Periodical References and Additional Reading: *New York Times* 6-10-93, p B10; *New York Times* 11-16-93, p D4; *Discover* 1-94, pp 60 & 93; *New York Times* 1-5-94, p D2; *New York Times* 2-4-94, p D5; *New York Times* 2-18-94, p A16; *New York Times* 4-2-94, p 1; *New York Times* 5-8-94, p III-7; *New York Times* 8-7-94, p III-9; *Scientific American* 5-95, p 32; *New York Times* 12-18-95, p D1)

TIMETABLE OF TECHNOLOGY AND INVENTION TO 1992

2,400,000 B.C.	Ancestors of human beings begin to manufacture stone tools.
750,000 B.C.	Ancestors of human beings learn to control fire.
23,000 B.C.	People in the Mediterranean regions of Europe and Africa invent the bow and arrow.
7000 B.C.	People in what is current day Turkey begin to make pottery and cloth.
5000 B.C.	Egyptians start mining copper ores and smelting the metal.
3500 B.C.	The potter's wheel and, soon after, wheeled vehicles appear in Mesopotamia (now Iraq, Syria, and Turkey).
2900 B.C.	The Great Pyramid of Giza and the first form of Stonehenge (with only three stones) are built.
2000 B.C.	Interior bathrooms are built in palaces in Crete.
522 B.C.	Eupalinus of Megara constructs a 3,600 ft (1,100 m) tunnel on the Greek isle of Samos to supply water from one side of Mt. Castro to the other.
290 B.C.	The Pharos lighthouse at Alexandria is built.
260 B.C.	Archimedes [Greek: 287–212 B.C.] develops a mathematical description of the lever and other simple machines.
200 B.C.	The Romans develop concrete.
140 B.C.	The Chinese start making paper, but do not use it to write on. (Early paper was used as a kind of cloth.)
100 B.C.	Water-powered mills in Illyria (now Yugoslavia and Albania) are introduced.
1	About this time the Chinese invent the ship's rudder.
100	The Chinese are beginning to use paper for writing on.
190	The Chinese develop porcelain.
600	The first windmills are built in what is now Iran.
704	Between 704 and 751, the Chinese start printing with woodblocks.
1040	The Chinese develop gunpowder.
1041	Between 1041 and 1048, Pi Sheng invents movable type in China.
1070	The Chinese begin to use the magnetic compass for navigation.
1190	The first known reference to a compass is made in Europe.
1267	A book written by Roger Bacon [English: 1220–1292] in 1267–1268 mentions eyeglasses to correct farsightedness.

1288	The first known gun, a small cannon, is made in China.
1310	Mechanical clocks driven by weights begin to appear in Europe.
1440	Johann Gutenberg [German: 1398–1468] reinvents printing with movable type about this time.
1450	Nicholas Krebs develops eyeglasses for the nearsighted.
1590	About this time, the compound microscope (using two lenses) is invented in Holland, probably by Zacharias Janssen.
1608	The telescope is invented in Holland, probably by Hans Lippershey [German/Dutch: 1570–1619].
1620	Cornelius Drebbel [Dutch: 1572–1633] builds the first navigable submarine.
1642	Blaise Pascal [French: 1623–1662] invents the first adding machine.
1643	Evangelista Torricelli [Italian: 1608–1647] makes the first barometer, and in the process produces the first vacuum known to science.
1654	Christiaan Huygens [Dutch: 1629–1695] develops the pendulum clock.
1658	Robert Hooke [English: 1635–1703] invents the balance spring for watches.
1671	Gottfried Wilhelm Leibniz [German: 1646–1716] builds a calculating machine that can multiply and divide as well as add and subtract.
1698	Thomas Savery [English: c. 1650–1715] patents the "Miner's Friend," the first practical steam engine.
1709	Gabriel Daniel Fahrenheit [German-Dutch: 1686–1736] invents the first accurate thermometer.
1733	John Kay [English: 1704–1764] invents the flying-shuttle loom, which, along with improvements in making iron and the steam engine, is a key to the start of the Industrial Revolution.
1751	Benjamin Huntsman [English: 1704–1776] invents the crucible process for casting steel.
1761	A clock called a marine chronometer, built by John Harrison [English: 1693–1776], is shown to be accurate within seconds over a long sea voyage, making modern navigation possible.
1764	James Hargreaves [English: 1720–1778] introduces the spinning jenny, a machine that spins from 8–120 threads at once.
1765	James Watt [Scottish: 1736–1819] builds a model of his improved steam engine.
1769	Richard Arkwright [English: 1732–1792] patents the water frame, a spinning machine that complements the spinning jenny.
1783	Joseph-Michel Montgolfier [French: 1740–1810] and Jacques-Etienne Montgolfier [French: 1745–1799] develop the first hot-air balloon.
	Jacques Charles [French: 1746–1823] builds the first hydrogen balloon.
1792	William Murdock [Scottish: 1754–1839] is the first to use coal gas for lighting.
1793	Eli Whitney [American: 1765–1825] invents the cotton gin, a machine for separating cotton fibers from seeds.

1807	Robert Fulton [American: 1765–1815] introduces the first commercially successful steamboat.
1822	Joseph Niepce [French: 1765–1833] produces the earliest form of photograph.
1825	George Stephenson [English: 1781–1848] develops the first steam-powered locomotive to carry both passengers and freight.
1837	Samuel Finley Breese Morse [American: 1791–1872] patents the version of the telegraph that will become commercially successful.
1839	Louis-Jacques Daguerre [French: 1789–1851] announces his process for making photographs, which come to be called daguerreotypes.
	Charles Goodyear [American: 1800–1860] discovers how to make rubber resistant to heat and cold, a process called vulcanization.
1856	Henry Bessemer [English: 1813–1898] develops the way of making inexpensive steel now known as the Bessemer process.
1859	Edwin Drake [American: 1819–1880] drills the first oil well in Titusville, Pennsylvania.
1876	Alexander Graham Bell [Scottish/American: 1847–1922] invents the telephone.
	Karl von Linde [German: 1842–1934] invents the first practical refrigerator.
1877	Nikolaus Otto [German: 1832–1891] invents the type of internal combustion engine still used in most automobiles.
1878	Louis-Marie-Hilaire Bernigaud [French: 1839–1924] develops rayon.
1879	Thomas Edison [American: 1847–1931] and Joseph Swan [English: 1828–1914] independently discover how to make practical electric lights.
1885	Karl Benz [German: 1844–1929] builds the precursor of the modern automobile.
1889	Gustave Eiffel [French: 1832–1923] builds his famous tower in Paris. At 993 ft (303 m) it is the tallest free-standing structure of its time.
1890	Herman Hollerith [American: 1860–1929] develops an electronic system based on punched cards, which is used in counting the United States census. Later, he founds the company that will become IBM.
1893	Rudolf Diesel [German: 1858–1913] describes the Diesel engine.
1903	Orville Wright [American: 1871–1948] and Wilbur Wright [American: 1867–1912] fly the first successful airplane.
1904	John Fleming [English: 1849–1945] develops the first vacuum tube.
1909	Leo Baekeland [Belgian/American: 1863–1944] patents Bakelite, the first truly successful plastic.
1929	Robert Goddard [American: 1882–1945] launches the first instrumented, liquid-fueled rocket.
1930	Frank Whittle [English: 1907–] patents the jet engine.
1931	Ernst Ruska [German: 1906–] builds the first electron microscope.
1937	John V. Atanasoff [American: 1903–] starts work on the first electronic computer.

Chester Carlson [American: 1906-1968] invents xerography, the first method of copying using ordinary, untreated paper.

1939   Paul Müller [Swiss: 1899-1965] discovers that DDT is a potent and long-lasting insecticide.

1941   John Rex Whinfield [English: 1901-1966] invents Dacron.

1948   William Shockley [English/American: 1910-], Walter Brattain [American: 1902-1987], and John Bardeen [American: 1908-1991] invent the transistor.

1957   Gordon Gould [American: 1920-] develops the basic idea for the laser, which he succeeds in patenting in 1986 after a long struggle.

1959   In a speech on December 29, Richard Feynman [American: 1918-1988] proposes that computers can be built from tiny transistors made by evaporating material away, that new products can be made by combining single atoms, and that circuits can be made from as few as seven atoms.

1965   John Kemeny [American: 1926-] and Thomas Kurtz [American: 1928-] develop the computer language BASIC, which for a time will become the main programming language for personal computers.

1970   The floppy disk is introduced for storing data on computers.

1971   The first computer chip is produced in the United States.

The first pocket calculator is introduced in the United States. It weighs about 2.5 pounds and costs about $150.

1975   The first personal computer (in kit form) is introduced in the United States. It has 256 bytes of memory.

1981   Gerd Binnig and Heinrich Rohrer invent the scanning tunneling microscope (STM).

The world's longest suspension bridge, 4,626 ft (1,410 m) long, opens over the Humber estuary in Great Britain.

1982   Compact disc players are introduced.

1985   Gerd Binning, Christoph Gerber, and Calvin Quate invent the atomic force microscope (ATM).

1986   The Chernobyl nuclear reactor number 4 explodes, killing dozens of people and making a large area around it uninhabitable.

1988   Computer chips based on reduced instruction set computing (RISC) are introduced by Motorola.

1989   On January 3, Japan initiates daily broadcasts of High Definition Television (HDTV) with a one-hour program featuring the Statue of Liberty and New York Harbor.

Japan's Ministry of International Trade and Industry establish the Laboratory for International Fuzzy Engineering Research to develop applications of fuzzy logic.

1990   On January 1, the volt and ohm are redefined in terms of atoms.

The Keck Telescope with a mirror of 36 segments, each 6 ft (1.8 m) in diameter, sees first light.

A railroad tunnel from England to France (dubbed the Chunnel) passes under the English Channel, with the two parts meeting on December 21.

<table>
<tr><td>1991</td><td>Woo Paik and coworkers produce the first prototype of digital HDTV.</td></tr>
</table>

1991   Woo Paik and coworkers produce the first prototype of digital HDTV.

On February 1, Philips Corporation demonstrates the digital audio cassette.

Jagdish Narayan, Vijay Godbole, and Carl White grow single-crystal diamond films on metal.

MAKING LIGHT

LIGHTING NEWS

Although we think of Thomas Edison as the inventor of *the* light bulb, there were successful bright lights powered by electricity well before Edison's invention. As early as 1650, Otto von Guericke demonstrated that electricity could produce light. By 1860, almost 20 years before Edison, incandescent lamps existed, the principle of fluorescent lights had been discovered, and electric arc lights were in regular use in lighthouses. Edison's accomplishment (like Ford's for the automobile) was to make electric lights practical and available. The electric incandescent light we use today is a far cry from the carbon-filament lamp invented by Edison, but he did make it all possible by building power plants and transmission lines as well as by selling the light bulbs.

A Better Light Bulb

Utility companies encourage their customers to abandon incandescent light bulbs. These bulbs, like Edison's original, produce light by using the resistance of a solid filament to an electric current to heat the filament sufficiently, thus causing electrons in the filament to become so energized that they shed higher-energy photons. That is, filaments in incandescent bulbs produce light as a by-product of heat. This is a wasteful way to light a room. For much less electricity, the electrons in a gas through which a current is passed can be induced to shed high-energy photons. These electrons are then absorbed by a solid coating inside the bulb, causing electrons in the coating to emit lower-energy photons that we perceive as light. Bulbs based on this principal are called fluorescent bulbs. Not only do they use less electricity, but they also last longer and produce less heat. For an office that is lighted throughout working hours, fluorescent bulbs producing the same amount of light as incandescent bulbs use only about two-thirds as much energy. The utility companies, which have found that it is cheaper to get customers to use less electricity than it is to build new power plants, encourage the use of fluorescent bulbs.

Consumers find fluorescent bulbs less satisfactory, in part because of a higher initial cost, but also because the quality of light is different from that of the incandescent light, since they use a different set of wavelengths. Also, until recent advances, fluorescent bulbs required special fixtures and had to be large and attached at two ends, unlike the simple screw-in incandescent bulb. Recent advances have resulted in a series of size reductions, down to what are called "compact" fluorescent lights which screw into the sockets of ordinary fixtures. Some of the new compact bulbs introduced in 1995 are very similar in size to the incandescent lights they can replace. Also in 1994, General Electric and Motorola teamed to produce a new and better ballast, the essential device in a fluorescent bulb that serves to start the arc through the gas which causes the discharge that makes the bulb glow.

For about twice the initial cost of a compact fluorescent bulb, and a cost of operation that is only slightly greater, the customer may be induced to buy an E-bulb.

The E-bulb is a different kind of fluorescent bulb that was invented in 1992. The main difference is that instead of an electric arc passing through the gas to cause it to shed high-energy photons, radio waves are used. After that, the same procedure follows, with the actual light originating in a phosphorescent inner coating on the bulb. One of the main advantages of the E-bulb is that it comes closer in size to the standard electric light bulb and will fit where the compact fluorescent bulb will not. Like the compact fluorescent, the E-bulb has a much greater life span and lower energy cost than the incandescent bulb. The E-bulb even has a much greater life-span than a fluorescent bulb. An E-bulb in ordinary use (typically four hours a day) could last 14 years without replacement—more than twice as long as a compact fluorescent, and more than 20 times as long as an average incandescent bulb.

In 1994, a truly different kind of bulb was developed by Fusion Lighting, Inc. Microwaves are used to induce sulfur to emit a bright yellow light. The new bulb is so bright that a small bulb the size of a golf ball could be used to produce as much light as hundreds of the high-intensity mercury-vapor lamps of the type now used primarily to light sporting events. Initially, the microwave light will most likely be restricted to lighting large outdoor spaces.

Or No Bulb at All

In science-fiction one often encounters lighting for which there is no bulb at all. A wall simply glows with light as needed. Making such a wall in some way other than putting bulbs behind a translucent panel has proved very difficult in the real world. Recently, the greatest progress has been toward the development of light-emitting diodes (see also "Display Progress," p 691). These would produce a white light, but this technology is far from the point of commercial availability.

White light, as Newton showed, is a mixture of all different wavelengths. A light-emitting diode, or LED, is normally set by its chemical composition to produce light of one particular wavelength only. However, as we learned in art class, white is also the combination of the three primary colors. The primary colors for combining light are different from those for pigments: while blue, green and red pigments combine to produce a muddy brown, combining these same colors of light produces white light. Using three layers of translucent organic LEDs, Junji Kido and coworkers at Yamagata University in Japan produced a small but working white-light LED. The actual LED is more complex than three layers, however, since there also need to be layers that inject holes, that inject electrons, and that partially block holes. The final device has eight layers, counting the glass on which the layers are placed. The white light that was produced was brighter than the light given off by a typical computer monitor and about half as bright as a fluorescent bulb. Use of a higher voltage and optimization of materials might bring the material to the luminance of a fluorescent bulb. The intention of these experiments is more along the lines of producing a display for computers than for lighting a room, but organic LEDs can be produced in virtually any shape. A problem with current organic LEDs, however, is that they break down too quickly with use.

Conventional LEDs can now also be combined to make white light, since they come in blue, green, and red. Used in the place of incandescent bulbs, these LEDs would have some advantages of lower power and longer life, although power requirements are greater than that of a compact fluorescent or an E-bulb producing the same amount of light. Colored LEDs are already used in some places for traffic signs.

(Periodical References and Additional Reading: *New York Times* 6-1-92, p A1; *Discover* 1-93, p 93; *New York Times* 3-2-94, p D4; *New York Times* 10-21-94, p A17; *Science* 1-6-95, p 51; *Science* 3-3-95, p 1262)

DISPLAY PROGRESS

The ability to change rapidly from one text or picture to another via mechanical, electrical, or electronic means is vital to the operation of all sorts of devices but is taken for granted much of the time by the consumer. The most primitive display devices include various dials for which the different possible outcomes (usually just numbers) are constantly visible, with a "hand" or arrow pointing to the one that is intended. Such devices began to emerge in the Middle Ages, along with clocks, although the first clocks did not have faces or any other visual display, merely ringing bells. Even before the clock face, however, there were clockwork or water-powered devices that showed the positions of astronomical bodies as a display. Earlier displays were often integral to the device itself, as for example the hourglass, in which the amount of sand that runs from one part to another is both the display and the mechanism behind the tool's function.

Even early calculators reported answers using dials, although in later models gears were used to produce displays of digits. The first computers used banks of lights as displays but soon moved to the teletypewriter for display. It was another long step before the typewriter keyboard could also be used as an input device. Banks of lights could also be used as displays by combining the individual lights from an array so that the pattern of its lights reproduced digits. Some later calculators also used this form of display.

In recent times, the display has been dominated by two different technologies, neither of which has been entirely satisfactory. The cathode ray tube (CRT), versatile enough for both text and pictures in many shades and tones (including the moving pictures of television), takes up too much space and may cause eyestrain if viewed long enough. If not properly adjusted, CRTs can also emit harmful radiation. Most desktop computers and almost all television sets today use cathode-ray technology, but scientists and inventors continually look for some other way to achieve an equivalent result without the disadvantages.

The other dominant technology today is the liquid crystal display (LCD), common on hand-held calculators, laptop computers, digital watches, and various appliances, from microwave ovens to VCRs. An LCD cannot produce as good a picture as a CRT and often cannot be read easily unless the light hits the screen at just the correct angle. On the other hand, LCDs are flat-panel displays, unlike the CRT (which must have several inches of depth to operate). The common LCD, called "passive," is made from two panes of glass enclosing a fluid that changes color when an electric current is applied. By applying a low voltage charge in the right places, the image appears. A more sophisticated LCD, used in better laptop computers and called "active," has its own transistor at each picture element, giving the device more flexibility as well as the ability to produce full color instead of just a single color on a different background.

Another type of display is called polyvision. Its advantage is that it uses a coating that turns black when it is electrified. By using a white background, polyvision can achieve a much greater contrast than that possible with a passive liquid crystal. Instead of using an electrode or transistor at each point on the display where color is needed, a polyvision picture element (or pixel) can be induced to turn black, or back to white, by selecting perpendicular coordinates from a grid, just as in drawing a graph in algebra. This technology has applications where visibility is important, such as displays for bathroom scales, meters, and calculators.

Light-Emitting Diodes

Before LCDs became common on handheld calculators, another form of display was used. This type of display still appears on some household appliances, especially those that simply display numbers. In its first form, this display produced an odd dull red glow, although more modern ones (while still appearing strangely radioactive) generally glow a brighter red, amber, orange, or light green. This display is called the light-emitting diode (LED), a solid-state cousin of the transistor. LEDs recently became available in blue, so they now span the color spectrum. As discussed below, experimental LEDs also produce white light.

A coded display that is somewhat out of the context of this discussion uses LEDs as the red brake lights in recent-model automobiles. Newly available green LEDs can be combined with red and amber to make traffic lights that are more energy efficient and require less maintenance, although at a higher initial cost.

An LED is a semiconductor device, a close cousin to the transistor. In semiconductors made from some solids (such as gallium arsenide or gallium nitride), two layers of material are prepared, one doped to have an excess of electrons and the other doped to have holes where electrons might otherwise be. When a current pushes electrons out of the first layer, they fall into the holes in the second and give up some energy in the form of a photon. The color of the light produced by an LED is normally determined by the basic material from which the semiconductor is built, as the material determines how far electrons have to fall to reach the holes.

In a suitably doped LED, the interaction can also cause nearby molecules to become excited. When the molecules fall back, they drop to a lower energy state, and the fall causes emission of a photon. The LED has been doped with the molecule, which is called a fluorescent dye. As before, the distance from one energy state to another decrees the wavelength of light that will be emitted, which can be controlled through the choice of the materials used to produce the electrons and holes or by the fluorescent dye molecule that is used as a dopant. In recent years, organic LEDs have been developed. These are easier to fabricate in thin films and can be made into a versatile display device.

In 1994 and 1995, two different groups of researchers produced organic LEDs that give off white light when an electric current is applied to the devices (*see also* "Lighting News," p 689). In 1994, Anath Dodabalapur and coworkers at AT&T Bell

Labs produced white light by embedding a diode that produces various wavelengths of light into a device that selectively reflects red, green, and blue toward the viewer and yellow and orange off to the side. The result is a combination of primary colors: white light. In 1995, a group led by Junji Kido of Yamagata University in Japan combined three different diode layers to make white light. Other techniques are being studied as well. Presumably, a white-light LED against a dark background would make a screen that is much easier to read than the present generation of LCD monitors.

The AT& T Bell Labs approach to making white light was just one application of a more general idea, which involves the insertion of small optical resonators possessing a wavelength or half a wavelength of a particular frequency of light. The resonator, called a microcavity, then promotes that single wavelength and prevents others. In the early 1990s, a number of groups built such microcavities with a variety of optically active materials in the wells. When microcavities are placed in an LED, the result can be engineered into a bright display device, or into a light source for use as a switch in optical communications (*see* "Optical Computing," p 715). The method increases light output by a factor of five in experiments, and in theory could do twice that well with some adjustments.

Another new technology for LEDs was developed by V. L. Colvin, M. C. Schlamp, and A. P. Allvisatos of the Lawrence Berkeley Laboratory in California. As in the use of microcavities, this one relies on a quantum effect. Resembling the layered method of producing white light, this LED achieves its effects by using different layers, one of which is a doped semiconducting polymer forming an organic LED on its own. The other working layer consists of very small crystals (nanocrystals) of cadmium selenide, which form an inorganic LED when layered with a hole donor like the organic polymer. The nanocrystals, which can be adjusted in size to produce different wavelengths of light in the red-yellow region of the spectrum (this is the quantum part), are deposited on the polymer, which acts as a green LED. Applying an electric current at a low voltage activates the nanocrystals while a higher voltage activates the polymer. Therefore, the same LED can be either red or green or a bit of both. This technique means that the color generated is a function of the level of electric current—a very desirable characteristic for a display.

(Periodical References and Additional Reading: *New York Times* 3-1-92, p III-7; *Nature* 8-4-94, p 354; *Science* 8-12-94, p 943; *Science* 1-6-95, p 51; *Science* 3-3-95, pp 1262 & 1332)

Laser UPDATE

The laser is now over 35 years old. If it were a person, it would be entering middle age. However, like an aging hippy, the laser continues to astound us with energy, oddity, and continual change. Although it might seem that lasers are already ubiquitous, new lasers and new uses for them are just over the horizon.

X-ray Lasers

The progenitor of the laser was the MASER invented by Charles Townes in 1953. Although essentially the same as a laser operating at microwave frequencies, it was

---

Background

The word **laser** is an acronym that describes exactly how the device works: light amplification by stimulated emission of radiation. The key part is the stimulation, although the whole phrase is needed for the device to work. If an atom or molecule has absorbed some energy from any source, the most likely way that this energy will be stored is in outer electrons. Unless energy is continually added (which is called pumping) the outer electrons spontaneously release the extra energy as light, which can be thought of as waves, or, perhaps more clearly in this instance, as the particles called photons. Normally this happens at random, and the energy emitted is absorbed by the nearby atoms. If the pumping has excited a lot of atoms, an incident light of the correct energy can cause the atoms to release their extra energy all at once. This burst of light, sometimes called lasing, has an unusual property in that all the photons are released at exactly the same time. The photons marching in step as they head away from the laser form what is called a coherent beam. The beam's unusual properties include the fact that all the light is the same wavelength, the waves all match their crests and valleys, and the photons tend to stick together and spread very little. As a result, the beam can be focused to carry all the energy released to a small point, a property that is used in many laser applications.

A semiconductor laser operates like a high-powered LED (*see* "Display Progress," p 691). Photons are released when electrons merge with holes rather than when electrons in atoms drop to a different state.

A single burst of light usually does not contain as much energy as is needed. Lasers nearly always have mirrors that bounce the light back and forth several times so that it can be used to stimulate other pumped atoms to give up their energy as well. If one of the mirrors allows some but not all of the light to escape, then the laser can be operated continuously or nearly so.

---

originally used as a powerful amplifier. Townes and others recognized that the same basic theory could be used to produce coherent electromagnetic waves at any wavelength from radio to gamma rays, but the physical materials needed to do so were not immediately apparent. Five years later, the physics was worked out for frequencies around red visible light. By 1960, the first ruby laser had been demonstrated. Subsequent work produced lasers made from various solids and gases that emit light at frequencies from infrared to ultraviolet. Just above the top of the ultraviolet is the bottom of the x-ray region. Although scientists and military planners could see great potential for an x-ray laser, producing one seemed just out of reach.

The main problem concerned the amount of energy it would take to pump enough atoms up to such a high level that when they descended to lower levels they would emit x rays. The energy need rises approximately as the fifth power of the wavelength. Since visible light is around 500 nanometers in wavelength as compared to less than 1 nanometer for an x ray, the energy needed for an x-ray laser is roughly 30 trillion times the amount needed for a light laser.

The U.S. Defense Department tried to tap the energy of nuclear explosions to produce x-ray lasers for the Strategic Defense Initiative ("Star Wars") program, but this is thought not to have been successful. Later, a powerful military laser called the alpha laser used chemical reactions similar to those in rocket fuel to achieve a high energy input. Even the laser at the Nova nuclear fusion experiment (*see* "Fusion Progress at Last," p 639) was tried as a power source, but it only produced low-frequency, or "soft," x-rays.

Finally, civilian scientists led by Charles K. Rhodes of the University of Chicago in Illinois found a kinder, gentler way to get a material substance to emit x-ray laser light. In 1994, they produced brief pulses of hard x rays from clusters of xenon atoms by exposing the clusters to extremely brief bursts of ultraviolet laser light. These atoms were ionized by the bursts in an unusual way. Instead of stripping off one or more of the outer electrons, the incident energy, with its wavelength carefully adjusted to match the amount needed for the purpose, caused the outer electrons to move back and forth. This oscillation in turn raised the energy levels of several electrons in an inner shell. If one pictures the atom in the old Bohr model, with electrons in several orbits about a nucleus, it is as if some of the electrons in inner orbits were raised to join those in the outer orbit. Since xenon already has eight electrons in its outer orbit it cannot accommodate any more, so some of these have to be occupied elsewhere in the cluster for this method to work. The researchers call the atom with inner electrons forced out a "hollow atom." It cannot, however, stay hollow for long, so the xenon electrons return with a rush to their original orbits, giving up their excess energy as an x ray in the 0.2 to 0.3 nanometer wavelength region.

If these hard x rays can be tamed as well as generated, they could be used for probing the inner secrets of complex molecules, just as softer x rays are used to image the interior of larger objects, such as human beings.

There may also be yet another way to obtain high-energy laser light (ultraviolet and x ray also) by inducing a different way for atoms to store and release energy. In the substances whose atoms or molecules are the workers in a laser, most of the atoms are normally in an unexcited state, with only a few of the atoms excited. When one of the few excited atoms relaxes by emitting a photon, the photon is usually absorbed by one of the previously relaxed atoms. Therefore, one excited atom is simply exchanged for another, with no photons emitted. Pumping the atoms with energy creates an unusual situation called a "population inversion," in which more of the atoms are excited than not. In this situation, an arriving burst of light simultaneously knocks all the atoms in the inversion into a relaxed state.

However, in a new approach pioneered by Martin O. Scully of Texas A&M University in College Station and coworkers, an entirely different notion is used. Instead of pumping up a population inversion, laser light is used to block absorption but not emission of photons. Those few atoms that are naturally excited then emit their photons, as when there is no stimulation, but now the previously relaxed atoms are constrained from getting excited. So the photons have to leave the substance as a coherent laser beam. Some pumping is needed to keep the absorption path from availability, but not as much as is necessary to cause a population inversion. Two lasers are used—a weak one to do the minimal pumping, and a strong one to block the absorption path.

The reason this new method might be useful in producing high energy lasers is that the large amount of pumping needed to produce a population inversion for ultraviolet or x-ray photons is eliminated. In the first two successful experiments with the new method, however, lasing was at lower wavelengths.

The same principles can also be used to produce effects of importance other than lasing. By using lasers to control optical properties in a gas, experimenters have been able to change the refractive index of the gas at will by preventing certain routes to absorption and emission of photons. In one experiment, transparency was induced for a specified wavelength. In another, the laser was used to produce a lens that could be adjusted to different situations by tuning the laser. In addition to using this effect to produce a lasing without population inversion, the method also shows promise for the creation of new kinds of microscopes, measuring instruments, and small-sized particle accelerators.

The Quantum Cascade Laser

The ability to deposit a few molecules in specific places in a controlled fashion has been among the main sources of new semiconductor technology in recent years. In January 1994, physicists at AT&T Bell Labs, led by Federico Capasso, were able to use the new technology to create a semiconductor laser that works by a relatively new principle without doping. The basic concept, however, was proposed as early as 1970 by Leo Esaki and Raphael Tsu, when both were with IBM. The concept was seen as a possible source of laser light the next year. Technology was not advanced enough to produce it effectively until the 1990s.

The quantum cascade layer consists of wells of indium gallium arsenide sandwiched between layers of aluminum indium arsenide. There are 500 separate layers, each built up a molecule at a time. The thin layers are quantum wells, so thin that they can only hold certain wavelengths of electrons. Each group of layers contains 20 layers of different thicknesses, alternating in composition, and producing three different-sized quantum wells. Despite the apparent complexity of the task, molecular beam epitaxy has already become commercially viable. With a properly programmed computer and the equipment for it to run, it is possible to turn out tens of thousands of quantum-well laser chips in only a few hours.

Applying a current to the chip causes electrons to pass from thinner into thicker wells, losing energy at each step of the way. As the electrons lose energy, they pour out photons, all of the same wavelength, creating a laser. The cascade is the electrons falling like a waterfall from one quantum well to another. One of the big differences between the quantum cascade laser and a conventional semiconductor laser is that there are no holes needed. All of the action is initiated by electrons, as in a gas laser or solid-state laser.

The advantage of the quantum cascade laser is that the size of the quantum wells controls the frequency of the laser light. A conventional semiconductor layer is not tunable in that way, since the frequency depends on the materials used in its construction. The main initial disadvantage of the quantum cascade laser is that it only works when cooled to near absolute zero, although the temperature needed has risen in later experiments to as high as 125K ($-325°F$ or $-150°C$). Furthermore, for the laser to work at all requires a commercially unacceptable amount of electric current. As a final disadvantage, it only fires in pulses, while most applications need a continuous beam. The difficulties are all interrelated, stemming from the high amperage needed for operation, which produces so much heat that the device needs to start very cool and operate in brief pulses. The physics of this laser is innovative, however, so later developments might take it anywhere.

If other problems can be resolved, the quantum cascade laser has several possible applications. This laser produces infrared light of a long wavelength that is not easy to obtain from other lasers. The atmosphere is largely transparent to those wavelengths unless polluted, so a quantum cascade laser device could be designed to monitor air pollution. Since its light normally travels almost unimpeded through air, applications that require communication in straight lines might also be developed, such as collision detectors for automobiles, or communication by two entities that are in sight of each other.

Applications UPDATE

When lasers were first invented, no one was quite sure what they could be used for. They were soon found to be suited for what seemed to be an endless array of applications. It is a rare American today who does not interact with a laser several times every day, although most of these interactions go unnoticed: code scanners at stores, CDs or CD-ROMs, fiber-optic telephone communication, and so forth. One does not see the laser, but it is the heart of the enterprise.

Scanning Interferometric Apertureless Microscope (SIAM)   In 1995, scientists of IBM's Thomas J. Watson Laboratory in Yorktown Heights, New York, led by Kumar Wickramasinghe, combined a laser source with a vibrating nanometer-meter silicon tip to process interference patterns from samples one-five-hundredth as small as can be imaged by visible light. One pass of the SIAM produces an interference pattern that can be used to obtain the spectrogram of the object, which has a laser focused on an extremely thin sample. Successive scans permit a tomographic analysis, similar to a CT scan, that allows a computer to put together a three-dimensional view of the object. Objects as small as five atoms were among the first successes. In principle, the SIAM can observe atoms and molecules. One expected application is in scanning ultracompact discs, which would use the same general technology as compact disks but with an incredibly high density for the stored information. The IBM scientists envision a diskette the size of a penny with a storage capacity equal to 30 of the new Digital Video Discs (DVDs), designed to store computer programs and motion pictures.

Fiber-Optic Connections   Fiber-optic systems use lasers to transmit messages over the fiber-optic cable and detectors at the other end. The system can transmit many messages at once by flicking the lights on and on at great speeds. By using lasers of different wavelengths, the volume of information over a single cable can be greatly increased. A strand of cable about as thick as a piece of damp spaghetti can carry as many as 100,000 telephone conversations at once if needed. A new type of laser is called a vertical cavity surface emitting laser (VCSEL), so called because the light is emitted perpendicular to that of other lasers. The VCSEL may be useful in fiber-optic light generation.

More and more businesses and even some homes are being directly connected to fiber-optic systems. In Japan, however, the telephone company plans to connect every home to a fiber-optic network was scaled back at the start of 1995, in part due to recognition of the limitations of fiber-optics, such as providing cable-free mobile communications.

Controlling Matter with Lasers   The use of lasers to tame matter in various ways is sweeping through vast reaches of science and technology. It is behind the methods of moving, holding, and changing the speed of individual atoms and molecules in a gas (*see* "The Newest Forms of Matter," p 645), as well as the brute-force transfer of energy that can even cause atoms to fuse (*see* "Fusion Progress at Last," p 639). A whole new technology is developing in the 1990s based on the use of a pair of lasers to adjust the energy levels of electrons in atoms (*see* "Data Storage and Retrieval UPDATE," p 708, and "Optical Computing," p 715). In this technique, the first laser is adjusted to match the level of energy needed to raise an electron into a specific excited state, leaving a "hole" behind. While the electron would, after a very short time, drop back down and fill the hole, in principle it can be kept from doing this by maintaining the laser. Minimally, in a group of electrons, enough would be kept in that state to make a device work. The second laser, however, can be set to be just out of phase with the first, so that the peaks of its waves come during the valleys of the first. When the second laser is added to the mix, it not only fails to hold the electrons in the higher state, but actively pushes the electron back to its original state. A more powerful version of this same technology is used to create the "hollow" atoms of x-ray lasing.

(Periodical References and Additional Reading: *New York Times* 6-30-92, p D5; *Science* 4-22-94, pp 503 & 553; *Physics Today* 7-94, p 20; *Nature* 8-25-94, pp 595 & 631; *New York Times* 8-25-94, p A14; *New York Times* 12-6-94, p C1; *Discover* 1-95, p 46; *New York Times* 1-10-95, p D2; *Nature* 4-20-95, p 679; *Physics Today* 5-95, p 9; *Physics Today* 9-95, p 19; *Science News* 9-30-95, p 223; *Science* 10-6-95, p 29; *New York Times* 12-19-95, p C11; *The Sciences* 1/2-96, p 9)

TRANSPORTATION AND COMMUNICATIONS TECHNOLOGY

TRAINS AND TUNNELS: BACK TO THE FUTURE

The history of transportation on land might be summed up very briefly with the following timetable (mya means "million years ago").

THE TIMETABLE OF LAND TRANSPORTATION

414 mya	Centipedes are known to be using legs for walking on land.
5 mya	Ancestors of human beings begin to walk upright.
7000 B.C.	Cattle are used to pull sledges.
5000 B.C.	Skis are in use in what is now northern Russia.
4000 B.C.	People in Central Asia learn to ride horses.
3500 B.C.	In what is now Iraq, wheels are added to sledges to make carts.
2000 B.C.	Wheels in current-day Iraq are made with spokes and copper rims. In Crete, the first paved roads for travel by wheeled vehicles are built.
100 B.C.	In Europe, the Celts invent a way to turn the front wheels of a wheel-and-axle pair independently of the back wheels, making carts easier to steer.
1350	In Europe, the suspension system for absorbing shocks is applied to carts.
1457	The passenger coach with a strap suspension for public travel is introduced in Hungary.
1630	Wooden rails for carts are installed at coal mines in England.
1769	Nicolas-Joseph Cugnot, a French military engineer, builds a steam carriage that can carry four passengers.
1779	The first version of the bicycle appear in Paris, France.
1804	In Britain, Richard Trevithick builds the first steam locomotive to run on iron rails.
1815	Scottish engineer John McAdam invents road paving with crushed rock.
1825	British engineer George Stephenson builds the first successful steam locomotive to be regularly scheduled and to carry both passengers and freight.
1863	Jean-Joseph-Etienne Lenoir builds the first horseless carriage to be powered by an internal combustion engine. In London, the first Underground (subway) system is started on January 10 using steam trains.
1869	The first transcontinental rail linkup in the United States is made on May 10 at Promontory Point, Utah.
1876	H. J. Lawson designs the first bicycle powered by pedals and a chain using gears.
1879	Werner von Siemens demonstrates the first electric railway in Berlin, Germany.
1885	Gottlieb Daimler invents the motorbike using an internal-combustion engine.

1889	Karl Benz and Wilhelm Maybach build a four-wheeled automobile powered by a gasoline engine, the prototype of the modern automobile.
1894	Electric automobiles begin to be sold commercially.
1895	The German firm Benz builds the first gasoline-powered passenger bus.
1896	The first electrically powered subway system, and the first on the continent of Europe, is opened in Budapest, Hungary.
1908	Henry Ford begins to manufacture the Model T, the first widely used automobile.
1909	Robert Goddard, better known for the development of liquid-fuel rockets, proposes a magnetically levitated train.
1934	The Chrysler Airflow is the first commercially available streamlined automobile.
1964	The first "bullet trains" begin operating in Japan, with an average speed of 100 mph (160 kph).
1968	James R. Powell and Gordon T. Danby obtain the first patent for a magnetically levitated train.
1975	Fiat builds the first prototype of a train that tilts on curves, enabling it to take these curves faster. The train has a top speed of 150 mph (240 kph).
1981	In France, the regular passenger run starts of the *Train à grande vitesse* (TGV), a train clocked that year at a record of 236 mph (380 kph). It eventually reaches 320.3 mph (515.7 kph), and regularly travels at 185 mph (300 kph) on passenger runs.
1990	Sweden introduces the X-2000 tilting railroad car that leans on curves, enabling it to take curves 30–40% faster. The train has a top operating speed of 155 mph (250 kph).
1993	The Metroliner service experiments for 3 months with Swedish X-2000 railway cars. Although use of the train would enable Amtrack to cut more than 40 minutes from the New York to Washington run, the experiment follows the regular schedule of 2 hours 55 minutes.
1995	The Metroliner begins working toward introducing tilting trains, expected to be in place between Boston and Washington, D.C., by 1997.

Today humans still travel about on land by walking, riding on animals, and riding in carts, carriages, passenger trains, automobiles, trucks, and buses. Although there have been some experiments with vehicles that walk on legs, especially for military purposes, it does not appear that these are going to have widespread use in the near future. Land transportation is possible with an air-cushion (or ground-effect) machine, but the few in use are nearly always employed at sea.

Can We Regain the Train?

Many critics of the transportation system that has emerged in industrialized nations (especially in the United States) believe that automobiles cause more problems than they solve. Transportation by automobiles is more dangerous, more wasteful of time, more polluting, and more of a drain on resources of all kinds than transportation by railroad trains.

It is difficult to keep Americans off the road by persuading them to ride the rails. With all forms of public transportation considered, the number of trips per person by public means has fallen from 114 in 1950 to 31 in 1990. Public transportation on the ground includes buses at every level, as well as trains called urban rapid transit within a city and suburban or interurban outside of a city.

This drop in public transportation usage has occurred despite a steady but gradual increase in the number of cities offering rail transportation. New systems that provide advanced versions of these subways and (sometimes) elevated trains are becoming available, although older systems are very slow to upgrade. A few large cities have good suburban rail systems as well as urban rapid transit, but even those cities have their roadways jammed with automobiles every rush hour. Among the new subway systems is one inaugurated in Los Angeles, California, in January 1993. Residents are reportedly skeptical of the L.A. subway. There are no fast interurban trains in the United States, although the moderately fast Metrorail serves the corridor from Washington, D.C., to Boston, Massachusetts.

It may take sociological changes to encourage Americans to give up their cars for the train, although technological progress continues. In many cases, however, new technology already in existence is not being utilized for one reason or another, usually economic. Trains of all kinds in all industrialized nations have tended to lose money since World War II. This has discouraged venture capitalists from financing new lines or improvements on old ones. Progress, therefore, is left to governments.

In 1960, James R. Powell and Gordon T. Danby invented the basic principles for the magnetically levitated train, or maglev, while working at Brookhaven National Laboratory on Long Island in New York. The United States and its local governments, however, have shown little interest in financing a maglev, although considerable test development has been done in both Japan and Germany. In 1994, the German Maglev was carrying passengers at 260 mph (420 kph) on test runs, and had runs as fast as 281 mph (450 kph). The German government at that time approved the construction of a maglev for regular service between Berlin and Hamburg, a distance of roughly 170 mi (275 km). It is expected to reduce the present trip of nearly three hours to about 66 minutes.

Although the German maglev train does not expose the passenger to elevated magnetic fields, the different design of the Japanese train causes the magnetic field inside cars to be 20 times that of the Earth's field. This, however, can be reduced or eliminated by shielding.

Conventional trains can be upgraded to be fast enough to compete with airplanes for medium-distance trips. Even the slow (nearly three-hour) Metroliner trip from New York City to Washington, D.C., can be faster downtown to downtown than a trip by air and taxi. Tilting trains, however, are being introduced, and are expected to replace the current Metroliners by 1997, reducing the New York to Washington run to two hours 11 minutes, and decreasing the Boston to New York trip from about five hours to about three. The tilting trains, already widely used in Europe since their commercial introduction in 1988, do not travel as fast as high-speed trains on a straightaway, but make up for that by taking curves at higher speeds. Passengers also find the tilting cars more comfortable than standard trains.

Another alternative for modernizing train travel in the United States is the monorail, which has become familiar to millions of Americans from exposure to the trains used at Disneyland and Disney World. Airports have become so large that they often have automated trains or automated monorails running from terminal to terminal. In the case of Newark Airport, the monorail system is scheduled to run right out of the airport to a connection to New Jersey transit. This is projected to occur in 1999, and will make it possible for the first time to travel from midtown New York City to one of its airports entirely on the train, although one still has to schlep bags from the conventional electric train to the monorail system.

All forms of train travel are limited to some degree by air resistance. Since trains travel along predetermined routes, it would be possible to cover the routes, or to put tracks into tunnels and then remove most of the air. This would greatly enhance speeds for runs of medium distances between cities. A straight, evacuated tunnel from New York, New York, to Los Angeles, California, could in theory reduce time for a maglev train trip between the cities to as little as an hour, considerably faster than air travel.

The Chunnel

Speaking of tunnels, one breakthrough in engineering in the early 1990s was quite literally a breakthrough. On December 1, 1990, at 11:21 a.m. Greenwich Mean Time, English and French workers on each side of a tunnel under the English Channel broke through and shook hands. They were 13.9 mi (22.4 km) from the English coast and 9.7 mi (20.3 km) from the French. The original plans for such a tunnel dated back to the late nineteenth century, but plans toward building such a tunnel were scrapped as a result of a political sentiment that discouraged closer ties between Britain and the European Continent.

The channel tunnel is known officially as the Eurotunnel, in English slang as the Chunnel, and in France as *Le Shuttle*. The tunnel is considered by many to have been the world's largest engineering project so far. Digging began in December, 1987. The Chunnel was officially inaugurated on May 6, 1994, when Queen Elizabeth II of Britain and then-President François Mitterand of France rode through the tunnel in the Queen's Rolls Royce, while the Rolls in turn rode on a specially designed train car for carrying automobiles. Full operation as a railroad tunnel did not begin until November 14, 1994, however. Since that time, three somewhat different systems have been in operation: a freight service to accommodate trucks and their drivers; an automobile service that is somewhat similar to the freight service in that the automobiles are loaded onto freight trains for the journey; and a regular passenger service connected to a high-speed surface line from London to Paris or Brussels. Actual passage through the Chunnel takes between about 30 and 45 minutes if all goes well—and so far it has.

The methods employed in construction and ventilation for the Eurotunnel are descendants of those methods first used nearly three-quarters of a century earlier for the building and ventilating of the Holland Tunnel under the Hudson River in 1927.

(Periodical References and Additional Reading: *New York Times* 3-3-92, p C1; *New York Times* 10-25-92, p III-6; *New York Times* 1-31-93, p I-34; *New York Times* 5-7-94, p 6; *New York Times* 6-13-94, p B1; *New York Times* 6-27-94, p B3; *New York Times* 10-13-94, pp A4; *New*

York Times 11-6-94, p I-6; *New York Times* 11-13-94, p III-10; *New York Times* 11-15-94, p A10; *Scientific American* 5-95, p 32; *New York Times* 7-23-95, p I-27)

REINVENTING COMMUNICATIONS

Aside from speech, the newest way of communicating is still the oldest: E-mail and the fax have brought back the written letter. The Internet is the great communications revolution of the early 1990s, but the Internet is essentially something new. The old forms of communication, however, are all being reinvented with new technology, just as next-morning delivery reinvented the postal system.

Consider the following.

Reinventing the Telephone

The telephone system today is vastly different from what it was a few years ago, although not what those who had attempted to project the future had anticipated. For example, prognosticators at the New York World's Fair in the 1960s offered as the phone of the future a booth from which one might see and speak to another person in a booth somewhere else. The "Picturephone" is now available (minus booth) as the Videophone, first marketed by AT&T in August 1992. Few seem to want it, even after the price was lowered to $1,000 per telephone.

Instead, the cellular telephone, or cell'phone, has become ubiquitous. Even at home, the idea that a telephone should be tied to one place in the house by a cord is old hat. The only phones with cords in the modern household are those speakerphones built into the new computers, which include the voice mail successor to the answering machine as well as the high-speed fax-modem needed to stay connected. The utility of the cell'phone was vastly improved in 1992 when a digital signal replaced the previously used analog version, tripling capacity and greatly improving reception. Several companies have been working on cell'phones that could work like a cordless telephone inside a building and like a cell'phone outside.

As the 1990s reached their midpoint, the mobility of the telephone continued to improve dramatically. Even on public transportation, where the cell'phone is inoperative, there are telephones available, sometimes at every seat. Cellular networks have become larger, and problems with echoes have decreased.

In 1993, the U.S. Federal Communications Commission (FCC) allocated several narrow bands of electromagnetic radiation in the region where the top frequencies of FM and VHS TV merge into lower microwaves. The bands were allocated for personal communications services (PCS)—wireless devices including telephones as well as various personal communications, faxes, and wireless computer communication. In 1994-1995, the FCC auctioned off the frequencies to various entities for nearly $9 billion.

A half-dozen or more projects have sprung up to make the telephone as available worldwide as the global positioning system (*see* "The State of Technology," p 679). Some of these projects are based on the same GPS technology of Earth-orbiting satellites. The starting gun for these projects was the agreement in March 1992 of the

120-nation World Administrative Radio Conference to allocate global radio frequencies to low-orbit satellites serving the telephone and paging industries. Among the plans for these frequencies are the following:

Iridium   The first satellite telephone system to make waves in the early 1990s was the 77-satellite Iridium system, later scaled back to 66. The plan is based on direct phone-to-satellite-to-phone conversations, with one low-orbit satellite passing the callers along to the next as the satellites flew over, or from satellite to satellite when the callers were too far apart for a single satellite to handle. On September 20, 1994, Motorola, the principal backer of Iridium with a 28% share, announced that the project was fully funded and would be in operation in 1998, only a year behind its original schedule. However, Motorola then found it had to withdraw a large debt offering because of lack of investor interest.

Globalstar   Most of the other satellite telephone systems are banking on a less ambitious system in which a few satellites would connect to ground stations which would connect to regular long-distance lines. While this kind of system is less expensive for everyone, including the consumer, it would not provide service for hand-held phones all over the Earth, which is the goal of Iridium. Globalstar appears to be the farthest along of several similar system ideas. If it goes into operation in 1998 as planned, it will launch only 48 satellites. Globalstar is focusing on customers in places around the world that are not served by existing wire-based telephone systems.

Teledesic   In March 1994, McCaw Cellular and Microsoft announced that the two companies planned to launch 840 communications satellites along 21 north-south orbits at an average altitude of 435 mi (700 km). The 1,700 lb (770 kg) satellites would each be able to handle 20,000 transmissions at once, relaying data to small low-powered stations with 18 in (45 cm) antennas on Earth and to other satellites in the system as needed. This system would enable a message from any point on Earth to be transmitted and received at any other point, although not by someone on the move. That is, one would need a fixed antenna and not just a handheld phone to use the Teledesic system. Going far beyond telephone needs, the system would also be able to transmit video and other data at a rate comparable to a fiber-optic system. According to a 1994 estimate, completion of the system would require $9 billion as well as approval from governments around the world, since this proposal would use frequencies beyond those allocated for telephone use. If all goes well, the system should be in operation by 2001.

Others   The International Maritime Satellite Organization (Immarsat) is a coordinating body for satellites used for telephone service at sea. It has started ICO Global Space Communications with the intention of broadening its telephone offerings. Odyssey is a proposal to launch a 12-satellite system.

If you can't wait ...   Available since 1993, the NEC Mlink-5000 uses a service available from geosynchronous Immarsat satellites. The Mlink-5000 is larger than a handheld telephone, weighing 29 lb (13 kg), and the size of a briefcase. It is set up for faxes as well as computer connections at a slow 4,800 baud. However, it only operates for an hour before it needs to be recharged. In 1993, it cost about $25,000 a set, and calls were billed at $5.50 an hour.

Telephone lines may themselves be upgraded or perhaps replaced by links from cable television suppliers in the future as well. Currently, there are a few options

DIGITAL VS. ANALOG

To a surprising extent, the history of mathematics, technology, and science in general can be interpreted as the interplay of digital and analog. This bipolar view leaves out a lot, but it also explains much that otherwise might be overlooked.

Mathematics, for example, began as a digital enterprise. In fact, a few mathematicians today still distrust mathematics that transcends the integers (positive and negative whole numbers). Early Greek mathematicians, however, proved that lengths are not digital but analog. For example, the diagonal of a square whose sides measure an integer cannot also be expressed in terms of integers. The calculus of Newton and Leibniz is the triumph of continuity and therefore analog over discreteness, the essence of digital.

In technology, early carpenters and masons simply fit wood or stone together, trimming until the fit was as perfect as possible. That is analog construction. Building plans, however, specify exact lengths in terms of a unit (such as a sixteenth of an inch), which is a digital concept.

The first calculating devices after the digital abacus and its relatives were slide rules, which are analog. About the same time in the seventeenth century, the first calculating machines based on gears were built, which were digital. The thermometer and the barometer were true analog devices, using the **analogy** of the height of mercury in a tube to represent temperature and air pressure. In the nineteenth century, most devices for interpreting power or current were analog dials similar to the thermometer and barometer. Babbage's partly built computing devices bucked the trend by returning to digital principals, following Jacquard's looms that used digital input (as holes punched in cards) to produce analog designs in cloth. The first computer of the twentieth century, Vannevar Bush's differential analyzer of 1930, used voltages to approximate numbers, and was therefore analog. When computers first became important after World War II, reference books felt that it was necessary to stress that these were **digital** computers, using a code based on the digits 0 and 1 to obtain results. This implied that analog computers were still being made, although this was not true.

Sound recording systems also reflect the war between digital and analog. The Edison phonograph and all other early recording systems were analog, using the analogy between sound waves and physical waves in a disk or electronic waves in a tube to record and reproduce the sounds. The compact disc system replaced these analogies with strings of digits recorded as pits in plastic that specify particular samples of a sound wave, although at some point the digital samples need to be returned to continuous sound waves to be heard. Today, the precision and reproducibility of digital information has caused it to be the system of choice not only for computers and compact discs but also for new tape systems, telephone systems, and in the near future radio and television.

available for up-grading the amount of data that can be transmitted from a home or office, options that are becoming increasingly popular because of the Internet. The integrated services digital network can be reached in most places via an ISDN line from local telephone suppliers, but that approximate quadrupling of data-transmission is not enough for applications of the near future. When combined with high-speed fiber optics, an approach called asynchronous transfer mode (ATM) may resolve that problem. It is an expensive option that can transfer data at a rate about 45,000 times that available with an ordinary telephone connection.

Reinventing Radio

In 1992, the French began transmitting the first experimental digital FM broadcasts from Paris. The following year, U.S. FM stations began to use the Radio Data System (RDS), also known as the Radio Broadcast Data System. This system had been in use in Europe since 1986. In RDS, digital data is transmitted along with an analog FM signal. The digital data does not replace the FM signal for broadcast of music or talk, but supplements it by enabling special receivers to display messages on a small auxiliary unit. The thought is that the stations can use the displays to insert call letters, format, the type of music that the station plays, song titles, or visual advertising. A person could even have the radio search the airwaves for a desired format, such as easy listening, classic rock, or traditional jazz. One of the more useful applications would be to provide emergency information, such as tornado warnings. Among the handier possibilities is finding the strongest signal from among all the NPR stations broadcasting the same show.

In the United States, a number of companies have been hoping to offer satellite-based digital radio systems that would be intended primarily for use in automobiles. If the FCC would provide the wavelengths from the electromagnetic spectrum to one of the companies and if the companies can raise the money, they might be providing the service toward the end of the millennium.

Reinventing Television

Early in the 1990s, the newspaper business pages were filled with reports of the advantages of High Definition Television (HDTV). The Japanese were the first to provide an expensive version of the new television, which they accomplished by the simple expedient of using an analog system in which the 525 lines of pixels (480 of which appear on the screen) were doubled. (Pixels are picture elements, or dots of varying shades.) Despite regular daily broadcasts in Japan that started as early as June 3, 1989, generally for an hour each day, the system, with its very expensive receivers, did not catch on. Analog HDTV was almost abandoned by Japan in 1994, but protests from electronics companies involved in the business won the day, and the government promised to continue it into the next millennium.

A far more successful product in Japan has been TV in the wide-screen format familiar in the United States from recent motion pictures. "Wy-doh terebi," as the wide-screen format is known in Japan, can be obtained in the cheapest version for just a tenth the cost of a Japanese-system HDTV set. The wide-screen sets became popular even before Japanese television stations began broadcasting in wide-screen formats in 1995, a system called "enhanced definition" television.

Japan also offered another alternative television set, although not a new broadcasting system. Flat-screen TV came to Japan in 1993, although at such a high cost per set that few if any are expected to be sold. One version simply uses hundreds of thousands of tiny cathode-ray tubes instead of one big one as a screen, while another displays what amounts to hundreds of thousands of tiny fluorescent lights as its pixels. It is believed that eventually some other technology will be required for successful flat-panel displays, although present LCD and LED technology does not seem to provide a bright enough picture (*see* "Display Progress," p 691).

In the United States, various manufacturers also worked on developing HDTV using systems similar to the Japanese. No one seemed to think that the digital system used for computers and compact music discs would carry enough data to replace the analog system used for radio and existing television. In 1990, however, new compression techniques were developed for digital transmission. In 1991, the relatively small General Instrument Corporation of San Diego, California, demonstrated that a digital version of HDTV is workable. That same year, the U.S. Federal Communications Commission began a search for the single system to which it would award sole possession of the U.S. HDTV airwaves.

By 1993, it had become apparent that the Japanese approach of simply improving analog television would not work. Japan, which had its version entered in the FCC competition, dropped out. In May 1993, the various other competitors decided to form an alliance to work on a single standard for digital HDTV. Various problems since that time were mostly associated with the familiar problem of making the old system compatible with the new, as well as the new problem of making television screens compatible with computer monitors. These difficulties delayed introduction of the new standard in the United States, although eventually the FCC accepted the alliance version. So HDTV is coming soon to a television set near you, probably as near as your living room.

In the meantime, millions of customers have signed up for direct-satellite digital TV, which is not high definition. The service, which began in the United States on September 19, 1994, uses a band of frequencies that can be interrupted by rainstorms, and a system that compresses data in such a way that the number of frames per second is reduced. In some systems, pixels are even visible on screen. Yet because of the data compression, a great many different channels are available.

Reinventing Miscellaneous Communications

The Fax In 1994, a new standard for faxing messages by computer was issued, making it easier to send faxes by computer. Still, most computers continue to use the old standard. The new standard is binary file transfer, which is somewhere between sending ASCII code for letters and sending a bit-mapped image. Without binary-file transfer, a computer user can send a message to a fax machine only in a bit-mapped form (as a picture of the document being faxed) but can send a file directly, without special formatting, to a computer.

The Photograph   As early as 1992, Kodak was making a digital camera system for $25,000, primarily so news photographs could be sent over telephone lines to home offices. From the beginning, the photos had about the same resolution as HDTV, more than sufficient for newspaper or most magazine use. That same year, Kodak also introduced the Photo CD system, in which photographs taken on file are preserved on CDs for display on television monitors. Customers of home photography were not thrilled. Among others, Sony and Canon had already introduced systems requiring special electronic cameras. These were not popular either. Kodak began to have some success, however, with various professional applications of the technology. By 1996, Kodak had moved toward all-electronic photography based upon the removable PCMI units for computers, bypassing the expensive CD player for television. The first cameras in the Kodak Digital Science line cost $1,000 each, although cameras priced as low as $300 were expected to be available by the end of 1996. Once a photograph has been captured in this system, it can be transmitted over the Internet or inserted into a report prepared by computer, although the images are not as sharp as with film.

(Periodical References and Additional Reading: *New York Times* 3-3-92, p D1; *New York Times* 3-18-92, p D1; *New York Times* 6-7-92, p III-1; *New York Times* 8-26p D4; *New York Times* 8-26-92, p D5; *New York Times* 9-23-92, p A1; *New York Times* 1-6-93, p D4; *New York Times* 1-8-93, p D4; *New York Times* 1-15-93, p A1; *New York Times* 2-17-93, p D2; *New York Times* 3-10-93, p D19; *New York Times* 3-31-93, p D4; *New York Times* 5-25-93, p A1; *New York Times* 5-26-93, p D1; *New York Times* 7-11-93, p III-9; *New York Times* 7-30-93, p D3; *New York Times* 9-17-93, p D3; *New York Times* 9-18-93, p D5; *New York Times* 10-24-93, p III-9; *New York Times* 12-26-93, p III-10; *Discover* 1-94, p 97; *New York Times* 1-24-94, p D1; *New York Times* 3-21-94, p A1; *New York Times* 9-15-94, p D1; *New York Times* 9-21-94, p D1; *New York Times* 9-22-94, p D1; *New York Times* 11-27-94, p III-8; *Discover* 1-95, p 100; *New York Times* 1-15-95, p III-7; *New York Times* 1-9-96, p 50; *New York Times* 1-10-96, p D2)

Data Storage and Retrieval UPDATE

Communications and data storage and retrieval tend to use similar technologies. Certainly both are involved with data. For the most part, any system that can be used to store or retrieve data can, without much adjustment, be used in communications. Thus, the history of the two are intertwined. In a sense, data storage and retrieval is simply communicating with the future first and the past later. It is an ancient ability of humans that has reached a level undreamt of even as recently as a few years ago.

TIMETABLE OF DATA STORAGE AND RETRIEVAL

30,000 B.C.	People in Europe begin making drawings and carvings, including tallies, simple devices for indicating numbers by collections of strokes.
15,000 B.C.	Elaborate paintings of animals are made in caves. The first known map is carved on a piece of mammoth bone.
10,000 B.C.	Clay tokens are introduced in the Middle East to record commercial transactions. These are the precursors of writing and numeration. If the presence or absence of a token is taken as a single bit of information, each bit

was the size of a modern bead, variable but often less than half an inch across in each direction. It might be more appropriate to relate a token to a byte (eight bits), the amount of space allocated to code a single letter or a digit in the decimal notation system.

3500 B.C.	Clay tokens in the Middle East are kept in spherical baked clay envelopes that hold on average of nine or so tokens. Envelopes are spheres about 1–1.5 in (3–4 cm) in diameter, so that a volume of 3–7 cu in (30–50 cu cm) is the space used to store about nine bytes, or 72 bits, of information.
3300 B.C.	In Egypt, people begin using the earliest form of hieroglyphic writing.
1500 B.C.	In what is now Syria, the first alphabets are developed, with about 30 signs.
600	The Chinese develop the system of printing whole pages using wood blocks.
870	The Chinese use wood-block printing to print an entire book.
900	In India, the concept of zero is developed, completing the Hindu-Arabic system of numeration now used.
1050	Movable type is invented in China.
1440	Johann Gutenberg and Laurenz Janszoon Koster invent a workable system of casting and printing with movable type on paper that becomes the basis of modern printing. A typical book contains about 1 million bits of information in a volume of 70 cu in (1,000 cu cm).
1805	Joseph-Marie Jacquard develops a method for storing patterns on punched cards that then direct the operation of looms.
1816	Joseph Niepce begins experimentation that lead to the first photograph.
1859	John Benjamin Dancer develops microdot photography, reducing book pages to dots on slides.
1877	Thomas Alva Edison records sound.
1888	Emile Berliner invents the flat-disk recording.
1893	Valdemar Poulsen develops the first form of magnetic recording of sound: the wire recorder.
1920	Lee De Forest invents recording of sound on motion-picture film.
1935	Tape recorders using magnetic recording are in use in Germany.
1937	Alec Reeves in England proposes a method for changing analog sound into digital and back.
1941	The ABC computer, the first of the modern computers, uses punched cards for program storage and input.
1944	J. Prosper Eckert develops a method of storing data as acoustic pulses in a tube of mercury, but also proposes using a magnetized disk to store data and instructions for a computer.
1947	Dennis Gabor develops the theory of holography, storing information as interference patterns of light.
1948	The magnetic drum is developed for storing information in computers. The long-playing phonograph record is invented by Peter Goldmark.
1953	The ferrite core memory for computers is invented by Jay Forrester.
1956	The first form of videotape recorder is introduced commercially. The first

forms of hard-drive disk storage are developed for computers. An early hard disk can hold about 2,000 bits per square inch.

1962   Phillips introduces the compact cassette, or audiocassette, for tape recordings.

1970   Intel develops a memory chip that can store 1,024 bits in a space about 0.2 sq in (1 sq cm) of negligible thickness, which replaces the ferrite core memory in computers. The first form of floppy disk is developed in the same year for storing data to be used by a computer.

1972   Phillips develops the first system for recording images and sound on a disk using a laser for input and reading.

1974   The bar code, read by a scanner, is used in a limited number of supermarkets.

1976   VHS videotape is introduced.

1978   Apple introduces the first disk drive for personal computers, using floppy disks that record data magnetically. It stores about 180,000 bytes, or 1,440,000 bits, in an area of 86 sq in (45,000 sq cm), or about 16,000 bits per square inch.

1982   The audio compact disc (CD) is introduced, using data stored as tiny pits that are read by a laser. It stores about 72 million bits per square inch.

1983   IBM introduces a computer that comes with a hard-disk drive memory, capable of holding 10 megabytes (10,240 bytes; a byte is eight bits) of information.

1984   The CD is adapted for computer use as the CD-ROM, an optical disc that can store large amounts of data. Computer memory chips with 256 kilobits (262,144 bits) are available.

1993   Fujitsu in Japan announces a 256-megabit (268,435,456 bits) memory chip.

Hitachi announces that they have developed a single-electron transistor that works at room temperature, although many other practical problems would have to be overcome before it can be used in applications.

1995   The best commercially available magnetic-storage hard disks reach 0.3 gigabits (322 million bits) of data per square inch. The best experimental system can store ten times as much per square inch.

A peculiarity of spelling is that because the floppy was originally called a "diskette," it is usually spelled *disk*. However, because the compact disc originated in the Netherlands (where people speak British English), the spelling follows standard English. Hard disks have followed the spelling for floppies. As a result, optical media use *discs*, but magnetic media use *disks*.

In early 1996, the typical new home computer contained 8-16 megabytes of random access memory; 840 megabytes to 2 gigabytes of memory on a hard drive that can store about a million bits per square inch of surface; and the capability of randomly accessing information from a floppy disk with 1.4 megabytes, or from a CD-ROM with 640 megabytes stored on it (which is the equivalent of 330,000 pages of printed text). An average household in the United States contained at least one VHS video recorder, using tapes that could record either two, four, or six hours of television pictures and sound with somewhat different quality. The home or car, or more likely both, would have a music CD player, playing CDs designed to hold and access 74 minutes of music. There might be a few audiocassettes around as well. This amount of information storage capability is in addition to any print media available in the household.

It does not seem to be enough.

The Video CD

Many technologies vie for the data-storage and retrieval business, and most fall by the wayside. Even data storage devices that are technically successful may not bring the revolution that their inventors no doubt expected. The compact cassette is a case in point. Its superiority to audio reel-to-reel tape for home use is so obvious that reel-to-reel vanished from the consumer market, although it remains an important part of commercial enterprises. While the audio cassette did not replace the long-playing record, neither did it vanish in the face of the CD. The compact cassette format was used for data storage and retrieval in the first personal computers, but failed when challenged by floppy disks.

In nearly all these cases, the main disadvantage of the compact cassette has been accessibility of data. One has to move the tape through the machine to get to the desired place, which takes too much time. Although the new Digital Audio Tape (the lineal descendant of the compact cassette) produces better sound quality and no tape hiss, it still cannot be accessed at random. VHS videotape, a bigger cassette, suffers from the same problem. Furthermore, digital videotape, which has been under development for years, cannot store nearly as much video as the audio format does. Cassettes of one sort or another, audio or video, may linger because they can easily record new data over old, but ultimately another format seems to have the edge. For example, the new standard for digital videotape, adopted in 1994 but not yet available, calls for cassettes similar in size to an audiocassette, using tape half the width of VHS tapes.

When data is stored in a long line that must be read sequentially, access is limited. If data is stored in two dimensions but accessed by a stylus or other thin and essentially one-dimensional probe, access can be as fast as it takes to move the probe a short distance across two dimensions. Even the earliest photograph recordings, in which the long line of data was in a spiral groove in two dimensions, could be roughly accessed at any point by lifting the needle and dropping it into the groove. Magnetic data stored on rapidly spinning disks can be accessed exactly in thousandths of a second at any point on the disk by modern computers. The user may not even realize that a modern hard-disk drive has taken some time to find data, and that it takes some time to read it. A small amount of data appears so fast that it seems to be instantaneous, although the wait for a large amount of data to be read is noticeable. The time required to find where the data is located is lower than casual perception of it.

The ability to find data on a CD is not that different from the use of other rotating discs, although the amount of data on the disc may be far greater. A CD uses small pits read by a laser beam to store and retrieve data. This format enables the CD to store somewhat more music than both sides of a single 12-inch LP on one side of a 4.7 in (12 cm) disc. Furthermore, with no change in technology, the same format can store the equivalent of a large encyclopedia set, with some music, animations, photographs, and even small sections of motion picture thrown in as well.

In August of 1995, IBM obtained an agreement from two contending groups of corporations that had each found ways to extend the storage capacity of a CD. Because the main goal of the extended storage has been home reproduction of motion pictures, the new technology has come to be known as a Digital Video Disc (DVD), although the same format can be used for music and as a CD-ROM in computers. The system depends largely upon placing the pits in the disc closer together, allowing for a lens to provide two different foci for the laser reader, and the option of two-sided discs. Depending upon the options taken, storage can range from 4.7 billion bits per DVD up to 19 billion bits per disc. As a device for video, the disc can store and replay a motion picture up to four hours long, using the MPEG 2 digital video system also used in small-antenna satellite TV and digital surround sound. Although this technology certainly would seem to supply enough data, progress in developing blue-light lasers could in short order result in a third-generation CD that would store three or four times as much data in the same space (*see* "Laser UPDATE," p 693). The virtue of a blue-light laser over the red-light laser now used is that the shorter wavelength means that the dot diameter can be decreased by half. This ideally means that four times as much data could be stored, although the need to put some space between dots reduces that advantage slightly.

Audiences might do well to get used to digital movies in any case. In 1994, the first tests were run on digital movies for theaters, with distribution to be by fiber-optic cable rather than delivery of physical objects to the theater.

The DVD has the option of writing two sets of information on a single side in two different layers. In 1994, IBM scientists announced that they had a method for expanding that to as many as 10 layers of data per side. IBM suggested that such a technology, when used with a blue laser, might reach as much as 240 billion bits per side in a CD format.

Other Ideas

At some point in the hundreds of billions of bytes possible for CD storage, the string runs out. There is no more space on a disc to pack any additional data on either side. **Bring Back Mechanical Recording** H. John Mamin and coworkers at IBM's Almaden Research Center in San Jose, California, have developed a system for recording and reading data based on the movement of a tiny needle in a soft plastic. To write data, a laser pulse heats the needle as it touches the swiftly rotating disk, creating pits in the plastic. The needle can then read the data by sensing the pits. The system is amazingly similar to the original method of recording sound mechanically on wax cylinders and discs. The experimental method is just slightly slower than standard hard disks using magnetism to record and read data.

Pack It In Various schemes use one of the existing technologies but try to squeeze in greater storage capability. One form of floppy disk, for example, utilized a very thin magnetic coating, with about 100 megabytes of data storage obtained mostly by pushing the technology. Most of the high-density disks similar to this effort have been expensive, unreliable, and unpopular.

Magnetic Pillars Stephen Y. Chou of the University of Minnesota in Minneapolis is using molecular beam lithography to create a material made from thin pillars, creating a sort of carpet pile. Each pillar can be magnetized or demagnetized separately via technology similar to that used in ordinary hard disks. The magnetic pillars could record 65 million bits per square inch. Individual bits could be tracked more easily, since each bit would be on a separate pillar.

From the Planet Krypton   In the Superman movies, data from the planet Krypton was stored in gleaming crystals. There are several experimental methods in which lasers store data in a transparent medium, typically a crystal. In the whole-image method, holography (*see* "Optical Computing," p 715) is used to store an entire image with about 320×230 pixels at each location within the volume of the crystal. A crystal only 1 in (3 cm) on a side can store 10,000 such holographic images, recorded and read by lasers. This is about 100 megabytes, or about 100 million times as much as the original data storage and retrieval system of tokens in clay envelopes in about the same volume. Of course, the equipment needed to retrieve the holograph data is complex and the clay envelope data can be read with the naked eye. Furthermore, early experiments have taken hours to store data, and the image has faded after being read several times. Theoretically, however, the holographic method could be used to store hundreds of billions of bytes of data in a small region, with the ability to retrieve a random billion or more bits per second after a search of less than 100 microseconds.

Plastic Instead of Crystal   Another version of holography uses newly developed polymers that undergo chemical changes when exposed to light. These produce read-only memories, as they cannot be reprogrammed (which can be done with crystals). In one experiment, researchers stored a thousand pages in a polymer film only 0.004 in (0.1 mm) thick. The data is more stable than in a crystal, and does not slowly vanish when accessed.

Reading the Waves   A newer experimental method of data storage in a crystal, called spatial-spectral selective optical memory, has been developed by Thomas Mossberg and coworkers at the University of Oregon. Two lasers are used to record data as waves in a crystal kept near absolute zero. Retrieval of data is obtained by a popular mathematical trick called an inverse Fourier transform. The data is recorded by preventing certain electron transitions (*see* "Laser UPDATE," p 693), and it vanishes in less than 200 microseconds after it is stored. Still, the method can store up to eight gigabits of information per square inch of the surface of the crystal.

Scanning the Dots   The near-field scanning optical microscope (NSOM) is near field because the tip of a tiny optical fiber, narrower than a wavelength of light, is placed extremely close to the object to be scanned, or viewed one bit at a time in the light of a laser. For data purposes, the same equipment can be used to write and then read tiny dots at densities of 45 gigabits per square inch.

Nanostorage   The atomic force microscope, or AFM (*see* "Nanotechnology UP-DATE," p 732), is the physicist's tool for looking at atoms, and has been shown capable of moving individual atoms from place to place. The Japanese corporation Matsushita Electric demonstrated that it can be used to change the phase of tiny locations on a crystalline disk. The same AFM is then used to read the data by observing which bits have been changed. Each bit of information with this system takes up a dot only a hundred thousandth of a millimeter. By crowding dots together, one might get a trillion bits into a square centimeter, which translates to 155 gigabits per square inch. It might happen in the future, but an AFM costs $10,000 or more, making the method impractical for home computers.

The AFM method is not even the tiniest dot. The honors for this may go to an AT&T Bell Labs experiment in 1992, in which they used a fiber-optic cable that had been pulled like taffy to focus laser light on a dot 60 nanometers across, about a thousandth of the diameter of an AFM dot. This is essentially the same principal as the NSOM.

Quantum Storage   Michael W. Noel and Carlos R. Stroud of the University of Rochester have what may be, at least for now, the smallest possible storage medium. They are using lasers to manipulate electrons in atoms, producing various energy levels controlled by combinations of lasers at different frequencies. They believe that they can encode as much as 900 bits of data in a single atom as the wave function of a particular electron state. Their goal is to use this to encode a color picture of the word "optics" on a single atom. There are numerous obstacles to be overcome, ranging from the instability of an atom with high electron energy levels to the quantum-mechanical dogma that observation of a small system, such as a single atom, will change its state.

Electron Trapping Optical Memory   A far more practical technology being developed for commercial introduction in the next few years is, like quantum storage and the spatial-selective optical memory, dependent on using two lasers to affect the energy levels of electrons. In this case, however, the electrons that are imbued with high energy become captured by ions of another element, where they stay for months, and possibly for years, if left undisturbed. A gentle jog at the correct frequency of laser light, however, dislodges them. They emit their own pulse of laser light, which is recovered and collected as a data bit. The method is "electrical trapping optical memory," so the result is termed an ETOM disc. There are two advantages to the ETOM discs over DVDs: not only can more data be recorded by using four levels instead of two on a side of the disc, but the ETOM disc can also be written and read over and over, as in magnetic data storage on a computer hard drive or floppy.

(Periodical References and Additional Reading: *New York Times* 8-2-92, p III-6; *New York Times* 8-6-92, p D1; *New York Times* 2-23-93, p C1; *New York Times* 3-7-93, p III-9; *New York Times* 7-2-93, p D3; *New York Times* 8-26-93, p D4; *New York Times* 12-8-93, p D5; *New York Times* 3-1-94, p A1; *New York Times* 3-21-94, p D2; *New York Times* 4-16-94, p 36; *New York Times* 5-13-94, p D1; *Science* 5-8-94, p 737; *New York Times* 9-10-94, p 39; *New York Times* 9-19-94, p D2; *New York Times* 12-17-94, p 41; *Physics Today* 1-95, p 17; *New York Times* 1-11-95, p D1; *Science News* 4-1-95, p 207; *Science News* 4-21-95, p 245; *Science News* 6-3-95, p 344; *Physics Today* 11-95, p 9; *Scientific American* 10-95, p 40; *Scientific American* 11-95, p 70; *Nature* 11-23-95, p 339)

COMPUTER TECHNOLOGY

OPTICAL COMPUTING

From time to time in the early 1990s, computer engineers have announced a prototype of an optical computer. The earliest was on January 29, 1990, when Bell Labs demonstrated their first optical computer, using a kind of optical transistor based on gallium arsenide. Three years later, in January 1993, electrical engineers from the University of Colorado claimed the first fully optical computer. The Colorado version could store, transmit, and process data entirely in the form of light, with no electron intermediaries.

Optical computers have several inherent advantages over those based on movement of electrons. The main advantage is not that the speed of light in a vacuum is the highest possible speed, although that does not hurt the case for optics as opposed to electronics. The main advantage is that photons, which carry light, are bosons, but electrons are fermions (*see* "The Newest Forms of Matter," p 645). Fermions cannot occupy the same location, but bosons easily crowd together without even bumping into each other. Thus, fiber-optic cables can transmit thousands of telephone conversations at once.

The Colorado optical computer was built over a five-year period by Vincent Heuring and Harry Jordan. The light used is generated by lasers, and is transmitted through fiber-optic cables. Data is stored by sending it around and around in a fiber-optic loop. Instead of using gallium-arsenide equivalents to transistors, which are really electro-optical devices, the Colorado computer works by using wave guides to switch light beams, an awkward way to change data. An improved switching system may enable more light beams to be switched at one time, thus also eliminating the need for transmission through fiber-optic cables.

Fundamentally, to produce an optical computer one needs

- methods of producing the light that will be used, which can come from lasers, light-emitting diodes (LEDs), or some other source more suited to switching;
- ways of routing the light through the computer, for which fiber-optic cables and/or beams of light using mirrors and lenses are well developed;
- switches that will be the equivalent of transistors;
- methods of connecting the switches to form integrated circuits;
- memory of various types.

Light-Emitting Adventures

Optically active chips are generally made from gallium arsenide, which can easily be induced to produce light from electrical stimulation, or electric currents from stimulation by light. Gallium, however, is hard to work with, and arsenic is famous for being poisonous. Various other means of obtaining light may be desirable for other reasons as well.

Spongy Silicon Superlattice   The electrical computer business is built on silicon, an easily worked material whose properties are well known and which is nontoxic. Furthermore, it is easier to combine two silicon devices with each other than to get a gallium arsenide device to work with silicon. Thus, development of silicon chips that emit light has been desirable.

Physicists discovered that endowing silicon with a spongy texture, filled with tiny holes, could cause it to emit light, although no one is entirely sure why this happened. Furthermore, the quantity of light has not been great.

In 1995, Z. H. Lu and coworkers at the National Research Council of Canada in Ottawa reported that they had found a way to use alternating thin layers of silicon and silicon dioxide (the compound in quartz or glass) to produce a higher-quality light whose frequency could be adjusted by the thickness of the layers. Each layer is laid down one atom or molecule at a time with molecular beam epitaxy. Quantum electrodynamics can easily explain why this process works, so the results of specific adjustments can be predicted and desired effects can be obtained. The resulting silicon structure is called a superlattice after the arrangement of the atoms in material, which is not regular enough to be termed a crystal. There is good reason to hope that superlattices may fill several previously unmet needs in optical computing and data transmission.

Quantum Dots   Sometimes it is useful to have light emitted at a specific frequency for one purpose, at another frequency for a second purpose, and so forth. Still, most lasers and LEDs emit light of one frequency only, dependent upon the chemical composition of the device (see "Laser UPDATE," p 693, and "Display Progress," p 691). Tiny dots of one semiconducting material embedded in a different one can become miniature light emitters as a result of quantum effects caused by the small size of the dots. The dots attract both electrons and holes, which then combine in the dots when activated with a weak laser. The exact wavelength a particular dot will emit is hard to predict or to measure, since methods of forming them produce thousands of dots in a small region. Still, progress is being made. If the dots are tamed, they could be an excellent source of tunable light. For example, a row of dots emitting known wavelengths could be lined up in a small region. Shining the laser from dot to dot would enable one to vary frequencies in much the way a one-fingered pianist picks out a tune.

Optical Switching Progress

There are various ways to turn a switch on and off with light, but not very many of them are practical for computers.

A fast switch can be achieved by using a laser to raise the energy in selected electrons (see "Laser UPDATE," p 693). This kind of switch can be turned on almost as fast as one feels the need. The problem arises if the electron is then allowed to relax in order to turn the switch off. This takes about a nanosecond (a billionth of a second), which seems short until one realizes that switching is conducted in computers in femtoseconds (a femtosecond is a millionth of a nanosecond). A group of Hitachi employees working at Cambridge University's Cavendish Laboratory have solved this switching problem with the addition of a second laser slightly out of phase with the first. Firing the second laser turns off the switch. The researchers have been able to demonstrate turning the switch on and off again in as little as one ten-thousandth of a nanosecond (100 femtoseconds). The problem is that the switch is made from expensive ultrapure gallium arsenide cooled almost to absolute zero. Still, now that the principal has been developed, perhaps it can be applied to a more practical material.

Turning a Corner with Light

Although fiber optics can be used to connect all the elements in an optical computer, some designers prefer to use beams of laser light traveling through open space. The geometry of the computer then becomes one based on the straight line, for that is how a photon travels if not disturbed by intervening matter or, very slightly, gravitational attraction. Neither of these methods is suitable for moving light from one switch to another when both switches are embedded in the same chip of a solid. The similar problem for carrying an electric current between transistors on a chip has been solved for a long time, with lithography being the method of choice today. A mask during chemical treatment allows narrow connecting parts of a silicon chip to become conductors, while preventing the chemical from reaching other parts of the chip.

Finding a similar concept for moving light from place to place in a gallium-arsenide based light-emitting chip has been difficult. One notion has been to simply etch out deep channels, which works well so long as the light is traveling in a straight line. If the inside of the channels are polished, the light will be reflected and transmitted somewhat as in a fiber-optical cable, but it is difficult to polish the inside of narrow channels sufficiently. A lot of light gets lost every time it has to make a corner, which happens frequently on a tiny integrated chip. In the April 10, 1995 issue of *Applied Physics,* a team from the University of Illinois in Urbana-Champaign and from the Swiss Federal Institute of Technology in Zurich announced what appears to be a solution to the problem. The type of semiconducting device they worked with consists of a layer of gallium arsenide between two layers of aluminum-gallium-arsenide. The key was a 1990 discovery that aluminum-gallium-arsenide-oxide conducts more light than aluminum-gallium-arsenide; that is, it has a lower index of refraction. A mask could be used to set up the channels and a chemical, heated damp air, used to oxidize the material, just as it oxidizes iron into rust. This process worked satisfactorily on the upper layer, but the oxidation process stopped at the middle layer where there was no aluminum. This problem was solved by using silicon to move some aluminum atoms out of the top layer into the middle one where they were needed. The result is a great improvement over trying to move light around corners with etched-out channels. There seems to be no reason why this light channeling method could not produce an all-optical microprocessor, the chip that is the "brains" of a computer.

Data Storage

One possibility for an all-optical system would involve the use holograms to store information. This could replace the common method used for memory storage on floppy disks or hard disks, which utilizes patterns of magnetic domains. While holography and CDs share the use of lasers, holography relies on an entirely different application of light. Magnetic storage and CDs store information one bit at a time using mechanical means. A holographic system can store information in complete images, somewhat like the three-dimensional images seen on credit cards, and can store and retrieve those images with no moving parts. Each image is stored as an interference pattern in a particular part of a crystal or a special transparent polymer. The interference occurs between a fixed reference beam produced by one laser and (in one holographic system at least) another beam produced by a tiny laser on a chip. One such chip can contain as many as 10,000 lasers. The even tinier lasers developed in 1991 by Sam McCall at Bell Labs are expected to see future use in making optical computers a practical reality in this and a number of other applications. Recovery of data stashed in holographic storage is nearly instantaneous, since the only require-ment is the activation of a particular laser. This could prove much faster than the mechanical searching necessary to recover data from even the fastest magnetic hard disk or CD.

Note that holographic storage does not need to wait for optical computers to become a reality. The method can be used with present electronic computers. In 1992, however, a principal barrier to the uses of holographic data storage involved the fact that although the actual storage location can be the size of a Scrabble tile, and the chips with the tiny laser on them are even smaller, the equipment used so far to operate the device is larger than most current laptop computers. This part of the operation needs to be miniaturized before it becomes practical. Costs are also high, although this is always the case for prototypes.

The Optico-Electrical Interface

Nearly all applications using light are dependent upon the interactions between electrons and photons; even firelight, the first optical tool of humans, comes from such interactions. Despite this intimate connection at the level of individual electrons and photons, the connection of the two at control levels is not especially simple. Ideally, at least in the immediate future, we will use machines that are powered by moving electrons, a technology that has been in place for a hundred years. If it is advantageous to operate a device with light, the initial power supply will come from moving electrons. Some switches may be electrons controlled by light, while others may be light controlled by electrons, in addition to pure electronic and photonic switches.

One difficulty in getting electronics and photonics to work together is that the silicon-based devices of conventional modern electronics fail to mesh comfortably with the gallium-based devices common to photonics. Because of the different crystal structure of silicon and gallium-based materials, one cannot simply grow a crystal of one on the other. Because of the tiny sizes used in modern switches, it is far too difficult to assemble devices by gluing one piece to another. John S. Smith of the

University of California, Berkeley, and graduate student Jay Tu have developed an unusual way of solving the problem. They make a slurry of tiny lasers that have been formed as truncated pyramids. They also separately produce a silicon chip in which the appropriate places for the lasers have been scooped out and lined with metal. This is essentially like a jigsaw puzzle, with the laser fitting in exactly one way. The slurry, lasers suspended in alcohol, is then swirled about over the silicon wafer. When a laser is positioned correctly, so that it fits in the hole, it stays there. Otherwise, it flows away with the slurry. When one thinks that all the lasers are in place, the chip is heated to melt the metal, gluing the lasers in place and providing at the same time a conductor for the electrons from the chip. When this method was presented to a Conference on Lasers and Electro-Optics in May 1995, not all the bugs had been worked out, although the principle seemed to have been established.

In April and June 1995, however, two groups announced separate sets of improvements discovered by each team. The improvements involved an entirely different way to blend optical and electrical materials into one, a method relying on chemical instead of mechanical joining. The basic method employs organic polymers called nonlinear optical (NLO) materials. (Some inorganic crystals are also NLO materials.) An NLO polymer changes its optical properties, such as the index of refraction, in response to an electric field. The polymer also can be manipulated by light waves. The main limitations of NLO materials have been that they are not very responsive to light, and that the polymer breaks down when it gets hot, a condition that occurs in most electronic devices after a short operating time. In April, however, Seth Marder and coworkers at CalTech in Pasadena, California, developed an NLO polymer that is more responsive than the inorganic lithium niobate crystals used in most NLO applications to date. In June, Don Burland and coworkers at IBM's Almaden Research Center in San Jose, California, announced that they had produced an NLO polymer that could be heated to temperatures as high as would be expected in electro-optical devices without breaking down. The remaining problem: the Pasadena NLO melts easily and the San Jose NLO is not very responsive. But it is not very far from one city to the other.

(Periodical References and Additional Reading: *Discover* 1-94, p 95; *Science* 12-12-94, p 1478; *Science* 12-16-94, p 1807; *Science* 5-26-95, p 1131; *Science* 6-16-95, pp 1570 & 1604; *Science* 6-23-95, p 1702; *Science* 10-6-95, p 29; *Nature* 11-16-95, pp 238 & 258)

NEW COMPUTERS AND CHIPS

The kind of device that remains the basis of nearly all computers today was first built in the 1940s, with commercial versions available starting in 1951. Since then, computers have been significantly improved at an astonishing rate. On average, every two years since their invention, the components that are at the heart of the computer have been halved in size. The result is that the transistorized regions on chips today are about 33 million times as small as the vacuum tubes used in the first commercial electronic computers. Furthermore, the speed of computers has doubled every two years, in part as a result of the reductions in component size. Thus, the best computer today is about 33 million times as fast as the 1951 UNIVAC.

Another change is equally significant. In 1995, the personal computer finally moved out of the office and into the home in a significant way. Figures for 1994 demonstrated that dollar sales of television sets and of personal computers were nearly even.

The Handwriting on the Computer

One problem limiting the spread of computers into the home and into many aspects of business is that many people do not know how to type. Entering any great amount of data using a keyboard is slow going if one is not a touch typist, or at least a skilled hunt-and-peck artist. The mouse, track-ball, and other devices for point-and-click, in combination with the pull-down menu, have made it easy to use a computer for most purposes with little or no typing. A numerical keypad on the keyboard solves the problem of entering numbers, although entering words generally continues to require typing.

Early in the 1990s, a number of developments were intended to change this dependence on keyboard entry. On May 29, 1992, Apple introduced the Newton, a computer of sorts that depended on handwritten instructions instead of keyboard entry. Cartoonists lampooned the odd words that Newton "read" from a handwritten entry. On April 16, 1992, the first IBM Thinkpad was based on handwritten data entry, although it was only available on a limited basis. In August, AT&T announced that it was increasing financial support of the small company that had developed the software IBM used in its pen-based Thinkpad. By October, however, it was clear that the handwriting may have been on the wall, but it was not on the computer pad. Two years later, several of the pen-based systems had failed completely, and Apple's Newton was nowhere near being the success expected. Sony tried the idea yet again, with new software in its Magic Link. Still, no dice.

Supercomputers

Early in the 1990s, it became apparent that for most data intensive purposes, the massively parallel version of the supercomputer would replace the vector supercomputer, familiar from various Cray supercomputer models of the 1980s. A massively parallel computer uses a large array of separate microprocessors to enable it to work on many small aspects of a problem at once. Older supercomputers, like almost all computers since the first, had been designed to perform all operations sequentially, so that each operation passed through one microprocessor, or sometimes a small number of microprocessors.

Progress has been rapid. On June 4, 1991, Thinking Machines of Cambridge, Massachusetts, claimed that its latest version of a massively parallel supercomputer had set a record for speed in calculation at 9.03 billion mathematical calculations a second. Two years later, Cray Research had also produced a massively parallel computer with about the same top speed. On November 7, 1994, the NEC Corporation claimed that its new massively parallel supercomputer had reached a trillion calculations a second, more than a hundred times as fast as the Thinking Machines computer's claim of just more than three years earlier. In 1995, the Intel Corporation, which pioneered massively parallel computers along with Thinking Machines, received a contract to build a computer using 9000 Pentium Plus chips. These would operate at 1.8 trillion calculations per second (technically known as 1.8 teraflops, a *flop* being a "floating point operation").

Japan began its own effort to develop a new approach to computers in 1982 when it began the Fifth Generation computer project. The earlier generations were thought to be as follows:

First   The giant computers that, after a brief try at using relay switches, based their operations on vacuum tubes. The first of these was the ABC computer in 1943.

Second   The same kind of giant computers made smaller, faster, and more reliable by replacing the vacuum tubes with transistors. Bell Telephone built the first of these in 1955, and a commercial version was available the next year.

Third   Computers that use integrated circuits instead of transistors. The first of these was available commercially in 1968.

Fourth   Computers that use very large scale integrated circuits. These circuits were invented in 1973, and began to work their way into computer architecture soon after.

The Fifth Generation was intended to be identified not as the earlier generations had been (by virtue of the type of switching circuits employed), but by the abilities of the new computer: speech, understanding, and intelligence. The project failed to produce anything with notable capabilities along these lines, however, and was abandoned in June 1992 after ten years of software development. In its stead, the Japanese began again on a project to further develop a powerful parallel processing computer, prototypes of which were the main benefit of the whole Fifth Generation experiment.

Other supercomputer plans also ran aground. Cray Computer, started by Seymour Cray after leaving his original company Cray Research, filed for bankruptcy. Stern Computing Systems in France went into receivership. Cray's supercomputer was difficult to produce and his investors lost confidence. Stern Computing Systems also could not produce the special-purpose computers they had designed. Neither were parallel computers, which are inherently easy to produce although difficult to program. Thinking Machines also got into trouble as a result of not progressing as fast as the competition.

The DNA Computer

Mathematician Leonard Adleman (best known as the "A" in RSA public key encryption) had an interesting idea that went far beyond mathematics. He recognized that DNA is an information storage and retrieval system as well as a program instructing a cell to produce specific proteins on a complex schedule. Thus, DNA is essentially the same as a computer. Instead of simply writing a paper discussing this insight, in 1994 he used biology to demonstrate that the DNA computer could be used to solve a difficult problem in mathematics. In terms of density of computer power, the DNA system ranked up near the top, since about one-fiftieth of a teaspoon of DNA was far more than enough to solve the problem. In terms of information storage, DNA is far more impressive than any of the artificial media so far (*see* "Data Storage and Retrieval UPDATE," p 708). Adleman calculated that DNA stores 1 bit per cubic nanometer (10^{27} bits per cubic centimeter), compared with 3×10^7 bits per cubic centimeter for holographic data storage.

One might think that Adleman merely developed the *idea* of using DNA as a computer, inveigling some biologist into working out the details. Instead, intrigued by modern biology, Adleman took a year off to study the subject in detail. When he recognized that DNA could be used as a computer, he spent six weeks working out the kind of problem for which it could be used; for this he needed a good grasp of what a biologist could actually do. He then went into the lab and used the techniques developed by molecular biologists to put DNA to work on the problem.

The problem Adleman chose was a version of the traveling salesman problem, well known to mathematicians as a difficult problem for which there is no good algorithm, and an important example of a whole class of difficult problems. The traveling salesman problem, however, is in a class of problems for which it is easy to check a "solution" to see if it is correct. The problem is also simple to state: plan a salesman's route through a given number of cities via predetermined one-way air routes between cities. The beginning and end cities are specified, and the salesman can only pass through each city one time. Adleman's problem stipulated seven cities and 14 air routes.

Mathematicians have used physical systems to solve mathematical problems in other situations. Perhaps the best known example is the use of soap bubbles to solve problems of least area for particular configurations. A wire figure is made representing the boundary of the figure containing the conditions that must be satisfied. When the wire frame is dipped into a solution of soapy water, the film which forms necessarily has the least possible area for a figure with its edges on the frame. Calculating such a solution without resorting to the bubble is a difficult problem in a branch of mathematics called the calculus of variations.

The essence of Adleman's method involved representing the parts of the problem as stretches of DNA and then allowing many copies of these parts to form combinations in a kind of DNA stew. He then used probes that fished out the DNA combination representing the solution to the problem. He made additional copies of the DNA solution using PCR, and then "read" the solution from the DNA. In practice, the steps were a bit more complicated, however.

The model of the problem was to use three different kinds of 20-base long DNA fragments. One set of seven fragments was randomly chosen to be the cities. These did not themselves need to be used, but were used to create 21 other DNA sequences. The second set of 14 fragments was created with the last 10 bases of a city from which the salesman could depart as its first 10 bases and the first 10 bases of a city at which the salesman could land as its last 10 bases. For example, if the seven cities are represented with a single letter of the alphabet as a sequence of 10 bases, then the cities would be AB, CD, EF, GH, IJ, KL, and MN. The 14 routes would be formed from different combinations of those same 10-base sequences, here represented using the same letters as the names of the cities:

BC DE FG HI JA JK LM LE EC BM LC DC BG HE

Thus, the combination LE represents a trip starting at KL and ending at EF. Finally,

Adleman built the 10-base complementary strands to the city strands. If a complement to GH is represented as G'H', that means that each base in GH is replaced with the base that fits it when two strands of DNA form a double helix. If G'H' is mixed with the 14 air-route bases, the part G' will connect to FG and BG, while the part H' will bind with HI and HE. Adleman then mixed all of the complements of the cities, except for the starting city AB and the end city MN (that is, the DNA sequences represented by C'D', E'F', G'H', I'J', and K'L') with the 14 air routes. The complements bound to pairs of air routes forming chains of various lengths. Then the fishing expedition used A'B' and M'N' to pull out only the strands that started with B at one end and ended with M at the other. Some of these would be the shortest path, but others might double back and be longer. Still others might take short cuts and miss some cities completely. So a second fishing expedition used C'D', E'F', G'H', I'J', and K'L' sequentially to make sure that each city was represented. Now the remaining DNA all represented paths that started at the beginning city, passed through all the cities at least once, and ended at the end city. Separating these molecules by weight showed one molecule with the least weight, which was the solution to the problem. There were also a few results suggesting that not all the chemicals combined according to plan, possibly as a result of some contamination of one type by another along the way.

The essence of this plan is that the DNA operates much like a massively parallel computer, with each strand acting on its own. One result is that the amount of energy expended is close to the theoretical minimum needed to make a calculation; that is, the amount expended by the DNA. Adleman expended a lot of his own energy as he worked for days to plan and perform the whole sequence of a half-dozen basic steps. The DNA may have "performed the calculation" in a very short time, but locating and separating out the answer took a week. On the other hand, many of the steps would have been the same even if the problem had involved many more than seven cities, so the increase in time for additional cities, which is very great when this problem is solved by conventional means, would not be large using DNA.

Quantum Computers

Quantum effects are beginning to be used to store and retrieve data and are about to be introduced in a commercial data-storage product (*see* "Data Storage and Retrieval UPDATE," p 708). Other researchers have built quantum-based switches, including a switch that operates by changing the state of a single electron. The idea of combining these advances into a quantum-based computer, which had been contemplated as a theoretical idea in the 1980s, looked like a practical feasibility in the 1990s. Furthermore, in 1993, Seth Lloyd of M.I.T. in Cambridge, Massachusetts, worked out a theory of quantum computing that would avoid the difficulty that plagued earlier notions. This difficulty involves the unpredictability of quantum states based on probability leading to undetectable errors. In 1994, Paul W. Shor of AT&T Bell Labs established that a quantum computer would be better at some tasks than other varieties of computer. By 1995, people were beginning to assemble the different parts needed to create a quantum computer.

In addition to the use of electron levels to store data, a computer needs logic gates. Some NOT, AND, and COPY gates have been developed based on photons and the spins of atoms, but these cannot be connected together practically. Another solution, which has been demonstrated to work for a few bits of information at a time, is based upon trapping isolated ions and representing information by a rocking motion imparted to the ions. In November 1994, a group from the University of Notre Dame in Indiana demonstrated that they could produce arrays of quantum dots in which electrons can influence each other without being in direct contact. These arrays could then be used to perform addition and other operations based upon quantum rules.

One virtue of even the simplest quantum computer cannot be duplicated on even the most sophisticated nonquantum computer. A quantum computer can produce strings of bits that are truly random. Random numbers are used in many important applications, and while it is easy to think that the "random numbers" generated by ordinary computers are really completely random, they are always produced as a result of some formula, and therefore fail mathematicians' strict tests for randomness. Quantum processes, however, are inherently random, so generating random numbers is the easiest operation of a quantum computer. Another advantage of quantum computers is that they may be set to work in parallel with various different atoms or electrons working all at once, as in the DNA computer. However, a particle in quantum theory is in all possible states it can occupy simultaneously (until it is observed). Thus, a collection of particles representing n bits of information is really in 2^n states all at once. Finally, there is the odd ability of two quantum states that are said to be entangled keeping in touch with each other over intervening space without any apparent means of communication. One part of a computer can affect another part without wires or fiber-optic cables or light guides, and such an effect is instantaneous, not even constrained by the speed of light in a vacuum.

Set against these advantages are three major difficulties. Since the effects being measured are all small compared to random noise, a quantum computer may make errors that it cannot detect or correct. Such errors could not be detected by the computer because quantum theory says that any observation of a state changes the state. The second problem concerns the fact that electrons would not necessarily stay in any higher energy state long enough to accomplish anything, in part because quantum states tend to change suddenly into classical states if any information about the state leaks out to the environment. Lastly, there are only a small class of problems for which quantum algorithms have been worked out, the most noticeable being factoring large numbers. In these algorithms, the trick is that the answer is converted from a measurement of individual states to the measurement of the period of a wave, which is easier to accomplish in practice.

So far, the only practical application of quantum states similar to those used in a computer is in transmitting public-key codes. Because a quantum state vanishes if it is observed, a coded message can detect whether or not anyone is attempting to eavesdrop on it. In 1995, scientists from Los Alamos National Laboratory, led by Richard Hughes, demonstrates that these methods can be used to transmit data over an ordinary fiber-optic cable 14 kilometers long. The hitch in this case is that a message sent in this fashion cannot be amplified, since any attempt to do so would alter the photons and therefore cause the message to evaporate.

Chip UPDATE

When people speak of computer "chips," they may mean the devices that store the random access memory of a computer, although most of the time the emphasis is on the microprocessor that serves as the central processing unit of a computer. This is the chip with which a computer becomes identified. A "Power Mac" is a Macintosh computer with the PowerPC chip that was introduced in March 14, 1994. This chip, also used in some IBM workstations, is notable because it is based on reduced instruction set computing (RISC architecture) in which there are fewer instructions, allowing for more speed. A fast chip does not guarantee success. A version of Digital Equipment Corporation's chip based on RISC architecture was introduced a few months after the PowerPC and was touted as being two to three times as fast. Lack of software, however, has kept the Alpha chip from having much of an impact. The most successful chip has been the one with the most software available for it, the Pentium, although it, too, had a problem early in its lifetime.

The Pentium Problem   One surprise of the 1990s was that a computer chip could have a systematic source of error programmed into it, and that the error might go undetected for a fairly long time, considering that the chip in question was from the beginning immensely popular with consumers.

Known as the Pentium, the chip is the lineal descendant of the Intel 8086 chip that took part in the personal computer revolution in 1978. Essentially the same chip, the 8088, was used in the IBM Personal Computer. As subsequent chips in this line were developed, the initial 8 was first dropped, so one spoke of a 286 or 486 chip. In 1993, with what would have been the 586, Intel switched from a nerdy number to a catchy name: the Pentium. The logical name for the next chip in line would have been the Sexium, but for some reason Intel decided to call it the Pentium Plus when it was introduced in 1995.

In the summer of 1994, Thomas Nicely from Lynchburg College in Lynchburg, Virginia, was checking the parameters in a number-theory problem when he discovered an odd error. At first he thought that only his computer made the error, but he eventually tested the computation on several computers and found that the same error occurred only on the Pentium-based computers. He told Intel that he had found a bug, but the company was not responsive. Finally, on October 30, 1994, Nicely E-mailed several of his colleagues to warn them that their computers might not be reliable. The word was really out by November 22, when it was broadcast by CNN on television. Intel had, as it emerged, already found the error on its own and was shipping revised chips to certain customers. Intel staff, however, felt that the average customer would not be handicapped when a division problem involving seven-digit numbers showed an answer that was off by 256.

Intel's marketing judgement was flawed, however, and Intel stock fell 5% in value. The media reported details of the fiasco, people made jokes about errors made by Pentium chips, and customers demanded replacements. Earlier errors found in other Intel chips and chips from other manufacturers had been quietly corrected, but this one was very noisily resolved. Before the issue was settled, IBM suspended Pentium-based computer sales for a time, and consultant firms recommended that their clients not purchase computers containing the Pentium. By the end of 1994, Intel was replacing chips for all who requested replacements and offering software patches via the Internet as well.

The Pentium contains 3.1 million transistors on a single chip, which suggests how complex it is. Since each chip can be either on or off, the total number of possible states for the system is $2^{3,100,000}$, a number much larger than the number of electrons in the universe. Checking all of these states for errors one at a time is clearly impossible. Various means have been developed, however, that are used to cull out unnoticed flaws. In mathematics, theorems since Euclid's time have been able to establish results for not just $2^{3,100,000}$ possibilities, but for an infinite number of situations. Therefore, one approach to testing a chip is to create a mathematical statement that in effect says "this chip works," and then to use mathematical techniques to prove that the statement is a theorem. More commonly, and with more chance for missing an error, a model of a working chip can be set up and the model tested. It is not clear that either of these methods would have been able to predict the Pentium problem. It is certain, however, that Intel and other manufacturers will make error detection a main part of chip development in the future.

Plastic Chips   Computer chips in personal computers are all made from doped silicon, while some supercomputers and optical computers use gallium arsenide or variations on gallium arsenide. These materials are inorganic crystals, meaning that they have the virtues and defects of inorganics. Just as plastic has replaced many inorganic devices throughout the world, plastic may come to replace at least some of those inorganic chips. Large sheets of flexible plastic could be used for solar cells or for displays. The reasons for using plastic are the same for chips as for other devices: plastic is cheaper and easier to work. Plastic materials may, however, be less durable, or be possessed with flaws that were not found in the inorganic they are displacing.

In 1994, there were two promising developments in plastic-chip technology. In September, a team from the French National Center for Scientific Research in Thiais, France, combined large sheets of thin organic polymers with conducting inks to print transistors that contain no metal at all. The main advantage of this approach is the use of the familiar techniques of printing to produce the sheets of transistors. The disadvantage is that devices produced this way are so large that they operate slowly due to the time it takes electrons to move from one place to another. Then in November, Y. Yang and A. J. Heeger of the UNIAX Corporation in Santa Barbara, California, developed a different plastic transistor whose structure is more like that of a vacuum tube. The polymer grid triode, as it is called, is speedy because electrons do not have to move very far when they are working. Polymer grid triode design can also be used as a light-emitting diode in optical applications.

Tiny Chips   One of the first improvements noticed about the transistor when compared to the vacuum tube was the smaller size of the transistor. In fact, the only thing people could think to do with a transistor in the late 1940s was to put it into hearing aids, since small size was a clear advantage there. Most of the general public became aware of the transistor when the Japanese began making what at the time seemed to be tiny radios—although the first transistor radios in 1954 were far from being as small as many radios commonly used today.

Smallness in a transistor is still desirable today. Among the approaches that might work for very small transistors are methods based on using the spins of electrons as gates instead of the electron's energy levels or their positions in a substance. Another more conventional approach that is already in commercial production involves making chips three-dimensional by adding layers or by boring into the silicon to establish a profile. The hard part is making two layers of transistors, although some progress has been made.

The smaller and smaller size of even conventional new chips and circuits has had some unexpected results. One of these was the collapse of a division of Alcoa called Alcoa Electric, which made ceramic holders for microprocessors. The ceramics were the firebricks of computer chips, dissipating the heat produced by all that artificial thought. While ceramic holders also provide support and protection from dirt or other damage, heat dissipation is why ceramics have been used instead of a cheaper material such as a plastic. Because smaller chips make less heat, Intel (the manufacturer of the Pentium and Pentium Plus chips) began in 1995 to switch from ceramics to plastic. Alcoa Electric ended up firing ("permanently laying off") half of its workers and announcing that the division might go out of business completely. Alcoa Electric was only one of the four suppliers of the ceramic holders.

(Periodical References and Additional Reading: *New York Times* 5-29-92, p D3; *New York Times* 4-17-92, p D1; *New York Times* 6-5-92, p D1; *New York Times* 10-6-92, p D4; *Discover* 1-93, p 94; *New York Times* 3-15-93, p D4; *New York Times* 5-19-93, p D5; *New York Times* 8-16-93, p D1; *New York Times* 9-27-93, p D1; *New York Times* 9-8-94, p D4; *New York Times* 9-16-94, p A24; *Science* 9-16-94, p 1684; *New York Times* 9-28-94, p D1; *Science* 11-11-94, pp 993 & 1021; *New York Times* 11-8-94, pp D1 & D7; *New York Times* 11-22-94, p C1; *Nature* 11-24-94, p 344; *New York Times* 11-24-94, p D1; *New York Times* 12-2-94, p D4; *Science News* 12-3-94, p 375; *New York Times* 12-13-94, pp A1 & C1; *New York Times* 12-21-94, pp A1 & D1; *New York Times* 1-7-95, p D39; *New York Times* 1-8-95, p III-7; *Science* 1-13-95, p 175; *Science* 1-20-95, p 332; *Nature* 3-30-95, p 394; *Science* 7-8-95, p 28; *Nature* 9-14-95, p 96; *Physics Today* 10-95, p 24; *Scientific American* 10-95, p 140; *New York Times* 12-22-95, p D16)

COMPUTER SPEECH AND INTELLIGENCE

In the dim beginning of computer science, someone noted that all the metal and glass comprising a physical computer was just "a bunch of hardware." Therefore, the logical jokey description of the programs enabling the hardware to function should be *software*. The name has stuck, and has even inspired other words associated with computers, including firmwear and wetwear. The availability of useful software has driven the computer revolution far more than any development of hardware. Personal computers were toys until spreadsheet software made them essential in business. Writers of all kinds, except for a few old-fashioned literary types, abandoned their typewriters when they discovered word processing. The first computers could beep, but high-quality sound along with good graphics seems to have been what it took to make the computer at home in the home.

Speak When Spoken To

For many people, the concept of an advanced computer is largely based upon Hal, the talking computer in the motion picture *2001*. If one could speak to a computer and be understood, and if the computer could speak the answer instead of displaying it, then there would be no need for input devices such as keyboards and mice. We are getting close to the year 2001 now, and while much of the movie's view of life in the next millennium is clearly not going to happen by then, the computer that can be spoken to is here.

IBM has been a leader in this field, introducing affordable speech recognition for its computers as early as the end of 1992, and offering many home computers today with speech recognition built in. Indeed, speech recognition is in many new devices, not just in our computers. The first encounter many people have with computerized speech recognition is a telephone answering device that recognizes a few responses spoken into the phone, often only words like *yes* and *no*. High-end automobiles have been speaking to their owners for years, telling them to take out their key or not to go so close to the curb. Today, some owners of expensive automobiles can speak back. An automobile alarm that can be shut down by the speech of the auto owner may be one feature. Other devices, such as some automatic teller machines, also recognize speech.

A computer requires several steps to recognize human speech. First, it has to change the sound wave from analog pressure differences into digital samples. Using the same approach as in making a CD audio recording, the computer samples the speech with thousands of air-pressure values taken every second. These are recorded as digital data points, from which such information such as pitch or volume can be deduced. The computer analyzes the data in terms of pitch and volume to get a rough notion of what it has heard, and then compares that set of data points with the digital data for various sounds that it has stored in its vocabulary. The closest match wins, whether it makes sense or not. Then the computer takes the action it is programmed to do, such as saving a file.

A similar approach is helping computers to see, a task at which they are not as skilled as they are at sound recognition. Instead of trying to take in a whole scene, a computer using a system developed by the David Sarnoff Research Center in Princeton, New Jersey, focuses on the one part of a scene that it has been programmed to recognize, most often anything that resembles a human face. The computer then has to digitize the picture and compare it with a picture stored in memory. As far as this software has been developed, if the picture in its memory is not exactly the same as the face presented, the computer won't make the connection. This means that the computer is less able to recognize people than sounds. On the other hand, this aspect of the software enables it to be used as a security device. A person must put his or her face in a specific attitude before the computer, which will then recognize the person and unlock the gate.

Translation

Related to the problems of speech recognition in many ways is the much studied minefield known as computer translation. One of the first two or three computers ever built was COLOSSUS, the British computer used for decoding German messages (decoding is translation from an artificial language). Natural-language decoding was tried soon after the development of the computer, and it has proved much more difficult.

One of the few successes of Japan's Fifth Generation effort at computer translation was an improvement in computer dictionary skills, enabling the computer to combine several special-purpose dictionaries before making a translation. Words in one dictionary, especially nouns, are arranged according to a taxonomic system, so that a given word can be identified by species, genus, family, and so forth. A footstool could be recognized as in the living room family of the furniture class, for example. Another dictionary identifies how concepts are related to each other, so that a footstool has legs, a leg is a part of furniture as well as a part of an animal, a living room is part of a house, and so forth. A different dictionary connects words that are simply used together, such as *footstool* and *move* or *on*. It would tend to eliminate constructions as "footstool spoke" or "speeding footstool." With the several dictionaries used together, the computer can produce sensible translations that are fairly close to what a human would produce.

Far less sophisticated translating software can be purchased for a home computer today for less than $100 in some cases. These relatively simple programs are likely to translate "out of sight, out of mind" into something that means "invisible, insane," but they do a fairly good job of translating simply written, non-idiomatic language. Each program has to be specific to the language being translated.

The personal-computer translation programs all need someone to input the words in the initial language to be translated via a keyboard. A more advanced computer can combine voice recognition with translation, although just barely. On January 28, 1993, for example, the first system for telephonic simultaneous computer translation was demonstrated. In Kyoto a Japanese researcher, Toshiyuki Takezawa, spoke the phrase *moshi moshi* into a telephone connected to a computer. The computer used voice recognition to produce a text version of the word. It then passed the word, still in Japanese, along to a second computer programmed to translate Japanese into English. The translating computer transmitted the data in digital form along a standard international telephone connection (via satellite) to another computer in Pittsburgh, Pennsylvania. That computer recognized the word in the dictionary in its memory, and used voice synthesis to produce the word *hello* in English. The process took 12 seconds from "moshi moshi" to "hello."

Machine Learning

Although the Japanese Fifth Generation computer project failed to develop artificial intelligence to the degree its founders had hoped, considerable progress continues to be made on the skills associated with the artificial-intelligence movement. This progress is not only in speech recognition and production and computer translation, but also expert systems. There is not much likelihood of a computer demonstrating rational thought, but most computer specialists today admit that artificial intelligence is not the same as real intelligence. Computers based on artificial intelligence are simply large computers, such as the one that watches what American Express customers are buying in hopes of preventing credit-card fraud, with software that enables them to make judgements.

The American Express computer works by using a form of neural network, a kind of imitation of the way that parts of the brain are supposed to be connected. Neural networks have become very popular in the 1990s. In 1993, for example, Intel introduced a computer chip, the Ni1000, that was based on the neural network concept for use in computers that learn skills of one kind or another. Not only have neural networks proven useful in making judgmental decision relating to finance, but they have also been used for developing machine vision and improved speech recognition.

The original neural networks employed machine learning simply by noting mistakes and avoiding their repetition. Later, machines were programmed to come closer and closer to an ideal. For example, a computer teaching itself to play the stock market might compare its performance with the Dow Jones Industrial Average and keep only those behaviors that produced better results for a given day than had been reported on the D-J ticker. Other tricks of successful teachers have been found to work on computers as well as or better than they do on human students. Giving hints and trying out hypothetical examples have been found to be especially effective.

At this stage, the two main problems with machine learning are a tendency to spend too much time and energy computing possibilities before taking action, and a tendency to repeat back what the teacher intended as an example, as if that were the only possible solution. Both problems may be familiar to teachers of human students as well.

It's Only a Game

One long-recognized test of artificial intelligence is the ability to play chess. Chess has such a powerful strategy component that it has long been treated as something more than a mere game. Yet computers have played chess almost since the invention of the computer. Furthermore, even moderately powerful computers have been able to beat indifferent human chess players for a long time. Still, no one expected computers to be able to beat the very best human players for a long time.

This "long time" was over in August, 1994. A computer program called Genius 2 won one match with the human world chess champion Garry Kasparov and played him to a draw in the second match. Later in the same tournament, however, chess grandmaster Vishy Anand defeated Genius 2, knocking it out of the tournament. Genius 2 runs on a computer based on a 166 mHz Pentium chip. In 1996 Kasparov played a six-game match with an IBM computer known as Deep Blue, programmed by Chung-Jen Tan and coworkers at IBM's Thomas J. Watson Research Center in Yorktown Heights, New York. Deep Blue won the first game, but Kasparov adjusted to the computer's style and won the match, winning games two, five, and six and drawing games three and four.

Bugs in the Programs

The problem with division in the original version of the Pentium chip was technically a hardware-based problem, although it originated in the way the hardware had been programmed. As computers become more complex, other problems are bound to develop. In the early 1990s, one such example came when several spacecraft got the wrong instructions because of software bugs. As a result, a Russian spacecraft fell silent just as it was approaching Mars, and a United States space probe that was intended to leave a successful survey of the Moon and head off to the asteroid Geographos went into a spin and could not complete its mission. Perhaps the best known 1994 software problem occurred at Denver's new airport, where software problems with the baggage handling system were so bad that they kept the airport from opening for nearly a year.

Software programmers have developed various methods intended to eliminate such bugs. In the early 1990s, one of the most useful has been a move toward programming languages in which the basic parts of programs already exist in a tested format. Among such programs, the one most likely to be encountered by the average computer user is called Java, used especially for making Internet applications.

Although most bugs appear in programs by accident, another sort of computer bug, the virus, is a deliberate creation of people intent upon disrupting computer operations—mostly for the fun of it. As a result, the creation of software intended to handle problems caused by viruses has become a respectable sub-industry. In July 1994, Jeffrey O. Kephart of IBM's High Integrity Computing Laboratory introduced a new method for dealing with computer viruses, which he based on how mammals combat biological viruses. The IBM system sets up decoys that are easily infected and then monitors for infection. The program notifies any computers to which it is attached in a network, and each of them begins a systematic search for infection and notifies any other computers to which they are themselves attached. IBM has been using this system itself since 1991 and found out how to recognize more than 2,500 viruses in just the first three years of operation.

(Periodical References and Additional Reading: *New York Times* 8-19-92, p D5; *New York Times* 11-17-92, p D5; *New York Times* 9-30-92, p III-9 *New York Times* 2-13-93, p 35; *New York Times* 7-18-93, p IV-6; *Scientific American* 9-94, pp 86 & 97; *Discover* 1-95, p 103; *Scientific American* 4-95, p 64)

DEVICES AND DEEDS

Nanotechnology UPDATE

Technically, the prefix *nano-* refers to a billionth of a measure, such as a nanosecond or nanometer. It has been appropriated by scientists who need a general prefix that is smaller than *micro-*, which had a long usage as "very small" even before it became the international prefix for a millionth of a measure (think of the microskirt). Thus, an oceanographer may refer to very small plankton as nanoplankton, even if they are several orders of magnitude larger than a nanometer. Nanoengineering, however, is somewhere in between the nanometer and micrometer range. Specifically, it is defined as the creation of devices smaller than 0.1 micrometer, and therefore smaller than 100 nanometers, which is the same measurement expressed a different way.

Sensing Molecules and Rearranging Atoms

Noel MacDonald, Susanne Arney, Jason Yao, and coworkers at Cornell are building the smallest possible scanning tunneling microscopes (STMs), but not for imaging. While the business end of an STM, the tungsten tip, consists of only a few atoms, the rest of the machinery used to control the tip's position over a sample is usually somewhere between the size of a VCR and a microwave oven. The Cornell team hopes eventually to get the entire tunneling device down to the size of the cut end of a human hair. Their goal is to use tiny tunneling devices to sense specific molecules or changes in light or motion in the environment. They plan to use two tips instead of one and to use piezoelectric crystals and tunneling to monitor the current between the tips. Changes in the current in a given environment could mean the presence of a specific molecule or some other change of interest. Several such devices could be placed in an array to detect a range of changes in the given environment.

Other teams of researchers are also working on miniature tunneling devices with various purposes. For example, William J. Kaiser and coworkers at the Jet Propulsion Laboratory in Pasadena are developing small tunneling sensors that can be used as parts of navigational systems for spacecraft. They also envision using tunneling devices as seismometers small enough to be carried on spacecraft to other planets. Small silicon-based seismometers could be utilized to monitor prospective earthquakes or volcanoes on Earth as well.

When the tip of a tunneling device is placed 30–50 nanometers (3–5 angstroms) from a silicon surface and an electric potential of about $+3$ volts is applied, some of the atoms of silicon form a small mound. The mound may contains a few silicon atoms, depending upon the exact distance between the tip and the silicon, the size of the voltage, and the sharpness of the tip. Moving the tip slightly closer to the mound causes the top layer of atoms to part from the surface and adhere to the tip. The number involved can be as small as one, or it can be of the order of 10 or so.

If the tip is then moved to a different place above the surface and a voltage of the opposite polarity but same amount (that is, about -3 V) is applied, the silicon atom or cluster of atoms drops off the tip and sticks to the other silicon atoms of the surface.

The inventors of this technique, In-Whan Lyo and Phaedon Avouris of IBM's Thomas J. Watson Research Center, envision using it to create very tiny semiconductor devices, a process they have named nanoelectronics. Another possible use for their technique would be in the development or manufacture of the small tunneling sensors that Noel MacDonald's team is developing, since these sensors are built largely of silicon.

In 1994, Calvin Quate of Stanford University in Pasadena, California, and coworkers used an AFM to oxidize a line in a silicon coating by pushing aside a protective coating of hydrogen, allowing oxygen to get to surface as a part of ambient water vapor. Oxidation then gave the silicon beneath the line protection while the remainder of the silicon coating was dissolved away with a solvent. With this technology, the Stanford group produced gates for metal-oxide-silicon field effect transistors (MOSFETs) as small as 0.2 micrometers long. Although this is not much better than existing MOSFETs made by conventional lithographic techniques, the Stanford method can be adapted to produce much smaller gates. Conventional methods, however, are already near the end of their potential. Furthermore, there is reason to think that an AFM system can be partly automated in such a way as to be cheaper than conventional means.

In addition to creating nanoelectonic devices, similar technology has been used in experimental data storage and retrieval systems (see "Data Storage and Retrieval UPDATE," p 708), as well as in experiments that might lead to new kinds of chips or even displays.

C. T. Salling and M. G. Lagally of the University of Wisconsin in Madison reported in July 1994 that they have used the scanning tunneling microscope to create straight-walled structures on silicon wafers that were just a single atom deep and two atoms wide. These structures stay stable at room temperature, but tend to disappear at the high temperatures used in fabricating modern electronic devices, which limits their applications at present.

A group of physicists from the University of Basel in Switzerland used a scanning force microscope (see box) to modify fast-reacting liquid crystals so that they contained nanoscale wave guides. This work could lead to use of liquid crystal devices in optical computers as well as possibly improving LCDs (liquid crystal displays).

An entirely different device has been developed by a group from the University of Hannover in Germany. They developed an atomic beam for the purpose of imaging nanometer sized objects but realized that the same device could be used to place atoms in exact locations as desired, making the atomic beam microscope yet another tool for nanofabrication. The beam uses lasers and magnets to control a beam of cesium atoms.

SCANNING TUNNELING, ATOMIC FORCE, AND THEIR RELATIVES

The basic method used by the scanning tunneling microscope (STM) is to maintain a specific level as the tip moves over a surface. The amount of electron tunneling (see below) between the tip and the surface varies with the height of the surface, increasing over a hill and decreasing over a valley. The electric current is then converted into an image on a computer screen.

Conversion of distances into an image of the surface is something like the process of converting a contour map (a map that shows the heights of each point as a numerical distance above sea level) into a relief map that shows approximate heights by using techniques to make a two-dimensional drawing look three-dimensional. A contour map can also be converted into actual relief by making a three-dimensional model of the surface described by the numbers on the map. With such a relief model, the surface can be sensed by touching it. One can even discern characteristics of the surface in the dark. If the surface is a real mountain chain on Earth, the scale is too great for one to envision it by feeling the actual mountains, but one can get an idea of the shape of the system by feeling a relief model of the mountains. The mountains need to be represented on a human scale.

Now about **tunneling**. In short, tunneling is passing through a barrier that is like the gospel-song description: "so high, you can't get over it; so wide, you can't get around it; so deep, you can't get under it." Some particles, however, get over, around, or under the barrier anyway. Tunneling was one of the early astonishing paradoxes of quantum theory, first explained by George Gamow in 1928. Any particle can tunnel, but as in most quantum effects, the smaller the particle, the easier the tunneling. Electrons are among the smaller particles, so they tunnel very easily.

Perhaps the easiest way to think of tunneling is to remember that everything has both a wave aspect and a particle aspect—and then forget the particle aspect, since tunneling is a function of the wave. It may also help to remember that, comparatively speaking, large particles are small waves, and small particles are large waves. Thus, a small particle like an electron is associated with a large wave.

Tunneling works with any kind of barrier, whether a barrier caused by the electromagnetic force (such as the barrier a positive charge effects on another positive charge), one caused by gaps (such as a deep ditch around a castle), or a solid wall (such as an insulator to an electric charge). A person cannot tunnel through a brick wall because a person is fairly large, making for a small wave, and a brick wall is rather thick. An electron can tunnel through a gap of several angstroms because the electron has a big wave, and several angstroms is not far.

Because it is a wave, the electron is not located exactly at any point in space. Some of the wave penetrates into the gap. If the gap is small enough, some of the wave appears on the other side of the gap. Here is where quantum weirdness comes in. The wave we are discussing is a probability wave, the probability of the electron being located (as a particle) at a particular point. So if some of the wave is on the other side of the gap, there is a definite probability greater than zero that the electron will appear on the other side of that gap, which it sometimes does. That's what tunneling is all about.

From a point of view that ignores quantum weirdness, tunneling is just the process by which an electron suddenly vanishes from one side of a barrier, at the same instant appearing on the other. Shades of "Beam me up, Scotty!"

The frequency of tunneling can be predicted very accurately by quantum mechanics. For a given source of electrons, the amount of tunneling depends just upon the size of the gap. Thus, an STM can use tunneling to measure quite exactly the distance of a tip to the individual hills and hollows of a surface, even when the hills are individual atoms.

In 1986, when Gerd Binnig and Heinrich Roher of IBM won a Nobel Prize in Physics for their 1981 invention of the scanning tunneling microscope, the device was primarily a way to examine surfaces in terms of individual molecules or atoms. By the early 1990s, however, the basic principles of the microscope were being applied in a multitude of ways, many of which did not involve imaging anything at all. It might be more appropriate to call the machines based upon the STM's main components (for example, very sharp needles, piezoelectronic controls, and tunneling electrons) scanning tunneling devices (STDs) instead of microscopes. Even that modification in name does not go far enough, given that some of the new applications of these devices do not even use scanning, and others do not use tunneling. (Furthermore, STD is already in use by medical workers to denote "sexually transmitted disease.")

Another coinage is "scanning probe microscopes," which includes many recent devices similar to the STM but which do not use tunneling. All of these devices scan surfaces with tiny tips, but the atomic force microscopes (AFMs) press a sharp tip against the surface; the cantilever-based scanning force microscopes (SFMs) use a tip in various ways; other devices use the tip to measure magnetic fields; and still others use ions or sound. Like the STM itself, some of these scanning nontunneling microscopes are available commercially. Probably the most versatile is the atomic force microscope, which rests the tip against the surface being studied, measuring variations in the height of the surface scanned. In biology, however, the scanning force microscope has had more applications, in part because it can be used by touching a surface, tapping it, or even just interacting with it in chemical or electronic ways of one kind or another. Another variation, the scanning polarization force microscope, is just a scanning force microscope with its tip coated with platinum, enabling it to observe and manipulate liquids.

Tiny Machines and Tools

One of the goals of nanotechnology is to produce very small versions of the machines, motors, and tools that humans have developed for the macroworld. Thus far, most of these tools have been crafted from silicon, because the techniques for working with very small silicon objects have been worked out by the electronics industry. In some cases, however, materials with different properties are needed. For example, some uses would require a substance harder than silicon (*see* "Diamond Discoveries," p 453).

Claims of "the smallest" depend to some degree on how objects are measured, or the description that the claimant prefers to use. With that caveat, here are some of the claims.

In 1992, Toshiba claimed the world's smallest electromagnetic motor, with a diameter of 0.03 in (0.8 mm), operating at 10,000 rpm. It does not have sufficient force to operate nanoequipment and is not the smallest motor of any kind—it is not even nanoengineering—but its virtue is that it is a real electromagnetic motor and the basic design is capable of improvement. The following year, Paul McWhorter at Sandia National Laboratories claimed the world's smallest steam engine, built from silicon using electronics techniques. Unlike the Toshiba motor, the Sandia steam engine is powerful enough to do useful work. The concept is that a tiny bit of water can be heated almost instantly to form water vapor, pushing a piston. When the heat is turned off an instant later, the water condenses. This is much like the first pre-Watt steam engines in principle, but more effective because the small size permits heat to dissipate quickly.

What kind of application might micromachines have? Although much has been made of their ability to travel through blood vessels for medical procedures, one possible use may be to influence the big forces that act on objects traveling through fluids at high speeds. Calculations show that the best way to reduce drag is by making many tiny changes in flaps as small as 0.2 millimeters. These would need to be raised and lowered by even smaller actuators, for which the tiny steam engine could be a candidate. Sensors would tell the flaps whether to raise or lower to adjust for the lowest possible drag.

Scientists at the National Institute of Standards and Technology developed a tunneling electron microrefrigerator. Hot (that is, energetic) electrons tunnel through the barrier, but cool electrons are channeled through a superconducting channel back to the application being cooled. The purpose of the refrigerator is to cool electronic measuring devices so that they will experience less noise while measuring such parameters as the cosmic-ray background. Variations of the idea using slightly different physics were also proposed by Sean Washburn of the University of North Carolina at Chapel Hill and other physicists.

Charles E. Martin, Matsuhiko Nishizawa, and Vinod P. Meno of Colorado State University in Fort Collins have created what may be one of the most useful new tools: an adjustable nanosieve. They found a way to insert tiny gold tubes through a plastic membrane. By putting an electric charge on the tubes, the sieve will let either positive or negative ions pass, as desired. By slightly changing the manufacturing procedures, the diameter of the tubes can be adjusted from as little as 1.6 nanometers up to about 20 nanometers.

Even tools in the nanoworld need to be built to precise measurements, which is not easy to do when one is at times working at distances shorter than the wavelength of light and at other times working at distances visible to the naked eye. In December 1995, Clayton Teague of the U.S. National Institute of Standards and Technology in Gaithersburg, Maryland, introduced a tool that will help solve the measurement problem. Their molecular measuring machine, called the M^3, uses a combination of an STM and laser interferometry to produce a multipurpose tool that can measure distances between points as far as a millimeter apart to within 20-40 nanometers. One of the first applications of the M^3 will be to correct the 600-micrometer grids used by semiconductor manufacturers to calibrate their own STMs and other measuring devices.

(Periodical References and Additional Reading: *New York Times* 9-25-92, p D3; *New York Times* 10-5-93, p C5; *Science* 7-22-94, pp 502 & 512; *Science* 10-28-94, p 543; *New York Times* 1-3-95, p B13; *Nature* 1-12-95, p 106; *Physics Today* 2-95, p 9; *Science News* 4-8-85, p 223;

Nature 5-18-95, p 214; *Science* 5-5-95, pp 639 & 700; *Physics Today* 12-95, p 32; *Science* 12-22-95, p 1920)

REFRIGERATION ADVANCES

A nano-refrigeration scheme for electrons (*see* "Nanotechnology UPDATE," p 732) is only one of many advances in a technology that extends well beyond the large white device found in most kitchens. Refrigeration technology includes the techniques needed to cool a thin gas of metals to the lowest temperature ever achieved, making it into a state of matter that probably never before existed in the universe (*see* "The Newest Forms of Matter," p 645). Other such technology is aimed at keeping machinery and other devices from becoming so hot that they fail to work properly.

Refrigeration and its close cousin, air conditioning, had to be rethought in the early 1990s after the nations of the world joined in an agreement to stop using chlorofluorocarbons, the backbone of cooling since World War II. The problem with chlorofluorocarbons is that although they are inert vapors in the lower atmosphere, they gradually migrate to the upper stratosphere where they break up into reactive chlorine and other gases when exposed to ultraviolet radiation. The chlorine acts as a catalyst to remove one oxygen atom from ozone, effectively destroying the layer of ozone that keeps excessive ultraviolet radiation from reaching the surface of the Earth. Powerful ultraviolet radiation can cause cancer and can kill small but essential organisms. When the gradual ban on chlorofluorocarbons started into effect at the end of the 1980s, there were no suitable substitutes. Not only was a chlorofluorocarbon used as the coolant liquid in refrigerators, but a different chlorofluorocarbon filled the bubbles of foam insulation in the refrigerator walls. The international ban was scheduled to be complete in 1996. In 1992, the manufacturers of refrigerators and air conditioners and the manufacturers of chlorofluorocarbons began to test different chemicals, as well as completely different approaches to refrigeration.

In 1992, for example, a group of electric utilities scattered across the United States went so far as to announce a $30 million prize for the developer of a refrigerator that would use at least 25% less power and no chlorofluorocarbons. Part of the motivation was the realization that, at that time, refrigerators and freezers between them consumed a fifth of all the electric power used in the average home. The utilities had learned that they could make more near-term profit by persuading customers to use less electricity, allowing capital investment in power plants to be kept low. For other reasons, the government also wanted refrigerators to use less electricity and, of course, no chlorofluorocarbons. Instead of offering a prize, the government mandated standards for new refrigerators in two phases, one starting on January 1, 1993, and a second lower standard scheduled for 1998. The 1993 standards called for a greater than 25% reduction in power requirements for a typical home refrigerator.

By the end of 1992, some 450 organizations and businesses had expressed some interest in the competition to produce what the utilities called the Super Efficient Refrigerator. Frigidaire and Whirlpool, two familiar names in household refrigeration, were selected by the utilities to be finalists. Manufacturers such as Owens-Corning were presenting ideas for new products to improve refrigeration as well. The Owens-

Corning idea was to replace the chlorofluorocarbon foam insulation with thin fiberglass thermos bottles, which would be much more insulating and take up about the same amount of space.

Meanwhile, the U.S. Environmental Protection Agency (EPA) was not waiting for the utilities and manufacturers to solve the problem. One big concern expressed by the refrigerator manufacturers was that the most promising replacement for chlorofluorocarbons was a chemical called hydrofluorocarbon 134a that might, over the long lifetime of a refrigerator (often more than 20 years), degrade the lubrication that protects the moving parts of a refrigerator, primarily the piston in the compressor that causes the refrigeration fluid to flow and carry away heat. The EPA therefore funded development of a new compressor that did not need oil as a lubricant and that also used 15% less power.

Refrigerator and compressor manufacturers were skeptical of all the new fixes. They worried that the Owens-Corning thermos bottles might leak. Some of the manufacturers had earlier tried out an idea similar to the EPA-funded compressor and found that it tended to break down disastrously.

Turning Heat into Sound, Cooling with Light

In the 1970s and 1980s, an effect discovered by accident by European glassblowers in the nineteenth century finally began to be understood in terms of basic physics. The glassblowers discovered that connecting a hot glass bulb to a cool glass tube caused the glass to make a wonderful sound as the tube warmed up. The effect ties together oscillation, heat, and sound in such a way that, aligned in one direction, it can be used as a simple engine that turns heat into movement. Aligned in the other direction, it can be used to remove heat from a particular environment with great efficiency. Thermoacoustic refrigeration is flexible enough to be of interest to scientists trying to cool experiments to cryogenic temperatures, as well as to manufacturers of refrigeration devices for use in home kitchens. One of the advantages of the thermoacoustic system is that the gas central to the cooling process is usually helium under pressure instead of chlorofluorocarbons. Helium is truly inert, whether at ground level or in the stratosphere, so it does not contribute to air pollution or ozone thinning. The basic concept of thermoacoustic refrigeration is to use an acoustically insulated loudspeaker to set up a train of sound waves in the helium. Helium is chosen because it transmits sound better than any other gas. The train passes the heat along from wave front to wave front like a bucket brigade. When the acoustic bucket brigade reaches the heat sink, it tosses the heat down the well.

In 1994, Stephen L. Garrett of the Naval Postgraduate School in Monterey, California, and coworkers developed the ThermoAcoustic Life Science Refrigerator for home refrigeration. This invention uses about one-third the energy of an efficient conventional refrigerator. Despite the use of a high-powered loudspeaker as the main cooling element, the unit is almost totally silent. The high cost of the prototype rules against it being the exact model for mass production, but Garrett is convinced that the essential design, which can be adjusted for everything from home air conditioning to cryogenic cooling, will eventually become an option for consumers.

Another approach to refrigeration involves cooling with light. The idea itself dates back to 1928, but only recently have lasers with the correct wavelengths and energy become available. The same principle is used for laser cooling of individual atoms (*see* "The Newest Forms of Matter," p 645), although the laser refrigerator cools a specially doped glass. As the glass glows in the laser light, it cools slightly. The amount of cooling is not very great at room temperature, but that same amount becomes important at temperatures around that of liquid nitrogen—a temperature at which a household refrigerator would not produce any cooling. Keeping liquid nitrogen cool could be important, as high-temperature superconducting devices need to be bathed in liquid nitrogen to work. If there are ever to be large-scale applications of high-temperature superconductivity, the laser refrigerator might be a part of the machinery.

(Periodical References and Additional Reading: *New York Times* 7-8-92, p D3; *New York Times* 8-30-92, p III-3; *New York Times* 10-22-92, p D5; *New York Times* 12-8-92, p D5; *New York Times* 6-27-93, p III-10; *Science News* 2-26-94, p 135; *Physics Today* 7-95, p 22; *Science News* 10-14-95, p 246)

TECHNOLOGY LISTS AND TABLES

THE NATIONAL INVENTORS HALL OF FAME

In 1973 the U.S. National Council of Patent Law Associations (now the National Council of Intellectual Property Law) began the practice of naming inventors who hold U.S. patents to the National Inventors Hall of Fame. Although many of those honored have several patents, a selection committee chooses one patent for each inventor as the occasion for the award, which they identify using the title of the original patent application (given in capital letters below). Note that holding a U.S. patent does not mean that the honoree is necessarily American. Some, such as Rudolf Diesel and Louis Pasteur, were lifelong citizens of other countries.

Alexanderson, Ernst F. W. [Swedish/American: 1878-1975]
HIGH FREQUENCY ALTERNATOR
This basic device makes it possible for radio and television to transmit voices and music, not just dots and dashes. Alexanderson was also a pioneer in television and many other kinds of electrical equipment, receiving 322 patents. (Inducted 1983)

Alford, Andrew [Russian/American: 1904-1992]
LOCALIZER ANTENNA SYSTEM
With the localizer and other inventions, Alford developed the radio system for airplane navigation and instrument landing systems. (Inducted 1983)

Alvarez, Luis Walter [American: 1911-1988]
RADIO DISTANCE AND DIRECTION INDICATOR
Despite his citation, Alvarez is far better known as the developer of specialized equipment for studying subatomic particles (which led to his 1968 Nobel Prize) and as a proponent—along with his son Walter—of the theory that dinosaurs became extinct as the result of the impact of a massive body on Earth 65 million years ago. (Inducted 1978)

Armstrong, Edwin Howard [American: 1890-1954]
METHOD OF RECEIVING HIGH FREQUENCY OSCILLATIONS
Armstrong's several inventions connected with radio broadcasting and reception made him rich and created the "radio days" of the 1920s through the 1940s. In 1939, he invented FM broadcasting and reception, which helped lead to another revolution in radio. (Inducted 1980)

Baekeland, Leo Hendrik [Belgian/American: 1863-1944]
SYNTHETIC RESINS
Baekeland's plastic (synthetic resin) that he named Bakelite was not the first plastic to be manufactured (celluloid was first), but it was the first to make people realize the potential of plastics in general. Baekeland also developed the first commercially successful photographic paper. (Inducted 1978)

Bardeen, John [American: 1908-1991]
TRANSISTOR
In the mid-1940s, Bardeen invented the transistor with William Shockley and Walter Brattain. The transistor is the essential semiconductor device used on microprocessors and other chips. (Inducted 1974)

Beckman, Arnold O. [American: 1900-]
APPARATUS FOR TESTING ACIDITY
Although there are many simple ways to determine acidity, a precise measuring instrument developed by Beckman became the foundation of the Beckman Instrument Co., a leading company in the scientific instrumentation field. The other precision instruments Beckman invented also contributed to the company's growth. (Inducted 1987)

Bell, Alexander Graham [Scottish/American: 1874-1922]
TELEGRAPHY
Despite the title, the invention was the telephone. Another important Bell invention was the disk phonograph. (Inducted 1974)

Bennett, Willard Harrison [American: 1903-1987]
RADIO-FREQUENCY MASS SPECTROMETER
A mass spectrometer is a device to separate particles according to their masses. Prior to the Bennett spectrometer, patented in 1955, heavy magnets were used for this task; the Bennett device uses easy-to-generate radio waves. In addition to his invention, Bennett is noted for pioneer work in plasma physics. (Inducted 1991)

Berliner, Emile [German/American: 1851-1929]
MICROPHONE AND GRAMOPHONE
Berliner's microphone made it possible to use Alexander Graham Bell's telephone over long distances. With the $50,000 patent rights payment he received from the Bell Telephone Company, Berliner developed the gramophone, the forerunner of the record player. (Inducted 1994)

Binnig, Gerd Karl [German: 1947-]
SCANNING TUNNELING MICROSCOPE
(With Heinrich Rohrer) The scanning tunneling microscope enables scientists to make images of objects even smaller than an atom. The innovative device also paved the way for a number of similar inventions that both observe and manipulate structures as small as individual atoms and molecules. (Inducted 1994)

Bird, Forrest [American: 1921-]
MEDICAL RESPIRATORS
The little green box known as "the Bird" prevented thousands from dying from cardiopulmonary failure. Its inventor also developed the Babybird respirator for low birth-weight babies, cutting mortality rates for infants with respiratory problems from 70% to 10%. (Inducted 1995)

Black, Harold Stephen [American: 1898–1983]
NEGATIVE FEEDBACK AMPLIFIER
The basic principle of the negative feedback amplifier—the use of feedback of information to control a process—has become fundamental to many other devices since Black's first use of it to control distortion. Black also invented pulse-code modulation, which remains an important concept in communications. (Inducted 1981)

Blumberg, Baruch [American: 1925–]
HEPATITIS B VACCINE
Since it was patented in 1972, this vaccine (invented with Irving Millman) has protected hundreds of millions of people worldwide against the hepatitis B virus, which is still one of the leading causes of death worldwide. (Inducted 1993)

Brattain, Walter H. [American: 1902–1987]
TRANSISTOR
(*See* Bardeen, John) (Inducted 1974)

Brown, Rachel Fuller [American: 1902–1987]
NYSTATIN
(With Elizabeth Lee Hazen) The world's first nontoxic antifungal antibiotic. Not only did this medicine cure many disfiguring and disabling skin, mouth, and throat infections, but it could be combined with other antibacterial drugs to balance their effects. Nystatin has been used to treat Dutch elm disease and to rescue water-damaged artwork from mold. (Inducted 1994)

Burbank, Luther [American: 1849–1926]
PEACH
Burbank holds 16 plant patents with numbers between 12 and 1041 (all issued posthumously). His work in developing more than 800 new varieties of plants was in part responsible for development of the plant patent program, which began in 1930. (Inducted 1986)

Burckhalter, Joseph H. [American: 1912–]
FITC
FITC, a yellow-green compound known to doctors as fluorescein isothiocyanate, discovered by Burckhalter and Robert J. Seiwald, has proved invaluable in identifying different antibodies. In addition to helping identify the causes of AIDS, FITC has been used to speed the diagnosis of leukemia and lymphoma. (Inducted 1995)

Burroughs, William Seward [American: 1857–1898]
CALCULATING MACHINE
Although a calculating machine had been built by Wilhelm Schickardt as early as 1623 (and other types by later inventors), Burroughs was the first to develop a practical device that could be mass produced and easily used. (Inducted 1987)

Burton, William Meriam [American: 1865-1954]
MANUFACTURE OF GASOLINE
The highlight of Burton's years in the oil business came when he developed the first commercially successful cracking process, a method that yields twice the amount of gasoline from crude oil as had previous methods. (Inducted 1984)

Camras, Marvin [American: 1916-1995]
METHOD AND MEANS OF MAGNETIC RECORDING
Before the tapes currently used to record sound and pictures, sound was recorded on the wire recorder that Camras invented in the 1930s. He went on to develop more than 500 inventions, most connected with improvements in recording methods. (Inducted 1985)

Carlson, Chester F. [American: 1906-1968]
ELECTROPHOTOGRAPHY
Carlson invented the dry copying method used in most offices today, which he named xerography. Although first patented in 1940, the dry copier did not reach the market until 1958, by which time Carlson had many patents on improvements in the process. (Inducted 1981)

Carothers, Wallace Hume [American: 1896-1937]
DIAMINE-DICARBOXYLIC ACID SALTS AND PROCESS OF PREPARING SAME AND SYNTHETIC FIBER
Despite the formidable title of his patent, Carothers's invention was simply nylon, not only an important fiber, but also a useful plastic. Carothers also developed the first commercially synthetic rubber. (Inducted 1984)

Carrier, Willis Haviland [American: 1876-1950]
APPARATUS FOR TREATING AIR
Not only did Carrier invent the first really workable air-conditioning system, he also invented many of the techniques used in modern refrigerators. (Inducted 1985)

Carver, George Washington [American: 1864-1943]
PRODUCTS USING PEANUTS AND SWEET POTATOES
Carver was a successful black scientist in Iowa when he accepted an invitation to return south to Tuskegee Institute in Alabama to help other African Americans advance. He recognized that traditional agricultural practices were destroying the soil. To encourage farmers to plant regenerative plants instead of cotton and tobacco, Carver developed over 300 new uses for the peanut and 118 sweet potato byproducts, taking no personal profits from any of these inventions. (Inducted 1990)

Colton, Frank B. [Polish/American: 1923-]
ORAL CONTRACEPTIVES
Colton not only pioneered the first birth control pill, in 1960, he also was an early developer of anabolic steroids. (Inducted 1988)

Conover, Lloyd H. [American: 1923-]
TETRACYCLINE
Before Conover created tetracycline (patented January 11, 1955), no one thought that a natural drug could be chemically modified to improve its action. Tetracycline remains the drug of choice to treat tick-spread diseases such as Rocky Mountain spotted fever and Lyme disease. (Inducted 1992)

Coolidge, William D. [American: 1873-1974]
VACUUM TUBE
The "Coolidge tube" is actually an x-ray generator. Among Coolidge's many other inventions is the modern tungsten-filament electric light. (Inducted 1975)

Cottrell, Frederick G. [American: 1877-1948]
ELECTROSTATIC PRECIPITATOR
Cottrell's invention (patented August 11, 1908) uses high-voltage electricity to capture the particulates, including fly ash, dust, and droplets of acids or other chemicals, found in smoke from the burning of fossil fuels and many industrial processes; in this way tons of waste can be removed from smoke instead of being spread around the countryside. (Inducted 1992)

Damadian, Raymond V. [American: 1936-]
APPARATUS AND METHOD FOR DETECTING CANCER IN TISSUE
Damadian was the first to realize that the nuclear magnetic resonance technique could be used on living creatures (it was already a success as a laboratory tool used by chemists) and that it could detect cancer cells. (Inducted 1989)

Deere, John [American: 1804-1886]
PLOW
Anyone who grew up near a farm knows the name John Deere from the company he founded, which still makes farm tools; few remember that a vastly improved plow was the start of his whole enterprise. (Inducted 1989)

DeForest, Lee [American: 1873-1961]
AUDION AMPLIFIER
Although he eventually acquired more than 300 patents related to radio, De Forest's invention of the triode was the key to modern radio and later developments in amplification of signals. (Inducted 1977)

Diesel, Rudolf [German: 1858-1913]
INTERNAL COMBUSTION ENGINE
The pressure-ignited heat engine he invented is known today as the diesel engine. The phrase "internal combustion engine" is more often used for a gasoline-powered engine in which the fuel is ignited by a spark plug rather than by pressure. (Inducted 1976)

Djerassi, Carl [Austrian/American: 1923-]
ORAL CONTRACEPTIVES
Djerassi has been a major influence on modern organic chemistry, although his interests within that field have been so broad, it is difficult to single out any one of them. (Inducted 1978)

Dow, Herbert Henry [Canadian/American: 1866-1930]
BROMINE
Besides new methods of extracting bromine and chlorine from naturally occurring salt deposits, Dow patented more than 90 inventions and founded the Dow Chemical Company. (Inducted 1983)

Draper, Charles Stark [American: 1901-1987]
GYROSCOPIC EQUIPMENT
Because a spinning gyroscope maintains the same relative position if it is free to do so, a gyroscope can be used to maintain a missile, a plane, or a ship on its course. In some ways a gyroscope is better than a compass; for example, it can be set to point true north and it is not affected by variations in magnetism. Draper's gyroscopic stabilizer helped both antiaircraft guns and falling bombs hit their targets in World War II. Later he developed systems for navigation and for guiding missiles. (Inducted 1983)

Durant, Graham J. [English/American: 1934-]
CIMETIDINE (TAGAMET®)
With John C. Emmet and C. Robin Ganellin, Durant developed the major drug used as an acid suppressor for stomach and intestinal ulcers. Introduced in the United Kingdom in 1976 and in the United States in 1977, by 1980 Tagamet® was the best-selling drug in America. (Inducted 1990)

Eastman, George [American: 1854-1932]
METHOD AND APPARATUS FOR COATING PLATES FOR USE IN PHOTOGRAPHY
Before Eastman's work, photographers had to use cumbersome wet plates. Eastman learned that it is possible to make a dry plate, and developed the necessary equipment to prepare the plates commercially. He also invented the transparent roll film that was the basis of the first Kodak box camera, and a stronger motion picture film for use in the newly invented cinema. (Inducted 1977)

Edgerton, Harold E. [American: 1903-1990]
STROBOSCOPE
Although the idea of using a flash of bright light to stop action in a photograph goes back to at least 1908, in the 1930s Edgerton created the special device called the stroboscope to produce such flashes on a regular basis. His classic photograph of the crown produced by a drop of milk falling into a bowl of milk dates from the 1930s also, but Edgerton went on to make many other analyses of events using the stroboscope to stop motion. He also contributed inventions to underwater photography. (Inducted 1986)

Edison, Thomas Alva [American: 1847–1931]
ELECTRIC LAMP
In addition to the carbon-filament electric light bulb, Edison patented a phonograph, the mimeograph, the fluoroscope, and motion picture cameras and projectors. (Inducted 1973)

Elion, Gertrude Belle [American: 1918–]
DNA-BLOCKING DRUGS
Working at the Burroughs Wellcome Company in the 1940s, Elion and her partner George H. Hitchings created artificial variations on the genetic material DNA that could block replication of cells. The chemicals became the basis for drugs to treat leukemia, septic shock, transplant rejection, herpes, and other diseases. Elion became the first woman to enter the National Inventors Hall of Fame. (Inducted 1991)

Emmet, John C. [English: 1939–]
CIMETIDINE (TAGAMET®)
(*See* Durant, Graham J.) (Inducted 1990)

Ericsson, John [American: 1803–1889]
PROPELLER
Ericsson's invention offered a highly efficient, difficult to damage, and easily maintained means of propulsion that sped the transition from wind power to steam power in shipping. (Inducted 1993)

Farnsworth, Philo Taylor [American: 1906–1971]
TELEVISION SYSTEM
While still a high-school student, Farnsworth conceived of an all-electronic television system. Crude pictures had been transmitted previously, but the most common system relied on spinning mirrors. Farnsworth patented many of the components of all-electronic television and was a pioneer in electronic microscopes, radar, use of ultraviolet light for seeing in the dark (a system used in World War II), and nuclear fusion. (Inducted 1984)

Fermi, Enrico [Italian/American: 1901–1954]
NEUTRONIC REACTOR
Fermi's nuclear reactor is the basis of nuclear power today. His many contributions to modern physics include basic theoretical work as well as experimental physics. (Inducted 1976)

Ford, Henry [American: 1863–1947]
TRANSMISSION MECHANISM
Many of Ford's ''inventions'' that revolutionized society, such as the automobile assembly line, vertical integration of manufacturing, inexpensive automobiles, and the $5-a-day wage (in 1914), were not patentable. Ford did, however, invent and patent numerous mechanisms used in automobiles. (Inducted 1982)

Forrester, Jay W. [American: 1918-]
MULTICOORDINATED DIGITAL INFORMATION STORAGE DEVICE
A pioneer in the development of electronic computers after World War II, Forrester's main invention was magnetic storage of information. Most computers today, from giant mainframes to tiny laptops, still use magnetic storage to retain data even when the computer has been shut off. (Inducted 1979)

Ganellin, C. Robin [English: 1934-]
CIMETIDINE (TAGAMET®)
(*See* Durant, Graham J.) (Inducted 1990)

Ginsburg, Charles P. [American: 1920-1992]
VIDEOTAPE RECORDER
The ubiquitous VCR, now known mostly from Japanese imports, was actually developed in the United States by an engineering team led by Ginsburg. (Inducted 1990)

Goddard, Robert Hutchings [American: 1882-1945]
CONTROL MECHANISM FOR ROCKET APPARATUS
The father of American rocketry, Goddard's experiments with liquid-fueled rockets between the two world wars were often derided, notably in a famous *New York Times* editorial. During both of the wars, however, the military accepted his offer of help, and he devised successful rocket weapons and rocket-assisted takeoff mechanisms for carrier-based airplanes. He obtained 214 patents on various aspects of rocketry. (Inducted 1979)

Goodyear, Charles [American: 1800-1860]
IMPROVEMENT IN INDIA-RUBBER FABRICS
Goodyear's discovery of how to make rubber that could withstand heat and cold is an illustration of the principle that people who find new ideas by accident are usually looking for that idea in the first place. While Goodyear was working on ways of making rubber more practical in 1844, he accidentally dropped rubber mixed with sulfur on a hot stove. The result, which Goodyear named vulcanized rubber, was what he had been seeking. Although Goodyear patented vulcanization and other ways to improve rubber, his patents were constantly infringed upon, and he died poor. (Inducted 1976)

Gould, Gordon [American: 1920-]
OPTICALLY PUMPED LASER
Although Gould developed the idea for the laser slightly before anyone else, he was not the first to get a patent filed. Despite this, after a 20-year battle, he finally obtained patent rights in 1977. By then lasers were already in use in a host of applications. He also holds patents in other fields, notably fiber optics. (Inducted 1991)

Greatbatch, Wilson [American: 1919–]
MEDICAL CARDIAC PACEMAKER
Greatbatch's pacemaker has helped millions of people who suffer from heart disease. He is also the inventor and manufacturer of batteries that can be implanted along with the pacemaker to keep the machinery running without the adverse effects sometimes caused by the chemicals in the battery. (Inducted 1986)

Greene, Leonard M. [American: 1918–]
AIRPLANE STALL WARNING DEVICE
In the 1950s, when Greene's stall warning device was just coming into use, more than half of all aviation accidents occurred because pilots failed to recognize that the plane was about to stall. Subsequently, Greene patented more than 60 air-safety features, including a device that warns of low-altitude wind shears. (Inducted 1991)

Hall, Charles Martin [American: 1863–1914]
MANUFACTURE OF ALUMINUM
After his college chemistry teacher remarked that discovering a way to make cheap aluminum (then selling at five dollars a pound—in 1886 dollars!) would make a person rich and famous, Hall set himself to the task. In only eight months, he found the method and indeed became rich and famous. That same year, the French metallurgist Paul-Louis-Toussaint Héroult discovered the same process. Patent litigation between the two independent discoverers was eventually resolved amicably. (Inducted 1976)

Hall, Robert N. [American: 1919–]
HIGH VOLTAGE HIGH POWER SEMICONDUCTOR PIN RECTIFIER
This invention greatly reduced the waste of power and dangerous heat buildup that accompanies large-scale power transmission. Hall also invented the first semiconductor laser, which is now commonly found in CD players. (Inducted 1994)

Hanford, W. E. "Butch" [American: 1908–1996]
POLYURETHANE
Hanford and his partner Donald F. Holmes developed their first form of polyurethane in 1937 while both were at E. I. du Pont Nemours & Co., but they are recognized for an improved form which they patented in 1942. Hanford later was involved in improving Teflon and in developing the first liquid detergent. (Inducted 1991)

Hazen, Elizabeth Lee [American: 1885–1975]
NYSTATIN
(*See* Rachel Fuller Brown) (Inducted 1994)

Hewlett, William R. [American: 1913–]
VARIABLE FREQUENCY OSCILLATION GENERATOR
The first invention of one of the founders of Hewlett-Packard was the audio oscillator in 1939 (not patented until January 6, 1942), a device for generating high-quality

audio frequencies that could be used for many different purposes. Among its first uses was the production of special sounds for the movie *Fantasia*. (Inducted 1992)

Higonnet, René Alphonse [French: 1902–1983]
PHOTO COMPOSING MACHINE
Along with Louis Marius Moyroud, Higonnet developed the first machine to set type by recording the images of letters on film. After its invention in 1946, film composition gradually became the standard way of setting type, replacing type set from hot metal. Today, however, type is often set by a laser printer and then photographed. (Inducted 1985)

Hillier, James [Canadian/American: 1915–]
ELECTRON LENS CORRECTION DEVICE
Although Hillier was not the first to make a microscope using electrons, his microscopes became the standard in the field. Electron microscopes can enlarge much smaller details than light microscopes because the wavelength of an electron is much smaller than the wavelength of a photon of visible light. (Inducted 1980)

Hollerith, Herman [American: 1869–1929]
STORAGE AND PROCESSING OF NUMERICAL DATA
The punched cards and readers Hollerith developed for the 1890 American census became the basis of data processing. The company he founded in 1896, along with three other companies, became IBM in 1924. (Inducted 1990)

Holmes, Donald Fletcher [American: 1910–1980]
POLYURETHANE
(*See* Hanford, W. E. "Butch") (Inducted 1991)

Houdry, Eugene J. [French: 1892–1962]
CATALYTIC CRACKING OF PETROLEUM
Houdry's patent recognition is for a process by which high-grade gasoline and airplane fuel is made from crude oil, but he also developed many other important catalytic processes and devices, ranging from a method of making synthetic rubber to the basic catalytic converter used in modern automobile mufflers. (Inducted 1990)

Julian, Percy F. [American: 1899–1975]
SYNTHESIS OF CORTISONE AND OTHER HORMONES
Despite a Ph.D. from the University of Vienna and an already brilliant record in chemistry, Julian was rejected for a professorship at DePauw University, presumably because he was black. In private industry, he developed many important industrial products based on soybeans before discovering that soybeans can be used as the basis for synthesizing cortisone, an important hormone with many medical uses. (Inducted 1990)

Keck, Donald B. [American: 1941-]
FUSED SILICA OPTICAL WAVEGUIDE
(With Robert D. Maurer and Peter C. Schultz) This optical fiber helped launch the information age: it carries 65,000 times more information than copper wire 30-50 times farther without boosting the signal. (Inducted 1993)

Kettering, Charles Franklin [American: 1875-1958]
ENGINE STARTING DEVICES AND IGNITION SYSTEM
Although one tends to think of Kettering as the genius behind many developments in automotive engineering at General Motors, his inventing career actually began with the electric cash register at National Cash Register. Kettering's Dayton Engineering Laboratories Co. (Delco) produced the self-starter for automobiles and the first small electric generator that could be used on isolated farms before rural electrification. After he sold Delco to General Motors, Kettering continued to run a research laboratory. In addition to automobile-related inventions, Kettering's laboratory developed diesel locomotive engines. (Inducted 1980)

Kilby, Jack S. [American: 1923-]
MINIATURIZED ELECTRONIC CIRCUITS
A number of people worked on putting several transistors and other solid-state electronic devices on a single chip, but the monolithic integrated circuit that Kilby developed for Texas Instruments in 1959 was the beginning of the modern integrated circuit. (Inducted 1982)

Kolff, Willem J. [Dutch/American: 1911-]
SOFT-SHELL, MUSHROOM-SHAPED HEART
Although Kolff's American patent that is cited is for an early version of an artificial heart, his most important work was the development of the artificial kidney dialysis machine while he was still in the Netherlands. (Inducted 1985)

Kwolek, Stephanie [American 1923-]
KEVLAR
This substance is five times as strong as steel and weighs 40% less than glass, making it ideal for use in aircraft construction, golf clubs, tennis rackets, fiber-optic cables, and perhaps most importantly, bulletproof vests. (Inducted 1995)

Land, Edwin Herbert [American: 1909-1991]
PHOTOGRAPHIC PRODUCT COMPRISING A RUPTURABLE CONTAINER
CARRYING A PHOTOGRAPHIC PROCESSING LIQUID
Land's first success was not the instant camera for which he became world famous, but the development of substances that polarize light and applications of those substances. He also made important contributions to the theory of color vision. (Inducted 1977)

Langmuir, Irving [American: 1881–1957]
INCANDESCENT ELECTRIC LAMP
The original Edison-Swan light bulbs relied on a vacuum to keep the filament from burning too fast, but in 1913 Langmuir realized that filling the bulb with a nonburning gas would result in a longer-lasting light. He also made many basic scientific discoveries, including those stemming from work with the chemistry of surfaces for which he won the 1932 Nobel Prize in chemistry. (Inducted 1989)

Lawrence, Ernest Orlando [American: 1901–1958]
METHOD AND APPARATUS FOR THE ACCELERATION OF IONS
Although Lawrence did not develop the very first particle accelerator (popularly known as an atom smasher), his 1930 cyclotron, with many refinements and contributions by other inventors, has set the basic pattern for the most successful and most powerful machines of its type since. (Inducted 1982)

Ledley, Robert S. [American: 1926–]
WHOLE-BODY CT SCANNER
When Ledley proposed his revolutionary device that would make three-dimensional images of living tissue, manufacturers were not interested—but radiologists were. Ledley had to form his own company to make the device, which soon became a successful part of the medical marketplace. (Inducted 1990)

Maiman, Theodore Harold [American: 1927–]
RUBY LASER SYSTEMS
There has been a lot of dispute about the invention of the laser, but it is clear that Maiman's ruby laser was the first to be recognized worldwide and to be commercially successful. (Inducted 1984)

Marconi, Guglielmo [Italian: 1874–1937]
TRANSMITTING ELECTRICAL SIGNALS
Marconi's patent was for using radio waves to carry coded messages—also known as wireless telegraphy. (Inducted 1975)

Maurer, Robert D. [American: 1924–]
FUSED SILICA OPTICAL WAVEGUIDE
(*See* Donald B. Keck) (Inducted 1993)

McCormick, Cyrus [American: 1809–1884]
REAPER
McCormick developed his machine for harvesting grain in 1831 and patented it in 1834, when he learned that there might be competition in the field. By 1847 his factory was turning out the machines that would be among the first part of the revolution in agriculture in the United States. (Inducted 1976)

Mergenthaler, Ottmar [German/American: 1854–1899]
MACHINE FOR PRODUCING PRINTING BARS
Mergenthaler's invention, known as the Linotype, constituted the first major improvement in setting type since Gutenberg began using movable type about 1440. The Linotype machine casts each line of type from melted metal, based on instructions from a keyboard. Modern versions of the Linotype are still in use, but most typesetting today is done by photographic processes or laser printing. (Inducted 1982)

Millman, Irving [American: 1923–]
HEPATITIS B VACCINE
(*See* Blumberg, Baruch S.) (Inducted 1993)

Morse, Samuel F. B. [American: 1791–1872]
TELEGRAPH SIGNALS
The telegraph that Morse developed was the first that was commercially successful. Joseph Henry [American: 1797–1878] was the genius behind the electronics, but Morse and his dot-dash code made instantaneous long-distance communication possible. (Inducted 1975)

Moyer, Andrew J. [American: 1899–1959]
METHOD FOR PRODUCTION OF PENICILLIN
World War II brought a great need for medicines that would halt infections. Although penicillin was known to achieve the desired results, no one knew how to make it in quantity. Moyer, a microbiologist at the American Department of Agriculture's Northern Regional Research Laboratory in Peoria, Illinois, solved the problem. Combining a strain of mold found on a rotting muskmelon with a new growing medium (a byproduct of the manufacture of corn starch) solved the problem. The basic method continues to be used today in the manufacture of many other antibiotics and substances produced by microorganisms. (Inducted 1987)

Moyroud, Louis Marius [French: 1914–]
PHOTO COMPOSING MACHINE
(*See* Higonnet, René Alphonse) (Inducted 1985)

Noyce, Robert N. [American: 1927–1990]
SEMICONDUCTOR DEVICE-AND-LEAD STRUCTURE
Noyce was at the center of development at two seminal semiconductor producers, Fairchild and Intel, both of which he helped found. Intel today makes the most widely used microprocessor chips for personal computers—those at the heart of various IBM models and their clones. (Inducted 1983)

Olsen, Kenneth H. [American: 1926–]
IMPROVED MAGNETIC CORE MEMORY
After leaving MIT, Olsen founded Digital Equipment Corporation (DEC) to manufacture computers based on his new memory devices. DEC, with Olsen as president,

soon became one of the leaders in electronic computing. Olsen also contributed extensively to the development of the minicomputer. (Inducted 1990)

Otis, Elisha Graves [American: 1811-1861]
IMPROVEMENT IN HOISTING APPARATUS
Modern skyscrapers would have been impossible without the safety elevator that Otis devised in 1853 when his employer asked him to build a hoist to lift heavy equipment. Eight years later the first Otis passenger elevators were being installed. (Inducted 1988)

Otto, Nikolaus August [German: 1832-1891]
GAS MOTOR ENGINE
While Otto's four-stroke engine of 1876 is the basis of the modern internal combustion engine, it ran on compressed natural gas instead of gasoline; the gasoline-powered engine was a development pioneered by Gottlieb Daimler and Wilhelm Maybach in 1889. (Inducted 1981)

Parker, Louis W. [Hungarian/American: 1906-]
TELEVISION RECEIVER
Parker not only invented the basic type of television receiver that is in common use today, but also the basic type of color television transmission and reception that is most commonly used. (Inducted 1988)

Parsons, John T. [American: 1913-]
NUMERICAL CONTROL OF MACHINE TOOLS
This 1958 patent introduced automation to manufacturing and design and promoted development of machines that can mill steel with an accuracy of one-tenth the thickness of newsprint. (Inducted 1993)

Pasteur, Louis [French: 1822-1895]
BREWING OF BEER AND ALE
Louis Pasteur is not usually thought of as an inventor, and his work on beer and ale is generally thought to have been unsuccessful, but he did invent vaccines for several diseases (although these very important inventions were not patented) as well as pasteurization, the process of protecting beverages and food from contamination by microbes through heating. (Inducted 1978)

Plank, Charles J. [American: 1915-1989]
CATALYTIC CRACKING OF HYDROCARBONS WITH A CRYSTALLINE ZEOLITE CATALYST COMPOSITE
Along with Edward J. Rosinski, Plank discovered in the early 1960s that zeolites (various aluminum silicates, a fairly common type of mineral) could be used to improve the production of gasoline and other petroleum products. (Inducted 1979)

Plunkett, Roy J. [American: 1910-1994]
TETRAFLUOROETHYLENE POLYMERS
In 1938, Plunkett discovered the tetrafluoroethylene polymer that we know as Teflon. He later developed many of the chlorofluorocarbons (Freons) that have caused considerable concern recently, since they erode the atmosphere's protective layer of ozone. (Inducted 1985)

Rines, Robert H. [American: 1922-]
CONTRIBUTIONS TO HIGH RESOLUTION IMAGE-SCANNING RADAR
Rines's patents were the basis for almost all the high-definition radar used to provide the armed forces with early warning radar during the Persian Gulf War. In peacetime, his inventions have been used in underwater archaeology and for locating the *Titanic* and the *Bismarck*. (Inducted 1994)

Rohrer, Heinrich [Swiss: 1933-]
SCANNING TUNNELING MICROSCOPE
(*See* Binnig, Gerd Karl) (Inducted 1944)

Rosinski, Edward J. [American: 1921-]
CATALYTIC CRACKING OF HYDROCARBONS WITH A CRYSTALLINE ZEOLITE CATALYST COMPOSITE
(*See* Plank, Charles J.) (Inducted 1979)

Rubin, Benjamin A. [American: 1917-]
BIFURCATED VACCINATION NEEDLE
Rubin's needle (patented July 13, 1965) enabled easy use of small amounts of smallpox vaccine, making it possible for vaccine supplies to be stretched and helping the World Health Organization to eliminate smallpox as a disease by May 8, 1980. (Inducted 1992)

Sarett, Lewis Hastings [American: 1917-]
THE PROCESS OF TREATING PREGNENE COMPOUNDS
Pregnene compounds are the predecessor chemicals of certain steroids. In 1944 Sarett found a way to produce cortisone as artificial steroid from its predecessor chemicals. By 1949, Sarett and his collaborators had learned to make cortisone from simple inorganic chemicals. Cortisone and closely related steroids are widely used in medicine today for treatment of conditions from arthritis to psoriasis. (Inducted 1980)

Schultz, Peter C. [American: 1942-]
FUSED SILICA OPTICAL WAVEGUIDE
(*See* Donald B. Keck) (Inducted 1993)

Seiwald, Robert J. [American: 1925-]
FITC
(*See* Burckhalter, Joseph H.) (Inducted 1995)

Semon, Waldo Lonsbury [American: 1898-]
SYNTHETIC RUBBER-LIKE COMPOSITION AND METHOD OF MAKING SAME;
METHOD OF PREPARING POLYVINYL HALIDE PRODUCTS
While working for B. F. Goodrich in the 1920s, Semon was assigned the problem of finding a better material for binding rubber linings to metal tanks used for corrosive chemicals. In 1928, while investigating vinyl chlorides, he found a simple method of making a polymerized form (polyvinyl chloride, or PVC) found to be waterproof, fire resistant, and insulating, all desirable properties. But the crash of the stock market in 1929 caused Goodrich to hold off on any new development. On his own, Semon developed a way to coat cloth with vinyl, which he later patented as Koroseal. Soon, Goodrich was producing vinyl-coated cloth, which was commercially used for raincoats, shower curtains, and umbrellas. Today, PVC is the world's second best-selling plastic. Semon also contributed to the development of polyurethane and synthetic rubber, and invented a bubble gum. (Inducted 1995)

Sheehan, John C. [American: 1915-1992]
SEMI-SYNTHETIC PENICILLIN
Sheehan's research developed a synthetic penicillin that could be produced cheaply and quickly, in contrast with the slow and expensive natural way of producing it. (Inducted 1995)

Shockley, William Bradford [English/American: 1910-1989]
TRANSISTOR
(*See* Bardeen, John) (Inducted 1974)

Sikorsky, Igor I. [Russian/American: 1889-1972]
HELICOPTER CONTROLS
Although known primarily today for his invention of the first commercially successful helicopter, Sikorsky designed and built many early airplanes that were widely used. In 1931 he made a critical breakthrough in helicopter design, which he had worked on for years; his continued developments led to the modern helicopter. (Inducted 1987)

Sperry, Elmer A. [American: 1860-1930]
SHIP'S GYROSCOPIC COMPASS
Sperry used the idea behind what was until then a child's toy to develop a compass that remains pointed in the same direction without constant correction for the difference between magnetic north and true north. In addition to the gyroscopic compass, he also invented mining and railroad equipment. (Inducted 1991)

Stanley, William [American: 1858-1916]
ELECTRIC TRANSFORMER
This invention allowed for the long-distance transmission of alternating current (AC), which was far less dangerous than transmitting direct current (DC) at very high voltages. (Inducted 1995)

Steinmetz, Charles Proteus [German/American: 1865–1923]
SYSTEM OF ELECTRICAL DISTRIBUTION
Steinmetz was an important theoretician as well as an inventor, and his most significant work was the development of the theory of alternating current (AC) that makes our present power grids possible. Among his inventions was a machine that produced "lightning in the laboratory," a device that made Steinmetz a household word near the end of his life. (Inducted 1977)

Stibitz, George R. [American: 1904–1995]
COMPLEX COMPUTER
Stibitz was one of several scientists who developed electromechanical computers in the late 1930s and during World War II. His innovations at Bell Telephone Laboratories and in the U.S. Office of Scientific Research and Development include floating decimal point arithmetic and taped computer programs. (Inducted 1983)

Tabern, Donalee L. [American: 1900–1974]
THIO-BARBITURIC ACID DERIVATIVES
In 1936, Tabern and Ernest H. Volwiler developed Pentothal, the anesthetic of choice for short surgical procedures and for use prior to the administration of a general anesthetic. Tabern later introduced the therapeutic use of radioactive chemicals. (Inducted 1986)

Tesla, Nikola [Croatian/American: 1857–1943]
ELECTRO-MAGNETIC MOTOR
Tesla's induction motor was simpler than previous electric motors and was powered by AC current, which can more easily be distributed over long distances than direct current. (Inducted 1975)

Tishler, Max [American: 1906–1989]
RIBOFLAVIN AND SULFAQUINOXALINE
In the late 1930s, Tishler developed an economical method for synthesizing riboflavin, also known as vitamin B_2. Later he and his coworkers developed a commercially useful method to produce sulfaquinoxaline, an antibiotic that prevents and cures a common disease of poultry. (Inducted 1982)

Townes, Charles Hard [American: 1915–]
MASERS
The maser, which preceded the better-known laser, is essentially a laser that works at microwave wavelengths instead of at the shorter wavelength of visible light. Masers are used in many applications. Townes, who was solely responsible for the maser, also contributed to the development of the laser. (Inducted 1976)

Volwiler, Ernest H. [American: 1893–1992]
THIO-BARBITURIC ACID DERIVATIVES
(*See* Tabern, Donalee L.) (Inducted 1986)

Wang, An [Chinese/American: 1920–1990]
MAGNETIC PULSE CONTROLLING DEVICE
Although best known for his state-of-the-art word processor of the 1960s and 1970s, Wang contributed many fundamental ideas to the development of electronic computers, including the principle upon which magnetic core memory is built. (Inducted 1988)

Westinghouse, George [American: 1846–1914]
STEAM-POWERED BRAKE DEVICES
Westinghouse specialized in improving rail transportation at the time of its greatest expansion, just after the U.S. Civil War. In 1869 he patented his first air brake for locomotives, his most important contribution to railroad safety. Later he worked on improving signals and switches, work that led him to form Westinghouse Electric Company in 1884, chiefly to implement the possibilities of alternating current. (Inducted 1989)

Whitney, Eli [American: 1765–1825]
COTTON GIN
By making it possible to remove seeds from cotton mechanically, the gin made large-scale cotton farming possible. Whitney also was a pioneer in the use of interchangeable parts—the beginning of mass production. (Inducted 1974)

Williams, Robert R., Jr. [American: 1886–1965]
ISOLATION OF VITAMIN B$_1$ (THIAMINE)
As the son of Baptist missionaries to India, Williams gained early familiarity with beriberi, the disease caused by thiamine deficiency. After World War I, Williams began vitamin research in his spare time, but it was not until 1933 that he isolated pure thiamine. Two years later he synthesized the vitamin, which almost immediately went into commercial production. Money from his patents has been used to encourage vitamin enrichment of foods as a way to stop deficiency diseases. (Inducted 1991)

Wright, Orville [American: 1871–1948] and **Wilbur Wright** [American: 1867–1912]
FLYING MACHINE
Not only did the Wright brothers invent the first airplane, they also popularized, manufactured, and sold the new machines. For the first few years after their first 1903 flight, people took little notice of their work. In 1908, however, Orville demonstrated a flight of an hour's duration that was widely reported. By World War I, airplanes were regularly used by the armed services of the combatants. (Inducted 1975)

Zworykin, Vladimir Kosma [Russian/American: 1889–1982]
CATHODE RAY TUBE
The cathode tube that Zworykin invented in 1928 is the kinescope, the basic picture tube used in modern television. Ten years later he developed the iconoscope, the first practical television camera. His later work on electron microscopes created the type of electron microscope that is commonly used today, although not the first electron microscope to be developed. (Inducted 1977)

Appendices

APPENDIX A

OBITUARIES: 1992-1994

Aleksandrov, Anatoly Petrovich [February 13, 1903–February 3, 1994] Soviet physicist who designed and advocated graphite-moderated reactors such as the one that exploded at the Chernobyl nuclear power plant in 1986, resulting in a tragedy for which he was blamed. During World War II, he developed a method of demagnetizing naval vessels to protect them from mines and assisted in developing the Soviet nuclear-powered fleet of surface vessels and submarines.

Anderle, Richard J. [October 8, 1926–January 28, 1994] American researcher who was known for his contributions to the field of satellite geodesy; he developed and used satellite Doppler techniques for determining geopotential models and for global geodetic positioning and was the first to use Doppler data for determining polar motion.

Arnon, Daniel Israel [November 14, 1910–December 20, 1994] American plant physiologist, born in Warsaw, who shed light on plant photosynthesis by helping to explain the synthesis of adenosine triphosphate, or ATP, which sends energy to living cells; his research showed how light was used to produce chemical energy and oxygen. He was awarded the National Medal of Science in 1973.

Asch, Timothy [1932–October 3, 1994] American anthropologist and University of Southern California professor who as an undergraduate was a teaching assistant to Margaret Mead. He was known for producing more than 70 documentary films about remote societies in South America, North America, Africa, Afghanistan, and Indonesia.

Asimov, Isaac [January 2, 1920–April 6, 1992] American popular-science writer, born in the Soviet Union, who wrote books and articles on an extensive range of non-fiction subjects but was best known for his science-fiction; he won five Hugo Awards for works including his *Foundation Trilogy* (written in the early 1950s) and *The Gods Themselves* (1972). In his short-story collection *I, Robot* (1950), he introduced his famous Three Laws of Robotics, which played a part in his 14 subsequent robot-related novels.

Auslander, Maurice [August 3, 1926–November 18, 1994] American mathematician whose theories in the study of abstract algebra have been used by others to solve practical problems; he also codeveloped with Idun Reiten a concept that was instrumental in the representation theory of Artin algebras.

Avigad, Nahman [1905–January 28, 1992] Israeli archaeologist whose important achievements included discovering the Herodian "Upper City" in the Old City section of Jerusalem, which had been destroyed by the Romans in A.D. 70, and in 1956 identifying the last of the seven 2,000-year-old Dead Sea Scrolls.

Bay, Zoltan Lajos [July 24, 1900–October 4, 1992] American physicist, born in Gyulavari, Hungary, whose achievements included research in 1946 with microwaves reflected from the Moon, which led to the beginning of radar astronomy. He invented electron multiplier devices used for counting particles and a method for producing light by burning carborundum. He was a staff member at the U.S. National Bureau of Standards, where he contributed to new measurements of the speed of light.

Bers, Lipman [May 22, 1914–October 29, 1993] American mathematician and Columbia University professor, born in Riga, Latvia, who was best known for his human rights activism; he established the Committee on Human Rights of the National Academy of Sciences and helped dissident mathematicians from the Soviet Union, such as Yuri Shikhanovich, Leonid Plyushch, and Valentin Turchin, to obtain exit visas to go to the United States and teach at Columbia.

Bessis, Marcel Claude [November 15, 1917–March 28, 1994] French physician, born in Tunisia, who was known for developing methods and instruments to isolate and study parts of a single live cell to better understand how the cell functions as a whole.

Beverage, Harold Henry [1893–January 27, 1993] American radio engineer who was a pioneer in communications technology, co-inventing the wave antenna and the diversity system for high-frequency technology. He received several awards, including the Lamme Gold Medal, given by the American Institute of Electrical Engineers in 1957.

Birch, Albert Francis [August 22, 1903–January 31, 1992] American geophysicist and Harvard University professor whose career focused on the study of how materials respond to high pressures found in Earth's interior, using measurements of sound velocities and other data; this helped other scientists understand the seismic waves of earthquakes. He also helped design and transport the Little Boy atomic bomb that was later dropped on Hiroshima, and he was awarded the National Medal of Science by President Johnson in 1968.

Boehm, George A. W. [1922–October 7, 1993] American journalist whose career as a science writer and editor included co-authoring *The New World of Math* (a book explaining new mathematical ideas) and serving as science editor for *Newsweek* and *Fortune* magazines and as a member of the board of editors of *Scientific American* during the 1950s and 1960s.

Bogert, Charles M. [1907–April 10, 1992] American scientist who studied reptiles and amphibians, focusing on their temperature control mechanisms as well as analyzing sounds made by toads and frogs. He was curator of the herpetology department at the American Museum of Natural History and had 21 species and subspecies named after him.

Bohm, David Joseph [December 20, 1917–October 27, 1992] American-born physicist and University of London professor who in 1951 wrote *Quantum Mechanics*, one of the most popular textbooks on the subject at the time. He was acquitted after being involved in a controversy that same year when the U.S. House Un-American Activities Committee accused him of being a Communist and of withholding information about possible Communist activity in a radiation laboratory at the University of California at Berkeley during World War II.

Bolton, John [1922–July 6, 1993] British-born researcher who was an expert in radio astronomy; in 1947, he discovered the first radio "stars," or galaxies that broadcast strong signals in radio wavelengths. He taught at the California Institute of Technology, starting a radio observatory there. He was also director of the Australian National Radio Astronomy Observatory.

Bovet, Daniele [March 23, 1907–April 8, 1992] Swiss-born pharmacologist who was known for his discovery of the first antihistamines in 1941, of muscle-relaxing drugs derived from curare, and of several sulfa drugs. He was awarded the Nobel Prize in Physiology and Medicine in 1957.

Braus, Harry [1914–March 1, 1993] American chemist who, while working for the United States Public Health Department in the 1950s, developed a method to detect pollutants, using activated carbon to filter contaminants from large quantities of air or water. After being forced out of government during Senator McCarthy's crusade because he was accused of being a communist, Braus worked for the U.S. Industrial Research Company, inventing and patenting more than 50 products and methods.

Broneer, Oscar T. [1894–February 22, 1992] Swedish-born archaeologist and professor at the University of Chicago who researched in Greece; in 1952, he discovered the site of the Temple of Poseidon, built by the Corinthians at Isthmia around the fourth century B.C. and the last of four Panhellenic shrines to be uncovered.

Bruce-Mitford, Rupert Leo Scott [June 14, 1914–March 10, 1994] British archaeologist involved in excavations at Sutton Hoo in southeastern England from 1965 to 1968 and who performed an official study for the British Museum; the items unearthed included the remains of a seventh-century Saxon ship which he believed could have been used for the burial of Raedwald, King of Anglia, who may have been cremated in the 620s.

Buchsbaum, Solomon Jan [December 4, 1929–March 8, 1993] American physicist, born in Poland, who served as senior vice-president in charge of technology systems at Bell Laboratories and also as a science adviser to five United States presidents; he was chairman of the White House Science Council under Presidents Reagan and Bush and adviser on the Strategic Defense Initiative (also known as Star Wars), among other posts. He was awarded the National Medal of Science by President Reagan.

Callen, Herbert Bernard [July 1, 1919–May 22, 1993] American physicist whose work in the area of statistical mechanics was highlighted by a paper he wrote with Theodore A. Welton, in which they proved the fluctuation-dissipation theorem that deals with charged particles. He was the author of a standard text, *Thermodynamics* (1960), and he was elected to the National Academy of Sciences in 1990.

Caminos, Ricardo Augusto [July 11, 1915–May 28, 1992] American Egyptologist, born in Buenos Aires, who was chairman of Brown University's department of Egyptology from 1972 to 1980 and was an expert in epigraphy (the study of ancient inscriptions) and paleography (the study of ancient documents).

Campbell, William J. [1930–November 20, 1992] American meteorologist who led important polar ice projects for the United States Geological Survey, using earth satellite sensors and other methods to interpret data. He was known for his expertise in sea-ice dynamics and remote sensing, and received several awards, including the United States Antarctic Medal in 1965.

Carpenter, Frank Morton [September 6, 1902–January 18, 1994] American zoologist and Harvard University professor who devoted his life to the study of insects, publishing *Superclass Hexapoda*, a two-volume history of fossil insects, in 1992. The Entomological Society of America honored him with the Thomas Say Award for his life achievements.

Clifford, Paul Clement [November 23, 1910–October 6, 1993] American mathematician and Montclair (New Jersey) State College professor who was a pioneer in applying statistics to manufacturing, starting from World War II when his methods reduced flaws and improved production of airplane parts made by Wright Aeronautical Company. His lifelong work in the area of quality control led to his receiving the highest award given by the American Society of Quality Control in 1965.

Closs, Gerhard Ludwig [May 1, 1928–May 24, 1992] American organic chemist, born in Wuppertal, Germany, who expanded the study of chemical reactions through his research in the magnetic properties of intermediate compounds formed in reactions. He is also known for his discovery that certain chemical reactions polarize atomic nuclei.

Cockburn, Sir Robert [March 31, 1909–March 21, 1994] British radar specialist who during World War II led a group of engineers to develop the Window — anti-radar technology designed to defeat the ability of German radar to spot British and American bombers. He was knighted in 1960.

Cooley, Austin G. [1900–September 7, 1993] American inventor who, starting in the 1920s, developed successful facsimile (or fax) machines, in which photographic negatives were translated into electrical signals and then transmitted by radio or telephone. He achieved the first successful transmission in 1935, containing photographs of survivors of the U.S. Navy dirigible *Macon,* which had exploded over the Pacific Ocean. He served as president of both the Times Facsimile Corporation, a subsidiary of the New York Times Company, and of Litton Systems, Inc.

Cressman, Luther S. [1897–April 4, 1994] American archaeologist who in 1938 led a dig at Fort Rock in south-central Oregon, where his team found 75 pairs of sandals made of grass and sagebrush bark, the first indication that humans had inhabited the Northwest 9,000 years ago instead of 4,500 as anthropologists previously believed. He also established the Oregon Museum of Natural History and was a founding director the Oregon State Museum of Anthropology.

Dahlberg, Albert A. [November 20, 1908–July 30, 1993] American dentist who was a leader in the field of dental anthropology; he trained others in evolutionary dentition, paleoanthropology, and primatology, and studied ancient civilizations in the American Southwest, Alaska, Greenland, Mexico, and Asia.

Dales, George F. Jr. [1927–April 18, 1992] American archaeologist who studied the ruins of the ancient Indus Valley, dating from 4000 B.C. to 2000 B.C, as well as Mahenjo Daro in what is now Pakistan. He attempted to decipher language fragments and identified evidence of trade routes among the civilizations of the region.

Davis, Bernard David [January 7, 1916–January 14, 1994] American geneticist and Harvard Medical School professor who demonstrated how genes are regulated in bacteria and who was one of the first researchers to use penicillin to isolate mutants, helping other scientists produce mutants which they could compare with normal bacteria in studying how each used chemicals for growth. He was senior author of a widely-used textbook, *Microbiology*, and was elected to the National Academy of Sciences and to the American Academy of Arts and Sciences.

Davydov, Aleksandr Sergeivich [December 26, 1912-1993] Ukrainian chemical physicist who established during the 1960s his "exciton theory" of quantum mechanics, which used the uncertainty principle and the role of exciton particles to explain a peculiar effect of light passing through crystals. His discoveries had a considerable impact on solid-state physics.

Dethier, Vincent Gaston [February 10, 1915-September 8, 1993] American biologist and University of Massachusetts professor who was known for his books about insect behavior, including *To Know a Fly* (1962) and *Crickets and Katydids, Concerts and Solos* (1992), for which he was awarded the John Burroughs Medal for Distinguished Nature Writing.

DuBridge, Lee Alvin [September 21, 1901-January 23, 1994] American physicist who served as president of the California Institute of Technology from 1946 to 1969, overseeing the school's increase in size, endowment, and resources. While he was a physics professor at the University of Rochester from 1934 to 1946, he led a team that built a cyclotron that produced the highest-energy proton to that date, and during World War II he headed a laboratory at the Massachusetts Institute of Technology, developing radar technology for the military.

Eagle, Harry [July 13, 1905-June 12, 1992] American medical scientist who was known for groundbreaking achievements in biological research, including his 1959 development of compounds necessary to sustain the reproduction of human and other mammalian cells in the laboratory, known as Eagle's growth medium, which aided future research on viruses and genetic defects and cancer. He also discovered that blood clotting is an enzyme process and developed a treatment for arsenic poisoning and a diagnostic test for syphilis. He was awarded the National Medal of Science by President Reagan in 1987.

Eisenhart, Churchill [March 11, 1913-June 25, 1994] American mathematical statistician who worked for the National Bureau of Standards, an agency of the United States Commerce Department, beginning in the late 1930s. He founded the Bureau's Statistical Engineering Laboratory in 1947 and served as its director until 1963.

Fairservis, Walter A. [1921-July 12, 1994] American archaeologist who had worked with the American Museum of National History in New York and was known for locating and exploring lost civilizations, including a city in Afghanistan he discovered while leading the first archeological expedition there in 1949 and a ceremonial complex of the prehistoric Harappan civilization of Pakistan.

Falicov, Leopoldo Maximo [June 24, 1933-January 24, 1995] American physicist and professor at the University of California at Berkeley, born in Buenos Aires, whose work in condensed matter physics resulted in the calculations of the definitive electronic structures of magnesium, zinc, cadmium, arsenic, and other metals. He was a member of the National Academy of Sciences.

Feinberg, Gerald [May 27, 1933-April 21, 1992] American physicist and Columbia University professor who in 1958 wrote a groundbreaking paper in which he explained that two kinds of neutrinos (a type of nuclear particle) existed instead of just one. Leon Lederman, Melvin Schwartz, and Jack Steinberger won the Nobel Prize in Physics for their experiments confirming his hypothesis.

Feld, Bernard T.   [December 21, 1919–February 19, 1993] American physicist and professor at the Massachusetts Institute of Technology who worked with Enrico Fermi to develop the atomic bomb but later advocated the end of the arms race. He worked with scientists who were concerned about the effects of nuclear war, serving as a leader of the Pugwash movement, co-founder of the Federation of American Scientists, president of the Albert Einstein Peace Foundation, and editor of the Bulletin of Atomic Scientists,

Feyerabend, Paul Karl   [January 13, 1924–February 11, 1994] A controversial philosopher of science, born in Vienna, who taught at the University of California at Berkeley and the Polytechnic Institute of Zurich and was known for his theories that scientists and scientific research are not objective, arguing that scientific objectivity is a myth.

Field, William Osgood   [January 30, 1904–June 16, 1994] American geologist who was considered one of the founding fathers of the field of glaciology, studying the fluctuations of about 200 glaciers in the western Andes, Greenland, and other places, serving as head of the World Data Center A for Glaciology at the American Geographical Society in New York, and writing a book, *Mountain Glaciers of the Northern Hemisphere*, published in 1975.

Fogel, Seymour   [September 27, 1919–September 24, 1993] American molecular biologist who specialized in using yeast to shed light on the genetics of human reproduction and evolution. In 1982, his research team developed a method to increase the output of useful chemicals produced by gene splicing.

Fosberg, F. Raymond   [1908–Sep 25, 1993] American botanist who was affiliated with the Smithsonian Institution's National Museum of Natural History, where he became known as an expert on tropical plant life; he published more than 600 papers and co-edited *The Flora of Ceylon*, and helped establish the Rachel Carson Council and the Nature Conservancy.

Fox, Arthur Gardner   [November 22, 1912–November 24, 1992] American physicist who worked for Bell Telephone Laboratories and held 53 patents in the microwave and quantum electronics fields. He helped invent the SCR 545 antiaircraft radar and a microwave beam switcher for submarines using radar, both used during World War II. Later he played a role in creating AT&T's first transcontinental microwave relay system for long-distance calls, and in developing the basic theory of light behavior in laser beams, which led to further innovations in laser technology.

Fuller, Calvin Souther   [May 25, 1902–October 28, 1994] American chemist and Bell Laboratories researcher who in 1954, along with Gerald L. Pearson and Daryl M. Chapin, invented the solar battery; now called the solar cell, this device converts solar energy into electricity and facilitated the development of space vehicles that could take advantage of available sunlight.

Gagge, Adolf Pharo   [January 11, 1908–February 13, 1993] American biophysicist who applied his knowledge of human body temperature at the John B. Pierce Laboratory at Yale University; his research included the development of the "met unit," which measured the combined effect of temperature and humidity on body heat regulation and comfort, and the "clo," which gauged the insulation value of clothing. These and other advances influenced the design of clothing, equipment, and heating and cooling systems to improve humans' comfort, health, and safety.

Gaymont, Stephen A. [1905–December 16, 1994] American bacteriologist, born in Hungary, who is credited with introducing yogurt to the United States and with inventing frozen yogurt; he started his own company, Gaymont Laboratories, Inc., which produced the yogurt culture for more than 500 companies in the United States, and he also introduced the first yogurt with fruit added to it.

Gelbart, Abraham [December 22, 1911–September 7, 1994] American mathematician who codeveloped the theory of pseudoanalytic functions, which explains the behavior of materials such as air floating around airplane wings. He was also the founding dean of the Belfer Graduate School of Science at Yeshiva University, established in 1959.

Gentry, Alwyn Howard [January 6, 1945–August 3, 1993] American biologist who was tragically killed in an airplane crash while attempting a tree-top survey of the Ecuadorean coast; a senior curator at the Missouri Botanical Garden, he was a world authority on the flora of Latin America, and collected about 70,000 specimens of plant life.

Gimbutas, Marija [January 23, 1921–February 2, 1994] American archaeologist, born in Vilnius, Lithuania, who was known for her thesis that goddesses were worshiped and societies were centered on women until about 6,000 years ago when the harmony between men and women was destroyed and warlike gods became the focus of worship. Her work led to the development of the field of archeomythology.

Ginsburg, Charles P. [1920–April 9, 1992] American engineer who was hired by the Ampex company to lead a research team in an industry-wide race to develop the first video recorder and video tape; the group, which included the sound technology pioneer Milton M. Dolby, was successful inventing the first such device in 1956. As a result, Ampex earned an Emmy award and in 1990 Ginsburg was inducted into the National Inventors Hall of Fame.

Glasse, Robert M. [1929–January 1, 1993] American anthropologist known for his work in New Guinea; his research of the deadly kuru neurological disease helped D. Carleton Gajdusek win a Nobel Prize, and he published a book, *Huli of Papua*, in 1968.

Gordon, Albert Saul [August 8, 1910–June 12, 1992] American physiologist and New York University professor whose blood disease research showed that the production of red blood cells is controlled by a factor that responds to the physiologic needs of the organism; when the factor was cloned, it provided the basis for later development of a drug that is widely used in the biotechnology industry.

Gorenstein, Daniel [1923–August 26, 1992] American mathematician who focused on Galois' Theory of Groups, establishing relationships among algebra, geometry, coding theory, quantum mechanics, and elementary particle physics. Gorenstein completed the classification of the finite simple groups, the lengthiest proof in mathematical history.

Grace, Virginia [1900–May 22, 1994] American anthropologist who specialized in studying ancient civilizations in Greece, often using the dates on the handles of amphoras (Greek jars or vases) to find information on the Mediterranean wine trade and other activities.

Gray, Truman Stretcher [May 3, 1906–November 7, 1992] American engineer and professor at the Massachusetts Institute of Technology who in 1931 invented the Photo-Electric Integraph, a machine that performed complex mathematical calculations in minutes by converting the problems into rays of light. He also wrote a standard text, *Applied Electronics*, published in 1954.

Griffin, John W. [1920–September 3, 1993] American archaeologist who worked for the State of Florida, studying the remains of a Spanish mission at San Luis near Tallahassee. In 1951 he and historian Mark Boyd wrote *Here They Once Stood*, considered a fine account of urban archeology.

Gurzey, Feza [1920–April 13, 1992] American physicist who, as a specialist in elementary particles, published more than 100 papers; in his research, he was credited for new ideas on the symmetries of weak and strong interactions, grand unified gauge theories, and other related projects.

Hafstad, Lawrence Randolph [June 18, 1904–October 12, 1993] American physicist who, in 1939 along with Richard Roberts and Merle Tuve, was the first in the United States to split atomic nuclei, paving the way for the development of the first atomic bomb in the 1940s. He also helped develop the first nuclear reactors and nuclear engines for submarines, and led a team that devised a variable-time fuse to explode artillery shells before they reached their targets, which was used successfully during the Battle of the Bulge during World War II.

Hanson, Earl Dorchester [February 15, 1927–October 26, 1993] American biologist who was best known for his academic career at Wesleyan University; he promoted ethical training for scientists and scientific teaching for nonscientists and received the Harbison Award for Distinguished Teaching in 1970. He was also chairman of the national Commission on Undergraduate Education in the Biological Sciences from 1965 to 1967.

Hanson, William B. [December 30, 1923–September 11, 1994] American physicist who was known for his research in aeronomy and ionospheric physics. His work included studies of the F region of the ionosphere; development of an ion potential analyzer for the OGO spacecraft and a spacecraft-borne anemometer, both used for global measurements; and assistance in forming the Atmosphere Explorer program.

Harden, Donald Benjamin [July 8, 1901–April 13, 1994] British archaeologist who was affiliated with several museums, including the London Museum where he served as director from 1956 to 1965, and was an expert on ancient glass from prehistoric times to the Medieval period. He wrote several books on the subject as well as essays, including the acclaimed "Glass of the Caesars," which concerned a 1987–88 exhibition at the Corning Museum of Glass in Corning, New York.

Harrington, Robert Sutton [October 21, 1942–January 23, 1993] American astronomer who worked at the United States Naval Observatory in Washington, D.C., and specialized in positional astronomy, an area that concerns the dynamics and distances of stars and the solar system. He was also involved in the search for "Planet X," a body once thought to exist beyond Pluto, and he had Minor Planet 3216 named for him.

Haury, Emil Walter [May 2, 1905–December 4, 1993] American archaeologist and anthropologist who specialized in the American Southwest; he played a major role in shedding light on two prehistoric cultures, the Mogollon Indian civilization which existed for several centuries before A.D. 1000, and the Hohokam culture, which thrived for approximately 1,000 years until about A.D. 1400.

Hawkins, Walter Lincoln [March 21, 1911–August 20, 1992] American chemist who, while at AT&T's Bell Laboratories, co-invented additives that extended the life of plastic coatings for shielding wire cables, and anti-oxidizing agents that gave plastic a life span of 70 years, thus helping the telecommunications industry save billions of dollars. One of the first successful African Americans in his field, he was awarded the National Medal of Technology by President Bush for his achievements in chemistry as well as for his efforts to encourage minorities to pursue scientific study.

Heiligenberg, Walter F. [1938–September 8, 1994] American scientist, born in Germany, who studied electric fish found in South America; their easily measurable signals shed light on how nerve networks respond to sensory information, thus leading to an increased understanding of animal behavior. He was also senior editor of the *Journal of Comparative Physiology* and a member of the American Academy of Arts and Sciences.

Hempelmann, Louis Henry [1914–June 30, 1993] American biologist and University of Rochester radiology professor who was best known for his role in the Manhattan Project to build the atomic bomb, for which he was director of the health division at the Los Alamos laboratory; he studied the effects of radiation exposure, which included the accidental exposure of workers to an uncontrolled fission reaction.

Henize, Karl Gordon [October 17, 1926–October 5, 1993] American scientist who was an astronomer at the Smithsonian Astrophysical Observatory and a professor at Northwestern University; he is credited for finding the third nova (an exploding new star) in the Magellanic clouds. He later worked for NASA as a member of the support crew of the 1971 *Apollo 15* mission, and at the age of 58 was involved in the *Spacelab 2* mission of the space shuttle *Challenger*, which distinguished him as the oldest American in space.

Herold, Edward William [October 15, 1907–June 30, 1993] American electrical engineer who, while working for the RCA Corporation, led a team that developed the first color television picture tube and performed RCA's first projects in the area of transistors. He was affiliated with RCA for 42 years, working his way up to serving as director of the corporation's David Sarnoff Research Center in Princeton, New Jersey.

Higinbotham, William Alfred [October 25, 1910–November 10, 1994] American physicist who was a group leader in electronics at Los Alamos, New Mexico, where the first atomic bomb was developed, then later helped form the Federation of American Scientists, a group advocating control of nuclear weapons. In 1958 while at the Brookhaven National Laboratory on Long Island, where he spent most of his career, he developed what is believed to be the first video game—a tennis game similar to the Pong game of the early 1970s, although he did not seek a patent for it.

Hiller, Lejaren [February 23, 1924–January 26, 1994] American chemist and composer who was the first to write music with a computer, using Princeton University's Illiac computer to compose the *Illiac Suite* for string quartet in 1956. He continued to work in both science and music, directing the electronic music studio at the University of Illinois, collaborating with composer John Cage on his well known *HPSCHD*, and cowriting books including *Experimental Music* (1959) and *Principles of Chemistry* (1960).

Hirst, George Keble   [March 2, 1909–January 22, 1994] American medical scientist who in 1941–42 found a way to detect influenza viruses in blood when he discovered that red blood cells clump together when mixed with such viruses. This led to his development of the hemagglutination assay method, which helps researchers estimate how much virus can be found in a sample and how much antibody is in a person's serum. He served as director of the Public Health Research Institute from 1956 to 1980 and was elected to the National Academy of Sciences and the American Academy of Arts and Sciences.

Hoard, James Lynn   [December 28, 1905–April 10, 1993] American chemist whose early work in using x-ray diffraction to study structures earned him a position with the Manhattan Project, where he studied the nature of uranium compounds. He later was known for his research on boron, an element that is important for plant growth and its use in industry, and in 1972 he was elected to the National Academy of Sciences.

Hodgkin, Dorothy Mary Crowfoot   [May 12, 1910–July 29, 1994] British chemist who was awarded the Nobel Prize in Chemistry in 1964; her achievements included using x-ray crystallography to shed light on the functions of biochemical compounds in living organisms and discovering the structures of penicillin, insulin, and vitamin B_{12}. She had taught at Oxford University and was chancellor of Bristol University from 1970 to 1988.

Holtzman, Eric   [May 25, 1939–April 6, 1994] American biologist and Columbia University professor whose research provided insight into how membranes function in nerve cells, how lysosomes break down harmful substances in cells, how cells produce certain proteins, how cell markers track down the movement of proteins within cells, and how membrane cycling occurs.

Holzer, Robert E.   [November 21, 1906–May 19, 1994] American geophysicist whose achievements included helping build a special analog computer which facilitated the study of cloud motion, publishing a paper that was essential in establishing that the fair weather ionosphere-to-earth current was caused by worldwide thunderstorms, and designing instruments for rocket flights into the ionosphere as well as instruments that monitored the ELF (Extremely Low Frequency) and VLF (Very Low Frequency) signals from the magnetosphere.

Hughes, George R.   [1907–December 21, 1992] American Egyptologist who was director of the Oriental Institute at the University of Chicago and an expert in Egyptian writing; he was involved in the development of a Demotic dictionary, covering the writing used from the seventh century B.C. to about A.D. 500, and in 1965 translated a Coptic prayer book that may have dated from the tenth or eleventh centuries.

Hurwitz, Henry Jr.   [December 25, 1918–April 14, 1992] American research physicist who worked on the hydrogen bomb at the Los Alamos National Laboratory in New Mexico and helped design and develop early nuclear power plants and the *Seawolf*, a nuclear submarine. He held several important positions at the Knolls atomic laboratory, operated by General Electric for the Atomic Energy Commission, as well as 15 patents.

Jerne, Niels Kaj   [December 23, 1911–October 7, 1994] London-born immunologist who was a corecipient of the 1984 Nobel Prize in Physiology or Medicine for groundbreaking theories that shed light on how immune defenses develop and fight disease. His positions included director of the Basel Institute of Immunology during the 1970s and chief medical officer for immunology for the World Health Organization in Geneva from 1956 to 1962.

John, Fritz   [June 14, 1910–February 10, 1994] American mathematician and New York University professor, born in Berlin, who was known as an expert on partial differential equations, and who worked on ways to mathematically describe actions such as the movement of waves in water, the bending of material like cardboard, and what occurs at the boundaries of oscillating objects. His work had important applications for physics and engineering.

Josten, Kurt   [1912–July 10, 1994] U.K. scholar, born in Neuss, Germany, who was a curator of the Museum of the History of Science in Oxford, England, from 1950 to 1964. He helped expand the museum's holdings of old astronomical and mathematical instruments and found the key to the written code of Elisas Ashmole, a seventeenth century British antiquary, scholar, and collector.

Kahng, Dawon   [May 4, 1931–May 13, 1992] American physicist, born in Seoul, South Korea, whose inventions included the first silicon transistor and the floating gate memory cell, essential for many forms of semiconductor memory devices. He was affiliated with Bell Laboratories until 1988, then became founding president of NEC Research Institute, known for its work in computer and communications technologies.

Kerst, Donald William   [November 1, 1911–August 19, 1993] American physicist and professor at the University of Wisconsin at Madison who in the 1940s developed the betatron, one of the first particle accelerators; although it was surpassed by higher-energy synchrotron type accelerators that emitted less waste energy, the betatron was considered an important step toward those more efficient designs. Kerst also worked on the Manhattan Project to develop the atomic bomb and was a member of the National Academy of Sciences, which awarded him its Comstock Prize in 1943.

Kildall, Gary   [1942–July 11, 1994] American computer scientist who in 1973 created Color Program/Monitor (CP/M), the first operating system that made it possible for a computer's central processing unit to easily store and retrieve information from a floppy disk drive.

Kimura, Motoo   [November 13, 1924–November 13, 1994] Japanese geneticist who was known for his evolutionary theory that challenged neo-Darwinism by suggesting that certain mutations that have no selective advantage can thrive in a population; his books included *Neutral Theory of Genetic Evolution*, published in 1968.

Kinzey, Warren Glenford   [October 31, 1935–October 1, 1994] American anthropologist who was an expert in primatology; his research of the wear patterns of the teeth of early hominids who lived in what is now Latin America provided insight into what their diets may have been.

Kleene, Stephen Cole [January 5, 1909-January 25, 1994] American mathematician who played an important role in establishing recursion theory, a mathematical version of the theory of computable functions that was the foundation of the field of theoretical computer science. He was awarded the National Medal of Science in 1990 and was elected to the National Academy of Sciences in 1969.

Kline, Morris [May 1, 1908-June 10, 1992] American mathematician and New York University professor who was known for his criticism of the way mathematics was taught as an isolated subject with little connection to the real world. His books included *Mathematics in Western Culture* (1953), *Mathematics: The Loss of Certainty* (1980), and *Why Johnny Can't Add: The Failure of the New Math* (1973).

Kolsky, Herbert [September 22, 1916-May 9, 1992] American physicist and Brown University professor, born in London, who invented the split-Hopkinson bar, a device used to measure permanent deformations in objects placed under sudden and heavy pressure that has been helpful in the development of aircraft design. He also published *Stress Waves in Solids* (1963), as well as 90 papers related to subjects such as conditions of metals and other materials.

Kolthoff, Izaak Maurits [February 11, 1894-March 4, 1993] American chemist, born in Almelo, the Netherlands, whose research on the emulsion polymerization process improved the quality of synthetic rubber used during World War II. He was professor and chief of the Division of Analytical Chemistry at the University of Minnesota; he co-authored with E. B. Sandell the *Textbook of Quantitative Inorganic Analysis*, a standard text; and he was awarded the William H. Nichols Medal, one of the highest honors given by the American Chemical Society, in 1949.

Kopal, Zdenek [April 4, 1914-July 23, 1993] American astronomer, born in Litomysl, Bohemia (now the Czech Republic), who was the astronomy department head at the University of Manchester in England from 1951 to 1981 but had also been affiliated with Harvard University and the Massachusetts Institute of Technology. He founded *Icarus*, an international journal on the solar system; and as a consultant to NASA he helped compile a detailed contour map of the Moon beginning in 1958.

Kotani, Masao [1905-June 6, 1993] Japanese physicist who served as president of Tokyo Science University and received awards in Japan for his research on the theory of magnetrons and microwave circuits, as well as his contributions to molecular physics and biophysics.

Kusch, Polykarp [January 26, 1911-March 20, 1993] American physicist, born in Germany, who shared a Nobel Prize in Physics with William E. Lamb for their work on the atom, including Kusch's determination of the magnetic moment of the electron. Kusch was a professor at Columbia University, where he received the Great Teacher Award in 1959 and served as provost and executive vice president in 1970-71, and later taught at the University of Texas in Dallas.

Layton, James P. [1920-December 2, 1993] American aeronautics engineer who was considered a pioneer in rocket technologies; in 1948, he directed the testing of the first large American rocket, the Viking series; he set up a rocket research laboratory at Princeton's Guggenheim jet propulsion laboratory, and conducted experiments using liquid ozone as a rocket propellant, among other projects.

Lozier, William Wallace [August 3, 1906-October 9, 1993] American physicist who worked for Union Carbide Corporation. He was awarded an Oscar from the

Academy of Motion Picture Arts and Sciences for his contributions to color photography, which included developing the lights that made color films possible in the 1940s and 1950s.

Luyten, Willem Jacob   [March 7, 1899–November 21, 1994] American astronomer born on Java, who in the 1960s developed a computerized version of a photographic-plate scanner. This device tracked stars on plates taken several years apart with the same telescope, thus facilitating the study of star motions and enabling Luyten to find more than 6,000 white dwarfs, a type of star that is extremely dense but gives off little light.

Lwoff, André Michel   [May 8, 1902–September 30, 1994] French-born biologist who shared the 1965 Nobel Prize in Physiology or Medicine with François Jacob and Jacques Monod; they discovered how some genes control the function of others to produce substances and regulate cell metabolism. His research on the role of vitamins as aids to enzymes and viruses that infect bacteria was a vital basis to modern biology.

MacDonald, Elizabeth G.   [1893–May 11, 1992] American inventor who marketed the Spic & Span household cleaner. With the help of her chemist aunt, she developed the product during the Depression, set up a company with her husband and friends to produce it, and in 1945 sold it to Proctor and Gamble.

Malkin, Myron Samuel   [August 6, 1924–October 24, 1994] American physicist who from 1973 to 1980 served as the first director of NASA's space shuttle program, playing a key role in getting the reusable spacecraft built on time for its first flight in 1981. Before then, he worked for the U.S. government as Deputy Assistant Secretary of Defense for Technical Evaluation.

Margolis, Stanley V.   [1943–November 7, 1992] American geochemist and University of California professor who helped establish authenticity of a 2,500-year-old Greek statue using x rays, electron microscopy, and a microprobe to examine the statue's magnesium. He also contributed to an understanding of how a collision of a meteorite or comet with the earth 65 million years ago could have led to the extinction of dinosaurs and other fauna.

Mark, Herman Francis   [May 3, 1895–April 13, 1992] American chemist, born in Vienna, whose groundbreaking research in polymers (giant long-chain molecules used in plastics and other materials) earned him many awards worldwide, including the National Medal of Science in the United States and the Humboldt Medal of Germany. He helped to define the structure of the polymer molecule in 1928 and to establish the kinetic theory of rubber elasticity in the 1930s, and led research into the development of polystyrene and other materials.

Martin, John Holland   [February 27, 1935–June 18, 1993] American oceanographer known for his theory that dumping iron into the ocean could reduce the increase of carbon dioxide in the atmosphere, thus easing global warming, and for his accurate measurements of heavy metals in the oceans.

Mason, Edward Allen   [September 2, 1926–October 27, 1994] American chemical physicist and Brown University professor who was known for his role in formulating the theory of transport phenomena, particularly the thermal conductivity of molecular masses, which brought a quantitative treatment of gas transport in porous media into use in engineering.

Mayall, Nicholas Ulrich [May 9, 1906–January 5, 1993] American astronomer who was former director of the Kitt Peak Observatory in Arizona and who in 1955 cowrote a report on research that concluded that the universe is 6 billion years old and three times the size previously estimated. Earlier in his career, while working for Lick Observatory in California, he took a photograph revealing 27 undiscovered bright-line stars.

McQueen, Robert G. [1924–October 18, 1994] American geophysicist specializing in the Earth's interior, who combined experimental and theoretical concerns in research using shock waves on minerals found inside the Earth.

Mitsui, Akira [January 25, 1929–May 31, 1994] American marine scientist, born in Shizuoka, Japan, whose research explored the possibilities of using blue-green algae in the production of new chemicals, food, medicines, and technologies in order conserve limited resources and use nonpolluting fuels, thus preserving the environment.

Moishezon, Boris G. [1938–August 25, 1993] Ukrainian-born mathematician who was one of many Soviet Jews who felt shut out of top positions in his country; after gaining his exit visa in 1972, he taught at Tel Aviv University and Columbia University, establishing a reputation as a world authority in algebraic geometry.

Montgomery, Deane [September 2, 1909–March 15, 1992] American mathematician and Princeton University professor who focused on topology, which deals with the intrinsic features of objects; his advances included working with Andrew Gleason and Leo Zippin to solve a question about Euclidean topological groups known as Hilbert's Fifth Problem. Montgomery was a member of the National Academy of Sciences and 1988 recipient of the American Mathematical Society's Steele Award.

Moore, Ian David [1951–September 28, 1993] Australian agricultural engineer who performed research in the areas of erosion mechanics, hydrological modeling, dry-land salinization, and water quality, focusing on problems such as land degradation and calling for more interdisciplinary science and the development of a field of environmental management.

Morgan, William Wilson [January 3, 1906–June 21, 1994] American astronomer who made the significant discovery of the shape of Milky Way galaxy, which is characterized by two spiral arms comprised of stars and interstellar gas. An expert in astronomical morphology, his research shed light on the structures of stellar populations, Earth's distance to remote stars, and the existence of super-giant galaxies.

Neuringer, Leo J. [November 20, 1928–May 4, 1993] American researcher who served as a scientist at M.I.T.'s Francis Bitter National Magnet Laboratory and who was best known for his developments in magnetic resonance imaging (MRI), which creates cross-section pictures of body organs. He developed technologies used in cardiac surgery and the treatment of brain disease, diabetes, and cancer.

Newell, Allen [March 19, 1927–July 19, 1992] American scientist who was considered one of the four pioneers in the field of artificial intelligence; he developed programs for complex information processing, including the Soar software system in the 1980s, was founding president of the American Association for Artificial Intelligence, and wrote or cowrote 10 books, including *Unified Theories of Cognition* (1990), and 200 articles. One month before Newell's death, he was awarded the National Medal of Science by President Bush.

Nier, Alfred Otto Carl   [May 28, 1911–May 16, 1994] American geophysicist who used a mass spectrometer to measure the isotopes of lead, which helped the establish the age of the Earth at 5 billion years, and to identify uranium 235, which led to the evolution of the atomic bomb. He also worked in Oak Ridge, Tennessee, on the Manhattan Project to develop the first such bomb.

Nolan, Thomas Brennan   [May 21, 1901–August 2, 1992] American geologist who served as director of the United States Geologic Survey from 1956 to 1965, establishing and leading projects to evaluate the effects of underground nuclear tests and carry out a photographic mapping of the moon, as well as other programs for researching naturally occurring chemicals, earthquakes, volcanoes, and other developments.

Northrup, Filmer Stuart Cuckow   [November 27, 1893– July 22, 1992] American philosopher and Yale University professor who was known for his work as a scholar of Einstein's theory of relativity and other scientific advancements of the twentieth century; he believed that such events played a larger role in foreign relations than political and ideological forces.

O'Neill, Gerald Kitchen   [February 6, 1922–April 27, 1992] American physicist who taught and cofounded the Space Studies Institute at Princeton University and invented the storage ring principle, used to produce colliding particle beams in high energy physics research. He was also an advocate of space colonization, drafting ideas for self-supporting solar-powered space stations. He served on the National Commission for Space under President Reagan starting in 1985 and was awarded an honorary degree from Swarthmore College in 1978.

Oort, Jan Hendrik   [April 28, 1900–November 5, 1992] Dutch astronomer whose major advances in the field included gathering evidence of the rotation of the Milky Way galaxy and locating the solar system about 30,000 light years away from its center. His research into galactic structure and dynamics helped reveal the existence of the "missing mass"—undetected matter that may make up 90% of the universe and could explain the clustering of stars into galaxies. He is most famous for his theory of the Oort Cloud, the likely source of comets that enter the Solar System.

Page, Robert Morris   [1903–May 15, 1992] American physicist and researcher at the Naval Research Laboratory whose radar technology inventions, designed to detect targets and missile launchings, included the pulse radar system, the planned position indicator (or PPI), and Project Madre, a system used during the cold war. He received citations from four presidents, including Dwight D. Eisenhower, who presented him with the Presidential Award for Distinguished Civilian Service in 1960.

Paine, Thomas Otten   [November 9, 1921–May 4, 1992] American engineer who spent most of his career working for General Electric but was Deputy Administrator and Administrator at NASA from 1968 to 1970, overseeing the first seven Apollo missions into space, including lunar expeditions during which four astronauts walked on the Moon's surface.

Parker, Theodore A. III   [1953–August 3, 1993] American ornithologist who perished in an airplane crash while surveying the Ecuadorean coast; a well-known expert on birds, he could identify nearly 4,000 species by their sounds alone. He served as senior scientist for Conservation International, an organization which focused on finding ways that people and nature could coexist.

Paul, Wolfgang [August 10, 1913–December 6, 1993] German physicist who shared a Nobel Prize in Physics in 1989 for his work in the 1950s and 1960s, when he developed a method of isolating ions and electrons from outside influences such as temperature and pressure; this "Paul trap" enabled scientists to measure particles with great precision and led to the development of the cesium clock. He also served for 10 years as president of the Alexander von Humboldt Foundation, an organization which coordinates international academic exchanges.

Pauling, Linus Carl [February 28, 1901–August 19, 1994] American chemist and peace advocate who won the 1954 Nobel Prize in Chemistry for his work on the chemical bond, about which he also wrote a classic textbook, and the 1962 Nobel Peace Prize for his antinuclear and antiwar efforts. In his later years Pauling also became famous for his advocacy of vitamin C for preventing both the common cold and cancer, and for the development of a promising approach to using vaccination as a cure for AIDS.

Petty, Olive Scott [April 15, 1895–March 2, 1994] American geophysicist who in 1925 invented an electrostatic seismograph detector that was used to find deposits of petroleum through vibrations in the earth, and who started one of the first seismic service companies in the oil industry. He was later known for exploring commercial oil fields England and India in the 1930s.

Platt, John Radec [June 29, 1918–June 17, 1992] American physicist and biophysicist who was known for research in his fields as well as for his studies of social trends in science. In 1971, he and two political scientists unveiled the results of a study that indicated that most of the social research in the latter part of this century has been conducted by scholars in major intellectual centers such as universities; because of this he believed that more funding and attention should be given to social research in regional centers.

Pollack, James Barney [July 9, 1938–June 13, 1994] American research scientist who was affiliated with NASA's Ames Research Center at Moffett Field, California, and who authored, with Carl Sagan and three others, an article that introduced the concept of nuclear winter, a global chilling effect they believed would be caused by nuclear war.

Pollara, Luigi Zummo [1914–July 29, 1994] American physical chemist who, while at the Stevens Institute of Technology in Hoboken, New Jersey, served as dean for graduate and professional studies and provost, and in 1982 founded the Polymer Processing Institute, which conducted chemical industry-sponsored research.

Pollister, Arthur Wagg [1903–October 18, 1994] American zoologist who in 1949 led a team that developed the micro-spectro-photometer, an electronic device that used visible and invisible light to examine and determine the quantities of a cell's components; this invention led to developments in research on DNA, cancer, and anemia, among other fields.

Ponnamperuma, Cyril Andrew [October 16, 1923–December 20, 1994] American chemist, born in Ceylon (now Sri Lanka), who believed that there could be a chemical explanation to the origins of life.

Pontecorvo, Bruno Maksimovich [August 22, 1913–September 24, 1993] Italian-born physicist who during the 1930s worked with Enrico Fermi in Rome on experiments in which they found ways to produce radioactive isotopes of certain elements, then later emigrated to France, the United States, Canada and England, where he was employed by the Atomic Energy Research Laboratory. When he and his family defected to the Soviet Union in 1950, there was speculation that he intended to provide secrets on building a hydrogen bomb; however, he stated that he believed that Russia provided the best academic atmosphere for pursuing nuclear research for peaceful purposes.

Popper, Sir Karl Raimund   [July 28, 1902–September 17, 1994] British philosopher, born in Vienna, whose writings about science concerned his opposition to Marxism and arguments for more rigorous testing of scientific theories; his books included *The Open Society and Its Enemies* (1945). He was a professor at the London School of Economics from 1949 to 1969.

Qian, Sanquiang   [October 16, 1913–June 28, 1992] Chinese physicist who received his advanced education in France but returned to China to lead a team in the 1950s that successfully worked on an atomic bomb. An expert on uranium fission, he received the French Academy's Henry de Parville Award for Physics in 1949, and in China he served as vice president of the Academy of Sciences and president of Zhejiang University.

Rainey, Froelich G.   [1907–October 11, 1992] American archaeologist whose unearthing of bone shafts used to kill animals supported the theory that humans migrated across the Bering Strait to America; he also created a Peabody Award-winning television program called *What in the World* and was director of the Museum of Archaeology and Anthropology at the University of Pennsylvania from 1947 to 1976.

Ringwood, Alfred Edward   [April 19, 1930–November 12, 1993] Australian geologist whose extensive research concerning the Earth's mantle led to his 1962 pyrolite model for its bulk composition. He also found that the siderophile (iron-loving) elements in the mantle are overabundant, which raised questions about the Earth's history and core formation; and he developed a theory based on his experiments with basalts that suggests a genetic connection between the Earth and the Moon.

Roller, Paul S.   [April 2, 1902–May 21, 1993] American chemist and inventor who had his own company, the Liquids Process Company in Washington, and held more than a dozen patents, for a particle size analyzer, water purification and desalination devices, and other innovations for manufacturing and other industries.

Rossi, Bruno Benedetto   [April 13, 1905–November 21, 1993] American physicist, who was born and began his career in Italy, becoming a pioneer in cosmic-ray physics. His contributions included the Rossi coincidence circuit, an instrumental technique that recorded the simultaneous occurrence of electrical pulses and was later used for high-energy physics and computers; other discoveries related to the nature of cosmic rays; and his role in the construction of the *Explorer X* satellite in 1961, which discovered the magnetopause (the space boundary beyond the powerful region of the Earth's magnetic field). He was awarded the National Medal of Science, among other honors.

St. Joseph, John Kenneth Sinclair   [1912–March 11, 1994] British geologist and Cambridge University professor whose pioneering use of aerial photography revealed the locations of over 200 Roman forts and led to his conclusion that the Roman invasion of Scotland during the first century under Gnaeus Julius Agricola had a greater impact on the island of Great Britain than previously believed.

Sands, George Dewey [June 16, 1919–November 2, 1994] American chemist who, while a project scientist at NASA, was known as the "Voice of Viking" for fielding questions from scientists and reporters about the spacecraft's two missions to Mars, in which over 1,400 photographs of the planet were transmitted back to Earth.

Schaefer, Vincent Joseph [July 4, 1906–July 25, 1993] American chemist whose inventions included cloud "seeding" and the first artificially induced snow and rainfall; although these weather-controlling innovations were never used on a grand scale, they have been adopted by some countries to clear clouds over airports and for other purposes. Schaefer was also founder and director of the Atmospheric Sciences Research Center at the State University of New York.

Schelkunoff, Sergei A. [1896–May 2, 1992] American inventor, born in Samara, Russia, who codeveloped the coaxial cable that could transmit television or up to 200 telephone circuits and is still used widely in television today. He was employed at Western Electric (which became Bell Laboratories) and taught at Columbia University.

Schevill, William Edward [1906–July 23, 1994] American biologist who was affiliated with the Woods Hole Oceanographic Institute in Massachusetts and who was known as an expert on whale sounds. In 1949, he and his wife, Barbara Schevill, made the first recording of whales by using a dictating machine to capture the sounds made by beluga whales in the Saguenay River in Quebec.

Schwinger, Julian Seymour [February 12, 1918–April 16, 1994] American physicist who shared the 1965 Nobel Prize in Physics for his work in the quantum field theory of electromagnetism, particularly his mathematical formulations that shed light on the interaction between charged particles and a field. He was a professor at Harvard University and the University of California at Los Angeles, and he received the National Medal of Science from President Johnson in 1964, as well as the first Albert Einstein Prize, which he shared with mathematician Kurt Gödel in 1951.

Seacord, Daniel Freeman [August 26, 1921–November 5, 1994] American nuclear physicist who was known for his specialty in theoretical hyperdynamics (measurement and analysis of shock pressures) and for his ability to provide calculations necessary to build shielding for nuclear processing plants. He participated in the atomic tests at Los Alamos, New Mexico, in the 1950s.

Sheehan, John Clark [September 23, 1915–March 21, 1992] American chemist who was the first to develop synthetic penicillin, a drug that had previously taken months to generate from molds, and discovered ampicillin, a semi-synthetic penicillin that could be taken orally. He also served as a professor at the Massachusetts Institute of Technology and as a scientific adviser to Presidents Kennedy and Johnson, and published a book, *The Enchanted Ring: The Untold Story of Penicillin*, in 1982.

Shore, Arthur Frank (A. F.) [1924–November 27, 1994] British scholar who was an expert on Egypt during the period of Roman domination from about 30 B.C. to A.D. 642; his writings included *Portrait Painting from Roman Egypt*, which concerned mummy portraits unearthed in the Fayum area of Upper Egypt, and he was an authority on Coptic (a form of the Egyptian language used in the early centuries of the Christian era) as well an ancient Egyptian writing system.

Shriner, Ralph Lloyd   [October 9, 1899–June 7, 1994] American organic chemist and Indiana University professor who was known for his system of identifying organic compounds and for a standard textbook, *The Systematic Identification of Organic Compounds*, which was first published in 1935. In 1962, the American Chemical Society presented him with the James Flack Norris Award for outstanding achievement in the teaching of chemistry.

Shrock, Robert Rakes   [August 27, 1904–June 22, 1993] American geologist and Massachusetts Institute of Technology professor who was known for his researching and indexing of fossils, including one that was named for him: *Kentlandoceras shrocki*, an invertebrate which lived during the Ordovician period 500 million years ago.

Shugg, Carleton   [1899–January 23, 1992] American engineer known for his innovations in submarines; he helped develop the rescue diving bell used when the *Squalus* sank in 1939, and a water-filled tower used for underwater escape practice. He and Hyman G. Rickover developed the first nuclear submarine, the *Nautilus*, in 1955.

Slayton, Donald Kent (Deke)   [March 1, 1924–June 13, 1993] American astronaut who was one of the Mercury Seven, the first group of astronauts from the United States. Because of a heart problem, he never participated in the Mercury flights, but did complete an Apollo mission in 1975, in which a docking with a Soviet *Soyuz* spacecraft occurred, and he served as chief of flight operations at the Johnson Space Center in Houston, training and selecting crews for the Apollo flights to the Moon and other missions.

Smith, Alan Paul   [March 31, 1945–August 26, 1993] American plant ecologist who had been based in Panama, where he was known for his research of plant and animal life in the tropical forest canopy, or treetop layer, and for his design of cranes that allowed scientists to study the area in safety.

Smith, Cyril Stanley   [October 4, 1903–August 25, 1992] American metallurgist and Massachusetts Institute of Technology professor, born in Birmingham, England, whose research in metals resulted in his development of structural theory; he also led the preparation of fissionable metal used for the atomic bomb at Los Alamos, which resulted in his receiving a Presidential Medal for Merit, He combined his interests in metallurgy and history to write such books as *A History of Metalography: The Development of Ideas on the Structure of Metals to 1890* (1988) and to establish the Institute for the Study of Medals at the University of Chicago and the Laboratory for Research in Archeological Materials at M.I.T.

Smith, Harold Hill   [April 24, 1910–October 19, 1994] American geneticist who, while at Brookhaven National Laboratory on Long Island, performed important research on the genetic basis of the formation of tumors and the genetic effects of irradiation in plants; in 1976, his research team became the first to produce the first fusion of human cell with a cell from a plant.

Smith, Ora   [April 13, 1900–February 4, 1993] American plant physiologist and Cornell University professor who was known as the world's leading expert on potatoes; his achievements included the discovery of a chemical used to inhibit the sprouting of potato eyes during storage and an additive that preserved the color of cooked potatoes, and the invention of the potato hydrometer for measuring water content. His innovations paved the way to mass production of french fries, potato chips, and similar foods.

Smithson, Evelyn Lord   [1923–March 9, 1992] American anthropologist and classics professor who devoted her summers to working on the ancient marketplace excavations in Greece in order to shed light on the culture of the Homeric Age.

Sperry, Roger Wolcott   [August 20, 1913–April 17, 1994] American neurobiologist who shared the 1981 Nobel Prize in Physiology or Medicine for helping to reveal the function of the brain's corpus callosum, a bundle of about 200 nerve fibers, as a passageway for information between the two hemispheres; this shed light on why patients who had an epilepsy operation in the early 1960s developed "split brains," resulting in problems with coordination and other functions.

Standen, Anthony   [1906–June 22, 1993] American chemist, born in Great Britain, who was executive editor of the 22-volume second edition of the *Kirk-Othmer Encyclopedia of Chemical Technology*, and who wrote books that challenged the inflated egos of scientists, including the popular *Science is a Sacred Cow*, published in 1950.

Starr, Richard Francis Strong   [1899–March 9, 1994] American archaeologist who from 1929 to 1931 was leader of an expedition to Nuzi, Iraq, where discoveries included the world's oldest map (dated to 2500 B.C.) and armor, as well as cuneiform tablets that were the first evidence of the Horites, who are mentioned in the Bible. He later worked for the U.S. State Department and Central Intelligence Agency as a research specialist on the Middle East.

Stern, Arthur Cecil   [March 14, 1909–April 17, 1992] American scientist who was one of the first to warn about the effects of air pollution and to advocate for controlling it; he served as director of air pollution research for the United States Public Health Service beginning in 1955, after working at the New York State Labor Department, where he supervised an air quality survey in the 1930s and drafted New York City's 1949 air pollution control law. He was involved in several pollution organizations and wrote the three-volume *Air Pollution*.

Steward, Frederick Campion   [June 16, 1904–September 13, 1993] American botanist and Cornell University professor, born in London, who in 1958 found that plants could be regenerated from one cell; this discovery meant that cuttings were no longer necessary to produce hybrids and mutations and also defined the field of plant molecular biology. He was also known for other insights on plant metabolism, protein synthesis, and hormone regulation.

Stoker, James Johnston Jr.   [March 2, 1905–October 19, 1992] American mathematician who served as director of New York University's Courant Institute of Mathematical Sciences from 1958 to 1966, known for research involving water flow and flood waves. He used high-speed digital computers to calculate the flow of an Ohio River flood and similar occurrences, which helped engineers find ways to alleviate flood-wave problems.

Storms, Harrison Allen Jr.   [1916–July 11, 1992] American aeronautical engineer who, while working for Rockwell International (formerly North American Aviation), helped design and develop B-25 bombers and P-51 Mustang fighters used in World War II, the F-86 fighter used in the Korean War, the F-100 Super Sabre, and the X-15 space research rocket plane. He received the International von Karman Wings Award for Lifetime Achievement, which is given by the California Museum of Science and Industry, and the Air Force Meritorious Civilian Service Award.

Strong, John Donovan   [January 15, 1905–March 21, 1992] American scientist who in the 1930s played a role in developing a vacuum evaporation process used for coating glass with aluminum to make telescope mirrors more durable and sensitive; he applied this coating to the 200-inch diameter telescope at Mount Palomar, California, the largest of its kind in the 1940s. The process is still in use today. He was also the first to discover water vapor in the atmosphere of Venus, using instruments mounted on a piloted balloon.

Swinton, William Elgin   [September 30, 1900–June 12, 1994] Scottish-born paleontologist who had been a Fellow of the Royal Society of Canada, professor of zoology at the University of Toronto, and director of life sciences at Ontario Museum. While he was with the Natural History Museum in London, he tutored Queen Elizabeth II when she was a child; he also wrote several textbooks, including *The Dinosaurs* (1934), as well as works for children.

Temin, Howard Martin   [December 10, 1934–February 9, 1994] American cancer researcher who shared the 1975 Nobel Prize in Physiology or Medicine with Renato Dulbecco and David Baltimore for their discovery of reverse transcriptase (an enzyme that helps certain viruses destroy the genetic machinery of the cells they infect). This development proved his theory that the genetic information of some viruses could be carried in the form of RNA and passed on to DNA, contrary to the previous notion that only DNA transmitted information to RNA, and led to the production of drugs like human insulin and the discovery of retroviruses such as H.I.V.

Theremin, Leon   [1896–November 3, 1993] Russian inventor, born Lev Sergeyevich Termen in St. Petersburg, who was best known in the United States for his introduction in 1927 of an electronic musical instrument called the theremin. RCA marketed and sold about a thousand sets of the instrument and Theremin moved to New York, where he set up a laboratory and developed more musical inventions, before returning to the Soviet Union in 1938. He remained there, serving in prison for anti-Soviet propaganda but continuing to work in electronics, developing an eavesdropping device for the KGB. He also taught at the Moscow Conservatory.

Timby, Elmer K.   [1906–October 28, 1992] American civil engineer who taught at Princeton University and played key roles in the design and construction of several important projects in the United States, including the Golden Gate and Bronx-Whitestone Bridges, the New Jersey (and other) Turnpikes, and the Miami International Airport. He was particularly known for his detailed working metal models, which were used to study stress problems and make design changes.

Turnbull, Colin M.   [1925–July 28, 1994] British-born anthropologist who wrote several well-known books, including *The Forest People*, an account of pygmy hunters and gatherers in Zaire which became a standard text, and *The Mountain People*, which depicted the plight of the dwindling Ik civilization in Uganda. He also worked as a professor at George Washington University and as an associate curator at the American Museum of Natural History in New York.

Van Gelder, Richard G.   [December 17, 1928–February 23, 1994] American scientist who was a curator at the American Museum of Natural History in New York and an expert on taxonomy and the evolution of animals; in 1968 he oversaw the museum's installation of Whale II, a 94-foot long blue whale, which replaced a 76-foot version that had been on display for 60 years. He wrote books and presented exhibitions to children and campaigned to save endangered species.

Vine, Allyn Collins   [June 1, 1914–January 4, 1994] American oceanographer who advocated for and helped develop manned submersible vessels; the first American vessel, named Alvin in his honor, was christened in 1964 and is credited for recovering a lost hydrogen bomb that had sunk to the floor of the Mediterranean, finding wreckage of the Titanic in 1986, and discovering new life forms in the deep sea. During World War II, Vine's revised design of an instrument called the bathythermograph found layers of temperature change in the sea that made it easier for submarines to hide from enemy sonar, thus saving lives and equipment.

Walker, John Charles   [July 6, 1893–November 25, 1994] American agricultural scientist and University of Wisconsin professor who was the first to demonstrate the chemical nature of disease resistance in plants; he used his findings to develop several varieties of vegetables, save cucumber crops from spot rot, and cure boron deficiency in beets during the 1940s.

Walton, Richard R.   [1908–June 24, 1993] American inventor who never earned a college degree and worked out of his basement. His innovations, mostly in fabrics, included a device that could pick up a single layer of cloth from a stack used in automated production; the Pak-nit process that alleviated shrinkage in knit fabrics; agitating devices for automatic clothes washers; and a hand-operated clothes washer used in developing countries.

Watson, William Weldon   [September 14, 1899–August 3, 1992] American physicist who played a role in developing the atomic bomb as director of the Metallurgical Laboratory of the University of Chicago. He was a professor at Yale University, where he also served as physics department chairman from 1940 to 1961 and oversaw the construction of particle accelerators. He was also a trustee of the Brookhaven National Laboratory. Watson was also a pioneer in suggesting that nuclear energy could be used for aircraft and space satellites.

Webb, James Edwin   [October 7, 1906–March 27, 1992] American administrator whose leadership at NASA from 1961 to 1968 led to several major successes of the space program: John Glenn made the first orbital flight; Alan B. Shepard Jr. completed the first U.S. manned space flight; and Neil Armstrong walked on the moon in 1969 ahead of schedule, among other achievements. Webb had also worked for the U.S. government in several capacities, including serving as budget director under President Truman from 1946 to 1949.

Weick, Fred E.   [1900–July 8, 1993] American engineer, affiliated with NASA's predecessor, the National Advisory Committee for Aeronautics; his pioneering innovations in aviation design were often adapted to larger aircraft. He improved airplane aerodynamics and reduced drag and turbulence, developed stable landing gear that reversed the traditional two sets of wheels in front and one in back to the opposite, designed the first tailspin-proof plane, and helped construct the first large wind tunnel for testing aircraft.

Whitten, Charles Arthur   [October 2, 1909–July 12, 1994] American researcher who was Chief Geodesist at the U.S. Coast & Geodetic Survey and whose achievements included the application of modern computing techniques to geodetic adjustments, starting in 1951. His analyses of crustal movements made scientists better able to describe seismic mechanisms and possibly able to predict earthquakes.

Wick, Gian Carlo [October 15, 1909–April 20, 1992] Italian physicist whose research in particle physics began when he collaborated with Enrico Fermi in the 1930s. He came to the United States in 1951, where he worked at Columbia University and other institutions. In 1951 he presented a mathematical system for the quantum theory of electromagnetic radiation, which was helpful in theoretical physics, and theoretical description of certain particles was widely used in the 1960s.

Wiesner, Jerome Bert [May 30, 1915–October 21, 1994] American engineer who served in President Kennedy's administration as Special Assistant to the President for science and technology, during which he worked on a treaty to ban all but underground nuclear tests; the treaty was signed in 1963. He later became president of the Massachusetts Institute of Technology, a post he held for nine years.

Wigglesworth, Sir Vincent Brian [April 7, 1899–February 12, 1994] British biologist who was considered an authority on insect hormones; he wrote several acclaimed books, including *Insect Hormones* (1970), *Insect Physiology* (1934), and *The Principles of Insect Physiology* (1939). His research revealed interesting features about insects, including how their feet stick to surfaces and how the gnatlike midge can flap its wings more than 1,000 times per second.

Wigner, Eugene Paul [November 17, 1902–January 1, 1995] American physicist and Princeton University professor, born in Budapest, who won the Nobel Prize in Physics in 1963 for his groundbreaking research in quantum mechanics, which included discovering the actions of electrons as they jump from one quantum to another, and establishing what are known as Wigner crystals, Wigner theorems, Wigner energy, and Wigner rules, related to the symmetry and order of subatomic particles. He was also credited as one of three scientists who forewarned Albert Einstein and the United States that Hitler might build an atomic bomb, which resulted in the United States government's creation of the Manhattan Project to build its own.

Wilson, Edgar Bright [December 18, 1908–July 12, 1992] American chemist and Harvard University professor who cowrote some standard textbooks, including *Introduction to Quantum Mechanics* with Linus Pauling, and was one of a group of prominent scientists who met secretly with President Lyndon Johnson to urge him to bring an early end to the Vietnam War. He was awarded the National Medal of Science by President Ford in 1976.

Wilson, John Tuzo [October 24, 1908–April 15, 1993] Canadian geophysicist and University of Toronto professor whose work supported the continental drift hypothesis, now known as plate tectonics; he cited similarities in fossils found in Norway and Scotland with American fossils as evidence that the Atlantic Ocean may have closed up and reopened several times in occurrences that have been named Wilson cycles.

Wormington, Hannah Marie [September 5, 1914–May 31, 1994] American archaeologist who was one of the first women to enter the field. A specialist of the paleo-Indian period in North America, she was curator of archeology at the Denver Museum of Natural History from 1937 until she became curator emeritus in 1968, the same year that she was elected the first woman president of the Society for American Archaeology.

Wu, Shien-Ming   [October 28, 1924–October 28, 1992] American engineer, born in Chekiang, China, who developed the "dynamic data system," which used mathematical formulas and computerized data to analyze production in order to find ways to improve precision, quality, and efficiency in mass production. The system was adopted by the three major automobile manufacturers in the United States. He taught manufacturing technology at the University of Michigan and served as director of the National Science Foundation's Industry-University Cooperative Research Center.

Wyckoff, Ralph Walter Graystone   [August 9, 1897–November 3, 1994] American chemist whose achievements in electron microscopy included playing a role in the development of the ultracentrifuge (a high-speed centrifuge that helped determine the sizes and weights of small particles) and purifying the virus that causes equine encephalomyelitis, which paved the way for a vaccine for horses and later a program for producing a typhus vaccine. He published more than 200 books and 400 papers, and was a member of the National Academy of Sciences and other organizations.

Yegorov, Boris B.   [November 26, 1937–September 12, 1994] Russian physician who was the first in his field to fly in space; he participated in the 24-hour, 16-orbit *Voskhod 1* mission, launched on October 12, 1964, where he observed astronauts' reactions to microgravity, drew their blood samples, and performed experiments on plants and fruit flies.

Zener, Clarence Melvin   [December 1, 1905–July 2, 1993] American physicist who worked as a director of research for the Westinghouse Electric Corporation and as a professor at the Carnegie Mellon University. His achievements in solid-state physics included a 1934 paper on the breakdown of electrical insulators, which led to the development during the 1950s of a widely used voltage regulator known as the Zener diode; advances in the areas of internal friction and ocean thermal energy conversion; and the development of better alloys based upon mathematical theories.

Zetlin, Lev   [July 14, 1918–December 4, 1992] American civil engineer, born in Russia, who helped design and construct buildings and other structures, such as the Roosevelt Island Tramway; the New York State Pavilion, or Tent of Tomorrow, for the 1964 World's Fair; and Disney's Epcot Center. He was also an expert on structural disasters, investigating tragic events such as the collapse of the Hyatt Regency's suspended walkway in Kansas City, which killed 110 people.

Zhou, Peiyuan   [1902–November 24, 1993] Chinese physicist who was a student of Albert Einstein while at Princeton University from 1935 to 1936, and who worked on launching torpedoes from aircraft and other research projects with the U.S. Navy between 1943 and 1947. In China from 1952 to 1981, he was Beijing University's dean of students before serving as its vice-president and, ultimately, president.

Zorn, Max August   [1906–March 9, 1993] American mathematician and Indiana University professor, born in Krefeld, Germany, who during 1934–35 developed a proposition known as Zorn's lemma (equivalent to the Axiom of Choice in set theory), which states that a set can be defined by choosing one item from each of an infinite number of sets; it serves as an important premise of algebraic set theory.

Zuckerman, Lord Solly   [May 30, 1904–April 1, 1993] British scientist, born in Cape Town, South Africa, who wrote 28 books about animal anatomy and primate behavior and served as a military advisor and researcher under Winston Churchill and other prime ministers. His activities included persuading military leaders to change their attack strategies during World War II; participating in talks in 1960 concerning the banning of nuclear tests; and supervising an oil cleanup of the Cornwall coast in 1967.

Zumberge, James Herbert   [December 27, 1923–April 15, 1992] American geologist who led several polar expeditions to the Antarctic, where Cape Zumberge and the Zumberge Coast are named after him. He was also president of the University of Southern California from 1980 to 1981, where he was successful in raising funds for the school and in improving its academic reputation, which had previously been secondary to athletics.

Zygmund, Antoni   [December 26, 1900–May 30, 1992] American mathematician and University of Chicago professor, born in Warsaw, Poland, who was known for his work in harmonic analysis (used to analyze periodic functions) and Fourier analysis (a treatment of mathematical functions that are applied to partial differential equations). He was awarded the National Medal of Science in 1986.

APPENDIX B

UNITS OF MEASURE

There are two measurement systems that are widely used. Most of the world uses a system often still called the metric system, a simplification of the International System of Units (abbreviated SI from *Système International*, its name in French). The true metric system was the French ancestor of SI. The United States continues to use a system that is called U.S. customary measure. Scientists in all countries, however, nearly always use SI.

THE INTERNATIONAL SYSTEM

The International System of Units had its origins in the rational idealism of the French Revolution. It represents a serious effort to make people measure in a rational and consistent fashion; but people often refuse to be rational or consistent, so there are various accommodations that have to be made.

The idea of SI is to have a limited number of basic units that can then be combined with a fixed set of prefixes to indicate multiplication or division by various powers of ten, giving a reasonably sized unit for every measurable quantity. Some of the problems that arise in developing this concept include the choice of a base unit that already has a prefix (such as the kilogram, kept as a base unit in SI because it was a base unit for the French metric system); the use of units that derive from nature (such as the speed of light or the mass of the electron); and the use of units that derive from SI units in such a complicated fashion that they must be used with special names. Furthermore, the official SI units are often awkward in size for the object being measured, whereas a former metric or customary unit may be more convenient.

Because it is an international system, the official way of writing whole numbers of more than three digits is to use a thin space at each period of three digits instead of a comma. In Europe, the comma is often used as a decimal point instead of the period in unofficial usage. The thin spaces are also inserted for numbers smaller than 0.001. In that case, the periods of three units at taken from the decimal point, but moving to the right. Thus, the numeral 1234567890.0987654321 would be written as 1 234 567 890.098 765 432 1.

The seven official base units are as follows:

Quantity to be measured	Base unit (abbreviation)	Definition
Time	second (s)	Time it takes hot cesium atoms to vibrate 9 192 631 770 times.
Length	meter (m)	Distance light travels in a vacuum in 1/299 792 458 of a second.
Mass	kilogram (kg)	Mass of a platinum-iridium cylinder kept in Sèvres, France, called the International Prototype Kilogram.
Electric current	ampere (A)	A current that if maintained in two infinite, parallel conductors of negligible cross-section placed 1 meter apart would produce a force between them equal to 10^{-7} newton.
Temperature	kelvin (K)	1/273.16 of the thermodynamic temperature of the triple point of water.
Amount of substance	mole (mol)	The amount of a substance that contains 6.0225×10^{23} (Avogadro's number) of basic units, such as atoms or molecules.
Luminous intensity	candela (cd)	1/600 000 of the light produced by a 1-square-meter cavity at the temperature of freezing platinum (2,042K).

These units are maintained by diplomatic body termed the Conférence Générale des Poids et Mesures (CGPM, or the General Conference of Weights and Measures), which in 1960 established the system now in use. CGPM continues to refine the system.

For reasons known only to CGPM, measures for plane and three-dimensional angles have been termed supplementary, although they are at least as basic as the seven base measures. The radian (rad) for plane angles is most easily thought of as the measure of the central angle of a circle that intercepts an arc equal to the radius of the circle. A solid angle is measured in steradians (sr), a three-dimensional central angle that intercepts a region measuring $\pi/4$ square units on a sphere whose radius is one unit. In October 1995, CGPM moved the two angle measures to another category—derived units. Unlike the other derived units, however, such as speed (meters per second) and the newton (kilogram meters per second per second), radians and steradians are not derived from SI units.

These units are modified by a set of standard prefixes to produce smaller or larger measures. For example, centi- is the prefix for 1/100 (10^{-2}) and is abbreviated c, so 1/100 of a second is a centisecond (cs) and 1/100 of a candela is a centicandela (ccd). Since the base unit for mass, the kilogram, already uses the prefix that means 1 000 (10^{3}), prefixes for measures of mass are attached to the gram, so a mass of 1 000 kilograms is not labeled a kilokilogram, but is instead a megagram (Mg).

Factor	Scientific notation	Prefix	Symbol
1/1 000 000 000 000 000 000 000 000	10^{-24}	yocto-	y
1/1 000 000 000 000 000 000 000	10^{-21}	zepto-	z

Factor	Scientific notation	Prefix	Symbol
1/1 000 000 000 000 000 000	10^{-18}	atto-	a
1/1 000 000 000 000 000	10^{-15}	femto-	f
1/1 000 000 000 000	10^{-12}	pico-	p
1/1 000 000 000	10^{-9}	nano-	n
1/1 000 000	10^{-6}	micro-	μ
1/1000	10^{-3}	milli-	m
1/100	10^{-2}	centi-	c
1/10	10^{-1}	deci-	d
10	10^{1}	deka-	da
100	10^{2}	hecto-	h
1 000	10^{3}	kilo-	k
1 000 000	10^{6}	mega-	M
1 000 000 000	10^{9}	giga-	G
1 000 000 000 000	10^{12}	tera-	T
1 000 000 000 000 000	10^{15}	peta-	P
1 000 000 000 000 000 000	10^{18}	exa-	E
1 000 000 000 000 000 000 000	10^{21}	zetta-	Z
1 000 000 000 000 000 000 000 000	10^{24}	yotta-	Y

In the tables below, the most commonly used SI units are defined in terms of each other instead of strictly in terms of the base units. Also included are derived SI units and units commonly used by scientists that are not SI units. Following this listing is a brief table of U.S. customary units. At the end there are tables converting SI or related units into U.S. customary units and vice versa.

SI Length, Area, and Volume

Despite the best efforts of the SI police, many scientists persist in using metric measures that are no longer part of the system. In computer science, for example, measurements are often stated in *microns* instead of the SI equivalent of *micrometers*. Throughout the sciences, the *angstrom* is used instead of appropriate measurements in parts of a micrometer or in nanometers.

One angstrom (Å) = 10 nm = 0.000 1 micrometer = 0.000 000 1 $\overline{mm}$

Length or Distance

decimeter (dm) = 10 centimeters = 0.1 m

centimeter (cm) = 0.01 m

millimeter (mm) = 0.1 cm = 0.001 m

micrometer (μ) = 0.001 mm = 0.000 1 cm = 0.000 000 1 m

dekameter (dam) = 10 m

hectometer (hm) = 10 dekameters = 100 m

kilometer (km) = 10 hectometers = 100 dekameters = 1 000 m

Area

1 square millimeter (mm^2) = 1 000 000 square micrometers

1 square centimeter (cm^2) = 1000 mm^2

1 square meter (m^2) = 10 000 cm^2

1 square kilometer (km^2) = 100 ha = 1 000 000 m^2

An unofficial unit often used in measurement of land is the *hectare* and its associated measure the *are*.

1 are (a) = 100 m^2

1 hectare (ha) = 100 a = 10 000 m^2

Volume

1 cubic centimeter (cm^3) = 1 000 cubic millimeters (mm^3)

1 cubic decimeter (dm^3) = 1 000 cm^3

1 cubic meter (m^3) = 1 000 dm^3 = 1 000 000 cm^3

The unit *cubic centimeter* is sometimes unofficially abbreviated *cc* and is used in fluid measure interchangeably with the unofficial measurement milliliter (mL).

Fluid volume measurements are given in cubic measure in SI, but the use of a separate group based upon the liter is tolerated. A *milliliter* of fluid occupies a volume of 1 cubic centimeter (cm^3 or cc). A *liter* of fluid (slightly more than the customary quart) occupies a volume of 1 cubic decimeter or 1 000 cubic centimeters. In SI the letter L is the official abbreviation for liter, although most dictionaries still use l, which is easily confused with the numeral 1.

1 centiliter (cL) = 10 milliliters (mL)

1 deciliter (dL) = 10 cL = 100 mL

1 liter (L) = 10 dL = 1 000 mL

1 dekaliter (daL) = 10 L

1 hectoliter (hL) = 10 daL = 100 L

1 kiloliter (kL) = 10 hL = 1 000 L

SI Mass and Weight

Mass and weight are often confused. Mass is a measure of the quantity of matter in an object and does not vary with changes in altitude or in gravitational force (as on the Moon or another planet). Weight, on the other hand, is a measure of the force of gravity on an object, so it does change with altitude or gravitational force.

The International System generally is used to measure mass instead of weight. The early metric system's basic unit for measurement of mass was the gram, which was originally defined as the mass of 1 milliliter (= one cubic centimeter) of water at 4 degrees (4°) Celsius (about 39° Fahrenheit). Today, the official standard of measure is the kilogram (1 000 grams), a platinum-iridium object 4 cm (1.5748 in.) high and 4 cm in diameter kept at the Bureau International des Poids et Mesures in France that reflects the older water standard. In 1995 it remained the only SI standard of measure based upon a single macroscopic object, and physicists predicted that soon it would be replaced by some atom-based standard.

1 centigram (cg) = 10 milligrams (mg)

1 decigram (dg) = 10 centigrams = 100 milligrams

1 gram (g) = 10 decigrams = 100 centigrams = 1 000 milligrams

1 kilogram (kg) = 10 hectograms (hg) = 100 dekagrams (dag) =
    1 000 grams

1 metric ton (t) = 1 000 kilograms

SI Time

In 1967 the International System adopted a *second* that is based on the microwaves emitted by the vibrations of hot cesium atoms. A second is the time it takes the atoms to vibrate exactly 9 192 631 770 times. This became the U.S. standard in 1970. In the customary measure of time, a second is 1/86 400 of one rotation of Earth, or one day. For most nonscientific purposes, an International System second (s) and a customary second can be treated as the same. Since Earth's rotation is gradually slowing, scientists since 1972 have also periodically added a second to one day in the year to keep astronomical time in step with SI clocks. The most recent such addition was on December 31, 1995, just before midnight (universal—formerly Greenwich—time; just before 8:00 p.m. EDT). It was the 19th time that such a second had been added.

In the United States and around the world, the official time is kept by a Master Clock consisting of 54 different cesium atomic and hydrogen maser clocks located around the world whose timekeeping is merged to prevent drift cause by minor environmental changes. These clocks consult each other every 100 seconds and bring laggards or speedsters into line with the majority—truly democracy in action. Comparison with the natural time caused by movements of Earth in space is made by using a network of 462 very distant quasars as a coordinate system. The rotation of Earth is slowing down slightly, mostly as a result of interactions with the Moon, but the rotation also is affected by random events, especially wind currents.

In SI, decimal fractions of time are used to measure smaller time intervals, typically the following:

$$\text{millisecond (ms)} = 0.001 \text{ second } (10^{-3})$$

$$\text{microsecond } (\mu s) = 0.000\,001 \text{ second } (10^{-6})$$

$$\text{nanosecond (ns)} = 0.000\,000\,001 \text{ second } (10^{-9})$$

$$\text{picosecond (ps)} = 0.000\,000\,000\,001 \text{ second } (10^{-12})$$

Although not part of SI, longer intervals are measured in minutes (60 seconds) or hours (3600 seconds) or days of 24 hours (86 400 seconds).

A calendar year, technically called the *tropical year*, is 365.242 2 days (31 556 926 seconds). Astronomers also use the *sidereal year*, based on the position of the Sun with respect to the background of stars. It is 365.256 4 days (31 558 153 seconds).

SI Temperature

The International System measures temperature with the Kelvin scale (sometimes called the absolute scale) denoted by the symbol K. However, for temperatures in the range at which humans live, many scientists still use the Celsius (formerly centigrade) system, whose symbol is C. No degree sign is used to represent Kelvin units, but Celsius units (and Fahrenheit units; see below) use a small raised circle to indicate degrees along with the letter abbreviation for the unit. When only one system is being used, the designation C (or F) is dropped and just the degree sign is used. Since Kelvin does not use a degree sign, the K must always be used for temperature measurements.

In the original Celsius scale, water freezes at 0°C and boils at 100°C. The temperature change denoted by 1 degree is the same size in the Kelvin and Celsius scales, but the zero point for Kelvin is set at *absolute zero*. Starting from absolute zero and defining 1 degree K (and therefore 1 degree C) in terms of molecular motion changes slightly the relationship of temperature to properties of water. In the revised system, adopted in 1990, water still freezes at 273.15K or 0°C, but the boiling point is shifted from 100°C to 99.97°C. A third temperature system, called Fahrenheit, is commonly used in the United States. Its symbol is F, and it has a degree 5/9 the size of a Kelvin degree. The equivalences between Kelvin, Celsius, and Fahrenheit are as follows:

	Kelvin	*Celsius*	*Fahrenheit*
Absolute zero	0	− 273.15	− 459.7
Freezing point, water	273.15	0	32
Normal human body temperature	310.15	37	98.6
Boiling point, water	373.15	100	212

SI Force, Work/Energy, and Power

In physics, compound measurements of force, work or energy, and power are essential. There are two parallel systems using metric units, but only one is official in SI.

The *meter/kilogram/second* (mks) system is the official SI system. Many scientists, however, are used to working the in the *centimeter/gram/second* (cgs) system, especially if they regularly measure small quantities. In both systems, units of measure are treated as if they were numbers, being both multiplied and divided. Thus, the newton meter (often unofficially written newton-meter) is the product of a newton and a meter. Similarly, a speed of a meter per second is usually written meter/second or abbreviated as m/s.

Measurement of Force

mks unit: newton (new)	The force required to accelerate a mass of 1 kg 1 m/s^2
cgs unit: dyne (dy)	The force required to accelerate a mass of 1 g 1 cm/s^2

Measurement of Work or Energy

mks unit: joule (j)	The newton meter; i.e., the work done when a force of 1 newton produces a movement of 1 m (= 10 000 000 ergs)
cgs unit: erg	The dyne centimeter; i.e., the work done when a force of 1 dy produces a movement of 1 mc

Measurement of Power

mks unit: watt (w)	The joule/second, a rate of 1 joule per second (= 10 000 000 erg seconds)
cgs unit: erg/second	A rate of 1 erg per second

Heat energy is also measured using the *calorie* (cal), which is defined as the energy required to increase the temperature of 1 cubic centimeter (1 mL) of water by 1°C. One calorie is equal to about 4.184 joules. The *kilocalorie* (Kcal or Cal) is equal to 1 000 calories and is the unit in which the energy values of food are measured. This more familiar unit, also commonly referred to as a Calorie, is equal to 4,184 joules.

SI Electrical Measure

The basic unit of quantity in electricity is the *ampere* (defined above), but in the older metric system it was the *coulomb*. A coulomb is equal to the movement of 6.25 × 10^{18} electrons past a given point in an electrical system. In SI, where the ampere is the base unit, the coulomb is defined as an ampere second. When the coulomb is used as a base unit, the ampere is equal to a coulomb second (i.e., the flow of 1 coulomb per second). The ampere is thus analogous in electrical measure to a unit of flow such as gallons per minute in physical measure.

The unit for measuring electrical power in SI is the *watt* (W), defined as one joule per second. Since the watt is such a small unit for many practical applications, the *kilowatt* (= 1 000 watts) is often used. A kilowatt hour is the power of 1 000 watts over an hour's time.

The unit for measuring electrical potential energy is the *volt*, defined as one watt per ampere. This can also be translated into 1 joule/coulomb (i.e., 1 joule of energy per coulomb of electricity). The volt is analogous to a measure of pressure in a water system.

The unit for measuring electrical resistance is the *ohm*, whose symbol is Ω (capital omega). The ohm is the resistance offered by a circuit to the flow of 1 ampere being driven by the force of 1 volt, which can also be interpreted as 1 volt per ampere. It is derived from Ohm's law, which defines the relationship between flow or current (amperes), potential energy (volts), and resistance (ohms).

OTHER INTERNATIONAL MEASURES

Measures of Angles and Arcs

Angles are measured by systems that are not exactly part of either the customary or International systems, although degree measure might be described as the "customary system" for most people most of the time. Arcs of a circle can be measured by length, but they are also often measured by angles. In that case, the measure of the arc is the same as the measure of an angle whose vertex is at the center of the circle and whose sides pass through the ends of the arc. Such an angle is said to be subtended by the arc.

The most commonly used angle measure is *degree measure*. One degree is the angle subtended by an arc that is 1/360th of a circle. This is an ancient system of measurement that was probably originally developed by Sumerian astronomers. These astronomers used a numeration system based on sixty ($60 \times 6 = 360$), as well as a 360-day year. They divided the day into twelve equal periods of thirty smaller periods each ($12 \times 30 = 360$), and used roughly the same system for dividing the circle. Even when a different year-length and numeration system were adopted by later societies, astronomers continued to use a variation of the Sumerian system.

$$1 \text{ degree } (1°) = 60 \text{ minutes } (60') = 3600 \text{ seconds } (3600'')$$

$$1 \text{ minute } = 60 \text{ seconds}$$

When two lines are perpendicular to each other, they form four angles of the same size, called *right angles*. Two right angles make up a line, which in this context is considered a *straight angle*.

$$1 \text{ right angle } = 90°$$

$$1 \text{ straight angle } = 180°$$

While the degree system is workable for most purposes, it is artificial. Mathematicians discovered that using a natural system of angle measurement produces results that make better sense in mathematical and scientific applications. This system is called *radian measure*. The SI system accepts this, but calls radian measure "supplementary" instead of basic. One radian is the measure of the angle subtended by an arc of a circle that is exactly as long as the radius of the circle.

1 radian is about 57° 17′ 45″

The circumference, C, of a circle is given by the formula $C = 2\pi r$, where π is a number (approximately 3.14159) and r is the radius. Therefore, a semicircle whose radius is 1 is π units long, which implies that there are π radians in a straight angle. Many of the angles commonly encountered are measured in multiples of π radians.

$0° = 0$ radians

$30° = \pi/6$ radians

$45° = \pi/4$ radians

$60° = \pi/3$ radians

$90° = \pi/2$ radians

$180° = \pi$ radians

$270° = 3\pi/2$ radians

$360° = 2\pi$ radians

To convert from radians to degrees, use the formula t radians $= (180/\pi)t°$. To convert from degrees to radians, use the formula $d° = (\pi/180)d$ radians.

The U.S. artillery uses the *mil* to measure angles. A mil is the angle that is subtended by an arc that is 1/6 400 of a circle.

1 mil $= 0.05625° = 3′ 22.5″$

1 mil is almost 0.001 radian

Astronomical Distances

One very large measure of distance useful in astronomy is the *light year*. It is defined as the distance light travels through a vacuum in a year (approximately 365¼ days). Light travels though a vacuum at the rate of about 186,250 mi per hour (exactly 299 792 458 m/s, exact because the meter is defined in terms of the speed of light in a vacuum). It is approximately equivalent to 5,880 billion miles (9 460 billion km).

Astronomers also use a measure even larger than the light year, the *parsec*, equal to 3.258 light years, or about 19,180 billion miles (30 820 billion km). A parsec is the distance a star would be if it appeared to shift by an angle of 1 arc second over the course of half a year. The apparent shift is caused by viewing the star from opposite ends of Earth's orbit.

A smaller unit, for measurements within the solar system, is the *astronomical unit*, which is the average distance between Earth and the Sun, or about 93 million mi (150 million km).

U.S. CUSTOMARY MEASURES

Length

1 foot (ft) = 12 inches (in)

1 yard (yd) = 3 ft = 36 in

1 rod (rd) = 5 1/2 yd = 16 1/2 ft

1 furlong (fur) = 40 rd = 220 yd = 660 ft

1 mile (mi) = 8 fur = 1,760 yd = 5,280 ft

An International Nautical Mile has been defined as 6,076.1155 ft.

Area

1 square foot (sq ft) = 144 square inches (sq in)

1 square yard (sq yd) = 9 sq ft

1 square rod (sq rd) = 30 1/4 sq yd = 272 1/2 sq ft

1 acre (A.) = 160 sq rd = 4840 sq yd = 43,560 sq ft

1 square mile (sq mi) = 640 A.

1 section = 1 mile square (that is, a square region 1 mi on a side)

1 township = 6 miles square, or 36 sq mi

Note that the U.S. customary acre is abbreviated with a period (A.), while the SI ampere has no period (A). In practice it is easy to tell the two apart from context.

Volume and capacity

1 cubic food (cu ft) = 1,728 cubic inches (cu in)

1 cubic yard (cu yd) = 27 cu ft

A gallon is equal to 231 cu in of liquid or capacity.

1 tablespoon (tbs) = 3 teaspoons (tsp) = 0.5 fluid ounce (fl oz)

1 cup = 8 fl oz

1 pint (pt) = 2 cups = 16 fl oz

1 quart (qt) = 2 pt = 4 cups = 32 fl oz

1 gallon (gal) = 4 qt = 8 pt = 16 cups

1 bushel = 8 gal = 32 qt

Weight

In customary measure, it is more common to measure weight than mass. The most common customary system of weight is the *avoirdupois*.

1 pound (lb) = 16 ounces (oz)

1 (short) hundredweight = 100 lb

1 (short) ton = 20 hundredweights = 2,000 lb

1 long hundredweight = 112 lb

1 long ton = 20 long hundredweights = 2,240 lb

A different system called *troy weight* is used to weigh precious metals. In troy weight, the ounce is slightly larger than in avoirdupois, but there are only 12 ounces to the troy pound.

Other U.S. Customary Measures

Customary measure is often said to include the Fahrenheit scale of temperature measurement (abbreviated F). In this system, water freezes at 32°F and boils at 212°F. Absolute zero is −459.7°F.

The *foot/pound/second* system of reckoning includes the following units:

poundal = fundamental unit of force

slug = the mass to which a force of 1 poundal will give an acceleration of 1 ft per second (= approximately 32.17 lb)

foot-pound = the work done when a force of 1 poundal produces a movement of 1 ft

foot/pound/second = the unit of power equal to 1 foot-pound per second

Another common unit of power is the *horsepower*, which is equal to 550 foot-pounds per second.

Thermal work or energy is often measured in *British thermal units* (Btu), which are defined as the energy required to increase the temperature of a pound of water 1°F. The Btu is equal to 0.778$^+$ foot-pound.

CONVERSIONS

From time to time, the U.S. government has taken steps to change from the customary system to the International System, but these efforts have failed. Metric measure is legal in the United States, but nearly everyone continues to use the customary system. One result is that there is a need in the United States to be able to convert back and forth from one system to another. The following tables include methods of making conversions for each of the major forms of measure.

In 1959 the relationship between customary and SI measures of length was officially defined as follows:

0.0254 meter (exactly) = 1 inch

0.0254 meter × 12 = 0.3048 meter = 1 *international* foot

This definition, which makes many conversions simple, defines a foot that is shorter (by about 6 parts in 10,000,000) than the *survey foot*, which had earlier been defined as exactly 1200/3937, or 0.3048006, meters.

Using the *international foot* standard, the following are the major equivalents:

Length

1 inch = 2.54 cm = 0.0254 m

1 foot = 30.48 cm = 0.3048 m

1 yard = 91.44 cm = 0.9144 m

1 mile = 1609.344 m = 1.609344 km

1 centimeter = 0.3937 in

1 meter = 1.093613 yd = 3.28084 ft = 0.00062137 mi

Area

1 square inch = 6.4516 cm^2

1 square foot = 929.0304 cm^2 = 0.09290304 m^2

1 square yard = 8361.2736 cm^2 = 0.83612736 m^2

1 acre = 4046.8564 m^2 = 0.40468564 hectares

1 square mile = 2,589,988.11 m^2 = 258.998811 ha = 2.58998811 km^2

1 square centimeter = 0.1550003 sq in

1 square meter = 1550.003 sq in = 10.76391 sq ft = 1.195990 sq yd

1 hectare = 107,639.1 sq ft = 11,959.90 sq yd = 2.4710538 A. = 0.003861006 sq mi

1 square kilometer = 247.10538 A. = 0.3861006 sq mi

Volume

1 cubic inch = 16.387064 cm^3

1 cubic foot = 28,316.846592 cm^3 = 0.0028316847- m^3

1 cubic yard = 764,554.857984 cm^3 = 0.764554858- m^3

1 cubic centimeter = 0.06102374 in^3

1 cubic meter = 61,023.74 in^3 = 35.31467 ft^3 = 1.307951 yd^3

1 fluid ounce = 29.573528 mL = 0.02957 L

1 cup = 236.588 mL = 0.236588 L

1 pint = 473.176 mL = 0.473176 L

1 quart = 946.3529 mL = 0.9463529 L

1 gallon = 3785.41 mL = 3.78541 L

1 milliliter = 0.0338 fl oz

1 liter = 33.814 fl oz = 4.2268 cup = 2.113 pt = 1.0567 qt = 0.264 gal

1 pint, dry = 33.600 cu in = 0.551 L

1 quart, dry = 67.201 cu in = 1.101 L

Mass and Weight

Since mass and weight are identical at standard conditions (sea level on Earth), grams and other International System units of mass are often used as measures of weight or converted into customary units of weight. The following conversions are correct under standard conditions.

1 ounce = 28.3495 g

1 pound = 453.59 g = 0.45359 kg

1 short ton = 907.18 kg = 0.907 metric ton

1 milligram = 0.000035 oz

1 gram = 0.03527 oz

1 kilogram = 35.27 oz = 2.2046 lb

1 metric ton = 2,204.6 lb = 1.1023 short ton

Temperature

The simplest means of converting is by formula.

To convert a Fahrenheit temperature to Celsius on a calculator, subtract 32 from the temperature and multiply the difference by 5, then divide the product by 9. Alternatively, subtract 32 and divide the difference by 1.8. The formula is

$$C = 5/9 \times (F - 32) = (F - 32)/1.8$$

To convert a Celsius temperature to Fahrenheit on a calculator, multiply the temperature by 1.8, then add 32; the formula is often given with the fraction 9/5 instead of the equivalent decimal, 1.8.

$$F = 9/5\ C + 32 = 1.8C + 32$$

To convert to Kelvin, find the temperature in Celsius and add 273.15.

$$K = C + 273.15$$

Measurement of Force

1 poundal = 13,889 dynes = 0.13889 newton

1 dyne = 0.000072 poundal

1 newton = 7.2 poundals

Measurement of Work/Energy

1 foot-pound = 1356 joules

1 British thermal unit = 1,055 joules = 252 gram calories

1 joule = 0.0007374 foot-pounds

1 (gram) calorie = 0.003968 Btu

1 (kilo) Calorie = 3.968 Btu

Measurement of Power

1 foot-pound/second = 1.3564 watts

1 horsepower = 746 watts = 0.746 kilowatts

1 watt = 0.73725 ft-lb/sec = 0.00134 horsepower

1 kilowatt = 737.25 ft-lb/sec = 1.34 horsepower

SIMPLIFIED CONVERSION TABLE (ALPHABETICAL ORDER)

To convert	into	multiply by
centimeters	feet	0.03281
centimeters	inches	0.3937
cubic centimeters	cubic inches	0.06102
cubic feet	cubic meters	0.02832
degrees	radians	0.01745
feet	centimeters	30.48
feet	meters	0.3048
gallons	liters	3.785
gallons of water	pounds of water	8.3453
grams	ounces	0.03527
grams	pounds	0.002205
inches	centimeters	2.54
kilograms	pounds	2.205
kilometers	feet	3280.8
kilometers	miles	0.6214
knots	miles per hour	1.151
liters	gallons	0.2642
liters	pints	2.113
meters	feet	3.281
miles	kilometers	1.609
ounces	grams	28.3495
ounces	pounds	0.0625
pounds	kilograms	0.4536

(Periodical References and Additional Reading: *Natural History*, 2-95, p 72; *Science* 5-12-95, p 804; *Physics Today* 8-95, p BG15)

Index